ORDINARY CHONDRITES

Ordinary chondrites, the most abundant meteorites, constitute about 80% of meteorite falls and are essential to our understanding of cosmochemistry. They provide important information about planetary accretion, the early Solar System, and the geological history of asteroids, including such processes as thermal metamorphism, shock metamorphism, and aqueous alteration. This comprehensive guide begins with meteorite classifications and useful definitions, followed by a discussion of fall phenomena and terrestrial weathering. It provides a detailed overview of the three main ordinary-chondrite groups, which include the most primitive, least-processed meteorites known. Compositional differences among these samples furnish clues to the nature of processes operating in the solar nebula 4.5 billion years ago. These rocks also disclose information on the nature and origin of chondrules, matrix material, and metallic iron-nickel grains. This book is a valuable resource for graduate students and research professionals interested in meteorites and planetary science, as well as amateur meteorite enthusiasts.

Alan E. Rubin is a world-leading expert on meteorites. He is now semi-retired but previously served as an adjunct professor and research geochemist in the Department of Earth, Planetary, and Space Sciences at the University of California, Los Angeles. He is a fellow of the Meteoritical Society and winner of the Nininger Meteorite Award and seven Griffith Observer science writing awards. He is the namesake of the garnet mineral rubinite and the main-belt asteroid 6227 Alanrubin. He has published about 230 peer-reviewed papers on the subject, as well as four previous books, including the well-received *Meteorite Mineralogy*, published in our Cambridge Planetary Science series in 2021.

Cambridge Planetary Science

Series Editors:
Jim Bell, Leigh Fletcher, David Jewitt, Ralph Lorenz, Francis Nimmo, David Vokrouhlický

Books in the Series:

1. Jupiter: The Planet, Satellites and Magnetosphere †
   Edited by Bagenal, Dowling and McKinnon
   978-0-521-03545-3
2. Meteorites: A Petrologic, Chemical and Isotopic Synthesis †
   Hutchison
   978-0-521-03539-2
3. The Origin of Chondrules and Chondrites †
   Sears
   978-1-107-40285-0
4. Planetary Rings †
   Esposito
   978-1-107-40247-8
5. The Geology of Mars: Evidence from Earth-Based Analogs †
   Edited by Chapman
   978-0-521-20659-4
6. The Surface of Mars †
   Carr
   978-0-521-87201-0
7. Volcanism on Io: A Comparison with Earth †
   Davies
   978-0-521-85003-2
8. Mars: An Introduction to its Interior, Surface and Atmosphere †
   Barlow
   978-0-521-85226-5
9. The Martian Surface: Composition, Mineralogy and Physical Properties
   Edited by Bell
   978-0-521-86698-9
10. Planetary Crusts: Their Composition, Origin and Evolution†
    Taylor and McLennan
    978-0-521-14201-4
11. Planetary Tectonics†
    Edited by Watters and Schultz
    978-0-521-74992-3
12. Protoplanetary Dust: Astrophysical and Cosmochemical Perspectives†
    Edited by Apai and Lauretta
    978-0-521-51772-0
13. Planetary Surface Processes
    Melosh
    978-0-521-51418-7
14. Titan: Interior, Surface, Atmosphere and Space Environment
    Edited by Müller-Wodarg, Griffith, Lellouch and Cravens
    978-0-521-19992-6
15. Planetary Rings: A Post-Equinox View (Second edition)
    Esposito
    978-1-107-02882-1

16. Planetesimals: Early Differentiation and Consequences for Planets
    Edited by Elkins-Tanton and Weiss
    978-1-107-11848-5
17. Asteroids: Astronomical and Geological Bodies
    Burbine
    978-1-107-09684-4
18. The Atmosphere and Climate of Mars
    Edited by Haberle, Clancy, Forget, Smith and Zurek
    978-1-107-01618-7
19. Planetary Ring Systems
    Edited by Tiscareno and Murray
    978-1-107-11382-4
20. Saturn in the 21st Century
    Edited by Baines, Flasar, Krupp and Stallard
    978-1-107-10677-2
21. Mercury: The View after MESSENGER
    Edited by Solomon, Nittler and Anderson
    978-1-107-15445-2
22. Chondrules: Records of Protoplanetary Disk Processes
    Edited by Russell, Connolly Jr. and Krot
    978-1-108-41801-0
23. Spectroscopy and Photochemistry of Planetary Atmospheres and Ionospheres
    Krasnopolsky
    978-1-107-14526-9
24. Remote Compositional Analysis: Techniques for Understanding Spectroscopy, Mineralogy, and Geochemistry of Planetary Surfaces
    Edited by Bishop, Bell III and Moersch
    978-1-107-18620-0
25. Meteoroids: Sources of Meteors on Earth and Beyond
    Edited by Ryabova, Asher and Campbell-Brown
    978-1-108-42671-8
26. Meteorite Mineralogy
    Rubin and Ma
    978-1-108-48452-7
27. Vesta and Ceres: Insights into the Dawn of our Solar System
    Edited by Marchi, Raymond and Russell
    978-1-108-47973-8
28. Ganymede
    Edited by Volwerk, McGrath, Jia and Spohn
    9781108832953
29. Martian Impact Craters: A Morphological Perspective
    Mouginis-Mark and Boyce
    978-1-107-15491-9
30. Ordinary Chondrites: The Most Common Meteorites
    Rubin
    978-1-009-65613-9

† Reissued as a paperback

ORDINARY CHONDRITES

The Most Common Meteorites

ALAN E. RUBIN
University of California, Los Angeles

CAMBRIDGE
UNIVERSITY PRESS

CAMBRIDGE
UNIVERSITY PRESS

Shaftesbury Road, Cambridge CB2 8EA, United Kingdom
One Liberty Plaza, 20th Floor, New York, NY 10006, USA
477 Williamstown Road, Port Melbourne, VIC 3207, Australia
314–321, 3rd Floor, Plot 3, Splendor Forum, Jasola District Centre, New Delhi – 110025, India
103 Penang Road, #05–06/07, Visioncrest Commercial, Singapore 238467

Cambridge University Press is part of Cambridge University Press & Assessment, a department of the University of Cambridge.

We share the University's mission to contribute to society through the pursuit of education, learning and research at the highest international levels of excellence.

www.cambridge.org
Information on this title: www.cambridge.org/9781009656139

DOI: 10.1017/9781009656122

When citing this work, please include a reference to the DOI 10.1017/9781009656122

First published 2026

Cover credit: Image courtesy of Larry Atkins and CosmicConnectionMeteorites.com.

A catalogue record for this publication is available from the British Library

A Cataloging-in-Publication data record for this book is available from the Library of Congress

ISBN 978-1-009-65613-9 Hardback

Dedicated to

Harold Urey and Harmon Craig
(for defining the H and L groups)

and

Klaus Keil and Kurt Fredriksson
(for defining the LL group)

and

Randall Van Schmus, John Wood, Robert Dodd, Derek Sears, and Jeffrey Grossman
(for characterizing the petrologic types and subtypes of ordinary chondrites)

and

John Wasson, Gregory Kallemeyn, Hugo Wiik, and Eugene Jarosewich
(for determining the bulk chemical compositions of ordinary chondrites)

and

Robert Clayton, Toshiko Mayeda, Jitendra Goswami, and Edward Olsen
(for determining the bulk oxygen-isotopic compositions of ordinary chondrites)

and

Robert Dodd (again), Dieter Stöffler, Klaus Keil (again), and Edward Scott
(for characterizing the principal shock features of ordinary chondrites)

and

John Wood (again)
(for determining metallographic cooling rates of ordinary chondrites)

and

William Cassidy, Ralph Harvey, John Schutt, Harvey Nininger, and Glenn Huss
(for developing, implementing, and inspiring programs for the systematic collection of meteorites, thereby bringing numerous ordinary chondrites to the attention of the research community)

and

Meteorite hunters (too many to list)
(for doing the arduous work of discovering finds and retrieving falls, most of which are ordinary chondrites)

and

George Prior, Oliver Farrington, Max Hey, Andrew Graham, Alex Bevan, Robert Hutchison, Monica Grady, Jörn Koblitz, and Jeffrey Grossman (again)
(for cataloging meteorites – mostly ordinary chondrites – listing their properties, and facilitating global dissemination of the information)

and

Numerous scientists (friends and colleagues)
(for developing fireball camera networks, for studying ordinary chondrites and their components, thereby determining modal abundances, textures, physical properties, shock and alteration histories, mineral chemistries, organic constituents, and chemical and isotopic compositions, for constructing theoretical models of ordinary-chondrite formation in the Early Solar System, and for determining the mechanisms that deliver ordinary chondrites to Earth)

Other Books by Alan E. Rubin

Disturbing the Solar System: Impacts, Close Encounters, and Coming Attractions. Princeton University Press, 2002, Princeton, New Jersey. 376 pp.

Son of Man: A Personal Memoir by Yehoshua ben Yahweh, Translated and Edited by Arthur Melton and Monica Wheatley. A novel. CreateSpace, 2015. 147 pp.

With Chi Ma: *Meteorite Mineralogy*. Cambridge University Press, 2021, Cambridge. 404 pp.

Surface/Volume: How Geometry Explains Why Grain Elevators Explode, Hummingbirds Hover, and Asteroids Are Colder than Ice. Springer, 2023, Cham, Switzerland. 182 pp.

sic itur ad astra

Virgil, *The Aeneid*

The Fall

Some distance from a quaint old town, meteoroids came hurtling down.
Residents heard claps of thunder; piercing shockwaves split the sky.
Falling stones in black were shrouded; 'neath the crust, chondrules were crowded.
Fine-grained matrix can't be doubted; discov'ring cosmic clues was nigh.
An H3 fall with lots of metal, and iron-sulfide right nearby.
Research interest was quite high.

Oh, so clearly I recall; it was a cloudy day in Fall
When a section of that fallen chondrite came before my eager eye.
Some chondrules had fine rims of metal, outstretched like a flower's petal.
Opaque veins were like a nettle; 'though hard to see, I had to try.
And native copper, next to sulfide, within iron grains did lie.
But I could not find a C.A.I.

Well-formed chondrules, ripe for study; in crossed nicols, blue and ruddy.
In some rims, microchondrules could I now spy.
Most were made of pyroxene with radial bars both straight and lean
With mesostasis that was clean. Their parent melt was very dry.
For fine details of smaller features, into which I craved to pry,
The S.E.M. I need now ply.

The bulk rock's graphical position marks its isotopic composition:
Just above the T.F. line, the plotted points so clearly lie.
Richer in O-seventeen than terrestrial rocks have ever been,
At least among those I have seen, and I've seen many with my eye.
Since heavy oxygen in Type-IA chondrules is not high;
In other components it must lie.

When I looked about for alteration, I saw a rounded aggregation:
Carbide-oxide-sulfide-metal intergrowths I could espy.
Smectite grains in low abundance (montmorillonite redundance).
Surely there was no reluctance to find the whole rock was not dry.
The histories of ord'nry chondrites are thus revealed by and by.
Through careful study – our ally.

AER

Contents

Preface

In Plato's *Phaedrus*, Socrates posits that the art of writing will provide its practitioners with "not truth, but only the semblance of truth; they will be hearers of many things and will have learned nothing; they will appear to be omniscient and will generally know nothing; they will be tiresome company, having the show of wisdom without the reality." This characterization comports with John Maynard Keynes' disparagement of academic scribblers. These admonitions notwithstanding, I wrote this book; it is focused on ordinary chondrites – the most abundant meteorites. Ordinary chondrites are of enormous importance in cosmochemistry, and no single person has a full understanding of their myriad characteristics.

They constitute about 78% of well-classified observed falls and 85% of all well-classified meteorites (not taking possible pairings into account). In his influential 1794 monograph on meteorites, Ernst Chladni listed 18 observed falls from antiquity to modern times. Many are ordinary chondrites: for example, Eichstädt, Bavaria, in 1785 (H5); Tabor, Bohemia, in 1753 (H5); Pleskowitz (officially Ploschkovitz), Bohemia, in 1723 (L5); Albareto, Italy, in 1766 (L/LL4); Lucé, France, in 1768 (L6); and Aire-sur-la-Lys, France, in 1769 (unclassified meteorite, now lost, reported by the Academy of Sciences in Paris to contain "small metal and pyritic parts" and to be "covered with a black crust"; probably an ordinary chondrite). After Chladni's monograph was published, there was a spate of widely witnessed ordinary-chondrite falls that helped convince the scientific community that rocks really fell from space: Siena, Italy (1794, LL5 chondrite); Wold Cottage, England (1795, L6 chondrite); Portugal, from the town of Évora Monte (1796, unclassified stone, possibly an ordinary chondrite, which was lost); Salles, France (1798, L5 chondrite); Benares (a), India (1798, LL4 chondrite); L'Aigle, France (1803, L6 chondrite); Weston, Connecticut, USA (1807, H-chondrite regolith breccia). In addition, the Barbotan, France fall (1790, H5 chondrite) came to the attention of scientists only after Chladni's book came out.

The oldest extant meteorite fall is an ordinary chondrite – the Nogata L6 chondrite (472 g), which plowed into soft soil in Fukuoka Prefecture, Japan on the night of May 19, 861. It has been preserved in a Shinto shrine for more than a thousand years. The next-oldest properly classified meteorite is also an ordinary chondrite: LL6 Ensisheim. This single 127-kg stone gouged a meter-deep hole in a field in Alsace, France on the morning of November 16, 1492. The meteorite was dug up and brought to Ensisheim parish church. A Latin inscription was attached: *De hoc lapide multi multa, omnes aliquid, nemo satis*, loosely translated as "There are many things about this stone, all have said something, but it is not enough." It is likely the fall of Ensisheim inspired Jules Verne's posthumously published 1908 novel *The Chase of the Golden Meteor*. The Ensisheim meteorite is currently on display in the Palace of the Regency, an impressive sixteenth-century structure in the center of town.

Ordinary chondrites have even been cited in modern science fiction. In the disaster novel *Impact!*, written by Ron Fodor and Jeff Taylor in 1979, there is a spate of meteorite falls, 28% of which are L chondrites. These projectiles were dislodged from an 11-km asteroid that imperiled the Earth. (This novel has no sequel.)

The first meteorite falls whose orbits were determined through photographic evidence are all ordinary chondrites: H5 Pribram in 1959 (Ceplecha, 1961); H5 Lost City in 1970 (McCrosky et al., 1971); L5 Innisfree in 1977 (Halliday et al., 1978). Their orbits (fig. I-9 of Wasson, 1985) resemble those of Apollo asteroids (Wetherill, 1979, 1985).

Throughout the nineteenth and twentieth centuries, numerous ordinary chondrites were recovered and examined. For example, in the 1980s alone, there were dozens of published descriptions of ordinary-chondrite finds from the United States (e.g., Sipiera et al., 1980, 1983, 1987; Keil et al., 1981; Rubin et al., 1981a; Povenmire, 1984; Recca et al., 1986; Struempler and Kaelin, 1987; Ehlmann and Keil, 1988; White, 1989), Brazil (e.g., Gomes and Keil, 1980; Fredriksson and Wlotzka, 1985), other countries (e.g., Fredriksson et al., 1981; Keil et al., 1986; Bevan and Binns, 1989), and Antarctica (e.g., McSween and Wilkening, 1980; McKinley et al., 1981; Scott et al., 1982).

Chondritic meteorites were reviewed by Van Schmus (1969) and Zolensky et al. (2018a). The enormous increase in our understanding in the (nearly) half-century between these reviews was made possible by the development of new instrumentation and the examination and analysis of tens of thousands of meteorites recovered from cold and hot deserts. My own research on chondrites postdates Van Schmus's review and spans more than four decades, but it is vastly overshadowed by the burgeoning literature on these fascinating rocks. Numerous scientists have contributed to our current understanding of ordinary chondrites. This book reflects their efforts, but it was not practicable to cite every important paper.

Because of their high abundance, ordinary chondrites are the least expensive meteorites to acquire for personal collections and classroom instruction: eBay commonly offers unclassified ordinary-chondrite finds from Northwest Africa for about $1.00 or so per gram (plus modest shipping costs). (*Caveat emptor*: A few "meteorwrongs" are sometimes also on offer.) There is no better way to introduce students to the wonders of planetary science than firsthand experience with celestial samples. Meteorites bring students into the realm of "tactile astronomy" where (metaphorically à la William Blake) they can hold infinity in the palm of their hand. Nininger (1936a), in an article entitled "Importance of Meteorite Collections," affirmed that "meteorites are an extremely important class of objects; probably the most important of all natural history objects from the standpoint of the information which they hold for serious students regarding the earth and the universe." Today, there are more meteorite collectors than ever; books written for this community include Bagnall (1991), Carman (1995), Kichinka (2005), Norton and Chitwood (2008), Notkin (2011), Stinchcomb (2011), Croning (2018), and Bryant (2018).

I am grateful to numerous colleagues who answered my questions, provided data, and discussed research results. Many images in this book were made available through the courtesy of friends and colleagues and by the permission of research journals and their publishers and editors. Other images are licensed under Creative Commons: https://creativecommons.org/licenses/by/4.0/.

One final point. Advances in science often come from youthful investigators. Einstein's *annus mirabilis* occurred in 1905, the year he turned 26. In that year, Einstein published papers on special relativity, mass-energy equivalence, the photoelectric effect, and Brownian motion. It is good to remember "Planck's principle," commonly paraphrased as "science progresses one funeral at a time." The principle is well-illustrated by the reluctance of an elderly Lord Kelvin (1824–1907) to accept an ancient age for the Earth. His calculations, made as a young man in 1844 and 1846, based on thermodynamics and data from geophysics and experimental petrology, yielded an age for the cooling Earth of a few tens of millions to a few hundred million years. Although the relevance of these calculations was brought into question by the 1896 discovery of radioactivity by Henri Becquerel, Kelvin never accepted a terrestrial age of billions of years. (Radioactive decay within the Earth provided a long-lived heat source, facilitating much slower global cooling, and permitting a much older age.) Planck's principle pervades every field of science; it has more than anecdotal support (Azoulay et al., 2019). As my generation of researchers is beset with joint effusion, edentulism, and dyspepsia, I hope this book proves useful to statistically vigorous early career scientists.

Introduction

Meteorite researchers include astronomers, physicists, astrophysicists, planetary scientists, organic and inorganic chemists, materials scientists, metallurgists, and even some industrious collectors and dealers, but the largest cohort comprises geoscientists of various stripes (mainly petrologists, mineralogists, crystallographers, and geochemists). Nevertheless, most geoscientists are practical folk, working in the mining, petroleum, geological engineering, agricultural, meteorological, and environmental industries. According to the online recruitment services provider Zippia, only 6% work in education. And among university-level geo-educators, most have little experience with meteorites. The website of the geology department at the University of Illinois at Urbana-Champaign (where I took an introductory geology course) provides a list of faculty research areas: Earth Materials (Mineral Science), Sedimentary and Earth Surface Processes, Geobiology, Hydrogeology, Seismology, Geochemistry/Petrology, Geophysical Fluid Dynamics, Paleoclimate, and Tectonics/Structural Geology. Absent are any stated interests in meteoritics, cosmochemistry, or planetary science.

If they snag an interview, meteorite researchers applying for a job at such an institution may be asked about the relevance of their field. In such moments, there is no need for apprehension or reticence. As far back as 1858, the German chemist Karl Reichenbach noted that meteorites are simultaneously "cosmological, astronomical, physical, geological, chemical, mineralogical, and meteorological" objects (Burke, 1986). Pejovic (1982) pointed out that meteorites are of special interest to chemists, metallurgists, nuclear physicists, biologists, and aerodynamicists.

And more than a century ago, Farrington (1915) presented three cogent reasons that meteorites should be a topic of particular interest:

1. They are our only tangible sources of knowledge regarding the universe beyond us.
2. They are portions of extraterrestrial bodies.
3. They are a part of the economy of Nature.

These points are valid today, but with the advent of sample-return space missions, meteorites are no longer the sole source of tangible extraterrestrial material. Farrington's list can be augmented. Here are 25 succinct modern reasons (cosmochemical, astrophysical, geological, anthropological, and sociological) why meteorite research is important and relevant:

1. Meteorites are our most abundant source of extraterrestrial material.
2. Meteorites formed under a wide range of redox conditions and contain many minerals not found on Earth.
3. The study of meteorites extends the range of known petrological and geochemical processes.
4. Meteorites serve as concrete examples of shock metamorphism of natural materials.
5. Many meteorites contain the most ancient examples of aqueous alteration of minerals that can be studied in the lab.
6. Meteorites are the building blocks of the planets.
7. The O-isotopic composition of the upper layers of Earth's atmosphere can be measured by analyzing cosmic spherules.
8. CI carbonaceous chondrites provide the cosmic abundances of most elements (excluding noble gases and a few other elements).
9. Meteorites derive mainly from meter-to-decameter-sized bodies that were bombarded by cosmic rays

in interplanetary space and thus can provide information about the interactions of cosmic rays with solid materials.

10. Meteorites likely delivered raw materials (e.g., water and organics from carbonaceous chondrites; phosphorus from irons) to the Early Earth, possibly helping to facilitate the origin of life.
11. Carbonaceous chondrites provide the most ancient organic compounds (including hundreds of amino acids) that can be studied in the lab.
12. Collisions of asteroids with the Earth changed the course of biological evolution.
13. Impact-crater formation by asteroidal collisions is the predominant geomorphological process in the Solar System.
14. The existence of iron meteorites supported the idea that the Earth has an iron core.
15. Iron meteorites provide otherwise inaccessible samples of planetesimal cores.
16. Meteorites provide clues to the geological history of asteroids.
17. Lunar meteorites come from the entire lunar globe, enhancing our understanding of the geology of the Moon.
18. Martian meteorites allow us to study the geological history of Mars.
19. Ancient refractory inclusions in chondritic meteorites yield the age of the Solar System.
20. Chondrules, which occur only in chondritic meteorites, were the most abundant discrete objects in the Solar System during the epoch when planetesimals accreted.
21. The relative numerical abundances of different varieties of meteorites reflect the relative sizes of parent bodies and source regions and the durability of meteoroids/meteorites in interplanetary space and the terrestrial environment.
22. Presolar grains in chondritic meteorites permit the in situ examination of materials from stars that existed long before the birth of the Solar System.
23. Meteorites were occasionally worshipped by ancient peoples.
24. For some technologically primitive peoples, iron meteorites served as the sole source of metallic iron for making tools.
25. Meteorite collecting and meteorite museums promote public appreciation of planetary science.

The components of primitive chondrites reveal vital information about the early Solar System. Unequilibrated, type-3 ordinary chondrites contain about 70–75 vol.% chondrules – the most abundant igneous objects formed in the solar nebula. The fine-grained matrix of these (and other) chondrites contains presolar grains, small particles that condensed in the atmospheres of dying stars. Also present in these meteorites are rare calcium-aluminum inclusions – the oldest objects formed in the Solar System. The differences among the H, L, and LL groups in metal abundance and metal composition provide evidence for chemical fractionations in the nebula.

A panoply of complex parent-body processes can be studied in the ordinary chondrites, including thermal metamorphism, shock metamorphism, and aqueous alteration. These processes may have operated in tandem (Lewis et al., 2022) and caused fluid-assisted metamorphism, hydrothermal alteration, metasomatism during thermal metamorphism, shock-induced annealing, and shock heating of ice and mobilization of liquid water. Such processes could also have operated sequentially: for example, multiple shock events, post-shock annealing, post-shock alteration, and finally post-alteration annealing.

Some iron meteorites were formed from carbonaceous-chondrite-like parent asteroids, others from noncarbonaceous asteroids (e.g., Kruijer et al., 2017a, 2017b, 2017c; Poole et al., 2017). But the only chondrite clan that can be tied specifically to iron meteorites is that of ordinary chondrites: IIE irons may be a fourth ordinary-chondrite group, more reduced than H (e.g., Bild and Wasson, 1977; Rubin, 2022a); IVA irons are related to L or LL chondrites (Clayton et al., 1983); and the Guin ungrouped iron is related to LL chondrites (Rubin et al., 1986).

Ordinary chondrites have been bombarding the Earth throughout geological time. After the L-chondrite parent asteroid was disrupted by a massive collision ~470 Ma ago (Korochantseva et al., 2007), L-asteroid fragments were dispersed throughout the Inner Solar System (Schmitz et al., 2001). Numerous L-chondrite samples from this event were recovered as fossil meteorites in a quarry of Ordovician limestone in Sweden (e.g., Thorslund and Wickman, 1981). The Lockne-Målingen doublet crater (7.5 km, 0.7 km) in Central Sweden formed 458 Ma ago; it may have been produced by a small L-chondrite asteroidal fragment derived from the L parent-body breakup.

Ordinary chondrites are still falling. In the five-year inclusive period of 2019–2023, 56 meteorite falls were recovered, analyzed, and approved (so far); as of this writing, an additional H-chondrite fall from France (on September 9, 2023) is going through the classification and approval process. Including this French fall, 46/57 (81%) are ordinary chondrites (22 H, 20 L, 1 L/LL, 3 LL). The other falls include six carbonaceous chondrites, one enstatite chondrite, two HEDs, and two aubrites.

Most terrestrial impact craters with accompanying meteorites are associated with irons.[1]

The most famous example is Meteor Crater, Arizona and the Canyon Diablo IAB iron. Three craters have yielded intact ordinary-chondrite fragments: (1) A 25-cm, ~750-g, LL6 chondrite breccia (Morokweng) was found within the 145-Ma-old, ~70-km-wide Morokweng crater in South Africa (Maier et al., 2006). (2) A 900-kg H5 chondrite (Kunya-Urgench) was recovered from a 6-m-wide, 4-m-deep crater (or impact pit) after a large mass crashed in Turkmenistan at 5:25 PM local time on June 20, 1998 (Alexeev et al., 2001). An additional 100–200 kg of material was later retrieved. (3) A 342-g H4-5 chondrite (Carancas) was recovered from the ~14-m-wide Carancas crater in Peru (Kenkmann et al., 2009). The collision that produced this crater occurred at 11:45 AM local time on September 15, 2007. Carancas is the youngest known impact crater on Earth. (There is a younger and far smaller circular depression [7 cm wide, 3 cm deep] dubbed a "crater" – with the requisite quotation marks – by Bartoschewitz et al. [2017]. It occurs within a pebble-concrete step in front of a house in Braunschweig-Melverode, Germany. The crater, which has an ejecta blanket and prominent rays, formed at 2:07 AM local time on April 23, 2013. Straddling the crater is the 214-g main mass of the Braunschweig L6 chondrite. The meteorite shattered into hundreds of pieces upon impact; a total mass of 1.3 kg was recovered.)

Toward the other end of the size scale is the ~100-km-diameter Popigai impact crater in Northern Siberia. Analysis of Pt-group elements in samples of impact melt from the crater indicate a meteoritic contaminant of ~0.2 wt.% (Tagle and Claeys, 2005). The Ru/Rh, Pt/Pd and Pd/Ir ratios of the impact melt show the contaminant to be an ordinary chondrite, probably L. This finding is in accord with ordinary chondrites' having made up a large fraction of the impactors that created >1.5-km-diameter craters since the beginning of the Cambrian Period 538.8 Ma ago (McDonald et al., 2001; McDonald, 2002) (Table I.1). These bodies would likely be S-type asteroids in Earth-crossing orbits.

In his 1923 collection of New Hampshire poems, Robert Frost included "A Star in a Stoneboat" (the "star" being a meteorite and the stoneboat being a handcart used for hauling stones). The first stanza reads:

> Never tell me that not one star of all
> That slip from heaven at night and softly fall
> Has been picked up with stones to build a wall.

Although Frost lacked firsthand knowledge, he appeared certain that farmers must have encountered meteorites as they cleared their fields and stacked the meteorites along with terrestrial rocks to build stone walls at the borders of their property. An approximation of this scenario came to

[1] Impact craters associated with iron meteorites include Agoudal, Morocco (IIAB); Boxhole, Northern Territory, Australia (IIIAB); Camp del Cielo, Chaco, Argentina (IAB); Barringer (Meteor Crater), Arizona (Canyon Diablo IAB); Gebel Kamil, East Uweinat Desert, Egypt (ungrouped iron); Henbury, Northern Territory, Australia (IIIAB); Hickman, Western Australia (IAB); Kaalijärv, Estonia (IAB); Monturaqui, Antofagasta, Chile (probable IAB); Morasko, Poznan, Poland (IAB); Odessa, Texas (IAB); Sikhote-Alin, Primorskiy kray, Russia (IIAB); Veevers, Western Australia (IIAB); Wabar, Ash Sharqiyah, Saudi Arabia (IIIAB); Whitecourt, Alberta, Canada (IIIAB); and Wolfe Creek, Western Australia (Wolf Creek IIIAB). Two craters are associated with stony-iron meteorites: Dalgaranga, Western Australia (with a mesosiderite of the same name) and Haviland, Kansas (with the Brenham pallasite). There are two other major collisions associated with identifiable meteoritic material: (1) A 2.5-mm fossil carbonaceous chondrite is probably a fragment of the 10-km asteroid that crashed into the Yucatán Peninsula ~66 Ma ago. This collision formed the Chicxulub impact structure and wiped out the non-avian dinosaurs. The fossil meteorite was found ~9000 km west of the crater in a deep-sea drill core at the Cretaceous-Paleogene (K-Pg) boundary in the North Pacific Ocean (Kyte, 1998; Shukolyukov and Lugmair, 1998). (2) The Eltanin mesosiderite occurs as about two dozen small basaltic breccia fragments within Ir-rich sediments from deep-sea drill cores from the Bellingshausen Sea in the southeastern Pacific Ocean. The meteorite fragments were derived from a kilometer-size asteroid that slammed into the Pacific in the Late Pliocene, about 2.5 Ma ago (Kyte, 2002). No crater has been identified.

Table I.1 *Ordinary-chondrite crater-forming projectiles from the Phanerozoic.*

Crater	Diameter (km)	Age (Ma)	Projectile type
Pingualuit (Canada)	3.44	1.4±0.1	L
Brent (Canada)	3.8	450±30	L-LL
Rio Cuarto (Argentina)	4.5	<0.1	H
Bosumtwi (Ghana)	10.5	1.03±0.02	OC
Rochechouart (France)	23	214±8	IIE
Clearwater East (Canada)	26	290±20	L, LL
Morokweng (South Africa)	70	145±0.8	L, LL
Popigai (Russia)	100	35.7±0.2	L

Pingualuit is from the Inuit word for "pimple." This impact structure was previously called Chubb Crater and New Quebec Crater.
Data for this table are from Tagle and Claeys (2005) with references therein.

pass with an ordinary chondrite in 1996, when a farmer found the 22-kg main mass (plus 11 smaller pieces) of Korra Korrabes (H3) in a dry riverbed in Namaland, Namibia. Eventually 120–140 kg of the meteorite were collected with the aid of metal detectors. The main mass was placed in a garden wall. Other ordinary chondrites have been identified in rock gardens – Clifford (L6), 11.36 kg; and Kramer Creek (L4), 2.30 kg (as enthusiastically reported by Schmidt, 1973). An ordinary chondrite that may have been retrieved from a rock garden is Roosevelt County 075 (H3.10; 258 g), one of the least equilibrated H chondrites known. It was found on the concrete driveway of a home near Portales, New Mexico after a rock fight the night before (McCoy et al., 1993).

The first meteorite discovered by the indefatigable Harvey Nininger in a region not previously known to harbor meteorites was an ordinary chondrite. On the evening of May 17, 1944, Nininger had stopped for lunch along U.S. Highway 85 in Socorro County, New Mexico when he spotted a small, dark, 7.673-g pebble. Closer inspection revealed it to be a meteorite – the Puente-Ladron L6 chondrite (Nininger, 1944). The only other previously unknown meteorite discovered by Nininger himself is also an ordinary chondrite: in 1955 he spied the 800-g cucumber-shaped Cottonwood H5 chondrite residing in a rock-strewn field in Yavapai County, Arizona (Nininger, 1972). Through his meteorite hunting, trading, and extensive public outreach, Nininger had assembled one of the largest meteorite collections in the world by 1946. It contained an unprecedented proportion of finds and included many ordinary chondrites.

Ordinary chondrites span the Earth: from north to south – Greenland (L6 Ella Island; L5 Ryder Gletcher) and Norway (e.g., H3-6 Oslo; L/LL6 Trysil) to Antarctica (e.g., L5 Adelie Land; LL6 Yamato 003259); from east to west – Japan (e.g., L5 Fukutomi; H4/5 Yonozu) to Alaska (LL6 Hope Creek); from A to Z – Afghanistan (L6 Kandahar) to Zimbabwe (e.g., H3-5 Magombedze; L6 Nkayi); from large countries to small ones – Russia (e.g., LL5 Chelyabinsk; H3.8 Raguli) to Mauritius (LL5-6 Mauritius); and from urban areas to rural ones – cities (e.g., H6 Peekskill; H5 New Orleans) to the Australian outback (e.g., L4 Biduna Blowhole 001; H5 Murrili). There is even one ordinary chondrite from the United States in my very small personal collection (L4 Gold Basin).

Although they commonly elude mention in geology or astronomy textbooks, more than 65,000 ordinary chondrites have currently been classified. Thousands more have been recovered and await classification. Despite their relative abundance, ordinary chondrites are extraordinary objects.

Let's learn about them.

1

Approaches to Experimental and Observational Science

All things are implicated with one another, and the bond is holy; and there is hardly anything unconnected with any other thing. For things have been coordinated, and they combine to form the same universe. For there is one universe made up of all things ...

Marcus Aurelius, *Meditations*

Test everything; retain what is good.

1 Thessalonians 5:21

Whene'er the mist, that stands 'twixt God and thee,
Sublimates to a pure transparency,
That intercepts no light and adds no stain–
There Reason is, and then begins her reign!

Samuel Taylor Coleridge, "Reason"

Meteorites don't come with labels. Before these rocks reveal their secrets, they must be analyzed. Theoreticians can construct models of the solar nebula and offer plausible geological and dynamic histories of planetary bodies, but without hard data at their fingertips, theoreticians would have little to do. In *An Enquiry Concerning Human Understanding*, David Hume (1748) wrote that "causes and effects are discoverable, not by reason, but by experience." As Bertrand Russell (1935) pointed out, "Science is the attempt to discover, by means of observation, and reasoning based upon it, first, particular facts about the world, and then laws connecting facts with one another and (in fortunate cases) making it possible to predict future occurrences."

Before engaging in any scientific endeavor, researchers make four (usually tacit) metaphysical assumptions: (1) the rules of logic are universally applicable, (2) there are universal laws that govern the behavior of all materials and their interactions (e.g., matter, dark matter, energy, dark energy, fields, fundamental forces, spacetime), (3) these universal laws are invariant everywhere (with the possible exception of the interiors of black holes) and everywhen (at least since an instant after the Big Bang), and (4) all universal laws are ultimately discoverable, comprehensible, and expressible by humankind. Paul Dirac (1938) questioned the third assumption, speculating that some physical constants may have changed over time. However, no such changes have yet been discerned. Albert Einstein (1936) endorsed the fourth assumption, remarking that "the most incomprehensible thing about the world is that it is comprehensible." Scottish biologist J. B. S. Haldane (1927) disagreed with this notion, asserting that "the universe is not only queerer than we suppose, but queerer than we *can* suppose" (emphasis added). And, in a different context, Ludwig Wittgenstein (1922) claimed that some things are unsayable (ineffable) – they cannot be put into words. These four assumptions are akin to Euclid's postulates and common notions. Although they are not falsifiable like scientific theories (cf. Popper, 1963) and have likewise not been proven, they are necessary starting points. As a practical matter, scientists adhere to the principle of methodological naturalism: Only naturalistic explanations are sought for phenomena. If no such explanations are found, they speculate (and openly disclose this) or admit ignorance.

Formally, experimental science is restricted to fields amenable to direct experiments such as mixing chemicals at the lab bench, placing material in a diamond anvil cell, or forcibly colliding high-energy, fast-moving beams in a particle accelerator. Observational science encompasses fields such as cosmology and paleontology, which treat subjects that are not commonly open to experimentation (but permissive of computer simulations). There are three basic approaches to experimental and observational science, illustrated here with some historic cases and examples from ordinary-chondrite studies. These approaches are not mutually exclusive: Many research projects combine more than one method, and some studies also include theoretical modeling.

Historians of science can trace the development of the post-Aristotelian scientific method to Francis Bacon (1561–1626), Lord High Chancellor of England under James I. In his 1620 book *Novum Organum* (*New Method*), Bacon argued that scientific knowledge could be achieved through a methodical, iterative process involving inductive reasoning and careful observation of natural phenomena.

1.1 Hypothesis Testing/Problem Solving

The heart of the scientific method is hypothesis testing: formulating conjectures, making predictions, and running experiments to test those predictions.

As professional skeptic Michael Shermer is fond of pointing out (e.g., Shermer, 2001), Charles Darwin wrote in 1861, "How odd it is that anyone should not see that all observation must be for or against some view if it is to be of any service." This is an endorsement of hypothesis testing. In statistics, hypothesis testing is an inferential method applied to analytical data to determine the probability that a given assumption is true – the method is used to check whether the results of a research study support a previously advanced theory about a population. More colloquially (and as affirmed by Darwin), hypothesis testing is an experimental approach to gathering data to evaluate a hypothesis.

An excellent example of this approach is the saga of the discovery of fossils of *Tiktaalik* – a bony fish that lived 375 Ma ago and was transitional between fish and land animals. The creature had a flat head, eyes on top, and a mobile neck separating the head from the shoulder. It had gills and scales like most fish, but its fins were unusual: They were shaped like cleavers. In addition to the typical thin ray bones of most fish, *Tiktaalik*'s fins contained sturdy bones corresponding to the upper arm, forearm, and a functional wrist joint, permitting the fish to push up against the streambed. Adult members of this species may have reached 2.5 m in length. Edward Daeschler and his colleagues were searching for just such a fish – the link between normal lobe-finned fish and tetrapods (four-limbed land vertebrates). They examined geological maps looking for exposed sedimentary rocks of the right age and type. This research led them to the Canadian Arctic, where, after several field seasons, they unearthed the first intact *Tiktaalik* specimens in 2004 (Daeschler et al., 2006).

Hypothesis testing also occurs in meteorite studies. Many FeO-rich porphyritic chondrules contain olivine phenocrysts a few hundred micrometers in size. If these grains crystallized during a single melting/cooling event, the chondrules would likely have remained hot for tens of minutes. Jones (1990) concluded that the FeO-rich chondrules in the Semarkona LL3.01 chondrite cooled at rates on the order of 1000 K hr^{-1}. Chondrules cooling at these moderate rates would likely exhibit distinctive geochemical traits. For example, Yu et al. (2003) studied the behavior of Na and K in synthetic chondrules produced in a vacuum furnace. They found that alkalis were lost by evaporation and that K isotopes were significantly fractionated in the synthetic chondrules; these objects were depleted in alkalis and enriched in δ^{41}K. It was an experimental observation ripe for hypothesis testing.

Alexander and Grossman (2005) measured 28 Semarkona chondrules with different textures and that possessed a wide range of bulk K/Al ratios. They found no systematic variations in K-isotopic composition and concluded that real chondrules probably did not form under the same conditions as those pertaining to the synthetic chondrules in the Yu et al. (2003) experiment. Real chondrules thus did not behave as relatively slow-cooling, open systems surrounded by gas at very low pressure. One might therefore conclude that (1) chondrules were not open systems, (2) the surrounding gas pressure was high (leading to solid-gas exchange), (3) chondrules cooled too quickly for K-isotopic fractionation to have occurred, and/or (4) subsequent parent-body alkali metasomatism altered the initial chondrule K-isotopic composition (which may have been fractionated). More experiments are needed to choose which alternatives are viable.

1.2 Serendipity

In an 1854 speech at the University of Lille, France, Louis Pasteur proclaimed that "in the fields of observation, chance only favors the mind which is prepared." There are numerous occasions on which experienced researchers, conducting an otherwise unrelated examination of a specimen or a dataset, have happened to notice an unusual feature that prompted further analysis. Exclamations akin to "That's odd" entered their thoughts and may have escaped their lips. A novice (or an unprepared mind) may have overlooked such anomalies.

Among the most famous examples of scientific serendipity is Wilhelm Röntgen's accidental discovery of X-rays in 1895. While attempting to determine if cathode rays could pass through glass, he noticed that an incandescent green light emitted by a cathode tube (shrouded in black paper) lit up a fluorescent screen. Röntgen embarked on a series of experiments and demonstrated that these unknown (or "X") rays would pass through many materials and produce shadows of dense solids. He received the first Nobel Prize in Physics in 1901.

On March 1, 1896, a few months after the discovery of X-rays, Henri Becquerel discovered (actually, rediscovered) radioactivity. Becquerel had been studying phosphorescence and wanted to test the idea that uranium salts could absorb sunlight and emit X-rays. He wrapped black paper over photographic plates to block the sunlight, put uranium salts on top of them, and placed the plates outside. When he developed the plates, Becquerel noticed the outline of the salt crystals; he deduced incorrectly that the uranium salts had absorbed sunlight and emitted X-rays or a similar type of penetrating radiation. However, during an extended period of cloudiness that precluded further experiments, he put everything away in a dark drawer. After developing this set of wrapped photographic plates, Becquerel found that the salt crystal outlines were clearly visible even though the uranium had not been in the sun. He realized that uranium salts emit radiation without external stimulation. Additional experiments showed that, unlike X-rays, uranium radiation was affected by electric and magnetic fields. By 1898, Marie and Pierre Curie had isolated other radioactive elements (polonium, thorium, radium) and dubbed the phenomenon "radioactivity." (Becquerel had been beaten to the punch, nearly four decades earlier, by photographer Claude Félix Abel Niépce de Saint-Victor. In 1857, Niépce de Saint-Victor found that, even when shrouded in complete darkness for six months, uranium salts could expose photographic emulsions. He stated in 1861 that these salts emitted a form of invisible radiation and reported his findings to the French Academy of Science [Niépce, 1861].)

Another momentous discovery occurred in 1928, when Alexander Fleming noticed that mold (*Penicillium rubens*) was growing on a Petri dish of *Staphylococcus* bacteria. The mold produced a chemical – an antibiotic – that killed the bacteria, constructing a bacteria-free annulus around itself. Fleming named the antibiotic substance "penicillin." He received the 1945 Nobel Prize in Physiology or Medicine.

Serendipity also plays a role in research and development. In the 1830s, self-taught chemist Charles Goodyear was trying to improve the rubber tube tires then in use. His experiments had yielded inconsistent results, until one day in 1839 he accidentally dropped a mixture of rubber and sulfur into a hot frying pan and discovered that the rubber became harder and more durable. He worked out a formal, replicable process for this hardening and called it vulcanization in honor of the Roman god of fire.

Serendipitous discovery occurs in all branches of science. While looking through the petrographic microscope at the Piancaldoli LL3.10 chondrite, I noticed a clast containing ~100 radial pyroxene microchondrules, 0.2–64 μm in apparent diameter (Rubin et al., 1982a). These objects ranged to far smaller sizes than any chondrules previously observed (the upper limit for the diameter of microchondrules was subsequently set at 40 μm [Rubin, 1989]). Later work showed the Piancaldoli "clast" was part of a rim around a normal-sized chondrule that had been largely excluded from the thin section during preparation. Microchondrules within chondrule rims were formed by the flash-melting of small portions of pyroxene phenocrysts at the inside chondrule margin (Krot and Rubin, 1996; Krot et al., 1997a; Bigolski et al., 2016).

Caillet Komorowski et al. (2012) were studying opaque aggregates in ordinary chondrites by field emission secondary electron microscopy when they happened upon a unique assemblage consisting of submicrometer-sized grains of cinnabar (HgS), Cu sulfides, and metallic Hg associated with fine-grained metallic Cu in the matrix of the Tieschitz H/L3.6 chondrite. They concluded that this complex assemblage had formed at low temperatures near the asteroid surface.

Experimental and theoretical studies have shown that the most abundant phase in the Earth's mantle is an $(Mg,Fe)SiO_3$ orthorhombic mineral with the perovskite crystal structure. But because no naturally occurring examples of this phase were in hand, the mineral could not be formally named. Tschauner et al. (2014) were examining shock melt veins in the Tenham L6 chondrite when they came across the first natural occurrence of the phase. They named it bridgmanite after Nobel laureate physicist Percy Bridgman (1882–1961), who had studied high-pressure effects on an assortment of materials.

As Alexander Graham Bell said in a 1914 speech, "Anyone, if he will only observe, can find some little thing he does not understand as a starter for an investigation." Chakraborty (2025) pointed out that although serendipity "has always been a major driver of scientific advance … one cannot rely on that." Other approaches are necessary.

1.3 Using New Analytical Tools or Examining New Samples

Axioms in philosophy are not axioms until they are proved upon our pulses.

John Keats, Letter to J. H. Reynolds, May 3, 1818

In a 2018 interview, NASA Apollo Sample Curator Ryan Zeigler noted that "Every decade since the Apollo samples came back has seen significant advances in instrumentation that have allowed samples to be analyzed at higher levels of precision, or smaller spatial resolution." Some lunar samples remained sealed for 50 years, awaiting analysis by more sophisticated techniques. The characterization of new samples can also be described as "fishing."

A month after Röntgen reported the discovery of X-rays in December 1895, physicians used X-rays to examine the human skeleton and internal organs. Within six months, they were treating skin rashes by radiation. These were the beginnings of diagnostic imaging and radiology.

Perhaps the most rapid deployment of a new tool that resulted in significant scientific breakthroughs relates to the invention of the telescope. Hans Lipperhey (c. 1570–1619) is often credited with inventing the telescope; although he filed a patent in 1608, he did not receive it due to competing claims. Galileo heard about the invention in 1609 and built his own astronomical telescope to scan the heavens. Over the next few years, he found imperfections in heavenly bodies, observing mountains and valleys on the Moon and dark spots on the Sun. These sunspots showed that the Sun rotates about once a month. This undermined the geocentric model of the Solar System: If the Sun could rotate on its axis, why not the Earth? Galileo found that Jupiter has four large moons revolving around it; this was also detrimental to the geocentric model because it demonstrated that some objects orbited a body other than Earth. As Jupiter moved through the sky, its moons were not left behind. (Proponents of the geocentric theory had reasoned that if Earth actually orbited the Sun, it would have left the Moon behind, stranded alone in space.) Galileo observed that Venus showed phases like the Moon's, consistent with its being an interior planet orbiting the Sun, as Copernicus had suggested in his heliocentric model. Galileo also observed "ears" or "handles" on Saturn (deduced several decades later by Christian Huygens to be rings). After turning his telescope to the Milky Way, Galileo found it studded with stars. This indicated that the stars were quite distant; thus, their parallactic shift would be too small to measure. (Many contemporary scholars believed the stars were not that far away. Proponents of the geocentric view pointed out, quite correctly, that if the Earth orbited the Sun, nearby stars should show a parallactic shift if their positions were measured six months apart when the Earth was on the other side of the Sun.)

Within a few years of the development of the electron microprobe, meteorite researcher Klaus Keil and his colleagues used this instrument to analyze numerous phases in ordinary chondrites and other meteorites. Keil and Fredriksson (1964) analyzed olivine and orthopyroxene in 86 ordinary chondrites and found that each chondrite group had narrow ranges in the FeO/(FeO+MgO) ratios of these two minerals; they also identified LL chondrites as an ordinary-chondrite group distinct from L (only H and L ordinary chondrites had been recognized previously; see Urey and Craig, 1953). Bunch et al. (1967) and Snetsinger and Keil (1969) analyzed chromite and ilmenite, discovering systematic compositional variations in these phases among H, L, and LL chondrites, commensurate with whole-rock changes in oxidation state.

The category of new or underutilized analytical tools is not confined to instruments; it could also include statistical analysis. Benford's law is a scale- and base-invariant probability distribution that has been applied to many physical phenomena (e.g., surface areas of rivers and islands, distances between galaxies, durations of volcanic eruptions, speeds of seismic P waves in the upper mantle, reversals of the geomagnetic field). Deviation of a population from Benford's law may suggest the sample distribution has been modified (Sambridge et al., 2010). Rubin (2021) applied Benford's law to ordinary-chondrite mass distributions and found it pertained to the set of individual objects that survived intact on the Earth's surface after atmospheric disruption of meteoroids. Deviations from the law among different data sets resulted from terrestrial weathering and from the nonsystematic collection of samples by meteorite hunters influenced by commercial considerations.

New meteorite falls and finds also offer opportunities for investigators to use extant analytical tools to determine what the new samples have to offer. For example, two H-chondrite regolith breccias that fell in 1998 (Zag in the Western Sahara; Monahans (1998) in Texas) were found to contain millimeter-sized aggregates of blue-to-purple halite (NaCl); the halite grains in Monahans (1998) included crystals of sylvite (KCl) (e.g., Zolensky et al., 1999a, 1999b; Rubin et al., 2002). The halite crystals in both meteorites are ~4.5 Ga old and contain numerous inclusions of aqueous fluids, solids, and organic material. The H5-an chondrite, Sidi El Habib 001, which was recovered in Algeria in 2021 (and may be related to a fireball observed in the area on December 20, 2020), also contains blue-to-purple grains of halite and minor sylvite. Zolensky et al. (2016) suggested the halite grains in these meteorites originated on the asteroid Ceres. (Indeed, sylvite and halite of likely extraterrestrial origin were reported among particles retrieved from asteroid 25143 Itokawa [Noguchi et al., 2014]. This is an S-type asteroid of LL-chondrite composition.)

Because CAIs are moderately abundant in carbonaceous chondrites but very rare (and small) in ordinary chondrites (e.g., MacPherson et al., 1988), Russell et al. (1996) employed optical and electron microscopy to hunt down a few ordinary-chondrite CAIs. McKeegan et al. (1998) used an ion microprobe to determine the oxygen-isotopic compositions of these ordinary-chondrite (OC) calcium-aluminum-rich inclusions (CAIs) and found them similar to carbonaceous-chondrite CAIs. These researchers concluded that most CAIs formed in a small region of the solar nebula before they were heterogeneously distributed by aerodynamic processes.

1.4 Mélange

Theoretical work, hypothesis testing, use of new instruments, and serendipity all played roles in the development of Big Bang cosmology. Serendipity: While examining variable stars on photographic plates, astronomer Henrietta Swan Leavitt found in 1912 that one type of star, called a Cepheid variable, exhibits a strong positive correlation between intrinsic luminosity and the period between consecutive episodes of peak brightness. This relationship can be used as a "standard candle," allowing astronomers to estimate distances to galaxies and nebulae from the periods of the Cepheid variables inside them. Use of New Instruments: Using the new 100-inch (2.5 m) Hooker Telescope at Mount Wilson, California in 1924, astronomer Edwin Hubble identified Cepheid variables in the Andromeda Nebula and Triangulum Nebula, demonstrating that these nebulae are actually distant galaxies, not part of the Milky Way.

Theory and Modeling: In 1927, Georges Lemaître, a cosmologist (and Catholic priest), published a solution to the equations of general relativity and predicted the expansion of the universe. Observational Science: Hubble measured the distances to 24 galaxies in 1929 and found that the farther these galaxies were from the Milky Way, the faster they were receding. Theory and Modeling: From these observations, Lemaître proposed that the universe had originated at a finite time in the distant past from an extraordinarily dense region of small spatial extent (a "primeval atom"). The explosion of this "atom" caused the expansion of the universe.

Theory and Modeling: In 1953, theoretical physicist George Gamov realized that the universe must have cooled as it expanded; he predicted that there would be a cosmic background temperature a few degrees above absolute zero. His estimate was 7 K. Hypothesis Testing: More than a decade later, astronomer Robert Dicke and his associates independently predicted the existence of the cosmic microwave background (CMB) radiation and began to build a microwave radiometer to search for it. (Shortly after the Big Bang, the universe was extremely hot and consisted of a quark-gluon plasma – a dense soup of subatomic particles. Photons interacted with electrons and could not travel far before being absorbed; the universe was effectively opaque. After about 380,000 years, as the universe continued to expand, it had cooled sufficiently to allow protons to combine with electrons and form neutral atoms [mainly hydrogen]. Photons could then travel freely; the universe became transparent. The CMB is the remnant echo of this epoch.)

Use of New Instruments and Serendipity: Before Robert Dicke could construct his radiometer, he received word that the CMB had been discovered accidentally by two employees of Bell Labs: physicist Arno Penzias and radio-astronomer Robert Wilson. Penzias and Wilson had been experimenting with a new, extremely sensitive, 6-m horn antenna. After they eliminated all recognizable sources of interference from their receiver (including pigeon droppings), there remained a uniform noise emanating from across the sky with an estimated temperature of 3.5 K (the modern value is 2.725 K). This was the CMB. In 1978, Penzias and Wilson were awarded the Nobel Prize in Physics for their discovery.

Use of New Instruments: Three spacecraft, the Cosmic Background Explorer (COBE) (launched in 1989), the Wilkinson Microwave Anisotropy Probe (WMAP) (launched in 2001), and the Planck satellite (launched in 2009) mapped the CMB and found it remarkably homogeneous. Temperature differences between warmer and cooler spots were at the level of only about one part per million. John Mather and George Smoot shared the 2006 Nobel Prize in Physics for the COBE discoveries. Theory and Modeling: This high degree of homogeneity in the CMB can be accounted for by the theory of cosmic Inflation. This is the idea that, at the earliest stages of universal history, starting at 10^{-36} seconds after the Big Bang and ending sometime between 10^{-33} and 10^{-32} seconds after the Big Bang, the universe increased immensely in volume (by a factor of $>10^{78}$), smoothing out its internal structure. Tiny quantum fluctuations in the inflation field were greatly magnified during cosmic inflation and ultimately produced the modern universe – a structure consisting of enormous rotating galaxy-laden filaments (hundreds of millions of lightyears in length) built upon strands of dark matter, clusters of galaxies stationed at filament intersections, and huge interstitial voids with few galaxies within them.

Meteorite studies incorporate a mélange of scientific approaches. Observations are explained by theoretical models; models are subject to hypothesis testing; new instruments are used to examine specimens in ever finer detail; and occasionally, a strange new sample or an unexpected observation comes to light that demands investigation.

Observations and Data Analysis: Using published data, Warren (2011a, 2011b) showed that carbonaceous chondrites (CC) and noncarbonaceous chondrites (NC) (mainly ordinary and enstatite chondrites) plot in widely separated regions in diagrams of O-, Ti-, Cr-, and Ni-isotopic composition. He also found that IIIAB irons, main-group pallasites, HED meteorites, mesosiderites, angrites, and ureilites are all isotopically related to NC chondrites, whereas Eagle-Station pallasites are isotopically related to CC chondrites. Use of New Instrumental Techniques: Budde et al. (2016, 2017, 2018) used MC-ICP-MS (multicollector inductively coupled plasma mass spectrometry) to show that CC and NC chondrites have distinct ^{95}Mo and ^{94}Mo concentrations. Theory and Modeling: Warren (2011a, 2011b) concluded that CC and NC chondrites formed within different isotopic reservoirs of the solar nebula and suggested that CC chondrites formed outside the orbit of Jupiter. Use of New Instrumental Techniques: By examining the Mo-isotopic compositions of iron meteorites, Budde et al. (2016, 2018), Kruijer et al. (2017a, 2017b, 2017c), and Poole et al. (2017) determined that iron meteorites could be divided into two classes – those derived from carbonaceous-like precursors (CC irons) and those from noncarbonaceous precursors (NC irons). Among magmatic iron groups (i.e., those formed via fractional crystallization within the well-mixed molten metal cores of differentiated asteroids), CC irons include groups IIC, IID, IIF, IIIF, and IVB, while NC irons include groups IC, IIAB, IIIAB, IIIE, and IVA. Theory and Modeling: There is an evolving consensus that the CC magmatic irons formed in the cores of differentiated asteroids residing in the Outer Solar System beyond Jupiter, while the NC magmatic irons formed from differentiated asteroids in the Inner Solar System. Kruijer et al. (2017a, 2017b, 2017c) suggested that proto-Jupiter accreted rapidly between the CC and NC regions and exceeded 20 Earth masses ($M_\oplus$) within ~1 myr after CAI formation. Proto-Jupiter functioned as a barrier between the CC and NC reservoirs by impeding the radial drift of small particles toward the Inner Solar System. These small particles likely included CAIs, which appear to have concentrated in dust-rich nebular regions (Rubin, 2011) located outside Jupiter's orbit (Gerber et al., 2017; Scott et al., 2017). Proto-Jupiter continued to grow by accreting gas onto its core; by 3–4 Ma after CAI formation, it had grown to ~50 $M_\oplus$ and opened a gap in the nebular disk (Morbidelli et al., 2016), allowing particles to drift inward.

Hypothesis Testing: Although CAIs occur in all chondrite groups, their modal abundance is much higher in carbonaceous-chondrite groups (0.6–4 vol.%, excluding CI chondrites) than in noncarbonaceous-chondrite groups (0.01–0.04 vol.%). Many CAIs are enriched by factors of 10–20 in refractory siderophile elements (RSEs) (e.g., Re, Os, Ru, Ir, Pt) relative to bulk CI chondrites. A significant fraction of RSEs in CAIs is present in 0.05–5 μm-sized refractory metal nuggets that consist of pure metals or metal alloys. These nuggets appear to have crystallized within CAIs. A reasonable hypothesis is that CC iron cores inherited RSEs from CAIs bearing melted refractory metal nuggets when their carbonaceous-chondrite-like precursors melted during asteroid differentiation. Consistent with this model, Rubin (2018a) found that, among magmatic iron groups, the median Ir/Ni contents of CC irons (1.22×10^{-4}) is three times higher than that of NC irons (0.402×10^{-4}).

Hypothesis Testing: Because carbonaceous chondrites are more oxidized than noncarbonaceous chondrites and have higher bulk FeO/Fe^0 ratios, one might expect that the cores of differentiated CC asteroids would be proportionately smaller than those of differentiated NC asteroids. In general, magmatic CC iron groups should have significantly fewer members than magmatic NC iron groups. This expectation is borne out: The February 23, 2023 edition of the Meteoritical Bulletin Database shows that the five CC magmatic iron groups have an average of 13.8 ± 9.7 members each, whereas the five NC magmatic iron groups have an average of 122 ± 136 members each. These numbers exclude undocumented and discredited samples and do not take possible meteorite pairings into account. Wiley and Fitzwater Pass were omitted in this calculation from groups IIC and IIIF, respectively, because recent data show significant isotopic differences from other members of their assigned groups.

2

Meteorite Classification

Although meteorite classification has been reviewed several times in recent years (e.g., Krot et al., 2005a; Weisberg et al., 2006; Rubin and Ma, 2021), the system becomes more complex as more meteorites are recovered and analyzed. Unique specimens are matched with congeners, and grouplets evolve into groups (e.g., Bischoff, 2001; Burbine et al., 2002).

2.1 Taxonomy

Meteorite classification is the systematic organization of meteorites into levels of increasing complexity: unique sample, grouplet, subgroup, group, clan, class, superclass, and domain – the latter term corresponds to the degree of differentiation of the parent body. These categories encompass all meteorites, but only certain meteorites belong to a grouplet, subgroup, or superclass (Fig. 2.1).

The fundamental unit of meteorite classification is the *group*. Convention (stemming from Wasson, 1985) dictates that a group should include at least five members. Members of a single meteorite group are assumed to have originated on the same parent body (although this assumption can be questioned). Group members span limited ranges in their texture, mineralogy, mineral chemistry, and bulk chemical and isotopic composition. *Texture*: For example, the majority of the chondrules in LL3 chondrites has an apparent diameter within a factor of two of the group average (Nelson and Rubin, 2002). The difference between the highest and lowest values in the mean apparent diameter of chondrules (excluding microchondrules) in different LL3 chondrites is ~90 μm (Nelson and Rubin, 2002), whereas the difference among all major chondrite groups is ~750 μm (Rubin, 2010a; Friedrich et al., 2015). *Mineralogy*: The principal minerals in all type-3 ordinary-chondrite falls are olivine, low-Ca pyroxene (major clinopyroxene and in some cases minor, accessory, or trace orthopyroxene), low-Ni metallic Fe (kamacite), taenite, and troilite. (Kamacite and taenite both derive their mineral names from Greek words describing their shapes: *kamas* [pole or shaft] and *tainia* [ribbon].) Chondrules in the type-3 OC contain glassy and microcrystalline mesostases that may include Ca-pyroxene crystallites, pigeonite, and/or rare tiny grains of merrillite (in addition to major olivine ± low-Ca pyroxene, and in many cases, some metallic Fe-Ni and/or troilite). Minerals that form during thermal metamorphism in ordinary chondrites include orthopyroxene, Ca-pyroxene, plagioclase, phosphate (merrillite and chlorapatite), chromite, ilmenite, and rutile. *Mineral chemistry*: The difference between the highest and lowest values for the mean olivine Fa content among typical equilibrated chondrites in individual OC groups is ~3–6 mol%, whereas among all typical equilibrated H, L, and LL chondrites it is ~15 mol% (Rubin, 1990). (The merrillite in all three OC groups is of similar composition [Lewis and Jones, 2016].) *Bulk chemical composition*: The difference between the highest and lowest values for the CI-normalized Ir/Ni ratios of H-chondrite falls (0.17), L-chondrite falls (0.22), and LL-chondrite falls (0.26) is appreciably less than that of all OC falls (0.56) (Kallemeyn et al., 1989). *Bulk isotopic composition*: The difference between the highest and lowest values of bulk $\Delta^{17}O$ among equilibrated chondrites in each OC group is ~0.3–0.5‰, but

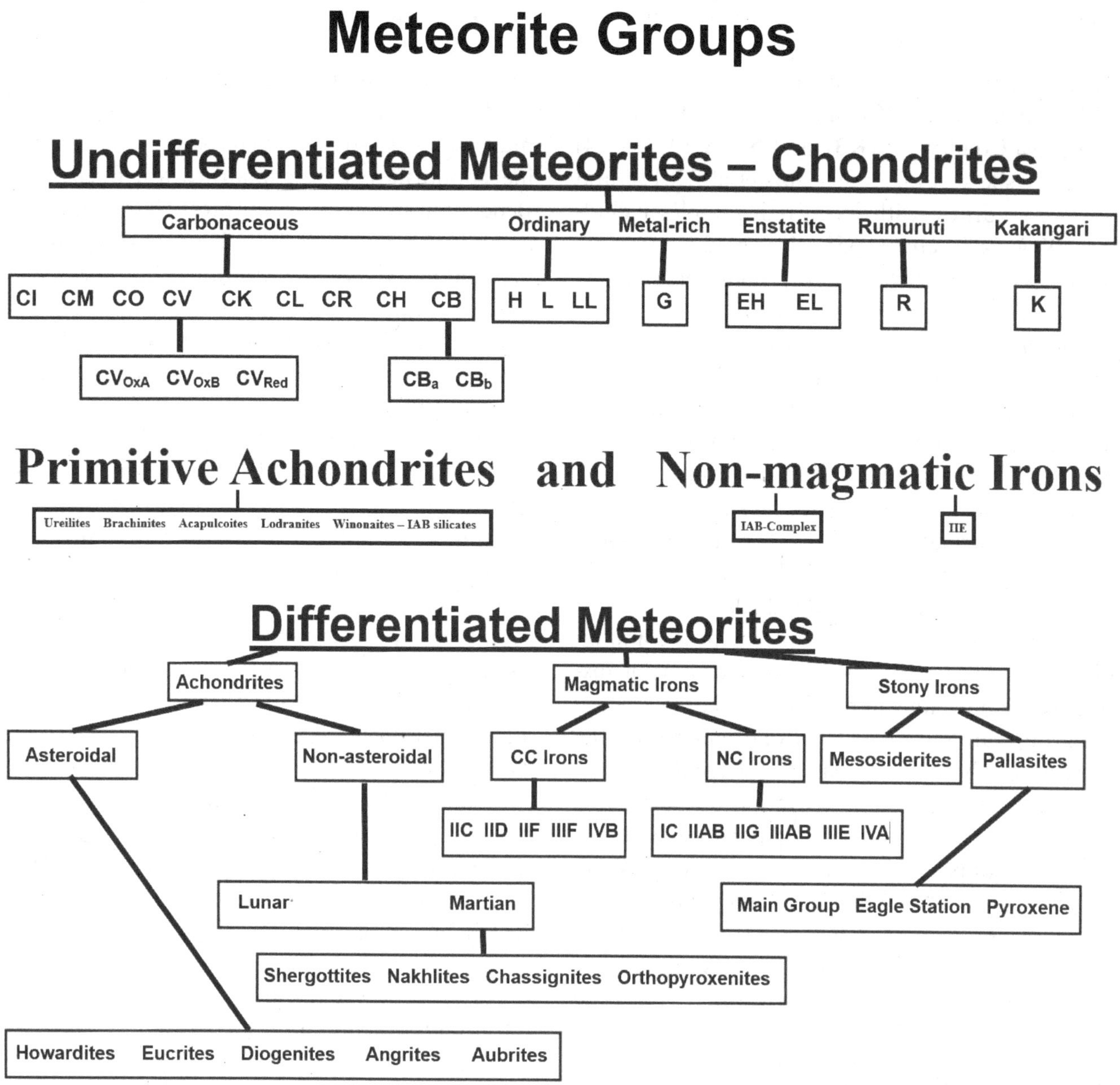

Fig. 2.1 Classification of meteorite groups. Meteorites can be divided into three broad categories: undifferentiated meteorites (chondrites), primitive achondrites and non-magmatic irons, and differentiated meteorites (achondrites, magmatic irons, stony-irons).

among all equilibrated H, L, and LL chondrites it is $\sim$0.9‰ (Clayton et al., 1991; Greenwood et al., 2017).

The May 20, 2025 Meteoritical Bulletin Database lists 63 ordinary chondrites designated H/L or H(L). Individuals with this classification could be anomalous H chondrites, anomalous L chondrites, or members of an intermediate group derived from a separate H/L asteroid. The list also includes 162 ordinary chondrites designated L/LL or L(LL); individual samples could be anomalous L chondrites, anomalous LL chondrites, or members of an intermediate group derived from a separate L/LL asteroid.

A collection of two to four meteorites that spans a limited range of petrologic and chemical properties is a *grouplet*. Examples of grouplets are (a) the South Byron Trio of iron meteorites (comprising

South Byron, Babb's Mill (Troost's Iron), and Inland Forts 83500) (Hilton et al., 2019; McCoy et al., 2019); (b) Kakangari chondrites, currently with four members: Kakangari (K3), LEW 87232 (K3), Lea County 002 (K3), and NWA 10085 (K4) (e.g., Weisberg et al., 1996; Utas et al., 2017); and (c) G chondrites, with metal-rich samples including NWA 5492, GRO 95551, NWA 12379 (and its paired specimen, NWA 12273), and Sierra Gorda 009 (e.g., Ivanova et al., 2019, 2020).

Unique and ungrouped meteorites tend to join congeners and become grouplets, while grouplets tend to turn into groups as more meteorite specimens are recovered. And they are recovered in great numbers. For example, the Meteoritical Bulletin Database lists 2141 new meteorites that were classified and formally approved for the three-year inclusive period 2020–2022; the list includes 58 ungrouped or anomalous individuals.

A *subgroup* is a set of samples with a common group designation, but with each set having distinctive textural and mineralogical characteristics. Subgroups are presumed to have originated on the same parent body but to have subsequently experienced different modes of alteration on that body. Two carbonaceous-chondrite groups have recognized subgroups: (1) CV chondrites (CV_{OxA}, CV_{OxB}, CV_{Red}) and (2) Bencubbin-like CB chondrites rich in metallic Fe-Ni (CB_a, CB_b).

CV_{OxA} chondrites are oxidized CV chondrites of the Allende subgroup. They contain very little bulk H_2O and only rare phyllosilicate grains in the matrix. Common secondary phases include Ni-rich metallic Fe (tetrataenite and awaruite), pentlandite, magnetite, nepheline, sodalite, and hedenbergite. $CV3_{OxB}$ chondrites are oxidized CV chondrites of the Bali subgroup. They contain several phyllosilicates: Fe-bearing saponite, saponite, Na phlogopite, Al-rich serpentine, and Na-K mica. Also present are secondary magnetite, carbonates, Ca-phosphates, Ni-rich sulfides, fayalite, Ca-Fe-rich pyroxene, and andradite. $CV3_{Red}$ chondrites are reduced CV chondrites; their matrix regions contain major olivine and minor-to-accessory pyrrhotite, low-Ca pyroxene, metallic Fe-Ni, and hydroxylapatite. Some CV chondrites are breccias that comprise more than one subgroup.

CB_a chondrites are coarse-grained Bencubbin-like CB chondrites characterized principally by centimeter-sized ellipsoidal nodules rich in metallic Fe-Ni with minor-to-accessory troilite, and multimillimeter-to-centimeter-sized skeletal olivine chondrules. CB_b chondrites are finer-grained Bencubbin-like CB chondrites characterized by compositionally zoned and unzoned grains of metallic Fe-Ni, common refractory inclusions, and hydrated lithic clasts containing magnetite, sulfide (pentlandite and pyrrhotite), complex Fe-Mn-Mg-Ca carbonates, and phyllosilicate (serpentine and saponite).

The classification order higher than the group is the *clan*. Clans typically consist of two or more groups derived from materials that formed in close proximity. For chondrites, this proximity refers to separate asteroids accreting in the nebula; for achondrites, primitive achondrites, and irons, this proximity typically applies to distinct regions of the same differentiated or partially differentiated body. Clan members share bulk chemical, bulk isotopic, and mineralogic and/or textural characteristics. The chondrite clans comprise five noncarbonaceous sets: (1) H, L, and LL ordinary chondrites (along with H/L and L/LL samples, and I argue, IIE irons), (2) Rumuruti chondrites (R chondrites), (3) EH and EL enstatite chondrites, (4) G chondrites, and (5) Kakangari chondrites (K chondrites). Chondrite clans also include four carbonaceous sets: (1) CM-CO, (2) CV-CK-CL, (3) CR-CH-CB, and (4) CI. Primitive achondrite clans include: (1) ureilites, (2) acapulcoites and lodranites, (3) brachinites and related ungrouped meteorites (NWA 6112, MIL 090206, Divnoe), and (4) winonaites and IAB silicate inclusions (and perhaps tissemouminites; Stephant et al., 2023). Achondrite clans include: (1) Martian meteorites, (2) howardite–eucrite–diogenite (HED) samples, (3) lunar meteorites, (4) aubrites, and (5) angrites. Each magmatic iron group is its own clan except for the IIAB-IIG clan (chemical groups that were likely derived from the same asteroidal core; Wasson and Choe, 2009). The non-magmatic irons consist of two clans: (1) the IAB Complex (which might hail from more than a single asteroid; Wasson and Kallemeyn, 2002) and (2) IIE (which are related to ordinary chondrites; Rubin, 2022a).

Many clans include ungrouped meteorites. For example, Ningqiang (Rubin et al., 1988a; Kallemeyn et al., 1991; Wang and Hsu, 2009) is a member of the CV-CK-CL clan; Acfer 094 (Greshake, 1997; Choe et al., 2010) is a member of the CO-CM clan; and Flensburg (Bischoff et al., 2021) and LEW 85332 (Rubin and Kallemeyn, 1990; Brearley, 1997) are members of the CI clan.

The order directly above clan is *class*, comprising one or more clans that formed in the same general region of the Solar System. The members of a class may resemble one another in one or more of their whole-rock properties, such as texture, mineralogy, inclusion abundance, oxidation state, bulk chemistry, and isotopic composition. There are four major chondrite classes: (1) ordinary chondrites (including chondritic inclusions in IIE irons), R chondrites, and G chondrites, (2) enstatite chondrites, (3) carbonaceous chondrites, and (4) K chondrites. There are two major classes of iron meteorite: (1) those derived from carbonaceous chondrites (CC irons) (IIC, IID, IIF, IIIF, IVB) and (2) those derived from noncarbonaceous chondrites (NC irons) (IAB, IC, IIAB, IIE, IIG, IIIAB, IIIE, IVA). Pallasites are a class consisting of (1) main-group pallasites, (2) Eagle Station pallasites, and (3) pyroxene pallasites. Mesosiderites constitute a separate meteorite class (although they may be related to a few dunite-rich achondrites such as NWA 2968 and NWA 3329 [Greenwood et al., 2015]).

Rubin and Ma (2021) introduced the term "superclass," an order higher than class. Members of a superclass are derived from different asteroids that presumably accreted in the same general locale of the Solar System.

The first superclass is that of enstatite meteorites: (1) EH and EL chondrites, (2) aubrites (enstatite achondrites) originating from more than one parent body, and (3) an assortment of ungrouped and anomalous enstatite chondrites, reduced metal-rich meteorites, and reduced impact-melt breccias and impact-melt rocks (Rubin, 2015a).

The second superclass contains differentiated meteorites with similar bulk O-isotopic compositions: (1) HED achondrites (eucritic basalts, diogenitic orthopyroxenites, and fragmental howardite breccias composed mainly of eucrite and diogenite clasts), (2) mesosiderites (metal-rich samples containing eucritic and diogenitic clasts), (3) rare dunite-rich samples, (4) main-group pallasites (consisting of metallic Fe-Ni and cumulate magnesian olivine), and (5) IIIAB iron meteorites. All these samples formed from material derived from similar O-isotopic reservoirs in the nebula.

There is a consensus that the majority of HED achondrites were derived from asteroid 4 Vesta (mean diameter 525 km). This is based on several factors:

(1) HED meteorites share spectral similarities to (a) Vesta (e.g., McCord et al., 1970; Consolmagno and Drake, 1977; McSween et al., 2013), (b) numerous small, spectrally related asteroids nicknamed "vestoids" (Binzel and Xu, 1993; Binzel, 2001; Burbine et al., 2001), and (c) some Near-Earth Asteroids (NEAs) including 3551 Verenia, 3908 Nyx, and 4055 Magellan (e.g., Cruikshank et al., 1991).
(2) The Dawn mission showed Vesta's characteristics to be consistent with its being the parent body of HED samples. These features include Vesta's geologic diversity, extent of space weathering, impact history, chronology of surface units, presence of carbonaceous material, and inferred core dimensions (McSween et al., 2013).
(3) The Rheasilvia crater (diameter 505 km) in the southern hemisphere of Vesta is a plausible source for the excavation and ejection of the vestoids (e.g., McSween et al., 2013).
(4) The orbits of vestoids extend between the orbit of Vesta and gravitational escape hatches (such as the Jovian 3:1 resonance at ~2.5 AU and the ν6 resonance) that can perturb vestoids toward the Inner Solar System (Wisdom, 1985; McSween et al., 2013).
(5) The computed orbits of meteorite falls (including the anomalous basaltic meteorite Bunburra Rockhole; Bland et al., 2009) resemble those of Earth-crossing asteroids (e.g., Borovička et al., 2015).

Although mesosiderites contain eucrite-like and diogenite-like clasts, these metal-rich breccias are probably not from the same parent body as the HEDs. While about 40% of large silicate clasts in mesosiderites are gabbros with extremely fractionated rare earth element (REE) abundances (Mittlefehldt et al., 1992), this rock type has not been reported in HED meteorites (Rubin and Mittlefehldt, 1992, 1993). Because Vesta is still intact, the IIIAB irons and main-group pallasites (which have similar, but not identical, bulk O-isotopic compositions; e.g., Greenwood et al., 2015) must have been derived from one (or more likely, two; Yang et al., 2010) separate differentiated asteroids that were collisionally disrupted hundreds of millions of years ago (based on their cosmic-ray exposure [CRE] ages; e.g., Eugster, 2003).

The only higher-order classification level consists of domains of meteorites that experienced different degrees of differentiation: (a) differentiated meteorites (achondrites, iron meteorites, stony-irons), (b) undifferentiated meteorites (chondrites), and (c) partially differentiated meteorites (primitive achondrites). The origins of primitive achondrites are controversial. For example, chondrule-bearing samples of some acapulcoites (Yanai and Kojima, 1991; McCoy et al., 1996; Rubin, 2007) and winonaites (Benedix et al., 2003) texturally resemble chondritic impact-melt breccias. They may share such an origin (Rubin, 2007) or be products of partial differentiation in internally heated asteroids (e.g., McCoy et al., 1996; Hunt et al., 2017; Keil and McCoy, 2018).

As of this writing, there are seven groups of meteorites that can be assigned to specific parent bodies: (1) lunar meteorites from the Moon (e.g., Keil and Papike, 1983), (2) Martian meteorites from Mars (e.g., Treiman et al., 2000), (3) HED meteorites from asteroid 4 Vesta (e.g., McSween et al., 2013), (4) more than 50 specimens of the Almahata Sitta polymict ureilite from asteroid 2008 TC_3 (fell October 7, 2008) (Jenniskens et al., 2009), (5) about two dozen specimens of the Motopi Pan howardite from asteroid 2018 LA (fell June 2, 2018) (Jenniskens et al., 2021), (6) more than 30 specimens of the Saint-Pierre-le-Viger L5-6 chondrite from asteroid 2023 CX_1 (fell February 13, 2023) (e.g., Antier, 2023; Bischoff et al., 2023), and (7) a few dozen samples of the Ribbeck aubrite from asteroid 2024 BX_1 (fell January 21, 2024) (Bischoff et al., 2024a).

In addition, it is possible that H chondrites come from asteroid 6 Hebe (mean diameter 195 km) (Farinella et al., 1993a; Migliorini et al., 1997; Gaffey and Gilbert, 1998; Bottke et al., 2010), but 3D models of Hebe may be inconsistent with this hypothesis (Marsset et al., 2017). Brož et al. (2024) proposed that H chondrites hail from members of the Karin and $Koronis_2$ subfamilies of the large Koronis family (~6000 members). L chondrites could be from the Gefion asteroid family (e.g., Nesvorný et al., 2009) or, more likely, the Massalia asteroid family (Marsset et al., 2024). LL chondrites could be derived from 8 Flora (mean diameter 146 km) (Vernazza et al., 2008; Dunn et al., 2013).

A minority of micrometeorites (especially those with ultracarbonaceous compositions) is derived from comets. Some individual micrometeorites can be linked with high probability to specific comets. Messenger (2002) noted that cometary outbursts could produce high concentrations of particles entering the Earth's atmosphere. He predicted that these particles would have very short space-exposure ages; they would neither contain noble gases implanted by the solar wind nor bear particle tracks from solar flares. NASA subsequently conducted "timed collections" to recover stratospheric particles from comets 26P/Grigg-Skjellerup (orbital period 5.31 years) and 73P/Schwassmann-Wachmann (orbital period 5.44 years) (Trigo-Rodríguez and Blum, 2022). The properties of these particles matched Messenger's predictions.

Although the abundances of most non-volatile elements in CI chondrites are close to those in the solar photosphere (Fig. 3.1), meteorites do not come from the Sun. The pre-Socratic Greek philosopher Anaxagoras of Clazomene (c. 500–c. 428 BCE) was said to have predicted that, on a particular day, a stone would fall from the Sun, but Pliny the Elder (23–79 CE) was skeptical of the claim. He wrote in book 2, chapter 60 of the *Natural History* that "our knowledge of the nature of things, and indeed everything else, would be thrown into confusion were we to suppose either that the Sun is itself composed of stone, or that there was even a stone in it" (Bostock and Riley, 1855).

Even though tiny presolar grains within primitive chondrites originated outside the Solar System, no meteorites are known to have such an origin. Nevertheless, a fireball (dubbed CNEOS 2014-01-08) produced by a 500-kg meteoroid near Papua New Guinea on January 8, 2014 appears to have entered Earth's atmosphere on a hyperbolic trajectory at about 44 km s^{-1}; this exceeds the escape velocity of the Solar System at 1 AU. In 2023, Harvard astrophysicist Avi Loeb and his expedition team recovered hundreds of small metallic spherules from the ocean floor off the coast of Papua New Guinea. A few of these spherules are rich in Be, La, and U. As of this writing, most researchers doubt that these objects are extraterrestrial, let alone interstellar. In any case, with uncertainties about the trajectory, velocity, and tensile strength of the meteoroid and the limitations of available reduction tools for meteor observations, Vaubaillon (2022) concluded that evidence was insufficient to confirm interstellar provenance of the meteoroid.

No meteorites from Mercury, Venus, the moons of Mars, the moons of the giant planets, or Kuiper Belt objects are currently known.

Ancient giant impacts must have blasted terrestrial rocks off the Earth and onto the Moon. There may be up to 200 kg of terrestrial material per km^2 on the lunar surface (Armstrong et al., 2002), but no such samples have yet been identified. Although Bellucci et al. (2019) suggested that a 1.8-g felsite clast within Apollo 14 breccia 14321 was derived from a terrestrial meteorite launched from Earth ~4 Ga ago, Warren and Rubin (2020) pointed out that the sample is not heavily shocked (as would be expected if it had been launched from Earth) and contains concentrations of volatile trace metals (Zn, Ga, Ge) and refractory elements (e.g., Ta, Lu) typical of highly evolved lunar felsites and generally much lower than those in terrestrial rocks. They concluded that the small felsite clast was lunar after all.

Debaille et al. (2022) speculated that the ungrouped achondrite NWA 13188 could be terrestrial impact ejecta that made it into Earth orbit before coming back down. The rock is a basaltic andesite with a fusion crust. Its O-isotopic composition is close to the terrestrial fractionation line ($\Delta^{17}O = -0.02‰$) and its REE pattern and Nd-isotopic composition are in the terrestrial range. However, NWA 13188 is not heavily shocked, casting doubt on the possibility that it was blasted into space from the Earth (Rubin, 2015c). Its ^{10}Be concentration (0.028 dpm kg^{-1}) is very low for a meteorite (commonly 10–20 dpm kg^{-1}), suggesting that if this rock was ever in orbit, it remained there for no more than a few thousand years (unless it was heavily shielded). Such a brief sojourn implies a large terrestrial cratering event in historical times. Modeling by O'Keefe and Ahrens (1985) shows that a crater with a 50-km diameter on an atmosphereless Earth would be necessary for 17-m ejecta blocks to reach escape velocity. (This is close to the pre-atmospheric size of the Chelyabinsk meteoroid [Popova et al., 2013].) There is no known multi-kilometer impact crater on Earth with such a young age. If NWA 13188 is indeed a meteorite (rather than a terrestrial rock with a darkened surface), it is more likely from a heretofore unsampled differentiated asteroid.

2.2 Methods of Meteorite Classification

Before meteorites are officially approved and listed in the Meteoritical Bulletin, they must be properly classified. The steps involved in classification require answering more than a dozen specific questions, although not every question is pertinent to every sample: (1) Is the sample a meteorite? (2) Is it a find or a fall? (3) How many paired individuals were recovered? (4) What is the total known mass? (5) Is the meteorite a chondrite, primitive achondrite, achondrite, stony-iron, or iron? (6) Does the sample fit into a well-established group? (7) If it is a member of a group, is it compositionally and/or texturally anomalous? (8) If not a group member, is the sample compositionally related to any other meteorites? (9) Is the sample brecciated? If so, what are the included lithologies? (10) Does the sample contain rare gases implanted by the solar wind? (11) What is the extent, if any, of aqueous alteration? (12) To what degree has the sample been thermally metamorphosed? (13) What is the extent of shock metamorphism? and (14) What is the degree of terrestrial weathering?

The techniques involved in meteorite characterization should include field data collection, naked-eye and stereomicroscopic inspection of whole-rock samples, and, in most cases, optical microscopy (transmitted and reflected light) and mineral identification, and mineral analysis (commonly by electron microprobe [EPMA]). Another possible classification scheme for ordinary chondrites is based on the relative areas of Mössbauer spectral components (Goryunov et al., 2025).

Initial characterization could be done using magnetic susceptibility measurements, gauging a sample's response to a magnetic field. Terrestrial rocks generally contain no free metal and different meteorite varieties have distinct average metal contents (e.g., Rochette et al., 2003, 2008a, 2009, 2012; Consolmagno et al., 2006; Gattacceca et al., 2004). Uehara and Gattacceca (2023) developed a handheld device that measures magnetic susceptibility and electrical conductivity, useful for the identification of meteorites in the field or at outreach events.

Nath et al. (2025) showed that nondestructive reflectance spectroscopy coupled with a deep-learning model was able to classify meteorites with a 93% accuracy rate. They examined a plethora of spectral features including the characteristics of absorption features and any associated inflections along with the slope of the continuum. However, at the current state of model refinement, ordinary chondrites and ureilites are lumped together, as are CM, CR, and C2 chondrites, CO and CV chondrites, and HED samples.

Although there are handheld x-ray fluorescence (XRF) devices that have been used in attempts to measure meteorite bulk compositions, most of these instruments are designed (or programmed) to analyze metal ores. Their analyses commonly omit the lithophile elements that constitute the major components of silicate-rich rocks (including stony meteorites). As Randy Korotev (Washington University, St. Louis) has pointed out, in many cases, these instruments detect the K_β *or* L_α emission lines of common elements and incorrectly identify them as the principal K_α emission lines of trace elements (Korotev, n.d.). Although the devices can detect Fe, Ni, and Co in iron meteorites, the reported compositions typically are inaccurate. Handheld XRF devices are not recommended for meteorite identification.

Oxygen isotopes are often determined during meteorite characterization, mainly via laser-assisted fluorination, secondary ion mass spectrometry (SIMS), or UV laser ablation. Great precision can be achieved with the use of ultra-high resolution mass spectrometers – as pointed out by Greenwood et al. (2017), samples as small as 0.5 to 2 mg can be analyzed with a precision of at least ±0.08‰ for $\delta^{17}O$, ± 0.16‰ for $\delta^{18}O$, and ± 0.05‰ for $\Delta^{17}O$ (2σ).

Additional sophisticated techniques can be brought to bear in meteorite characterization; their use depends on the nature of the sample and the type of information being sought. As detailed in McSween and Huss (2010), such techniques include wet chemical analysis, x-ray fluorescence (XRF), neutron activation analysis (both instrumental [INAA] and radiochemical [RNAA]), scanning electron microscopy (SEM), transmission electron microscopy (TEM), auger electron microscopy (AEM), proton-induced x-ray emission (PIXE), inductively coupled plasma atomic emission spectroscopy (ICP-AES), x-ray diffraction (XRD), micro-XRF, scanning transmission x-ray microscopy (STXM), x-ray absorption near-edge spectroscopy (XANES), several varieties of mass spectrometry (including thermal-ionization [TIMS], inductively coupled plasma [ICPMS], and isotope dilution [IDMS]), Raman spectroscopy, gamma-ray and neutron spectrometry, alpha particle x-ray spectrometry (APXS), and Mössbauer spectroscopy.

About 80% of meteorites in our collections are ordinary chondrites; to date, I have been involved in the classification of about 120 approved OC samples. Here is the typical procedure for getting an ordinary-chondrite sample properly characterized:

(1) Field data are obtained (as much as practicable). This information includes (a) fireball observations, (b) whether the sample is a find or fall, (c) what day the first specimen was recovered, (d) the numbers of paired specimens, the dates of their collection, and their individual and collective masses, (e) the size and shape of the strewn field, (f) the nature of the ground surface, (g) whether late-recovered specimens of falls were rained upon, and (h) whether falls are "hammerstones" (i.e., whether individual stones pelted a man-made structure).

(2) A name should be proposed for the sample. If there are numerous geographic features near the recovery site (such as towns, streams, canyons, and mountains), a name can be chosen that is distinct from

previously approved meteorite names. If the sample is from a dense collection area (e.g., Northwest Africa, Allan Hills, Roosevelt County, Coyote Dry Lake, or Alatage Mountain), the Nomenclature Committee will assign the next available number upon approval (e.g., NWA xxxxx). However, this can potentially lead to confusion. For example, Mundrabilla is a well-known IAB-ungrouped iron with a total mass of 24 tons found in Western Australia in 1911. But there are more than 20 other meteorite finds (typically <100 g) that bear the names Mundrabilla 002, 003, 004, etc., most of which are equilibrated H and L chondrites. Names can be unusual: Nothing is a IID iron found 4.5 km north of the uninhabited ghost town Nothing, in Mohave County, Arizona. There are also meteorite names of higher numerical value, like Lone Star (Texas; H4), Twodot (Montana; H6), Three Little Hills (New Mexico; H5), Four Corners (New Mexico; IAB-ung), Almahata Sitta (Nahr an Nil, Sudan; ureilite-an), Dix (Nebraska; L6), and Twentynine Palms (California; L6).

(3) Whole-rock specimens are examined with the naked eye and with a stereomicroscope. Features to be investigated include the presence and nature of the fusion crust, evidence for brecciation, the proportions and dimensions of clasts (if any are visible), and the extent and nature of surficial weathering.

(4) A chip of the rock is removed from the sample for analysis; either it is cut off using a wafering saw with diamond-tipped blades or broken off mechanically. If a saw is used, the sample can be cut dry at a low speed. If a coolant is necessary, it should not be water (because water can dissolve or alter several meteoritic phases) or kerosene (because it can contaminate the sample with organic material); the best choice is pure ethanol. However, because ethanol is hygroscopic (apt to absorb moisture from the air), newly opened bottles should ideally be used once and then discarded (or utilized for a different purpose) after each sample cut.

(5) A doubly polished thin section is prepared for transmitted and reflected optical microscopy. (a) The first step is to determine the mineralogy: ordinary chondrites should contain abundant olivine and low-Ca pyroxene; all but the most weathered will have metallic Fe-Ni grains and troilite. Chondrule textural types and sizes should be noted. (b) The second step is to ascertain the OC group, although confirmation may have to await electron microprobe analysis. Average chondrule size increases from H (450 μm) to L (500 μm) to LL (550 μm). The average abundance of metallic Fe-Ni in unweathered samples decreases from H (17.8 wt.%) to L (8.33 wt.%) to LL (3.56 wt.%). Point-counting of phases can be used to determine the modal abundances of metallic Fe-Ni, troilite, chromite, and silicates. (c) The third step is to estimate the petrologic type. If the sample is not an impact-melt rock or an impact-melt breccia, it should contain recognizable chondrules. Even the most highly recrystallized type-6 OC will retain a few identifiable (albeit recrystallized) chondrules. The determination of petrologic type is based on petrographic properties and mineral chemistry (Van Schmus and Wood, 1967). Details are presented in Table 2.1 and in Chapter 13. In general, type-3 OC have very sharply defined chondrules. Mesostasis with isotropic glass is most abundant in subtypes 3.0–3.3. Chondrule outlines become increasingly blurred and integrated with the matrix in types 5 and 6. A fine-grained silicate matrix is evident in type-3 samples; a recrystallized matrix can be discerned in type-4 samples. Ordinary chondrites of types 5 and 6 have no fine-grained matrix; their matrix grains appear coarsened and integrated to some extent with chondrule boundaries. The low-Ca pyroxene grains in type-3 samples (particularly in chondrules) are monoclinic; they are characterized by polysynthetic twin lamellae on (100) and inclined optical extinction (evident in crossed polars). The low-Ca pyroxene in unshocked samples of types 5 and 6 is orthorhombic and exhibits parallel extinction. Type-4 OC may contain both monoclinic and orthorhombic low-Ca pyroxene. Coarse plagioclase grains (≥50 μm in size) are indicative of type 6. Coarse grains of steel-gray chromite (and to a lesser extent, ilmenite, which is gray with a pinkish tinge) are present in the matrix of samples of petrologic type ≥3.8. Further refinement of petrologic type must await electron microprobe analysis. (d) The fourth step of petrographic analysis is to determine the weathering stage using the weathering index in Table 9.1 (from Rubin and Ma, 2021; adapted from Zurfluh et al., 2016). Fresh falls should be W0, but some specimens recovered days after the fall (particularly if they were rained on) could reach W1. Old observed falls that have resided in old-fashioned museum cabinets for decades may have reacted with Earth's atmosphere and also reached W1. (e) The fifth step involves shock stage determination following the criteria listed in Table 2.2 (see Section 13.2.1 for more information).

Table 2.1 *Properties of ordinary-chondrite petrologic types.*

–	Type 3	Type 4	Type 5	Type 6
Homogeneity of olivine and low-Ca pyroxene compositions	>5% mean deviations	<5% mean deviations to uniform	Uniform	Uniform
Structural state of low-Ca pyroxene	Mainly monoclinic	Abundant monoclinic crystals	Orthorhombic	Orthorhombic
Development of secondary feldspar	Absent	Mainly microcrystalline aggregates	Mainly microcrystalline aggregates	Clear interstitial grains ≥50 μm in size
Chondrule mesostasis	Clear isotropic chondrule mesostases	Turbid or devitrified glass	No isotropic glass	No isotropic glass
Chondrule recognizability	Very sharply defined chondrules	Well-defined chondrules	Readily delineated chondrules	Poorly defined chondrules
Texture of fine-grained silicate matrix	Opaque matrix	Translucent to recrystallized matrix	Matrix recrystallized and integrated with chondrules	Matrix recrystallized and integrated with chondrules
Bulk carbon content	0.2–1.0 wt.%	<0.2 wt.%	<0.2 wt.%	<0.2 wt.%

Modified from Van Schmus and Wood (1967).

Table 2.2 *Shock stage determination mainly as a function of type of extinction and fracture in common silicate minerals in chondritic meteorites.*

Shock stage	Description	Olivine	Low-Ca pyroxene	Plagioclase
S1	Unshocked	Sharp optical extinction; irregular fractures	Sharp optical extinction; irregular fractures	Sharp optical extinction; irregular fractures
S2	Very weakly shocked	Undulose extinction; irregular fractures	Undulose extinction; irregular fractures; some planar fractures	Undulose extinction; irregular fractures
S3	Weakly shocked	Undulose extinction; irregular fractures; planar fractures	Undulose extinction; irregular fractures; planar fractures; clinoenstatite lamellae on (100)	Undulose extinction
S4	Moderately shocked	Weak mosaic extinction; planar fractures	Weak mosaic extinction; planar fractures; clinoenstatite lamellae on (100)	Undulose extinction; partly isotropic grains
S5	Strongly shocked	Strong mosaic extinction; planar fractures	Strong mosaic extinction; planar fractures	Maskelynite
S6 (applies to restricted local regions in or near melt zones)	Very strongly shocked	Solid-state recrystallization; yellow-brown staining; ringwoodite; melting	Grain melting; majorite	Shock melting and formation of normal glass
S7	Whole-rock melting	Whole-rock melting	Whole-rock melting	Whole-rock melting

Reproduced from Rubin and Ma (2021) and adapted from Stöffler et al. (1991, 2018), Rubin (2015a), Rubin et al. (1997), and Izawa et al. (2011).

In addition to the overall shock stage, other shock features should be noted, if present. The most common features are opaque shock veins, glassy shock veins, metallic Cu grains at troilite-metal boundaries, small irregular troilite grains within metallic Fe-Ni, chromite-plagioclase assemblages, chromite veins and veinlets, and polycrystalline troilite. If the thin section is etched with Nital (a dilute solution of ethanol and nitric acid), metallic Fe-Ni grains can be examined in reflected light; a search should be made for martensitic grains and polycrystalline kamacite. Shocked samples (≥S3) of all petrologic types will contain low-Ca pyroxene grains with clinoenstatite lamellae on (100). If the sample is a breccia, the sizes, modal abundances, and varieties of clasts should be determined.

(6) The thin section should be carbon-coated and placed in the electron microprobe for quantitative mineral analysis. (a) Group designation: For well-equilibrated samples, only olivine is necessary to establish the OC group, but it is good practice to analyze olivine, low-Ca pyroxene, kamacite and, if present, plagioclase and chromite. (During probe analysis of metallic Fe-Ni, care must be taken to correct the device-reported Co concentration for any overlap of the Fe Kβ peak with the Co Kα peak. Care must also be taken in the analysis of chromite to correct for any overlap of the Ti Kβ peak with the V Kα peak [Snetsinger et al., 1968].) The compositions of olivine, low-Ca pyroxene, and kamacite will help determine the proper OC group: H (Fa 17.3–20.2 mol%; Fs 15.7–18.1 mol%; kamacite Co: 4.4–5.1 mg/g); L (Fa 23.0–25.8 mol%; Fs 18.7–22.6 mol%; kamacite Co: 7.0–9.5 mg/g); LL (Fa 26.6–32.4 mol%; Fs 23.2–25.7 mol%; kamacite Co: 14.2–370 mg/g Co). (1 mg/g = 0.1 wt.%) H/L and L/LL chondrites will either have some intermediate values or some values characteristic of two different main groups. Chromite, ilmenite, and plagioclase also have different average compositions in H, L, and LL chondrites (Gomes and Keil, 1980), but these values are somewhat variable and are not often used as classificatory parameters of OC whole-rocks. (b) Petrologic type: Because petrologic type is a function of whole-rock equilibration, the variation in olivine Fa and low-Ca pyroxene Fs (as outlined in Table 2.1) correlates with the petrographic properties; their percent mean deviation (PMD) should be determined (see Table 13.1). During thermal metamorphism, chondrule mesostases devitrify (evident in type 3.7–3.9) and albite crystals nucleate and grow. Type-5 OC contain 2–10-μm plagioclase grains and type-6 OC contain many plagioclase grains exceeding 50 μm in maximum dimension. With increasing metamorphic grade, Ca-pyroxene increases in size from submicrometer grains in type-4 OC, to 2–5-μm grains in type-5 OC, to grains tens of micrometers in size in type-6 OC. Olivine grains with >2 wt.% FeO can be used to determine subtypes among the least equilibrated type-3 OC (Grossman and Brearley, 2005). These olivine compositions are sensitive to the lowest levels of thermal metamorphism. Electron microprobe data can be plotted in Fig. 2.2 to derive the subtype. If the rock is a breccia, minerals in the clasts should be identified and analyzed.

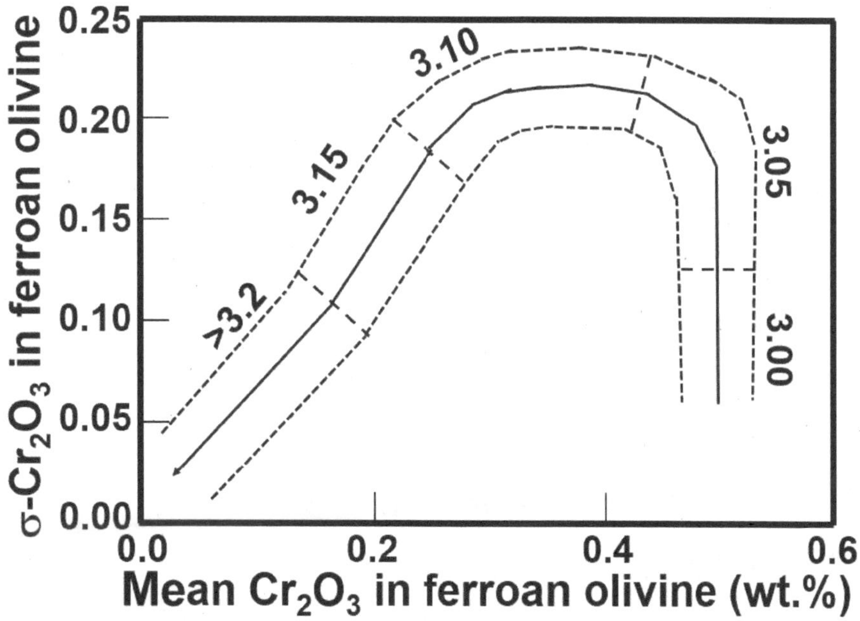

Fig. 2.2 Diagram of σ-Cr_2O_3 vs. mean Cr_2O_3 in ferroan olivine grains.
Note: This is useful for determining the petrologic subtype of type-3 ordinary chondrites (adapted from Grossman and Brearley, 2005).

(7) In some cases, meteorite chips can be analyzed for the rock's bulk O-isotopic composition. This property is also diagnostic of OC group. Mean group compositions relative to Standard Mean Ocean Water (SMOW) are: H4-6 ($\Delta^{17}O = 0.73‰; \delta^{18}O = 4.08‰; \delta^{17}O = 2.85‰$), L4-6 ($\Delta^{17}O = 1.07‰; \delta^{18}O = 4.70‰; \delta^{17}O = 3.52‰$), and LL4-6 ($\Delta^{17}O = 1.26‰; \delta^{18}O = 5.04‰; \delta^{17}O = 3.88‰$). The ranges in O-isotopic composition on the standard three-isotope diagram for the three major OC groups are shown in Fig. 3.4.

(8) If the sample is a fragmental breccia and has a light/dark structure (i.e., angular light-colored clasts set within a dark-colored fine-grained matrix), the meteorite might be a regolith breccia containing noble gases implanted by the solar wind, acquired at the surface of the parent asteroid. The gases, if present, would occur in the fine-grained dark matrix (representative of the asteroid's "soil"), not within the light-colored (metamorphosed) clasts. Analyses of the noble gas content should be made for ordinary chondrites with such a texture.

(9) When sufficient data are obtained for a new meteorite, the information should be submitted to the editor of the Meteoritical Bulletin for approval by the Meteorite Nomenclature Committee. The committee will evaluate the proposal and may request additional data before granting approval. Once approved, the information will be published in the online Meteoritical Bulletin Database. The site is updated frequently.

3

A Few Essential Definitions

Definitions too fussy are hard to define.

E. J. Olsen (1989), "The Chondrule"

As pointed out by Nininger (1936b), "Clear and concise use of language is essential in any body of literature if it is to be effective. Consistency and uniformity in the use of technical terms is equally essential in any field of science." In reference to the field of meteoritics Nininger found "a wide variation in the use of certain terms by various writers in the field." This remains the case today, even though The Society for Research on Meteorites (the predecessor of the Meteoritical Society) appointed a committee on terminology in 1932. To address this longstanding problem, precise definitions of common terms are offered here.

Farrington (1915) introduced the term **meteoritics** for the scientific field of meteorite study. Previous suggestions included "astrolithology" (Shepard, 1846) and "aërolitics" (Story-Maskelyne, 1870). Farrington's term caught on (e.g., Nininger, 1933), and Leonard (1953) introduced volume 1, number 1 of the journal *Meteoritics*, the successor to *Contributions of the Meteoritical Society* (1947–1951) and *Contributions of the Society for Research on Meteorites* (1935–1946). Meteorite researchers are meteoriticists; like Gatsby, they are "boats against the current, borne back ceaselessly into the past." Most scientists who study meteors are meteor astronomers, not meteoriticists.

Since the late eighteenth century, when the concept that rocks fell from space was accepted by the broader scientific community (e.g., Burke, 1986; Marvin, 1996, 2006; Rubin, 2022b), different definitions of **meteorite** have been offered. There are important constraints on a proper definition: (1) Because man-made objects have been falling to Earth since January 1958 (when Sputnik 1, the first artificial satellite, reentered the atmosphere), the designation of meteorite should be restricted to a natural object. (2) Because natural objects can be dumped by orbiting spacecraft, meteorites should be launched by natural means from a natural body. (3) Because meteorites can land on other celestial bodies (e.g., asteroids, the Moon, and Mars) or penetrate spacecraft, the definition of a meteorite must be expanded beyond Target Earth. (4) An object launched by a collision into interplanetary space from a target body can eventually land back on its home world as a meteorite. However, objects blasted by impacts into suborbital trajectories (such as the australite-tektite buttons [e.g., Barnes, 1958], or projectiles that produce secondary impact craters after being ejected from a large primary crater [Shoemaker, 1962]) never reach interplanetary space and should be excluded from the definition. (5) There should also be size limits.

The quasi-Talmudic discussions of Rubin and Grossman (2010) and Rubin and Ma (2021) can be modified slightly to define a meteorite as *a natural solid object, between 2 mm and 1 m in size, derived from a celestial body (e.g., asteroid, moon, comet, planet), that was transported by natural means (e.g., impact excavation) from the body on which it formed to a region outside the dominant gravitational influence of that body (i.e., to interplanetary space) and that later collided with a natural or artificial body larger than itself (e.g., Earth, Moon, Mars, an asteroid, a comet, or the International Space Station).*

According to the *Oxford English Dictionary*, there are two occurrences of the word "meteorite" per million words in modern written English.

Micrometeorites are *small meteorites, 1 μm to 2 mm in size.*

A **meteoroid** is *a natural solid object, 1 μm to 1 m in size, moving in interplanetary space.* Meteoroids can consist predominantly of a variety of materials (e.g., silicate, metallic Fe-Ni, sulfide, phosphide, carbide, or ice). If it survives collision with a larger body (e.g., lands as a solid body on Earth), a meteoroid becomes a meteorite.

Micrometeoroids are *small meteoroids, 1 μm to 2 mm in size.*

Dust grains are natural particles larger than a molecule and less than *1 μm* in size.

An **asteroid** is a meteoroid larger than 1 m in diameter (and smaller than 1000 km) that is orbiting its central star (and not orbiting a planet or moon) and is incapable of appreciable outgassing. Larger objects orbiting a central star are planets.

A **comet** is a natural volatile-rich solid subplanetary object, at least 1 m in size, in orbit around its central star and that exhibits (or would exhibit) extensive outgassing and develops (or would develop) a coma and tail upon close approach to the star; a comet is also a body inferred to have been rich in volatiles and to have undergone extensive outgassing in the past. A **dark comet** lacks a coma and tail and does not appear to be outgassing; nevertheless, it exhibits nongravitational accelerations that cannot be ascribed to anisotropic emission of thermal photons (the Yarkovsky effect) or to the scattering of solar radiation and emission of thermal radiation (the Yarkovsky–O'Keefe–Radzievskii–Paddack, or YORP, effect). Objects smaller than 1 m are meteoroids, micrometeoroids, or dust grains.

Rubin and Ma (2021) defined a **fall** as a recovered meteorite that was (a) responsible for a fireball observed by people, fireball-network cameras, Doppler radar, seismometers, and/or satellites, (b) responsible for thunderous sonic booms, whistling or hissing sounds (probably produced by very low frequency radio waves, i.e., electrophonic sounds), or noise produced when the object strikes a structure or the ground, (c) inferred to have fallen recently from the damage it caused to structures, the ground, or living creatures, or (d) found on a surface where recent prior observation had not indicated its presence. Individual samples of a fall may be recovered years after reaching the ground. Most falls encompass more than one observational category.

A **find** is a meteorite discovered at the surface of a target body without it (or samples paired with it) having been observed to fall.

A **chondrule** is defined as a natural rounded igneous body, of submicrometer-to-multi-centimeter size, that formed in the solar nebula from precursor melt-bearing droplets prior to the agglomeration of chondrites. The average modal abundance of chondrules outside the CI group (which is heavily altered and essentially chondrule-free) varies from ~15 vol.% in CK3 to ~70–75 vol.% in type-3 ordinary chondrites. Mean chondrule diameters range from 150 μm in CO chondrites to 900 μm in CK and CV chondrites. Although most of the chondrules in any chondrite group have diameters within a factor of two of the mean, chondrules in ordinary chondrites range in diameter from 0.8 μm (the smallest microchondrule reported in LL3.10 Piancaldoli; Rubin et al., 1982a) to ~5 cm (a fragmented macrochondrule in L3.2 Gunlock; Prinz et al., 1988). Most chondrules have discernable igneous textures, indicating that they formed from a melt or a melt-crystal mush. Most unbroken chondrules tend to be spheroids or triaxial ellipsoids; their melt-bearing droplet precursors assumed such rounded shapes due to surface tension. Although millimeter-sized impact-melt spherules with igneous textures have been found in lunar regolith breccias and called "lunar chondrules" (e.g., Keil et al., 1972; King et al., 1972a; Symes et al., 1998), most researchers distinguish these objects from the chondrules in chondrites (e.g., Taylor et al., 1983). Genuine chondrules are hypothesized to have formed in the solar nebula prior to chondrite agglomeration; they predate the asteroids that their present chondritic host rocks were dislodged from. Many chondrules may have been melted multiple times (e.g., Rubin, 2010a, 2013a, 2017a, 2017b; Rubin and Baecker, 2021). Compound chondrules also occur (e.g., Wasson et al., 1995; Weyrauch and Bischoff, 2012): In some cases, these objects formed by small droplets colliding with and adhering to another droplet or a preexisting solid chondrule. In other cases, chondrules became trapped within dustballs; the assemblages were then heated sufficiently to melt most (or all) of the dust but not much of the primary chondrule itself.

A **CAI (calcium-aluminum inclusion or refractory inclusion)** is defined as *a natural submillimeter-to-multi-centimeter object rich in refractory phases.* Calcium-aluminum inclusions (CAIs) are solid objects rich in refractory minerals, most commonly Al-Ti diopside (fassaite), anorthite, corundum, forsterite, grossite, hibonite, melilite, perovskite, and spinel. The largest CAIs are 2–3 cm in size and occur exclusively in CV3 chondrites; most CAIs outside the CV group are <1 mm. The mean modal abundance of CAIs (including CAI fragments) varies across chondrite groups (vol.%): CI, <0.01; CM, 1.2; CO, 1.0; CV, 3.0; CR, 0.12; CK, 4; R, 0.04; H, 0.01; L, 0.01; LL, 0.01; EH, 0.01; EL, 0.01 (Table 7.1 of Rubin and Ma, 2021; Dunham et al., 2023). Many CAIs have been divided into categories (A, B, C) and subcategories that include objects differing in average shape, size, mineralogy, and mineral chemistry.

Different chondrite groups have different CAI populations: Type-B CAIs are restricted to CV chondrites. Most CAIs in CM chondrites are enriched in spinel, spinel + pyroxene, spinel + pyroxene + olivine, pyroxene, and pyroxene + olivine. A few CM CAIs contain hibonite.

Compact Type-A CAIs (CTAs) are spheroidal and consist predominantly of zoned melilite enclosing spinel, hibonite, perovskite, and noble-metal nuggets; some CTAs are nearly pure melilite. Many CTAs are surrounded by Wark-Lovering rims – a series of mono- or bi-mineralic layers a few tens of micrometers thick, consisting of melilite, spinel and diopside along with some hibonite and perovskite (Wark and Lovering, 1977).

Fluffy Type-A CAIs (FTAs) are irregularly shaped; many appear to be aggregates of numerous melilite-rich nodules; many nodules resemble individual CTAs with Wark-Lovering rims.

Type-B CAIs occur exclusively in CV chondrites as multi-millimeter- to multi-centimeter-sized coarse-grained spheroidal igneous objects consisting mainly of melilite, Al- and Ti-rich diopside, spinel, and minor anorthite. Type-B1 inclusions consist of a coarse-grained melilite mantle surrounding a core of pyroxene, spinel, anorthite, metallic Fe-Ni, and noble-metal nuggets. Type-B2 inclusions have no melilite mantle and resemble B1 cores. Type-B3 inclusions contain coarse forsterite grains along with pyroxene, spinel, and melilite.

Type-C CAIs are igneous objects containing major spinel, calcic pyroxene and anorthite exhibiting poikilitic or ophitic textures.

One variety of CAI is the set of "FUN" inclusions exhibiting Fractionation and Unidentified Nuclear isotopic effects, exemplified by unusual isotopic compositions of Mg, O, Ba, Ti, Ca, and other elements (e.g., Wasserburg et al., 1977). These inclusions probably formed during evaporation of isotopically heterogeneous material (Krot et al., 2014a).

CAIs are not restricted to chondrites: about 1% of the rocky material recovered from Comet Wild 2 by the *Stardust* mission consists of CAIs and CAI fragments.

CAIs were the among the first solids to form in the Solar System, appearing 4.56730 ± 0.00016 Ga ago (Amelin et al., 2002; Bouvier and Wadhwa, 2010; Connelly et al., 2012). They are rich in ^{16}O (e.g., Clayton et al., 1973) and formed, in part, from material derived from a ^{16}O-rich reservoir of composition: $\delta^{18}O \approx \delta^{17}O = -50‰$ to $-60‰$ relative to Standard Mean Ocean Water (SMOW) (e.g., MacPherson, 2005). Nevertheless, CAIs from CB chondrites are younger and depleted in ^{16}O; they likely formed within an impact-generated plume on an asteroid later in Solar System history, ~4.563 Ga ago (Krot et al., 2009).

An **amoeboid olivine aggregate** (also known as an **amoeboid olivine inclusion**) is a natural fine-grained (typically 1–20 µm), irregularly shaped object consisting of major olivine, major-to-minor calcic pyroxene, minor anorthite, accessory metallic Fe-Ni and perovskite, and in some cases, accessory spinel ± low-Ca pyroxene ± melilite; many have porous interiors. AOAs are about as abundant as CAIs in carbonaceous chondrites and are very rare in type-3 ordinary chondrites. The mean AOA size varies among chondrite groups: LL3, 200 µm; CM2, 225 µm; CO3, 260 µm; CR2, 600 µm; CK3, 900 µm; CV3, ~3000 µm (Rubin, 2013b).

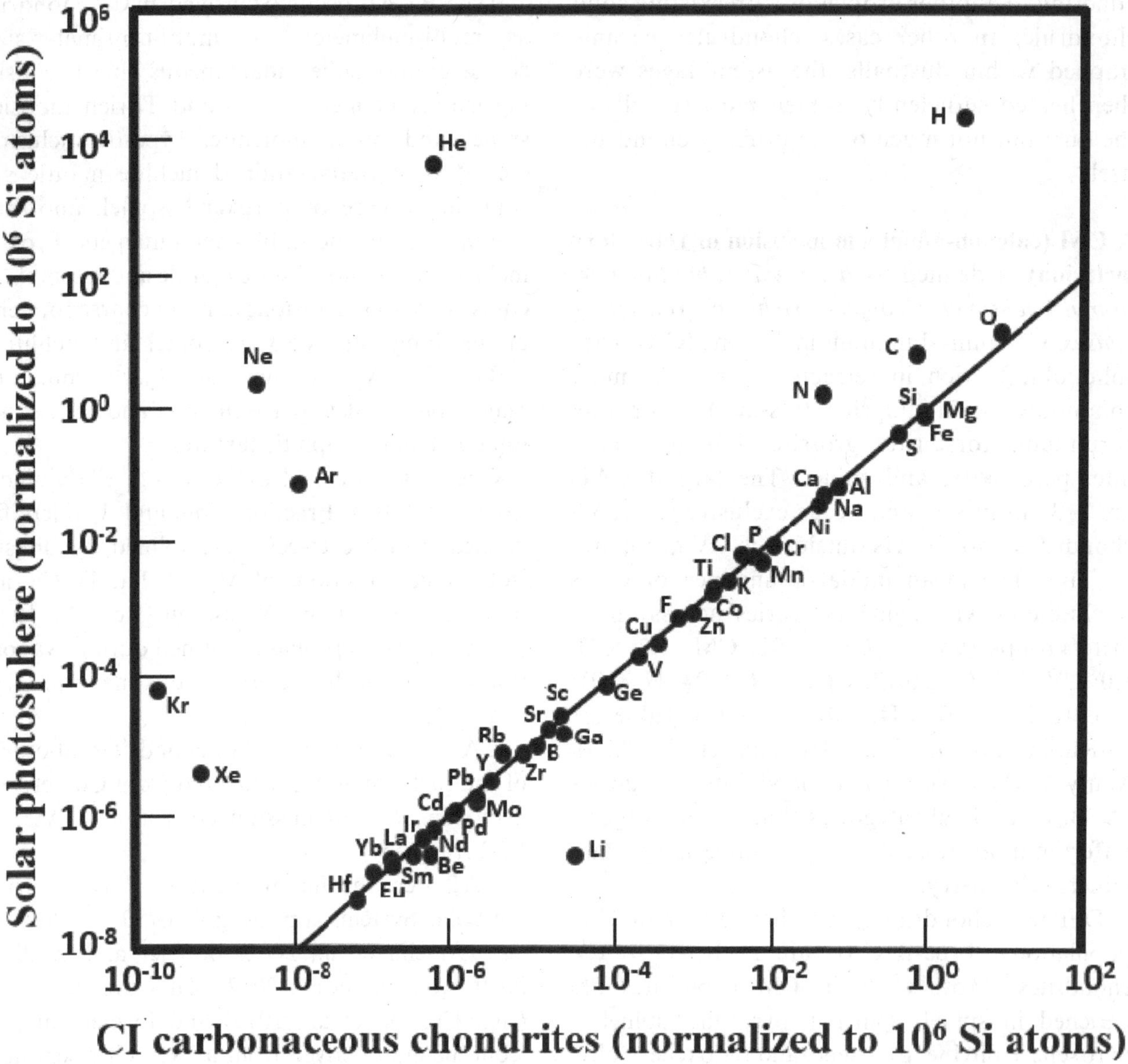

Fig. 3.1 Log-log plot of selected elemental abundances in the solar photosphere vs. those in CI carbonaceous chondrites. The photosphere/CI-chondrite ratios for most elements are close to unity; these elements include refractory, common, and volatile elements with lithophile, siderophile, and chalcophile affinities. Noble gas abundances are much higher in the Sun than in CI chondrites because they did not quantitatively condense into solids. Diagram redrawn from Rubin and Ma (2021), modified from McSween and Huss (2010) and Dye et al. (2015).

Many elements in CI carbonaceous **chondrites** have abundances very similar to those of the solar photosphere (Fig. 3.1). These abundances are considered "solar," "cosmic," or "chondritic" (although small differences between CI chondrites and the photosphere may depend on the 50% condensation temperatures of individual elements; Jurewicz et al., 2024). When normalized to CI chondrites and to a common lithophile element (such as Mg or Si), chondrites have "unfractionated" (i.e., relatively flat) refractory-lithophile-element abundance patterns (e.g., Ca, Al, Ti, Sc, REEs) (Fig. 3.2). Chondrites are sedimentary agglomerates that formed more than 4.5 Ga ago as mechanical mixtures of primitive nebular constituents (e.g., chondrules, refractory inclusions, fine-grained matrix, metallic Fe-Ni, and sulfide grains) and small quantities of presolar grains. Chondrites were derived from undifferentiated asteroids (bodies that escaped extensive internal melting). Metallic Fe-Ni and silicate are assumed to be more-or-less homogeneously distributed in these asteroids; the bodies have not differentiated into distinct layers (silicate crust, silicate mantle, and a metallic-Fe-Ni-rich core). Chondrites did not form with igneous textures, although many chondrites contain inclusions (e.g., chondrules and CAIs) that were themselves largely or completely melted in the solar nebula. Some chondrites also contain impact-melt-rock clasts – inclusions with igneous textures that formed by localized impact melting in the near-surface regions of asteroids.

Many chondrites experienced extensive whole-rock thermal metamorphism (involving grain coarsening, chondrule-glass devitrification, secondary mineralization, and volatile loss). Many were subjected to various levels of shock metamorphism, ranging from negligible to extensive; some rocks were impact melted (involving the development of igneous textures, localized metal-silicate separation, and volatile loss). One subgroup of CV carbonaceous chondrites (the CV_{OxA} chondrites) underwent alkali-halogen-Fe metasomatism. Nevertheless, in most cases, these changes had only minor effects on the abundance pattern of whole-rock refractory lithophile elements; the pattern remained relatively flat.

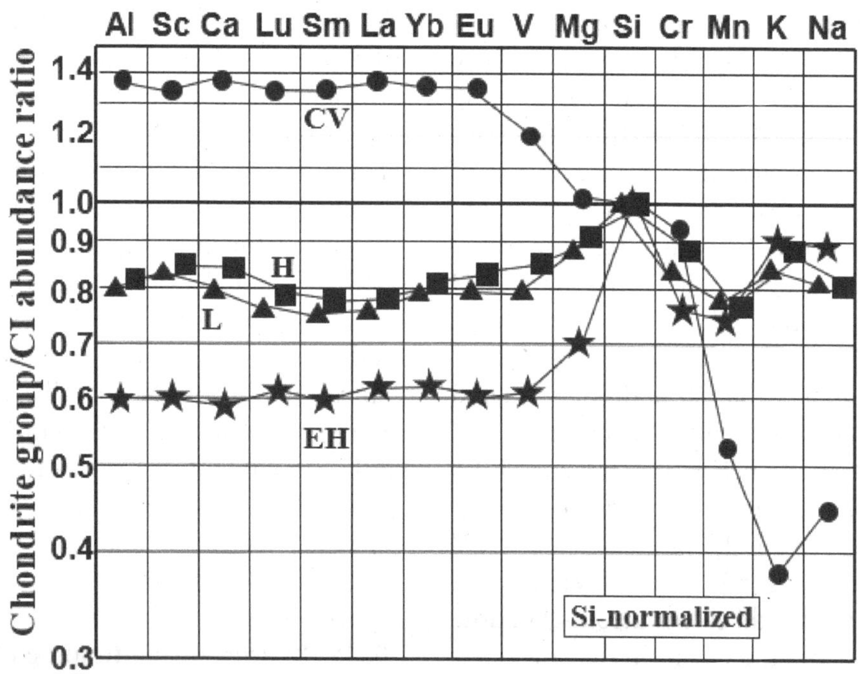

Fig. 3.2 Logarithmic plot showing the bulk lithophile-element abundances of several chondrite groups normalized to CI chondrites and to Si. Elements are arranged from left to right in order of increasing volatility. Chondrite groups tend to have relatively flat (unfractionated) patterns for their refractory lithophile elements. Only four groups are shown (CV carbonaceous chondrites, H and L ordinary chondrites, EH enstatite chondrites). CI chondrites form a horizontal line at 1.0. The abundances of refractory lithophiles decrease from carbonaceous to ordinary to enstatite chondrites. Diagram modified from Wasson and Kallemeyn (1988).

From these constraints, a chondrite can be defined as *a rock with a relatively flat CI- and Mg-, Si- or Cr-normalized refractory-lithophile abundance pattern, formed in a natural manner as a sedimentary agglomerate from primitive nebular constituents.* In many cases, these constituents include chondrules, refractory inclusions, amoeboid olivine inclusions, fine-grained silicate-rich matrix material, grains of metallic Fe-Ni and sulfide, and a smattering of presolar grains. The texture, mineralogy, mineral composition, and bulk chemical and isotopic compositions of chondrite whole-rocks may have been modified by a combination of thermal metamorphism, aqueous alteration, metasomatism, shock metamorphism, and weathering on the body where the chondrite landed.

There are currently nine well-established groups of **carbonaceous chondrites** (Fig. 2.1), far more groups than for ordinary chondrites (three), enstatite chondrites (two), and R chondrites (one). In addition, about 70% of ungrouped chondrites are carbonaceous. It is clear that our present inventory of meteorites includes samples from many carbonaceous-chondrite parent asteroids. (Although a CY carbonaceous-chondrite group had been proposed [Ikeda, 1992; King et al., 2019; Suttle et al., 2021, 2023], it seems more likely these samples are a mixed group of heated CI and CM chondrites [Schrader et al., 2025].) Carbonaceous-chondrite falls tend to contain more bulk C and chemically bound H_2O (i.e., H_2O^+) than other chondrite clans (e.g., Jarosewich, 1990).

Figure 3.2 shows that the CI- and Si-normalized refractory lithophile abundances of ordinary and enstatite chondrites fall below the horizontal CI carbonaceous-chondrite line. One carbonaceous-chondrite group (other than CI) is plotted in Fig. 3.2: CV chondrites form a fairly flat line appreciably above that of CI chondrites. In fact, all carbonaceous-chondrite groups plot on or above the CI line in such diagrams (e.g., Wasson and Kallemeyn, 1988).

Figure 6.2 shows that carbonaceous chondrites also have distinctive abundances of various stable isotopes, setting them apart from noncarbonaceous materials (ordinary, enstatite, and R chondrites, Earth, Moon, Mars, main-group pallasites, mesosiderites, and howardites-eucrites-diogenites [HED samples], probably from Vesta), angrites, aubrites and ureilites.

A carbonaceous chondrite is defined as *a chondritic meteorite with CI- and Si- (or Mg- or Cr-) normalized refractory lithophile abundances ≥1 and with bulk stable-isotopic compositions* ($\Delta^{17}O$, $\varepsilon^{54}Cr$, $\varepsilon^{50}Ti$, $\varepsilon^{62}Ni$) *falling within the carbonaceous fields* (e.g., Fig. 6.2). Most groups will contain significant amounts of bulk C and chemically bound water.

Ordinary chondrites (often abbreviated "OC") constitute ~80% of observed meteorite falls and ~85% of meteorite finds. There are three major groups: H (high total iron), L (low total iron), and LL (low total iron, low metallic iron) that have similar textures and mineral assemblages. Nevertheless, the groups vary systematically in numerous mineralogical, chemical, physical, and isotopic properties. Most of these differences result from their distinct oxidation states. (1) Mean metallic Fe-Ni-Co decreases from H to L to LL (Jarosewich, 1990) as increasing proportions of iron change from Fe^0 to Fe^{2+}. In going from H to L to LL there are (2) increases in the FeO contents of olivine, low-Ca pyroxene, chromite, and ilmenite (e.g., Mueller and Olsen, 1967; Bunch et al., 1967; Snetsinger and Keil, 1969; Gomes and Keil, 1980; Rubin, 1990), (3) increases in the TiO_2 content of chromite (Bunch et al., 1967), likely due to higher Ti^{4+}/Ti^{3+} ratios in the set of chondrule olivine grains (e.g., Simon et al., 2008) with increasing bulk oxidation state, (4) increases in the Co and Ni concentrations in low-Ni metallic Fe (i.e., kamacite) (Sears and Axon, 1976; Rubin, 1990), (5) increases in the Ge content of the metal fraction (Chou and Cohen, 1973; Chou et al., 1973; Florin et al., 2020), (6) decreases in the kamacite/taenite modal abundance ratios (Afiattalab and Wasson, 1980), (7) higher olivine/low-Ca-pyroxene ratios (Hutchison, 2004), (8) decreases in total Fe (Jarosewich, 1990), (9) decreases in siderophile/lithophile element ratios (e.g., Ni/Mg) (Kallemeyn et al., 1989), (10) decreases in bulk density (Consolmagno et al., 2006), (11) lower refractory-siderophile/common-siderophile ratios (e.g., Ir/Ni) (Kallemeyn et al., 1989), (12)

Fig. 3.3 Portion of a slab of the NWA 2833 L3.8 ordinary chondrite showing abundant chondrules surrounded by dark-colored recrystallized silicate matrix material. Image courtesy of John Kashuba.

lower refractory-siderophile/volatile-siderophile ratios (e.g., Ir/Au) (Kallemeyn et al., 1989), (13) heavier O isotopes (Clayton et al., 1991; Greenwood et al., 2017), (14) heavier Ge isotopes (Florin et al., 2020) and (15) increases in mean chondrule size (Rubin, 2000; Nelson and Rubin, 2002; Friedrich et al., 2015).

Some ordinary chondrites are designated H/L and L/LL; any or all of them could either be anomalous members of one of the two designated groups or members of new groups (derived from separate asteroids) with intermediate properties.

Because the bulk chemical (e.g., Ir/Ni, Ir/Au, Au/Ni, Co/Ni) and bulk O- and Ge-isotopic compositions of IIE iron meteorites lie along extensions of LL-L-H trends, Rubin (2022a) suggested that these irons (and the chondrule-bearing silicate inclusions some contain) constitute a fourth major OC group, appreciably more reduced than H chondrites. By the definition below, iron meteorites such as Netschaëvo can be considered ordinary chondrites because they are meteorites containing ordinary-chondrite components.

All unweathered ordinary chondrites contain olivine, low-Ca pyroxene, and Fe-rich sulfide (troilite in almost all cases, troilite plus pentlandite or, possibly, pentlandite alone in the most oxidized samples). All but the most oxidized samples contain metallic Fe-Ni grains. Type-3 ordinary chondrites contain ~70 vol.% unrecrystallized chondrules (Fig. 3.3).

An ordinary chondrite is defined as *a chondritic meteorite, or a meteorite containing chondritic components, unweathered samples of which contain olivine, low-Ca pyroxene, and Fe-rich sulfide, belonging to one of the H, L, and LL groups, an intermediate group, or a group with numerous mineralogical, bulk chemical, and bulk isotopic properties lying along extensions of the systematic mineralogic, textural, and geochemical H-L-LL trends.*

R (Rumuruti) chondrites are more oxidized than ordinary chondrites; the olivine Fa compositional distributions of equilibrated R chondrites have peaks that cluster between 37 and 40 (Kallemeyn et al., 1996), higher than those in type 4–6 H (17.3–20.2), L (23.0–25.8), and LL (26.6–32.4) chondrites (Rubin, 1990). All R chondrites are olivine rich. Low-Ca pyroxene is the principal pyroxene phase in unequilibrated samples, but in equilibrated R chondrites, Ca-rich pyroxene dominates, and low-Ca pyroxene may be absent. Most R chondrites contain pentlandite and/or pyrrhotite and little or no troilite. Metallic Fe-Ni is very rare;

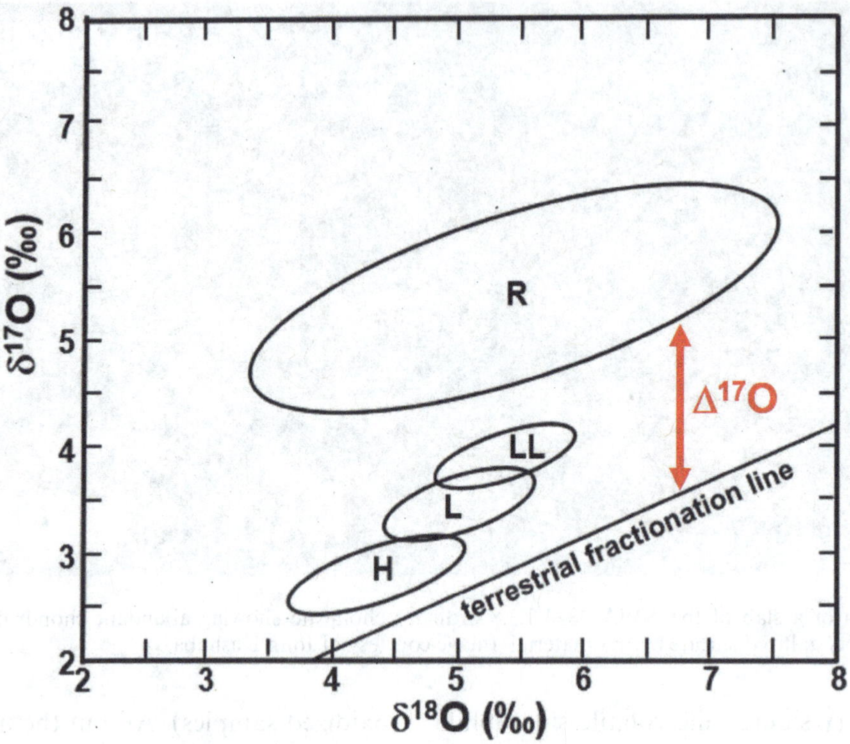

Fig. 3.4 Standard three-isotope diagram for oxygen showing that bulk R chondrites are enriched in ^{17}O relative to the three principal ordinary-chondrite groups. The R chondrites and the ordinary chondrites are richer in ^{17}O and plot above the terrestrial fractionation (TF) line. The parameter $\Delta^{17}O$ is the vertical displacement from the terrestrial fractionation line. Diagram based on data from Bischoff et al. (2011).

where present, most grains are awaruite (Ni_3Fe). Ca-Al-rich inclusions are also rare. The mean CI- and Mg-normalized abundances of refractory lithophile elements is ~0.95 × CI (Bischoff et al., 2011). The bulk O-isotopic compositions of R chondrites are enriched in ^{17}O compared to those of ordinary chondrites (Fig. 3.4).

An R chondrite (or Rumuruti chondrite) is defined as *an oxidized olivine-bearing chondritic meteorite, enriched in bulk* ^{17}O *relative to ordinary chondrites, and with CI- and Mg-normalized refractory lithophile abundances <1.*

Enstatite chondrites are the most reduced chondritic meteorite groups. In other chondrite groups, a significant fraction of the bulk iron occurs as FeO in silicates and oxides, but in enstatite chondrites, almost all the bulk iron occurs in Fe-bearing sulfide, low-Ni metallic Fe (kamacite), phosphide (schreibersite) and, in EH3 chondrites, silicide (perryite). The predominant silicate mineral in enstatite chondrites is enstatite with very low FeO (e.g., Wasson et al., 1994); kamacite containing metallic Si in solid solution is also abundant (e.g., Weisberg and Kimura, 2012). An assortment of sulfide minerals is present; many of these phases contain cations that, in more oxidized chondrite groups, are largely restricted to silicates and oxides: Na, Mg, K, Ca, Ti, Cr, Mn (e.g., Keil, 1968). There are two principal groups: EH (high siderophile-element abundances) and EL (low siderophile-element abundances). EH and EL chondrites also differ in chondrule size, sulfide mineralogy, and degree of reduction. The bulk O-isotopic compositions of enstatite chondrites (and enstatite achondrites) are very similar to those of the Earth and Moon (e.g., Clayton et al., 1984). An enstatite chondrite can be defined as *a highly reduced chondrite, rich in enstatite, sulfide, and low-Ni metallic Fe.*

Achondrites (stony meteorites with igneous textures) and planetary basalts (from the Earth, Moon, and Mars) exhibit fractionated (i.e., non-flat) CI-normalized refractory-lithophile

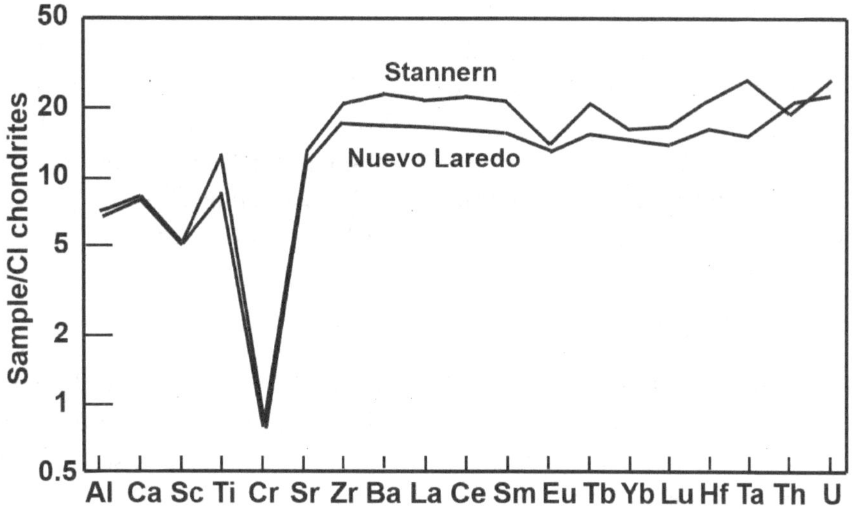

Fig. 3.5 Logarithmic plot showing the average, CI-normalized, bulk chemical compositions of the Stannern and Nuevo Laredo eucrites. These asteroidal basalts are enriched in refractory lithophile elements and have fractionated (non-flat) abundances compared to CI chondrites. Diagram based on Mittlefehldt and Lindstrom (1997).

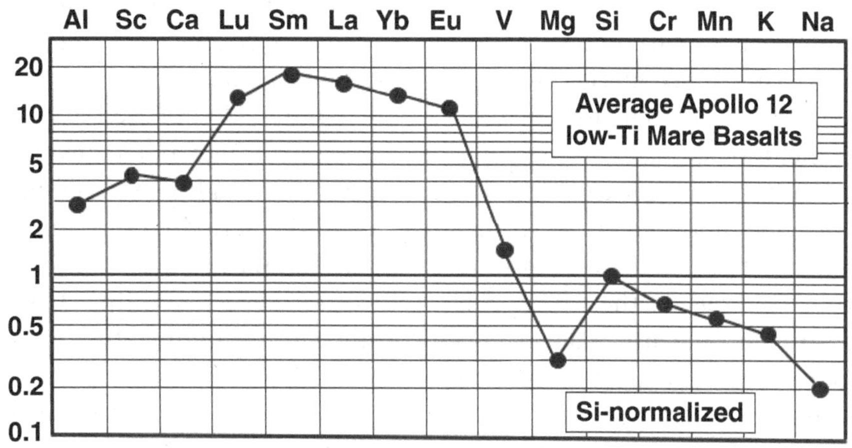

Fig. 3.6 Logarithmic plot showing the average bulk lithophile element composition of Apollo 12 low-Ti mare basalts normalized to CI chondrites and to Si. Elements are arranged from left to right in order of increasing volatility. These basalts are igneous rocks that crystallized from molten magma. They have fractionated (i.e., non-flat), non-chondritic abundance patterns. CI chondrites form a horizontal line at 1.0. Data compiled by P. H. Warren (UCLA) from numerous literature sources.

abundance patterns (e.g., Figs. 3.5, 3.6); these rocks formed within parent bodies (e.g., differentiated asteroids, Moon, Mars) that underwent wide-scale or global melting. Established asteroidal achondrite groups include angrites, aubrites (enstatite achondrites), and HEDs (howardites-eucrites-diogenites); non-asteroidal achondrites are lunar and martian meteorites.

An achondrite is defined as *a meteorite formed by large-scale melting processes in a parent body. It lacks chondrules, has an igneous or brecciated texture, and has fractionated (non-chondritic) abundances of refractory lithophile elements.*

A **primitive achondrite** is a natural rock that is intermediate in petrographic and/or geochemical properties between chondrites and achondrites; it formed on an asteroid by incomplete melting of chondritic material. It has an igneous, granoblastic, or brecciated texture but has retained some chondritic characteristics. These characteristics (which are found in some specimens) include chondritic mineralogy with moderately abundant metallic Fe-Ni (e.g., McCoy et al., 1996; Rubin, 2007), relict chondrules (e.g., Yanai and Kojima, 1991), and planetary-type rare gases (e.g., Stauffer, 1961; Weber et al., 1971; Wilkening and Marti,

1976). Primitive achondrite groups include (a) acapulcoites and lodranites, (b) winonaites and silicate inclusions in IAB irons, (c) brachinites, and (d) ureilites.

A **nonchondrite** is a meteorite that is not a chondrite; this category includes achondrites, primitive achondrites, iron meteorites, and stony-iron meteorites. It is an uncommon term.

Iron meteorites consist mainly of two iron-nickel minerals – low-Ni metallic Fe (kamacite, with a body-centered-cubic structure) and taenite (with a face-centered cubic structure). Additional relatively common minerals include other metallic Fe-Ni phases (martensite, tetrataenite), sulfide (troilite, daubréelite), phosphide (schreibersite, nickelphosphide), carbide (cohenite, haxonite), native carbon (graphite, diamond), native copper, nitride (carlsbergite), oxide (chromite), and phosphate (merrillite, chlorapatite). Some iron meteorites contain silicate inclusions.

The irons have been assigned to 13 major chemical groups (each symbolized by a Roman numeral and a one- or two-letter combination, e.g., IAB, IVA, IIIAB) based on their bulk concentrations of siderophile elements. The compositions of irons from a single chemical group tend to form tight clusters on element-element diagrams and are assumed to have been derived from the same unique asteroid. (The IIG group is an exception; it may have formed as a P-rich magma within the metallic core of the IIAB asteroid [Wasson and Choe, 2009].) Irons can be divided into "magmatic" groups (11 groups) that formed by fractional crystallization within a well-mixed molten metallic core on a differentiated asteroid, and "non-magmatic" groups (IAB and IIE), modeled as having formed by impact-melting processes on a chondritic asteroid. Based on their W-, Mo- and N-isotopic compositions, iron-meteorite groups can also be categorized as carbonaceous or noncarbonaceous (Fig. 6.3); these two populations probably formed at different heliocentric distances. Groups that formed from carbonaceous-chondrite-like precursors (CC irons) include IIC, IID, IIF, IIIF and IVB; groups that formed from noncarbonaceous precursors (NC irons) include IAB, IC, IIAB, IIE, IIG, IIIAB, IIIE and IVA (e.g., Kruijer et al., 2017a, 2017b, 2017c; Poole et al., 2017). Most ungrouped magmatic irons are carbonaceous (Rubin, 2018a).

An **iron meteorite** can be defined as *a metallic-Fe-Ni-rich meteorite that formed on an asteroid from a metal-rich melt.*

There are two principal groups of **stony-iron meteorites**: pallasites (rich in metallic Fe-Ni and olivine) and mesosiderites (composed mainly of metallic Fe-Ni and basaltic and diogenitic silicate clasts).

The pallasites are divided into several categories: main-group pallasites (PMG, ~82%), Eagle Station pallasites (PES, ~7%), and ungrouped pallasites (~11%), a few of which are commonly called "pyroxene pallasites." PMG pallasites are noncarbonaceous materials; PES pallasites are isotopically related to carbonaceous chondrites. Although pallasites are generally considered to be roughly 50% olivine and 50% metallic Fe-Ni, the actual modal abundance of olivine in PMG samples ranges from ~35 to 85 vol.% (e.g., Buseck, 1977). Phillips County (PMG-an) is unique in containing abundant troilite; its troilite/(metallic-Fe-Ni) ratio ranges from 5 to 10.

Mesosiderites are stony-iron breccias averaging ~50 wt.% metallic Fe-Ni, troilite, and schreibersite (with an actual range of 18 to 90 wt.% metallic Fe-Ni, 0.6 to 14 wt.% troilite, and 0 to 0.1 wt.% schreibersite) and ~50 wt.% silicate (with an actual range of 28 to 80 wt.%). Silicates include major orthopyroxene and plagioclase, minor Ca-pyroxene, olivine, and tridymite, and minor-to-trace amounts of merrillite, chromite, and ilmenite. Most of the silicate material is basaltic or diogenitic in mineralogy and composition.

A stony-iron meteorite is *a nonchondritic asteroidal meteorite rich in metallic Fe-Ni (± troilite) and silicate.*

"Nebula" is the Latin word for "fog." The **solar nebula** refers to the disk-shaped cloud of gas and dust rotating around the proto-Sun and early Sun within which the planets, asteroids, and comets accreted. It is the remnant of the parental molecular cloud that had either contracted under its own gravity or been compressed by shock waves from a nearby supernova.

A lexicographer, as defined by Samuel Johnson (1755), is "a writer of dictionaries; a harmless drudge, that busies himself in tracing the original, and detailing the signification of words." A **meteorite lexicographer** is *a didactic drudge, obsessed with meteoritics, who attempts to craft a set of unambiguous, non-conflicting definitions of meteorite-related terms* (see casuistry).

In Drudge Johnson's great 1755 work, *A Dictionary of the English Language*, he defined science as: "(1) knowledge; (2) certainty grounded on demonstration; (3) art attained by precepts, or built on principles; (4) any art or species of knowledge; (5) one of the seven liberal arts, grammar, rhetorick, logick, arithmetick, musick, geometry, astronomy." Johnson offered quotations illuminating each meaning.

4

Ordinary-Chondrite Abundances in Different Mass Ranges

As of this writing, the Meteoritical Bulletin Database lists six ordinary chondrites known to have delivered at least a metric ton of recovered material to the Earth's surface: H5 Jilin (fall), 4 tons (Fig. 4.1); L3-6 NWA 869 (find), 2 tons; L5 Ghubara (find), 1.75 tons; L5 Tsarev (find), 1.23 tons; H5 Kunya-Urgench (fall), 1.1 tons; LL5 Chelyabinsk (fall), 1 ton. (Most meteorites with masses ≥1 ton are metal-rich, i.e., irons and pallasites.) At the other end of the mass spectrum are 19 individual Antarctic finds with masses of ~10 mg each: eighteen L4, L5, and L6 chondrites from Grove Mountains and one H5 (Yamato 8333). The mass range of recovered ordinary chondrites thus spans more than eight orders of magnitude.

Carbonaceous-chondrite masses span more than seven orders of magnitude. There is only one recovered sample that exceeds a metric ton (CV3 Allende, 2 tons); four samples have masses of ~100 mg each (CO3 ALH 83026; $CV3_{OxA}$ Khatyrka;[1] CM2 MIL 13065; CM2 MIL 13233). No smaller individual carbonaceous chondrites have been reported.

Ordinary and carbonaceous chondrites together account for 98.5% of all chondritic meteorites (finds and falls). The mean carbonaceous-chondrite/ordinary-chondrite (CC/OC) number ratio (for all samples) is 0.05. Although only small numbers of samples are involved, the CC/OC number ratio for samples ≥1 metric ton is (1 / 7) = 0.17 (i.e., 3.4× larger than the ratio for all samples). This suggests that there may be significant differences in the CC/OC number ratio at different mass intervals. Such differences could result from processes affecting OC and CC precursor meteoroids in interplanetary space, differential winnowing of meteoroids traversing the Earth's atmosphere, differences in sample friability, variations in resistance to terrestrial weathering, and meteorite collection inefficiencies and biases. (Collection biases among ordinary chondrites are illustrated by the high proportion of type-3 samples among NWA finds [mainly collected by private individuals] compared to those among observed falls and Oman and Antarctic finds [mainly collected by scientific expeditions] [Rubin, 2021].)

In contrast to the low value (0.05) of the mean CC/OC number ratio, the ratio is very high for the smallest objects encountered by the Earth. Unmelted stratospheric interplanetary dust particles (IDPs) are typically ~10 μm in size (e.g., Brownlee et al., 1995). For the purpose of calculation, Rubin (2018b) assigned IDPs a density of ~3 g cm^{-3} (Table 2 of Kohout et al., 2014) and derived an average IDP mass of ~10^{-12} kg. Nearly all IDPs are carbonaceous: they have CI- or CM-like bulk chemical compositions (e.g., Schramm et al., 1989; Bradley, 2005) even though many differ from CI and CM chondrites mineralogically and isotopically (e.g., Bradley et al., 1988; Bradley, 2005). The CC/OC ratio of 100 for IDPs shown in Fig. 4.2 is an arbitrary lower limit; it is likely much higher.

Micrometeorites are typically 50–500 μm in size (e.g., Engrand and Maurette, 1998). Rubin

[1] The small Khatyrka $CV3_{OxA}$ find is associated with Al-rich quasicrystals: icosahedrite ($Al_{63}Cu_{24}Fe_{13}$) and $Al_{62}Cu_{31}Fe_7$; decagonite ($Al_{71}Ni_{24}Fe_5$)), metallic Al alloys (khatyrkite ($CuAl_2$); cupalite (CuAl); stolperite (AlCu); hollisterite (Al_3Fe); kryachkoite ($(Al,Cu)_6(Fe,Cu)$); steinhardite (Al,Ni,Fe); Fe-poor steinhardite; Al-bearing taenite (γ-(Fe,Ni))), Al-bearing sulfide ($Fe_{0.84}Al_{0.04}S_{1.12}$), Al-oxide (corundum ($Al_2O_3$)) and Al-bearing oxide (trevorite ($NiFe_2O_4$)) (e.g., Bindi et al., 2011, 2015, 2016; Hollister et al., 2014; Lin et al., 2017; C. Ma et al., 2017; Andronicos et al., 2018). A natural origin for many of these phases remains controversial (e.g., MacPherson et al., 2016).

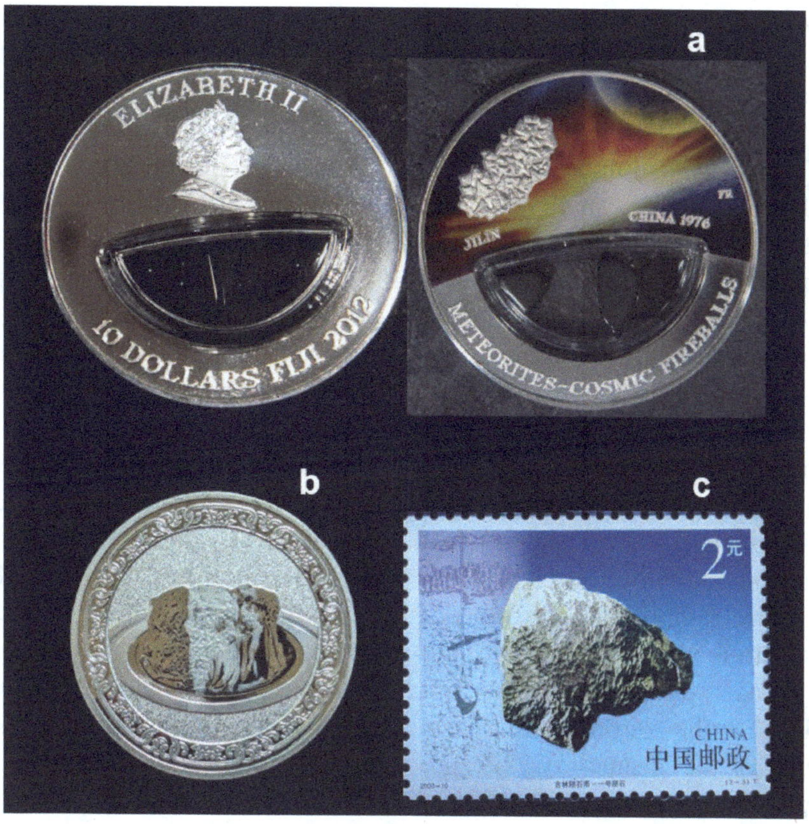

Fig. 4.1 Memorabilia of the Jilin H5-chondrite fall in China on March 8, 1976. The meteorite shower yielded ~4 metric tons of recovered samples. (a) Ten-dollar silver proof coin from the Republic of Fiji, dated 2012, depicting the Jilin H5 fall. (left) Obverse featuring a portrait of Queen Elizabeth II. (right) Reverse has a rendering of a Jilin specimen and a partially colored fireball accompanying the meteoroid. A small piece of the actual Jilin meteorite is encased in the reverse of the coin. A total of 999 coins were minted. (b) Obverse of a 44-mm-diameter metal token of the largest piece of Jilin (1770 kg), unofficially known as Jilin No. 1. The meteorite is on permanent display at the Jilin City Meteorite Museum. (c) Government-issued two-yuan postage stamp from China showing an artist's rendering of the same large piece of Jilin. The inscription at the bottom center of the stamp reads: Jilin Meteor Shower No. 1 Meteorite. The stamp was issued on June 21, 2003.

(2018b) assumed the average micrometeorite to be a sphere with a radius of 100 μm and mass of ~10^{-8} kg. Walter et al. (1995) showed that of the 262 unmelted micrometeorites in their sample set, 259 are carbonaceous-chondrite-like and three are ordinary-chondrite-like (most closely resembling H chondrites). This yields a CC/OC ratio of 86. As pointed out by Jonker et al. (2024), the proportion of micrometeorites derived from ordinary chondrites appears to increase with particle size: ~7% (<500 μm), 38% (>500 μm), ~70% (>800 μm) (van Ginneken et al., 2017; Suavet et al., 2011).

The CC/OC ratio is also very high for the largest objects currently encountered by the Earth. Fireball meteoroids range from ~0.2 to 10 m in size (Ceplecha et al., 1998); they are much larger than the estimated average diameter of observed ordinary- or carbonaceous-chondrite recovered falls (~0.11 m; Rubin, 2018b). Meteoroids become luminous at 80–90 km as they plunge through the increasingly dense layers of the Earth's atmosphere. More friable meteoroids break up more readily; they have shorter trajectories and higher end heights. Ceplecha and McCrosky (1976) divided fireballs into three categories based on end height: ordinary-chondrite-like (stony), carbonaceous-chondrite-like, and cometary. Ceplecha (1994) listed the percentages of fireballs in these categories. Rubin (2018b) calculated model masses for the larger fireball meteoroids: 1 m – 1.4×10^3 kg; 2 m – 1.1×10^4 kg; 5 m – 1.7×10^5 kg; 10 m – 1.4×10^6 kg. The derived CC/OC ratios for fireball meteoroids are: 1 m–4.8; 2 m–7.8; 5 m–14; 10 m–30.

In the mass range of most normal-sized meteorite falls (0.01–20 kg), the CC/OC number ratio is low (0.04–0.05); the ratio increases at greater mass ranges: at ≥200 kg, the ratio is 0.09; at ≥500 kg, the ratio is 0.20. The CC/OC number ratio also increases from 0.05 to 0.16 for small meteorite finds (10^{-3} to 10^{-4} kg). Thus, a diagram of CC/OC number ratio versus object mass forms a quasi-U-shaped curve (Fig. 4.2).

Different mechanisms are responsible for bringing extraterrestrial material of different sizes to the Earth: Poynting-Robertson drag affects particles such as IDPs and micrometeorites. The Keplerian orbit of a particle around the Sun causes solar radiation to appear to originate from slightly ahead of the particle; this process is analogous to the aberration of starlight. The radiation force impedes the particle's orbital motion, slowing it down and causing its semimajor axis to decrease (Wyatt and Whipple, 1950). This is the *Poynting-Robertson effect*. A recent study of the cosmic-ray exposure (CRE) ages of micrometeorites collected from urban areas and Antarctica shows that a significant fraction originates in the Outer Solar System (Feige et al., 2024). Poynting-Robertson processes affect particles mainly in the range of micrometers to centimeters. Its efficiency is a function of particle size and density – a 1-cm particle would take 10,000 times longer to spiral into the Sun than a 1-μm particle of the same bulk density.

Carbonaceous chondrites appear to be associated with C-type asteroids (Hiroi et al., 1996). The high CC/OC number ratio among small particles (and large objects) reflects the predominance of C asteroids beyond 3 AU (e.g., Fig. 1a of Bell et al., 1989; Fig. 9 of DeMeo and Carry, 2013). It is probable that the IDPs, ejected from C asteroids, spiral into the Inner Solar System via Poynting-Robertson drag. Large numbers reach Earth before being destroyed by collisions or swept out of the Solar System by planetary gravitational perturbations.

Most normal-sized meteorites (derived from meteoroids with pre-atmospheric diameters of ~2 to 20 cm) as well as 1- to 10-m-sized fireball meteoroids are affected mainly by Yarkovsky forces. The *seasonal Yarkovsky effect* is a thermal radiation force that exerts its greatest influence at different points in a body's orbit around the Sun. In simple cases (where the spin axis of a body is within the orbital plane), the Sun preferentially heats one hemisphere (let's call it the "northern" hemisphere) during northern summer and heats the other ("southern") hemisphere during northern winter. Because of thermal inertia (the resistance of a substance to temperature change), the body experiences the maximum radiative force after a delay, by which time it has moved some distance along its orbit. There is a net drag force that drives the body inward toward the Sun (e.g., Bottke et al., 2006).

The scattering of solar radiation off an asteroid and the emission of thermal radiation from an asteroid affect the spin rate and the obliquity of an asteroid's rotational axis. These changes, dubbed the Yarkovsky–O'Keefe–Radzievskii–Paddack (YORP) effect, modify the radial drift of an asteroid caused by Yarkovsky forces.

About 25% of Near-Earth Asteroids (NEAs) are <100 m in diameter (Fig. 5 of Stuart and Binzel, 2004), well within the size range of objects significantly affected by Yarkovsky forces (Farinella et al., 1998). These bodies were likely transferred to the Inner Solar System from the asteroid belt (where C asteroids predominate) by such forces. The predominance of carbonaceous-chondrite-like asteroids among NEAs (Stuart and Binzel, 2004) can account for the high CC/OC number ratio among the meter-sized fireball meteoroids (Fig. 4.2).

Bodies in the 0.03–0.3-m size range should spend appreciably more time in interplanetary space than larger objects due to the diminished influence of the seasonal Yarkovsky effect on smaller bodies. Because of collisions, a longer sojourn favors less friable objects. (Evidence that CI and CM carbonaceous chondrites, for example, are more friable than ordinary chondrites [e.g., Mason, 1963] includes their much shorter cosmic-ray exposure ages [0.05–2 Ma vs. 3–60 Ma; Eugster, 2003].)

It thus seems likely that the low CC/OC number ratio among most normal-sized meteorites is due

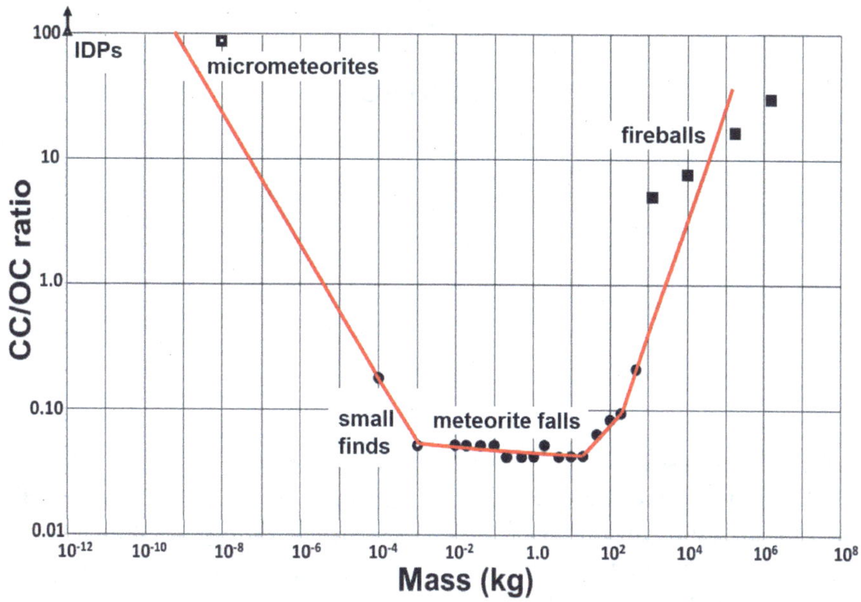

Fig. 4.2 Log–log plot of the number ratio of carbonaceous to ordinary chondrites (the CC/OC ratio) for different masses. The ratio is very high for the smallest objects – interplanetary dust particles (IDPs) and micrometeorites; it is moderately high for 1- to 10-m-sized fireball meteoroids. The x-axis values for meteorite falls are for objects with the indicated mass and all greater masses; the x-axis values for small meteorite finds are for objects with the indicated mass and all smaller masses. Literature data show that almost all unmelted stratospheric IDPs are carbonaceous chondrites; an arbitrary lower limit of 100 is plotted with an arrow for the IDPs. Micrometeorite data are from the sample set of Walter et al. (1995); data for meteorite falls and small finds are from the Meteoritical Bulletin Database; fireball data are from Table 4 of Ceplecha (1994). The masses of the fireball meteoroids are rough estimates based on an assumed density of 2.65 g/cm^{-3}. Diagram modified from Rubin (2018b).

to preferential collisional destruction of carbonaceous bodies during the long time objects in this size range spend in interplanetary space. The increase in the CC/OC number ratio among larger meteorite falls (Fig. 4.2) reflects the increased influence of the seasonal Yarkovsky effect on these larger objects. There was thus a tendency for their meteoroid precursors to have spent less time in interplanetary space, where friable CC materials are preferentially destroyed by collisions.

5

Asteroids

> The time will come when diligent research over long periods will bring to light things which now lie hidden. A single lifetime, even though entirely devoted to the sky, would not be enough for the investigation of so vast a subject ... And so this knowledge will be unfolded only through long successive ages. There will come a time when our descendants will be amazed that we did not know things that are so plain to them... Many discoveries are reserved for ages still to come, when memory of us will have been effaced.
>
> Seneca, *Natural Questions*

5.1 The Discovery of Asteroids

In 1766, Professor Johann Daniel Titius of the University of Wittenberg devised a simple geometric progression that approximated the distances from the Sun of the known planets (Mercury, Venus, Earth, Mars, Jupiter, Saturn). He inserted this mathematical formulation into a footnote in a French astronomy book (Charles Bonnet's *Contemplation de la Nature*) that he was translating into German. A version of the progression begins with the numbers 0, 3, 6, 12, 24, 48, 96 (each successive number after 3 being twice the preceding number). After adding 4 to each number and dividing the sum by 10, one gets 0.4, 0.7, 1.0, 1.6, 2.8, 5.2, 10.0. If 2.8 is omitted from the sequence, these numbers closely match the heliocentric distances (in astronomical units) of the planets: Mercury – 0.39, Venus – 0.72, Earth – 1.00, Mars – 1.52, Jupiter – 5.20, Saturn – 9.54. Johann Elert Bode of the Berlin Observatory noticed that footnote and in 1772 mentioned the formulation in an unsourced footnote in an astronomy book (entitled *Anleitung zur Kenntniss des gestirnten Himmels*). In subsequent editions of the book, Bode credited the progression to Titius. However, few astronomers paid serious attention to this mathematical gimmick.

Then, on March 13, 1781, using a large telescope he had constructed, William Herschel came across an object in the constellation Gemini. It had a slightly extended disk and did not exhibit scintillation. Herschel followed the object for several weeks, and after consulting with several astronomers, concluded that he had discovered a new planet. Johann Elert Bode of the Berlin Observatory suggested the name Uranus, after the god of sky and heaven in Greek mythology. Uranus has a mean apparent magnitude of 5.68, within the range of naked-eye observability (which extends to magnitude 6). Bode was excited by the discovery and gathered previous observations of Uranus by astronomers who had mistaken it for a star. Using these observations and Herschel's data, Bode was able to determine Uranus' orbit. Kepler's third law (the square of an orbital period is proportional to the cube of its semimajor axis) showed that the new planet resided 19.19 AU from the Sun. Bode pointed out this value is quite close to the next number in the Titius sequence: 19.6. This validation of the sequence immediately suggested that there might be a heretofore undiscovered planet lurking at 2.8 AU, between the orbits of Mars and Jupiter. (As early as 1596 Kepler had commented on "the huge gap between Jupiter and Mars.") Titius' mathematical gimmick came to be known as the Titius-Bode law or more commonly, Bode's law.

In 1796, Bode and Baron Franz Xavier von Zach of Seeberg began organizing a group of two dozen astronomers to search for "the missing planet" orbiting between Mars and Jupiter. Von Zach called the group the "celestial police" and assigned each astronomer a different section of the ecliptic to observe.

On January 1, 1801, the first night of the nineteenth century, Guiseppe Piazzi, director of the Palermo Observatory, was recording stellar positions in the constellation Taurus while compiling a star catalogue. He repeated his observations the following night and noticed that an eighth-magnitude "fixed star" had moved 4 arcminutes retrograde. It moved a similar amount on both the third and fourth nights. Piazzi was certain he had discovered a "new star" (i.e., a planet), but, acting conservatively, called the body a comet. Nevertheless, he wrote to a colleague that: "I have announced this star as a comet, but since it is not accompanied by any nebulosity and, further, since its movement is so slow and rather uniform, it has occurred to me several times that it might be something better than a comet" (Herschel, 1781). By January 12, 1801, the object had reversed direction and began moving prograde. Its movements were characteristic of a superior planet at opposition. However, unlike the known planets (including distant Uranus), the new object maintained a pinpoint starlike telescopic image. When Piazzi received Von Zach's letter inviting him to join the celestial police, he realized he had already discovered the missing planet. He proposed the name "Ceres Ferdinandea," honoring both the Roman goddess of grain and King Ferdinand IV of Sicily and Naples. The name was later shortened to Ceres.

Piazzi's observations of Ceres were complete until February 11, 1801, when illness forced him into bed, away from the telescope. He had made only about 20 observations covering just 3° of the sky. Piazzi's letter announcing the discovery of the missing planet did not reach Bode until March 20, 1801. By that time, Ceres was lost in the Sun's glare. Astronomers despaired of it ever being recovered. The twenty-four-year-old German mathematician Carl Friedrich Gauss took on the challenge of determining Ceres' orbit from the scant observations then available. He developed a new method to calculate an orbit based in part on Kepler's second law (an orbiting object sweeps out equal areas in equal times). Only three orbital positions were required. These calculations aided Ceres' subsequent recovery by Von Zach on December 7, 1801. With additional observations, it became clear that Ceres' semimajor axis is 2.77 AU, nearly identical to the position indicated by Bode's Law: 2.8 AU.

But Ceres was not alone in the Jupiter–Mars gap. On March 28, 1802, astronomer Heinrich Wilhelm Matthäus Olbers was trying to find Ceres from his home observatory when he stumbled upon a second body at 2.77 AU; he named it Pallas, after Pallas Athena. Astronomer Karl Ludwig Harding discovered a third body, Juno, at 2.67 AU on September 1, 1804. Olbers found Vesta (2.36 AU) on March 29, 1807 and gifted the naming rights to Gauss. Gauss named the new body after the Roman goddess of home and hearth.

No new asteroids were discovered until 1845. Prior to that time, the four new bodies were referred to simply as planets. Now, there were eleven of them. Olbers proposed in 1803 that Ceres and Pallas were remnants of an exploded planet; the subsequent discovery of Juno and Vesta supported this hypothesis. The exploded-planet model was endorsed by a few later astronomers (e.g., Newcomb, 1861; Ovenden, 1972, 1973; Van Flandern, 1978).

However, the diversity of asteroidal spectral characteristics (Section 5.3) and the great textural, compositional, and isotopic variability of meteorites (e.g., Davis, 2005) demonstrate that the roughly 800,000 (and counting) numbered asteroids and their more numerous unnumbered congeners are not remnants of a single disrupted body. (Nevertheless, individual Hirayama families [groups of asteroids with similar orbital elements, e.g., semimajor axis, eccentricity, inclination] probably originated from a single large asteroid that was collisionally disrupted [e.g., Hirayama, 1918, 1928, 1933].)

Instead of all asteroids being derived from a single body, 10–100-km asteroids accreted separately within the first few million years of Solar System history from smaller planetesimals experiencing inelastic collisions at low relative velocities (e.g., Johansen et al., 2014).

5.2 Links between Meteorites and Asteroids

One of the earliest suggestions for the provenance of meteorites (including those we now call ordinary chondrites) was the Moon. Olbers proposed in 1795 that meteorites may come from lunar volcanoes (Burke, 1986). Pierre-Simon Laplace suggested independently in 1802 that meteorites could be lunar volcanic ejecta. The idea was quickly endorsed by Laplace's protégé, Jean-Baptiste Biot. Also on board was the French mathematician Siméon Denis Poisson, who examined Laplace's hypothesis quantitatively and came out in favor of a lunar origin for meteorites.

There are, of course, meteorites from the Moon, but they constitute just 0.9% of all officially approved meteorites in our collections. Martian meteorites constitute 0.5%. The vast majority of meteorites comes from asteroids; ordinary chondrites make up ~85% of known meteorites (excluding possible pairings). Numerous links exist between meteorites and asteroids that pertain to size, orbital characteristics, and physical properties. These links confirm Humboldt's (1849) assertion that meteorites are "the smallest of all asteroids."

1. The metallographic cooling rates of many stony and iron meteorites are in the range of 1–100°C/Ma (e.g., H4 Kesen, 20 K Ma^{-1}; H5 Allegan, 15 K Ma^{-1}; L4 Bjurböle, 2 K Ma^{-1}; LL3.2 Krymka, 1 K Ma^{-1}; Taylor et al., 1987). This suggests that the meteorites were derived from bodies a few hundred kilometers in size (equivalent to that of many large asteroids).
2. The presence of noble gases implanted by the solar wind in meteorite regolith breccias indicates that these rocks come from the surfaces of bodies too small to retain significant atmospheres (i.e., subplanetary bodies). Examples include the Dimmitt (Rubin et al., 1983a) and Abbott (Fodor et al., 1976) H chondrites.
3. The relatively high concentrations of gas from the solar wind in meteorite regolith breccias (e.g., the Fayetteville H chondrite; Wieler et al., 1989a) are consistent with implantation at a distance of 2–3 AU from the Sun (overlapping the location of the asteroid belt). This is because the solar-wind flux falls off with the square of the distance from the Sun.
4. The old formation ages (~4.56 Ga; e.g., Fig. 9.11 of McSween and Huss, 2010) of most meteorites, including ordinary chondrites, indicate that they were derived from small, subplanetary bodies that cooled very early in the history of the Solar System.
5. The gravitational influence of Jupiter is expected to perturb some main-belt asteroids into resonances that facilitate transfer to the Inner Solar System.
6. The cosmic-ray exposure (CRE) ages of many stony meteorites are tens of millions of years; for example, L chondrites have CRE peaks at ~28 Ma and ~40 Ma (Fig. 4 of Eugster et al., 2006). These ages are consistent with the time inferred for bodies in the asteroid belt to have been perturbed into Earth-crossing orbits.
7. The orbits of more than a thousand fireballs (including dozens that yielded recovered meteorites) were determined to be very similar to those of Earth-crossing asteroids. These objects include H5 Lost City (McCrosky et al., 1971; Ceplecha and Revelle, 2005); L5 Park Forest (Brown et al., 2004) and LL6 Stubenberg (Spurný et al., 2016). By 2025, there had been 75 meteorites with photographic orbits and known orbital parameters; samples include 60 ordinary chondrites (H, L, LL, L/LL), six carbonaceous chondrites, two enstatite chondrites, six asteroidal achondrites, and an iron.
8. The spectral reflectivities of some asteroids (e.g., 6 Hebe; Fig. 4.21 of Burbine, 2017) closely match those of meteorites measured in the lab (e.g., H5 Allegan; Burbine et al., 2003) if space weathering is taken into account (e.g., Sasaki et al., 2001).
9. The brecciated nature of many meteorites (such as the H-chondrite regolith breccia Plainview [1917] [Fodor and Keil, 1976a], the LL5 breccia Paragould [Roy and Wyant, 1955], and the LL3-6 fragmental breccia Adzhi-Bogdo [Bischoff et al., 1993a]) shows that such meteorites suffered numerous collisions, consistent with the extensive cratering evident on asteroids.
10. Material returned from asteroid 25143 Itokawa matches that of LL chondrites (e.g., Nakamura et al., 2011, 2012; Nakashima et al., 2013).
11. Samples returned from asteroid 162173 Ryugu resemble CI carbonaceous chondrites (e.g., Yokoyama et al., 2022).
12. Samples returned from asteroid 101955 Bennu resemble CI and some ungrouped C2 carbonaceous chondrites (e.g., Lauretta et al., 2024).
13. The Dawn spacecraft's measurements of the composition and mineralogy of the surface of asteroid 4

Vesta match the characteristics of HED (howardite-eucrite-diogenite) achondritic meteorites.

14. Asteroid 2008 TC3, which crashed into the Nubian Desert in Sudan on October 7, 2008, ~19 hours after its discovery, yielded the Almahata Sitta polymict breccia two months later. Roughly 1000 fragments were (officially and unofficially) recovered. Although most examined fragments are ureilitic (~67%), also present are enstatite chondrites (~26%), ordinary chondrites (mainly H, but also one L and one LL fragment) (~4%), and rare specimens of carbonaceous chondrites, R-like chondrites, and a unique aubrite (Bischoff et al., 2022).
15. Asteroid 2018 LA, which crashed into Botswana on June 2, 2018, about eight hours after its discovery, yielded the Motopi Pan howardite (24 pieces with a total recovered mass of 214.5 g).
16. Asteroid 2023 CX1, which crashed into Normandy, France on February 13, 2023, less than seven hours after its discovery, yielded more than 30 individual pieces of the Saint-Pierre-le-Viger L5-6 chondrite.
17. Asteroid 2024 BX1, which crashed into northern Germany on January 21, 2024, about 90 minutes after its discovery, yielded several dozen specimens of the Ribbeck aubrite.

5.3 Spectral Properties of Ordinary Chondrites and Potentially Related Asteroids

5.3.1 Spectra of Common Minerals

Most of the radiation coming from the Sun is in the range of visible and infrared light with wavelengths of ~0.4 to 2.5 µm. Radiation that strikes an asteroid is either absorbed, reflected directly from the surface, or refracted through near-surface particles and scattered toward the detector. Many meteoritic minerals absorb light within the Sun's peak wavelength range. Particular absorption features are characteristic of specific minerals; they are also a function of mineral composition and particle size. In common practice, the reflectance spectrum of an asteroid is plotted as the reflectance (normalized to unity at a particular wavelength) versus wavelength.

Olivine has an asymmetric absorption feature near ~1 µm. It is composed of three separate features at ~0.9 µm, ~1.1 µm, and ~1.25 µm caused by the transitions of 3d electrons in Fe^{2+} in two distorted octahedral crystallographic sites (M1 and M2). There is also a strong absorption feature below ~0.7 µm (extending into the ultraviolet) formed by an oxygen-(ferrous iron) charge transfer. The

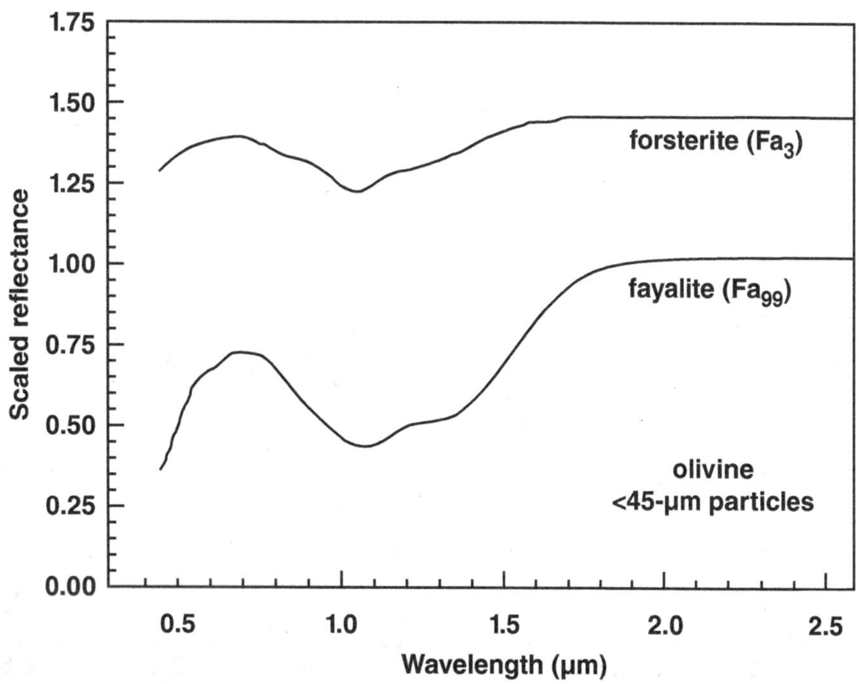

Fig. 5.1 Reflectance spectra of olivine (<45-µm-sized particles) covering almost all the compositional range from Fa_{99} (nearly pure fayalite) to Fa_3 (close to forsterite). Spectra are scaled to unity at 1.9 µm. Image modified from Sunshine and Pieters (1998).

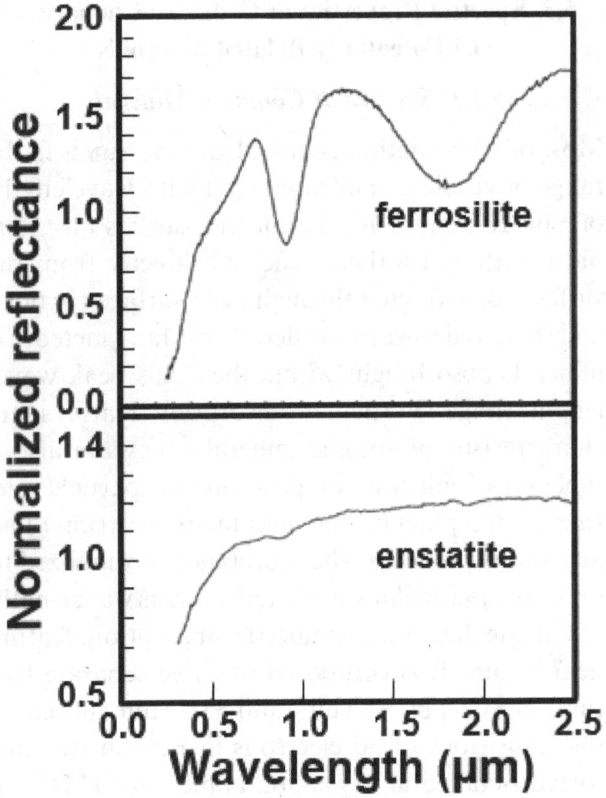

Fig. 5.2 Reflectance spectra in the visible and near-infrared of low-Ca pyroxene as represented by its two compositional endmembers – ferrosilite ($FeSiO_3$) and enstatite ($MgSiO_3$). Spectra are normalized to unity at 0.55 μm. Diagram modified from Burbine (2017).

overall reflectance spectrum of magnesian olivine (forsterite; Mg_2SiO_4) is shallower (flatter) than that of ferroan olivine (fayalite; Fe_2SiO_4) (Fig. 5.1).

Low-Ca pyroxene is a solid solution of two end-member compositions – enstatite ($MgSiO_3$) and ferrosilite ($FeSiO_3$). Ferrosilite has two prominent absorption features due to Fe^{2+}: one feature (Band I) is at ~0.9–1.0 μm, due to transitions of 3d electrons in the M1 and M2 sites; the second feature (Band II) is at ~0.9–2.0 μm, due to transitions of 3d electrons in the M2 site alone. Enstatite lacks ferrous iron and does not show either absorption feature; it has a relatively flat, featureless spectrum (Fig. 5.2).

The major metallic Fe-Ni minerals in meteorites are low-Ni iron (kamacite) with $\lesssim$7.5 wt.% Ni and taenite with ~20–40 wt.% Ni. Neither phase exhibits absorption features. The reflectance spectrum of powdered metallic Fe-Ni from the Odessa IAB iron meteorite illustrates this featureless spectrum, although it is slightly red (Fig. 5.3).

5.3.2 *Ordinary-Chondrite Spectra*

The minerals olivine, low-Ca pyroxene, and metallic Fe-Ni together constitute 84 vol.% of average H chondrites, 80 vol.% of average L chondrites,

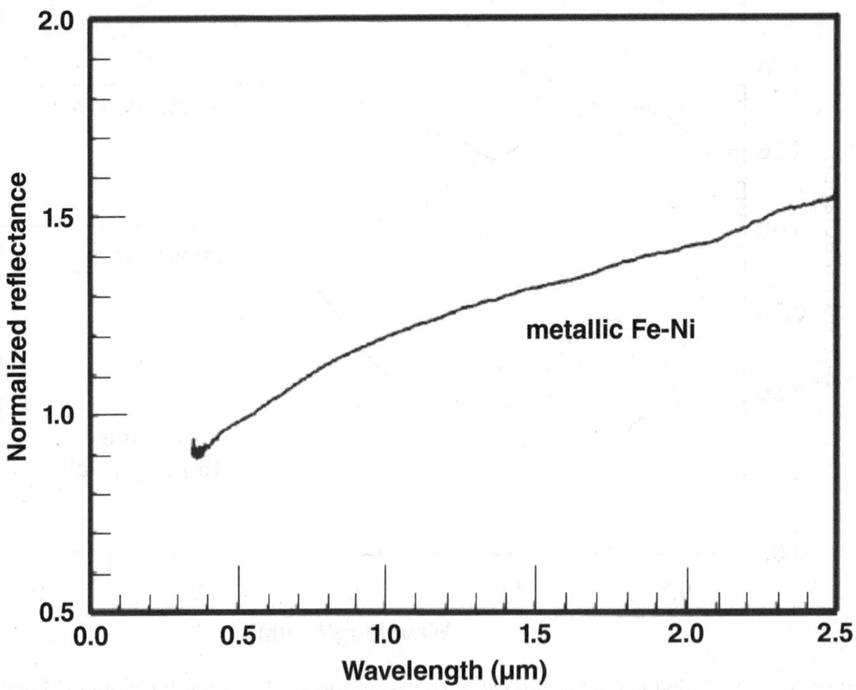

Fig. 5.3 Reflectance spectrum in the visible and near-infrared of metallic Fe-Ni derived from the Odessa IAB iron meteorite that was ground to a powder with a grain size of <45 μm. The spectrum is normalized to unity at 0.55 μm. Absorption features are absent. Diagram modified from Burbine (2017).

and 78 vol.% of average LL chondrites (Table 5 of Jarosewich, 1990; Table 5.1 of Hutchison, 2004). To a large extent, the spectrum of ordinary chondrites is a combination of the absorption features of these three phases. However, metallic Fe-Ni is featureless and serves only to decrease the depths of the olivine and low-Ca pyroxene absorption features. Thus, the spectra of ordinary chondrites in Fig. 5.4 were determined from powdered samples from which metallic Fe-Ni was sieved and removed.

The principal absorption features in ordinary chondrites are due to olivine and low-Ca pyroxene (e.g., Gaffey, 1976; Dunn et al., 2010a, 2010b). The deep absorption feature at ~0.9–1.0 μm is caused by the overlap of absorptions in olivine and Band I of low-Ca pyroxene. The shallow feature at ~1.9 μm is due to Band II of low-Ca pyroxene. In going from H to L to LL, there is a systematic increase in the olivine/(low-Ca pyroxene) modal ratio (H: 1.4; L: 2.0; LL: 3.6) (Table 5.1 of Hutchison, 2004). There is also a systematic increase in the proportion of ferrous iron in olivine and low-Ca pyroxene as exemplified by the increasing mean fayalite contents of olivine and ferrosilite contents of low-Ca pyroxene in equilibrated OC (H4-6: Fa18.8, Fs17.2; L4-6: Fa24.7, Fs21.3; LL4-6: Fa29.4, Fs24.1; Table 1 of Rubin, 2022a). These two characteristics account for the greater depths in L and LL chondrites than in H chondrites of the ~0.9–1.0-μm absorption feature relative to the ~1.9-μm feature.

5.3.3 Asteroid Spectra

Th'ayre shows such Meteors, as none can see,
Not onely what they meane, but what they bee.

John Donne, *The First Anniversarie. An Anatomie of the World*

Asteroids were initially classified by spectral reflectivities into two major groups (Fig. 5.5): C (resembling carbonaceous chondrites) and S (resembling silicaceous materials or stony or stony-iron meteorites) (Zellner, 1973; Chapman et al., 1975). S-type asteroids show two strong absorption features – one below about 0.7 μm (due to olivine), the other near 0.9 μm (due to the overlap between

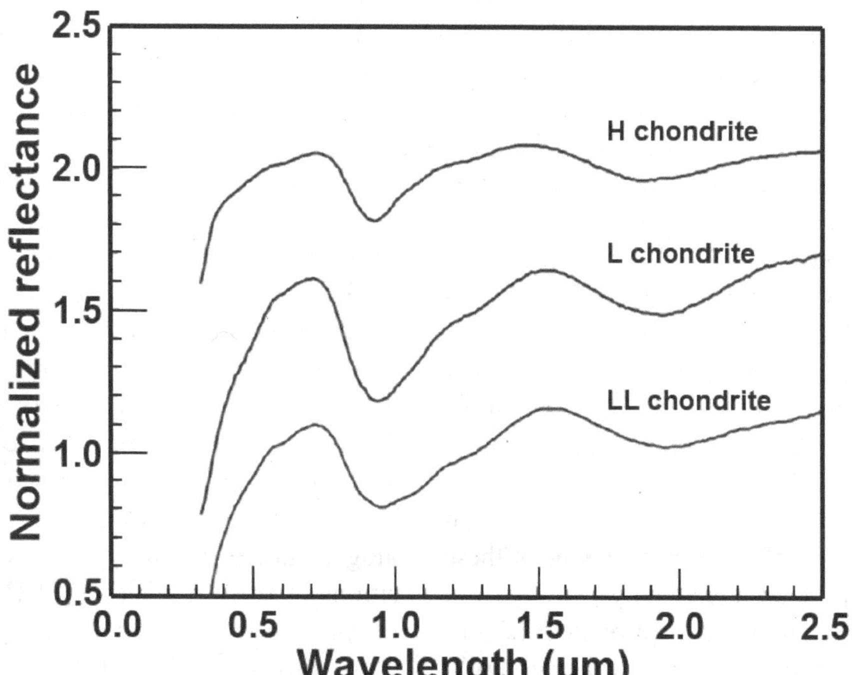

Fig. 5.4 Reflectance spectra in the visible and near-infrared of ordinary chondrites: H5 Allegan, L6 Girgenti, and LL4 Greenwell Springs. The spectra are normalized to unity at 0.55 μm and offset in reflectance. Bulk FeO increases from H to L to LL as do the widths of the troughs at about 1 and 2 micrometers. The spectra are determined on powdered samples from which the metallic Fe-Ni was sieved and largely removed. Diagram modified from Burbine (2017).

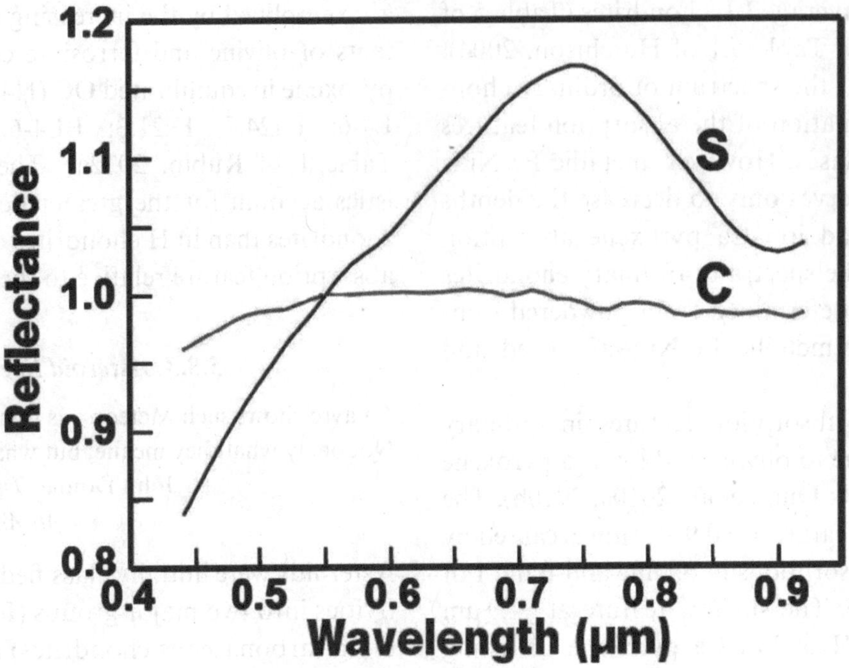

Fig. 5.5 Reflectances in the visible light range of the two main classes of asteroids (C and S), mainly in the visible light range determined by the Small Main Belt Asteroid Spectroscopic Survey (SMSS). The C spectra are relatively flat. Diagram modified from Delbo et al. (2012).

absorptions in olivine and low-Ca pyroxene); they have redder spectra and higher albedos than C asteroids. C-type asteroids have relatively flat spectra with weak or nonexistent absorption bands; they also have lower albedos than S asteroids.

Additional taxonomic classes of asteroids were introduced by Zellner and Gradie (1976) including M-types (presumed to consist mainly of metallic Fe-Ni) and E-types (presumed to contain mostly enstatite). Gaffey et al. (1993) divided the S class into seven subtypes – S(I), S(II), S(III), S(IV), S(V), S(VI), S(VII) – based on differences in their absorption features thought to represent distinct olivine/(low-Ca pyroxene) modal ratios. The most common asteroid taxonomy in current use is the Bus–DeMeo system (DeMeo et al., 2009); it divides asteroids into 25 classes and subclasses based on their visible and near-infrared spectra. One of these classes is the Q type, characterized by a strong 1-μm absorption feature, a weak absorption near 1.3 μm, and a 2-μm absorption feature of varying depth. Binzel et al. (2004) created a Q-complex class, folding in the Q types and the closely related Sq types from the Bus–DeMeo system.

Many workers have pointed out that S-complex asteroids are the most abundant types in the inner portion of the asteroid belt and that the most common meteorites observed to fall are ordinary chondrites. This suggests a statistical connection (e.g., Chapman, 1996). In addition, the spectra of S asteroids and ordinary chondrites are of similar shape – both have prominent absorption features near 1 μm and 2 μm. It is tempting to conclude that S asteroids are the parent bodies of ordinary chondrites, but their spectra are not identical.

This is well illustrated by the comparison of LL chondrites to samples returned from near-Earth asteroid 25143 Itokawa (S-type; 0.535 × 0.294 × 0.209 km). The Hayabusa spacecraft visited this asteroid in 2005 and returned a few thousand particles with a total mass of <1 mg. The mineralogy, mineral chemistry, and O-isotopic compositions of these particles most closely match those of LL5 chondrites (e.g., Nakamura et al., 2011; Yurimoto et al., 2011). It is clear that Itokawa is an LL-chondrite asteroid, but the spectrum of Itokawa is redder and has shallower absorption features than powdered LL chondrites (Fig. 5.6).

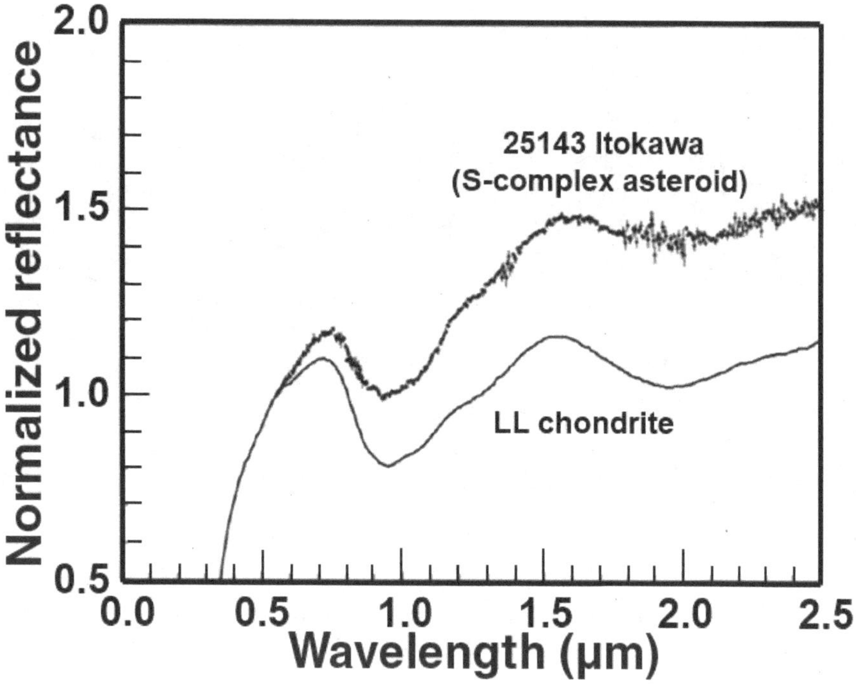

Fig. 5.6 Reflectance spectrum in the visible and near-infrared of LL4 ordinary chondrite Greenwell Springs compared to that of the S-complex asteroid 25143 Itokawa. The two spectra are normalized to unity at 0.55 μm. The spectra have similar shapes but different slopes. The spectrum of Greenwell Springs was determined on a powdered sample from which the metal was sieved and removed. Diagram modified from Burbine (2017).

Because Itokawa is composed of LL-chondritic material, some process must have affected the spectral characteristics of this asteroid that did not significantly affect LL chondrites that fall to Earth. Examination of Itokawa particles showed they are enriched in solar-wind-implanted noble gases, indicating derivation from a surficial regolith. Some LL chondrites (e.g., Holman Island; St. Mesmin) are regolith breccias, also enriched in gases from the solar wind.

The process that affected Itokawa is space weathering; it affects all substantial bodies in the Solar System that lack an atmosphere. The effect was first noticed on the Moon – young, fresh craters have bright ejecta deposits and prominent rays, while such features are absent around older craters (Fig. 5.7). Bright lunar surface material darkens with time. Among Apollo samples, fine-grained soils have redder spectra, lower albedos, and weaker absorption features than rock fragments. The potential causes of space weathering include micrometeorite bombardment (producing nanophase iron particles, abbreviated as $npFe^0$), sputtering of nanophase iron atoms (caused by the implantation of solar wind particles into the regolith), and the striking of the surface by solar radiation and galactic and solar cosmic rays. Samples returned by the Chang'e 5 lunar mission include pyroxene grains affected by space weathering. These grains consist of three distinct zones – an outer splash melt coating overlying quenched melt derived from the underlying pyroxene grain (Xian et al., 2024). Particle size is also important: MacLennan et al. (2024a) found that OC powders consisting of coarser particles produce a lower spectral slope in the 1-μm absorption band that mimics space-weathering effects.

Although Itokawa is of LL-chondrite composition, it has a redder spectrum and shallower absorption bands than LL chondrites (Fig. 5.6), consistent with the asteroid's having been subject to space weathering. In fact, among the Itokawa samples are nanophase iron particles and Fe-sulfide particles that occur within a thin layer at many grain surfaces. Some particles also have amorphous or vesicular rims.

Enhanced-color images of asteroid 243 Ida show that fresh craters are brighter and bluer than old

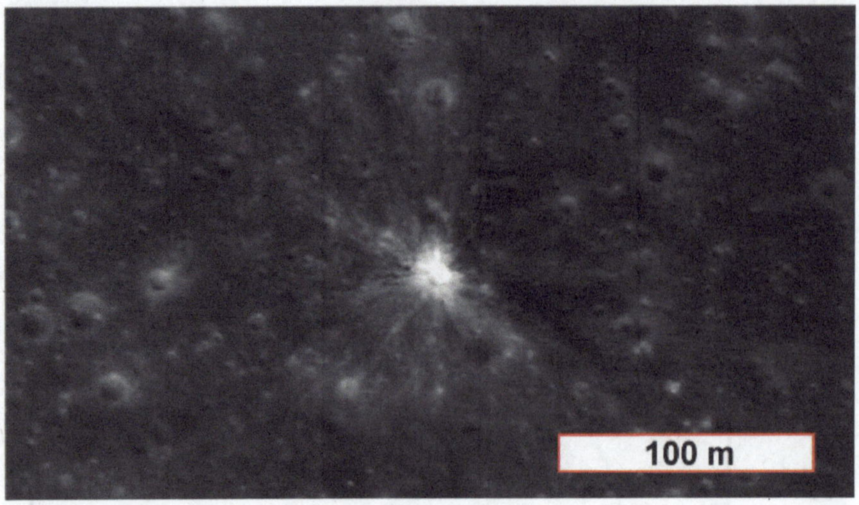

Fig. 5.7 This ~10-m-diameter impact crater formed on the Moon between August 1971 and September 2009. Bright subsurface material was excavated, forming ejecta deposits and crater rays. The ejecta are more than three times as reflective as the surrounding basalt. Older craters in the image presumably formed with bright rays and ejecta deposits, but they currently lack such features due to subsequent space weathering. This is a portion of an image taken by the Lunar Reconnaissance Orbiter Narrow Angle Camera (LROC NAC image M108971316L; NASA/GSFC/Arizona State University).

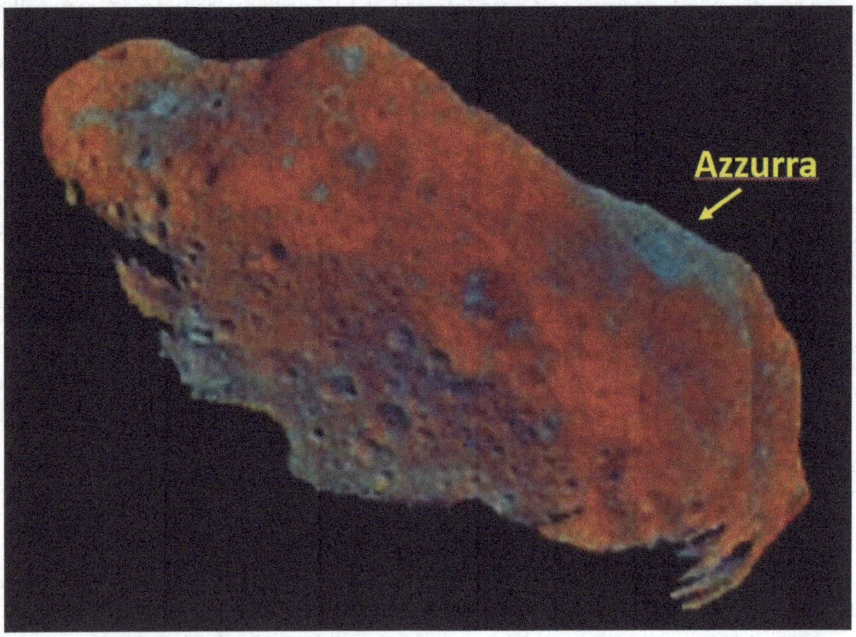

Fig. 5.8 Asteroid 243 Ida. Enhanced false-color view produced from the violet, 0.76 μm and 0.99 μm images of JPL photograph P-44131. Blue regions include small fresh craters and the ejecta surrounding the 7-km Azzurra crater; older, regolithic regions are red.

terrains (Fig. 5.8), consistent with the general process of space weathering on asteroids. The spectra of various features on Ida show that the younger features are less red and have deeper absorption bands than older features (Fig. 5.9). The youngest, bluest features with the deepest absorption bands are fresh craters; the Azzurra impact crater and its surrounding ejecta have somewhat shallower absorption bands. Old, weathered terrains are red and have shallow absorption features. In contrast, L chondrites are far less red and have prominent absorption bands.

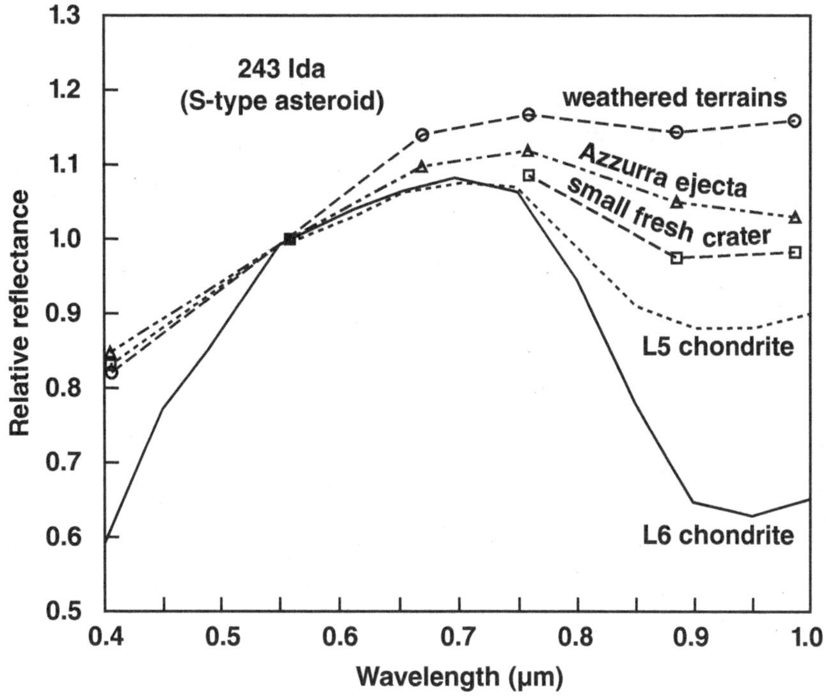

Fig. 5.9. Reflectance spectra for regions of asteroid 243 Ida. Spectra for weathered terrains are flatter and reddish; spectra for a small fresh crater and ejecta from the Azzurra crater show absorption features and are bluer. Also shown are typical spectra for L5 and L6 chondrites. Image modified from Chapman (1996).

Several workers have linked S asteroids, particularly of type S(IV), to ordinary chondrites (e.g., Wetherill, 1985, 1987). Asteroid 6 Hebe (205 × 185 × 170 km) was proposed to be the parent body of H chondrites (Farinella et al., 1993a; Gaffey and Gilbert, 1998; Bottke et al., 2010). The spectrum of Hebe is of similar shape to H chondrites but is redder and has shallower absorption features (Fig. 5.10), consistent with space weathering of H-chondrite material. Two arguments in favor of H chondrites being derived from Hebe are: (1) the location of Hebe near the 3:1 mean motion and ν_6 secular resonances (consistent with efficient transport of material knocked off of Hebe to the Inner Solar System) and (2) the distribution of CRE ages in H chondrites (consistent with derivation from Hebe). If Hebe is indeed the parent body of H chondrites, then it is currently supplying 32% of all meteorites to the Earth.

Q-complex asteroids have spectra similar to those of S asteroids, but Q asteroids are less red and have stronger absorption features. They resemble ordinary-chondrite-like material that has experienced only minimal space weathering. For example, the spectrum of Q-type asteroid 1862 Apollo (diameter 1.5 km) closely resembles that of LL chondrites (Fig. 5.11) with only a somewhat shallower 2-μm absorption feature. A small asteroid such as Apollo has very low surface gravity and is thus less likely to maintain a substantial regolith than a much larger asteroid (e.g., Housen et al., 1979); this may inhibit space weathering, making the spectral match much closer. In general, Q asteroids constitute a much larger proportion of small asteroids than of large asteroids (Fig. 6 of Binzel et al., 2019). Their small sizes inhibit regolith production, thereby diminishing the effects of space weathering.

MacLennan et al. (2024a, 2024b) investigated OC powders and found that linear-discriminant analysis is a useful tool for minimizing non-compositional parameters including grain size, thereby allowing equilibrated H, L, and LL chondrites as well as unequilibrated OC to be distinguished among asteroid spectra.

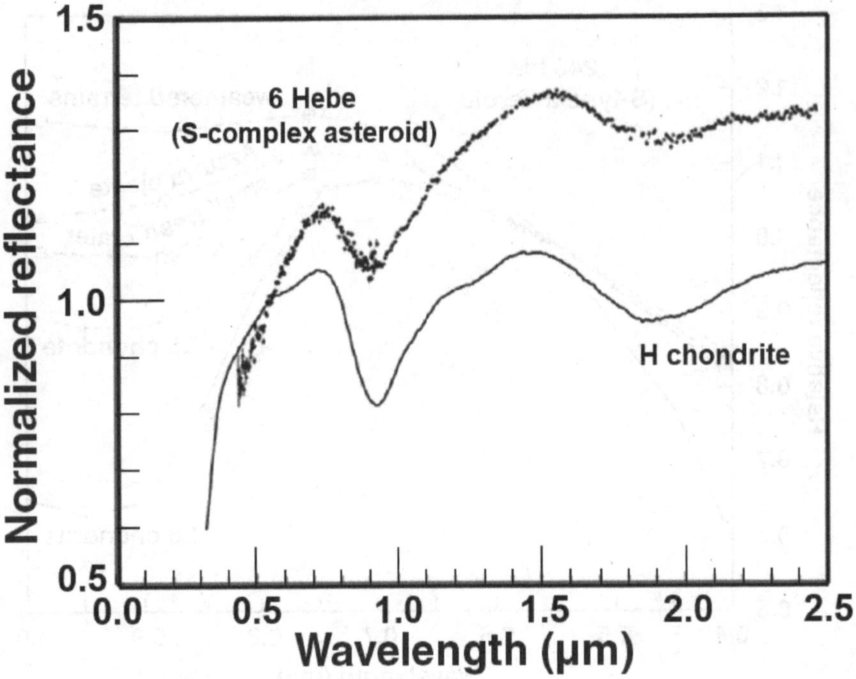

Fig. 5.10 Reflectance spectrum in the visible and near-infrared of an H chondrite (H5 Allegan) and of the S-complex asteroid 6 Hebe. The two spectra are normalized to unity at 0.55 µm. The spectrum of Allegan was determined on a powdered sample from which the metal was sieved and removed. Diagram modified from Burbine (2017).

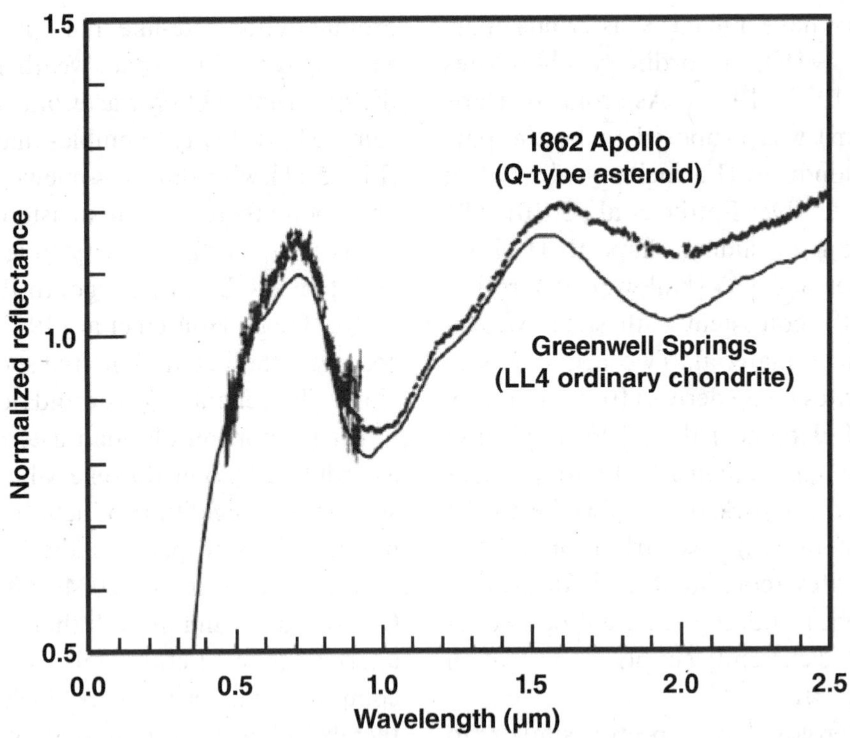

Fig. 5.11 Reflectance spectrum in the visible and near-infrared of LL4 ordinary chondrite Greenwell Springs compared to that of the Q-type asteroid 1862 Apollo. The two spectra are normalized to unity at 0.55 µm. The spectrum of Greenwell Springs was determined on a powdered sample from which the metal was sieved and removed. Diagram modified from Burbine (2017).

5.4 Internal Structures of Ordinary-Chondrite Parent Asteroids

Ordinary-chondrite parent asteroids likely exceeded 100 km in diameter (Wood, 1967); most estimates are on the order of 200–300 km (e.g., Trieloff et al., 2003; Kleine et al., 2008; Henke et al., 2012; Blackburn et al., 2017; Edwards and Blackburn, 2020). Lithostatic pressures within these bodies are very low. The internal pressure of a typical chondritic asteroid with a radius of 100 km is ~10 bars (i.e., ~1 MPa) (Weidenschilling and Cuzzi, 2006), a pressure reached on Earth at a depth of only 10 m. Even Vesta (the second-largest asteroid, with a radius of ~285 km) has a central lithostatic pressure of only ~800 bars (~80 MPa) (Ryan and Melosh, 1998). Ceres (the largest asteroid, also a dwarf planet, with a mean radius of ~470 km) has a central pressure of $\lesssim$2 kb ($\lesssim$200 MPa) (Zolotov, 2009).

There are two basic models for the structure of OC parent bodies (Fig. 5.12):

(1) *The onion-shell model* (Fig. 5.12a) posits that ordinary-chondrite asteroids originally consisted of nested layers in which higher petrologic types (i.e., more highly metamorphosed samples) occurred in progressively deeper layers (e.g., Heyes, 1978; Dodd, 1969, 1981; Anders, 1978; Minster and Allègre, 1979; Pellas and Storzer, 1981; Miyamoto et al., 1981; Trieloff et al., 2003; and Skirak et al., 2025). About 80–90 vol.% of the OC bodies would consist of deeply buried type-5 and type-6 materials (Bennett and McSween, 1996; Ciesla et al., 2013). This model is consistent with their experiencing internal heating (e.g., via the decay of ^{26}Al, $t_{1/2}$ = 717 ka). It is possible that an onion-shell asteroid could have been punctured by collisions during the epoch of thermal metamorphism, causing some deeply buried materials to be excavated, deposited near the surface, and quickly cooled (e.g., Scott et al., 2011). It is also possible that an onion-shell body experienced collisional disruption and gravitational reassembly before the body had cooled through ~500°C (Taylor et al., 1987), at which point diffusion in taenite becomes so sluggish that metallographic cooling rates are locked in. It is likely there was heating of different portions of the near-surface regions of the asteroid by stochastic collisions (e.g., Rubin, 2004b, 2005b; Friedrich et al., 2014a). Various combinations of these models (e.g., Ruzicka et al., 2015; Ruzicka and Hugo, 2018) are also viable. An extension of the onion-shell model hypothesizes that chondrite parent bodies grew to ~400 km in diameter within 1.5 Ma of CAI formation and developed a molten interior overlain by primitive undifferentiated material (e.g., Elkins-Tanton et al., 2011). If this model were correct, one might expect some chondritic breccias to contain abundant achondritic fragments (in contrast to impact-melt-rock clasts or rare calcic-plagioclase-bearing chondrule-like objects). None have been reported (see Section 14.6).

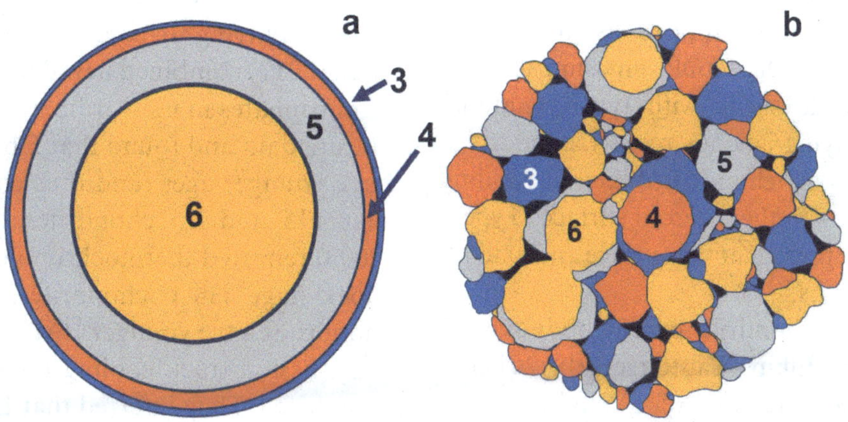

Fig. 5.12 Schematic cross-sectional diagrams of two distinct models for the structure of ordinary-chondrite parent asteroids. (a) The onion-shell model, positing internal heating after accretion, produced a stratified structure in which higher petrologic types were more deeply buried. The proportions of the petrologic types are based on those of H-chondrite falls (Scott and Rajan, 1981). (b) The metamorphosed-planetesimal model in which peak metamorphic temperatures occurred in small planetesimals before the $\gtrsim$100-km-diameter parent asteroid accreted. Planetesimals of different metamorphic grades are all jumbled up. Diagram based on Scott and Rajan (1981).

(2) *The metamorphosed-planetesimal model* (Fig. 5.12b) posits that ordinary-chondrite asteroids were initially composed of small (<10-km-radius) planetesimals in which peak metamorphic temperatures were reached prior to the accretion of the primary >100-km-diameter asteroid (e.g., Monnereau et al., 2013). A scenario consistent with this model is one in which most accretion occurred after much of the initial ^{26}Al had decayed (Scott and Rajan, 1981). Hugo et al. (2020) used optical microscopy, electron backscatter diffraction (EBSD), and transmission electron microscopy (TEM) to study two LL6 breccias (Saint-Séverin and Elbert). They found that Saint-Séverin was deformed by shock while the rock was already hot and was then quenched. In contrast, Elbert was shocked while it was cold and was then cooled within a locally warm region of the LL parent body. Macente et al. (2025) used EBSD and x-ray computed tomography (XCT) to find that the Ozerki L6 chondrite was still warm when it suffered impact-induced deformation. These results suggest that small planetesimals, heated mainly by the decay of ^{26}Al prior to accretion, were shocked and rapidly cooled after they accreted into a large asteroid; other materials were shocked after the asteroid had already cooled. One problem with the metamorphosed-planetesimal model is that these small planetesimals, isolated in interplanetary space, would have cooled so quickly (due to their high surface-area/volume ratios) they probably could not have reached high metamorphic temperatures.

There is a third model, suggested only occasionally, which posits that chondrites are derived from the outer layers of differentiated asteroids (e.g., Kleine et al., 2005). One could envision a series of nested chondritic shells with type-6 material immediately overlying igneous rocks (achondrites) and type-5 and type-4 chondritic material residing progressively closer to the asteroid surface. Type-3 material would be present at the surface and extending to shallow depths.

The model of a chondrite parent body with a molten core (e.g., Elkins-Tanton et al., 2011) is supported by the detection of a unidirectional magnetization in some CV chondrites (Butler, 1972; Nagata and Funaki, 1983; Carporzen et al., 2010; Weiss et al., 2010) consistent with an internal dynamo (Funaki and Wasilewski, 1999). A similar model was proposed for ordinary-chondrite parent asteroids; Bryson et al. (2019) measured magnetic remanence within nanostructures in metal veins of H6 Portales Valley and concluded that a magnetic field was present on the H-chondrite parent body about 100 Ma after CAIs. The field lasted for tens to hundreds of years. They found the magnetic record in H6 Portales Valley to be inconsistent with external fields and concluded that the H-chondrite asteroid possessed an advecting metal core. However, alternatives exist to the generation of a magnetic field by a dynamo: For example, Muxworthy et al. (2017) found that low-velocity impacts into porous bodies bearing a fine-grained matrix can generate brief temperature excursions in matrix grains, producing a magnetic fabric. Uehara et al. (2011) studied the magnetic microstructures of metal grains and found that the remanent magnetization of tetrataenite in equilibrated OC does not indicate the presence of an internal dynamo during the first 10–50 Ma after peak metamorphism. Furthermore, the occurrence in individual OC regolith breccias of taenite grains with M-shaped profiles that appear to have cooled at vastly different rates (to be discussed) is inconsistent with ~200-km-sized OC parent bodies having molten interiors.

There are studies offering support for the onion-shell model: (1) Göpel et al. (1994) found a negative correlation between petrologic type and phosphate Pb–Pb ages in a suite of seven H4 to H6 chondrites, consistent with later U–Pb closure times occurring at greater depths. (2) Blackburn et al. (2017) combined new data on the Pb–Pb ages of phosphates in nine ordinary chondrites with literature data and found that H6 and L6 chondrites have younger ages (and thus cooled more slowly) than H5 and L5 chondrites. (3) Trieloff et al. (2003) reported thermochronological data on low-shock-stage H4–6 chondrites, showing that H6 chondrites have younger ^{40}Ar/^{39}Ar ages and slower ^{244}Pu fission-track cooling rates than H4 and H5 chondrites. They inferred that H6 chondrites were more deeply buried than H4 and H5 chondrites. (4) Pellas and Storzer (1981) determined the ^{244}Pu fission-track cooling rates of a total of six H4, H5, and H6 chondrites; they found that higher petrologic types cooled more slowly. (5) Walton et al.

(2023) determined U–Pb ages of apatite in nine L chondrites and posited early parent-body fragmentation at >4530 Ma and cooling of an onion-shell structure until 4500 Ma, followed by collisional heating from 4480 to 4460 Ma.

However, whereas Trieloff et al. (2003) plotted nine H chondrites to find an inverse correlation between petrologic type and ^{244}Pu fission-track cooling rate, Taylor et al. (1987) determined the metallographic cooling rates of 12 H chondrites as well as five equilibrated H-chondrite clasts from regolith breccias and found no significant correlation with petrologic type. Scott et al. (2014) studied 30 H3–6 chondrites, also finding that H-chondrite cooling rates do not decrease systematically with increasing petrologic type (Fig. 5.13). Because metallographic and fission-track cooling-rate studies of the same meteorites yield consistent results (Fig. 4 of Taylor et al., 1987), the larger data sets should take precedence. In that case, one could reasonably conclude that there is no correlation between petrologic type and cooling rate. These studies do not support the onion-shell model.

But the controversy continued. Harrison and Grimm (2010) used literature data to show that an H-chondrite parent asteroid 150–260 km in diameter with an onion-shell structure and a thermally insulating regolith could account for 95% (20 out of 21) of the closure age data and 87% (62 out of 71) of the cooling rates. They concluded that the H asteroid was not collisionally disrupted in its early history.

Grimm (1985) attempted to reconcile these apparently contradictory results by proposing that the H-chondrite asteroid initially had an onion-shell structure but was collisionally disrupted prior to cooling below ~500°C (the temperature at which diffusion in taenite grains becomes too sluggish to preserve cooling-rate information). If disruption of an OC asteroid was followed by random mixing of fragments and gravitational reassembly within a period of a few days, the resulting metallographic cooling rates of ordinary

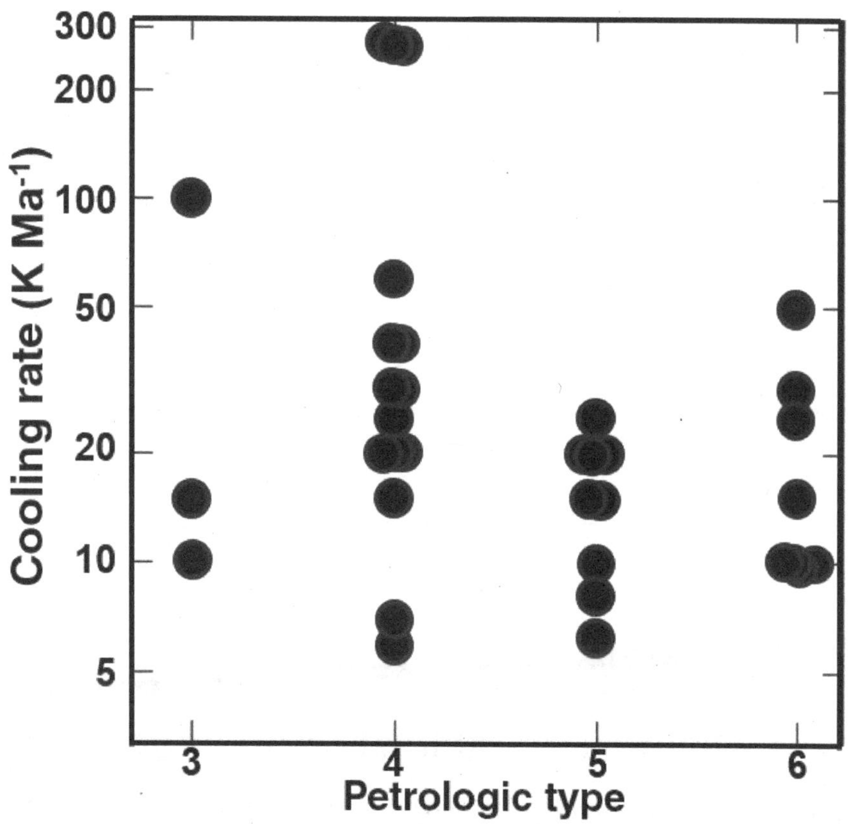

Fig. 5.13 Metallographic cooling rate vs. petrologic type for 35 H3–6 chondrites showing no correlation. Thirty samples are from Scott et al. (2014); five are from the literature. Diagram based on Scott et al. (2014).

chondrites would reflect the depths of their taenite grains (and surrounding silicate material) in the hot reassembled bodies. A similar model was proposed by Ganguly et al. (2013).

Blackburn et al. (2017) tried to harmonize Pb–Pb results and the metallographic cooling-rate data by proposing that: (1) The Pb-phosphate system recorded an onion-shell structure in the H and L asteroids >30 Ma after CAI formation, (2) at 60 ± 10 Ma, these bodies were catastrophically disrupted and subsequently reaccreted into rubble piles, and (3) some hot, deeply buried type-5 and type-6 chondrites were brought close to the surface and cooled quickly, destroying any inverse correlation between petrologic type and metallographic cooling rate. Edwards and Blackburn (2020) determined the $^{207}Pb/^{206}Pb$ dates of phosphate grains in LL chondrites and concluded that the LL parent asteroid had a total duration of cooling ≥75 Ma.

There are petrologic data supporting the hypothesis that OC asteroids were disrupted and reassembled. Measurements of individual taenite grains in OC regolith breccias indicate that these grains cooled at rates ranging between 1 and ~1000°C Ma^{-1} (Rubin, 1982; Fig. 5.14), consistent with burial depths ranging from about one kilometer to several tens of kilometers. Collisions energetic enough to excavate material from 10 to 30 km on

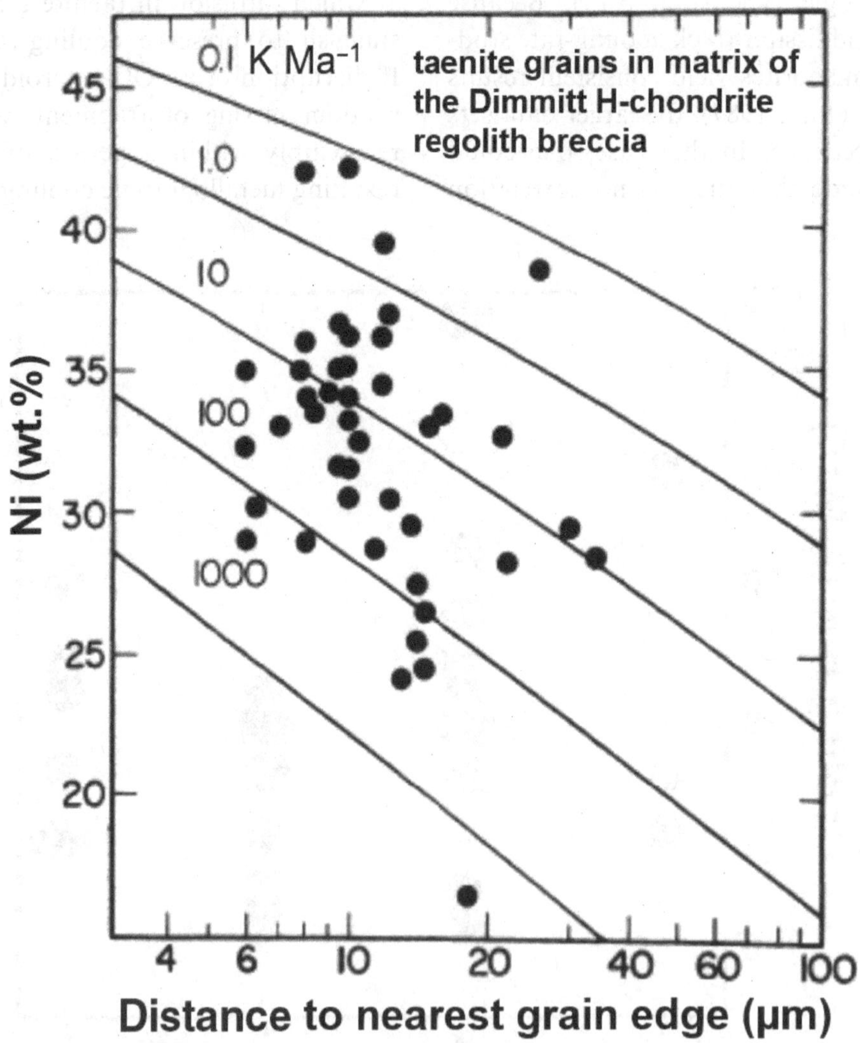

Fig. 5.14 Metallographic cooling rates of compositionally zoned taenite grains in the matrix of the Dimmitt H-chondrite regolith breccia. Each taenite grain has an M-shaped profile indicative of slow cooling. The grains record cooling at rates ranging from ~1 to 1000 K Ma^{-1}, consistent with their derivation from a variety of depths in the parent asteroid. Diagram modified from Rubin et al. (1982b).

Fig. 5.15 Cartoon illustrating the model of asteroid collisional disruption, fragmentation, and gravitational reassembly. Many large asteroids may have undergone this process. Diagram modified from Rubin et al. (1982b).

100–300-km-diameter bodies would tend to disrupt those bodies. It therefore seems plausible that the OC parent asteroids were collisionally disrupted, jumbled up, and gravitationally reassembled to create bodies wherein materials originating at different depths were all mixed together (e.g., Rubin et al., 1982b; Fig. 5.15). Impact gardening of the reassembled blocks brought small OC fragments together and scattered them across the rubble-pile surface. Theoretical modeling suggests this fate may be common among large asteroids (e.g., Fujiwara et al., 1977; Greenberg et al., 1978; Hartmann, 1978, 1979; Davis et al., 1979).

Theoretical studies of the collisional evolution of asteroids (e.g., Davis et al., 1979; Hartmann, 1979) indicate that, across a broad range of impact energies, collisions can disrupt asteroids while failing to impart sufficient kinetic energy to most of the fragments for permanent dispersal. Much of the disrupted material does not reach escape velocity and falls back together to form a rubble-pile asteroid that efficiently retains much of the heat generated in the collision. Y. Li et al. (2025) argued that just such a process affected the LL parent body.

There is increasing astronomical evidence that many asteroids are rubble piles with no effective tensile strength and with densities in the range of 1.3 g cm^{-3} (e.g., Cheng and Barnouin-Jha, 1999; Bottke et al., 1999; Veverka et al., 1999; Britt and Consolmagno, 2001). (1) For example, asteroid 253 Mathilde appears to be a rubble pile of low density (1.3 ± 0.2 g cm^{-3}; Veverka et al., 1997, 1999; Yeomans et al., 1997) and high porosity (~50%; Cheng and Barnouin-Jha, 1999). (2) Harris (1996) identified an abrupt cutoff at the high end of the distribution of asteroidal rotation rates and concluded most asteroids are rubble piles with no tensile strength. (3) Bottke et al. (1999) examined the shape of Near-Earth Asteroids (NEAs) and concluded that some may have been molded by planetary tides; if so, they must be rubble piles with no effective tensile strength

This possibility is supported by radar images of some NEAs in which such bodies appear to be aggregated boulders (e.g., Ostro et al., 1995; Benner et al., 2015). As discussed by Lauretta (2024), recently visited Near-Earth Asteroids (101955 Bennu [carbonaceous], 162173 Ryugu [carbonaceous], and 25143 Itokawa [LL ordinary chondrite]) are also rubble piles (e.g., Fig. 5.16).

A snapshot of individual planetesimals that have accreted into a larger body can be seen in the primitive Kuiper Belt object (486958) Arrokoth (Fig. 5.17). Spencer et al. (2020) superimposed a geomorphological map on this image, showing the approximate borders of different geological units. If such planetesimal conglomerates were assembled long after the decay of short-lived radionuclides, they may have retained substantial amounts of primitive volatile ices (Birch and Umurhan, 2024). The surface of Arrokoth contains methanol ices; when exposed to galactic cosmic rays, some

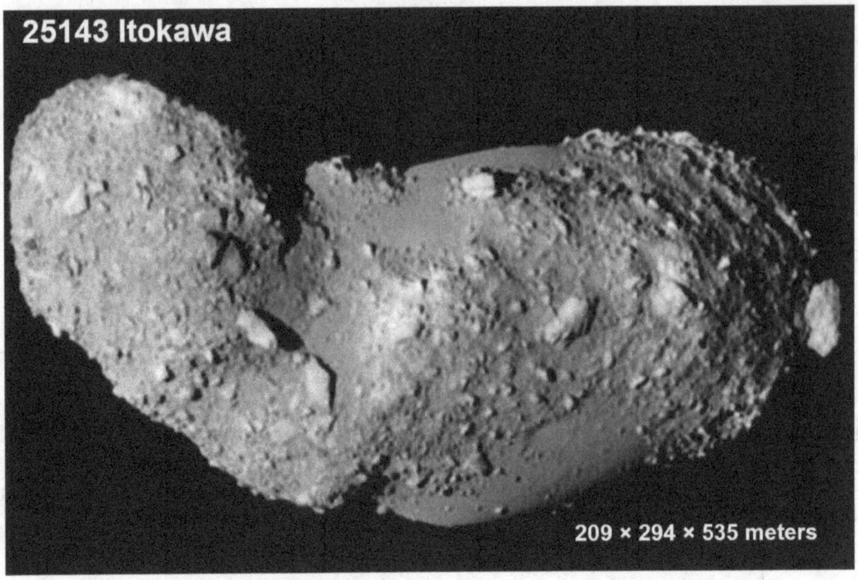

Fig. 5.16 Asteroid Itokawa, a sub-kilometer rubble-pile Near-Earth Asteroid of LL-chondrite composition. It was visited by the Japanese spacecraft *Hayabusa*, which collected particles in 2005 and returned them to Earth in 2010. Image courtesy of JAXA, the Japanese Aerospace Exploration Agency.

of the methanol is converted into sugars (including ribose and glucose) and polycyclic aromatic hydrocarbons (PAHs) (C. Zhang et al., 2024).

Several moons in the Outer Solar System (e.g., Saturn's Mimas, Enceladus, and Tethys, and Uranus' Miranda; Fig. 5.18) may also have been catastrophically disrupted by giant impacts and gravitationally reassembled (e.g., Movshovitz et al., 2014).

Lucas et al. (2020) used a recently developed thermometer involving rare earth elements (REE) in pyroxene to obtain cooling rates for 18 ordinary chondrites (mainly of petrologic type 6). They found that OC cooling rates through the Ca-in-olivine closure temperatures (~700–800°C) were three to six orders of magnitude faster than cooling rates sensitive to lower-temperature (≤500°C) regimes (i.e., metallographic, $^{40}Ar/^{39}Ar$ ages, ^{244}Pu fission-track methods). Lucas et al. suggested that OC parent asteroids initially had onion-shell structures but were later disrupted and rapidly reassembled into asteroidal rubble piles.

Information about the structure of OC parent bodies can also be gleaned from the study of shock effects. There is a positive correlation between petrologic type and mean shock stage among ordinary chondrites (Stöffler et al., 1991; Rubin, 1995a, 2004b; Bischoff et al., 2019). Rubin (2004b) compiled data from Grady (2000) on 1651 OC (not taking possible pairings into account) and found that mean shock stage increased with petrologic type. Rubin (2025) updated this analysis with data on 20857 OC (H, L, LL, H/L, L/LL) from the Meteoritical Bulletin Database compiled on January 11, 2024 (Fig. 5.19): type-3, mean shock stage = S2.18 ± 0.87 ($n = 1901$); type-4, S2.23 ± 0.92 ($n = 3944$); type-5, S2.49 ± 1.01 ($n = 8496$); type-6, S2.93 ± 1.17 ($n = 6516$). (For this correlation, $r = 0.945$, $n = 4$, $2\alpha = 0.075$, significant at the 92.5% confidence level.) The difference in mean shock stage between type-3 and type-4 OC is not great, but type-5 and type-6 OC are progressively and significantly more shocked. (There is no a priori reason that the correlation should be linear; there are no linear relationships between shock stage and shock pressure or between petrologic type and peak metamorphic temperature.)

The correlation is well illustrated by the distribution of differences in shock stages between type-3 and type-6 OC (Fig. 5.20): type-3 (18.0% S1; 57.1% S2; 15.8% S3; 7.6% S4; 1.4% S5; 0.1% S6) and type-6 (9.4% S1; 31.3% S2; 27.7% S3; 22.2% S4; 7.5% S5; 1.9% S6). A one-sided, two-sample t test shows that the set containing type-6 OC has

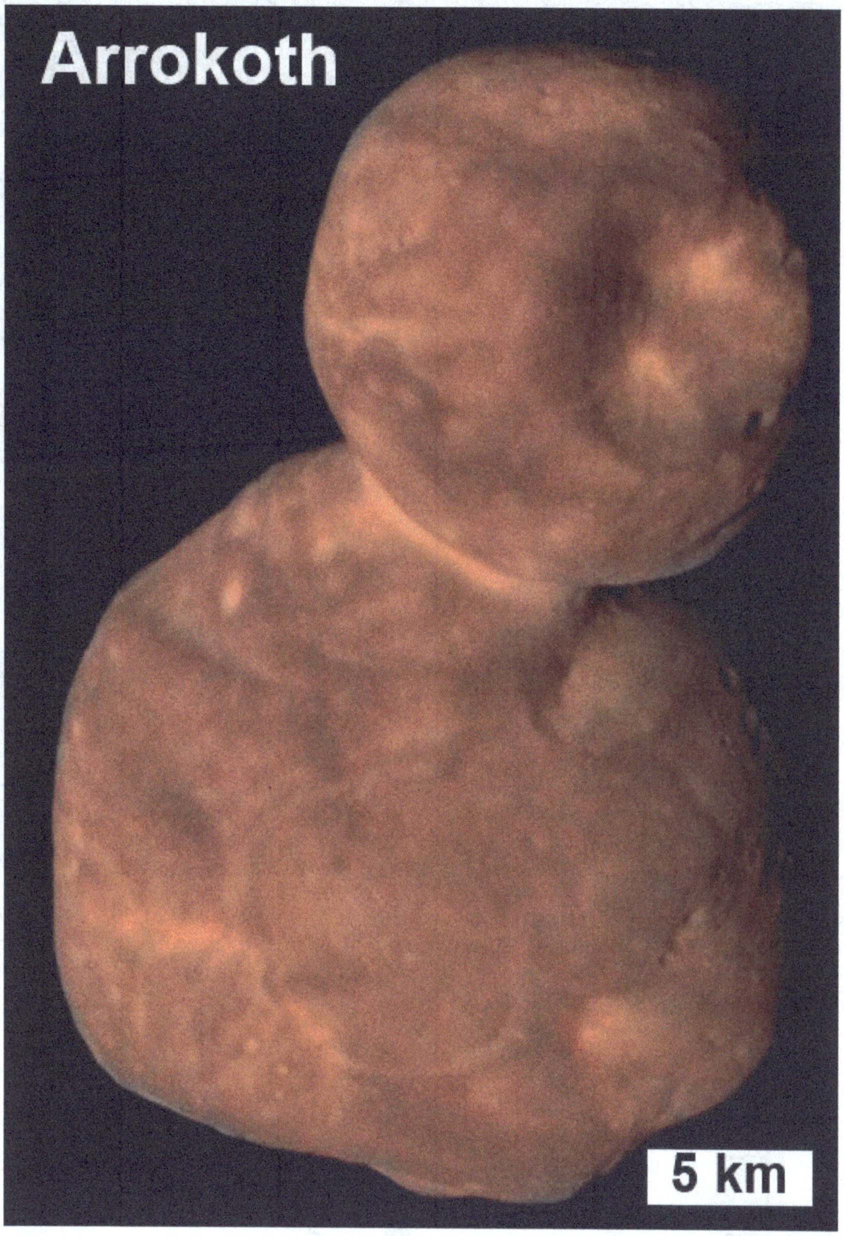

Fig. 5.17 Image of the Kuiper Belt object Arrokoth taken with the narrow-angle Long-Range Reconnaissance Imager (LORRI) camera in January 2019. The body is a bilobed binary composed of distinct geomorphological units of different shapes. Image courtesy of NASA/Johns Hopkins University Applied Physics Laboratory.

a higher mean shock stage than the set containing type-3 OC at a confidence level much greater than 99.99% ($t = 14.9$).

Does the positive correlation between petrologic type and mean shock stage hold for individual OC groups? This question is important because the three main groups are from different asteroids that experienced distinct thermal and shock histories: there are differences among these groups in their distributions of petrologic type (Section 13.1.1), shock stages (Section 13.2.1), cosmic-ray exposure ages (Eugster, 2003; Herzog, 2005), Ar–Ar ages (Bogard, 1995, 2011), and U/Th–He ages (Wasson and Wang, 1991). The three OC groups also have different bulk chemical compositions (Kallemeyn et al., 1989; Jarosewich, 1990) and distinct bulk O- and Ge-isotopic compositions (Clayton et al., 1991; Florin et al., 2020).

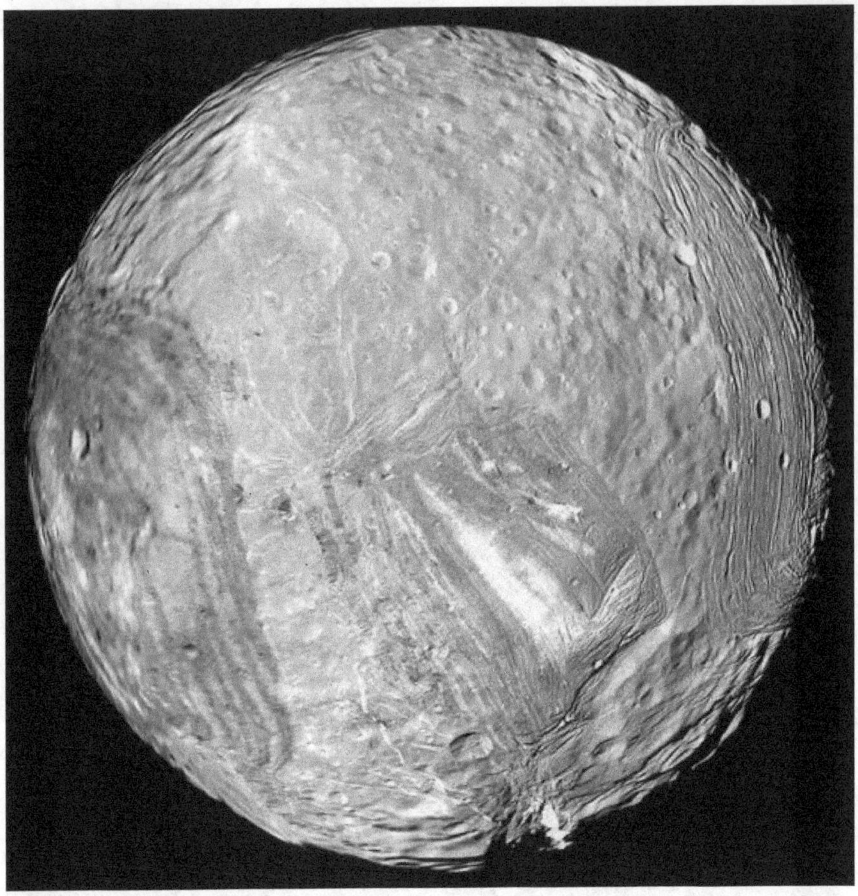

Fig. 5.18 Uranus' icy moon Miranda, imaged by the Voyager 2 spacecraft on January 24, 1986. The moon has dimensions of 465.8 × 468.4 × 480 km. The chaotic jumble of surface materials with distinct geomorphological characteristics may reflect a history involving catastrophic collision and rapid reassembly. NASA image.

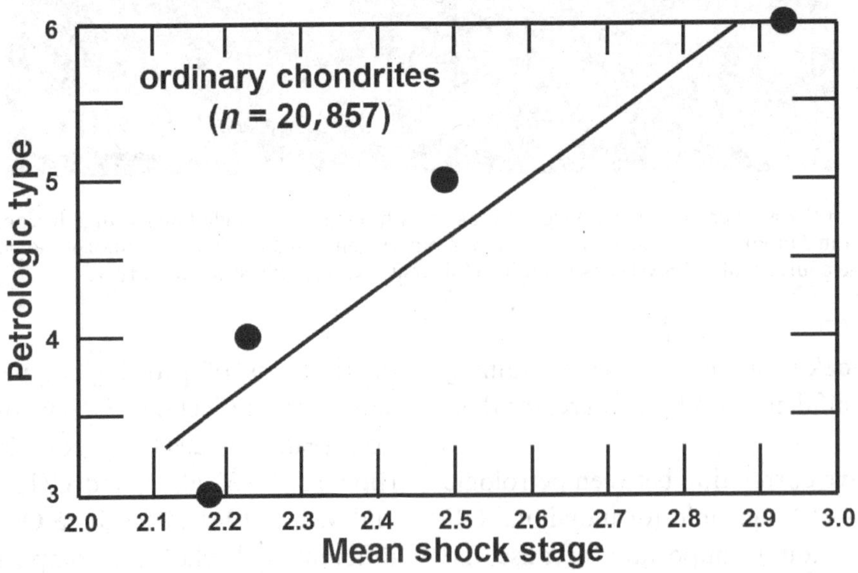

Fig. 5.19 Positive correlation between petrologic type and mean shock stage for ordinary chondrites (Rubin, 2025). Although it cannot be assumed the relationship is linear, a straight line is plotted for simplicity.

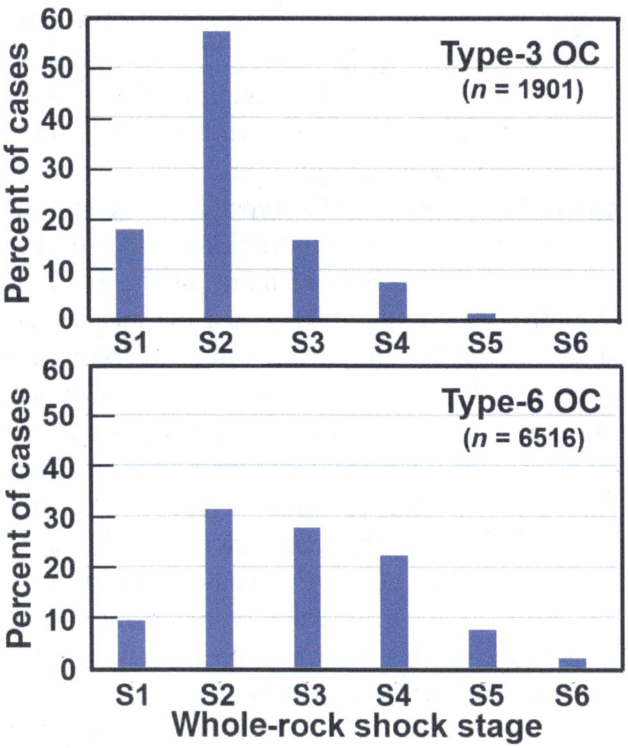

Fig. 5.20 Distributions of shock stages in type-3 and type-6 ordinary chondrites (Rubin, 2025). The type-6 OC distribution is skewed toward higher shock stages.

Even though different OC groups have somewhat different physical properties, varying mainly in their metal/silicate ratios (e.g., Flynn et al., 2018), each individual OC group shows a positive correlation between petrologic type and shock stage (Fig. 5.21). The same trend is weaker but present among the set of type-3 ordinary chondrites. Type 3.7–3.9 H, L, and LL chondrites have higher mean shock stages than those of lower subtypes: H3.7–3.9 (2.03), H3.0–3.6 (1.92); L3.7–3.9 (2.52), L3.0–3.6 (2.09); LL3.7–3.9 (2.31), LL3.0–3.6 (2.03). There are significant differences in the correlations between petrologic type and mean shock stage among the three OC groups (Table 5.1).

H chondrites have the smallest range of mean shock stages (2.07–2.32) with only a negligible difference between the mean shock stages of H3 and H4 chondrites (2.07 and 2.08, respectively).

L chondrites are generally the most shocked OC. Type-3, type-4, and type-6 L chondrites are more shocked than their H and LL counterparts. L5 chondrites have about the same mean shock stage as LL5 chondrites (2.68 and 2.70, respectively); both shock values are much higher than

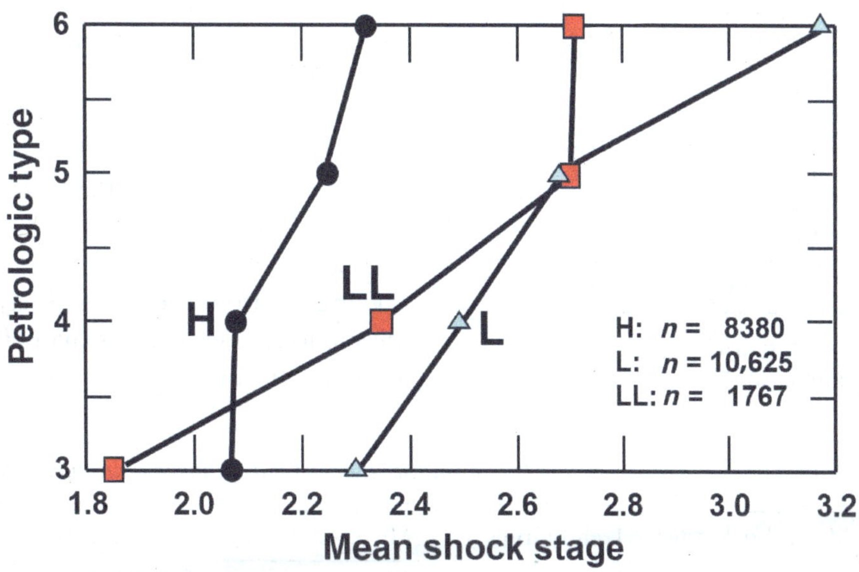

Fig. 5.21 Positive correlations between petrologic type and mean shock stage for each of the three main ordinary-chondrite groups (Rubin, 2025). In each group, the mean shock stage increases progressively from type 3 through type 6. H chondrites have the smallest range in mean shock stage. L chondrites are generally the most shocked OC. There are only insignificant increases in mean shock stage between H3 and H4 chondrites and between LL5 and LL6 chondrites.

that of H5 chondrites (2.25). L3 to L6 chondrites have the highest range in mean shock stage (2.30–3.18) and show the most regular progressive increase in mean shock stage with increasing petrologic type.

LL chondrites have the second-greatest range in mean shock stage (1.85–2.71). LL3 chondrites are by far the least shocked ordinary chondrites (1.85). LL5 and LL6 chondrites differ from each other only slightly in mean shock stage (2.70 and 2.71, respectively).

The negligible difference in mean shock stage between H3 (2.07) and H4 (2.08) could be attributable to the misclassification of a significant number of H3 and H4 chondrites. Van Schmus and Wood (1967) pointed out that "Petrologic type 4 is defined by relatively ambiguous characteristics" and that "many of the characteristics are not too explicit…" If misclassification of some type-3 and -4 H chondrites had occurred, one might expect there to be a large proportion of H3 chondrites classified as having a high petrologic subtype (i.e., 3.7–3.9). Some of these H3.7–H3.9 chondrites might actually be H4. Similarly, some H4 chondrites might be misclassified H3.7-H3.9 samples.

Table 5.1 *Mean shock stages of individual petrologic types for each ordinary-chondrite group.*

H chondrites	L chondrites	LL chondrites
H3: 2.07 ($n = 674$)	L3: 2.30 ($n = 695$)	LL3: 1.85 ($n = 451$)
H4: 2.08 ($n = 2397$)	L4: 2.49 ($n = 1282$)	LL4: 2.35 ($n = 280$)
H5: 2.25 ($n = 3743$)	L5: 2.68 ($n = 4343$)	LL5: 2.70 ($n = 385$)
H6: 2.32 ($n = 1566$)	L6: 3.18 ($n = 4305$)	LL6: 2.71 ($n = 651$)

Source: Data from the Meteoritical Bulletin Database (January 11, 2024 compilation).

To test this hypothesis, Rubin (2025) looked at the Meteoritical Bulletin (May 9, 2024) compilation for type-3 OC and found that 69% of H3 chondrites are classified as type 3.7–3.9, compared to only 36% of L3 chondrites and 33% of LL3 chondrites (Table 5.2). If many of the H3.7–3.9 chondrites are actually H4, then the mean shock stage of true H3 chondrites would be lowered to ~1.92 (Table 5.2), appreciably lower than that of H4 chondrites (2.08) (Table 5.1). The positive correlation between petrologic type and mean shock stage for H chondrites would then be strengthened.

It is not clear why many more H4 chondrites than L4 and LL4 chondrites would be misclassified as type 3.7–3.9. One possibility is that the smaller chondrules in H chondrites appear less recrystallized under superficial scrutiny.

There are discrepancies between the values of the shock stages for H3, L3, and LL3 chondrites in Table 5.1 and those of the shock stages for the subtypes listed in Table 5.2. These discrepancies arise from the fact that relatively few type-3 OC have shock-stage data *and* petrologic subtype listed in the Meteoritical Bulletin Database. Many samples with shock-stage data are classified only as H3, L3, or LL3; such samples are included in Table 5.1, but not in Table 5.2.

The fact that L chondrites tend to be the most shocked samples (Fig. 5.21) is likely due to the catastrophic disruption of the L parent body ~470 Ma ago (e.g., Korochantseva et al., 2007; Nesvorný et al., 2009; Swindle et al., 2014), consistent with the higher proportion of maskelynite-bearing samples among L chondrites than H or LL chondrites (Rubin, 2015c).

Table 5.2 *Proportions of subtypes and mean shock stages of unequilibrated ordinary chondrites.*

H chondrites	Mean shock stage	L chondrites	Mean shock stage	LL chondrites	Mean shock stage
H3.0–3.2 ($n = 22$; 6%)	1.67 ($n = 6$)	L3.0–3.2 ($n = 62$; 13%)	2.16 ($n = 25$)	LL3.0–3.2 ($n = 74$; 31%)	2.10 ($n = 42$)
H3.3–3.6 ($n = 90$; 25%)	2.00 ($n = 18$)	L3.3–3.6 ($n = 241$; 51%)	2.00 ($n = 21$)	LL3.3–3.6 ($n = 87$; 36%)	1.93 ($n = 30$)
H3.7–3.9 ($n = 247$; 69%)	2.03 ($n = 118$)	L3.7–3.9 ($n = 169$; 36%)	2.52 ($n = 75$)	LL3.7–3.9 ($n = 81$; 33%)	2.31 ($n = 42$)

Among type-3 OC, LL chondrites have the highest proportion of low petrologic subtype samples: 31% of LL3 chondrites are subtype 3.0–3.2 compared to 6% of H3 and 13% of L3 chondrites (Table 5.2). Most of these samples have a low shock stage, accounting for the overall low mean shock stage of the complete set of LL3 chondrites (1.85), which would likely include many LL3.0–3.2 chondrites currently classified only as LL3.

It is difficult to explain the near-identical values of mean shock stage in LL5 and LL6 chondrites (2.70 and 2.71, respectively) except as a coincidence.

Rubin (2004b) suggested that episodes of post-shock annealing tended to completely heal olivine crystal lattices, causing them to acquire sharp optical extinction and an apparent shock stage of S1. If this is correct, post-shock annealing of a shocked ordinary chondrite would not cause a decrease in shock stage from S5 to S3.

During post-shock annealing, prior shock-induced damage of olivine and plagioclase crystal lattices can be healed (Bauer, 1979). Experimental annealing of the shock-stage S5 Kyushu L6 chondrite for 90 hours at 900°C and 1000°C caused fractures in olivine grains to heal (Ashworth and Mallinson, 1985). Post-shock annealing of some shocked OCs could change some samples designated S3 or S4 (for example) to S1 (Rubin, 2002b, 2004b). If each petrologic type was affected more-or-less equally by post-shock annealing, similar proportions of each petrologic type would become S1. This would not destroy a preexisting correlation between petrologic type and mean shock stage (e.g., Fig. 5.21); the trend would just be shifted to the left (toward lower mean shock stages) (Fig. 5.22).

Unless they are impact-melted, chondrites that experience multiple shock episodes (without accompanying annealing) will likely just exhibit additional shock features; their assigned shock stage will either remain the same or increase. To a first approximation, if each petrologic type of an OC group was affected more-or-less equally by late-stage shock events, a preexisting correlation between petrologic type and shock stage (Fig. 5.21) would not be significantly affected; the trend would just shift to the right (toward higher mean shock stages) (Fig. 5.22).

Statistically, post-shock annealing and late-stage shock events probably affected the different petrologic types to similar extents, thereby modifying the trend of shock-stage versus petrologic

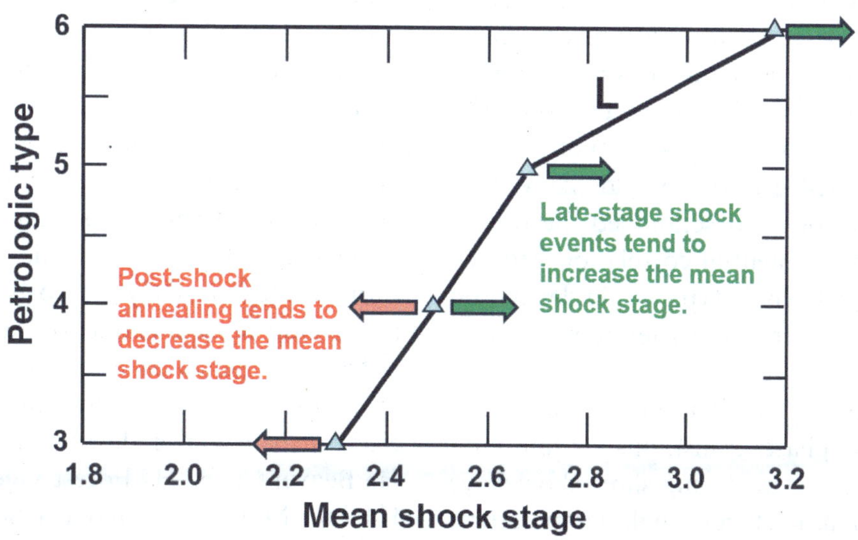

Fig. 5.22 The positive correlation between petrologic type and mean shock stage for L chondrites, showing how it would have been affected by two distinct late-stage processes (Rubin, 2025). Significant post-shock annealing of some shocked individuals would tend to cause a shift of the trend to the left (toward lower shock stages). Unless a sample was impact-melted, late-stage shocks would tend to cause the rocks to exhibit more-intense shock effects; they would either be classified at the same shock-stage level or at a higher level, causing a shift of the trend to the right (toward higher shock stages). These two processes might modify the trend to minor degrees but would tend to preserve a preexisting correlation.

type to only a minor degree. The occurrence of a positive correlation between mean shock stage and petrologic type in all three OC groups strongly suggests it is a preexisting feature; late-stage processes likely only shifted the trends to the right or left and did not modify them appreciably. It can be concluded that the correlations between petrologic type and mean shock stage for each ordinary-chondrite group (Fig. 5.21) date back to the early history of OC asteroids.

A possible explanation for the positive correlation between petrologic type and shock stage would be that a normal intact onion-shell asteroid suffered an enormously energetic impact and that deeply buried materials were excavated. The most deeply buried type-6 materials became more shocked because they required much higher shock energies for excavation. However, the energy necessary to excavate deeply buried materials from a large intact asteroid would likely form an impact crater with a diameter greater than the asteroid's radius (Fig. 5.23). For bodies with low microporosity ($\lesssim$0.21) (as in an intact asteroid), the largest impact crater that could form without disrupting the body would have a diameter equivalent to ~80% of the body's radius (Burchell and Leliwa-Kopystynski, 2010). In fact, collisional evolution models show it is statistically unlikely that a 200-km-diameter OC asteroid would experience an impact event energetic enough to excavate type-6 material (Davison et al., 2013). A collision sufficiently energetic to excavate type-6 chondrites from an intact onion-shell asteroid would likely have destroyed the asteroid and scrambled the onion-shell structure. Any potential correlation between cooling rate and petrologic type would depend on the internal temperature of the asteroid when collisional disruption occurred.

Processes with high activation energies (such as reconstructive phase transitions within mineral grains) cannot occur during short shockwave pulses at cold temperatures; high temperatures are required (e.g., in proximity to a shock melt). Langenhorst et al. (1992) reported that shock and deformation effects of target rocks depend strongly on their pre-shock temperature. Other experimental studies arrived at similar conclusions:

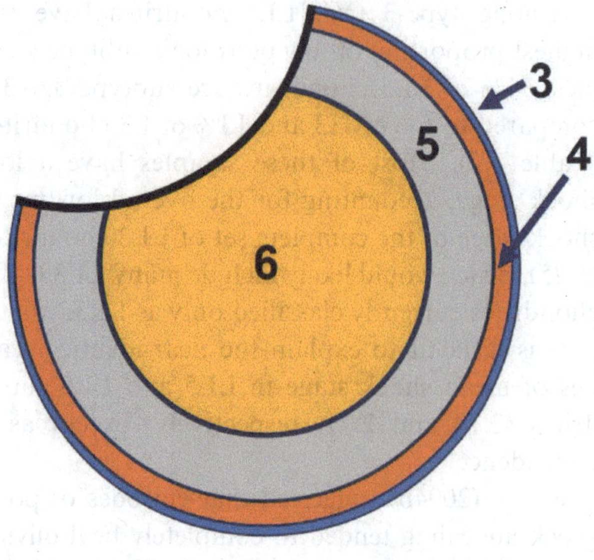

Fig. 5.23 Hypothetical cross-sectional structure of an onion-shell asteroid with an impact crater that penetrates the central type-6 region to one-third of its thickness. Proportions are based on the relative abundances of H-chondrite falls as shown in Scott and Rajan (1981). The diameter of the crater is two-thirds the diameter of the asteroid. An intact (non-rubble-pile) asteroid would be disrupted by this collision and could not support such a large crater. In contrast, 253 Mathilde, a carbonaceous-chondrite-like rubble pile, has two craters (Ishikari, 29.3 km; Karoo, 33.4 km) with diameters exceeding the mean radius of the asteroid (26.4 km) (e.g., Veverka et al., 1999).

(1) Nakamura et al. (2000) showed that CV3 chondrite samples shocked to 21 GPa at 300°C and 600°C exhibited more-intense shock features than room-temperature (~25°C) samples shocked to the same extent. (2) Schmitt (2000) found that olivine, orthopyroxene and plagioclase grains in hot samples of H6 Kernouvé developed shock features at lower shock pressures than in cold samples. Specifically, at ~650°C, sodic plagioclase transformed into maskelynite, orthopyroxene began to develop mosaic extinction, and olivine recrystallized or melted at lower shock pressures than in charges shocked at low temperatures. In their electron backscatter diffraction (EBSD) study of H5 Beaver Creek, H4 Forest Vale, H4 Quenggouk, H4 Ste. Marguerite, and H4 Sena, Goudy et al. (2023) found that all five samples were shocked while they were hot (due to thermal metamorphism) and were subsequently annealed.

If an OC asteroid were collisionally disrupted, chaotically mixed, and gravitationally reassembled while it was still hot (e.g., Grimm, 1985), the different petrologic types would be randomly scattered across the asteroid surface. Projectiles could not have preferentially targeted OC materials of higher petrologic type to produce a positive correlation between petrologic type and shock stage.

Instead, Rubin (2025) suggested that the positive correlation between mean shock stage and petrologic type (Figs. 5.19, 5.21) can be explained by stochastic impacts on a rubble-pile asteroid into randomly situated ordinary-chondrite materials that were already hot (presumably due mainly to the decay of ^{26}Al). Type-3 OC had peak temperatures of ~260–600°C, type-4 OC of ~600–700°C, type-5 OC of 700–750°C, and type-6 OC of ~820–930°C (e.g., Onuma et al., 1972; Dodd, 1981; Wlotzka, 1985, 1987; McSween and Patchen, 1989; Sears et al., 1991; Harvey et al., 1993; Huss and Lewis, 1994). Ordinary chondrites that were deformed while still warm include LL6 Saint-Séverin (Hugo et al., 2020) and L6 Ozerki (Macente et al., 2025).

If hot ordinary chondrites of all petrologic types, residing in the near-surface regions of a rubble-pile asteroid, experienced a similar range of random collisions of varying intensity, then, on average, the highest-temperature OCs (i.e., those of petrologic type 6) would exhibit the most prominent shock effects. This is because hotter materials record more-intense shock features from collisions of the same energy. In contrast, the merely warm type-3 OC, with their lower peak temperatures, would, on average, exhibit fewer and less-intense shock effects. This is consistent with the positive correlation in OC between the width of shock-induced melt veins and petrologic type (Miyahara et al., 2020).

Even within the set of type-3 chondrites (Table 5.2), type 3.7–3.9 OC experienced higher peak temperatures than type 3.0–3.6 OC (Sears et al., 1980); the higher subtypes manifested more-intense shock effects even though all the OC subtypes in a given group likely experienced a similar statistical collisional history.

Nakamura et al. (2000) found dramatic differences in the intensity of shock features between ~25°C samples and 300°C samples shocked to 21 GPa. Although the initial temperature differences between the petrologic types and subtypes are modest (e.g., ~200–400°C between type 4 and 3; ~100–150°C between type 5 and 4; and ~100–250°C between type 6 and 5), the differences in mean shock stage among petrologic types are also rather small (0.05 shock stage between type 4 and 3; 0.26 shock stage between type 5 and 4; and 0.44 shock stage between type 6 and 5).

It seems plausible that onion-shell OC asteroids formed early in Solar System history and were then disrupted, jumbled up, and rapidly reassembled while still hot (e.g., Grimm, 1985), perhaps while they had peak temperatures between ~800°C (the temperature at which Pb retention in phosphate begins; Blackburn et al., 2017) and ~650°C (a temperature at which Ni diffusion in taenite is still fairly rapid). This would have produced a random distribution of petrologic types in the near-surface regions of these bodies. Type-6 materials would have generally remained hotter than materials of lower petrologic type. Subsequent collisions into the hotter rocks would have produced more-intense shock features (i.e., features that would normally be associated with colder rocks that experienced higher levels of shock). This would have produced the observed correlation between shock stage and petrologic type (Fig. 5.19). Because H, L, and LL chondrites all show a positive correlation between petrologic type and mean shock stage (Fig. 5.21), it seems probable this scenario applies to all three OC parent bodies. Temperature differences could have been maintained for hundreds of thousands of years in the jumbled OC blocks if they had been buried at depths of ~1 km (Rubin, 2025).

It does not seem remarkable that all three OC parent asteroids were collisionally disrupted and reassembled at similar epochs in Solar System history. The ordinary chondrites form a clan exhibiting numerous similarities in their petrologic, mineralogic, chemical, and isotopic properties. Their parent bodies probably accreted in close proximity (e.g., Desch et al., 2018) and were thus subjected to a similar flux of projectiles. Bogard (1995) found that several H, L, and LL chondrites have Ar–Ar and Rb–Sr ages of ~3.5–4.0 Ga, possibly indicating times of disruption and reassembly. Nevertheless, each OC body had a distinct accretionary, metamorphic, and collisional history.

H chondrites have the smallest range of mean shock stages (2.07–2.32). If the H asteroid was disrupted and reassembled after it had already cooled more than the L or LL asteroids had, there would be only relatively small differences in the residual temperatures of H6, H5, H4, and H3 materials. The higher petrologic types would manifest shock features only a little more intense than those of lower petrologic types. Although consistent with the observations, the interpretation that the H-chondrite asteroid was disrupted later than the L or LL bodies is speculative.

As has been discussed, L chondrites are generally the most shocked OC. This is probably a result of their parent asteroid's catastrophic disruption ~470 Ma ago. The initial correlation between petrologic type and mean shock stage was shifted toward higher shock stages (Fig. 5.22). This late-stage disruption event is unrelated to the disruption/reassembly event that occurred early in Solar System history.

The most unusual feature of LL chondrites is the large proportion of type-3 samples (16%) (Section 13.1.1); 31% of those are of low petrologic subtype (3.0–3.2) (Table 5.2). One possible way to account for the high proportion of type-3 chondrites on the LL parent asteroid is to model the body as a relatively small onion-shelled structure with type-3 samples confined to an outer spherical shell extending from the surface to a depth of only a few kilometers. A model 250-km-diameter asteroid would have a surface-area/volume ratio of 0.024 km^{-1} and a 7-km-thick spherical shell (that would include 16% of the asteroid's volume and encase all the LL3 chondrites). (Type-3 individuals constitute 16% of LL chondrites by number; it is assumed this is equivalent to 16 vol.% of the LL asteroid.) In contrast, a 100-km-diameter body would have a greater surface-area/volume ratio (0.060 km^{-1}) and a 3-km-thick outer spherical shell of LL3 chondrites. The smaller body would cool much faster; its surface-area/volume ratio would be 2.5 × larger than that of the larger asteroid and its type-3 samples would be, on average, 2 km closer to the surface. In a small onion-shell body, there would be less metamorphic heating of materials close to the surface.

There is additional evidence supporting a relatively small size for the LL parent body: (1) It is consistent with the small proportion of LL chondrites (13.2%) within the set of all (stony) ordinary chondrites (H, L, LL, H/L, L/LL; falls and finds). LL chondrites constitute just 11.3% of OC falls. (2) The limited data of Blackburn et al. (2017), which include up to four samples per analyzed meteorite, suggest that LL6 chondrites have somewhat older Pb–phosphate ages than H6 or L6 chondrites: H6, 4505.94 Ma, $n = 1$; L6, 4508.70 Ma, $n = 3$; LL6, 4517.40 Ma, $n = 1$. This is consistent with the LL body having cooled faster to the closure temperature of Pb in phosphate due to the body's smaller size and higher surface-area/volume ratio.

Another potential line of evidence for a small LL parent asteroid involves its proportion of exogeneous contaminants – a small body with a high surface-area/volume ratio should have more contaminants than a large body subject to the same flux. For example, ~1 to 2% of lunar maria soils consist of a meteorite contaminant resembling CM chondrites (Wasson et al., 1975). There are also CM-chondrite clasts in OC regolith breccias (Section 11.13). Because CM-chondrite falls contain much higher mean bulk C and bulk H_2O^+ (chemically bound water) than OC falls (1.82 vs. <1.0 wt.% C; 8.73 vs. $\lesssim$ 2.0 wt.% H_2O^+; Jarosewich, 1990), a smaller parent OC body with a higher proportion of contaminants should be richer in bulk C and bulk H_2O^+. Because C and water abundances decrease during thermal metamorphism, it is best to compare intergroup concentrations among the type-3 OC. Among type-3 samples, median bulk C in LL chondrites is somewhat higher than in L chondrites, but only marginally higher than in H chondrites: LL3 (0.53 wt.%, $n = 5$), L3 (0.32 wt.%, $n = 3$), H3 (0.50 wt.%, $n = 2$) (Jarosewich, 1990). The same is true with mean H_2O^+: LL3 (1.18 ± 0.46 wt.%, $n = 4$), L3 (0.92 ± 0.18 wt.%, $n = 4$), H3 (1.09 ± 1.33 wt.%, $n = 2$) (Jarosewich, 1990). These few data are suggestive (but not conclusive) of greater contamination of the LL parent body (consistent with a smaller size).

The OC shock-stage data can be incorporated into the model of Blackburn et al. (2017), which attempted to reconcile Pb–phosphate data and metallographic cooling rates. This new composite model appears below:

1. The ordinary-chondrite parent asteroids were likely ~100–300 km in diameter (Wood, 1967; Trieloff et al., 2003; Kleine et al., 2008; Henke et al., 2012; Blackburn et al., 2017). The H and L bodies probably had diameters toward the upper end of the range, and the LL body toward the lower end (accounting for its high proportion of type-3 samples).
2. Each of the OC parent asteroids initially had an onion-shell structure, with type-6 material near the center of the body and type-3 material close to the surface. The principal heat source was the decay of ^{26}Al. Type-6 materials were well insulated and cooled slowly; consequently, the Pb–phosphate data revealed young closure ages. These bodies were collisionally disrupted at ~60 Ma or so after CAI formation (Blackburn et al., 2017), while temperatures were between ~800°C and ~650°C, locking in the Pb–phosphate record of their onion-shell structure. The H-chondrite parent body was collisionally disrupted somewhat later than the L or LL bodies and had cooled more than the L or LL bodies by that time.
3. Collisional fragments of each of the disrupted OC asteroids were jumbled up and rapidly reaccreted into rubble piles, perhaps within a few days (Grimm, 1985). The materials close to the surface of the rubble-pile asteroids cooled quickly. Because type-6 blocks located near the surface were now adjacent to type-3, -4 and -5 blocks, the different petrologic types cooled at similar rates through ~500°C (the temperature at which Ni diffusion in taenite is so sluggish that metallographic cooling rates are preserved). This precluded any correlation between petrologic type and metallographic cooling rate.
4. Shortly after the rubble piles formed, materials of higher petrologic type remained hotter than materials of lower petrologic type. All materials near the surface of each OC asteroid experienced a similar flux of impacting projectiles. The hotter materials recorded more-intense shock features, leading to the positive correlations in each OC asteroid between petrologic type and mean shock stage. The cooler materials near the surface of the H-chondrite rubble-pile body recorded lower-intensity shock effects, accounting for the small range of mean shock stages among H chondrites.

6

Carbonaceous and Noncarbonaceous Materials in the Solar System

It has long been known that the materials of the Solar System vary in their isotopic composition. Terrestrial rocks, moon rocks, and enstatite meteorites (chondrites and achondrites) can be plotted along the terrestrial fractionation (TF) line on the standard three-isotope oxygen diagram (Fig. 6.1). Ordinary chondrites and R chondrites are richer in ^{17}O and are plotted above the TF line. CI carbonaceous-chondrites plot near the TF line at values of $\delta^{18}O \approx 16‰$, $\delta^{17}O \approx 9‰$ (Clayton and Mayeda, 1989). Other carbonaceous-chondrite groups plot below the TF line.

The isotopic compositions of N, O, Mg, S, Ca, Ti, Cr, Fe, Ni, Zn, Sr, Mo, Ru, Nd, and Os in enstatite chondrites and aubrites resemble those of Earth (e.g., Clayton, 1993; Clayton et al., 1984; Javoy, 1995; Meisel et al., 1996; Javoy et al., 2010; Trinquier et al., 2007, 2009; Regelous et al., 2008; Warren, 2011a, 2011b; Moynier et al., 2012; Dauphas et al., 2004, 2014a, 2014b; Sikdar and Rai, 2020 and references therein). (Although the Earth and enstatite chondrites have different Si- and W-isotopic compositions [Fitoussi and Bourdon, 2012; Savage and Moynier, 2013; Lee and Halliday, 2000], these differences may not be primary. They could have resulted instead from core formation in the Earth [e.g., Trinquier et al., 2007].)

The parent asteroids of enstatite meteorites were modeled as having accreted at heliocentric distances of roughly 1 AU (e.g., Baedecker and Wasson, 1975; Sears, 1980; Wasson, 1985, 1988; Nakashima et al., 2006; Scott, 2007). Desch et al. (2018) pegged the formation locations at ~2 AU (see Section 14.1).

Because CI- and CM-chondrite falls contain ~11 and ~9 wt.% H_2O^+ (chemically bound H_2O), respectively (Jarosewich, 1990), it seemed likely that they formed in the Outer Solar System, accreting along with water ice beyond the snow line. (Closer to the Sun, most H_2O remained in the gas phase.) In contrast, ordinary- and enstatite-chondrite falls are rather dry ($\lesssim$0.5 wt.% H_2O^+) and probably formed in the Inner Solar System. It has long been apparent that the Inner Solar System differed in chemical and isotopic composition from the Outer Solar System.

But it was not until Paul Warren (UCLA) published two landmark papers in 2011 that the cosmochemical community more fully understood the implications of the positional/compositional structure of the Solar System. Warren (2011a, 2011b) found that carbonaceous chondrites (CC) and noncarbonaceous-chondrites (NC) plot in widely separated regions in diagrams of O-, Ti-, Cr-, and Ni-isotopic composition (i.e., ε^{50}Ti vs. ε^{54}Cr; Δ^{17}O vs. ε^{54}Cr; ε^{54}Cr vs. ε^{62}Ni) (Fig. 6.2). He also showed that IIIAB irons, main-group pallasites, mesosiderites, HED meteorites (presumably from Vesta), angrites, ureilites, terrestrial rocks, lunar samples, and martian meteorites are all related to NC chondrites, whereas Eagle Station pallasites are related to CC chondrites.

Because diverse types of meteorites exhibit distinguishable Mo-isotopic compositions (e.g., Burkhardt et al., 2011), Budde et al. (2016, 2017, 2018) were able to show that CC chondrites contain excess r-process Mo relative to NC chondrites. The two classes of chondrites plot along distinct s-process mixing lines on ε^{95}Mo versus ε^{94}Mo diagrams. The two isotopic reservoirs appear to have separated within ~1 Ma after CAIs formed (e.g., Budde et al., 2016; Kruijer et al., 2017a).

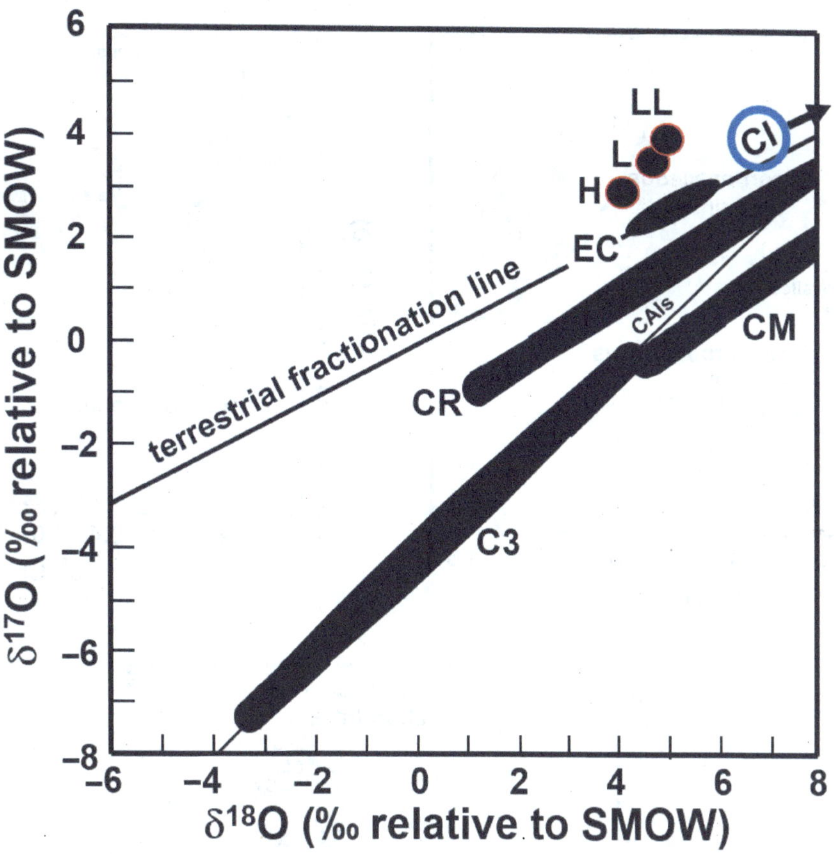

Fig. 6.1 Three-isotope plot showing the O-isotopic composition of chondritic materials. The ordinary-chondrite groups (H, L, LL) plot above the terrestrial fractionation (TF) line; enstatite chondrites (EC) plot on the TF line. Carbonaceous-chondrites plot on or below the TF line; CI chondrites plot on the TF line outside the limits of the diagram. C3 = type-3 CV and CO chondrites. Diagram modified from Clayton (1993) based on literature data.

CC and NC asteroids alike formed from chondrites or related primitive materials. Some of these bodies experienced large-scale melting, underwent differentiation, and produced iron cores. Iron meteorites derived from these cores can be dubbed CC irons and NC irons, respectively (e.g., Kruijer et al., 2017a, 2017b, 2017c; Poole et al., 2017). These so-called "magmatic" iron groups formed by fractional crystallization in well-mixed molten asteroidal cores. This ascribed origin is based on the substantial fractionation of refractory siderophile elements (RSEs) (e.g., Ir, Os, Re, Ru, Pt) from common (Ni, Co) and volatile (e.g., Au, Ga, Ge, As, Sb) siderophiles in some of these groups. For example, the concentration ranges of Ir, Re, and Os in group IIAB are ~4000 μg/g, ~6000 ng/g, and ~10,000 μg/g, respectively (Cook et al., 2004; Wasson et al., 2007). Fractionations arise because the solid/liquid distribution coefficient (D) is >1 for RSEs and <1 for many common and volatile siderophiles under ideal conditions of core crystallization (e.g., Wasson, 1985). (However, an important complication is that nonmetallic elements such as S, C, and P within the metallic liquid can affect the solid/liquid distribution coefficient significantly [e.g., Willis and Goldstein, 1982; Jones and Drake, 1983; Chabot, 2004; Walker et al., 2008].)

Particular magmatic iron-meteorite groups are related to different classes of chondrites. Clayton et al. (1983) used O-isotope data to show that IVA irons are related to noncarbonaceous chondrites; Shukolyukov and Lugmair (2006) used Mn–Cr systematics to show that IVB irons are related to carbonaceous chondrites; and Warren (2011a, 2011b) used O-, Cr-, and Ni-isotopic data to show that IIIAB irons are noncarbonaceous.

Additional associations of iron-meteorite groups with chondrites are based on Mo- and W-isotopic compositions (Budde et al., 2016, 2018; Kruijer et al., 2017a, 2017b, 2017c; Poole et al., 2017)

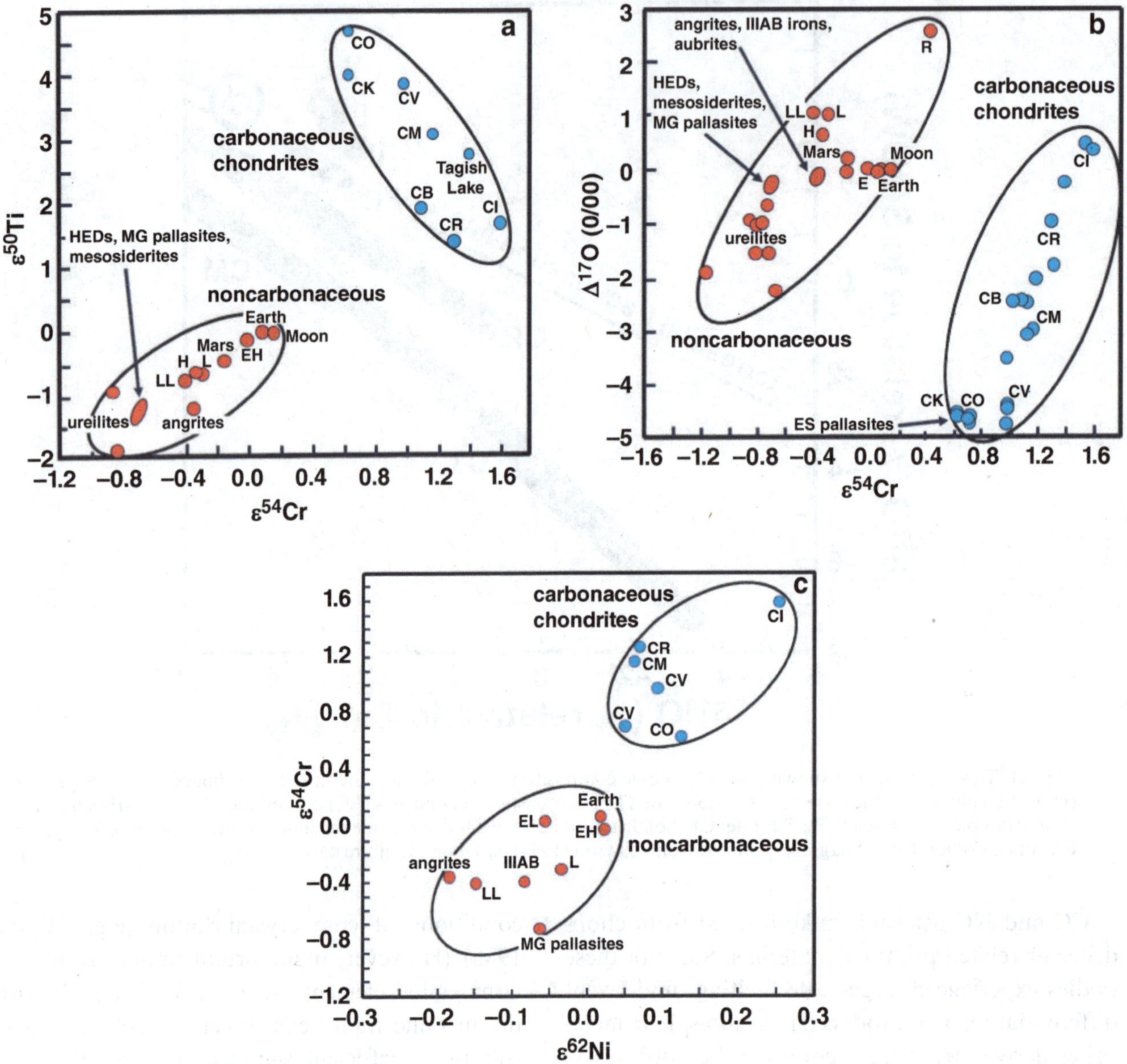

Fig. 6.2 Selected stable isotopes in various meteoritic and planetary materials. (a) ε^{50}Ti vs. ε^{54}Cr. (b) Δ^{17}O vs. ε^{54}Cr. (c) ε^{54}Cr vs. ε^{62}Ni. In all cases, carbonaceous-chondrites (light blue) form a field cleanly separated from noncarbonaceous material (red). Tagish Lake is an ungrouped C2 carbonaceous chondrite fall. HEDs are howardites, eucrites, and diogenites (all of which are presumed to have been derived from the asteroid 4 Vesta). MG pallasites refers to main-group pallasites; ES pallasites refers to Eagle Station pallasites (six samples have been described to date). In diagram (b), E stands for enstatite chondrites (including the EH and EL groups). All diagrams based on Warren (2011a).

(Fig. 6.3). It has been determined that the CC magmatic irons include groups IIC, IID, IIF, IIIF, and IVB, while the NC magmatic irons include groups IC, IIAB, IIIAB, IIIE, and IVA. There is a broad consensus that the CC magmatic irons formed in the cores of differentiated asteroids residing in the Outer Solar System beyond Jupiter, while the NC magmatic irons formed from differentiated asteroids in the Inner Solar System.

Both major groups of nonmagmatic irons (IAB and IIE) are noncarbonaceous (Budde et al., 2016; Kruijer et al., 2017a, 2017b, 2017c). These two groups have varied (and more radiogenic) bulk W-isotope signatures (Scherstén et al., 2006), consistent with late-stage formation by impact processes (e.g., Wasson et al., 1980; Rubin et al., 1986; Wasson and Wang, 1986; Wasson and Kallemeyn, 2002) that facilitated partial

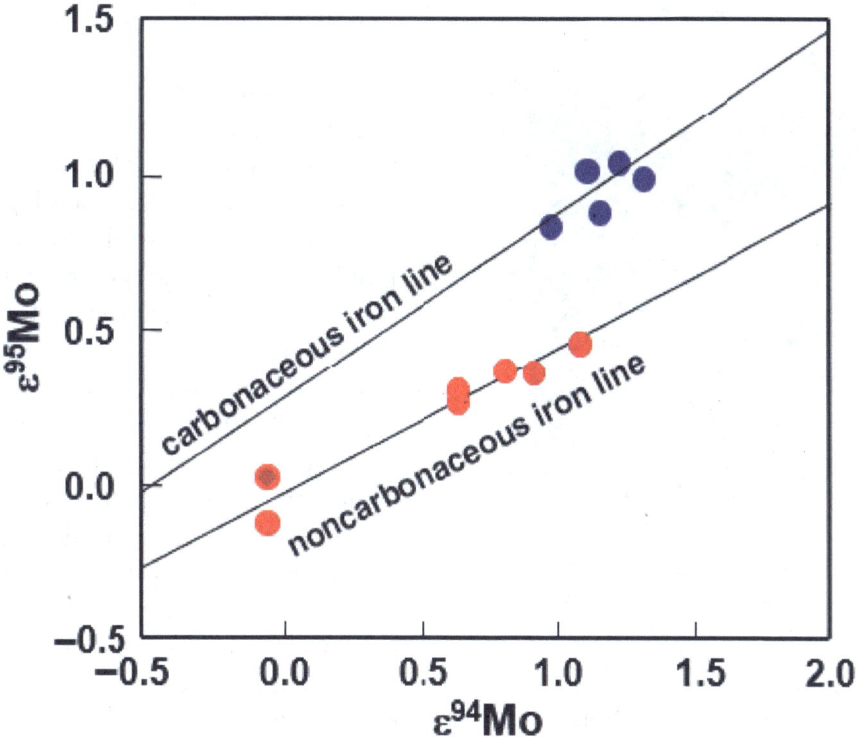

Fig. 6.3 Molybdenum-isotopic diagram showing that iron meteorites were derived from two different isotopic reservoirs – one identified with carbonaceous chondrites (blue points) and one with noncarbonaceous materials (red points). Diagram based on Budde et al. (2016).

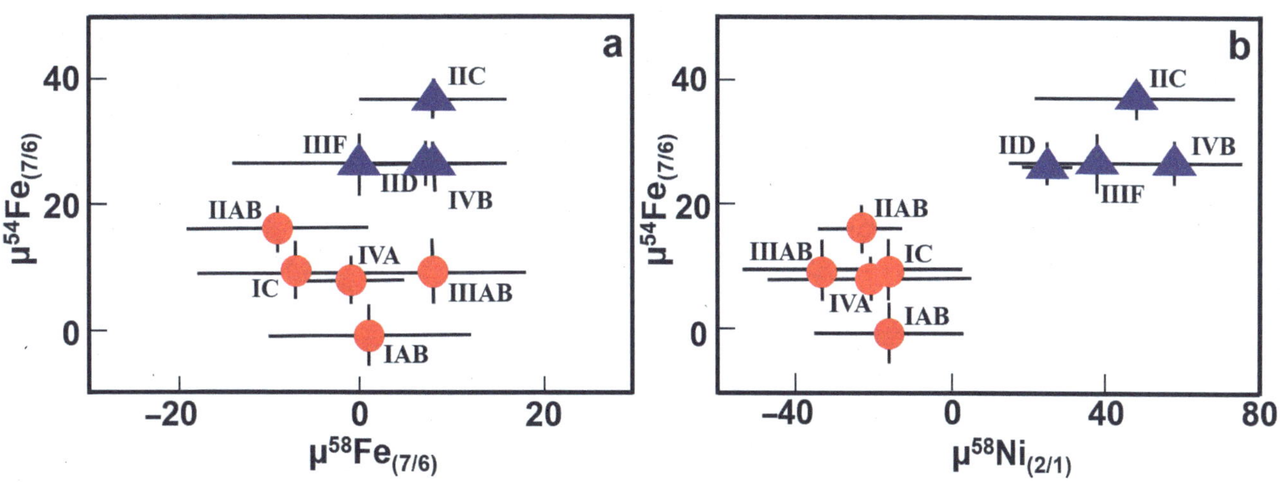

Fig. 6.4 The Fe- and Ni-isotopic compositions of CC irons (blue triangles) and NC irons (red circles) showing no compositional overlap. The symbol μ indicates parts per million. Data normalized to either the $^{57}Fe/^{56}Fe$ ratio (designated as $Fe_{(7/6)}$) or the $^{62}Ni/^{61}Ni$ ratio (designated as $Ni_{(2/1)}$). (a) $\mu^{54}Fe_{(7/6)}$ vs. $\mu^{58}Fe_{(7/6)}$. (b) $\mu^{54}Fe_{(7/6)}$ vs. $\mu^{58}Ni_{(2/1)}$. Diagram based on Hopp et al. (2022).

equilibration between the metal and younger (hence, more radiogenic) silicate.

The CC- and NC-iron groups differ significantly in their bulk W-, Mo-, N-, Ni- and Fe-isotopic compositions; CC irons have higher $\varepsilon^{182}W$, $\varepsilon^{100}Mo$, $\varepsilon^{95}Mo$, $\delta^{15}N$, $\mu^{54}Fe$, and $\mu^{58}Ni$ values (Prombo and Clayton, 1993; Budde et al., 2016; Kruijer et al., 2017a, 2017b, 2017c; Poole et al., 2017; Nanne et al., 2019; Hopp et al., 2022) (e.g., Figs. 6.3, 6.4). This indicates that the CC irons and

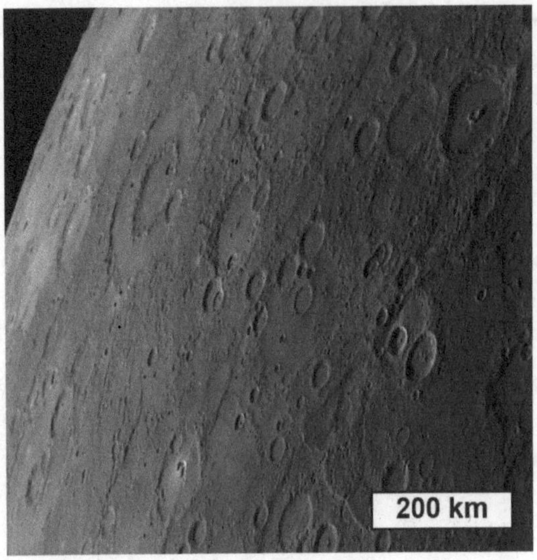

Fig. 6.5 Impact craters on Mercury. NASA Photo ID: PIA10936. Image courtesy of JPL, NASA, Johns Hopkins University Applied Physics Laboratory, and Carnegie Institution of Washington.

NC irons formed within separate isotopic reservoirs (e.g., Qin et al., 2011) very probably at different heliocentric distances.

Kruijer et al. (2017a, 2017b, 2017c) suggested that proto-Jupiter accreted rapidly between the CC and NC zones, exceeding 20 Earth masses ($M_\oplus$) within ~1 Ma after CAI formation. This nascent giant planet functioned as a barrier between the CC and NC reservoirs (Section 14.1), hindering the radial drift of small particles toward the Inner Solar System (e.g., Lambrechts et al., 2014). These small particles could include CAIs, which appear to have concentrated in dust-rich nebular regions (Rubin, 2011) likely located outside Jupiter's orbit (Gerber et al., 2017; Scott et al., 2017). Proto-Jupiter continued to grow by accreting gas onto its core; by 3–4 Ma after CAI formation, it had grown to ~50 $M_\oplus$, opening a gap in the nebular disk (Morbidelli et al., 2016). During this epoch, proto-Jupiter may have migrated (Walsh et al., 2011; Raymond and Izidoro, 2017), causing small CC bodies (including differentiated and undifferentiated objects) beyond its orbit to scatter. Many reached the Inner Solar System, colliding with the terrestrial planets (see, e.g., Fig. 6.5), the Moon, and the Sun.

7

Delivery of Meteorites to the Earth

7.1 Near-Earth Asteroids

The orbits of recovered meteorites closely resemble those of Near-Earth Asteroids (NEAs) (Fig. 7.1) and fireballs (Fig. 7.2). A particularly close orbital match is that of the 17–20-meter Chelyabinsk LL5 meteoroid (which entered the atmosphere over Russia on February 15, 2013 at 3:22 UT) and the Q-type asteroid NEA 86039 (1999 NC_{43}). This asteroid has a 2.33-year orbital period and is about 2 km in diameter. There are on the order of 1000 NEAs ≥ 1 km in size (Morbidelli et al., 2002; Mainzer et al., 2011), ~5000 with diameters between 0.3 and 1 km (data cited by Dai and Timmermann, 2025), and ~20,000 larger than 100 m (Mainzer et al., 2011). Multi-kilometer objects tend to have been derived from the central and outer main asteroid belt; sub-kilometer objects tend to come from the inner main belt (Nesvorný et al., 2023). Altogether, most NEAs have Apollo-like orbits (with perihelion distances less than 1.017 AU, semi-major axes greater than 1 AU, and orbital periods greater than 1 year); a minority has Amor-like orbits (with perihelion distances between 1.017 and 1.3 AU, and semi-major axes greater than 1 AU); and probably less than 10% have Aten-like orbits (with aphelion distances greater than 0.983 AU, semi-major axes less than 1 AU, and orbital periods less than 1 year) (e.g., Bottke et al., 2000).

A survey of NEA spectral types found that about 58% are Q/S-complex, 21% are B/C-complex, 10–11% are V types, and 10–11% are of miscellaneous taxonomic classes (Popescu et al., 2019). The Q-type NEAs have the lowest median perihelion value: 0.797 ± 0.244 AU. Ordinary chondrites resemble Q/S-complex objects that have experienced space weathering (e.g., Fig. 5.9). It is no coincidence that ordinary chondrites are the most abundant meteorites observed to fall and correspond spectrally to the most abundant Near-Earth Asteroids. Except for meteorites derived from ejecta blasted off the Moon and Mars, meteorites that impact the Earth are, for all practical purposes, derived from NEAs. And, of course, all meteorites – asteroidal, lunar, and martian – are derived from Near-Earth Objects (NEOs).

Jenniskens and Devillepoix (2025) reviewed orbital data based on video and photographic observations of 75 meteorite falls. They found that small meteoroids (0.1–1 m) tend to have different orbits than larger (1 km) asteroids of the same taxonomic class. (1) Some small H-chondrite meteoroids reached Earth along low-inclination orbits from a region slightly past the 5:2 resonance with Jupiter; others came from the inner main asteroid belt and have a ~35-Ma CRE age. However, S-class NEAs travel along high-inclination orbits associated with the 3:1 resonance; a few H chondrites (with distinct CRE ages of 6 and 18 Ma) arrived along similar orbits. (2) L-chondrite meteoroids reached Earth from the inner main belt by way of the ν_6 secular resonance, whereas L-chondrite-like NEAs arrive mainly via the 3:1 resonance. (3) LL chondrite meteoroids and LL-chondrite-like NEAs both arrive from the inner main belt along similar orbits.

7.2 Kirkwood Gaps and Resonances

The mean lifetimes of NEAs are only a few million years (Morbidelli et al., 2002); they crash into the Sun (or less commonly into a planet, moon, or

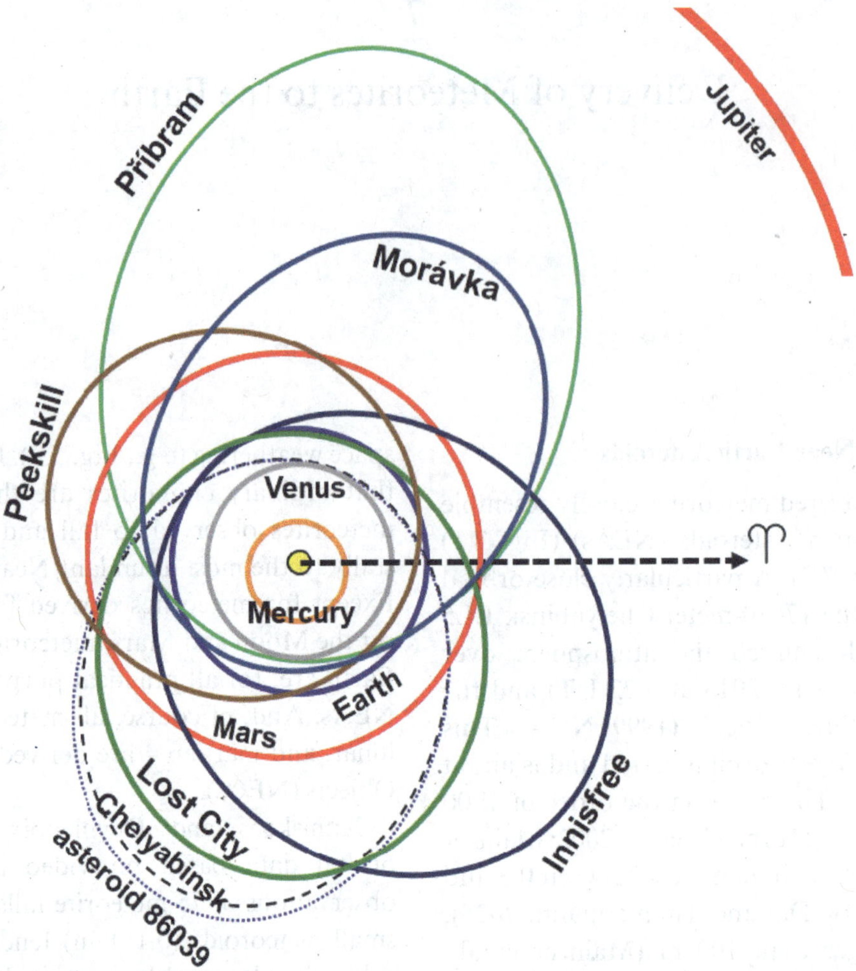

Fig. 7.1 The orbital paths of the recovered meteorite falls Příbram (H5), Lost City (H5), Innisfree (L5), Peekskill (H6), Morávka (H5), and Chelyabinsk (LL5) as well as that of the Near-Earth Asteroid 86039 (1999 NC_{43}). Also shown are the orbits of the inner planets and Jupiter. All orbits are projected onto the plane of the ecliptic. The orbits of Chelyabinsk and asteroid 86039 are very similar. The ram's horn symbol (♈) represents the vernal equinox (the "First Point of Aries," i.e., the intersection of the ecliptic and the celestial equator near the point where the Sun is located in the northern hemisphere during Spring). Diagram based on Borovička et al. (2015) and other sources.

other Near-Earth Object) or are ejected from the Solar System on hyperbolic trajectories. Over the course of the Solar System's history, NEAs must have been constantly replenished. The first strong evidence that the main asteroid belt was the source of NEAs was provided by American astronomer Daniel Kirkwood (Kirkwood, 1866, 1867). A histogram of the number of asteroids ordered by their semi-major axes reveals hiatuses (now known as "Kirkwood Gaps") at specific heliocentric distances (Fig. 7.3).

Each of the gaps reflects the ratio of the number of orbits that would be completed by an asteroid at that location relative to those completed by Jupiter in the same time period. For example, an asteroid at the 3:1 gap located at 2.5 AU (between the inner and central belts) would complete three heliocentric orbits in the time it would take Jupiter to complete a single orbit. An asteroid in the outer belt at about 2.82 AU (corresponding to the 5:2 gap) would orbit the Sun five times in the same time interval Jupiter would take to orbit the Sun twice. The most prominent Kirkwood Gaps are at 2.06 AU (corresponding to the 4:1 resonance), 2.50 AU (the 3:1 resonance), 2.82 AU (the 5:2 resonance), 2.95 AU (the 7:3 resonance), and 3.27 AU (the 2:1 resonance) (Fig. 7.3).

These orbital-period relationships are known as commensurabilities; each is caused by a mean-motion resonance. This is a common phenomenon

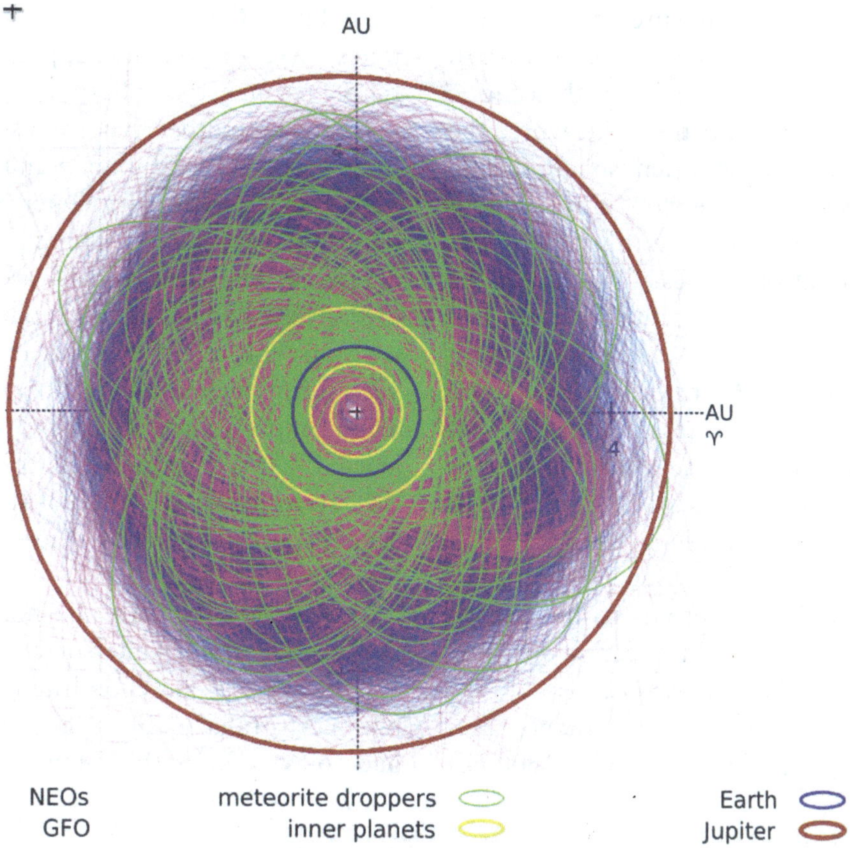

Fig. 7.2 Orbits of all known Near-Earth Asteroids (source: Minor Planet Center); the orbits of all fireballs detected by the Global Fireball Observatory and Desert Fireball Network; and the subset of fireballs that have delivered meteorites to the surface of the Earth (GFO/DFN and literature data). Image courtesy of Hadrien Devillepoix; GFO; Curtin University.

in celestial mechanics. In this case, the mean-motion resonance involves periodically repeating gravitational nudges on the asteroid by Jupiter. The cumulative effect of these nudges is an exchange of momentum and a shift in the asteroid's orbital elements. The asteroid is forced out of the gap. The process is analogous to an adult pushing a child on a swing; kinetic energy from the adult is added to the child+swing at the same point in the swing's oscillation. This "in-sync" motion causes an increase in the amplitude of the simple harmonic motion of the child+swing. The swing goes higher – closer to ±90° relative to the resting position.

Asteroids can enter Kirkwood Gaps due to Yarkovsky forces, gravitational perturbations, or via the transfer of momentum during collisions with other asteroids. Approximately 25% of NEAs are <100 m in diameter and would have been susceptible to Yarkovsky forces when they were in the main asteroid belt (Farinella et al., 1998).

Inside the 3:1 resonance, there is a broad region where asteroid eccentricity changes in a chaotic manner, allowing it to assume an Earth-crossing orbit in about ~1 Ma. There is also a narrow central region where asteroids evolve into Mars-crossers; these asteroids can be readily transferred to the broad region and become Near-Earth Objects (NEOs).

The median time it takes an asteroid at the 5:2 resonance to reach an Earth-crossing orbit is ~0.3 Ma (Morbidelli et al., 2002).

The υ_6 resonance is caused by Saturn (the sixth planet from the Sun, hence the subscript 6). It is among the most efficient mechanisms for delivering asteroids to the terrestrial planet region, thereby producing NEAs (Bottke et al., 2015). It occurs when there is a simple numerical ratio between the precession frequency of an asteroid's longitude of perihelion and that of the mean precession frequency of Saturn's longitude of perihelion. (The longitude of the perihelion [ϖ] represents the angular position of the perihelion of the orbit relative to the vernal equinox [the intersection of the ecliptic and the celestial equator near the point where the Sun is located in the northern hemisphere during Spring]. The precession is the slow movement of this position around the Sun due to gravitational interactions with multiple bodies.) This causes the eccentricity of the asteroid's orbit to increase. The υ_6 resonance delimits the inner edge of the main asteroid belt (Fig. 7.4a). Morbidelli et al. (2002) divided the υ_6 zone into two regions, each about 0.04 AU wide. In one region, asteroids greatly increase their eccentricities, achieving Earth-crossing orbits in ~0.5 Ma. Some become Venus-crossers; some collide with the Sun. The other υ_6 region is less powerful – asteroids in this region achieve Mars-crossing orbits, but can evolve further, allowing them to become NEOs.

Dynamical modeling shows that the υ_6 secular resonance supplies 83% of small (<300 m) Near-Earth Objects (~66% of which are S asteroids). The 3:1 resonance supplies 16% of NEOs and the 5:2 resonance supplies 1% (Sanchez et al., 2024).

There is another type of asteroidal orbit that resembles those of Jupiter Family Comets (JFCs). Some comets from the Outer Solar System that happen to pass close to Jupiter have had their orbits altered so that their aphelia are appreciably lowered. These bodies are the Jupiter Family Comets, a group with relatively short orbital periods (<20 years), low inclinations (<40°), and direct orbits. The group was first noted by Roller (1870), who listed 13 comets with aphelia between 4.1 and 6.2 AU, all within ~1 AU or so of Jupiter's mean heliocentric distance (5.2 AU). There are now more than 800 known JFCs.

It is possible there are also non-cometary objects in JFC-like orbits. Anderson et al. (2024) described a rather typical H5 chondrite (Arpu Kuilpu) that fell in Australia in 2019; its meteoroid progenitor had followed an orbit similar to those of JFCs:

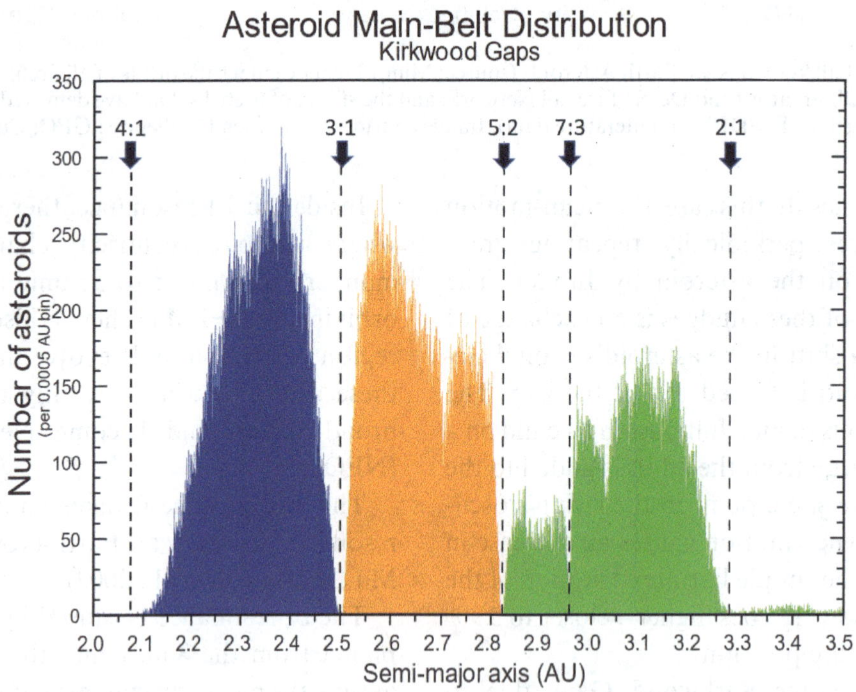

Fig. 7.3 Number of asteroids vs. semi-major axis (based on the JPL database). The prominent gaps at 3:1, 5:2, 7:3, and 2:1 reflect the ratios of the number of orbits completed by an asteroid at the corresponding heliocentric distance relative to the number completed by Jupiter in the same time interval. For example, an object at the 3:1 gap 2.5 AU from the Sun would complete three heliocentric orbits in the time it would take Jupiter to complete one orbit. This relationship is known as a mean-motion resonance and can cause significant changes in an asteroid's orbital elements. Diagram modified from Wikipedia (https://en.wikipedia.org/wiki/Kirkwood_gap), based on a plot by Alan Chamberlain (JPL/Caltech) (http://ssd.jpl.nasa.gov/images/ast_histo.ps).

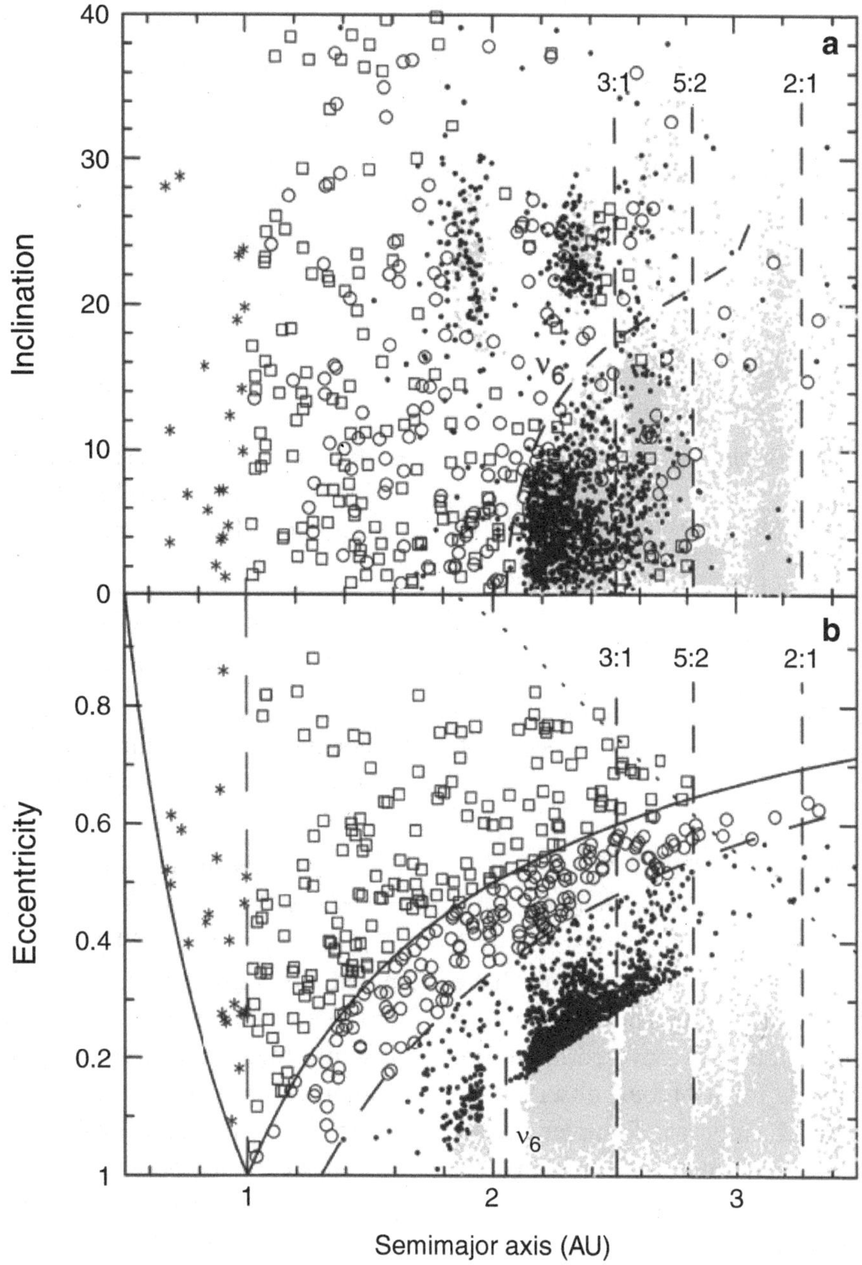

Fig. 7.4 The distribution of Near-Earth Objects (NEOs). (a) Orbital inclination vs. semi-major axis. (b) Orbital eccentricity vs. semi-major axis. The first 10,000 main-belt asteroids are light gray. Open squares are Apollo asteroids; black points are Mars-crossers; open circles are Amors; asterisks are Atens. The solid V-shaped curve with an inflection point at 1 AU bounds the Earth-crossing region. The dashed curve intersecting the x-axis at 1.3 AU bounds the extreme of the Amor region. The dashed vertical line intersecting the x-axis at 1 AU separates the Aten and Apollo NEOs. The υ_6 resonance is a function of the inclination – as shown in (a) –but is largely independent of the eccentricity – it is shown in (b) for an inclination of zero degrees. Diagram modified from Morbidelli et al. (2002).

($a = 2.75 \pm 0.03$ AU; $e = 0.671 \pm 0.003$; $i = 2.03 \pm 0.01°$; $T_J = 2.97 \pm 0.02$, wherein T_J is Tisserand's parameter with respect to Jupiter; Shober et al., 2022). The pre-atmospheric diameter of the Arpu Kuilpu meteoroid was probably only 2–10 cm. Thus, it seems possible that many similarly sized sub-meter bodies in JFC-like orbits are asteroidal fragments and not comets (Anderson et al., 2024).

8

Fall Phenomena

Now slides the silent meteor on, and leaves
A shining furrow …
Alfred Tennyson, "Now Sleeps the Crimson Petal"

… the strange huge meteor procession, dazzling and
clear, shooting over our heads,
(A moment, a moment long, it sail'd its balls of
unearthly light over our heads,
Then departed, dropt in the night, and was gone) …
Walt Whitman, "Year of Meteors (1859–60)"

8.1 What Is an Observed Fall?

An "observed fall" may have been witnessed by people and/or detected by one or more instruments; alternatively, the meteorite may only have been inferred to have fallen recently. As defined in Chapter 3, a meteorite fall is a recovered meteorite that was (a) responsible for a fireball observed by people, fireball-network cameras, Doppler radar, seismometers, and/or satellites, (b) responsible for thunderous sonic booms, whistling or hissing sounds, or noise produced when the object strikes a structure or the ground, (c) inferred to have fallen recently from the damage it caused to structures, the ground, or living creatures, or (d) inferred to have fallen recently because it was found on a surface where recent prior observation had not indicated its presence. A newly recovered meteorite specimen is commonly considered a fall if it is from a historic shower and is paired with an observed fall – even if the new specimen was found decades after its congeners hit the ground. For example, close to 100 new specimens of Holbrook (L/LL6) were recovered over a three-day meteorite-hunting trip on the 100th anniversary of the Holbrook, Arizona fall of July 19, 1912 (Verish, 2012). Altogether, about 180,000 individual Holbrook specimens have been recovered with a total mass of ~220 kg.

8.2 Fireballs, Bolides, Sonic Booms, and Electrophonic Sounds

And the second angel sounded, and as it were a great
mountain burning with fire was cast into the sea: and
the third part of the sea became blood …
Revelation 8:8; KJV

Every day about 54 tons of extraterrestrial material reach the Earth (G. Drolshagen et al., 2017; E. Drolshagen et al., 2020). Evatt et al. (2020) estimated the annual flux at 17,600 meteoroids with masses exceeding 50 g. About 71% of the material falls into the ocean. Meteor- and fireball-producing meteoroids strike the top of the Earth's atmosphere with average velocities of 16–18 km s^{-1} and median velocities of 15–17 km s^{-1} (E. Drolshagen et al., 2020), equivalent to Mach 44–53. At a height of 80–100 km (near the base of the thermosphere), the atmospheric density is sufficiently high to cause appreciable frictional drag on the meteoroid. The loss of kinetic energy during deceleration is compensated for by the production of thermal and electromagnetic radiation. The meteoroid becomes luminous, producing a meteor.

Rubin and Ma (2021) defined a **meteor** as the light phenomenon produced by a solid object transiting an atmosphere at high velocity. To date, meteors have been observed on Earth (e.g., Kronk, 1988; Arai et al., 2019) and Mars (Selsis et al., 2005). Collisions with atmospheric molecules cause the outer layers of the incoming object to sputter away, typically producing a vapor of metal atoms. Light is emitted when excited electrons in the orbits of these atoms fall back to lower-energy states. The color of the light is dependent on the element: Na – orange-yellow; Fe – yellow; Mg – blue-green; Ca^+ ions – violet; and Ni – green. Ionized oxygen and nitrogen atoms in the atmosphere can produce red light.

A **fireball** is a very bright meteor. The International Astronomical Union (IAU) defined a fireball to be a meteor brighter than absolute visual magnitude –4 (at a distance of 100 km), comparable in apparent brightness to Venus. Some fireballs can be much brighter; for example, the Žďár nad Sázavou fireball reached a magnitude of –15.26 (more than six times brighter than the full Moon) as it passed over the Czech Republic on December 9, 2014. A few small individual L3.9 ordinary chondrites were recovered (Spurný et al., 2020). In some cases, a fresh meteorite find can be matched with a fireball previously recorded by a camera network. Ischgl, for instance, a ~1-kg LL6 chondrite found in Tyrol, Austria in June 1976, seems to be associated with the bright fireball EN241170 observed visually and by camera networks on November 24, 1970 (Gritsevich et al., 2024). A **bolide** is a fireball-producing meteoroid that breaks apart in the atmosphere. A **superbolide** is a bolide brighter than absolute visual magnitude -17 (equivalent to the apparent magnitude at a distance of 100 km); this is about 40 times brighter than the full Moon. (The Chelyabinsk superbolide of February 15, 2013 had a maximum absolute magnitude of approximately –27 at an altitude of ~30 km [Brown et al., 2013; Popova et al., 2013]. It was thus momentarily brighter than the Sun [–26.74].)

The supersonic speed of an incoming meteoroid produces shock waves that can generate high amounts of sound energy reminiscent of thunderclaps or cannon fire. The sonic boom is produced continuously until the meteoroid slows below the speed of sound.[1] Typically, the boom is heard by observers whose position along a narrow strip on the ground intersects an expanding conical region of air in the projectile's wake.

Meteoroids can produce electrophonic sounds characterized by witnesses as "crackling," "cracking," "clicking," "fizzing," "hissing," "sparking," "sizzling," "popping," "whistling," "whizzing," "rustling," "swishing," "whirring," "rushing," "squeaking," and "faint buzzing" (e.g., Olmsted, 1834; Udden, 1917; Wegener, 1917; Sellards, 1929; Lamar and Romig, 1964; Nield, 2011; Finnegan, 2015; Keay, 1992). The sounds were discussed in the eighteenth century by Edmond Halley (1714) and Blagdon (1784). A plausible explanation for electrophonic sounds is that they are produced by the plasma trail associated with a large fireball generating several kilowatts of extremely low to very low frequency (ELF/VLF) radiation (i.e., 1 to 10 kHz). When appropriate metal or dielectric objects are in the vicinity of witnesses, audible sounds are generated by the direct conversion of some of the electromagnetic energy emanated by the meteoroid into acoustic energy (Keay, 1980, 1992). Kelley and Price (2017) suggested that when the glowing region (known as the coma) around the descending meteoroid crosses Earth's magnetic field lines, it produces a potential difference (via the Hall effect) and an electromagnetic wave. As proposed by Nininger (1934, 1939), the electromagnetic wave can interact with structures (particularly metallic ones) and transform into true sound waves. An efficiency as low as 0.1% of the conversion of electrophonic energy to sound would suffice for humans to see meteors and hear electrophonic sounds almost simultaneously.

8.3 Time of Day

Ordinary chondrites fall at all hours of the day and night: for example, H5 Košice 00:24; H4 Ourique 01:50; H6 Orvinio 05:15; L6 Andila 10:00; L5 Chajari 13:00; LL3 Semarkona 15:00; L6 Ruhobobo 16:30; L/LL5 Qidong 17:45; LL6 Stubenberg 22:36; H5 Juancheng 23:23. They enter the atmosphere with different geocentric velocities (averaging 16–18 km s^{-1}).

The escape velocity of the Earth is 11.2 km s^{-1} (determined by $v_{escape} = \sqrt{[(2GM_\oplus)/R_\oplus]}$ where G is the universal gravitational constant, $M_\oplus$ is the

[1] The speed of sound (c) in the atmosphere varies with altitude, mainly as a function of air temperature (which is inversely related to air density): c increases from ~276 m s^{-1} at 80–90 km to ~328 m s^{-1} at ~50 km, decreases to ~296 m s^{-1} between 20 and 11 km, and then increases steadily to 343 m s^{-1} near the ground.

mass of the Earth, and $R_{\oplus}$ is the radius of the Earth). An object launched from Earth on a ballistic trajectory must reach this minimum velocity to escape the planet's gravitational influence. In the absence of other gravity-producing bodies, an object at rest an infinite distance from Earth and beginning its fall toward Earth would collide with our planet (after an infinite time) at 11.2 km s^{-1}. All substantial bodies hitting the Earth have at least this minimum velocity (neglecting the decelerating effects of the atmosphere). (An exception: Spacecraft orbiting the Earth can have impact speeds as low as 4 km s^{-1} due to the orbital motion of the spacecraft [E. Drolshagen et al., 2020].)

The Earth travels around the Sun along the ecliptic at an average Keplerian velocity of 29.8 km s^{-1} relative to the Sun (Fig. 8.1). The escape velocity of the Sun at the Earth's distance of 1 AU is 42.1 km s^{-1}. If an object moving at this speed with respect to the Sun is in a retrograde Earth-like orbit and collides with Earth head-on, its relative velocity can be determined by adding the Earth's heliocentric velocity to the Sun's escape velocity at 1 AU: $42.1 \text{ km s}^{-1} + 29.8 \text{ km s}^{-1} = 71.9 \text{ km s}^{-1}$. This is the maximum impact velocity of a Solar-System object striking the Earth; it could pertain, for example, to projectiles originating at a great distance (such as comets from the Oort Cloud). Interstellar objects not bound to the Sun could impact Earth at even higher relative velocities; a rough upper limit to their collision velocity – assuming they are not intergalactic – would be the escape velocity of the Galaxy at the Sun's distance from the galactic center ($v_{escape} \approx 580 \text{ km s}^{-1}$; Monari et al., 2018).

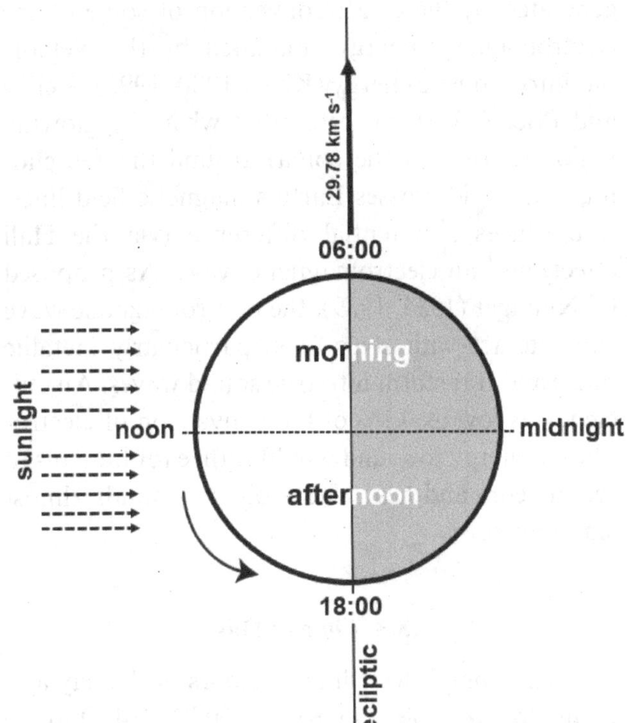

Fig. 8.1 Schematic view of Earth from the North Celestial Pole. Collisions of meteoroids and asteroids after local noon (bottom of circle) are in prograde orbits and generally have low geocentric velocities. Some meteoroids and asteroids colliding with Earth in the morning (top of circle) are in evolved prograde orbits and near their aphelion. Rare bodies in retrograde orbits would also impact in the morning. Arrows show direction of orbital motion.

A Near-Earth Asteroid (NEA) moving prograde (counterclockwise when viewed from the North Celestial Pole) in an orbit very similar to that of the Earth and overtaking the Earth (due to the Earth's gravitational attraction) would impact our planet in the afternoon (between local noon and midnight) at a relative velocity of 11.2 km s^{-1} (again, neglecting the decelerating effects of the atmosphere). The asteroid's heliocentric velocity and the Earth's heliocentric velocity cancel each other out:

$$v_{impact} = \sqrt{[(v_{\oplus escape})^2 + (v_{\oplus Kep} - v_{astKep})^2]}$$

$$v_{impact} = \sqrt{\begin{matrix}[(11.2 \text{ km s}^{-1})^2 \\ + (29.8 \text{ km s}^{-1} - 29.8 \text{ km s}^{-1})^2]\end{matrix}}$$

$$v_{impact} = 11.2 \text{ km s}^{-1}$$

where v_{impact} is the relative velocity of the asteroid when it hits the Earth, $v_{\oplus escape}$ is the escape velocity of the Earth, $v_{\oplus Kep}$ is the average Keplerian orbital velocity of the Earth around the Sun (i.e., essentially equivalent to the heliocentric velocity), and v_{astKep} is the orbital velocity of the asteroid. Most meteorite-dropping meteoroids are in prograde (a.k.a. direct) orbits (e.g., Newton, 1888; Pickering, 1910).

An NEA moving retrograde (clockwise) in an Earth-like orbit at the same speed as the Earth (29.8 km s^{-1}) would collide head-on with the leading side of our planet in the morning (between midnight and noon) at a relative velocity of 60.6 km s^{-1}:

$$v_{impact} = \sqrt{[(v_{\oplus escape})^2 + (v_{\oplus Kep} - v_{astKep})^2]}$$

$$v_{impact} = \sqrt{[(11.2 \text{ km s}^{-1})^2 + (29.8 \text{ km s}^{-1} - (-29.8 \text{ km s}^{-1}))^2]}$$

$$v_{impact} = \sqrt{[(11.2 \text{ km s}^{-1})^2 + (29.8 \text{ km s}^{-1} + 29.8 \text{ km s}^{-1})^2]}$$

$$v_{impact} = 60.6 \text{ km s}^{-1}$$

Moving at such high speeds through the atmosphere disrupts projectiles. Bodies in retrograde orbits follow the Dylan Thomas maxim and "do not go gentle into that good night."

For a meteoroid traversing Earth's atmosphere, the drag equation can be expressed as:

$$F_d = \frac{1}{2}\left(\rho \cdot v^2 \cdot c_d \cdot A\right),$$

where F_d is the drag force (frictional force), ρ is the mass density of the air at the meteoroid's altitude, v is the velocity of the meteoroid, c_d is the drag coefficient (dependent on both the Reynolds number and the Mach number), and A is the orthographic projected area of the meteoroid (approximated by the largest cross section). Because the drag force on a projectile plunging through the atmosphere is proportional to the square of the velocity, high relative velocities can lead to projectile disruption. Nininger (1933) noted that enhanced friction with the atmosphere encountered by retrograde meteoroids would disrupt these bodies, resulting in relatively small average recovered masses. The upper stability limit for incoming meteoroids likely lies between 22 km s^{-1} (Levin and Simonenko, 1969) and 30 km s^{-1} (ReVelle, 1979).

However, asteroids in retrograde orbits (e.g., 2007 VA85; 2009 HC82) are rare: ~99.9% of NEAs with absolute magnitudes in the range H = 17–25 (roughly corresponding to bodies with diameters of ~40 to ~1400 m) are in prograde orbits (Greenstreet et al., 2012a, 2012b; Granvik et al., 2018). Greenstreet et al. (2012a) suggested that the rare retrograde bodies were derived from main-belt asteroids that developed retrograde orbits while in the 3:1 resonance with Jupiter. These retrograde NEAs can have short lifetimes, as short as a few thousand years. The close encounters of retrograde NEAs with planets are high-energy events, more likely leading to a collision with the Sun (or in some cases Earth or Venus) or to ejection from the Solar System (e.g., Gladman et al., 2000).

There is a large single peak between 2:00 PM (14:00) and 6:00 PM (18:00) local time in the number of observed falls (Farrington, 1915; Leonard and Slanin, 1941; Heide and Wlotzka, 1995). These results may be influenced by sociological factors (Alekseev, 1988). People who are asleep are unlikely to witness fireballs.

Halliday and Griffin (1982) used camera-network data to show that meteorites fall at a fairly constant hourly rate in the afternoon and evening, between noon and midnight; however, falls exhibit a large dip in midmorning, around 06:00, exactly at the apex of the Earth's motion around the Sun (Fig. 8.1). (The dip becomes shallower at higher geographic latitudes.) The (midafternoon [18:00]/midmorning [6:00]) ratio of Halliday and Griffin (1982) is ~2. This deviation from unity is mainly a dynamic effect, not a sociological one (e.g., Wetherill, 1968).

There are reported differences in the frequency of falls during *daylight* hours between morning (defined as 06:00 until 12:00 noon) and afternoon (12:00 until 18:00). Hughes (1981) found that the afternoon/morning ratio of 644 meteorite falls (all meteorite groups) is ~2 (data from Hey, 1966 and Hutchison et al., 1977). Dudorov and Eretnova (2020) also found an afternoon/morning ratio of ~2 for 926 confirmed meteorite falls (all groups). Wetherill and Chapman (1988) showed that, among ordinary chondrites that fall during these hours (under statistically very similar near-ideal visual conditions), about twice as many OCs fall in the afternoon as in the morning. Similarly, Alekseev (1988) concluded that the afternoon/morning fall ratio of chondrites (all groups) is 2.3.

Afternoon (PM) meteorite falls are in prograde orbits and collide with Earth generally at low geocentric velocities. Morbidelli and Gladman (1998) showed that some morning (AM) falls are in evolved prograde orbits (such as those of the Aten asteroids). These AM meteoroid falls tend to be near their aphelia and are moving at relatively low heliocentric velocities (following Kepler's second law). The Earth has a higher heliocentric velocity than these meteoroids and ploughs into them with the planet's leading hemisphere.

8.4 Possible Variation with Latitude

Several studies have indicated that there are more meteorites per unit area impacting the equator than regions at higher latitudes (e.g., Halliday, 1964; Halliday and Griffin, 1982; Le Feuvre and Wieczorek, 2008). Most asteroids orbit the Sun near the ecliptic plane. Data from the Minor Planet Center website shows that the mean inclination of NEAs is 11.9 ± 10.6°; NEAs with high inclinations are rare (e.g., Greenstreet and Gladman, 2013). Most of these bodies are more-or-less aligned with the ecliptic and would be more likely to impact the Earth in an equatorial region than a polar region. If it is assumed that asteroids in ecliptic orbits are confined to a zone thicker than the Earth's diameter, then those bodies approaching Earth on orbits slightly above the ecliptic could be deflected toward the center of the Earth (Fig. 1 of Halliday, 1964); they would impact the Earth's surface at a lower latitude than they would have done had there been no gravitational focusing. This effect, however, is significant only for the low-velocity tail of the impact flux (Le Feuvre and Wieczorek, 2008). Evatt et al. (2020) used satellite observations of fireballs to show that the fireball frequency per unit area decreases with increasing latitude; they extended their analysis to smaller meteoroids. The modeled deficits of polar meteorites relative to equatorial meteorites range from ~4% (Le Feuvre and Wieczorek, 2008) to ~15% (Halliday and Griffin, 1982) to ~35% (Evatt et al., 2020). Although Antarctica may be the best place to find meteorites, these models indicate that fewer tend to land there than at lower latitudes.

However, Robertson et al. (2021) used updated models for the distribution of NEOs and precession models of gradual changes in the orbits of these bodies to predict just the opposite. They suggested that the flux of polar impacts is 22% greater than that near the equator. The principal contributing factor to latitudinal differences in the meteorite flux appears to be the radiant distribution of projectiles, particularly those with high relative velocities.

The bottom line is that there may be small differences in the flux of meteoroids with terrestrial latitude, but the magnitude of such differences depends on the model. As more data on NEOs are acquired and the observational biases of these data are accounted for, the models of terrestrial impacts will become more accurate.

8.5 Radar Observations

Most modern weather radars are pulse-Doppler systems that can gauge the distance to a target by using ~10^{-6}-s pulse transmissions of microwave radiation (with wavelengths of 1–10 cm) and evaluating the Doppler effect of those portions of the signals that are reflected. Between pulses, the radar system acts as a receiver for a duration of ~10^{-3} s; this is sufficient time for the transmitted pulse to reach the target and return, even at target distances of a few hundred kilometers. This technique can yield the target object's position and velocity; it can determine the motion of rain droplets and distinguish rain, snow, and hail.

Meteorite falls can be identified by their positions and velocity vectors. Clouds, airplanes, and birds tend to move more-or-less horizontally. Rain, snow, and hail fall mostly vertically (from heights on the order of 3–5 km), but meteorites fall mostly vertically from far greater heights (~80 km). Radar can also pinpoint sites where meteorites have landed, enabling more precise ground searches.

If eyewitness reports of a fireball occur in areas covered by weather radar, the recorded radar signals can be checked for objects falling from heights of tens of kilometers. Meteoroids of different sizes fall at different speeds; this is because drag affects small objects (with their higher surface-area/volume ratios) to a greater extent than large objects. Individual radar signatures are commonly attributable to meteoroids of a restricted size range. The radar data can provide the number of falling meteoroids, their sizes, and the locations where they struck the Earth's surface.

Here is an example: The Hamburg H4 chondrite fell in Michigan on January 16, 2018 at 8:08 PM local time. A fireball traveling east to west was observed by hundreds of people from seven states; it was also recorded by several security cameras. Mike Hankey of the American Meteor Society calculated the trajectory of the fireball and Mark Fries of Johnson Space Center (NASA) used weather

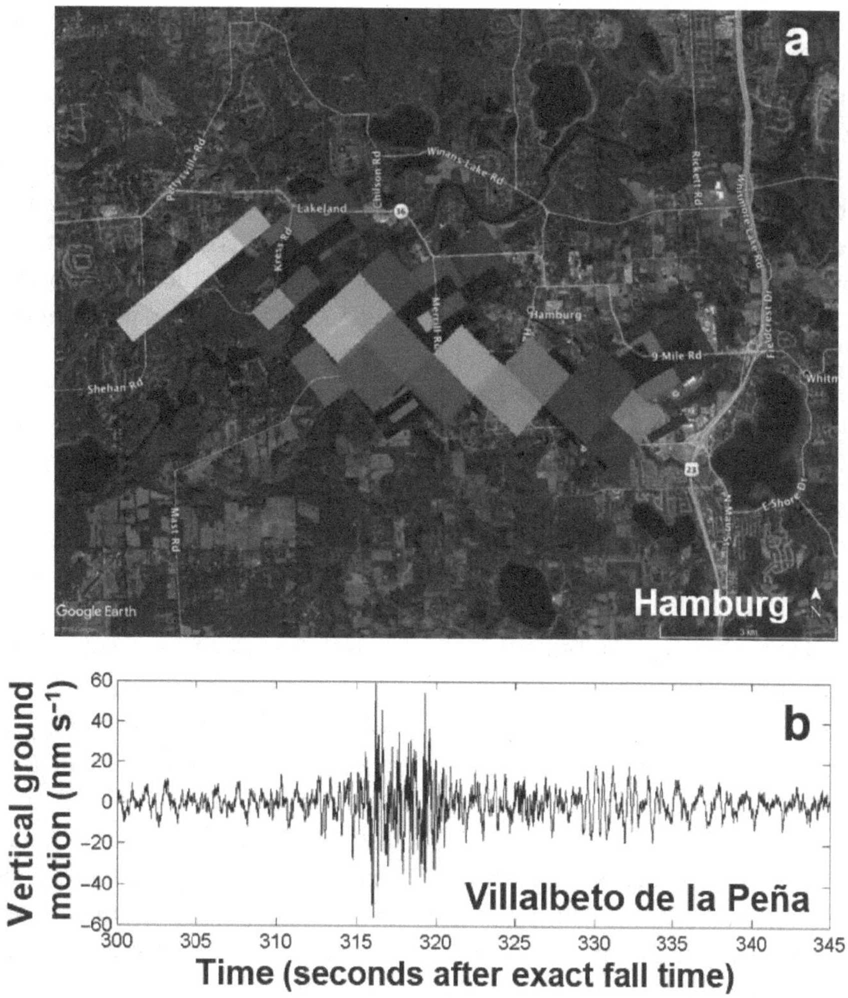

Fig. 8.2 Fall phenomena. (a) Radar signatures (blue/gray polygons) of falling individuals of the Hamburg H4 chondrite superimposed on a hybrid map from Google Earth. This is a composite image combining data from two different radar systems. In the NEXRAD weather radar network operated by NOAA, the KDTX radar outside Detroit recorded the signatures of falling meteoroids. The Terminal Doppler Weather Radar (TDWR) short-range weather radar for the Detroit airport also detected falling meteoroids. Image courtesy of NASA. (b) Filtered (1.25–40 Hz) vertical ground motion data of the fireball of the Villalbeto de la Peña L6 chondrite (January 4, 2004) recorded at a seismic station in Spain. Diagram from Llorca et al. (2005).

radar data to produce a map of the strewn field (Fig. 8.2a). The first six pieces were discovered two days after the fall; 13 additional pieces were found within two weeks. A total of ~1000 g was recovered. Loud sonic booms were heard, and the pressure wave from the meteoroid was recorded as a magnitude 2.0 earthquake.

In many cases, fireballs produce airwave signals that cause seismic ground motions. For instance, the fireball of the Villalbeto de la Peña L6 chondrite generated seismic signals; the vertical component of these signals is shown in Fig. 8.2b (Llorca et al., 2005). There are two distinct peaks 2.9 s apart, recording vertical ground motions of 59.0 and 54.6 nm s^{-1}, respectively. There is negative ground-motion velocity prior to the maxima of both peaks, probably caused by overpressure from an airwave.

8.6 Fusion Crust

As discussed by Rubin and Ma (2021), during the traverse of a chondritic meteoroid through Earth's atmosphere, the meteoroid's surface typically reaches temperatures of 1180–1410°C; the rock melts to depths of ~0.3–1.0 mm due to frictional heating with the surrounding air. Although the interior of the meteoroid remains cold, the surface ablates away, exposing a new surface that also melts and ablates away; this process can be repeated several times. An individual stony meteoroid that does not break into fragments could lose >95% of its original mass through ablation by the time it reaches the Earth's surface. A 1-meter-diameter meteoroid would be expected to lose 90% of its initial mass before reaching the ground (if it somehow avoided being torn apart by aerodynamic stress) (Melosh, 2011).

Chelyabinsk (LL5) broke apart in the atmosphere, but the main mass that plunged into Lake Chebarkul weighed ~640 kg. This is only about 0.005% of the original mass of the original meteoroid, which underwent both ablation and explosive fragmentation. Toward the other extreme, H5 chondrites from the Otway Massif, Antarctica pairing group (paired with GRO 03001) were derived from a 200–400-kg meteoroid that lost only 50–75% of its initial mass during atmospheric descent (Welten et al., 2024). (Based only on the refractive index of its olivine crystals, GRO 03001 was misclassified as L5 in the Meteoritical Bulletin Database [Righter, 2018].)

Most incoming meteoroids break into fragments as aerodynamic stresses increase during their descent. For objects travelling at 20 km s^{-1}, stresses rise from 10 MPa at an altitude of 30 km to 100 MPa at 15 km (Melosh, 1989). This exceeds the crushing strengths of many stony meteorites (which range from ~6 to ~500 MPa and average ~200 MPa).

The surfaces of the meteoroid fragments melt and ablate away until the meteoroids have decelerated sufficiently so that the temperatures reached during frictional heating are below the sample's melting temperature. Meteorite ablation spherules separate from the melt at the surface of the incoming meteoroid and fall as independent bodies. Many spherules are rich in silicates, but some are Fe-rich, formed by separated immiscible metallic melts (Genge and Grady, 1999).

The final fusion crust of the average meteoroid is a ~1–2-mm-thick coating (e.g., Figs. 8.3, 8.4a), commonly consisting of three zones: an outer zone composed of vesicular black opaque glass containing submicrometer magnetite, an intermediate zone containing partly melted silicate grains, and an inner zone containing unmelted silicate and rapidly solidified metal-troilite intergrowths. A detailed study of the fusion crust of the Auckland L6 fall (Scott et al., 2023) showed that the three zones consist of: (1) a vesicular, ~100-μm-wide outer zone that had largely melted and undergone sulfide volatilization, now consisting of vesicular glass (with embedded BSE-bright crystals, probably magnetite and/or wüstite) and ~50-μm-sized elongated olivine crystals, (2) a middle ~500-μm-thick zone that had

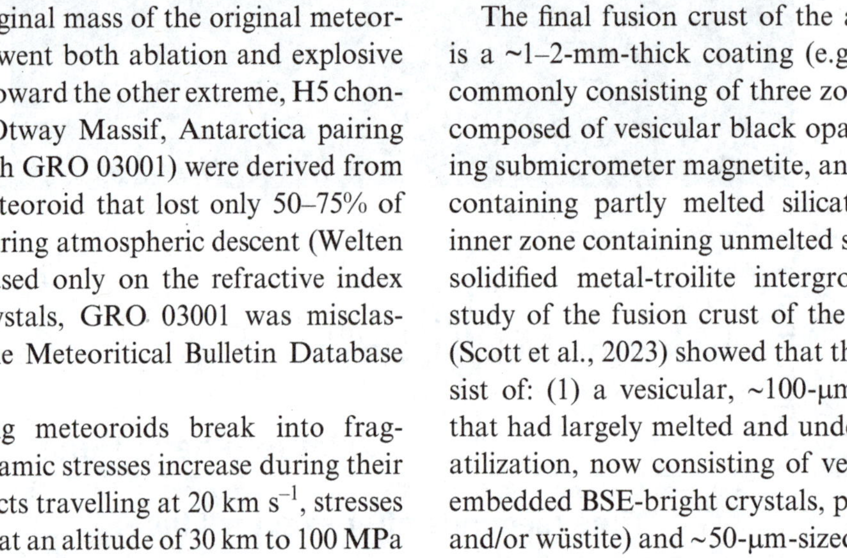

Fig. 8.3 Fusion crusts with irregular polygonal fractures on ordinary chondrites. (a) The Nuevo Mercurio H5-chondrite fall from Zacatecas, Mexico. Image courtesy of John Kashuba. (b) A 362-g specimen of the Hammadah al Hamra L6 chondrite, a possible fall from Libya. Image courtesy of Miguel Contreras (meteolovers.com).

been partially melted and is now composed of feldspathic silicate glass embedding immiscible blebs of metal and troilite as well as unmelted relict silicate grains, and (3) an inner 1.5–2-mm-thick zone exhibiting thin troilite-rich veinlets in pre-existing shock-induced fractures within olivine, orthopyroxene, and chromite grains. These chromite grains are typically a few micrometers in size; they tend to be smaller and have more variable MgO/(MgO+FeO) ratios than interior chromite grains (Bellesi et al., 2020). Similar chromites were reproduced with plasma flow experiments on the H5 chondrite El Hammami (Pittarello et al., 2019).

Oxygen isotope analyses show that up to 19% of the O_2 in the fusion crust is from the terrestrial atmosphere, incorporated during descent. (In fact, the O-isotopic composition of the upper layers of the Earth's atmosphere can be traced by analyzing cosmic spherules [e.g., Clayton et al., 1986; Pack et al., 2017].)

Some individual meteorites display more than one generation of fusion crust. One side may be a newly fragmented surface that was melted just before the meteoroid had decelerated below the point where frictional melting ceased; the fusion crust layer on this side may be thin and patchy. Another side of the individual may have a thicker fusion crustal layer generated a little higher in the atmosphere (thus experiencing a somewhat longer melting duration). In some cases, parts of the fusion crust adhere less strongly to the meteoroid interior and can flake off during deceleration.

The Tamdakht H5 fall in Morocco is unique in possessing a fusion crust up to 12 mm thick in some specimens (Zennouri et al., 2025). The meteoroid underwent several fragmentation events in the atmosphere, exposing new surfaces to frictional melting. Rotation of some of the fragments during descent may have caused large amounts of melt to accumulate at their surfaces (Ramdohr, 1967a).

Many fusion crusts on stony meteorites are covered with irregular polygonal cracks (e.g., Fig. 8.3). These are contraction cracks caused by thermal stress engendered by uneven cooling. Shortly after melt solidification, the outer surface of the crust cools rapidly while undergoing compression; the underlying meteorite cools much more slowly (and is in a state of tension). Similar processes cause cracking on the glazed surfaces of ceramics (e.g., Chen et al., 2023).

Fresh fusion crusts of ordinary chondrites have a matte black appearance (Fig. 8.4a). During terrestrial weathering, the fusion crust tends to turn rusty brown (e.g., Fig. 8.4b); the color change is not uniform across the surface. Moisture can penetrate the interior of the meteorite via cracks in the fusion crust.

Minerals that form in the fusion crusts of chondritic meteorites include eskolaite (Cr_2O_3), formed

Fig. 8.4 Ordinary chondrites with regmaglypts. (a) Largest specimen (7.01 kg) of the La Criolla L6 meteorite shower (Argentina, 1985) showing a black fusion crust covering a medium-gray interior. Also present are a few 1–3-cm-wide regmaglypts (more abundant on the right side of the sample). Image modified from a photograph taken by G. Natzler. UCLA specimen 1179. (b) Specimen of the Plainview (1917) find, an H5 regolith breccia found in Hale County, Texas. Although much of the fusion crust is rusty brown in color, some faded black patches remain. Thermoluminescence data indicate the meteorite has a terrestrial age of 225–300 years (Sears and Mills, 1974). Image courtesy of John Kashuba.

by the reduction and breakdown of chromite ($FeCr_2O_4$); hercynite ($(Fe,Mg)Al_2O_4$), formed possibly by the melting and oxidation of mafic silicates and plagioclase ($(Na,Ca)(Si,Al)_3O_8$); pseudobrookite (Fe_2TiO_5), formed by the breakdown of ilmenite ($FeTiO_3$), magnetite (Fe_3O_4), wüstite (FeO), and nanoparticles of goethite (α-FeO(OH)) (all formed by oxidation of metallic Fe); magnesioferrite ($Mg(Fe^{3+})_2O_4$), formed by the combustion of metallic Fe-Ni and silicate; Ni-rich sulfide, intergrowths of metallic Fe-Ni and troilite (FeS), formed by the melting and quenching of these phases; and olivine ($(Mg,Fe)_2SiO_4$), formed as overgrowths on relict grains and as new grains crystallized from the melt (e.g., Ramdohr, 1967a; Genge and Grady, 1999; Rubin and Ma, 2021; Petrova et al., 2024). The fusion crust of Tamdakht is depleted in metallic Fe-Ni-Co, orthopyroxene, clinopyroxene, and troilite and enriched in olivine relative to the interior (Goryunov et al., 2025). Some olivine grains in OC fusion crusts show reverse or oscillatory compositional zoning, indicative of complex redox effects. Most meteorite fusion crusts do not contain all these phases.

Oxidation of metallic Fe-Ni during atmospheric passage can cause a decrease in the saturation magnetic moment of ordinary-chondrite fusion crusts (Maksimova et al., 2020). (This is a property of ferromagnetic and ferrimagnetic materials wherein an applied external magnetic field cannot cause further increases in the magnetization of the material.)

Heating caused by atmospheric passage can produce back transformations of high-pressure phases to low-pressure ones. The fusion crust of the H6 chondrite Y-75267 transects a shock vein containing ringwoodite (Mg_2SiO_4), majorite-pyrope$_{ss}$ ($Mg_3(MgSi)Si_3O_{12} - Mg_3Al_2(SiO_4)_3$) and $NaAlSi_3O_8$ hollandite. However, portions of the vein near the fusion crust (where temperatures of ~1400°C were reached) contain no high-pressure phases – only olivine, low-Ca pyroxene ($(Mg,Fe)SiO_3$), and plagioclase glass (e.g., Fukimoto et al., 2020).

The bulk chemical composition of a fusion crust cannot be assumed to represent that of the meteorite as a whole. Heterogeneities in bulk composition and grain size in the near-surface regions of a meteoroid can cause variations in chemical composition of the resulting fusion crust (Thaisen and Taylor, 2009). Most OC fusion crusts are depleted in siderophiles relative to the bulk meteorite host (Genge and Grady, 1999). During atmospheric descent, meteoroids are continuously ablated. The fusion crust thus migrates inward into unfragmented specimens, melting regions that had been chemically modified a few seconds prior by frictional heating.

After a meteorite lands, its surface may still be warm (McSween, 1987); a prime example is the 1933 fall of LL6 Cherokee Springs (Cressy, 2016). An eyewitness to the 1917 fall of the Troup L6 chondrite recounted that the sample was "smoking" after it hit the ground (Udden, 1921). If this account is accurate, it is highly unusual (cf., Cressy, 2016).

8.7 Regmaglypts

Regmaglypts (also known as thumbprints) are dimples, pits, imprints, or (most commonly) shallow quasi-hemispherical depressions that form a scalloped surface within the fusion crusts of individual meteorites (Figs. 8.4, 8.5). Swindel and Jones (1954) refer to them as "piezoglyphs." Although regmaglypts are present in all types of meteorites (e.g., Lin and Qun, 1986), they are more common in irons than stones and occur much more frequently in large meteorites than small ones (Nininger, 1977; Lin and Qun, 1986). These concave structures are elliptical to almond-shaped in outline and commonly 1–3 cm wide (although some are much larger). (On large iron meteorites, regmaglypts range up to 20 cm in diameter.)

Regmaglypts form during the passage of a meteoroid through the Earth's atmosphere as localized vortices of hot turbulent gas scour the meteoroid surface. Their production is a function of numerous factors including the meteoroid's composition and friability as well as its orientation, velocity, and inclination during atmospheric passage. For meteorites that land on extraterrestrial bodies, the density of the atmosphere also plays a role in regmaglypt formation. (Regmaglypts have been reported on the surface of iron meteorites identified by rovers on Mars [Johnson et al., 2014].)

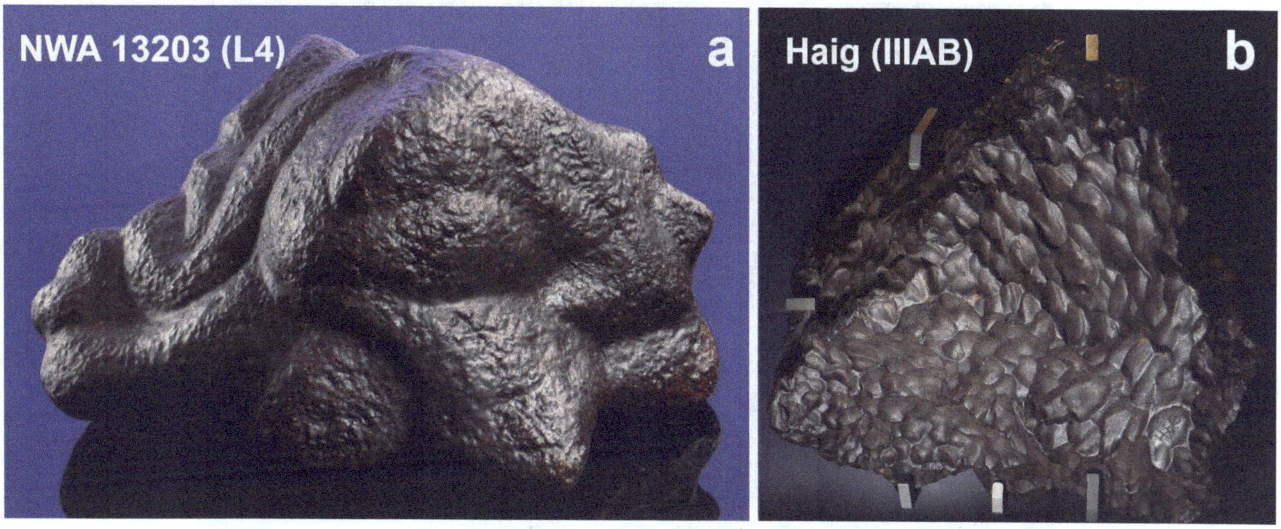

Fig. 8.5 Well-developed regmaglypts. (a) NWA 13203 is a 6.976-kg L4 chondrite with draping kielbasa-shaped regmaglypts. Photograph by Richard Goodbody and Darryl Pitt, courtesy of Christie's. (b) Haig is a 503-kg IIIAB iron found in Western Australia. The entire surface is covered with scallop-shaped depressions. Image courtesy of the Western Australian Museum.

8.8 Oriented Specimens

Most meteoroids tumble as they transit the atmosphere. When stresses exceed their crushing strength, they tend to break apart into angular fragments. A minority of meteoroids maintain a fixed orientation during descent; atmospheric ablation sculpts each of these samples into a conical shape with a blunt-nosed anterior and a moderately flat posterior (e.g., Fig. 8.6). The shapes have been described as shields or nose cones. Many specimens of meteorites (both large and small, and including irons, stony irons, chondrites, and achondrites) have oriented shapes (e.g., Nininger, 1981; Blood, 2013).

At least some of the sacred stones enshrined in temples in ancient times may have been oriented meteorites. Antoniadi (1939) mentioned several such stones, including one described by the Roman historian Tacitus as "a rounded block, larger at the base, and narrowing up to the summit, like a cone" stored in the temple of Aphrodite in Cyprus. The Roman historian Herodian wrote about "a very large stone, round at the base, and ending in a point" venerated at a shrine in Emesa, Syria in the third century CE (quoted from Antoniadi, 1939). This "Stone of Emesa" was associated with the Sun-God Sol Elagabalus (Bellemare, 1996) and brought to Rome by the young emperor Marcus Aurelius Antoninus (also known posthumously as Elagabalus), who demanded its public worship (Burke, 1986; Bellemare, 1996). An individual sacred stone was known as a betyl (also spelled baetyl; ultimately from the Semitic for "house of god") or an omphalos (from the Greek for "navel" or more generally, "center," as in the notion that Delphi was the omphalos, or center, of the world). Numerous ancient coins depicted these stones (e.g., Burke, 1986; Blood, 2019).

Amin et al. (2019) demonstrated that oriented meteorites achieve their characteristic shapes due to ablation, which sculpts meteoroids into the specific shapes that can maintain aerodynamic stabilization during descent. They conducted laboratory experiments on conical objects moving through water. They found that slender cones tumble, broad objects "rock and flutter," but intermediate forms (including those with stereotypical oriented-meteorite shapes) can achieve stable flight with the anterior of the specimen leading (Fig. 8.7).

A significant fraction of oriented meteorites exhibits flow lines. These typically form a radiating pattern of thin linear ridges extending from the apex of the meteoroid toward the posterior (e.g., Fig. 8.8). The flow lines form from melted surface material; the force of the surrounding air pushes the melt away from the apex. The phenomenon is somewhat akin to wind pushing small raindrops up

Fig. 8.6 Oriented individual of the Chelyabinsk LL5 chondrite. The specimen is conical with a blunt-nosed anterior and a broad, moderately flat posterior. The fluted fusion-crusted sample maintained a constant orientation with the anterior pointing downwards during atmospheric descent. Imaged modified from one provided courtesy of Darryl Pitt.

Fig. 8.7 Orientation and its development in meteorites. (a) Oriented individual of the Gao-Guenie H5 chondrite showing some flow lines. Image courtesy of John Kashuba. (b) Schematic diagram representing an experiment showing the fixed orientation of an aluminum body (1.2 cm in maximum width) falling through water. The shape of the body was machined to mimic that of an oriented meteorite. Time increases from top to bottom. Image based on Amin et al. (2019).

a windshield on a moving car when the wipers are turned off (e.g., Hooshanginejad and Lee, 2022).

In some cases, the melted material on a meteoroid, pushed away from the apex by rushing air, forms a lip – a roll-over rim near or atop the posterior of the specimen (e.g., Fig. 8.9). This feature is also well illustrated by specimens of Nuevo Mercurio (H5) and Juancheng (H5) in Blood (2013). During oriented flight, a partial vacuum develops on the posterior of the meteoroid. This facilitates

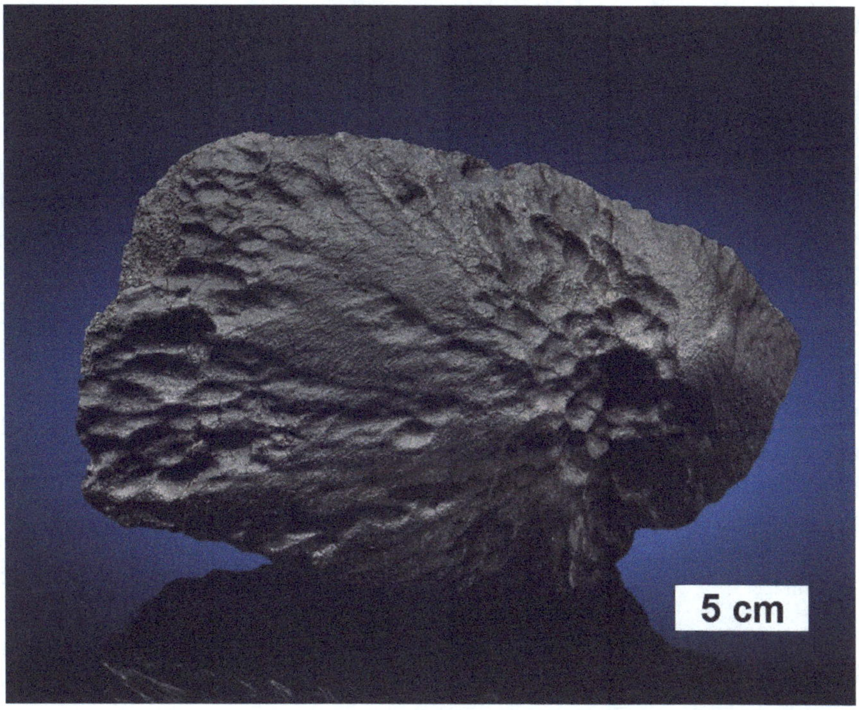

Fig. 8.8 Oriented shield-shaped individual of the NWA 13737 H5 chondrite showing sets of prominent flow lines. Image courtesy of Darryl Pitt, the Macovich collection, and Christie's.

Fig. 8.9 Specimen of the Gao-Guenie H5 chondrite with lipping – a roll-over layer of melt from the anterior onto the flat posterior of the specimen. Image courtesy of John Kashuba.

degassing and can cause the preservation of vesicles in the final fusion crust.

Many have noted that the shapes of oriented meteorites resemble those of space reentry vehicles, which are designed as blunted cones (e.g., Fig. 8.10). The principal difference is that such space vehicles plunge through the atmosphere with the apex pointing up, going the opposite direction of falling oriented meteorites. Reentry vehicles generate detached shock waves that spread the frictional heat over a large surface area. About 90% of frictional heat is dissipated this way, minimizing the heating rate.

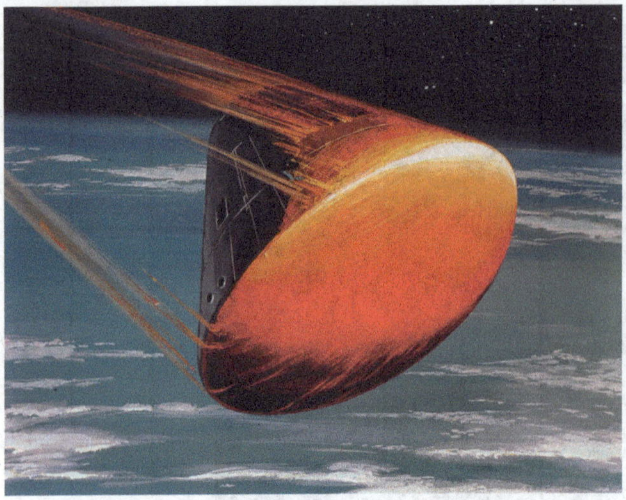

Fig. 8.10 Artist's rendition of an Apollo Command Module during reentry. It is oriented with the blunt end forward. NASA image.

8.9 Strewn Fields

When a descending meteoroid has decelerated sufficiently and loses its cosmic velocity, it can no longer heat the surrounding air to incandescence. It then begins dark flight, reaching a constant terminal velocity of about 300 m s^{-1} (about 25% faster than a commercial airplane at cruising altitude). At this point, the force of the wind and the freefall of the meteoroid are of comparable velocity. This results in changes of direction for many individual fragments (particularly the smaller ones) of a meteorite shower.

Individual fragments that reach the ground produce an elliptical pattern known as a strewn field. These areas typically range from a few kilometers to tens of kilometers in length; most strewn fields of stony meteorites are 2 to 4 kilometers wide. The strewn field of the Gold Basin L4 chondrite has an area of 225 km^2 (Kring et al., 2001); that of the Kumtag H5 chondrite is much smaller (62 km^2) (Du et al., 2011). Ideally, a comprehensive account of the mass-distribution of meteorite fragments after an observed fall can be gleaned from recovered masses combined with recorded trajectory data (Gritsevich et al., 2014). Monte Carlo models can aid in the discovery of the strewn fields of fresh falls (Moilanen et al., 2021). The practical difficulties of searching a strewn field are discussed by Notkin (2011).

After the Chelyabinsk LL5 ordinary-chondrite meteoroid exploded at an altitude of 27 km, approximately 500,000 individual meteorite fragments reached the ground; they formed a strewn field about 1 km wide and 60 km long. The combined mass of recovered meteorites was only about 1 metric ton, less than 0.008% of the initial meteoroid mass. The largest remaining fragment (the main mass) plunged into Lake Chebarkul, punching a 6-meter-wide hole in the ice. When recovered, the mass weighed 654 kg. A plot of the distribution of the recovered masses (Fig. 8.11) shows a strong tendency for larger masses to occur farther downrange than smaller masses along the meteoroid's trajectory (Rubin, 2023). The main mass sits at the far end of the strewn field; a dense concentration of small samples (<10 grams each) occurs several kilometers to the southeast (in the direction from which the meteoroid came). Tens of thousands of sparsely scattered small samples (many less than 1 gram) occur farther southeast along the trajectory.

The correlation between meteorite mass and location in strewn fields is a common one (e.g., Gnos et al., 2025). The largest masses tend to occur near the terminus of the field (the direction toward which the meteoroid was traveling); smaller individuals tend to be recovered at the opposite end of the strewn field.

Small samples, with their high surface-area/volume ratios, are preferentially slowed by the wind. Larger samples retain more of their pre-atmosphere momentum and travel the farthest; their small surface-area/volume ratios leave them less affected by drag. Nevertheless, the specific breakup dynamics of tumbling meteoroids during atmospheric passage complicate the distribution of individual fragments on the ground.

Relatively small strewn fields could result from a meteoroid entering the Earth's atmosphere at a steep angle (e.g., >60º) and/or breaking up at a low altitude (e.g., <20 km) (Welten et al., 2024). The determination of the true size of an individual strewn field could be complicated by erosion, varied terrain, or the overlap of several strewn fields resulting from the breakup of two or more meteoroids of the same group and petrologic type over a multi-century time period (e.g., Rubin et al., 2000).

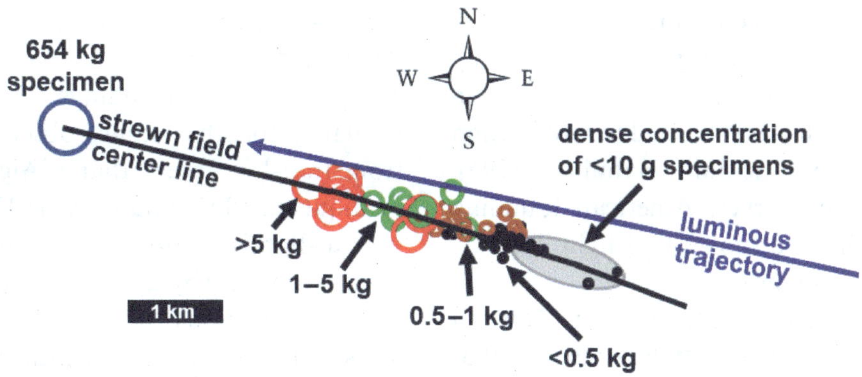

Fig. 8.11 Highly simplified map of the strewn field of the Chelyabinsk LL5 ordinary chondrite that exploded over Siberia on February 15, 2013. This map is based on a trajectory projection map by Svend Buhl (Buhl and Wimmer, 2013). Although there is significant overlap, individual meteorite specimens tend to be distributed according to their mass. The main mass (654 kg) is denoted by a blue circle; specimens >5 kg, red; specimens 1–5 kg, green; specimens 0.5–1 kg, brown; specimens <0.5 kg, black. The gray ellipse outlines a dense concentration of small (<10 g) specimens. The final portion of the luminous trajectory is shown; beyond the arrow, the meteoroid proceeded in dark flight until the main mass punched through the ice on Lake Chebarkul. Diagram from Rubin (2023).

8.10 Some Historic Falls and Their Role in the Recognition That Rocks Fall from the Sky

> And it came to pass, as they [the Amorites] fled from before Israel, and were in the going down to Bethhoron, that the Lord cast down great stones from heaven upon them unto Azekah, and they died…
>
> Joshua 10:11; KJV

Eighteenth- and nineteenth-century naturalists were aware of reports of rocks falling from the sky, but these were mixed in with fantastic tales of a wide variety of falling objects – flesh, blood, milk, wool, bricks, paper, money, and gelatin (Burke, 1986). There have also been stories of localized rains of ants, fish, frogs, mice, and rats (e.g., Anonymous, 1874). It could be challenging to determine which reports comported with reality.

One of the most important developments leading to the scientific community's broad acceptance of the idea that rocks fell to Earth from space was the spate of well-documented meteorite falls in the late eighteenth and early nineteenth centuries. All (or almost all) of the falls were ordinary chondrites. They include Barbotan, France, in 1790 (H5 chondrite); Siena, Italy, in 1794 (LL5 chondrite); Wold Cottage, England, in 1795 (L6 chondrite); Portugal, from the town of Évora Monte, in 1796 (unclassified stone that was lost); Salles, France, in 1798 (L5 chondrite); Benares, India, in 1798 (LL4 chondrite now officially designated Benares (a)); L'Aigle, France, in 1803 (L6 chondrite); and Weston, Connecticut, USA, in 1807 (H-chondrite regolith breccia).

The following accounts of historic falls are taken from Rubin (2022b):

Siena (LL5): Eyewitness accounts of the 1794 Siena fall were compiled by Abbé Ambrogio Soldani, professor of mathematics at the University of Siena, and discussed in a 288-page book, *On a Shower of Stones that Fell on the 16th of June at Siena* (Soldani, 1794). After a series of explosions, a shower of small stones fell from the sky; the largest weighed 3.5 kg. Soldani proposed that the stones were generated within the dust-laden clouds from which they fell.

William Thomson, an English mineralogist working in Naples, analyzed a sample of Siena and made the first magnetic separation of metallic iron from silicate in a meteorite.

Because Mount Vesuvius had erupted just 18 hours before the Siena fall, several scientists proposed that the Siena stones could have come from the volcano, having been propelled ~380 km. Other scientists questioned how any stones could have remained aloft for 18 hours. William Hamilton (English Ambassador to the Court of Naples) failed to find similar-looking stones on the flanks of the volcano, although it seemed possible they were buried beneath "a thick bed of fine ashes." He suggested that the Siena stones might have formed by the accumulation and hardening of volcanic ash in the atmosphere, which was subjected to vitrification by the action of an electric fluid (Hamilton, 1795). (More than 40 years earlier, American founding father Ben Franklin had shown that lightning was an electrical phenomenon.)

Lazzaro Spallanzani, Professor of Natural History at Pavia in Lombardy, Italy, suggested that the Siena stones had been set aloft by a local hurricane and scorched by electrical forces in the clouds.

As pointed out by Marvin (1996), Siena was a modern European city with nearly 30,000 residents. Many people had witnessed the fall and the rocks had been studied by scientists throughout Italy. Even though the origin of the stones was disputed, it was becoming more difficult to deny the phenomenon of rocks falling from the sky.

Wold Cottage (L6): After a series of explosions, a single 25.4-kg stone fell in Wold Cottage, England, on the afternoon of December 13, 1795. It was observed by several witnesses, one of whom, a young ploughman named John Shipley, saw the stone bury itself about 50 cm into the soil only eight meters from where he stood. As of this writing, Wold Cottage remains the largest stone known to have been observed to fall in the British Isles. The landowner, Major Edward Topham, exhibited the stone in London and in 1799 erected a brick monument on the spot where it fell. Topham was well known as a soldier, playwright, newspaper publisher, dog breeder, and fashion trendsetter. He was a captain in the army (automatically assuming the title of Major upon retirement); King George III had singled him out as his "tip-Top Adjutant" (Pillinger and Pillinger, 1996). Topham's notoriety and aura of integrity led many to accept the eyewitness reports of his workmen as honest accounts.

L'Aigle (L6): The fall of 2000 to 3000 stones in L'Aigle, France, in 1803 was witnessed by dozens of people from all walks of life: "children, aged men and women, peasants, laborers, war veterans, estate managers, and ecclesiastics" (Burke, 1986). The largest stone weighed 9 kg; the total mass recovered (from an area of $\sim 4 \times 10$ km) was about 37 kg. Less than two months after the fall, the French chemists Nicolas-Louis Vauquelin and Antoine-François de Fourcroy had already analyzed several stones and presented their findings. They noted L'Aigle's similarity to other stones observed to have fallen from the sky, except that L'Aigle was brecciated (Vauquelin, 1802; Fourcroy, 1803).

Jean-Baptiste Biot interviewed numerous witnesses from nearby farms and hamlets, stating that "All these persons, so diverse in professions, customs and opinions, having little or no relations one with another, agree all of a sudden to testify a same fact that they have no interest to suppose." The details in the report of Biot (1803) lent an aura of verisimilitude to eyewitness accounts of meteorite falls. Biot noted the physical effects of the meteorite shower – broken tree branches and holes punched into the soil by small stones. He compared numerous specimens and found them to resemble each other, but to differ from rocks in the surrounding countryside and from the products of local foundries or slag from nearby mines. Biot mapped the meteorite strewn field and made an estimate of the meteoroid's atmospheric trajectory. His observations did not support the lunar origin hypothesis he had endorsed just one year earlier.

Sears (1975) characterized Biot's report as "dramatic and exciting," contrasting with the "dry details of analytical techniques" offered by Vauquelin, Fourcroy, Edward Charles Howard, and other workers. Gounelle (2006) concluded that the engaging style of Biot's account led directly to large portions of the report being published in scientific journals as well as daily and monthly newspapers. Many authors credit Biot's account of the L'Aigle fall with firmly establishing the fact that meteorites are extraterrestrial gatecrashers.

But acceptance of falling rocks was not universal. In France, Guillaume-Antoine de Luc (1803) dismissed the Wold Cottage reports as the fantasies of mere "labourers" and remained skeptical that rocks could fall from the sky (Pillinger and Pillinger, 1996).

Skepticism also reigned in America. The 1807 fall of Weston (H-chondrite regolith breccia) in Connecticut was the first report of a meteorite falling in America. The chemists Benjamin Silliman of Yale and James Woodhouse of the University of Pennsylvania analyzed the stones (e.g., Silliman and Kingsley, 1809). Reports of the fall reached President Thomas Jefferson, but there is no evidence (e.g., Burke, 1986; Norton, 1998; Kichinka, 2005; Prince, 2011) he ever uttered the sentence often attributed to him (e.g., Pejovic, 1982) that "It is easier to believe that two Yankee professors would lie than that stones would fall from heaven." In a well-documented February 15, 1808 response to a letter from a finder offering to send a sample of Weston to Congress for official examination, Jefferson wrote: "We certainly are not to deny whatever we cannot account for… It may be very difficult to explain how the stone you possess came into the position in which it was found. But is it easier to explain how it got into the clouds from whence it is supposed to have fallen? The actual fact however is the thing to be established" (Burke, 1986).

8.11 Modern Falls

Alien they seemed to be; no mortal eye could see the
intimate welding of their later history,
Or sign that they were bent by paths coincident on
being anon twin halves of one august event…
Thomas Hardy, "The Convergence of the Twain"

8.11.1 Hammerstones

There have been some notable ordinary-chondrite falls over the last century (e.g., Roy and Wyant, 1951). Some have hit structures and been dubbed "hammerstones" by collectors. (There is some inconsistency about whether the "hammerstone" label applies only to those individual stones from a meteorite shower that strike a structure or to all specimens from that shower. Of course, meteorites other than ordinary chondrites also hit structures: e.g., a house and a doghouse were pelted by pieces of Aguas Zarcas [CM2] in April 2019.)

Ordinary-chondrite hammerstone targets have included:

houses (e.g., Baxter, L6 – 1916; Bloomington, LL6 – 1938; Kiel, L6 – 1962; Barwell, L5 – 1965; Andreevka, L3 – 1969; Suchy Dul, L6 – 1969; Wethersfield (1971), L6 – 1971; Wethersfield (1982), L6 – 1982; Juancheng, H5 – 1997; Dunbogan, L6 – 1999; Dongyang, H5 – 2002; San Michele, L6 – 2002; Thuathe, H4/5 – 2002; New Orleans, H5 – 2003; Park Forest, L5 – 2003; Auckland, L5 – 2004; San Juan de Ocotán, L5 – 2007; Cali, H/L4 – 2007; Mahadevpur, H4/5 – 2007; Xinglongquan, L3 – 2008; Cartersville, L5 – 2009; Draveil, H5 – 2011; Soltmany, L6 – 2011; Chelyabinsk, LL5 – 2013; Parauapebas, H4-5 – 2013; Wolcott, L5 – 2013; Famenin, H/L3 – 2015; Komaki, L6 – 2018; Mangui, L6 – 2018; Viñales, L6 – 2019; Santa Filomena, H5-6 – 2020; Narashino, H5 – 2020; Gatuto, L6 – 2020; Zhob, H3-4 – 2020; Golden, L/LL5 – 2021; Punggur, H7 melt-breccia – 2021; Longde, L5 – 2022; Ponggo, H3-5 – 2022; Matera, H5 – 2023),
a cabin (Oslo, H3-6 – 2012),
a hut (Sinnai, H6 – 1956),
sheds (Centerville, H5 – 1956; Ash Creek, L6 – 2009; Mifflin, L5 – 2010; Broek in Waterland, L6 – 2017),
greenhouses (Thika, L6 – 2011; Jinju, H5 – 2014),
commercial buildings (Barwell, L5 – 1965; Denver, L6 – 1967; Hiroshima, H5 – 2003; Lorton, L6 – 2010),
a fire station (Park Forest, L5 – 2003),
a mailbox (Claxton, L6 – 1984),
garages (e.g., Benld, H6 – 1938; Canon City, H6 – 1973; Worden, L5 – 1997; Novato, L6 – 2012),
cars (e.g., Benld, H6 – 1938; St. Louis, H4 – 1950; Barwell, L5 – 1965; Peekskill, H6 – 1992; Neagari, L6 – 1995; Turtle Lake, L5 – 1996, Park Forest, L5 – 2003),
a truck (Grimsby, H5 – 2009),
a boat (Chiang Khan, H6 – 1981),
a tiled courtyard (Ejby, H5/6 – 2016),
a man-made pond (Buzzard Coulee, H4 – 2008),
a cooking pot (Juancheng, H5 – 1997), and
a houseplant (Barwell, L5 – 1965).

This is an incomplete list; additional ordinary-chondrite hammerstones are awaiting approval.

Three OCs have struck living creatures: one cow (Valera, L5 – 1972) and two people (Sylacauga, H4 – 1954; Mbale, L5/6 – 1992) (Noda, 2024; Swindel and Jones, 1954; Jenniskens et al., 1994). Photographs of the stricken people appear in Smith et al. (2018). (The widely circulated report that the New Concord [L6] fall of May 1, 1860 killed a colt [e.g., Prior and Hey, 1953; Hey, 1966; Graham et al., 1985; Grady, 2000] is not supported by contemporary accounts [Greg, 1860; Horejsi, 2011].) La Paz (1951) attempted to calculate the probability of humans being injured by meteorite falls, concluding that between 2100 and 3300 CE, the odds are 99 to 1 that at least one meteorite fall will hit the world's human target area (HTA). Recent estimates indicate that ~100–1000 annual deaths are expected to be caused by asteroid impact, when averaged over very long time scales (e.g., Chapman, 2004a, 2007). One oft-mentioned estimate for the chances of a person being killed by an asteroid impact over a lifetime is about 1 in 1 million (0.0001%), when averaged over 100 Ma.

There are many examples of people injured or killed by volcanic bombs (e.g., Baxter and Gresham, 1997), but the most famous story of a person being struck by a terrestrial rock hurtling through the air is from 1 Samuel 17:49 (likely penned in the sixth century BCE): "And David put his hand in his bag, and took thence a stone, and slang it, and smote the Philistine in his forehead, that the stone sunk into his forehead; and he fell upon his face to the earth" (KJV). Goliath's fate was not unique in the Hebrew Scriptures; e.g., if it were proved that a woman in Israel was not a virgin when she married her husband, the judicial punishment was for her to be stoned to death at the door of her father's house (Deut 22:20–21). (Sacrificing one's children to Molech was also proscribed [Lev 20:2].)

8.11.2 Descriptions of a Few Modern Falls

A fictionalized account of a large LL3 fall in a small American town appears in the short story "Falling Stars" (Tobin, 2023). Many aspects of this fall are portrayed realistically; the details comport with those of Simms (2021) who discussed the fall of L5 Barwell in Leicestershire, England on December 24, 1965. It was the largest observed fall in U.K. history; a total of 44 kg was recovered. Cressy (2016) described falls that occurred in the United States between 1807 and 2016. He also listed 21 USA falls that landed within 10 feet (3 meters) of a person between 1859 and 2010; 16 of these are ordinary chondrites. Three particularly noteworthy OC falls are described below.

Peekskill (H6): At 7:50 PM local time on Friday, October 9, 1992, a bright fireball with a greenish hue was seen traveling in an ENE direction over West Virginia, crossing over Pennsylvania and into New York state (e.g., Beech et al., 1995; Ceplecha et al., 1996). It was captured by at least 16 videographers, many of whom were avid sports fans (and proud parents) recording local high-school (American) football games (cf. Bissinger, 1990). The luminous flight lasted more than 40 seconds, corresponding to a ground path of 700–800 km. At a height of ~41.5 km, the main meteoroid (with an initial size of 50–100 cm and an initial mass ~2–15 metric tons) broke into more than 70 fragments. The fireball reached a peak brightness of apparent magnitude −13, similar to that of the full Moon (−12.7).[2]

The Peekskill meteoroid produced sonic booms and electrophonic sounds. One witness of the fireball described a "crackling sound like that of a sparkler" lasting ~10 seconds after the first large fragmentation event (Beech et al., 1995).

At a height of 30 km, the meteoroid had slowed to less than 3 km s^{-1} and was no longer able to produce light via frictional heating of the surrounding air. It continued in this dark-flight regime across a ground distance of 50 km before a 12.37-kg fragment struck the trunk of a parked car (a red 1980 Chevy Malibu) in suburban Peekskill, New York at a vertical velocity of ~80 m s^1. Michelle Knapp, the owner of the car, described the sound of the impact as akin to a "three-car crash." She called the police who, after a brief investigation, concluded that criminal mischief was afoot. Smith et al. (2018) showed a picture of the meteorite and the dented car; Howgego (2025) listed additional details.

Peekskill is an H6-chondrite monomict breccia consisting of centimeter-sized light-colored angular H6 clasts surrounded by thin lenses of dark-colored

[2] Apparent magnitude is a function of the intrinsic luminosity of the light source, distance from Earth, and the diminution of light (i.e., extinction) caused by absorption of particles in space situated between the source and the Earth. The magnitude scale is reverse logarithmic – brighter objects have lower magnitudes. A difference in brightness by a factor of 100 corresponds to 5 units of magnitude. Thus, an object with an apparent magnitude number of 1.0 (similar to that of Spica, the brightest star in Virgo) is $\sqrt[5]{100}$ (or about 2.512 times) brighter than an object of magnitude 2.0 (similar to that of Polaris – the North Star, the brightest star in Ursa Minor).

H6 material. Dark-colored shock veins traverse many of the light-colored clasts. Some portions of the rock are highly fragmental, consisting of angular light-colored 140–720-μm H6 clasts embedded in dark H6 material.

Park Forest (L5): At about 11:50 PM local time on March 26, 2003, a ~1.7-m-wide meteoroid with a mass of $8.6 \pm 1.9 \times 10^3$ kg entered the atmosphere at an altitude of 82 km over the U.S. Midwest (Brown et al., 2004). The fireball reached a maximum brightness of –21.7 visual magnitude as the meteoroid traveled at 19.5 ± 0.3 km s^{-1} from the SSW to the NNE. One of the witnesses to the fireball was my late uncle, Robert Fine, a Chicago resident. Sonic booms were heard, loud enough to awaken some sleepers in the suburbs south of Chicago. The sonic booms produced infrasonic waves in the inaudible acoustic wave spectrum below ~20 Hz; these waves were detected in Maryland and Manitoba, Canada. Seismic waves were generated at points along the fireball trajectory. Dark flight commenced below an altitude of 18 km.

About 45 kg of meteorite fragments rained down on Park Forest, Illinois, a suburb ~50 km south of downtown Chicago. Some individual fragments broke windows, some broke through roofs or hit cars. The local fire station was also hit. The meteorite fragments (ranging in size from a few grams to 5.26 kg) formed a strewn field 9.5 km long. This was the most densely populated region on Earth to be the site of a meteorite shower in modern times (Simon et al., 2004). Although meteorite-researcher Steve Simon lived in Park Forest at the time, no meteorite fragments landed on his property; some were recovered a few hundred meters away to the NNE, SW, and E (Fig. 1 of Simon et al., 2004).

Park Forest is an L5 chondrite composed of light-colored clasts situated within a dark matrix. The meteorite is shocked (shock stage S5) – small grains of maskelynite are present as well as thin glassy shock veins containing small (~2 μm) blebs of metallic Fe-Ni and troilite. Troilite-rich veins pervade the dark matrix; this is a typical example of silicate darkening (also known as shock darkening) (Rubin, 1992), commonly observed in so-called "black chondrites" (Britt and Pieters, 1994).

Chelyabinsk (LL5): The largest airburst on Earth in more than a century occurred at 9:22 AM local time on February 15, 2013 over the southern Ural region of Russia. The meteoroid had an initial diameter of 19.8 ± 4.6 m and was first recorded at an altitude of 97 km traveling at 19.16 ± 0.15 km s^{-1}. It entered the atmosphere at a shallow angle: $18.3 \pm 0.2°$ relative to the horizon. It began to break apart at ~83 km; the disintegration rate increased at 54 km, and the final airburst occurred at 27 km. The estimated amount of total kinetic energy was equivalent to 590 ± 50 kT of TNT (Popova et al., 2013), about 39 times greater than that released by the atomic bomb dropped on Hiroshima. The peak brightness of the Chelyabinsk bolide was -27.3 ± 0.5 apparent magnitude, about 40% brighter than the Sun (apparent magnitude =–26.74). Electrophonic sounds were also reported.

The shock wave reached the ground 2–3 minutes after the primary explosion, causing extensive damage to buildings and numerous injuries. In the city of Chelyabinsk, 3613 apartments had shattered windows, and the roof of a zinc factory collapsed. In Yemanzhelinsk, a town ~50 km south of Chelyabinsk, windows broke, window frames bent inwards, and suspended ceilings fell. In total, more than 7000 buildings were damaged.

Below the fireball's trajectory, the shock wave knocked people to the ground. The intense radiation caused dozens of people to report sunburns; some had retinal burns and flash blindness. Most injuries were due to shattered window glass; about 1500 people sought medical attention.

Chelyabinsk is an LL5 breccia (shock stage S4) composed principally of mildly-to-moderately shocked light-colored clasts and moderately shocked and subsequently annealed dark clasts. Abundant black microcrystalline shock veins traverse both lithologies. The veins contain relict clasts of the meteorite groundmass as well as blebs of metallic Fe-Ni and troilite (Badyukov et al., 2015). At least eight pre-terrestrial impact events affected the portion of the parent asteroid (or asteroid fragment) from which Chelyabinsk was derived (Righter et al., 2015): these events occurred ~4.53 Ga, ~4.45 Ga, ~3.73 Ga, ~2.81 Ga, ~1.46 Ga, ~852 Ma, ~312 Ma, and ~27 Ma ago. The latest shock event likely caused Chelyabinsk to develop a strong petrofabric consisting of preferentially aligned metal grains (Popova et al., 2013).

Chelyabinsk has a cosmic-ray exposure (CRE) age of ~1.23 Ma (Righter et al., 2015), appreciably shorter than almost all LL chondrites (which form a peak at ~17 Ma) (Eugster, 2003).

8.12 Fossil Ordinary Chondrites

As of this writing, there are more than 1260 well-documented observed meteorite falls. The oldest preserved falls are Nogata (an L6 chondrite that fell in Kyushu, Japan in 861 CE), Narni (an unclassified stone that fell in Umbria, Italy in 961 CE), and three meteorites from the fifteenth century: Rivolta de Bassi (an unclassified stone that fell in Lombardia, Italy in 1491), Ensisheim (an LL6 chondrite that fell in Alsace, France in 1492) and Valdinoce (an unclassified stone that fell in Emilia-Romagna, Italy in 1496). Meteorite finds date back much farther: for example, two Antarctic ordinary chondrites (H5 ALH 88019 and L4 LEW 86360) have terrestrial ages $\gtrsim$2 Ma (Scherer et al., 1997; Welten et al., 1997).

A question addressed by some researchers in the mid twentieth century was whether ancient meteorites existed. Paneth (1956) and Heide (1964), for example, noted that meteorites had never been identified among the tens of billions of tons of coal that had been mined since the eighteenth century. Because most of these coal deposits date from the Carboniferous Period (359.2–299 Ma ago), it seemed "quite possible that during the early history of the Earth there were no meteorites in existence" (Paneth, 1956). In other words, these researchers were suggesting that falling meteorites might be a strictly contemporary phenomenon.

There are several objections to this inference: (1) It would have been exceedingly difficult for miners (or anyone else) to recognize highly altered meteorite fragments encased in coal. Nevertheless, one relict iron meteorite (probably IIIAB) named Gove, with a terrestrial age of 2.5–23 Ma, was excavated from a bauxite mine in Northern Territory, Australia in 1979 (Bevan et al., 2019). (2) The craters on the Moon had previously been modeled as having formed by impacting meteoritic/asteroidal projectiles (e.g., Baldwin, 1942, 1949; Dietz, 1946). Because the Earth is a larger target (and has a much greater effective gravitational cross section), it must have received an even greater share of projectiles passing through the 1-AU neighborhood over the Solar System's history. However, astronomy textbooks from the mid twentieth century (e.g., Baker, 1959; Abell, 1964) treated the origin of lunar craters as an unresolved issue. (3) Fossil meteorites are now known from the Middle Ordovician (470–458.4 Ma ago) (e.g., Schmitz et al., 2001), as discussed in this section. (3) Meteorites have also been identified among returned lunar samples: the Bench Crater carbonaceous chondrite from Apollo 12 (Wood et al., 1971; McSween, 1976; Zolensky, 1997), the Hadley Rille EH chondrite from Apollo 15 (Haggerty, 1972; Rubin, 1997a), and a possible LL4 chondrite from Luna 16 of shock-stage S3 (Demidova et al., 2020, 2022). These meteorites all impacted the Moon many millions of years ago. (4) Meteorites have also been identified on Mars (11 irons and four stony irons, according to the Meteoritical Bulletin Database from September 18, 2025).

The most abundant meteorite groups observed to fall in modern times are the three ordinary-chondrite groups: L chondrites (37%), followed by H chondrites (32%), and LL chondrites (8.7%). (HED achondrites constitute only 5.7% of falls.) The fact that ordinary chondrites constitute 78% of observed falls implies that their immediate parent asteroids are presently in locations (e.g., near the 3:1 mean motion or ν_6 secular resonances) that favor efficient transport of dislodged rocks to the Inner Solar System (e.g., Farinella et al., 1993b).

The high abundance of ordinary chondrites among meteorite falls appears to date back hundreds of millions of years. So far, about 130 highly altered L chondrites (e.g., Fig. 8.12) have been found as fossilized meteorites in a quarry of Middle Ordovician limestone in Sweden (e.g., Schmitz et al., 2001; Schmitz, 2013). Also present within sedimentary rocks of this age are numerous isolated grains of chromite and Cr-rich spinel from Sweden, China, and Russia with L-chondrite chromite compositions (e.g., Heck et al., 2010, 2024; Rout et al., 2017). It is clear that these chromite grains

Fig. 8.12 The arrowhead-shaped Middle Ordovician fossil L-chondrite Österplana 018 adjacent to a fossil nautiloid (genus *Orthoceras*). All common mineral phases (except chromite and Cr-rich spinel) in the meteorite were replaced during terrestrial diagenesis. Concentric alteration zones form a 50 × 60-mm halo around the meteorite within the surrounding marine limestone from the Thorsberg quarry in southern Sweden. Image modified from Rout et al. (2017).

were liberated from L-chondrite micrometeorites because the chromites contain high concentrations of solar-wind-implanted Ne and He, ions that penetrate mineral grains to depths of only a few nanometers (Schmitz et al., 2019). During their time of residence in interplanetary space, these chromite grains must have been at the surface of the micrometeoroids. The progenitors of all these Ordovician L-chondrite meteorites and micrometeorites were ejected from the L-chondrite parent asteroid after it suffered collisional disruption 470 ± 6 Ma ago (e.g., Bogard, 1995; Bogard et al., 1976, 1995; McConville et al., 1988; Turner, 1988; Schmitz et al., 2003; Korochantseva et al., 2007; Nesvorný et al., 2009; Swindle et al., 2014). Dust derived from this break-up event is found in a marine limestone section in central China (T. Zhang et al., 2024). This dust is associated with a positive Ir anomaly (Schmitz et al., 1996, 1997) and could potentially be used as a ~460 ± 10-Ma-old stratigraphic marker across the Earth; it may also occur on Mars (and possibly the Moon and large asteroids).

Within the sequence of stacked Ordovician sedimentary rocks in Sweden and Russia, the younger strata (those higher up in the stack) contain extraterrestrial chromite grains (derived from micrometeorites) with older CRE ages than the comparable chromite grains in older strata (lower in the sedimentary stack) (Heck et al., 2008; Meier et al., 2014). Chromites from the lower strata spent little time in interplanetary space. After the break-up of the L-chondrite parent body, these grains had been directly injected into a resonance and transferred into the Inner Solar System in about 10^5 years (Heck et al., 2004, 2008). In contrast, the chromite grains higher in the stack took longer to reach the Earth; their sojourn in interplanetary space involved orbital decay induced by Poynting-Robertson drag.

There is an order-of-magnitude excess of terrestrial impact craters formed during the Ordovician (Fig. 8.13) within about 30 Ma after the disruption of the L-chondrite parent asteroid (Schmitz et al., 2022). This excess has been dubbed the Ordovician impact spike. Many of the extant structures (preferentially preserved on continental crust) are in North America and the Balto-Scandinavia region. These structures include the Lockne-Målingen doublet crater (7.5 km, 0.7 km) in Central Sweden (458 Ma old) (Ormö et al., 2014)

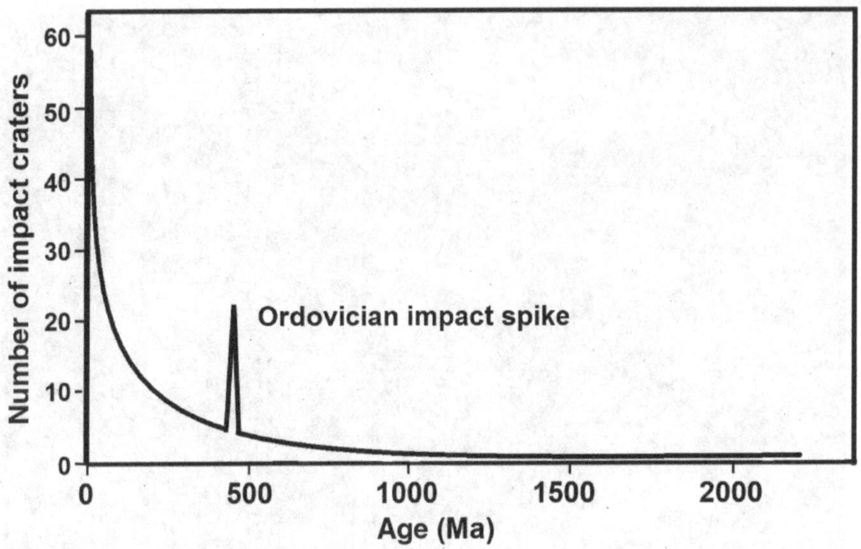

Fig. 8.13 Smoothed curve approximating the age distribution of 202 impact craters on Earth. The number of recognizable craters decreases with time due to erosion and sedimentation. However, there is a large spike during the Ordovician about 450–470 Ma ago. Diagram based on Schmitz et al. (2022).

and the ~26-km-diameter Clearwater East impact structure in northwestern Québec, Canada (~465 Ma old) (Schmieder et al., 2015). Five of the Balto-Scandinavian craters (Hummein, Sweden; Ilyinets, Ukraine; Kärdia, Estonia; Målingen, Sweden; Tvären, Seden) are appreciably smaller: 0.7–4.5 km in diameter (Schmitz et al., 2022). Another possibly associated crater is the Slate Islands structure in Ontario, Canada; it is 32 km in diameter and contains zircon grains that were reset 456.1 ± 6.9 Ma ago (Parisi et al., 2024). All the impact craters in the Ordovician impact spike lie within 30° of the equator, a region that contained only ~30% of the Earth's continental crust (Tomkins et al., 2024). This led to the speculation that an L-chondrite asteroid passed within Earth's Roche limit, was pulled apart by tidal forces, and formed an equatorial ring of L-chondrite material. The impact craters may have formed from asteroidal fragments that dropped to Earth from the ring.

During the Middle Ordovician subperiod, there was widespread (global) formation of coarse sedimentary megabreccias at continental margins. These breccias, which commonly contain meter-sized boulders embedded in a fine-grained matrix, were caused by marine debris flows initiated by impact-induced tsunamis and earthquakes (Parnell, 2009).

Just above the stratigraphic horizon in the mid Ordovician limestone where abundant L-chondrite material first appears, there is a marked increase in the relative abundance of brachiopods (e.g., Pedersen and Rasmussen, 2019). This feature has been dubbed the Great Ordovician Biodiversification Event (GOBE) (e.g., Rasmussen et al., 2016). Because brachiopods are shallow-water species that replace deeper-water marine organisms, it is likely that sea level was falling during this period. This marine regression was triggered by the onset of glaciation caused by global cooling (Pedersen and Rasmussen, 2019), itself plausibly attributable to increased L-chondrite dust in the atmosphere (Schmitz et al., 2008, 2019) or, conceivably, shading from a temporary ring (Tomkins et al., 2024). However, because the GOBE may not have been a single event (Servais et al., 2021), its indirect association with the breakup of the L-chondrite parent asteroid is uncertain.

Sediments that formed about 1 Ma before the L-chondrite asteroid disruption event also acquired H, L, and LL chromite grains derived from OC micrometeorites. The three OC groups are about evenly mixed; after the disruption event, more than 99% of the chromite grains are from L chondrites (Schmitz et al., 2019). Over the past 500 Ma (and likely much longer), the meteorite flux has been dominated by ordinary chondrites (Terfelt and Schmitz, 2021); this is ~11% of our planet's history.

Brunflo was the first fossil L chondrite to be discovered (Thorslund and Wickman, 1981; Thorslund et al., 1984). Except for relict grains of chromite ($FeCr_2O_4$) (a mineral highly resistant to alteration), all primary phases in Brunflo were replaced (Nyström and Wickman, 1991). As summarized by Rubin and Ma (2021), secondary phases include carbonate (major calcite), sulfate (barite), a disordered illite-type phyllosilicate, rare grains of elemental Ni, sulfarsenides (cobaltite, gersdorffite), arsenides (rammelsbergite, safflorite, nickeline, maucherite, orcelite), sulfides (chalcopyrite, sphalerite, galena, pyrite, and several poorly characterized Cu sulfides), TiO_2 (probably anatase or brookite), quartz, and carbonate-fluorapatite (francolite). Alteration phases in Österplana 001 (which was found in the same limestone quarry as Brunflo) include calcite, Cr-bearing layered silicate, barite, apatite, and TiO_2 (Nyström et al., 1988).

Wiltshire (also known as Lake House), Britain's largest meteorite (92.75 kg; 35 × 38 × 55 cm), is an H5 chondrite that was moderately shocked (S4). Its chondrules have indistinct boundaries and some contain coarse recrystallized mesostasis. The rock is dark brown and extremely weathered (W5); a network of weathering veins up to 2 mm in thickness occurs throughout the sample (C. Pillinger et al., 2011). According to news reports, Colin Pillinger (Open University) speculated the meteorite may have landed on or near a glacier in the vicinity of Salisbury Plain, near Stonehenge, about 30,000 years ago. Thousands of years after the Ice Age ended, the meteorite was picked up by Druids and used in the construction of a chalk burial mound. The reducing environment inhibited the oxidation of metallic Fe-Ni, allowing the meteorite to remain intact. It appears to have been excavated by archaeologists about 200 years ago and later used as an occasional doorstop at a large home (Lake House) near Wilsford-cum-Lake, Wiltshire.

A more recent fossil ordinary chondrite is Danebury, a ~30-g H5 chondrite pebble that fell 2350 ± 120 years ago. It was excavated from a grain pit dug into chalk by prehistoric peoples at the site of an Iron Age hillfort in Hampshire, England (Pillinger et al., 2014). The stone contains well-delineated porphyritic, barred olivine, radial pyroxene, and cryptocrystalline chondrules, a few of which include turbid mesostasis. Its olivine (Fa18.3 ± 0.9; $n = 17$) and low-Ca pyroxene (Fs16.5 ± 1.7; $n = 13$) compositions are in the range of H-group chondrites: Fa17.3–20.2; Fs15.7–18.1 (Rubin, 1990; Gomes and Keil, 1980). Although the interior of the stone is not appreciably weathered (W1/2), there is a 100–200-μm-thick weathering rind on the sample consisting of (1) lepidocrocite and goethite that experienced minor Ni substitution and (2) fragments of chalk derived from prehistoric infilling of the pit. The remarkable preservation of the meteorite is most likely due to its location amidst a compacted mixture of chalk, clay, and silt (Pillinger et al., 2014). Water percolating through this material would have had a high pH, incapable of rapid chemical alteration of metallic Fe-Ni.

Other fossil meteorites include: (a) Al Haggounia 001, a ≥3-metric-ton fossil EL chondrite impact-melt rock with a ^{14}C age of 23,000 ± 2000 years from Morocco (e.g., Chennaoui-Aoudjehane et al., 2009; Rubin, 2016), although its fossil status is disputed (Leili et al., 2024), (b) Gove, a relict iron from an Australian bauxite mine (Bevan et al., 2019), (c) Eltanin, a unique, metal-poor mesosiderite found amidst 2.5-Ma-old Ir-rich sediments within deep-sea sediment piston cores from the southeastern Pacific Ocean (Kyte, 2002), (d) Lake Murray (IIAB iron), encased in a thick rusty crust and found embedded within the Antlers sandstone of Oklahoma (which formed about 113 Ma ago) (Buchwald, 1975), (e) Muonionalusta (IVA iron), which fell roughly 1 Ma ago (e.g., Chang and Wänke, 1969) and likely experienced several episodes of glaciation, and (f) a ~2.5-mm fossil carbonaceous chondrite (from a deep-sea drill core at the Cretaceous-Paleogene [K-Pg] boundary) that was probably derived from the ~10-km asteroid that produced the Chicxulub impact structure in the Yucatán ~66 Ma ago (Kyte, 1998) and exterminated the non-avian dinosaurs.

Nininger (1938) and Buchwald (1975) listed several meteorites that had been sequestered or intentionally buried by Native Americans. These include (a) Winona (the winonaite type specimen, 24 kg), found in Arizona in a stone cist similar to

Fig. 8.14 Obverse of a medal struck in the United States by Rob Wesel of Nakhla Dog Meteorites in 2007 in honor of H. H. Nininger. It is based on a photograph of Nininger by Orin Seeley of the Denver Post; it shows Nininger examining meteorite specimens at his desk at the Denver Natural History Museum (now the Denver Museum of Nature and Science). The photo was first published on the back cover jacket of Nininger (1952).

a child's coffin; (b) Navajo (IIAB iron, the 1500 kg mass), buried along with ornamental beads beneath talus at the base of a sandstone ridge in Arizona; (c) Mesa Verde Park (IAB Complex iron, 3.5 kg), located amidst terrestrial rocks at the Sun Shrine in Pipe Shrine House in Colorado (dating from the thirteenth century); (d) a specimen of Glorieta Mountain (pallasite), buried in a pottery vessel in New Mexico; (e) Casas Grandes (IIIAB iron, 1545 kg), wrapped in cloth and found in a temple ruin located in northern Chihuahua, Mexico; (f) the 61.5-kg Camp Verde specimen of Canyon Diablo (IAB iron), found in a stone cist within an ancient dwelling in Arizona; (g) Huizopa (IVA iron, four masses totaling 140 kg), found in a ruin in Chihuahua; and (h) pieces of the Brenham pallasite, buried in Hopewell Mounds, Ohio after having been transported ~1500 km from Kansas. Nininger (1938) also reported the discovery of four new (unnamed) stony meteorites on old Native American campsites in eastern Colorado and western Kansas. (In honor of Nininger's contributions to meteoritics, a commemorative medal [a piece of exonumia, not a legal-tender coin] was struck in 2007 with his likeness [Fig. 8.14].)

8.13 Impact Craters with Ordinary-Chondrite Remnants

Ordinary chondrites that reach the Earth come in all sizes, from micrometeorites a few hundred micrometers in size and recovered from ice, snow, moraines, rock traps, deep-sea sediments, or rooftops in urban areas, to the multikilometer asteroids that produce enormous impact craters.

On September 15, 2007, a bright fireball with a smoky tail was observed moving toward the northeast over the Puno Region of Peru, near the border with Bolivia. The shock wave shattered windows 1 km away. The meteoroid impacted the clay soil near the village of Carancas in Peru, producing a 13.3×13.8-m crater, elongated in an east-west direction. There were reports of "fetid, noxious" gases and boiling water escaping from the crater. Some villagers who approached the crater felt ill, possibly because they inhaled vapors emanating from arsenic-contaminated groundwater.

Numerous small meteorite fragments were recovered from the vicinity of the crater and along the sides of the crater. Although only 342 g were reported officially, the amount of material actually collected is far higher (e.g., Fig. 8.15). The meteorite (named Carancas) is an H4-5 chondrite breccia of which most specimens lack fusion crust. Its matrix is highly recrystallized and relatively few relict chondrules are evident. Carancas is shock stage S3 and is replete with black shock veins; its olivine grains exhibit undulose extinction and contain irregular and planar fractures.

The 145-Ma-old Morokweng impact structure in South Africa is more than 70 km in diameter (e.g., Koeberl et al., 1997). It has a deep-seated, 870-m-thick impact-melt sheet with high concentrations of refractory siderophile elements, indicative of a significant meteoritic component. The contents of a borehole at a depth of ~766 m included a 25-cm meteorite fragment – an LL6 chondrite breccia with a 1-mm-thick brown alteration rind (Maier et al., 2006). Many other subcentimeter-sized fragments,

Fig. 8.15 A 345-g specimen of the Carancas H4-5 chondrite along with a 1-cm cube. The specimen appears altered and exhibits pitting and corrosion. Photograph courtesy of Peter Utas.

recrystallized to various degrees, also occur within the melt sheet (e.g., Hart et al., 2002). The olivine (Fa30-33) in the 25-cm fragment is at the upper extreme of oxidized LL chondrites. Additional phases include orthopyroxene (En_{69-73}), plagioclase (mostly An_{33-50}), apatite, pyrrhotite, and pentlandite; troilite and metallic Fe-Ni are absent. The chondrules are recrystallized but quite discernable; textural types include porphyritic pyroxene, radial pyroxene, and barred olivine.

Both the olivine and orthopyroxene exhibit undulose extinction, indicating that the sample has a shock stage of ≥S2. Although Maier et al. (2006) reported mosaicized olivine indicative of shock stage S4-S5, these olivines are actually individual assemblages consisting of numerous small, relatively unstrained, olivine grains, many with 120° triple junctures (Fig. 1e of Maier et al., 2006). These olivine assemblages resemble those in some ureilites that were shocked and annealed (e.g., ALH 81101) (Fig. 4c of Rubin, 2006). They are also similar to the coarse olivine aggregates within melt regions of the Chico and Rose City OC impact-melt breccias. Rubin (1995a) modeled these latter grains as having once been shocked-and-mosaicized olivine grains that were subsequently annealed by the surrounding silicate melt. These olivine aggregates presently consist of numerous 10–20-μm-sized, essentially unstrained, individual faceted olivine crystals with 120° triple junctures.

It seems likely that the Morokweng chondrite was annealed during its contact with the melt sheet; dislocations within the mosaicized olivines were healed, resulting in the formation of numerous small unstrained olivine crystals. This is consistent with the findings of Bauer (1979) and Ashworth and Mallinson (1985) who reported that shock-induced damage of the olivine crystal lattice can be repaired during annealing at subsolidus temperatures.

I suggest that the meteoroid may have been a relatively unshocked and unequilibrated (and not especially oxidized) LL chondrite (i.e., LL3-4, S1-2) prior to crashing into South Africa. It is plausible that the remnants of the chondritic projectile were shocked to S5 levels during impact with the Earth (when an enormous amount of energy was generated) rather than during its pre-terrestrial history on its parent asteroid. Contact with the melt sheet may have annealed the chondrite to ~S1 levels, healing dislocations in the mafic silicates. The present undulose extinction in some crystals may be due to later-stage, small-scale deformation (or to

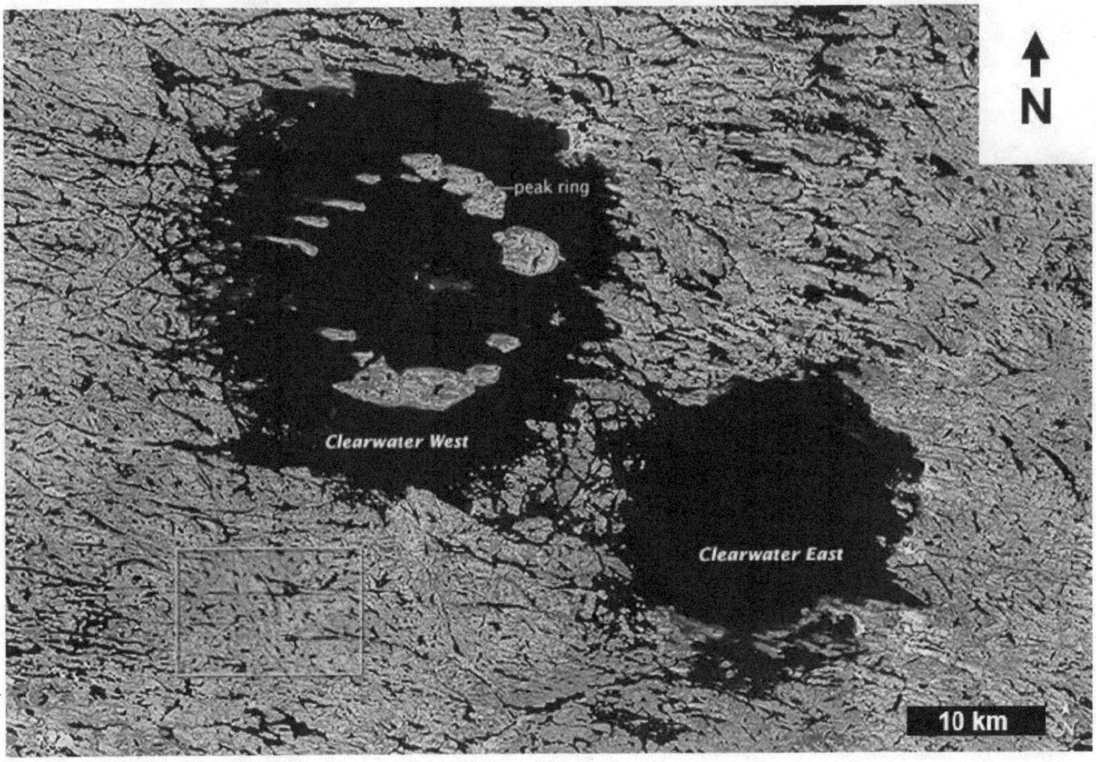

Fig. 8.16 Adjacent impact craters in northwestern Québec, Canada: Clearwater West (32 km) and Clearwater East (22 km). A series of islands in the Clearwater West crater are the remnants of a peak ring. Image from the Operational Land Imager on Landsat 8 on June 29, 2013. Ice sheets covering the area retreated 20,000 years ago after scouring the landscape. Image courtesy of NASA.

incomplete annealing). As has been pointed out by Hart et al. (2002), the chondritic fragments likely reacted with the more oxidized melt sheet to form the unusually oxidized metal-free silicate and sulfide assemblages.

The adjacent impact structures at Clearwater Lakes, northwestern Québec (Fig. 8.16) appear to have formed in different geological periods: Clearwater West (≥36 km) formed 286.2 ± 2.2 Ma ago in the Early Permian; Clearwater East (~26 km) formed ~465 Ma ago in the Middle Ordovician (Schmieder et al., 2015). Impact-melt rocks from the Clearwater East crater are rich in refractory siderophile elements and have interelement ratios (e.g., Ir/Os, Ir/Pd, Ir/Ni, Ni/Cr) comparable to those of ordinary chondrites, plausibly L chondrites (Palme et al., 1979; McDonald, 2002).[3] The relevant chromium isotope data are consistent with such an ordinary-chondrite source (Shukolyukov and Lugmair, 2000). The age of the Clearwater East crater is compatible with it having formed shortly after the break-up of the L-chondrite parent asteroid ~470 Ma ago (e.g., Korochantseva et al., 2007; Nesvorný et al., 2009; Swindle et al., 2014).

The eroded Popigai impact structure in northern Siberia is the remnant of a ~100-km-diameter crater that formed 35.7 ± 0.2 Ma ago during the Late Eocene (Bottomley et al., 1997). During the impact, graphite within target rocks near the center of the crater was transformed into diamonds; most crystals are 0.5–2 mm in size (Deutsch et al., 2000). Interelement ratios of refractory siderophile elements within impact-melt rocks from the crater show that the projectile was an ordinary chondrite (and probably an L chondrite) (Tagle and Claeys, 2005).

Koeberl et. al. (2007) analyzed Cr isotopes in samples from four impact craters: Bosumtwi

[3] Chondritic meteorites have higher abundances of refractory siderophile elements than terrestrial crustal rocks by several orders of magnitude (e.g., Barnes et al., 1985; Wasson and Kallemeyn, 1988; Wedepohl, 1995).

(Ghana), Clearwater East (Canada), Lappajärvi (Finland), and Rochechouart (France) and found that ordinary chondrites are the likely projectiles. Lappajärvi probably formed from an H chondrite; Clearwater East could have been formed by either an H or L chondrite.

The Australasian tektites hail from the largest tektite strewn field, covering ~15% of the Earth's surface. They are also the youngest tektites (0.78 Ma old). No known crater is associated with these objects, although Wasson (1991) suggested there were multiple kilometer-size craters and Wasson (2003) proposed a large aerial burst, possibly leaving no crater at all. In any case, the Australasian tektites contain an identifiable extraterrestrial chemical component. Folco et al. (2018) studied Australasian microtektites; based on Cr, Co, and Ni interelement ratios, they determined the most likely impactor to be an LL chondrite. Nickel–Mg mixing calculations by Folco et al. (2023) indicated that the chondritic component in Australasian tektites and microtektites is up to 4% and up to 6%, respectively.

8.14 Ordinary-Chondrite Micrometeorites

About 98% of micrometeorites >50 µm across appear to have an asteroidal source; about 50% of those <50 µm across are probably cometary (Genge et al., 2020). Most of the asteroidal particles were derived from carbonaceous chondrites.

One early study of micrometeorites is by Walter et al. (1995), who examined 303 Antarctic micrometeorites derived from ice. The particles range in size from 100 to 400 µm and consist of cosmic spherules (13.5%), scoriaceous micrometeorites (44.9%), dehydrated phyllosilicates (21.5%), objects composed mainly of phyllosilicates (3.6%), and coarse-grained crystalline objects (16.5%). Most of the crystalline micrometeorites are related to carbonaceous chondrites; only three of the coarse-grained crystalline objects appear to have been derived from ordinary chondrites. The paucity of ordinary chondrites among Antarctic micrometeorites (~1%) in this dataset is comparable to that among Greenland micrometeorites (e.g., Beckerling and Bischoff, 1995) as well as to the complete set of micrometeorites (Fig. 1 of Rubin, 2018b).

One 200-µm particle (M6) from the Walter et al. study is porphyritic with olivine ($Fa_{18.7}$) and low-Ca pyroxene ($Fs_{16.6}Wo_{0.6}$) grains of H-chondrite composition. A 410-µm-long particle (AM1) is porphyritic with olivine grains ($Fa_{16.2}$) surrounded by a fine-grained, clinopyroxene-bearing matrix. The third particle (Mc7/10; 150 µm long) has a granular texture with intergrown olivine ($Fa_{17.6-19.8}$), low-Ca pyroxene ($Fs_{14.1-16.3}Wo_{\sim 1.2}$), and plagioclase ($An_{14}$). The presence of plagioclase in Mn7/10 suggests that it was derived from an equilibrated H chondrite. The FeO/MnO ratios in the olivine grains in these three micrometeorites are 33–49, similar to those of equilibrated H chondrites (e.g., Brearley and Jones, 1998). Instrumental neutron-activation analysis (INAA) of two of the particles showed their trace element compositions to resemble those of ordinary chondrites (Kallemeyn et al., 1989).

In contrast to the low proportions of ordinary chondrites among 100–400-µm-sized micrometeorites, fractions of larger size contain greater proportions of ordinary chondrites. For example, van Ginneken et al. (2012) found that 62% (five out of eight) "giant" micrometeorites (in the 300–1100-µm-sized range) are related to unequilibrated or equilibrated ordinary chondrites. Similarly, Noguchi et al. (2025) reported some >800 µm cosmic spherules related to OC. Rochette et al. (2008b) identified ordinary-chondrite-like particles within a large size fraction (800–2000 µm) of micrometeorites recovered from the top of eroded nunataks (mountaintops or ridges protruding from an ice field) in Victoria Land, Antarctica. Some of these particles show readily-to-poorly delineated chondrules (characteristic of petrologic type-4 and type-5 OC); others have granoblastic textures similar to those of some highly recrystallized type-6 OC. Olivine (Fa_{18}–Fa_{26}) and low-Ca pyroxene (Fs_{17}–Fs_{22}) compositions in these micrometeorites are generally within the ranges of those in H ($Fa_{17.3-20.2}$; $Fs_{15.7-18.1}$) and L ($Fa_{23.0-25.8}$; $Fs_{18.7-22.6}$) chondrites (Rubin, 1990; Gomes and Keil, 1980). Other phases in these micrometeorites include Ca-pyroxene, chromite, Fe-sulfide, oligoclase, and Fe-Ni oxide grains, all typical of weathered equilibrated OC. These results are all consistent with

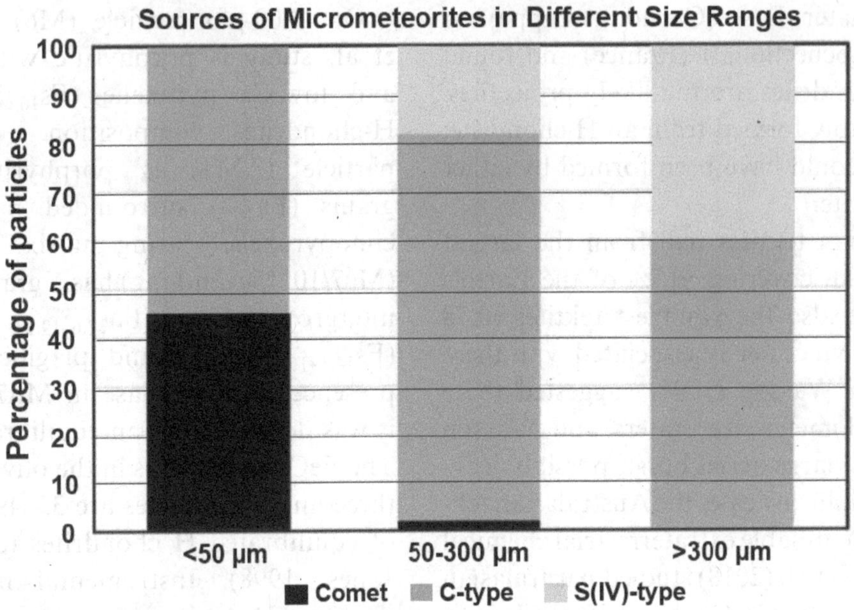

Fig. 8.17 The percentage of micrometeorites related to ordinary chondrites (S(IV) types) increases with size interval. About 40% of the >300-µm-sized particles are ordinary chondrites. Image based on Genge et al. (2020).

those of Genge et al. (2020), who found that the proportion of S(IV) asteroidal materials (which resemble ordinary chondrites) among micrometeorites increases with size interval (Fig. 8.17).

The proportion of ordinary chondrites among micrometeorites is correlated with petrologic characteristics. For example, equilibrated ordinary chondrites contain ~0.6 wt.% chromite (Dodd, 1981), and many type-3 OC contain low-FeO (Type-I) chondrules with relict ferroan olivine grains. These relict grains appear "dusty" in plane-polarized light due to the presence of low-Ni metallic Fe blebs formed by reduction of olivine FeO (e.g., Nagahara, 1981; Rambaldi, 1981; Jones, 1996; Rubin and Krot, 1996; Jones and Danielson, 1997). Chromite and dusty olivine grains are less abundant in carbonaceous chondrites than in ordinary chondrites. Thus, micrometeorites that contain chromite and/or dusty olivine grains are more likely to be derived from ordinary chondrites.

Prasad et al. (2015) investigated 481 micrometeorites from deep-sea sediments from the Indian Ocean and identified 15 with mineralogic and textural features indicative of their provenance. Seven of these particles contain chromite with characteristic ordinary-chondrite compositions; seven others have dusty olivine grains, and the remaining one is a POP chondrule. All 15 micrometeorites have OC-like bulk chemical compositions.

Fernandes et al. (2024) analyzed 128 FeO-rich relict olivine grains from 53 cosmic spherules collected from Antarctica and deep-sea sediments in the Indian Ocean. They found that ~21–23% of the spherules were derived from ordinary chondrites, and ~17–36% from carbonaceous chondrites; the remainder have compositional characteristics overlapping those of both ordinary and carbonaceous chondrites.

Furthermore, the oxygen isotopic compositions of most micrometeorite chromite grains from deep-sea sediments and from Antarctica are similar to those from ordinary chondrites (Rudraswami et al., 2019).

8.15 Pairing

There are several types of pairing among meteorites.

Pairs derived from a single meteoroid constitute the most common type of pairing. Such paired samples were separated when a single meteoroid fragmented during atmospheric passage and produced a strewn field. The number of recovered stones can be enormous: for example, ~14,000 (L/LL6 Holbrook), ~180,000 (H5 Pultusk; Fig. 8.18), and possibly

Fig. 8.18 Five-dollar silver proof coin (partially plated with Pd) from the Cook Islands, dated 2008, depicting the fall of the Pultusk H5 chondrite in Poland in 1868. About 250 kg of the meteorite were recovered. Obverse featuring a portrait of Queen Elizabeth II (left). Reverse shows a map of Europe with the site of fall indicated (right). Also shown are renderings of Pultusk meteorite specimens; there is also a small piece of the actual Pultusk meteorite embedded in the reverse of the coin. A total of 2500 coins was minted.

>500,000 (LL5 Chelyabinsk). Smaller, less friable meteoroids may break into relatively few pieces during atmospheric descent (e.g., L5 Cartersville; LL5 Tuxtuac; LL6 Karakol). Atmospheric fragmentation decreases the total mass of a meteorite fall that can actually be collected at the Earth's surface. Small meteoroids, on the order of 10 g, are not likely to break up at all during descent (Huss, 1991).

Cassidy (1983) estimated there are 10 individual pieces per average meteorite fall. Data from Hey (1966) indicate a specimen-to-fall ratio for recovered samples of 4 to 10 (Scott, 1984). Rubin (2021) reported that, among the 123 OC observed falls for which relevant data were listed in the August 8, 2020 edition of the Meteoritical Bulletin Database, the median number of recovered specimens was 3.

In dense collection areas, it is common for such pairs to receive separate numbers. For example, McKinley et al. (1981) paired ALH 77011 with 34 other individually named L3 chondrites (e.g., ALH 77015, ALH 77047, ALH 78038, ALH 79045) from the 1977–1979 collection from Allan Hills, Antarctica. The pairing of these samples was assigned because they all contain a few volume-percent C-rich aggregates, unlike most other Antarctic L3 chondrites.

Meteorites can also be paired outside dense collection areas on the basis of similar compositions and textures. For example, Plainview (e) and Plainview (f) are paired with the Plainview (1917) H5 regolith breccia from Texas. Nonetheless, some potential pairing remains uncertain. The Keysville L6 find (Georgia, 2018) might be paired with the Thomson L6 fall from 1888 (39 km to the northwest). The Zulu Queen L3.7 find (California, 1976) was found within 15 km of the Dale Dry Lake L3.7 find (California, 1957) and is probably paired with it (Verish, 2003a).

Some pairs were derived from a single meteorite that fragmented after landing. Meteorites can break apart from stress caused upon landing on a hard surface. They can also fragment long after landing due to terrestrial weathering, commonly by a type of mechanical weathering known as frost wedging. When meteorites with fractures are rained on, their fractures may fill with water. Water expands by 9.2% when it freezes, hence water-filled fractures exert appreciable internal pressure in a rock when temperatures dip somewhat below 0°C. This pressure widens the fractures; multiple freeze–thaw cycles can eventually split meteorites apart.

Individual OC finds may break into numerous pieces that remain near to each other on the ground. In some cases, many such fragments were recovered and the original meteorite partially reconstructed (Fig. 8.19). Collectors occasionally refer to these reconstructions as "puzzle chondrites." Nininger and Nininger (1950) called them complementary individuals.

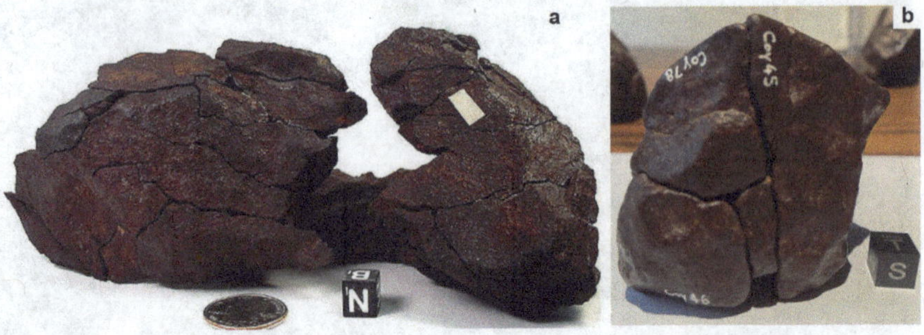

Fig. 8.19 Puzzle chondrites (complementary individuals). (a) More than 15 separate fragments that fit together from Stewart Valley 001, an H6 chondrite find from Nevada. The cube is 1 cm across. Also shown is a United States quarter coin (24.26 mm in diameter). Image from Verish (2003b), courtesy of Robert Verish. (b) An individual Coyote Dry Lake H5 chondrite find pieced together from four fragments found within ~15 m of one another. The total mass of this reconstituted individual is 296.6 g. The stone is probably paired with the officially approved H5 chondrite Coyote Dry Lake 054; some other H5 chondrite finds from this site are also likely paired. The cube is 1 cm across. Image courtesy of Peter Utas.

Launch pairs were derived from a single impact-excavation event on a parent body. About 45% of H chondrites, mostly those of petrologic types 4 and 5, plot within a CRE-age cluster peaking at 7–8 Ma (Graf and Marti, 1995). Almost all of these meteorites are launch pairs derived from the same parent asteroid during a major collision 7.5 Ma ago. Anders (1978) estimated that this event excavated material with a volume on the order of 1 km^3. Even though the samples were launched at the same time from the same crater, their paths diverged in interplanetary space. They fell to Earth in different places, in different months, and in different years: for instance, H4 Akbarpur (CRE age = 7.7 Ma) fell in India on April 18, 1838; H4 Ochansk (CRE age = 7.7 Ma) fell in Russia on August 30, 1887; and H5 Changde (CRE age = 7.6 Ma) fell in China on March 11, 1977. Some fell in the remote past: H5 ALHA79046 (CRE age = 7.6 Ma) fell in Antarctica, perhaps ~300,000 years ago (Benoit and Sears, 1993).

8.16 Meteorite Concentration Mechanisms in the Terrestrial Environment

For at least the last 50,000 years, meteorites have been falling more-or-less randomly over the globe at a fairly constant rate of 76 ± 40 falls exceeding 10 grams per 10^6 km^2 yr^{-1} and 109 ± 50 falls exceeding 20 grams per 10^6 km^2 yr^{-1} (Bland et al., 1996). Nonetheless, falls are about 1.5× more frequent in equatorial than polar regions (see Section 8.4). There are certain places on Earth where meteorite concentration is much higher than average; the two most fecund regions are portions of Antarctica (Cassidy et al., 1977) and Northwest Africa (e.g., Bischoff and Geiger, 1995; Schlüter et al., 2002). Another important region is the set of dry lakes (playas, hardpans, or flats) in the Southwestern United States (mainly in Southern California and Nevada, along with some in Arizona, Utah, and New Mexico). Other dense collection areas include Oman, the Australian outback, United Arab Emirates, Atacama Desert (Chile), northwestern China, and Roosevelt County (New Mexico). The high abundances of meteorites in these locations are due to natural concentration mechanisms.

Antarctica. According to the Meteoritical Bulletin Database (July 30, 2023), about 93% of characterized and approved Antarctic meteorites are ordinary chondrites. Hence, to a first approximation, a description of concentration mechanisms operating in Antarctica (Cassidy, 2003) is a story about ordinary chondrites. The Antarctic OC range in size from fewer than two dozen 10-mg samples (H5, L4, L5, L6) up to two meteorites exceeding 100 kg (H5 LEW85320, 110.2 kg; L6 ALHA76009, 407 kg). The Antarctic OC comprise approximately 45% H, 39% L, and 16% LL chondrites.

East Antarctica and West Antarctica are separated by the Transantarctic Mountain Range. Most recovered Antarctic meteorites are from East Antarctica – a vast ice plateau with a thin veneer of snow. Meteorites fall on the ice essentially at the same rate they fall everywhere on Earth and are eventually buried by falling snow. (The average rate of precipitation in Antarctica is 170 mm a^{-1}; Van de Berg et al., 2006.) As the years go by and more snow accumulates, snowflakes are compressed by the overlying snowpack; trapped air is squeezed out of interstices and the meteorites become encased in ice.

The highest point in East Antarctica is the Argus Dome, situated at an elevation of 4093 m. It is also the coldest naturally occurring place on Earth with temperatures dipping to lows of −90 to −98°C. From the dome, ice flows downhill toward the edge of the continent and the shore of the Southern Ocean. Also pouring downhill are cold high-density air currents (known as katabatic winds) that pick up speed over the barren landscape as they rush toward the sea. In places where the flowing ice encounters nunataks, it flows around them through narrow straits. Patches of blue ice, laden with meteorites, are swept clean of snow by the fierce winds. As the thick ice flows downslope, it acts like a conveyer belt, inching the entombed meteorites forward (Fig. 8.20).

At depth, subsurface ice flows over and around buried mountain ridges. As Ralph Harvey, long-time leader of ANSMET (The Antarctic Search for Meteorites) put it: "Those meteorites become essentially sprinkled throughout the ice sheet like the canonical raisins in a pudding." Where flowing ice abuts a mountain barrier or nunatak, the ice is deflected upward; meteorites trapped within the ice are brought toward the surface. Their mean surface residence time is on the order of 7000 years (Evatt et al., 2020). Ice on the windward side of the barrier is ablated at rates averaging 5–6 cm a^{-1} (Cassidy, 2003); ice is also lost through sublimation. Previously buried meteorites are exhumed. Adding to the accumulation of meteorites are those weighing less than ~100 g that are pushed downwind and concentrated at the base of the nunataks.

The ablation zones, where meteorite concentrations can be quite high, are referred to as meteorite stranding surfaces (or, less commonly, as meteorite traps). The Lewis Cliff Ice Tongue, for example, has a meteorite surface density (without considering potential pairing) of roughly 100 meteorites per km^2 (Cassidy et al., 1992), about a million times higher than the average at the Earth's surface. According to Ralph Harvey, there are several portions of the Davis–Ward icefields (e.g., Fig. 8.21) where an area equivalent to that of a standard tennis court (roughly 200 m^2) would have at least 50 meteorites (and vast numbers of terrestrial rocks)

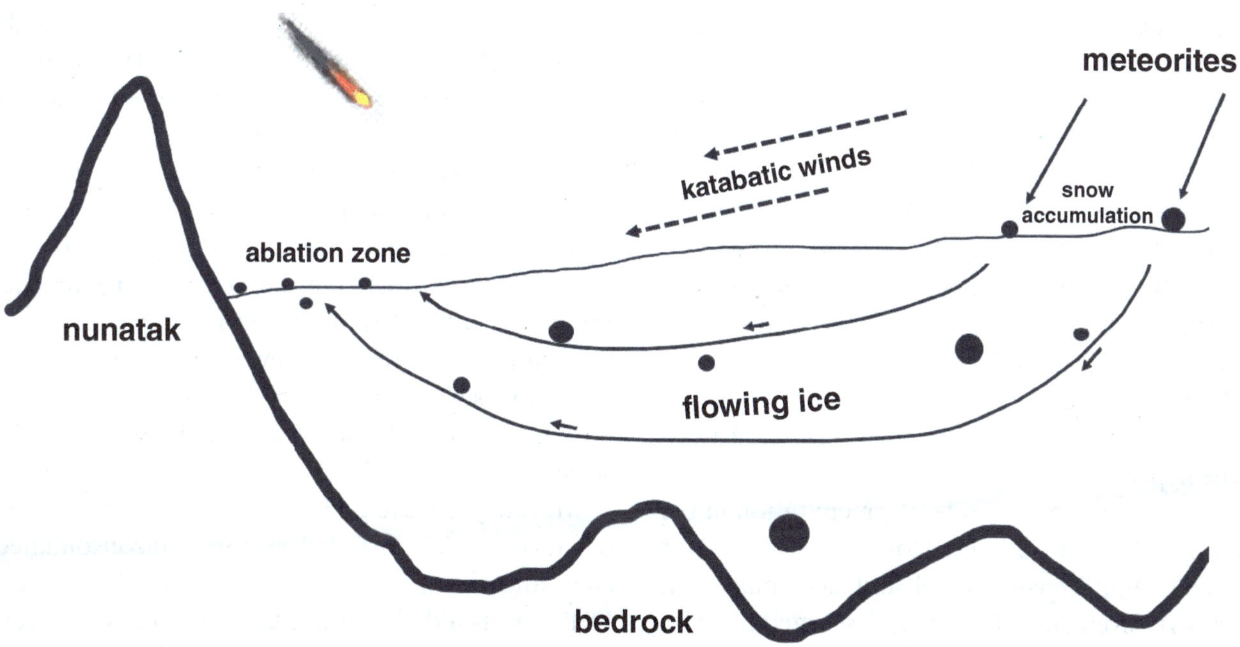

Fig. 8.20 Cross section of an ice sheet illustrating how meteorites can be concentrated in an ablation zone near a nunatak. The diagram shows vertical exaggeration of about 20–30 times. Meteorites (depicted as filled black circles and ellipsoids) are not to scale.

Fig. 8.21 A spot dubbed "The Beach" at the Davis–Ward icefields in Antarctica. Each red flag marks the position of a meteorite discovered by the search team on this ice-free gravelly surface. Image courtesy of Ralph Harvey and ANSMET.

upon it. Certain spots have meteorite concentrations approaching 5 per m^2 (equivalent to 5 million meteorites per km^2, about 50,000 times higher than at the Lewis Cliff Ice Tongue). Most of these meteorites are centimeter-sized and are scattered among millions of similarly sized terrestrial rocks.

Northwest Africa. In August 2023, the Meteoritical Bulletin Database showed that Northwest Africa (NWA) had yielded more than 12,000 meteorites, about 57% of which are ordinary chondrites.[4] NWA meteorites range in size from ~0.5 g (L5 NWA 8299 and martian shergottite NWA 8116) to 2 metric tons (L3-6 NWA 869). The NWA OC comprise approximately 34% H, 45% L, and 21% LL chondrites.

The two principal concentration mechanisms operating in the Sahara are time (tens of thousands of years) and wind deflation (an erosive process that clears fine-grained dry materials from the desert floor, leaving coarser rocks behind). These are aided by the hyper-arid climate, which acts to preserve old meteorites. (The average rate of precipitation in the Sahara is 100 mm a^{-1}; Le Houérou, 1980.) Over the millennia, meteorites fall and accumulate on the desert pavement; ^{14}C dating of selected samples yields terrestrial ages of 400 years to more than 40,000 years, with most samples being less than 20,000 years old (Aboulahris et al., 2019). There are only small numbers of terrestrial rocks and little summer vegetation at most sites; in many areas, the dark-colored NWA meteorites contrast with the reddish-brownish soil and pale limestones and sandstones. Meteorite concentrations can be high. For example, the meteorite density for Tunisia is 1.8 per km^2 before pairing and 1.5 per km^2 after pairing (Aboulahris et al., 2019).

Dry Lakes. The importance of dry lakes in the Southwestern United States as sites for high meteorite concentrations was discussed by Rubin et al. (2000). Since the end of the Pleistocene, many dry lakes partly fill with water every year (typically to depths of a few centimeters) during the rainy season and then dry out again (Oriti, 1965); this may happen multiple times in a single season.

Most dry lakes (e.g., Fig. 8.22) are subject to wind deflation, which lifts and removes unconsolidated clay- and silt-sized particles from the lake surface. Deflation is aided by the lack of protective vegetation and the occurrence of fine-grained sediments (Shaw and Thomas, 1997). Subsequent deposition of these particles forms clay and silt dunes near

[4] Most NWA meteorites are from Morocco, Algeria, Tunisia, Libya, Egypt, Mauritania, and Mali.

Fig. 8.22 Coyote Dry Lake, California in November 2022. The hills in the background mark the lake margin. Picture courtesy of Peter Utas.

many dry lakes. The erosive effect of wind produces ventifacts (sharp-angled rocks sculpted by wind erosion) on many dry lakes, and yardangs (wind-carved landforms of weakly cemented sediments) near a few dry lakes (Ward and Greeley, 1984; Meek, 1994).

Small meteoroids reach a terminal velocity of ~300 m s^{-1} (Wasson, 1985) and are unlikely to penetrate the lakebed when the bed is dry. Although such lakes are partly filled with water during the rainy season, little of the water penetrates more than ~30 cm below the lakebed because of the low permeability of the clay (Oriti, 1965). When wet, the playas contain some suspended clay particles, but generally not enough to cause meteorites to become deeply buried after the clay settles.

On dry lakes that experience subfreezing temperatures in winter, ice rafting can transport meteorites, terrestrial rocks, bones, and sticks toward the shoreline in the typical local windward direction (Gessler et al., 2002). Stanley (1955) found many parallel tracks on a playa surface and concluded that "wind-blown ice floes (were) dragging protruding stones." Direct observations by Norris et al. (2014) show the protruding rocks moving at 2–5 m min^{-1} under light winds. As they are pushed across the lake, small rocks, sticks, and bones leave shallow furrows on the lakebed (Fig. 8.23). Meteorites are concentrated on the windward shoreline, forming a stranding zone.

The Meteoritical Bulletin Database (August 21, 2023) shows that at 56 dry-lake sites in the Southwestern United States, there are 364 separate approved entries for meteorites (including some paired specimens as well as individual fragments that might or might not be paired with other recovered fragments) (e.g., Fig. 8.24). About 93% of approved dry-lake meteorites are ordinary chondrites. These dry-lake OC comprise 61% H, 35% L, and 4% LL chondrites.

Table 8.1 shows the large variations among falls and find sites in the percentages of ordinary chondrites among total meteorites and in the percentages of individual groups among the ordinary chondrites.

It is assumed that the percentages of individual meteorite groups among samples that reach the Earth's surface have been more-or-less constant over the last few tens of thousands of years (e.g., Wetherill, 1987; Bland et al., 1996). If this is true, then the significant differences in the proportions of ordinary-chondrite groups among falls and different collection sites of finds must be due to natural terrestrial processes (e.g., weathering, wind erosion, burial) and/or sociological effects (e.g., biases

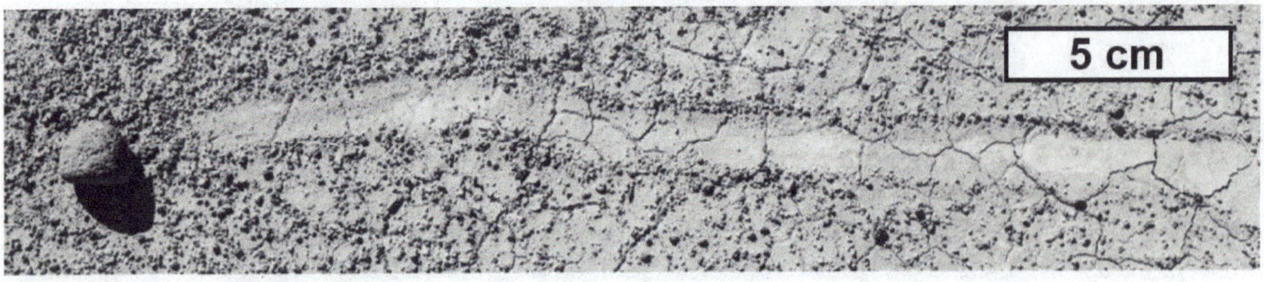

Fig. 8.23 Furrow left in the lakebed of Superior Dry Lake in Southern California by ice rafting of the pebble at left. Photograph by Robert Matson; image modified from Gessler et al. (2002), reproduced with permission.

Fig. 8.24 A quasi-cubic-shaped 46.76-g ordinary chondrite found at Coyote Dry Lake in November 2022. Head of magnetic pick-up tool at right side of image. Picture courtesy of Peter Utas.

Table 8.1 *Abundances of ordinary chondrites and of individual OC groups.*

–	Falls	Antarctic finds	NWA finds	Dry lake finds
Percentages of OC among total meteorites	84%	93%	57%	93%
Percentages of H chondrites among OC	41%	45%	34%	61%
Percentages of L chondrites among OC	48%	39%	45%	35%
Percentages of LL chondrites among OC	11%	16%	21%	4%
(%L)/(%H) ratio	1.17	0.87	1.32	0.57
(%LL)/(%L) ratio	0.23	0.41	0.47	0.11
(%LL)/(%H) ratio	0.27	0.36	0.62	0.07

in meteorite collection or reporting, willingness of researchers to classify specimens, eagerness of the Meteorite Nomenclature Committee to approve specimens).

Observed falls are generally considered to be the most representative set of the OC flux to the Earth (e.g., Dohnanyi, 1972; Huss, 1991; Cassidy, 2003). However, there are four principal sampling biases in the mass distribution of observed falls. (1) Over much of the land surface, vegetation and dark soil hinder the recognition of recently fallen meteorites, particularly small individuals. (2) There is an optimal mass range where samples can be readily identified and efficiently collected: ~0.1–1000 kg. Smaller samples are difficult to recognize under nonideal conditions. Some samples are collected but not officially reported. Specimens exceeding 1000 kg can get buried in the soil so deeply that they are unlikely to be found unless the impact pit is identified and excavated. (3) Many meteoroids, particularly large and friable ones, break up in the atmosphere and produce meteorite showers. (4) The total number of observed falls is relatively small (<1300 as of this writing); this is a result of the limited timespan over which meteorite falls have been recorded – only four OC falls in our collections fell prior to 1700 CE. This short period of time renders it unlikely that the full mass range of OC falls has been recorded.

Antarctic finds are highly visible on the ice; samples with masses less than a few grams can be spotted readily (Huss, 1990). In many ice fields, there are no terrestrial rocks to hamper meteorite recovery, and meteorites with masses exceeding ~10 g are efficiently collected. There are four principal sampling biases in the mass distribution of Antarctic finds: (1) Many samples <100 g are blown away by fierce wind gusts (Schutt et al., 1986; Harvey and Cassidy, 1989; Huss, 1990), but some are concentrated in traps (e.g., firn, moraines, snow bridges, ripples, and suncups on blue ice; Harvey, 2003) where they can be collected. (2) Samples dislodged from the ice are exposed to liquid water that can penetrate deep fractures. This results in increased fracture porosity (Spargur and Gooding, 1992; Bland et al., 1995) and subsequent mechanical disintegration (Huss, 1991). Small fragments produced by this process are likely to be blown off the ice. (3) In some regions, meteorites on blue ice areas adjacent to nunataks and moraines are scattered among terrestrial rocks, hindering recognition (Harvey et al., 2015). (4) Because the time–area integral for Antarctica is much smaller than that of the total land surface of Earth, the set of Antarctic finds does not contain samples approaching the mass of the largest meteorites among observed falls (Huss, 1991). Although collection efficiencies of Antarctic meteorites vary from season to season due to the vagaries of weather and drifting snow (and fluctuations in the visual acuity, search spacing, and persistence of searchers), the *entire* Antarctic collection reflects the time-integrated average product. Because the proportions of different petrologic types among Antarctic OC finds are roughly similar to those of falls (Table 8.2), it is evident that Antarctic meteorites have been collected fairly systematically.

Table 8.2 *Abundances of ordinary chondrites of different petrologic types.*

–	Falls	Antarctic finds	NWA finds	Dry lake finds
Type-3	5.7%	3.7%	18.3%	3.9%
Type-4	11.4%	15.9%	27.0%	18.8%
Type-5	35.2%	42.0%	27.2%	34.5%
Type-6	47.7%	38.4%	27.5%	42.8%
Total	100.0%	100.0%	100.0%	100.0%

Note: For this determination, breccias (e.g., L4-6) and transitional types (e.g., H3,4) were counted as the lowest petrologic type represented. Type-7 chondrites, impact-melt breccias, and impact-melt rocks were omitted.

NWA finds are subject to four principal sampling biases: (1) The color of the soil (red to brown) and the presence of some terrestrial rocks hinder the recognition of small, weathered meteorites. (2) Many searchers intentionally ignore, discard, or do not submit for analysis small meteorites, common meteorites (e.g., equilibrated OC specimens), and rusty meteorites (e.g., weathered H chondrites) because these samples are deemed to be of limited commercial value (e.g., Gilmer, 2020). (3) Many NWA finds are appreciably weathered; older specimens tend to disintegrate, making their fragments difficult to identify. (4) Because the time–area integral for Northwest Africa is much smaller than that of the total land surface of Earth, the set of NWA finds does not contain samples approaching the mass of the largest meteorites among observed falls.

The bias among collectors of NWA samples is revealed by the much higher proportions of type-3 and type-4 ordinary chondrites than among observed falls (Table 8.2). The type-3 and -4 OC tend to have recognizable chondrules and greater commercial value. There appears to be a tendency to either leave more-equilibrated OC in the field or not to submit them for classification. Several meteorite hunters have told me that they think it unlikely that equilibrated OC are intentionally left in the field. I can confirm that not all meteorite researchers are eager to analyze and classify weathered equilibrated OC.

Dry lake finds are subject to one principal sampling bias. Relatively few meteorites collected from dry lakes have been formally approved so far (364 as of this writing).[5] Because the time–area integral for dry lakes is infinitesimal relative to that of the total land surface of Earth, the set of dry-lake finds does not contain representative numbers of rare specimens or any samples approaching the mass of the largest meteorites among observed falls. The fact that the proportions of petrologic types among dry-lake finds are similar to those of falls (Table 8.2) indicates the dry-lake ordinary chondrites have been collected more-or-less systematically. Equilibrated samples were representatively collected and submitted for classification. Some meteorite hunters use a grid to aid systematic collection.

As pointed out by Rubin (2021), the degree of terrestrial weathering varies among the sets of falls and finds. Observed falls are essentially unweathered (median = W0, mean ≈ W0). The few falls with recorded weathering stages >W0 were all oxidized after reaching the Earth's surface; oxidation can occur even inside museum cabinets (e.g., Kichinka, 2005). Antarctic OC finds are somewhat more weathered than falls (median = W1.5, mean = W1.4); W2 samples constitute ~26% of the total, and ~4% of Antarctic OC finds have weathering stages ≥ W2.5. NWA OC finds tend to be even more weathered (median = W2, mean = W2.0); W2 samples constitute ~34% of the total, and ~28% of NWA OC finds have weathering stages ≥ W2.5. Dry-lake finds are the most weathered (median = W3, mean = W3.2); W2 samples constitute only ~19% of the total, and ~75% of dry-lake OC finds have weathering stages ≥ W2.5.

The H/L/LL ratio varies among falls and different find sites (Table 8.1). The H/LL ratio even varies between two neighboring dense collection areas (Catalina and El Médano) in the Atacama Desert, Chile: 3.0 and 13.4, respectively (Sadaka et al., 2025). The key difference among the OC groups is oxidation state. This is most clearly reflected by the mean bulk abundance of metallic Fe-Ni-Co, which decreases among falls from H (17.8 wt.%) to L (8.33 wt.%) to LL (3.56 wt.%) (Jarosewich, 1990).

The differences in intergroup ratios between Antarctic and NWA OC finds probably reflect differences in weathering environment. The (%L)/(%H), (%LL)/(%L), and (%LL)/(%H) ratios among NWA OCs are appreciably higher than those of Antarctic OCs or falls. These results show that the higher the mean modal abundance of metallic Fe-Ni-Co in an OC group, the more likely the samples are to weather away and disintegrate. In other words, the more extensive weathering of NWA samples relative to Antarctic samples causes more NWA H chondrites to be destroyed than L chondrites, and more L chondrites to be destroyed than LL chondrites.

Although dry-lake finds tend to be much more weathered than NWA finds, it appears that terrestrial weathering tends to fragment (but not destroy) dry-lake ordinary chondrites as a function of their bulk metal content (as attested by meteorite hunters). More dry-lake H chondrites are fragmented than L chondrites, and more L chondrites are fragmented than LL chondrites, accounting for the lower (%L)/(%H) and (%LL)/(%L) ratios in dry-lake OCs than in falls (Table 8.1). This fragmentation among dry-lake finds leads to numerous fragments that fit together like pieces of a 3D puzzle being found within a few meters of each other (Fig. 8.19) (e.g., Verish, 2003b).

Why are the more weathered OCs from dry lakes better preserved than the less weathered OCs from Northwest Africa? This apparent anomaly is probably a function of terrestrial age. Welton et al. (2004) found that the terrestrial ages of 74 ordinary chondrites from the Sahara typically range from ~0 to ~30 ka. The median age is ~8 ka and the mean age is ~11 ka. Dry-lake finds are likely much younger: some dry lakes held permanent water at the end of the Pleistocene (~11,000 years ago), whereas others contained shallow water at various times during the Middle Holocene (3500–4000 years ago) and Late Holocene (~400 years ago) (Enzel et al., 1992). Most recovered dry lake meteorites landed after the Pleistocene, most within the last few centuries. For

[5] Far more meteorites, however, have been collected. For example, more than 530 provisional meteorites were recovered from Tungsten Mountain, Nevada but have not been classified. Roach Dry Lake, Nevada has yielded nearly 100 provisional unclassified meteorites. Many meteorite samples (on the order of 3500) are estimated to have been recovered from dry lakes in California, Nevada, and Arizona but not reported.

example, Harlan et al. (2016) determined the terrestrial ages of two dry-lake LL5 finds near Stagecoach, Nevada – Misfits Flat 001 (326 g): 8100 ± 1300 years; Misfits Flat 002 (3.54 g): <300 years.

It is possible that the higher (%L)/(%H) ratio among falls than Antarctic ordinary chondrites (1.17 vs. 0.87) reflects one or more large H-chondrite showers among the Antarctic samples. For example, Huss (1991) inferred that >300 paired H5 chondrites from the Allan Hills region constitute an otherwise-unrecognized meteorite shower.

The most dramatic differences shown in Table 8.1 are in the percentages of ordinary chondrites among total meteorites: 84% among falls, 57% among NWA finds, and 93% among Antarctic and dry-lake finds.

The low abundance of OCs among NWA meteorite finds relative to falls is likely due to three factors: (1) terrestrial weathering causing higher percentages of metal-bearing OCs to disintegrate than other stony meteorites (which, except for enstatite chondrites, contain less metal), (2) ordinary chondrites preferentially being ignored, discarded, and/or not submitted to the Meteoritical Society's Nomenclature Committee by meteorite hunters (probably mainly the latter), and (3) meteorite researchers being reluctant to analyze and classify ordinary chondrites. (Meteorite hunter Bob Verish told me it was a "dark day" several years ago when he learned that meteorite researchers were no longer going to classify ordinary chondrites routinely.) A consequence of this third factor is the offering of small unclassified NWA meteorites for sale for about US $1.50 per gram on several websites.

The high abundance of OCs among Antarctic finds relative to falls is due to two factors: (1) fragmentation of metal-bearing OC samples coupled with near-quantitative collection of the fragments in the Antarctic environment and (2) the general lack of pairing of individual samples among Antarctic finds, resulting in single meteorite showers being counted as numerous separate meteorites. (In contrast, specimens of observed falls are routinely paired under the same meteorite name.)

The high abundance of OCs among dry-lake finds is due to three factors: (1) the systematic collection of specimens, (2) the relatively young terrestrial ages of the meteorites (leaving them insufficient time to disintegrate), and (3) the general lack of pairing of individual samples (resulting, as in Antarctica, in single meteorite showers commonly being counted as numerous separate meteorites). Nevertheless, there are exceptions: for example, the 104 individual fragments of H5 Primm collected at Roach Dry Lake in Clark County, Nevada were officially paired and counted as a single meteorite.

8.17 Lack of Evidence for Ordinary-Chondrite Meteoroid Streams

An asteroidal meteoroid stream can form when rocks are impact-excavated from an asteroid and dispersed within the asteroid's orbit. Streams are produced because collisions tend to eject rocks from asteroids at velocities on the order of 10 m s^{-1}, much lower than the heliocentric orbital velocities of the asteroids (which are on the order of 10^4 m s^{-1}; e.g., Jopek et al., 2002). Pauls and Gladman (2005) made detailed numerical orbital evolution calculations and found that asteroidal meteoroid streams would become decoherent (i.e., essentially randomized with respect to the fall date) within 10,000 to 100,000 years. This is far shorter than the typical cosmic-ray exposure (CRE) ages of ordinary chondrites (5–50 Ma; Marti and Graf, 1992). Pauls and Gladman concluded that orbital destruction of meteoroid streams is very efficient and that long-lived streams should not exist. In fact, attempts to assign ordinary chondrites to meteoroid streams have been unsuccessful because the orbits of the proposed members had different radiants (points on the celestial sphere from which the meteorite trajectories appear to originate) and, in some cases, significantly different CRE ages (Rubin and Matson, 2008).

Nonetheless, it seems possible that some meteorite falls could be traced to the same meter-sized immediate precursor body (IPB). Morbidelli and Gladman (1998) defined an IPB as a "body on which the meteorite acquired its penultimate exposure stage." Pairs of meteorites recently derived from the same IPB should share orbital characteristics and have similar CRE ages, chemical and isotopic compositions, and metamorphic, alteration, and shock histories.

Rubin and Matson (2008) identified four pairs of equilibrated OC (L6 Nejo, Salem; L6 Perpeti, Vouillé; L6 Drake Creek, Forsyth; H5 Okabe, Kerilis) wherein each member of the pair could conceivably have been derived from the same IPB. The members of each pair are of the same chondrite group and petrologic type; they have statistically indistinguishable olivine Fa compositions; they have similar CRE ages (within 5%) and fell within 1 calendar day of each other (in different years).

It is difficult to estimate the probability that two fragments of the same meter-sized Near-Earth Asteroid (NEA) would result in recovered meteorites, but it is likely to be very low. First, the Earth would have to intersect the precise location in the orbit of a particular NEA from which a recovered meteorite was previously observed to fall. Second, the Earth would have to sweep up one or more of these fragments and have them reach the surface in view of an observer. Out of ~13,700 meteorites with masses >100 g that are estimated to reach the Earth's surface each year (Halliday, 2001), only five or six falls are typically observed and recovered (Hutchison, 2004). (Recovered falls tend to be much larger than 100 g. The mean mass of the 115 OC that fell between 1980 and 2006 for which weights are listed in the Meteoritical Bulletin Database is 28 ± 107 kg.) Recovered meteorites constitute only ~0.04% of the >100-g masses that fall to Earth. It is improbable that one of those few recovered objects would be from the same NEA from which a previous meteorite fall had been observed and recovered.

Dodd et al. (1993) and Wolf and Lipschutz (1995) nevertheless inferred the existence of closely related H-chondrite meteoroid streams that produced falls in late May in the Mid- to Late-nineteenth century. The H chondrites vary in petrologic type from 4 to 6, in calendar date of fall from May 20 to May 27, and in CRE age from 4.2 to 63.6 Ma (Schultz and Weber, 1996). The large differences in their CRE ages alone render it unlikely that they are all members of the same stream. Although Dodd et al. (1993) reported these H chondrites to have abundances of labile trace elements (Rb, Ag, Se, Cs, Te, Zn, Cd, Bi, Tl, In) that are distinguishable from H chondrites that are not members of the proposed streams, Rubin and Matson (2008) found that, for each of these elements, the arithmetic means of the proposed stream and non-stream H-chondrite sets fall well within one standard deviation of the higher mean value. A simple statistical analysis shows the probability is low that the two sets of H chondrites are from different populations.

The similarities in the properties of the four pairs of ordinary chondrites selected by Rubin and Matson (2008) are likely coincidental. The meteorites were chosen from a list of nearly 400 OC with known CRE ages and fall dates. In addition, many members of each OC group have similar mean olivine Fa compositions (Rubin, 1990). There is no compelling evidence for the existence of meteoroid streams.

9

Terrestrial Weathering

Among the hills a meteorite
Lies huge; and moss has overgrown,
And wind and rain with touches light .
Made soft, the contours of the stone.

Thus easily can Earth digest
A cinder of sidereal fire,
And make her translunary guest
The native of an English shire.

C. S. Lewis (1946), "The Meteorite,"

9.1 Formation of Secondary Phases and the Utility of Weathering Scales

The breakup of incoming meteoroids during atmosphere passage produces small fragments with high surface-area/volume ratios, rendering them more susceptible to chemical weathering after they reach the Earth's surface. Upon landing, meteorites are contaminated with fungi and bacteria as well as other organic and inorganic materials. After a while (depending on the continent and local climate), different species of lichens may colonize the meteorite surface (e.g., H5 Wiltshire, also known as Lake House; J. Pillinger et al., 2011). The ubiquitous presence of water (atmospheric moisture, precipitation, groundwater, ponds, lakes, rivers, and oceans) causes alteration of indigenous mineral phases. (Even in the driest places on Earth, the air contains some moisture. For example, on June 20, 2017, Safi-Abad Dezful in western Iran hit a temperature of 46.5°C [~116 °F] with a relative humidity of 0.35%, close to the world record for dryness.)

Freshly recovered falls often rust in museum cabinets. Velbel (2014) found that the "oxidation, hydrolysis, hydration, and product-forming phenomena documented to affect OC finds have been documented to continue in OC samples in curatorial and laboratory settings, producing mineralogical and textural effects at scales easily discernable by electron microscopy, on timescales of decades."

Individual exposed metallic Fe-Ni grains on fresh falls can develop thin red coatings on both slabs and whole-rocks; this phase is mainly goethite ($\alpha FeO(OH)$), formed via oxidation from atmospheric water. This reaction increases the volume of the metal grains by 67 to 80% (as the density of goethite is much lower than that of metallic Fe: 3.3–4.3 vs. 7.6–7.8 g cm^{-3}). Kamacite is more susceptible to weathering than taenite. In both falls and finds, weathering rinds around metal grains that appear light gray in reflected light are commonly referred to as "limonite" even though this is not an IMA-approved mineral. In samples that are more weathered, troilite also is partially transformed into Fe oxides and hydroxides. Weathering effects tend to decrease from the surface to the interior of the samples.

Bland et al. (1998a) used Mössbauer spectroscopy to show that metallic Fe in OCs from hot deserts weathers into two distinct mineral varieties: paramagnetic phases (such as akaganéite, lepidocrocite, and goethite) and magnetically ordered phases (mainly magnetite and maghemite). Barium sulfate ($BaSO_4$), carbonates, and Fe oxides also occur in some OC finds from the Acfer region of Algeria (Stelzner et al., 1999).

Many meteorite finds are extensively altered (e.g., Nininger, 1948; Buddhue, 1957). Metallic Fe-Ni and sulfide show progressive alteration with time. This is reflected by the weathering index for

Table 9.1 *Weathering index for ordinary chondrites.*

Weathering degree	Metal oxidation (vol.%)	Troilite oxidation (vol.%)	Comments
W0.0	0	0	Fresh; some iron hydroxide staining possible.
W1.0	<20	≥0	Minor oxide rims around metal and troilite; small iron oxides and iron hydroxide veins may occur.
W2.0	20–60	<20	Onset of veining with iron oxides and iron hydroxides.
W3.0	>60	<20	Strong oxidation of metal. Troilite shows minor alteration.
W3.3	>60	20–60	Strong oxidation of metal. Troilite moderately altered; some troilites completely oxidized.
W3.6	>60	>60	Strong oxidation of metal and troilite. Most troilites oxidized or have reduced reflectivity.
W4.0	>95	>95	Nearly complete oxidation of metal and troilite; some troilite remnants may be present.
W4.5	100	100	All metal and troilite oxidized; only minor remnants of metal and troilite as inclusions in silicates. Some silicate alteration (mainly olivine) possible.
W5.0	100	100	Metal and troilite 100% oxidized. Major alteration of silicates, mainly olivine.
W6.0	100	100	Massive replacement of silicates by clay and oxides.

After Zurfluh et al. (2016), used with permission.

ordinary chondrites devised by Wlotzka (1993) and refined by Zurfluh et al. (2016) (Table 9.1). At high degrees of weathering (approximately beginning at stage W4.5), mafic silicates are transformed into phyllosilicates.

Bland et al. (1998b) gauged the rapidity of alteration of ordinary chondrites in semi-arid environments. They analyzed specimens of L/LL6 Holbrook (which fell as a large shower of stones in northern Arizona in 1912) by ^{57}Fe Mössbauer spectroscopy to determine the proportion of total Fe that had converted to Fe^{3+}. Less than 1% of the total iron had been converted to ferric iron in samples recovered shortly after the fall in 1912; by contrast, in samples recovered in 1968, 9.7% of the total Fe had been converted to Fe^{3+}.

If a large meteorite is partially buried, groundwater can preferentially weather the buried portion, leaving the above-ground portion relatively unaltered (e.g., Fig. 9.1).

It seems likely that weathering is a two-stage process. The initial phase of rapid weathering lasts less than 500 years and typically does not destroy the rock's integrity. During this stage, metallic Fe is weathered and transformed into goethite. There is an appreciable increase in volume as goethite begins to fill pores and cracks in the meteorite. As the extent of weathering of the meteorite increases, the whole-rock porosity decreases toward zero. Falls with longer terrestrial residence times tend to have lower porosities; almost all OC finds have zero porosity (Britt and Consolmagno, 2003). At that point, the grain volume (the volume of only the solid material in a sample) equals the bulk volume (the entire volume of the sample including solid material and pore space) (Consolmagno et al., 2008). The second stage of weathering proceeds slowly because the remaining metallic Fe-Ni grains can be reached by oxygen and water mainly through solid-state diffusion through mineral grains. This process eventually leads to disintegration of the rock.

Antarctic meteorites are traditionally assigned weathering categories (A, B, or C) based on the degree of surface rustiness (Table 9.2). Samples with light-colored evaporite minerals at their surfaces are labeled Ae, Be, or Ce. The evaporites include Mg carbonates (nesquehonite, hydromagnesite, amorphous Mg carbonate), Mg sulfates (epsomite, starkeyite), Ca sulfate (gypsum) and additional, poorly characterized, K-, Fe- and Mg-sulfates. The original meteoritic minerals provide the cations for

Table 9.2 *Weathering categories for Antarctic meteorites.*

Weathering category	Description	Comments
A	Minor rustiness	Minor rust haloes on metal particles; minor rust stains along fractures.
B	Moderate rustiness	Thick rust haloes on metal particles; extensive rust stains on internal fractures.
C	Severe rustiness	Near-complete rust staining of metal particles.
e	Evaporite minerals	Evaporite minerals visible to naked eye.

After the Antarctic Meteorite Newsletters.

Fig. 9.1 Large (205 kg) H4, S1 chondrite from Mauritania in Northwest Africa on display at the UCLA Meteorite Gallery. The lighter-colored portion of the stone was below the soil surface, subject to extensive terrestrial weathering. Analysis showed this portion to be of weathering stage W5. The top portion of the stone was above the soil and appears darker and less weathered. Flow lines are evident across the entire surface of the meteorite. Photograph courtesy of Peter Utas.

these secondary phases (specifically, Mg, Na, K, Ca, and Rb derive from the interior of the specimens). Much of the CO_2 in the carbonates is likely derived from the terrestrial atmosphere.

As discussed by Rubin and Ma (2021), weathering of the fusion crusts of some Antarctic meteorites produced poorly ordered clay minerals similar in composition to smectite and mica. Insolation heating of dark fusion-crusted Antarctic meteorites on calm summer days can locally raise the temperatures beneath the exposed surfaces to a few degrees above freezing, allowing the presence of liquid water. Freezing of this water can cause evaporite minerals to precipitate.

Altogether, there are about 70 minerals identified as having formed in meteorites by terrestrial weathering. Those formed in ordinary chondrites include: (1) oxides and hydroxides formed from metallic Fe-Ni by oxidation and the incorporation of H_2O, Cl^{-1}, and CO_3^{-2} and (2) sulfates formed from troilite. Numerous secondary minerals were formed in fossil L chondrites (as discussed in Section 8.12).

Color images of weathered meteorites, including many ordinary chondrites, appear in Fodor (1976), Yanai (1981), Yanai and Kojima (1987), Killgore and Killgore (2002), Stinchcomb (2011), and Cressy (2016). Images of weathered ordinary chondrites for sale also appear regularly on eBay.

9.2 Chemical and Isotopic Changes Caused by Weathering

Jarosewich (1990) compiled the bulk chemical compositions of 241 stone meteorites (including 144 ordinary chondrites) and 36 irons determined by standard wet-chemical analytical methods using aliquants of ~10-g powdered samples. He compared the concentrations of major elements in observed falls to those in Antarctic and non-Antarctic finds.

As shown in Tables 9.3a, 9.3b, 9.3c there are several compositional changes caused by terrestrial weathering:

1. Finds have been contaminated with terrestrial water. More chemically bound water (H_2O^+) derived from the terrestrial environment has been absorbed by finds than by fresh falls.
2. Finds have been contaminated with terrestrial carbon. Data show that H- and L-chondrite finds have bulk C concentrations as high as or higher than those of falls. Antarctic LL finds are an exception to the trend, but with only four analyzed samples, this apparent anomaly may be spurious.
3. Metallic Fe has been oxidized to Fe_2O_3. The bulk concentration of metallic Fe is lower in finds than in falls in H and L chondrites. The sole apparent exception to the trend is in non-Antarctic LL finds, but with only two analyzed samples, the reported value is probably nonrepresentative. Note that the metallic Fe concentrations listed for finds in Tables 9.3a, 9.3b, 9.3c have been recalculated from the Fe_2O_3 concentrations (Jarosewich, 1990), so the actual concentrations of metallic Fe in finds are significantly lower than shown in the table. Some OC finds contain high concentrations of Fe_2O_3: for example, H3.8-an Willaroy (found in New South Wales, Australia in 1970) contains 22.50 wt.% Fe_2O_3 and only 1.35 wt.% remaining metallic Fe. H5 Oro Grande (found in New Mexico in 1971) contains 21.25 wt.% Fe_2O_3 and only 0.84 wt.% metallic Fe. (In contrast, H-chondrite falls average 15.98 wt.% metallic Fe and little or no Fe_2O_3.)
4. Nickel has been oxidized to NiO. The mean bulk Ni concentration of OC finds tends to be lower than that of falls. The only exception is for H-chondrite Antarctic finds, where the value is slightly higher (but well within uncertainty of being identical) to that of falls (1.75 vs. 1.74 wt.%). But again, with only four analyzed samples, the set of Antarctic H-chondrite finds may not be representative.

In addition, greater degrees of weathering in hot-desert environments appear to increase the uniformity of bulk abundances of highly siderophile elements (HSEs) (Almas et al., 2024). This effect may be due to the destruction of labile phases and the redistribution of their constituent HSEs.

The surfaces of some OCs in Antarctica (a cold desert) are greatly enriched in iodine relative to the interiors of these rocks (e.g., 11.2 μg/g vs. 0.5 ng/g in an L6 chondrite analyzed by Ebihara et al. [1989]). Pourkhorsandi et al. (2021) reported that most Antarctic ordinary chondrites have slightly lower concentrations of rare earth elements (REE) than mean falls (1 μg/g = 1 ppm; 1 ng/g = 1 ppb).

Stelzner et al. (1999) found there to be significant enrichments in Ba and Sr in the outer regions (up to 126 μg/g Ba and 45.5 μg/g Sr) of weathered H5 chondrites from the Acfer region of Algeria. The interior regions of these specimens average ~4 μg/g Ba and 9 – 10 μg/g Sr. For comparison, the Earth's crust averages 584 μg/g Ba and 333 μg/g Sr (Wedepohl, 1995).

Pourkhorsandi et al. (2021) reported depletions in Sr, Hf, and Th, and enrichments in Ba and U in hot-desert finds caused by weathering. The REE in these meteorites have lower CI-normalized abundances than falls; many weathered samples have fractionated patterns showing light rare earth elements (LREE) enrichments.

Kallemeyn et al. (1989) determined the bulk compositions of 66 ordinary chondrites by replicate instrumental neutron activation analysis (INAA). Although only a few finds were included, there was a tendency for the finds to have 5–10% lower concentrations of alkalis (Na and K) and, in some cases, lower Ni and Co. Whereas mean L-chondrite falls contain 7.02 mg/g Na, 858 μg/g K, 12.3 mg/g Ni, and 600 μg/g Co, L3.6 Julesburg (found in Colorado in 1983) contains 5.80 mg/g Na, 788 μg/g K, 12.4 mg/g Ni, and 534 μg/g Co (a relatively unweathered interior sample was analyzed), L4 Barratta (found in New South Wales, Australia in 1845) contains 6.66 mg/g Na, 738 μg/g K, 11.0 mg/g Ni, and 609 μg/g Co, and L3.5 ALHA77011 (found in Allan Hills, Antarctica in 1977) contains 6.04 mg/g Na, 730 μg/g K, 11.2 mg/g Ni, and 550 μg/g Co. The relatively low alkalis in the finds are likely

Table 9.3a *Mean concentrations of major chemical components in H-chondrite falls and finds determined by wet-chemical analysis.*

–	H-chondrite falls	H-chondrite Antarctic finds	H-chondrite non-Antarctic finds
Number of samples	26	4	19
H_2O^+ (wt.%)	0.32 ± 0.44	1.09 ± 0.82	1.57 ± 0.13
Metallic Fe (wt.%)	15.98 ± 1.13	14.85 ± 1.79	15.08 ± 1.93
Ni (wt.%)	1.74 ± 0.09	1.75 ± 0.07	1.62 ± 0.13
C (wt.%)	0.11 ± 0.18	0.16 ± 0.15	0.12 ± 0.08

Data from Jarosewich (1990). In this table, Fe_2O_3 in finds was converted to metallic Fe and normalized to 100%.

Table 9.3b *Mean concentrations of major chemical components in L-chondrite falls and finds determined by wet-chemical analysis.*

–	L-chondrite falls	L-chondrite Antarctic finds	L-chondrite non-Antarctic finds
Number of samples	54	11	12
H_2O^+ (wt.%)	0.37 ± 0.45	0.51 ± 0.61	0.84 ± 0.60
Metallic Fe (wt.%)	7.03 ± 0.95	5.59 ± 1.66	6.39 ± 1.50
Ni (wt.%)	1.24 ± 0.10	1.20 ± 0.10	1.20 ± 0.09
C (wt.%)	0.12 ± 0.10	0.12 ± 0.30	0.14 ± 0.12

Data from Jarosewich (1990). In this table, Fe_2O_3 in finds was converted to metallic Fe and normalized to 100%.

Table 9.3c *Mean concentrations of major chemical components in LL-chondrite falls and finds determined by wet-chemical analysis.*

–	LL-chondrite falls	LL-chondrite Antarctic finds	LL-chondrite non-Antarctic finds
Number of samples	12	4	2
H_2O^+ (wt.%)	0.51 ± 0.55	0.53 ± 0.43	1.72
Metallic Fe (wt.%)	2.44 ± 1.61	2.21 ± 0.90	5.53
Ni (wt.%)	1.07 ± 0.13	0.71 ± 0.29	0.90
C (wt.%)	0.22 ± 0.21	0.05 ± 0.07	0.35

Data from Jarosewich (1990). In this table, Fe_2O_3 in finds was converted to metallic Fe and normalized to 100%.

attributable mainly to the leaching of chondrule mesostasis in the type-3 and -4 samples. The low Ni and Co concentrations in some finds are due to the weathering of metallic Fe-Ni-Co grains.

Dennison and Lipschutz (1987) reported radiochemical neutron activation analysis (RNAA) data for 14 siderophile, lithophile, and chalcophile elements in 45 H4, H5 and H6 chondrites from Victoria Land, Antarctica. They found statistically significant differences between samples that experienced minor-to-moderate weathering (A and B) and those that experienced moderate to severe weathering (B/C and C). The more weathered samples have lower concentrations of 10 elements: Co, Au, Sb, Rb, Ce, Bi, In, Zn, Tl, and Cd (Table 9.4), probably caused by leaching of the meteorites at the surface of the ice sheet.

Maeda et al. (2023) studied a suite of H chondrites, ranging from fresh falls to highly weathered finds from hot deserts. They found fractures within the weathered specimens to be enriched in REE, Th, and U; fractures in severely weathered specimens also contain Ce. They interpreted these enrichments as having been caused by terrestrial weathering following chemical leaching.

Weathering affects the abundance of noble gases in two ways: (1) heavily weathered specimens tend to be depleted in light noble gases (e.g., He) and (2) samples from hot deserts can become contaminated with heavy noble gases (Kr, Xe) from the terrestrial atmosphere that are carried by surface water into the meteorite interior via fractures (Stelzner et al., 1999).

Terrestrial weathering has also changed the bulk isotopic compositions of ordinary-chondrite finds. For example, Clayton et al. (1991) found that OC finds tend to have heavier O isotopes (i.e., they are enriched in ^{17}O and ^{18}O) (Table 9.5). To restore their initial O-isotopic compositions (at least partially), the finds were treated with 6 M HCl at 70°C for two minutes to remove the iron-oxide and -hydroxide weathering products. Acid washing of OC finds tends to lower their $\delta^{17}O$, $\delta^{18}O$, and $\Delta^{17}O$ values, bringing them closer to the means of falls in their respective OC groups.

Table 9.4 *Differences in bulk composition between mildly weathered* (A + B) *and moderately to severely weathered* (B/C + C) *H4, H5, and H6 chondrites from Victoria Land, Antarctica.*

–	A + B	B/C + C
Co (μg/g)	**925** ± 110	**801** ± 140
Au (ng/g)	**217** ± 9	**192** ± 39
Sb (ng/g)	**87** + 25, − 20	**69** + 14, −12
Rb (μg/g)	**2.07** + 0.44, − 0.36	**1.81** + 0.51, − 0.40
Ce (ng/g)	**93** + 220, − 65	**24** + 69, − 18
Bi (ng/g)	**4.7** + 10, − 3.2	**0.55** + 1.3, − 0.39
In (ng/g)	**0.53** + 2.8, − 0.44	**0.12** + 0.33, − 0.09
Zn (μg/g)	**46.3** ± 4.4	**41** ± 11
Tl (ng/g)	**1.6** + 7.0, − 1.3	**0.30** + 0.57, − 0.20
Cd (ng/g)	**1.2** + 4.1, − 0.90	**0.55** + 1.1, − 0.36

The uncertainties are for normal distributions (one value listed) or log-normal distributions (two values listed). Mean values are in bold.
Data from Dennison and Lipschutz (1987).

Two type-3 OC finds can serve as an illustration of the effects of acid-washing on weathered OC (Table 9.6). LL3.4 Ragland was found in New Mexico in 1978 or 1982; L/LL3.4 Inman was found in Kansas in 1966.

Ash and Pillinger (1995) found that one hot-desert ordinary chondrite (Adrar 003, L/LL3.10) has low, terrestrial-like, δD values of −82 to −68‰ (caused by terrestrial contamination) compared to those of type-3 OC falls (e.g., LL3.4 Chainpur has δD values of +58 to +160‰).

In hot-desert OC finds, the initial Sm/Nd ratios were disturbed during weathering. Meteoritic components were contaminated by soil materials, resulting in lower $^{147}Sm/^{144}Nd$ ratios in the meteorites (Pourkhorsandi et al., 2021).

Iron isotopes were also affected by weathering. Saunier et al. (2010) studied hot-desert H5 and L6 chondrites and found that iron isotopes tend to become slightly heavier with increasing degrees of weathering. This is presumably due to the leaching of isotopically light Fe^{2+} from heavily weathered samples. The least weathered OC finds (W ≤ 2) have mean bulk $\delta^{57}Fe/^{54}Fe$ ratios as low as −0.017 ± 0.027‰, whereas W5 OC have $\delta^{57}Fe/^{54}Fe$ ratios as high as 0.098 ± 0.069‰.

Table 9.5 *Bulk O-isotopic compositions of equilibrated ordinary-chondrite falls and finds.*

–	Number of samples	$\delta^{18}O$ (‰)	$\delta^{17}O$ (‰)	$\Delta^{17}O$ (‰)
H4,5,6 fall	22	4.08 ± 0.22	2.85 ± 0.15	0.73 ± 0.09
H4,5,6 finds	8	4.38 ± 0.33	3.05 ± 0.19	0.77 ± 0.06
L4,5,6 falls	26	4.70 ± 0.24	3.52 ± 0.14	1.07 ± 0.09
L4,5,6 finds	11	4.77 ± 0.40	3.58 ± 0.25	1.11 ± 0.13
LL4,5,6 falls	20	5.04 ± 0.24	3.88 ± 0.16	1.26 ± 0.12
LL5,6-7 finds	2	5.18	3.92	1.22

Data from Clayton et al. (1991).

Table 9.6 *Differences in bulk O-isotopic composition between unwashed and acid-washed type-3 ordinary-chondrite finds.*

–	$\delta^{18}O$ (‰)	$\delta^{17}O$ (‰)	$\Delta^{17}O$ (‰)
Ragland unwashed	6.44	4.21	0.86
Ragland washed	5.29	3.49	0.74
Inman unwashed	6.01	3.98	0.85
Inman washed	5.44	3.65	0.82

Data from Clayton et al. (1991).

9.3 Changes in Bulk Physical Properties

9.3.1 Intrinsic Properties

Intrinsic properties of a rock are independent of sample size. There are many intrinsic physical properties including hardness, luster, color, taste, specific gravity, porosity, grain density, petrofabric, compressive strength, tensile strength, melting point, boiling point, speed of sound, Young's modulus, bulk modulus, shear modulus, specific heat capacity, thermal conductivity and diffusivity, magnetic susceptibility (at specified temperatures below the Curie point), and electrical conductivity and resistivity.

The first intrinsic property of a meteorite to change during weathering is porosity (ϕ). It decreases from moderate values in observed falls (7.0 ± 4.9% in H, 5.6 ± 4.7% in L, and 8.2 ± 5.5% in LL) to 0% in finds of all three OC groups (Consolmagno et al., 2008). Grain density (ρ_{grain}), defined as the (mass of the meteorite)/(grain volume), decreases as pore spaces are filled with oxide and hydroxide grains (Britt and Consolmagno, 2003). (Pore spaces do not contribute to grain volume.)

Magnetic susceptibility (X_m) is the degree to which a material becomes magnetized in the presence of an applied magnetic field. X_m increases in OC falls from LL to L to H chondrites, coinciding with increasing proportions of metallic Fe-Ni. The X_m value is often used in ordinary-chondrite classification (e.g., Rochette et al., 2003). But weathering lowers the modal abundance of metallic Fe-Ni in OCs as metal is transformed into oxides and hydroxides. This lowers the magnetic susceptibility and can result in H chondrites exhibiting L-chondrite-like X_m values, and L chondrites exhibiting LL-chondrite-like X_m values (Flynn et al., 2018).

Flynn et al. (2018) pointed out that weathering not only reduces the porosity of ordinary chondrites, but also affects other intrinsic properties. For example, terrestrial weathering lowers the speed of sound, thermal conductivity, compressive strength, and tensile strength of ordinary chondrites.

In contrast, there are no significant differences between falls and finds in average bulk density. This is due to the pores in pristine falls becoming filled with oxides and hydroxides. The average densities of H-chondrite falls and finds are both 3.4270 g cm^{-3}; the average densities of L-chondrite falls and finds are both 3.3770 g cm^{-3} (Consolmagno et al., 2008).

9.3.2 Extrinsic Properties

Extrinsic properties of a rock include size, mass, shape, and orientation in three-dimensional space. The terrestrial environment can affect all these properties. Chemical and mechanical weathering can break a meteorite into small fragments, changing its size, mass, shape, and orientation. Mass wasting (downslope movement of rock and soil under the influence of gravity) can change a meteorite's orientation and potentially subject it to fracturing and crumbling.

Sand blasting, a form of aeolian erosion in arid and semi-arid environments, can abrade rocks at the surfaces of Earth and Mars (e.g., Laity and Bridges, 2009; Herkenhoff et al., 2023). In the terrestrial environment, meteorites are pelted and degraded by airborne or saltating quartz grains in hot deserts and ice particles in cold deserts, a process known as corrasion. In arid regions, wind gusts are unimpeded by vegetation. Wind-sculpted rocks include ventifacts (also known as windkanters), which typically have three or four flat (or flattish) sides that meet at angular interfaces (Fig. 9.2). Many samples have been pitted, etched, or wind-polished, and some display subparallel grooves.

Kamp and Lowe (1982) described iron-meteorite ventifacts from Antarctica and Zurfluh et al. (2012, 2016) described ordinary-chondrite ventifacts from hot deserts in Oman. The Oman specimens are strongly polished and display tiny "monadnocks" – ridges of more wind-resistant material protruding from the meteorite surface (Fig. 9.3).

Another OC ventifact is the H5 chondrite NWA 15527 (Fig. 9.4). This massive specimen (32.34 kg),

Fig. 9.2 Ventifact of a terrestrial basaltic rock from the Mojave Desert near Barstow, California. Image from Wikipedia (Creative Commons) with permission from photographer Mark A. Wilson (Department of Geology, The College of Wooster).

Fig. 9.3 Ventifact of a fractured specimen of Ramlat as Sahmah 418 (L6 S5 W3.3-4.0), the dark rock at the right, along with terrestrial ventifacts in Al Wusta, Oman. Black and white rectangles on the scaled paper strip next to the meteorite are 1 cm in length. Image courtesy of Florian Zurfluh, modified from Zurfluh et al. (2012).

found by nomads in northwest Africa in 2020 or 2021, has partially faceted, low-slope, undulating sides, some of which meet at acute angles. Partly eroded regmaglypts are visible on the surface.

9.3.3 Shatter Cones

Shatter cones are conical rock structures with striated fractures having a horsetail-like appearance. The striations appear to radiate from the cone apex. These structures were first identified by Branco and Fraas (1905) in the Steinheim Basin in Germany (a 3.8-km-diameter impact crater formed ~14.3 Ma ago); they have since been found associated with many terrestrial impact craters (e.g., Dietz, 1947, 1959, 1963, 1968). Shatter cones generally occur in place within rocks beneath the crater floor, exposed by subsequent erosion or uplift (French, 1998).

Impact experiments (e.g., Roddy and Davis, 1977) have shown that in situ shatter cones point roughly toward the hypocenter (i.e., ground zero, the point on the ground surface at which the explosion was centered) and that the axes of the cones are oriented approximately normal to the shock front. Shatter cones are produced at relatively low shock pressures, perhaps 2–10 GPa.

Fig. 9.4 NWA 15527, an H5 chondrite from Northwest Africa. The stone is a ventifact; its shape was, in part, sculpted by the wind. Shallow regmaglypts occur on the front side. Image courtesy of Christie's; photo by Richard Goodbody and Darryl Pitt.

Osinski and Ferrière (2016) concluded that shatter cones form by tensional stresses produced by shock waves scattered by heterogeneities in the target rocks. Well-formed cones have been found in sandstone, carbonate, and various igneous and metamorphic rocks; the cones range in size from a few millimeters to >10 m (e.g., Sharpton et al., 1996). The existence of shatter cones in a natural geologic structure is diagnostic of its impact origin. Shatter cones have been inconclusively identified on the Moon (e.g., Ulrich et al., 1981) and Mars (Newsom et al., 2015). They have also been produced artificially in nuclear explosions (Bunch and Quaide, 1968) and large-scale conventional explosions (Roddy and Davis, 1977).

Dietz (1966) searched through several institutional meteorite collections and identified 10 ordinary chondrites (5 H, 4 L, 1 LL) sporting striated fractures that resemble shatter cone features (not slickensides caused by fault dislocations in space; Farrington, 1915). McHone et al. (2012) described two NWA ordinary chondrites as definitive whole-rock shatter cones (Fig. 9.5). The larger of these

Fig. 9.5 Two NWA ordinary chondrites that are whole-rock shatter cones, replete with characteristic horsetail structures. (Left) A 1.74-kg H4-5 chondrite. (Right) An 851-g L5 chondrite. Ruler calibrated in inches and centimeters. Photo courtesy of Kitty Killgore. Samples initially described by McHone et al. (2012).

ordinary chondrites (1.74 kg) is unofficially classified as H4-5, W1-2. About 20% of its surface has the horsetail appearance of a shatter cone; the cone structure extends into the meteorite interior. The smaller NWA stone (851 g) is unofficially classified as L5, S5-6, W3-4. It contains numerous shock veins; its shatter-cone structure also appears to extend into the rock's interior.

Ferrière et al. (2013) described well-developed multi-centimeter-sized shatter cones in H4 Ybbsitz and L6 Zavid. Ferrière et al. (2025) collated data from several meteorite collectors and reported that similarly-sized shatter cones are fairly abundant and occur mainly in ordinary chondrites (H, L, and LL), including L6 L'Aigle, H5 Pultusk, and LL5 Chelyabinsk. Four H5 chondrites are currently listed in the Meteoritical Bulletin Database as having shatter-cone or partial shatter-cone structures: Tocache (Peru, 1998 fall), NWA 16420 (Morocco, 2023 find), Degtevo (Russia, 2016 fall) and Montferré (France, 1923 fall).

Because terrestrial shatter cones are produced beneath the floors of impact craters, individual ordinary-chondrite shatter cones are plausibly fragments of preexisting crater-floor material that were excavated from an asteroid and launched into interplanetary space. At least two major collisions were involved in their production and ejection.

9.4 Bric-à-Brac and Collectibles

Another way ordinary chondrites can be physically and chemically modified on Earth is by their intentional fashioning into novelty items (memorabilia) for collectors and hobbyists. In addition to earrings, necklaces, keychains, pendants, and bracelets (often available on eBay), an 83 × 44 × 64-mm sculpture of a human skull carved from an unclassified NWA ordinary chondrite was offered in 2019 by Andrew Jones Auctions.

Small pieces of meteorites, including ordinary chondrites, have been chipped from larger masses and encased within legal-tender commemorative coins (e.g., Figs. 4.1a, 8.17) sold to collectors (e.g., Rubin, 2004c, 2013c).

Pegasus Distillery in Burgundy, France recently crafted "Shooting Star Vodka," aged for up to 18 months in a terracotta amphora within which a piece of the Huntsman (b) H3 chondrite is suspended (e.g., Diaz, 2024; Sauers, 2024). The meteorite was found in 1977 as a single 10.215-kg stone on a family farm in Nebraska; the rock has been strongly shocked (S5) but only slightly weathered (W1). Alcohol seeps through the porous matrix of the suspended type-3 chondrite and fills cracks near the meteorite surface. The vodka in turn is infused with soluble phases and small particles dislodged from the meteorite. Shooting Star Vodka was initially priced at $180 to $200 a bottle.

10
Ordinary-Chondrite Groups

10.1 Establishment of the Ordinary-Chondrite Groups

This brief history of meteorite classification and the recognition of ordinary chondrites is adapted from Rubin and Ma (2021). In 1862, Gustave Rose at the Mineralogical Museum of the University of Berlin distinguished stones from irons and divided stones into seven categories: (1) chondrites (specimens containing the well-known spherulitic objects he dubbed "chondrules"), (2) howardites, (3) chassignites (with only a single member at the time – Chassigny), (4) chladnites (now known as aubrites, also with a single member – Bishopville), (5) shalkites (now known as diogenites, again with only one member – Shalka), (6) carbonaceous chondrites, and (7) eucrites. Rose subdivided the chondrites into groups on the basis of color or structure.

Gustav Tschermak at the University of Vienna modified the Rose classification system in 1872 and 1883. Between 1885 and 1904, Aristides Brezina, also at the University of Vienna, created more than 70 meteorite categories, based mainly on minor distinctions. Secondary characteristics (e.g., those resulting from metamorphism or shock) were used to separate related samples from one another. At the beginning of the twentieth century, the classification scheme in wide use was the Rose–Tschermak–Brezina system. In 1929, George Merrill, the head curator of geology at the Smithsonian Institution, used this system to differentiate 11 distinct groups of chondrites: howarditic, white, intermediate, gray, black, spherulitic, crystalline, carbonaceous, orvinite, tadjerite, and urelite. Many of these groups were subdivided on the basis of texture. (Many classifications from this era are easily translatable into modern terms: Merrill would have described an H5 chondrite with shock veins as a veined crystalline bronzite chondrite, an L3 chondrite as a spherical hypersthene chondrite, and a brecciated LL6 chondrite as a breccia-like gray amphoterite crystalline chondrite.)

Refinements to the Rose–Tschermak–Brezina classification system were made by Oliver Farrington (Field Museum), George Prior (British Museum), Frederick Leonard (UCLA), and Brian Mason (American Museum, later the Smithsonian Institution). After the classification system had been moderately cleaned up, Harold Urey and Harmon Craig (both from the University of Chicago) compiled superior chemical analyses of 94 chondrites (Urey and Craig, 1953). They showed that the data could be arranged into two distinct groups that differed in total iron content and oxidation state. They called the groups H (high total iron) and L (low total iron) and concluded that they had probably originated on different parent bodies. Klaus Keil (University of California, San Diego) reproduced the two groups based on specific gravity and modal abundance of metallic Fe-Ni (Keil, 1962). Two years later, Keil and Kurt Fredriksson (also University of California, San Diego) analyzed olivine and orthopyroxene in 86 ordinary chondrites by electron microprobe (Keil and Fredriksson, 1964). They reported distinct differences between the H and L groups and identified a third OC group – LL (low total iron, low metallic Fe).

Through the 1980s, H, L, and LL chondrites were occasionally referred to as olivine-bronzite chondrites, olivine-hypersthene chondrites, and amphoterites, respectively (e.g., Heide and Brückner, 1968).

This reflected differences between H and L in the FeO/(FeO+MgO) ratio of orthopyroxene (hypersthene has a higher ratio than bronzite), following the recommendations of Prior (1916). (The boundary between bronzite and hypersthene was different in Prior's time [Fs_{20}] from today [Fs_{30}], when these terms are seldom used.) The term amphoterite was inherited from the Rose–Tschermak–Brezina classification system, where it was listed as a type of achondrite (with achondrite simply meaning a meteoritic stone poor in metallic Fe and having few or no chondrules). Prior (1916) used the term amphoterite for meteorites such as LL6 Jelica and LL6 Manbhoom (which he equated with the Soko-Banja chondrite type; Soko-Banja is currently classified as LL4). The term amphoterite is a metaphor derived from amphoteric chemical compounds – substances that can react either as an acid or a base. The Greek word *amphoteroi* (αμφότεροι) means "both" and is the root of the modern English "amphibole," "amphibian," and "amphitheater." (Its Latinized counterpart *ambi-* is present in words including "ambivalent," "ambiguous," and "ambidextrous.") Presumably, the rationale for calling LL chondrites amphoterites relates to the intermediate nature of such rocks: although they have chondritic structure, they contain low amounts of metallic Fe-Ni, similar to the abundances in igneous meteorites – the achondrites (which lack chondrules).

The terms involved in classification were used inconsistently in this era. For example, Hey's (1966) *Catalogue of Meteorites* described several OC (e.g., L/LL4 Albareto, LL4 Benares (a), LL6 Mangwendi, LL6 St. Mesmin, LL4 Soko-Banja) as "olivine-hypersthene chondrites (amphoterites)," regarding amphoterites as a subgroup of olivine-hypersthene chondrites. Dietz (1966) classified LL6 Vavilovka as "a monomict brecciated achondrite (amphoterite)." These works came out two years after LL chondrites had been defined as a separate group by Keil and Fredriksson (1964), indicating that the new group had not yet been universally adopted.

Mason (1962) created a new category of olivine-pigeonite chondrites, cobbled together from the ornansites and spherulitic carbonaceous chondrites from the Rose–Tschermak–Brezina system. Although Anders (1964) was skeptical of this group's validity, Hey (1966) adopted the category and included 15 meteorite falls now classified as LL3 (e.g., LL3.4 Chainpur, LL3.2 Krymka, LL3.6 Ngawi); H/L3 (H/L3.6 Tieschitz); CV3 (e.g., $CV3_{OxB}$ Bali, $CV3_{OxB}$ Grosnaja, $CV3_{OxB}$ Kaba, $CV3_{OxB}$ Mokoia, $CV3_{Red}$ Vigarano); CK4 (Karoonda); and CO3 (e.g., CO3.3 Felix, CO3.2 Kainsaz, CO3.5 Lancé, CO3.4 Ornans, CO3.7 Warrenton). These meteorites all contain highly visible chondrules. Except for Tieschitz, they all have noticeably higher olivine/low-Ca-pyroxene modal ratios, more-ferroan olivine, and lower abundances of metallic Fe-Ni than typical H (olivine-bronzite) chondrites.

(However, Mason [1962] had concluded the low-Ca pyroxene in these rocks [present at the ~5 vol.% level] to be pigeonite [$Wo_{5–20}$], not hypersthene [$Wo_{0–5}$], based solely on refractive index measurements. In fact, it is *not* the case that these rocks tend to contain pigeonite instead of the low-Ca pyroxene that lies within the enstatite-ferrosilite field: for example, low-Ca pyroxene in Krymka clasts averages $Wo_{0.83}$ [Semenenko et al., 2001]; low-Ca pyroxene in Chainpur averages $Wo_{\sim 0.8}$ [Keil et al., 1964]; low-Ca pyroxene phenocrysts in Ornans chondrules average $Wo_{1.9}$ [Rubin and Wasson, 1988]; and low-Ca pyroxene in a Bali-like clast in Vigarano has $Wo_{0.9}$ [Krot et al., 2000].)

The subsequent edition of the *Catalogue of Meteorites* (Hutchison et al., 1977) omitted the olivine-pigeonite chondrite category. Chainpur, Krymka, and Ngawi were listed as amphoterites; Tieschitz as an olivine-bronzite chondrite; Bali, Grosnaja, Kaba, Mokoia, and Vigarano as Type III-V (Vigarano subgroup) carbonaceous chondrites; and Karoonda as a Type III-O (Ornans subgroup) carbonaceous chondrite along with Felix, Kainsaz, Lancé, Ornans, and Warrenton. (Karoonda subsequently became the type specimen of CK chondrites; Kallemeyn et al., 1991.)

The labels olivine-bronzite, olivine-hypersthene, and amphoterite were permanently replaced by H, L, and LL in the 1980s. About the same time, the names of the well-established carbonaceous-chondrite groups were simplified to CI, CM, CO, and CV. The CK group was not formalized until Kallemeyn et al. (1991). The CR group was not formalized until the early 1990s: Ash et al.

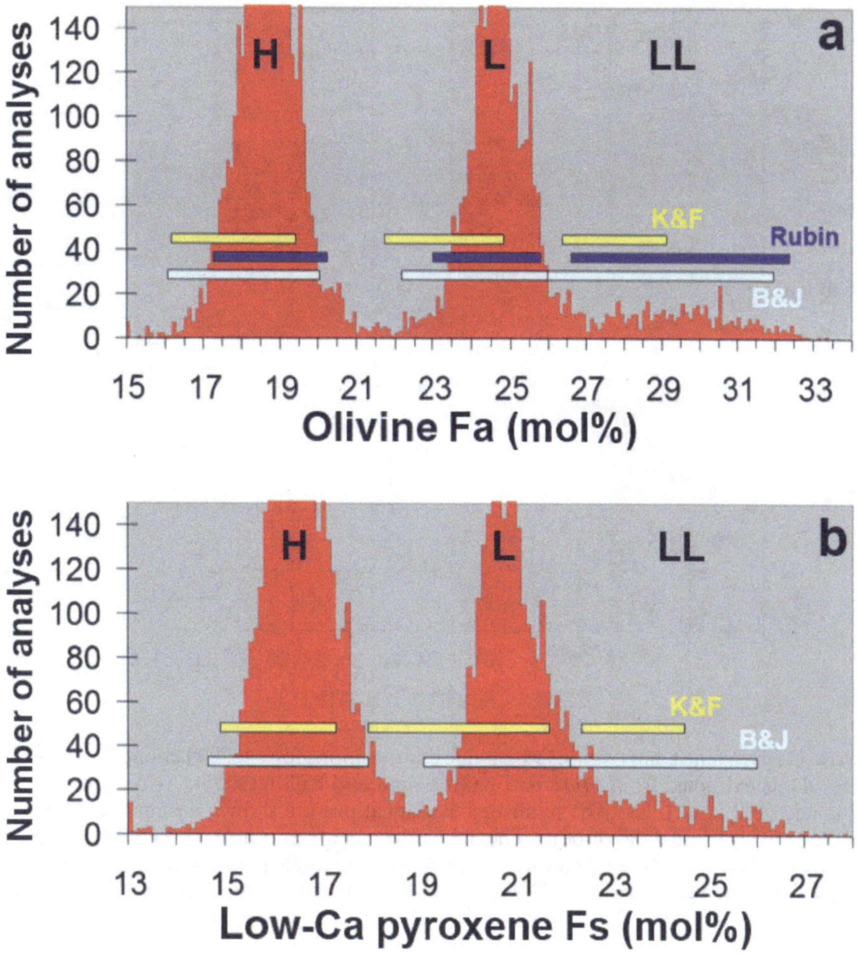

Fig. 10.1 Histograms of (a) olivine and (b) low-Ca pyroxene compositions derived from *MetBase* by Grossman and Rubin (2006). The dataset includes analyses of ~4600 H chondrites, ~2950 L chondrites, and ~570 LL chondrites; altogether, there are data for 8179 OC, 90% of which contain both olivine and low-Ca pyroxene analyses. Also shown are the ranges for the three OC groups recommended by Keil and Fredriksson (1964) = K&F, Rubin (1990) = Rubin, and Brearley and Jones (1998) = B&J. Rubin (1990) does not include data for low-Ca pyroxene. H and L chondrites are represented by two distinct tall peaks, but LL chondrites form a broad short peak. There is no clear boundary between L and LL chondrites. Diagram modified from Grossman and Rubin (2006).

(1991), Weisberg et al. (1992, 1993), Bischoff et al. (1993b), and Kallemeyn et al. (1994).

Rubin (1990) analyzed olivine (and kamacite) in 134 ordinary chondrites. Brearley and Jones (1998) compiled data from numerous sources for OC olivine and low-Ca pyroxene. Keil and Fredriksson (1964), Rubin (1990), and Brearley and Jones (1998) each constructed somewhat different ranges for the olivine and low-Ca pyroxene compositions of the H, L, and LL groups (Fig. 10.1). (Rubin [1990] did not report analyses of low-Ca pyroxene.)

Grossman and Rubin (2006) extracted data for 8179 OC from *MetBase* (the Meteorite Information Database and Bibliography of Meteoritics and Planetary Sciences, published by Jörn Koblitz) without regard for their reported classification. For both olivine and low-Ca pyroxene, the histograms (Fig. 10.1) show a prominent hiatus between H and L chondrites, but no clear boundary between L and LL chondrites. Potential sources of error include misclassified meteorites and systematic differences in analytical results.

Tables 10.1a, 10.1b give the ranges in olivine and low-Ca pyroxene for each OC group adopted by Rubin (1990) and Brearley and Jones (1998) along with values recommended by Grossman and Rubin (2006). Also presented are "proposed operational values," suggestions made to the Nomenclature Committee of the Meteoritical Society for use when considering the approval of new samples. There is

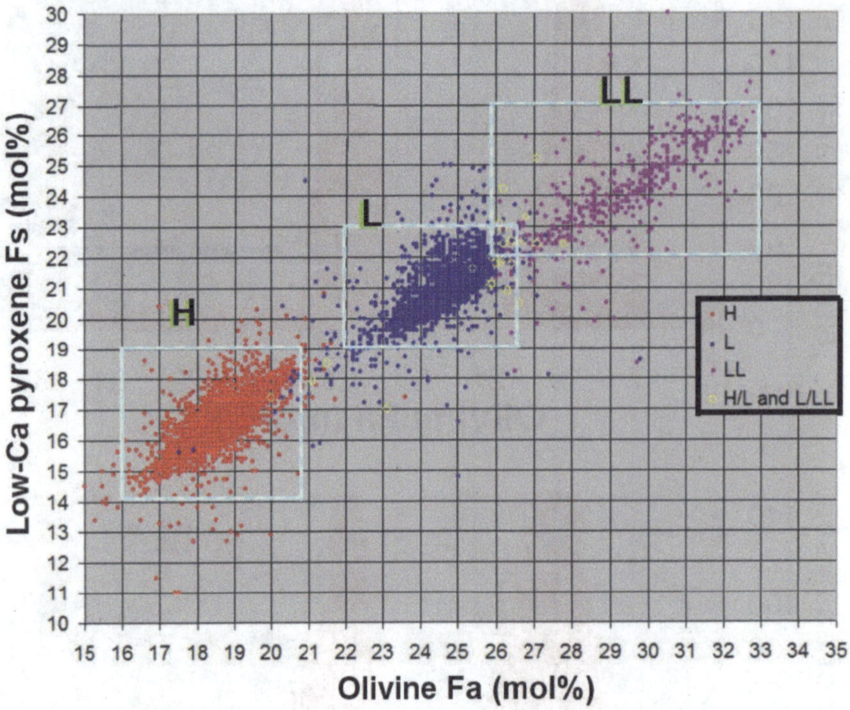

Fig. 10.2 Plot of low-Ca pyroxene compositions vs. olivine compositions for ~4600 H chondrites, ~2950 L chondrites, and ~570 LL chondrites. Data extracted from *MetBase* by Grossman and Rubin (2006). Also shown are OC with intermediate compositions labeled H/L and L/LL. The published classifications are shown; anomalously plotted points represent misclassifications (most likely), bad data, or both. The light-blue rectangles represent the proposed operational values for each group for use by the Nomenclature Committee when considering new meteorites. Data for H and L chondrites are distinct, but L- and LL-chondrite data overlap. Diagram modified from Grossman and Rubin (2006).

Table 10.1a *Ranges for olivine compositions for equilibrated (type 4–6) ordinary chondrites.*

–	Olivine Fa (mol%)		
Sources of data	H	L	LL
Rubin (1990)	17.3–20.2	23.0–25.8	26.6–32.4
Brearley and Jones (1998)	16.0–20.0	22.0–26.0	26.0–32.0
Recommended values	16.5–20.8	23.0–27.0	27.0–33.0
Proposed operational values	16.0–20.8	22.0–26.6	25.8–33.0

The recommended values are those that Grossman and Rubin (2006) deemed to be the best estimates for the actual ranges in mineral compositions based on the histograms of Fig. 10.1. The proposed operational values are suggestions for the Nomenclature Committee of the Meteoritical Society, to be used when considering the mineral chemistry data of new meteorites.

Table 10.1b *Ranges for low-Ca pyroxene compositions for equilibrated (type 4–6) ordinary chondrites.*

–	Low-Ca pyroxene Fs (mol%)		
Sources of data	H	L	LL
Rubin (1990)	–	–	–
Brearley and Jones (1998)	14.5–18.0	19.0–22.0	22.0–26.0
Recommended values	14.5–19.0	19.0–23.0	23.0–27.0
Proposed operational values	14.0–19.0	19.0–23.0	22.0–27.0

no hiatus between the L and LL ranges in Brearley and Jones (1998) or in the recommended ranges. The proposed operational ranges show an overlap in both olivine and low-Ca pyroxene between the L and LL groups. If these criteria were to be adopted, the group designation of new meteorites falling within the overlap would have to be based on other parameters (such as chondrule size, metallic Fe-Ni abundance, siderophile element content, and magnetic susceptibility).

Figure 10.2 is a plot of low-Ca pyroxene composition versus olivine composition for the data on equilibrated ordinary chondrites extracted from *MetBase* by Grossman and Rubin (2006). It is clear that there are three distinct OC clusters, despite a small overlap between the L and LL groups.

10.2 Inferred Proximity of the H, L, and LL Parent Bodies

The three most abundant meteorite groups among observed falls are the H, L, and LL ordinary chondrites, constituting 32.1%, 36.9%, and 8.7% of falls, respectively. Carbonaceous chondrites (all groups combined) comprise only 4.2% of falls; irons (all groups combined), only 4.0%. The high abundances of the three OC groups imply that their parent bodies must be near a resonance that facilitates the ready transfer of OC bodies (small asteroidal fragments and meteoroids) to the Inner Solar System. This is consistent with the observation that ordinary-chondrite spectra resemble those of Q/S-complex asteroids (which constitute 58% of NEAs) that have experienced space weathering. It is also consistent with the occurrence of numerous ordinary-chondrite chromite grains in sedimentary rocks from the Middle Ordovician; their presence indicates that ordinary chondrites have been reaching the Earth for roughly the last half-billion years. (And likely much longer.)

H chondrites comprise 44.8% of the three main OC groups (falls and finds together); L chondrites comprise 42.0%; and LL chondrites comprise 13.2%.

It is clear that the H, L, and LL groups all formed in the same restricted region of the solar nebula. This is indicated by the coherent mineralogic, geochemical, and textural trends among the three groups. As discussed in Chapter 3, in going from H to L to LL, there are systematic increases in (a) the FeO contents of olivine, low-Ca pyroxene, chromite, and ilmenite, (b) the Co and Ni concentrations in kamacite, (c) the Ge content of the metal fraction, (d) the olivine/low-Ca-pyroxene modal abundance ratio, (e) heavy oxygen isotopes, (f) heavy Ge isotopes, and (g) mean chondrule size. There are also systematic decreases in (a) total iron, (b) mean metallic Fe-Ni-Co, (c) the modal kamacite/taenite abundance ratio, (d) bulk siderophile/lithophile element ratios, (e) bulk density, (f) bulk refractory-siderophile/common-siderophile ratios, and (g) bulk refractory-siderophile/volatile-siderophile ratios.

10.3 Are There Additional Ordinary-Chondrite Groups?

10.3.1 H/L and L/LL

There are individual ordinary chondrites that either have some mineralogical properties intermediate between the established ranges of the major OC groups or have some properties characteristic of one group and other properties characteristic of another (e.g., Wang and Rubin, 1987; Kallemeyn et al., 1989; Rubin, 1990). Most of these pigeonhole-resistant samples have been designated H/L (e.g., Bremervörde; Tieschitz; Cali) or L/LL (e.g., Bjurböle; Holbrook; Qidong). H/L and L/LL chondrites constitute 0.4% and 1.2%, respectively, of ordinary-chondrite falls.

Table 10.2 shows some examples of L/LL chondrites that have olivine and kamacite compositions that do not both fit within the ranges of the same established group. (These ranges are: olivine Fa [L: 23.0–25.8 mol%; LL: 26.6–32.4 mol%]; kamacite Co [L: 7.0–9.5 mg/g; LL: 14.2–370 mg/g] [Rubin, 1990].) Qidong (which fell as a single 1275-g stone in China in 1982) is unusual in having olivine within the range of one OC group and kamacite within another (Wang and Rubin, 1987).

Table 10.2 *Mineral chemistry of selected L/LL chondrites.*

–	Type	Olivine Fa (mol%); characteristic group	Kamacite Co (mg/g); characteristic group
Bjurböle	4	26.2; L/LL	12.6; L/LL
Cynthiana	4	26.5; L/LL	10.9; L/LL
Holbrook	6	25.7; L	10.4; L/LL
Knyahinya	5	25.5; L	10.6; L/LL
Qidong	5	25.7; L	15.8; LL
Sultanpur	6	26.4; L/LL	9.4; L
Trysil	6	26.3; L/LL	9.0; L
Vera	4	25.7; L	9.8; L/LL
Xi Ujimgin	6	26.4; L/LL	8.0; L

Data from Rubin (1990).

It is not known whether any of the chondrites designated H/L or L/LL are from separate, otherwise-unsampled OC asteroids or are anomalous members of one of the three main established groups. Brown et al. (2023) studied the trajectory of the 2021 fall of L/LL5 Golden, compared it to those of previous L/LL orbits, and concluded that these meteoroids had originated predominantly from the inner main asteroid belt. Based on CRE ages and shock stages, Brown et al. inferred that L/LL chondrites hailed from several different parent bodies among the background S asteroids in the Hungaria region, 1.78–2.00 AU from the Sun. (434 Hungaria itself is an E-type asteroid.) It is unknown whether some or all of these parent bodies are fragments of a single large asteroid or were derived from several primary parent asteroids.

10.3.2 IIE Irons

Rubin (2022a) proposed that IIE irons and their chondritic inclusions constitute a fourth major ordinary-chondrite group. IIE irons have bulk chemical (e.g., Ir/Ni, Ir/Au, Au/Ni, Co/Ni) and bulk isotopic compositions (i.e., $\Delta^{17}O$ and $\delta^{74/70}Ge$) that lie along extensions of LL-L-H trends (Table 10.3). Chondrule-bearing silicate clasts in IIE Netschaëvo (Brezina, 1885; Zavaratskii and Kvasha, 1952; Buchwald, 1967; Olsen and Jarosewich, 1971; Bild and Wasson, 1977) have mineralogical and petrological characteristics that extend LL-L-H trends: Relative to H chondrites, Netschaëvo clasts have higher modal metallic Fe-Ni and lower values of olivine Fa, low-Ca-pyroxene Fs, and kamacite Co, as well as smaller mean chondrule diameters (Table 10.4).

The chondritic clasts in Netschaëvo are 1–2-cm-sized angular fragments (Bogard et al., 2000). They were informally designated as HH chondrites (very high total iron, very high metallic Fe) by Bild and Wasson (1977). Chondrule textural types in these clasts include BO, PO, PP, POP, and RP; constituent minerals include olivine, low-Ca pyroxene, Ca-pyroxene, sodic plagioclase (grains up to ~100 μm), merrillite, chlorapatite, chromite, troilite, schreibersite, and metallic Fe-Ni (Rubin, 1990; Schrader et al., 2017a).

Table 10.3 *Average bulk siderophile element concentration ratios and O-isotopic and Ge-isotopic compositions of IIE irons and ordinary chondrites.*

–	Ir/Ni	Ir/Au	Au/Ni	Co/Ni	$\Delta^{17}O$ (‰)	$\delta^{18}O$ (‰)	$\delta^{17}O$ (‰)	$\delta^{74/70}Ge_{metal}$ (‰)
IIE	61.5×10^{-6}	4.30	14.3×10^{-6}	53.7×10^{-3}	0.69	4.45	3.00	−0.58
H	48.5×10^{-6}	3.63	13.4×10^{-6}	51.0×10^{-3}	0.73	4.12	2.87	−0.51
L	41.6×10^{-6}	3.26	12.8×10^{-6}	48.8×10^{-3}	1.07	4.70	3.52	−0.31
LL	35.4×10^{-6}	2.91	12.2×10^{-6}	48.4×10^{-3}	1.26	5.04	3.88	−0.26

Oxygen isotope data for ordinary chondrites are for type 4–6 specimens.
$\delta^{74/70}Ge = ([^{74}Ge/^{70}Ge_{Sample} / ^{74}Ge/^{70}Ge_{Standard} - 1]\times1000)$. In the ordinary chondrites, $\delta^{74/70}Ge_{metal}$ is a proxy for $^{74/70}Ge_{bulk}$.
Data compiled by Rubin (2022a).

Table 10.4 *Mean mineralogic and petrologic properties of Netschaëvo chondrule-bearing clasts compared to those of equilibrated ordinary chondrites.*

–	Olivine Fa (mol%)	Low-Ca pyroxene Fs (mol%)	Kamacite Co (mg/g)	Kamacite Ni (mg/g)	Modal metallic Fe-Ni (vol.%)	Chondrule diameter (μm)
Netschaëvo	14.3	14.0	4.5	66.4	14	240
H4-6	18.8	17.2	4.7	69.2	7.5	450
L4-6	24.7	21.3	8.0	65.4	3.2	500
LL4-6	29.4	24.1	77.4	49.8	1.9	550

Data compiled by Rubin (2022a).

Three other IIE irons (Techado, Garhi Yasin, and Mont Dieu) contain highly recrystallized, petrologic-type-6, chondrule-bearing clasts (e.g., Van Roosbroek et al., 2015; McDermott et al., 2016). IIE Watson has a large, chondrule-free silicate inclusion with an H-chondrite-like bulk chemical composition (Olsen et al., 1994).

Many IIE irons contain alkali-rich silicate inclusions with nonchondritic bulk compositions (e.g., Bunch and Olsen, 1968; Wasserburg et al., 1968; Bunch et al., 1970; Olsen and Jarosewich, 1970; Ruzicka et al., 1999; Ruzicka and Hutson, 2010; Ruzicka, 2014). Some have high bulk K/Na ratios. These inclusions compositionally resemble alkali-rich clasts in OC breccias and melt-pocket glasses in shocked OC. All the alkali-rich objects are probably products of late-stage impact events that caused preferential melting of plagioclase (due to its low impedance to shock compression) (Rubin et al., 1986). In some cases, there was separation of (more volatile) K from Na during vapor transport.

10.3.3 Reduced Ordinary Chondrites

Yamato 982717 is a 10.27-g ordinary chondrite classified in the Meteoritical Bulletin as H4-an. However, its mean olivine ($Fa_{11.17\pm0.48}$) and low-Ca pyroxene ($Fs_{11.07\pm0.98}$) compositions (Yamaguchi et al., 2019) are far more magnesian than the mean values of equilibrated H chondrites ($Fa_{18.8}$; $Fs_{17.2}$). Bulk siderophile element abundances and whole-rock O-isotopic composition are in the H-chondrite range. Because Y-982717 lacks reduction textures, Yamaguchi et al. (2019) suggested that it is a bona fide H chondrite and, by implication, that the H-chondrite range extends to much more reduced samples. The widths of the established OC olivine Fa (H: 17.3–20.2 mol%; L: 23.0–25.8 mol%; LL: 26.6–32.4 mol%) ranges are relatively narrow: H (2.9 mol%), L (2.8 mol%), LL (5.8 mol%) (Rubin, 1990). However, if Y 982717 were a genuine H chondrite (i.e., from the same H parent asteroid), the new olivine Fa range width for H chondrites would be 9 mol% (i.e., 11.2–20.2 mol%), far greater than for any other OC group. It seems more likely that this meteorite and perhaps some H4-an congeners (e.g., Burnwell, $Fa_{15.8}$, $Fs_{13.4}$; LAP 04757, Fa_{13}, Fs_{13-16}) are either H chondrites that underwent reduction processes (e.g., Wasson et al., 1993) or are samples from one or more distinct, reduced, poorly represented OC asteroids (e.g., Russell et al., 1998).

Some rare varieties of ordinary chondrites are known only as clasts within breccias. Rubin et al. (2017) described a unique dark-colored, subrounded, 6.1 × 7.0-mm clast (Clast 6) which is more reduced than H chondrites or IIE chondrule-bearing clasts; it occurs within the NWA 10214 LL3-6 breccia. Clast 6 is a new kind of OC (of petrologic type 3.6) that consists of 60 vol% fine-grained matrix, 32 vol% coarse silicate grains, and 8 vol% coarse opaque grains (metallic Fe-Ni and troilite). (Most of the metal has been weathered into oxide and hydroxide.) The large chondrules and chondrule fragments are mainly Type IB (low-FeO PP); one small chondrule is Type IIA (high-FeO PO). Its mean chondrule size (330 ± 160 μm; $n = 28$) is within 1 standard deviation of that of H chondrules (450 μm). Also present are one 450 × 600-μm spinel-pyroxene-olivine CAI and one 85 × 110-μm AOA. Olivine (Fa13.5 ± 3.7; $n = 5$) is more magnesian than in equilibrated H chondrites. The clast resembles type-3 OC in its olivine and low-Ca pyroxene compositional distributions, and in the Fe/Mn ratio of ferroan olivine grains. The O-isotopic compositions of the chondrules in the clast are in the ordinary- and R-chondrite ranges.

So, how many ordinary-chondrite groups are there? There could be three (H, L, LL) or four (add IIE), or six (add H/L and L/LL) or seven (add Y-982717) or eight (add Burnwell) or nine (add Clast 6 from NWA 10214) or perhaps more than a dozen (add some of the reduced samples that plot within the low-FeO tail of H chondrites in Fig. 10.2, assuming the data are accurate). Some OC groups might be represented in our collections by only a single meteorite or a single clast in a polymict breccia. Each OC group likely hails from a separate asteroid.

It seems possible that the earliest accreted OCs were relatively rich in CAI (Rubin, 2022a). Extant samples may include Clast 6 within the NWA 10214 LL3-6 breccia as well as the three paired H3.2–3.4 chondrites Y-792947, Y-793408, and Y-82038 (which have an average CAI abundance of 0.19 vol.%; Kimura et al., 2002).

10.4 Ordinary Chondrites That Underwent Whole-Rock Reduction

Wasson et al. (1993) described three type-3 ordinary-chondrite finds (Willaroy, Suwahib [Buwah] and Moorabie) and one type-4 ordinary-chondrite find (Cerro los Calves) that are appreciably more reduced than other members of their particular OC group. As shown in Fig. 10.3, Moorabie and Suwahib (Buwah) have siderophile and chalcophile element abundance ratios within the ranges of L chondrites; they also have average chondrule sizes consistent with the L group. Willaroy and Cerro los Calvos have abundance ratios of siderophile elements that generally fall within the ranges of H chondrites; they also have average chondrule sizes consistent with the H group. However, the mean (rather uniform) olivine compositions of these meteorites are significantly lower than those of equilibrated members of their respective groups (Table 10.5).

Table 10.5 *Olivine compositions and chondrule sizes in four reduced ordinary chondrites compared to equilibrated H and L chondrites.*

–	Original chondrite group	Olivine Fa (mol%)	Mean chondrule diameter (μm)
Mean H chondrites	–	**18.8**	**450**
Willaroy	H	14.8 ± 1.6	320 ± 20 (n = 100)
Cerro los Calvos	H	12.8 ± 0.6	422 ± 22 (n = 373)
Mean L chondrites	–	**24.7**	**500**
Suwahib (Buwah)	L	14.4 ± 1.6	540 ± 30 (n = 100)
Moorabie	L	15.9 ± 1.5	650 ± 60 (n = 42)

Data from Wasson et al. (1993) and Table 10.4.

It seems likely that these four ordinary chondrites were reduced during thermal metamorphism. A plausible reducing agent is the set of C-rich aggregates and C-rich chondritic clasts found in some ordinary chondrites (Scott et al., 1988a). These objects are composed of fine-scale intergrowths of poorly graphitized C, amorphous C, metallic Fe-Ni, and minor chromite (Brearley, 1990); they range up to ~1 mm in size and constitute up to several volume-percent of some type-3 OC. These type-3 samples include H3.4 Sharps (which has more C-rich aggregates than fine-grained silicate-rich matrix material; Scott et al., 1981a) and L3.5 ALH 77011 and its paired meteorites (McKinley et al., 1981).

10.5 The Relationship between Ordinary Chondrites and R Chondrites

As of this writing, the major noncarbonaceous-chondrite clans are (in order of decreasing abundance): ordinary chondrites (~65,000 specimens,

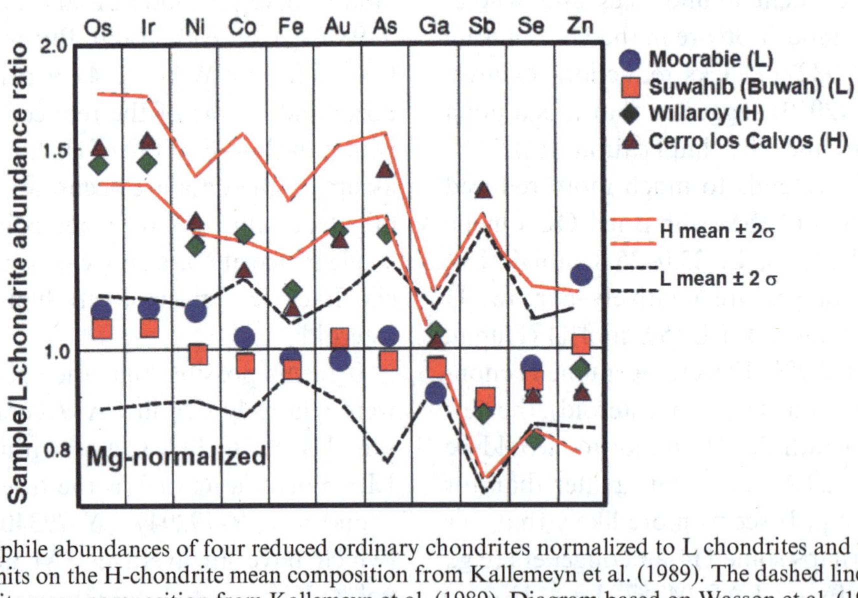

Fig. 10.3 Siderophile abundances of four reduced ordinary chondrites normalized to L chondrites and to Mg. The solid lines show 2σ limits on the H-chondrite mean composition from Kallemeyn et al. (1989). The dashed lines show 2σ limits on the L-chondrite mean composition from Kallemeyn et al. (1989). Diagram based on Wasson et al. (1993).

~980 falls), enstatite chondrites (~700 specimens, 17 falls), and R chondrites (also known as Rumuruti chondrites) (~330 specimens, 1 fall: Rumuruti, Kenya, 1934).

Most R chondrites are moderately metamorphosed rocks (mainly type 3.6–4); some are more-extensively metamorphosed (type 5–6). About half are regolith breccias, rich in solar-wind-implanted rare gases; many are genomict and contain clasts with a range of petrologic types (including some highly unequilibrated type-3 clasts).

R chondrites contain many of the same principal phases as ordinary chondrites. Major silicates in R chondrites include olivine, pyroxene, and sodic plagioclase. Opaques include rare metallic Fe-Ni (primarily awaruite with rare examples of grains with lower Ni), Fe-sulfide (mainly pyrrhotite, some monosulfide solid solution, and rare troilite), and, in some unequilibrated samples, magnetite. Other oxides in R chondrites include chromite, magnetite-chromite solid solution, chromite-hercynite-ulvöspinel solid solution, and ilmenite. Chondrules average 400 μm in diameter, similar to that of H chondrites (450 μm). As in ordinary chondrites, CAIs are rare (~0.04 vol.% in R chondrites, ~0.01 vol.% in type-3 OC).

There are major petrologic differences between R chondrites and OC:

(1) R chondrites contain a much higher average modal abundance of fine-grained silicate matrix material: 42 vol.% in R chondrites; 12 vol.% in type-3 OC.
(2) R chondrites are much more oxidized; this is reflected in their mean olivine Fa compositions: 37–40 mol% in R chondrites versus 29.4 mol% in LL, 24.7 mol% in L, and 18.8 mol% in H chondrites.
(3) Many R chondrites have a low modal abundance of low-Ca pyroxene. The abundance of low-Ca pyroxene is highly variable: it is abundant in unequilibrated R3 clasts, R3.6 chondrites (e.g., ALH 85151), and some R3.8–3.9 chondrites, but is rare to absent in other R3.8 chondrites and in many R5 and R6 lithologies. Other pyroxene phases in R chondrites include pigeonite, augite, and diopside.
(4) There are rare noble-metal-bearing phases in R chondrites but not in OC. These phases include metal alloys such as PtFe(Ir,Ni) alloy, Pt metal, Os with minor Re, Pt, Os-Ir-Ru-Pt alloy, native Au, electrum, ruthenosmiridim, rustenburgite, and PdPtSn; and very rare, small grains of noble-metal-rich sulfides, including erlichmanite, laurite, Os,Ru,Fe-rich sulfide, arsenides (sperrylite), sulfarsenides [irarsite ((IrPt)AsS) and tellurides (chengbolite, moncheite)].
(5) A few R6 chondrites contain amphibole (hornblende) and mica (phlogopite and titan-phlogopite) produced by hydrothermal alteration (McCanta et al., 2008a; Rubin, 2014).

There are similarities in bulk chemical composition between R chondrites and OC (reviewed by Bischoff et al., 2011). (1) Unlike carbonaceous chondrites, R chondrites and OC are not significantly depleted in Mn and Na relative to refractory lithophiles. (2) The total iron content of R chondrites (~24 wt.%) is between those of H- and L-chondrite falls (27.45 and 21.93 wt.%, respectively; Jarosewich, 1990). (3) The CI- and Mg-normalized refractory lithophile abundance ratios are ≥1.00 in carbonaceous chondrites, but <1.0 in non-carbonaceous chondrites (e.g., 0.974 in R, 0.899 in H, 0.904 in L, 0.890 in LL) (Rubin, 2011). (4) Whereas carbonaceous chondrites have bulk O-isotopic compositions that plot on or below the terrestrial fractionation (TF) line on the standard three-isotope graph, R chondrites and ordinary chondrites both have compositions above the TF line (Fig. 3.4).

Greenwood et al. (2000) observed a range of $\Delta^{17}O$ values in chondrule silicate grains from R chondrites: olivines with a composition in the range of Fa_{39-40} have $\Delta^{17}O$ values of –0.4 to +2.0‰. These are far lower than the whole-rock $\Delta^{17}O$ value of +2.9‰ in Rumuruti (the only R-chondrite fall). The similarity in $\Delta^{17}O$ between mafic silicates in R chondrites and ordinary chondrites suggests that both were derived from nebular components within the same O-isotopic reservoir.

There are two principal geochemical differences between R chondrites and ordinary chondrites: (1) R chondrites have significantly higher mean concentrations of trace elements, e.g., Zn (158 μg/g in R; 45–51 μg/g in OC) and Se (14.8 μg/g in R; 8.1–9.0 μg/g in OC). (2) R chondrites have appreciably higher proportions of bulk ^{17}O ($\Delta^{17}O$ = ~2.9‰) than OC (H: $\Delta^{17}O$ = 0.73‰; L: $\Delta^{17}O$ = 1.07‰; LL: $\Delta^{17}O$ = 1.26‰) (Fig. 3.4) or any other meteorite group.

Rubin and Ma (2021) defined a *meteorite class* as consisting of one or more meteorite clans that formed in the same general region of the Solar System. The members of a class may be related by one or more whole-rock properties including texture, mineralogy, inclusion abundance, oxidation state, bulk chemistry, and isotopic composition. Rubin and Ma identified four major chondrite classes, placing R chondrites and ordinary chondrites together: (1) ordinary chondrites (including chondritic inclusions from the IIE-an irons Netschaëvo and Mont Dieu and a unique clast in LL3 NWA 10214) and R chondrites, (2) enstatite chondrites, (3) carbonaceous chondrites, and (4) K chondrites. There may be a potential fifth chondrite class consisting of the metal-rich G chondrites such as NWA 5492, GRO 95551, NWA 12379 (and its paired specimen, NWA 12273), and Sierra Gorda 009 (Ivanova et al., 2019, 2020).

R chondrites are not restricted to individual meteorite falls. R-chondrite clasts (or R-like chondrite clasts) occur in Bencubbin-like breccias (Bencubbin and Weatherford) (e.g., Prinz et al., 1993), the Dar al Gani 310 ureilite (Downes and Mittlefehldt, 2006), and in the heterogeneous polymict breccias Kaidun (Brandstätter et al., 1996; Zolensky and Ivanov, 2003) and Almahata Sitta (Horstmann and Bischoff, 2010; Horstmann et al., 2010; Bischoff et al., 2010). (Although Isa et al. [2014] reported an R-chondrite clast in CM2 Murchison, further work by Kerraouch et al. [2019] showed the clast was a recrystallized CM-like chondrite.)

10.6 Physical Properties of Ordinary-Chondrite Groups

Flynn et al. (2018) and Ostrowski and Bryson (2019) compiled the measured values for many intrinsic physical and rheological properties of ordinary-chondrite whole-rocks. Compressive and tensile strengths of many meteorite groups were reviewed by Pohl and Britt (2020). The data in this section are culled mainly from Flynn et al. (2018).

First some definitions:

Bulk volume (V_{bulk}): volume of the entire sample, including solid material, liquid material and voids, commonly expressed in units of cm^3.

Bulk density (ρ_{bulk}): (mass of the sample)/(bulk volume), commonly expressed in units of g cm^{-3}.

Grain volume (V_{grain}): volume of only the solid material in a sample, commonly expressed in units of cm^3.

Grain density (ρ_{grain}): (mass of the sample)/(grain volume), commonly expressed in units of g cm^{-3}.

Porosity (ϕ): the amount of empty space in a sample (including voids, pores, and fractures), expressed as a percentage. It is equivalent to the percentage of the volume of a sample that consists of voids. In samples $\lesssim$1 meter in size, this property is often called microporosity. Meteorite parent asteroids (0.01–1000 km) also have macroporosity, consisting of large-scale voids, fractures, and crevasses. The total porosity of a parent body is the sum of the microporosity and the macroporosity.

Magnetic susceptibility (χ_m): a dimensionless proportionality constant that measures the degree to which a material becomes magnetized in the presence of an applied magnetic field. It is the ratio of magnetization to the applied magnetic field intensity; this ratio is called χ. Magnetic susceptibility is commonly expressed as log χ.

Electrical conductivity (σ): a measure of the ability of a substance to conduct an electrical current, commonly expressed in units of mS m^{-1} (milliSiemens per meter) or $(\text{ohm-cm})^{-1}$. The reciprocal of the Siemen (a unit of conductance) is the ohm (Ω) (a unit of resistance).

Thermal conductivity (κ): a measure of the ability of a substance to conduct heat; specifically, it is the rate at which heat is transferred by conduction through a unit area of material where there is a temperature gradient perpendicular to the unit area. It is commonly expressed in units of W m^{-1} K^{-1}.

Specific heat capacity (C_p): the amount of heat that must be added to a unit mass of a substance to make a unit change in the temperature of the substance. It is commonly expressed in units of J kg^{-1} K^{-1}.

The moduli are parameters that measure the amount of force per unit area required to achieve a given amount of deformation in a solid.

Young's modulus (E): a proportionality constant between the uniaxial stress (force per unit area) and strain (proportional deformation); it measures the compressive rigidity of a solid substance when a force is applied lengthwise. Materials with greater degrees of stiffness have higher values of E. The modulus is commonly expressed in units of GPa.

Bulk modulus (K): the measure of the resistance of a substance to bulk compression, defined as the ratio of the infinitesimal increase in pressure to the relative decrease in volume. It is commonly expressed in units of GPa.

Shear modulus (*G*): a measure of a substance's resistance to shear, defined as the ratio of shear stress to shear strain. It is commonly expressed in units of GPa.

Compressional wave velocity (v_p): a measure of the velocity of sound waves passing through a substance, commonly expressed in units of m s^{-1}. It is also known as the speed of sound.

Shear wave velocity (v_s): a measure of the velocity of shear waves passing through a substance; it is equivalent to $\sqrt{(G/\rho)}$ and is commonly expressed in units of m s^{-1}.

Compressive strength (*F*): the maximum compressive stress a material can withstand before it deforms; it is equivalent to the ratio of the maximum load to the cross-sectional area. It measures the ability of a substance to resist a reduction in size. It is commonly expressed in units of MPa.

Tensile strength (*s*): the maximum tensile stress a material can withstand before it deforms; it measures the ability of a substance to resist elongation. Tensile strength is generally lower than compressive strength for both brittle and ductile substances. It is commonly expressed in units of MPa.

Total iron and mean metallic Fe-Ni-Co in ordinary-chondrite falls decrease systematically from H to L to LL. Because Fe, Ni, and Co are the densest major elements in meteorites, the bulk density also decreases from H to L to LL (Table 10.6). Grain density likewise decreases from H to L to LL; because the porosities of the three OC groups are similar (8.0–9.5%), they do not much affect these systematic differences in density. Magnetic susceptibility depends on the amount of metallic Fe-Ni-Co; χ_m also decreases systematically from H to L to LL (Table 10.6).

Although OC falls have average porosities of 8.0–9.5% and average bulk densities of 3.52–3.71 g cm^{-3} (Table 10.6), some individual OC falls have much higher porosity and appreciably lower bulk density: e.g., Sena, H4 (26.6%; 2.61 g cm^{-3}); Eichstädt, H5 (26.6%; 2.51 g cm^{-3}) (Macke, 2010).

The electrical conductivity of the New Concord L6 fall was measured by Schwerer et al. (1971) and found to have values varying from ~10^{-5} $(\text{ohm-cm})^{-1}$ at ~30°C to $\sim 5\times10^{-3}$ $(\text{ohm-cm})^{-1}$ at ~700°C. As in lunar rocks, the electrical conductivity at 300 K, symbolized as $\sigma(300)$, is "extremely sensitive" to the meteorite's thermal history. (300 K ≈ 27°C.)

The thermal conductivities of ordinary chondrites vary by only a few percent between 100 and 300 K. Values are in the range of 1 to 3 W m^{-1} K^{-1} (Opeil et al., 2012). There is a strong positive correlation with porosity, particularly with fractures that are perpendicular to the direction of heat propagation. Weathered OC have low porosities and concomitantly lower thermal conductivities.

At relatively low temperatures (75–250 K), specific heat capacity strongly depends on temperature. At temperatures >400 K, this dependency decreases. The specific heat capacity of ordinary chondrites (of all three groups) in the temperature interval of ~300–370 K is roughly 730 ± 50 J kg^{-1} K^{-1}.

Values have been measured for the three moduli (Flynn et al., 2018). For ordinary chondrites, Young's modulus is 28.8 ± 2.4 GPa, the bulk modulus is 20.0 ± 1.8 GPa, and the shear modulus is 11.7 ± 1.0 GPa. Samples with lower porosity (e.g., weathered samples) have higher values for the bulk modulus and the shear modulus.

The mean values for the compressional wave velocity (i.e., the speed of sound), v_p, for

Table 10.6 *Some intrinsic physical properties of OC falls.*

–	Total Fe (wt.%)	Mean metallic Fe-Ni-Co (wt.%)	Bulk density (g cm^{-3})	Grain density (g cm^{-3})	Porosity (%)	Magnetic susceptibility log χ* (log 10^{-9} m^3 kg^{-1})
H falls	27.45	17.8	3.35 ± 0.01	3.71 ± 0.01	9.5 ± 0.4	5.30 ± 0.01
L falls	21.93	8.33	3.30 ± 0.01	3.58 ± 0.01	8.0 ± 0.3	4.87 ± 0.01
LL falls	19.63	3.56	3.18 ± 0.02	3.52 ± 0.01	9.5 ± 0.6	4.13 ± 0.05

Total Fe and mean metallic Fe-Ni-Co are from Jarosewich (1990). Other data are from the compilation of Flynn et al. (2018).

ordinary chondrites decrease systematically from H to L to LL chondrites: H (3634 ± 1060 m s^{-1}), L (3443 ± 1389 m s^{-1}), LL (3230 ± 1324 m s^{-1}). This trend is a function of the differences in density among the three groups (Gorshkov, 1973).

The mean values for the shear wave velocity, v_s, for ordinary chondrites decrease from H chondrites to L chondrites (although LL chondrites, with fewer samples, do not follow the trend): H (2109 ± 637 m s^{-1}), L (2028 ± 787 m s^{-1}), LL (2074 ± 920 m s^{-1}).

Porosity is anticorrelated with both v_p and v_s. Thus, weathered OC samples (with their very low porosities) have high values of v_p and v_s. In contrast, variations in porosity do not affect grain density, specific heat capacity, or magnetic susceptibility.

There are no statistically significant differences among the OC groups in mean compressive strength. As given in Flynn et al. (2018), the mean stress that can cause compressional failure in ordinary chondrites is 164 ± 102 MPa. Similarly, there are no statistically significant differences among OC groups in mean tensile strength. The mean stress that can induce tensional failure in OCs is 24 ± 11 MPa.

Determinations of the physical properties of ordinary chondrites (Yomogida and Matsui, 1983; Flynn et al., 2018), in concert with asteroid studies, have applications to the impact cratering and potential disruption of OC asteroids. Collisions into higher-porosity targets cause deeper craters, decreased ejecta velocities, and more locally restricted impact damage (e.g., Love et al., 1993; Michikami et al., 2007).

Melting temperature: Another important parameter is the temperature at which ordinary chondrites begin to melt. Heating experiments show that the approximate OC whole-rock solidus temperature is ~1100°C at an oxygen fugacity of IW-1 (1 log unit below the iron-wüstite buffer) (Jurewicz et al., 1995). (On oxygen-fugacity vs. temperature diagrams, the buffer curve represents the conditions under which the two phases coexist.) The lowest-temperature experiments (1170°C) show the presence of silicate melt and a depletion in low-Ca pyroxene. The metal-sulfide fraction of ordinary chondrites melts at a lower temperature: the Fe-FeS eutectic is 988°C (Brandes and Brook, 1998). With the addition of Ni, the eutectic temperature drops to ~940–950°C (e.g., Kullerud, 1963; Heymann, 1967).

10.7 Bulk Chemical Compositions of Ordinary-Chondrite Groups

There are two datasets for the bulk compositions of ordinary chondrites that are considered the most reliable:

1. Kallemeyn et al. (1989) analyzed elements by neutron-activation techniques. Sample sizes were ~250–300 mg and duplicate samples of each meteorite were analyzed. The standard deviations attributable to analytical error were ≤3% for all elements. This dataset can be supplemented by data determined by a variety of analytical techniques, as compiled by Lodders (2021) (Table 10.7). The compilation below supersedes that in Table 2 of Wasson and Kallemeyn (1988). In each OC group, there are no significant compositional differences among petrologic types in the abundances of major elements (Kallemeyn et al., 1989; Przylibski et al., 2023).
2. Jarosewich (1990) determined major elements by wet chemical techniques on aliquants of large (~10 g) powdered samples (Table 10.8). The potential problem of samples being heterogeneous at small size scales was greatly diminished.

In Table 10.9, the INAA data on ordinary-chondrite bulk chemical compositions from Kallemeyn et al. (1989) (Table 10.7) are compared to the wet chemistry data from Jarosewich (1990) (Table 10.8). All data are expressed in units of mg/g. Of the 30 comparisons, only five show a discrepancy above 5%; the highest discrepancy is 6.7%. Among the greatest discrepancies pertain to the elements K (which is of low concentration) and Ni (due to sample heterogeneities in metal abundances).

There are additional datasets that focus on volatile/mobile trace elements (e.g., Neal et al., 1981; Walsh and Lipschutz, 1982; Huston and Lipschutz, 1984; Lingner et al., 1987). Reasonable estimates for the mean concentrations of these elements in each OC group are given in Table 10.7.

The concentrations of mobile trace elements in OC tend to vary with metamorphic grade and degree of shock (Table 10.10). In going from H4 to

Table 10.7 *Bulk chemical compositions of ordinary chondrites.*

–	H	L	LL	–	H	L	LL
H (μg/g)	800	1260	1700	Rh (ng/g)	220	155	110
Li (μg/g)	1.7	1.9	2	Pd (ng/g)	845	620	510
Be (ng/g)	30	30	37	Ag (ng/g)	60	80	65
B (μg/g)	0.45	0.5	0.56	Cd (ng/g)	40	35	20
C (mg/g)	2.35	2.50	3.92	In (ng/g)	1	0.5	4
N (μg/g)	70	34	50	Sn (μg/g)	0.39	0.40	0.33
O (mg/g)	337.4	368.4	383.7	**Sb (ng/g)**	**67**	**68**	**58**
F (μg/g)	52	60	58	Te (μg/g)	0.35	0.4	0.39
Na (mg/g)	**6.26**	**7.02**	**7.09**	I (ng/g)	54	70	160
Mg (mg/g)	**142**	**148**	**151**	Cs (ng/g)	120	200	170
Al (mg/g)	**11.4**	**12.1**	**12.2**	Ba (μg/g)	3.7	3.8	4
Si (mg/g)	170.4	184.6	189.0	**La (ng/g)**	**303**	**320**	**320**
P (mg/g)	1.19	1.03	0.98	Ce (ng/g)	763	860	880
S (mg/g)	20	22	21	Pr (ng/g)	121	130	130
Cl (μg/g)	87	99	126	Nd (ng/g)	581	670	650
K (μg/g)	**782**	**858**	**857**	**Sm (ng/g)**	**191**	**198**	**198**
Ca (mg/g)	**12.4**	**12.9**	**13.1**	**Eu (ng/g)**	**74**	**77**	**76**
Sc (μg/g)	**7.88**	**8.39**	**8.34**	Gd (ng/g)	270	300	290
Ti (μg/g)	630	690	680	Tb (ng/g)	49	54	55
V (μg/g)	**73.3**	**76.8**	**76.6**	Dy (ng/g)	320	360	350
Cr (mg/g)	**3.67**	**3.79**	**3.80**	Ho (ng/g)	74	77	79
Mn (mg/g)	**2.32**	**2.57**	**2.63**	Er (ng/g)	213	230	230
Fe (mg/g)	**271**	**216**	**185**	Tm (ng/g)	33	33	35
Co (μg/g)	**831**	**600**	**489**	**Yb (ng/g)**	**208**	**220**	**225**
Ni (mg/g)	**16.3**	**12.3**	**10.1**	**Lu (ng/g)**	**31.9**	**32.8**	**33.1**
Cu (μg/g)	90	95	83	Hf (ng/g)	150	170	170
Zn (μg/g)	**45.5**	**50.8**	**46.5**	Ta (ng/g)	21	21	–
Ga (μg/g)	**5.78**	**5.69**	**5.16**	W (ng/g)	170	150	100
Ge (μg/g)	10	11	11	Re (ng/g)	74	55	32
As (μg/g)	**2.13**	**1.50**	**1.26**	**Os (ng/g)**	**848**	**546**	**387**
Se (μg/g)	**8.1**	**9.0**	**8.5**	**Ir (ng/g)**	**791**	**512**	**358**
Br (μg/g)	0.7	0.9	1	Pt (ng/g)	1580	1080	880
Rb (μg/g)	2.6	2.8	2.7	**Au (ng/g)**	**218**	**157**	**123**
Sr (μg/g)	8.7	11	11	Hg (ng/g)	–	2	22
Y (μg/g)	2	2.1	2	Tl (ng/g)	4	2	2
Zr (μg/g)	5.7	5.8	6.8	Pb (μg/g)	0.14	0.06	0.06
Nb (ng/g)	380	400	400	Bi (ng/g)	9	4	12
Mo (ng/g)	1400	1200	1100	Th (ng/g)	40	40	46
Ru (ng/g)	1100	750	550	U (ng/g)	13	13	13

Dashes signify the absence of reliable data.
Elements in bold determined by neutron activation analysis by Kallemeyn et al. (1989); elements in regular typeface are from Lodders (2021).

H5 or from H5 to H6, the concentrations of these mobile trace elements decrease. Selenium is the sole exception among the analyzed H chondrites; this may be because Se is concentrated in troilite (along with S) and troilite occurs mainly in coarse, heterogeneously distributed grains. The measured samples may therefore not have had representative concentrations of Se. In mildly shocked L chondrites, there are a few exceptions to the trend of decreasing abundances with increasing metamorphic grade: Au, Ag, and Bi are higher in mildly shocked L6 chondrites than in mildly shocked L4 and L5 chondrites. It is possible that the analyzed samples did not have representative concentrations of these elements. Nevertheless, it is apparent that, in general, thermal metamorphism causes a loss of mobile trace

Table 10.8 *Bulk chemical composition (wt.%) of ordinary-chondrite falls determined by wet-chemical methods.*

–	H falls	L falls	LL falls
SiO_2	36.60	39.72	40.60
TiO_2	0.12	0.12	0.13
Al_2O_3	2.14	2.25	2.24
Cr_2O_3	0.52	0.53	0.54
FeO	10.30	14.46	17.39
MnO	0.31	0.34	0.35
MgO	23.26	24.73	25.22
CaO	1.74	1.85	1.92
Na_2O	0.86	0.95	0.95
K_2O	0.09	0.11	0.10
P_2O_5	0.27	0.22	0.22
H_2O^+	0.32	0.37	0.51
H_2O^-	0.12	0.09	0.20
Fe (metal)	15.98	7.03	2.44
Ni	1.74	1.24	1.07
Co	0.08	0.06	0.05
FeS	5.43	5.76	5.79
C	0.11	0.12	0.22
Total	99.99	99.99	99.92
Fe (total)	27.45	21.93	19.63

H_2O^+ is chemically bound water; H_2O^- is water lost during analysis at 110°C.
Data from Jarosewich (1990).

Table 10.9 *Comparison of INAA data from Kallemeyn et al. (1989) and wet chemistry data from Jarosewich (1990).*

Element	Group	K (1989)	J (1990)	Difference
Na	H	6.26	6.4	2.2%
Na	L	7.02	7.0	0.3%
Na	LL	7.09	7.0	1.3%
Mg	H	142	140.3	1.2%
Mg	L	148	149.2	0.8%
Mg	LL	151	152.1	0.7%
Al	H	11.4	11.3	0.9%
Al	L	12.1	11.9	1.7%
Al	LL	12.2	11.9	2.5%
K	H	0.782	0.75	4.3%
K	L	0.858	0.91	6.1%
K	LL	0.857	0.83	3.3%
Ca	H	12.4	12.4	0.0%
Ca	L	12.9	13.2	2.3%
Ca	LL	13.1	13.7	4.6%
Cr	H	3.67	3.6	1.9%
Cr	L	3.79	3.6	5.3%
Cr	LL	3.80	3.7	2.7%
Mn	H	2.32	2.4	3.4%
Mn	L	2.57	2.6	1.2%
Mn	LL	2.63	2.7	2.7%
Fe	H	271	274.5	1.3%
Fe	L	216	219.3	1.5%
Fe	LL	185	196.3	6.1%
Co	H	0.831	0.8	3.9%
Co	L	0.600	0.6	0.0%
Co	LL	0.489	0.5	2.2%
Ni	H	16.3	17.4	6.7%
Ni	L	12.3	12.4	0.8%
Ni	LL	10.1	10.7	5.9%

All data in mg/g. The number of significant figures matches that of the original data sets.

elements. These elements tend to be volatile; almost all of their 50% equilibrium condensation temperatures (at 10^{-4} bar) are below 1000 K and several are $\lesssim$500 K (Table 10.10).

In every case, the concentrations of the mobile trace elements decrease in more-highly-shocked samples (Table 10.10). Shock-heating causes loss of these elements.

Because CI carbonaceous chondrites have many elemental abundances that closely match those of the solar photosphere (excluding noble gases, N, C, and Li) (Fig. 3.1), it is useful to compare ordinary-chondrite bulk chemical compositions to those of CI chondrites (e.g., Table 10.11; Figs. 10.4, 10.5). Such a comparison illustrates the effects of various processes that caused deviations (fractionations) from a presumed initial CI-like composition of OC precursor materials.

Lithophiles (Fig. 10.3): On diagrams plotted on a coarser scale (e.g., Fig. 1 of Wasson and Kallemeyn, 1988), the refractory lithophile abundances (Al, Sc, Ca, REE) of ordinary chondrites appear fairly flat. This is clear from the small differences among the normalized abundances of these elements in Table 10.11. The apparent differences in Fig. 10.3 are due to the fine scale of the diagram; many of these differences are slight or nonexistent (i.e., indistinguishable within analytical uncertainty). This indicates that the refractory lithophiles behaved mainly as a group – they were not quantitatively separated from each other during processing in the molecular cloud, the nebula, or on their respective parent asteroids. Sodium and Mn have similar abundances to those of refractory lithophiles in L and LL chondrites; in H chondrites, the Na and Mn abundances are lower than those of refractory lithophiles. In all three groups, K is enriched relative to Na, Mn, and refractory lithophiles.

Table 10.10 *Concentrations of some mobile and trace elements in H and L chondrites.*

–	50% condensation temperatures (K) at 10^{-4} bar	H4	H5	H6	L4, L5 (mildly shocked)	L4, L5 (strongly shocked)	L6 (mildly shocked)	L6 (strongly shocked)
Co (μg/g)	1354	–	–	–	650	529	630	540
Zn (μg/g)	704	–	53	41	58.0	–	–	47.2
Ga (μg/g)	1010	–	–	–	5.59	5.04	5.50	5.10
Se (μg/g)	701	–	8.0	8.86	7.0	–	–	9.7
Rb (μg/g)	752	–	2.52	1.82	2.46	–	–	2.10
Ag (ng/g)	699	–	–	–	109	65	141	66
Cd (ng/g)	502	10	3.4	–	–	13	70	–
In (ng/g)	492	–	0.46	0.24	2.08	0.73	1.78	0.55
Sb (ng/g)	890	81	65	–	–	–	–	–
Te (ng/g)	665	–	378	250	543	–	–	378
Cs (ng/g)	593	77	33	11	–	–	–	–
Au (ng/g)	967	236	174	–	179	160	190	156
Tl (ng/g)	365	–	–	–	–	1.8	19.1	–
Bi (ng/g)	480	–	–	–	7.0	2.7	9.3	4.5

Element concentration data from Huston and Lipschutz (1984) and Lingner et al. (1987). Equilibrium condensation temperatures from Wood et al. (2019).

Table 10.11 *CI- and Mg-normalized H, L, and LL chondrite bulk chemical compositions.*

–	H	L	LL
Al	0.906	0.922	0.911
Sc	0.928	0.948	0.924
Ca	0.921	0.919	0.915
Lu	0.889	0.877	0.868
Sm	0.876	0.871	0.854
La	0.877	0.889	0.871
Yb	0.894	0.907	0.909
Eu	0.903	0.901	0.872
V	0.910	0.915	0.895
Mg	1.000	1.000	1.000
Cr	0.946	0.937	0.921
Mn	0.834	0.887	0.889
K	0.954	1.004	0.983
Na	0.873	0.934	0.929
Os	1.182	0.730	0.507
Ir	1.175	0.729	0.500
Ni	1.041	0.753	0.606
Co	1.117	0.774	0.618
Fe	1.017	0.778	0.653
Au	1.034	0.715	0.549
As	0.791	0.534	0.440
Ga	0.403	0.381	0.338
Sb	0.299	0.291	0.244
Se	0.282	0.301	0.279
Zn	0.100	0.107	0.096

OC data from Kallemeyn et al. (1989). CI data from Wasson and Kallemeyn (1988).

Siderophiles and chalcophiles (Fig. 10.5): The refractory siderophiles (Os, Ir; but also Re, Ru, Pt, and Pd; Horan et al., 2003), common siderophiles (Ni, Co, Fe), and Au (a volatile siderophile) form fairly flat, unfractionated patterns. This indicates that these elements were not quantitatively separated from one another during processing in the molecular cloud, the solar nebula, or on the OC parent asteroids. Volatile siderophiles (Au, As, Ga, Sb) form a sloped trend – elements of higher volatility have progressively lower abundances. Chalcophile elements behave similarly; their volatilities are higher than those of the volatile siderophiles and their abundances are lower still. This trend is repeated in carbonaceous and enstatite chondrites (Wasson and Kallemeyn, 1988) and reflects nebular (or molecular-cloud) processes that depleted the more volatile species from accreting materials.

Plausible causes of these nebular fractionations include: (a) incomplete condensation of the more volatile elements from a gas of solar composition (e.g., Davis, 2006), (b) gas-solid interactions (e.g., in the Sun's parental molecular cloud, refractory elements resided mainly in dust grains while volatile elements were confined largely to the gas phase) (Yin, 2005) (c) nebular processing of dust and ice grains in the nebula caused volatile elements

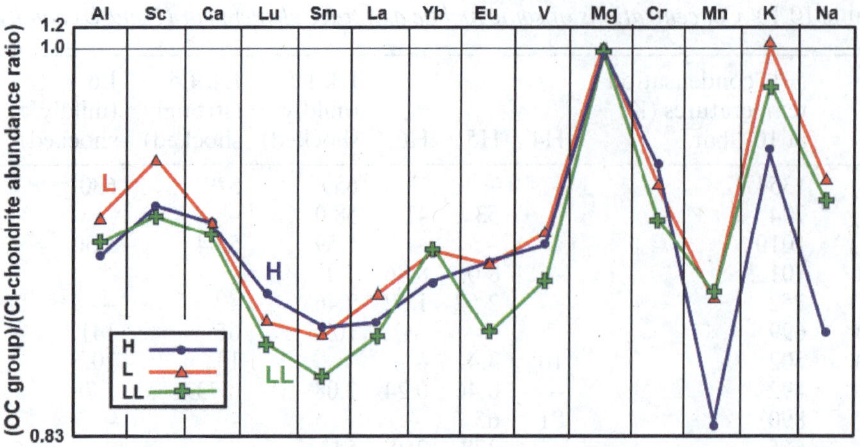

Fig. 10.4 The bulk composition of lithophile elements in the three ordinary chondrite groups normalized to Mg and to CI carbonaceous chondrites. H chondrites (blue circles), L chondrites (red triangles), and LL chondrites (green pluses) are represented. Elements are arranged from left to right in order of increasing volatility. The y-axis is shown at a very fine logarithmic scale to highlight slight-to-unresolvable compositional differences among the three groups. In broad view, all three OC groups have rather flat refractory lithophile patterns (Al to Eu). Elements in CI chondrites would plot along the 1.0 line.

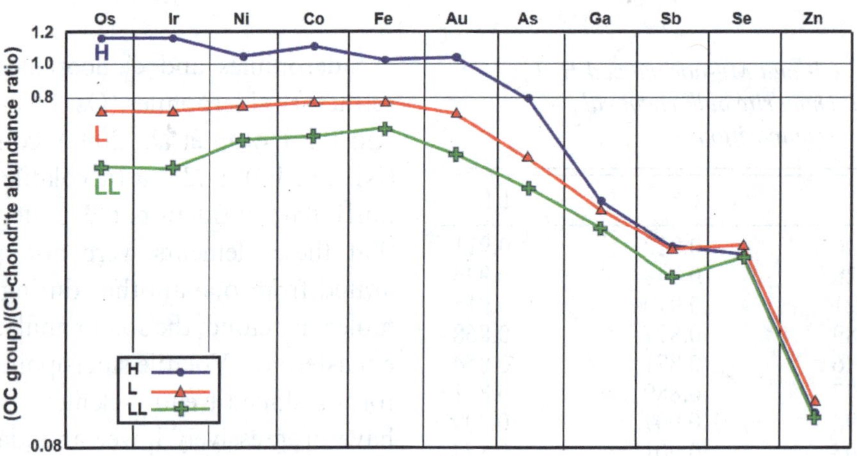

Fig. 10.5 The bulk composition of siderophile and chalcophile elements in the three ordinary chondrite groups normalized to Mg and to CI carbonaceous chondrites. H chondrites (blue circles), L chondrites (red triangles), and LL chondrites (green pluses) are represented. Elements are arranged from left to right in order of increasing volatility. The y-axis is shown at a logarithmic scale. All three OC groups have flat refractory (Os, Ir) and common (Ni, Co, Fe) siderophile patterns. Elements in CI chondrites would plot along the 1.0 line.

released from ice to recondense on dust grains as a function of their volatility (Huss et al., 2003; Huss, 2004), (d) unspecified redox reactions within nebular gas (mentioned by McSween and Huss, 2010), (e) differential settling of materials with distinct bulk chemical compositions to the nebular midplane prior to accretion (e.g., Wasson and Kallemeyn, 1988), and (f) loss of some volatiles and metal during chondrule formation (e.g., Rubin, 2022a). Various combinations of these processes are also possible.

10.8 Bulk Isotopic Compositions of Ordinary-Chondrite Groups

The three main OC groups have distinct bulk isotopic compositions. In this section, the focus is on oxygen, germanium, and hydrogen isotopes (Table 10.12).

There are three stable oxygen isotopes: ^{16}O (99.757%), ^{17}O (0.038%), and ^{18}O (0.205%). The O-isotopic compositions of meteorites are typically expressed in permille (‰) – parts per thousand;

Table 10.12 *Average bulk O-, Ge-, and H-isotopic compositions of ordinary chondrites.*

–	$\delta^{18}O$ (‰)	$\delta^{17}O$ (‰)	$\Delta^{17}O$ (‰)	$\delta^{74/70}Ge_{metal}$ (‰)	δD (‰)
H	4.12	2.87	0.73	−0.51	34.6 ± 230 ($n = 5$)
L	4.70	3.52	1.07	−0.31	-51.2 ± 102 ($n = 5$)
LL	5.04	3.88	1.26	−0.26	324 ± 548 ($n = 11$)

Oxygen isotope data for ordinary chondrites are for type 4–6 specimens and are relative to SMOW (Standard Mean Ocean Water). $\delta^{74/70}Ge = ([(^{74}Ge/^{70}Ge_{Sample}/^{74}Ge/^{70}Ge_{Standard})-1]\times 1000))$. In ordinary chondrites, $\delta^{74/70}Ge_{metal}$ is a proxy for $\delta^{74/70}Ge_{bulk}$. $\delta D = ([D/H_{measured}/D/H_{SMOW}-1)\times 1000$, with $D/H_{SMOW} = 155.76\times 10^{-6}$.
Deuterium data are for type-3 OC (Grant et al., 2023); O- and Ge-isotopic data are from the compilation of Rubin (2022a).

the values are relative to the arbitrary standard "SMOW" (standard mean ocean water).[1] Also common is the $\Delta^{17}O$ notation; this value is the vertical distance from a plotted point above the TF line (positive values) or below the TF line (negative values); it is calculated by the equation: $\Delta^{17}O = \delta^{17}O - 0.52(\delta^{18}O)$. Oxygen is sited predominantly in silicate minerals (mainly olivine, pyroxene, and plagioclase), chondrule glassy mesostases, and oxides (mainly chromite, ilmenite, and rutile).

Germanium has five stable isotopes: ^{70}Ge (20.5%), ^{72}Ge (27.4%), ^{73}Ge (7.76%), ^{74}Ge (36.5%), and ^{76}Ge (7.75%). Germanium is sited mainly in metallic Fe-Ni (kamacite, taenite, tetrataenite), to a lesser extent in silicates (depending on the redox state; e.g., Kegler and Holzheid, 2011), and to a minor extent (3–5%) in sulfide (troilite) (Chou and Cohen, 1973; Wai and Wasson, 1979).

Hydrogen has two stable isotopes: ^{1}H or H (protium) (99.9855%) and ^{2}H or D (deuterium) (0.0145%). Hydrogen in type-3 ordinary chondrites is sited primarily in water-bearing phases in the fine-grained matrix and, to a small extent, in organic materials.

[1] There have been several standards for mean ocean water. In 1961, Harmon Craig defined the composition of an arbitrary water standard based on measured values of D and ^{18}O in worldwide ocean samples. This was the first SMOW (standard mean ocean water) standard. In 1966, the Vienna SMOW (VSMOW) standard was prepared from distilled Pacific Ocean water mixed with minor amounts of other water samples. As VSMOW was gradually used up, a new standard (VSMOW2) with essentially the same isotopic composition was created in 1996. Because the isotopic differences between SMOW, VSMOW and VSMOW2 are tiny, it makes no practical difference which of the standards meteorite isotopic compositions are compared with.

10.8.1 Oxygen Isotopes

Along with texture, mineralogy, mineral chemistry, and bulk chemical composition, oxygen isotopes can serve as a fundamental meteorite classification tool. The importance of O isotopes is underscored by the fact that this element constitutes 34–38 wt.% of bulk ordinary chondrites (Table 10.7). The different chondrite groups occupy distinct regions of the standard three-isotope oxygen diagram (Fig. 10.6). Carbonaceous chondrites (except CI) and Kakangari chondrites (noncarbonaceous) plot below the TF line. CI chondrites tend to plot on the TF line or slightly above it. Enstatite chondrites plot on the TF line along with Earth materials (organic and inorganic) and Moon rocks. Ordinary chondrites plot above the TF line, forming somewhat overlapping fields. The L- and LL-chondrite fields overlap much more than those of H and L chondrites. R chondrites plot far above the TF line; they form the chondrite group richest in ^{17}O.

As has been pointed out by Clayton et al. (1991), there is greater variability among the OC $\delta^{18}O$ and $\delta^{17}O$ values than among the $\Delta^{17}O$ values because terrestrial contamination and isotopic fractionation in the lab move data points along a slope-½ mass-fractionation line (parallel to the TF line); this affects $\delta^{18}O$ and $\delta^{17}O$, but not $\Delta^{17}O$. The O-isotopic compositions of different minerals in OC also fall along a mass-fractionation line; variability in the modal abundances of these minerals will move the bulk O-isotopic compositions along this line.

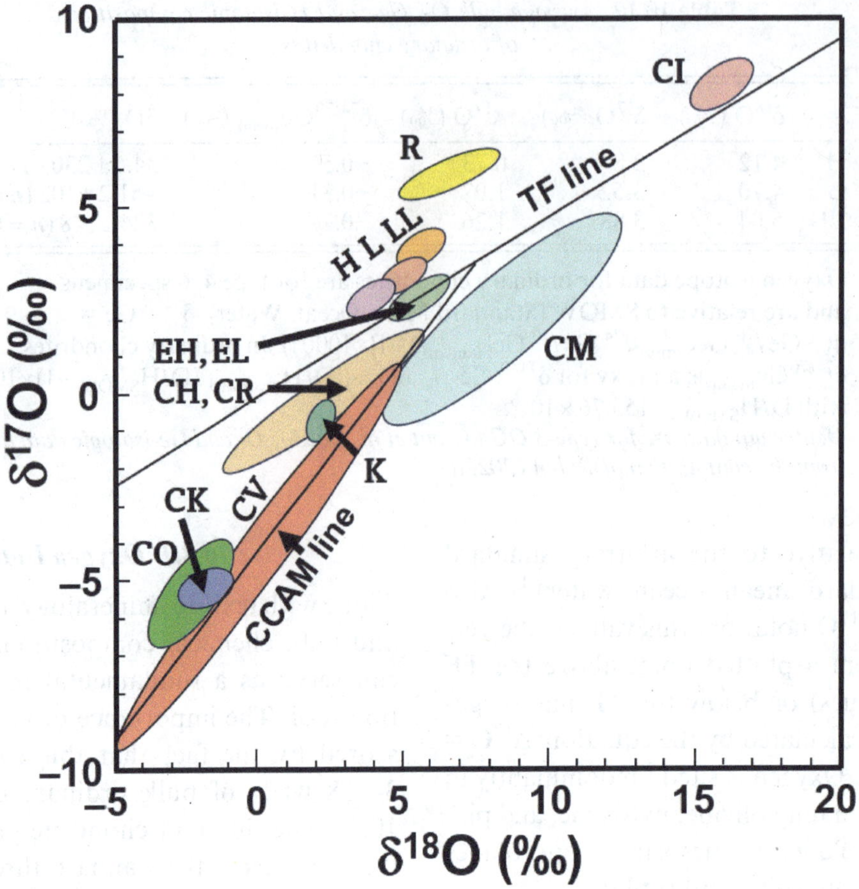

Fig. 10.6 The O-isotopic compositions of whole-rocks of the major chondrite groups. Data are in parts per thousand and are all relative to SMOW (standard mean ocean water). TF = terrestrial fractionation line; Earth materials and lunar rocks plot on the TF line. CCAM = carbonaceous-chondrites anhydrous minerals line; refractory inclusions plot along (or close to) this line. Data from Clayton (2005) and other sources.

These contingencies account for the spread of bulk OC points along mass-fractionation lines (Fig. 10.7a,b) and for the fact that the long axes of the OC data ellipses are subparallel to the TF line (which has a slope of 0.52) (Fig. 10.6). There are no significant differences in bulk O-isotopic composition in H chondrites or L chondrites between type-3 samples and type 4–6 samples. However, in the LL group, type-3 samples tend to have higher values of $\delta^{18}O$ and lower values of $\Delta^{17}O$ than type 4–6 samples.

Table 10.13 lists the mean bulk O-isotopic compositions of equilibrated OC separated by group and petrologic type (Clayton et al., 1991). Also listed are the mean olivine Fa values (Rubin, 1990). It is clear from this table and Fig. 10.8a that, in each OC group, the mean olivine Fa content increases from type-3 to type-4 to type-5 to type-6. Because olivine Fa is a proxy for oxidation state, it appears that the more metamorphosed OC are more oxidized (e.g., McSween and Labotka, 1993).

The same trend is evident in type-3 ordinary chondrites. Data for LL3 chondrites (Fig. 10.8b) show a strong positive correlation between petrologic subtype and mean olivine Fa. The correlation coefficient is 0.626 ($n = 78$) and $2\alpha < 0.0001$ – the correlation is significant beyond the 99.99% confidence level.

In the course of whole-rock oxidation during metamorphism, as more Fe^{2+} iron is produced at the expense of metallic Fe, the FeO/MnO ratio in olivine and pyroxene increases along with the FeO/(FeO+MgO) ratio (Gastineau-Lyons et al., 2002). Also increasing is the whole-rock olivine/low-Ca-pyroxene ratio, since in olivine, the divalent-cation/Si ratio is twice that of low-Ca pyroxene (2:1 vs. 1:1), as more divalent Fe becomes available.

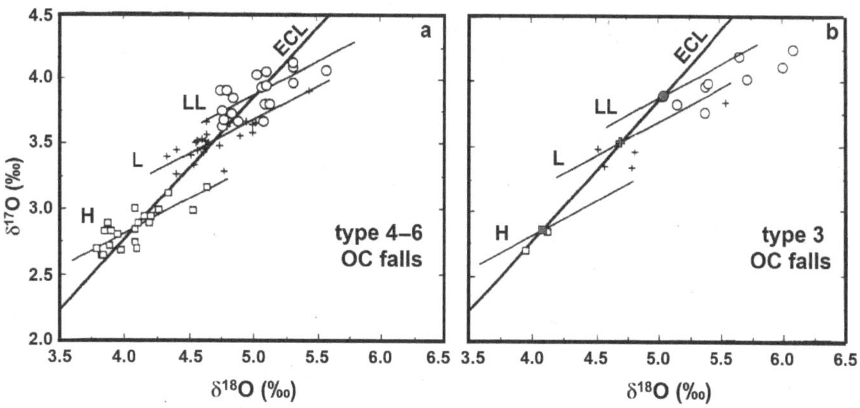

Fig. 10.7 O-isotopic compositions of ordinary-chondrite falls. (a) Type 4–6 samples. (b) Type 3 samples. H chondrites = squares; L chondrites = plus signs; LL chondrites = circles. ECL = Equilibrated Chondrite Line (this is the line drawn through the mean compositions of the three OC groups). Mass-fractionation lines of slope ½ are drawn through the three mean group compositions, which are shown in (b) as filled red symbols on the ECL. Type-3 H and L chondrites lie near their respective mass-fractionation lines, but LL3 chondrites tend to lie below the LL4-6 line. Diagram modified from Clayton et al. (1991).

Table 10.13 *Bulk O-isotopic compositions and mean olivine Fa contents of equilibrated ordinary chondrites listed by petrologic type.*

–	$\delta^{18}O$ (‰)	$\delta^{17}O$ (‰)	$\Delta^{17}O$ (‰)	Fa(mol%)
H4	4.13±0.24	2.88±0.13	0.74±0.09	18.3
H5	4.11±0.25	2.87±0.18	0.73±0.11	18.6
H6	3.97±0.16	2.79±0.12	0.72±0.05	19.4
L4	4.79±0.34	3.55±0.20	1.06±0.07	24.0
L5	4.75±0.18	3.57±0.09	1.10±0.07	24.7
L6	4.60±0.19	3.46±0.11	1.06±0.11	25.1
LL4	5.25±0.25	4.01±0.11	1.28±0.09	28.5
LL5	5.02±0.21	3.90±0.22	1.29±0.12	29.5
LL6	4.92±0.17	3.79±0.11	1.23±0.13	29.7

O-isotopic data from Clayton et al. (1991). Olivine Fa data from Rubin (1990).

It seems likely the principal oxidizing agent was water (which is ~89 wt.% oxygen). Aqueous alteration has been inferred for type-3 OC from the presence of phyllosilicates, Ni-rich sulfide, and smectite (Hutchison et al., 1987; Alexander et al., 1989a; Grossman et al., 2000), fayalite-silica associations (Wasson and Krot, 1994), carbide-magnetite assemblages (Krot et al., 1997b), and bleached radial pyroxene and cryptocrystalline chondrules (e.g., Grossman et al., 2000). If the oxidizing agent was indeed water and was associated with silicate, that agent could have been phyllosilicate.

During metamorphism, water is driven out of chondrites. Among OC falls, there is a progressive decrease in the mean concentration of H_2O^+ (chemically bound water) with petrologic type; nevertheless, 67% of type-6 OC falls still contain detectable H_2O^+ (0.1–1.2 wt.%) (Jarosewich, 1990). Type-3 OC falls average 1.06 ± 0.54 wt.% H_2O^+ (n = 10); type-4, 0.65 ± 0.72 wt.% (n = 17), type-5, 0.32 ± 0.36 wt.% (n = 26); and type-6, 0.27±0.33 wt.% (n = 43). [Lower limits of <0.1 wt.% H_2O^+ listed by Jarosewich (1990) are assumed here to be 0 wt.%.] These trends are evident in the individual OC groups as well: H3 falls contain 1.09 ± 1.33 wt.% H_2O^+ (n = 2), whereas H6 falls contain 0.20 ± 0.22 wt.% H_2O^+ (n = 5); L3 falls contain 0.92 ± 0.18 wt.% H_2O^+ (n = 4), whereas L6 falls contain 0.28 ± 0.37 wt.% H_2O^+ (n = 32); LL3 falls contain 1.18 ± 0.46 wt.% H_2O^+ (n = 4), whereas LL6 falls contain 0.26 ± 0.14 wt.% H_2O^+ (n = 5).

Clayton et al. (1991) stated that there was a "slight hint" that the $\delta^{18}O$ and $\delta^{17}O$ values of ordinary chondrites decrease with petrologic type. Rubin (2005a) pointed out that the correlation is, in fact, robust. Table 10.13 shows that the mean $\delta^{18}O$ and $\delta^{17}O$ values progressively decrease from type-4 to type-6 OC in all three groups. (The sole exception is L5, which has a slightly higher mean $\delta^{17}O$ value than L4.) The trend is quite evident in the $\delta^{18}O$ values (which are more precisely determined). Rubin (2005a) ran a series of Student's t-tests to compare the $\delta^{18}O$ values between type-4 and type-6 samples in each OC group: H4 and H6 (t = 1.39 with 11 degrees

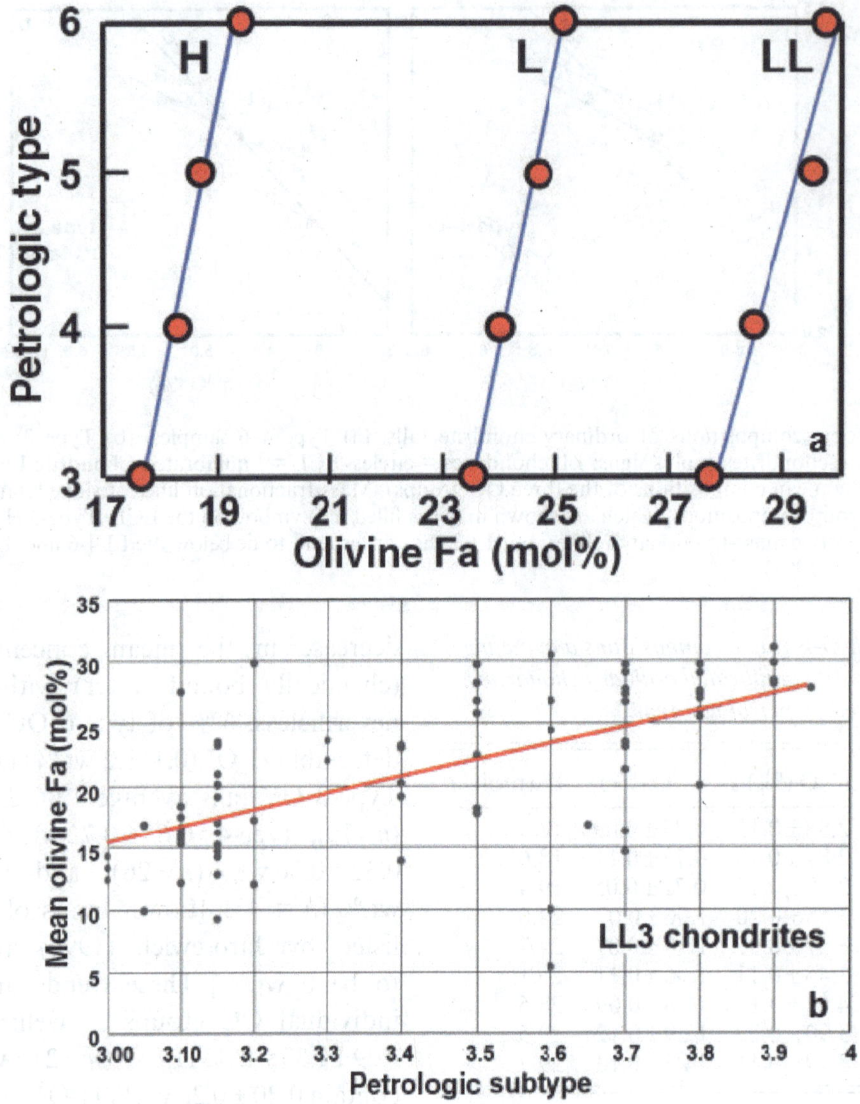

Fig. 10.8 Mean olivine Fa content vs. petrologic type for ordinary chondrites. (a) Data for ordinary chondrites of types 3–6. For each of the three groups, Fa values increase with increasing petrologic type. Trend lines are eyeball estimates. Data from Rubin (1990). (b) Data for all LL3 chondrites from the Meteoritical Bulletin (February 8, 2024) for which mean olivine data were listed and the petrologic subtype was defined (n = 78). Chondrites listed only as LL3 (with no listed subtype) were omitted. For both diagrams, breccias (e.g., L3-6) and transitional groups (e.g., H/L5; L/LL3) were excluded. Although there is some scatter, the LL3 chondrites show a strong positive correlation between mean olivine Fa and petrologic subtype. The trend line is an eyeball estimate.

of freedom, $2\alpha = 0.10$ for a one-sided test, significant at the 80% confidence level); L4 and L6 ($t = 1.48$ with 15 degrees of freedom, $2\alpha = 0.09$ for a one-sided test, significant at the 82% confidence level); and LL4 and LL6 ($t = 3.06$ with 13 degrees of freedom, $2\alpha = 0.005$ for a one-sided test, significant at the 99% confidence level). It seems clear that the mean O-isotopic composition of ordinary chondrites becomes lighter as petrologic type increases. This correlation is likely due to whole-rock metamorphism.

Although type-3 ordinary chondrites contain some phyllosilicates (Hutchison et al., 1987), phyllosilicates have not been reported in type 4–6 OC. This suggests that phyllosilicates suffered partial-to-complete dehydration during metamorphism and transformed into other phases. Nevertheless, it is possible that small amounts of phyllosilicates could have survived thermal metamorphism (e.g., Lipschutz et al., 1999). For example, saponite (a member of the smectite group of phyllosilicates) decomposes at 800 to 900°C (Akai, 1992); this is within the metamorphic temperature range of type-6 OC (i.e., 750–950°C; Dodd, 1981; Huss et al., 2006).

If indigenous water on OC asteroids had been enriched in heavy O isotopes (Choi et al., 1998), loss of this water during metamorphism (ignoring the limited exchange with O in silicates) would have caused OC of higher petrologic types to develop lower $\delta^{18}O$ and $\delta^{17}O$ abundances, consistent with observations (Table 10.13). (The inverse correlations between petrologic type and $\delta^{18}O$ and $\delta^{17}O$ are stronger than with $\Delta^{17}O$ because loss of water moves a point on the three-isotope diagram along a mass-fractionation line; this affects $\delta^{18}O$ and $\delta^{17}O$, but not $\Delta^{17}O$.)

Because olivine Fa composition is a proxy for oxidation state, the correlation between $\Delta^{17}O$ and olivine Fa for ordinary chondrites (Fig. 10.9) means that the bulk O-isotopic composition is also correlated with oxidation state. One potential cause is water released from ordinary chondrites during thermal metamorphism; another is the proportion of FeO-rich (Type-II) chondrules acquired by OC during accretion (see Section 10.11).

10.8.2 Germanium Isotopes

Another potential classificatory parameter for ordinary chondrites is their bulk Ge-isotopic composition. Florin et al. (2020) measured Ge in a suite of ordinary chondrites and found that $\delta^{74/70}Ge_{metal}$ serves as a proxy for $\delta^{74/70}Ge_{bulk}$ wherein:

$$\delta^{74/70}Ge = \left(\left[\left(\frac{{}^{74}Ge/{}^{70}Ge_{Sample}/}{{}^{74}Ge/{}^{70}Ge_{Standard}}\right)-1\right]\times 1000\right).$$

Each OC group (H, L, and LL chondrites as well as IIE irons) has a distinct mean $\delta^{74/70}Ge_{metal}$ value that increases from –0.58‰ in IIE to –0.51‰ in H to –0.31‰ in L to –0.26‰ in LL (Table 10.3).

The degree of oxidation increases from IIE to H to L to LL. This is illustrated by increasing mean olivine Fa values, decreasing modal abundances of metallic Fe-Ni, and increasing values of $\delta^{74/70}Ge_{metal}$ (Figs. 10.10, 10.11b). Because the bulk O-isotopic compositions of OC also correlate with oxidation state (e.g., Fig. 10.9), one would expect $\Delta^{17}O$ to correlate with $\delta^{74/70}Ge_{metal}$ as well. This is indeed the case (Fig. 10.12).

Ciesla et al. (2003) suggested that phyllosilicates could form in the nebula; Wasson (2008) posited that

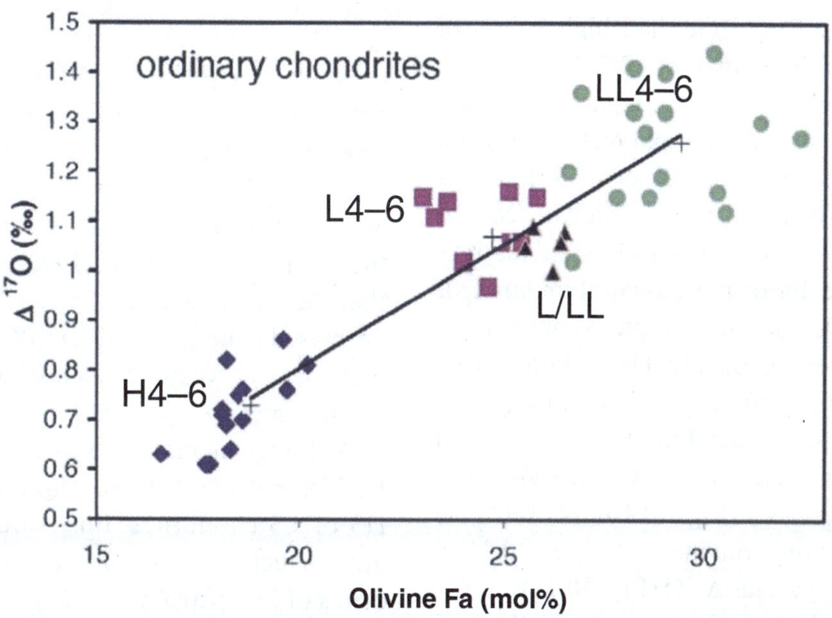

Fig. 10.9 $\Delta^{17}O$ vs. olivine Fa for equilibrated H, L, and LL chondrites. Olivine data from Rubin (1990); O isotope data from Clayton et al. (1991). The mean values for the three OC groups are shown with plus signs; the trend line is the best fit to the three group means. H4–6 chondrites (navy blue diamonds) plot close to the same line, exhibiting a positive correlation between $\Delta^{17}O$ and Fa. The L4–6 chondrites (plum-colored squares) and LL4–6 chondrites (sage-green circles) scatter around the trend line. Chondrites designated L/LL are shown as black triangles. Diagram from Rubin (2005a).

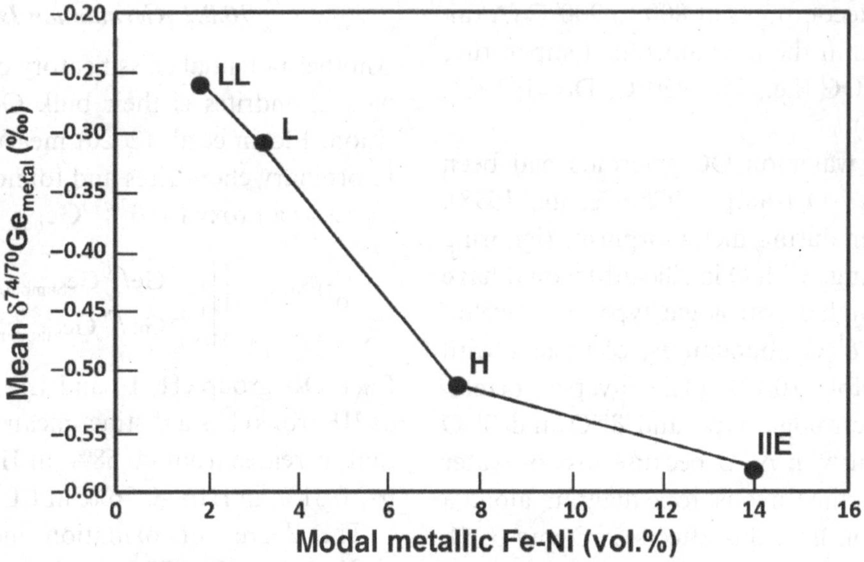

Fig. 10.10 Diagram of mean $\delta^{74/70}Ge_{metal}$ vs. modal metallic Fe-Ni in ordinary chondrites and silicate clasts in IIE irons, showing a correlation between Ge-isotope composition and metal abundance. The trend reflects increasing oxidation from LL to IIE. Diagram from Rubin (2022a).

volatiles derived from fine nebular dust grains that had evaporated during episodes of chondrule formation could react with water in surrounding nebular gas to produce small phyllosilicate grains. Florin et al. (2020) proposed that, in going from H- to L- to LL-chondrite parent bodies, there were increasing proportions of accreted hydrated phyllosilicates that contained 4 to 7 μg/g Ge, had high values of $\delta^{74/70}$Ge (roughly in the range of +2.53‰ to +3‰), and high values of Δ^{17}O (~ 7‰; e.g., Choi et al., 1998). During parent-body thermal metamorphism, the phyllosilicates dehydrated, releasing Ge that was incorporated into the metal phase. This can account for the correlations among oxidation state (as represented by olivine Fa or the modal abundance of metallic Fe-Ni), O- and Ge-isotopic compositions, and Ge abundance: specifically, (1) Δ^{17}O and oxidation state (olivine Fa) (Fig. 10.9), (2) $\delta^{74/70}$Ge and oxidation state (olivine Fa) (Fig. 10.11b), (3) $\delta^{74/70}Ge_{metal}$ versus oxidation state (modal abundance of metallic Fe-Ni) (Fig. 10.10), (4) bulk Ge versus oxidation state (olivine Fa) (Fig. 10.11a), and (5) $\delta^{74/70}Ge_{metal}$ versus Δ^{17}O (Fig. 10.12).

IIE iron meteorites conform to the H, L, and LL correlations (e.g., Tables 10.3, 10.4; Figs. 10.10, 10.13), consistent with their constituting a fourth, more reduced OC group (e.g., Florin et al., 2020; Rubin, 2022a).

10.8.3 Hydrogen Isotopes

The D/H ratio varies among Solar System materials by more than a factor of 25, ranging from $\sim 20 \times 10^{-6}$ for molecular hydrogen in the Sun (Geiss and Gloeckler, 2003) to $>500 \times 10^{-6}$ for ices in comets (Müller et al., 2022). Grant et al. (2023) measured bulk water and D/H ratios in a suite of type-3 ordinary chondrites. They showed that the bulk water content in these meteorites decreases with increasing petrologic subtype (Fig. 10.14a). The least-equilibrated OC falls contain about 1 wt.% H_2O (up to 1.14 wt.% in LL3.10 Piancaldoli; Marrocchi et al., 2020). This trend is consistent with the data of Jarosewich (1990), who demonstrated that the mean bulk H_2O^+ (chemically bound water) decreased from type-3 OC (1.06 wt.%) to type-6 OC (0.27 wt.%). Water was lost during parent-body thermal metamorphism.

Although it may appear from Fig. 10.14a that LL3 chondrites tend to contain more water than H3 or L3 chondrites, this is due solely to the relative paucity in our collections of H and L chondrites of low petrologic subtype. As of this writing, there are about 13 LL3 chondrites of subtype ≤3.05, compared to 6 H3 chondrites and 9 L3 chondrites. There are no subtype data available for the vast majority of type-3 ordinary chondrites.

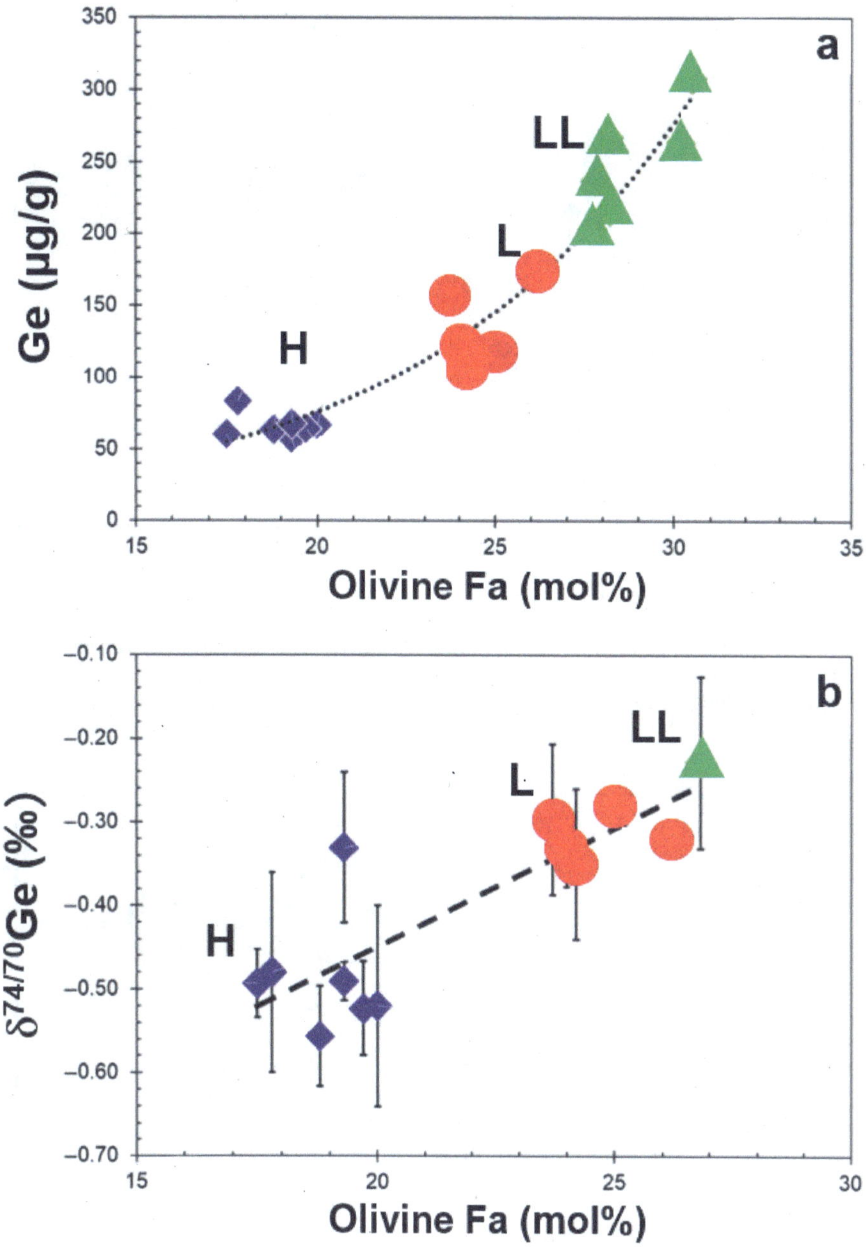

Fig. 10.11 Relationships between Ge and olivine Fa in ordinary chondrites. (a) The bulk Ge concentration vs. olivine Fa in ordinary chondrites, showing increasing concentrations from H to L to LL. (b) $\delta^{74/70}$Ge of the metal phase vs. olivine Fa in olivine in ordinary chondrites, showing increasing metal $\delta^{74/70}$Ge from H to L to LL. Error bars represent ±2 standard deviations. H chondrites (blue diamonds), L chondrites (red circles), and LL chondrites (green triangles) are represented. Diagram modified from Florin et al. (2020).

Type-3 finds tend to contain higher water contents than falls (Fig. 10.14a); this is probably due to contamination from terrestrial water. It is consistent with the data of Jarosewich (1990), who found that the total water contents of non-Antarctic finds exceeded those of observed falls: H (1.64 vs. 0.44 wt.%); L (1.11 vs. 0.46 wt.%); LL (2.13 vs. 0.71 wt.%).

Also anticorrelating with petrologic type is the mean δD value of ordinary chondrites (Fig. 10.14b). It is high in the least equilibrated samples (subtype 3.00–3.15), although there is appreciable variation among different samples of the same meteorite. The highest average values in individual type 3.00 OC are ~1500‰ (Alexander et al., 2012; Table 1 of Grant et al., 2023). Some water released from bulk

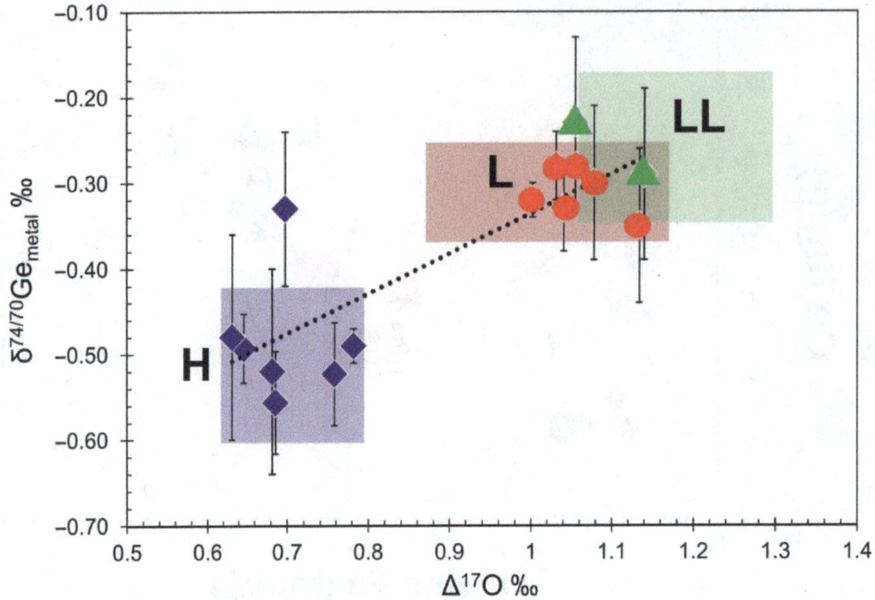

Fig. 10.12 The $\delta^{74/70}$Ge value of the metal phase vs. bulk Δ^{17}O in type 3–6 ordinary chondrites: H (blue shades), L (red shades), LL (green shades). Both parameters increase from H to L to LL. Points represent individual meteorites; rectangles represent the range of values for each OC group. The dotted line is a linear regression. H chondrites (blue diamonds), L chondrites (red circles), and LL chondrites (green triangles) are represented. Diagram modified from Florin et al. (2020).

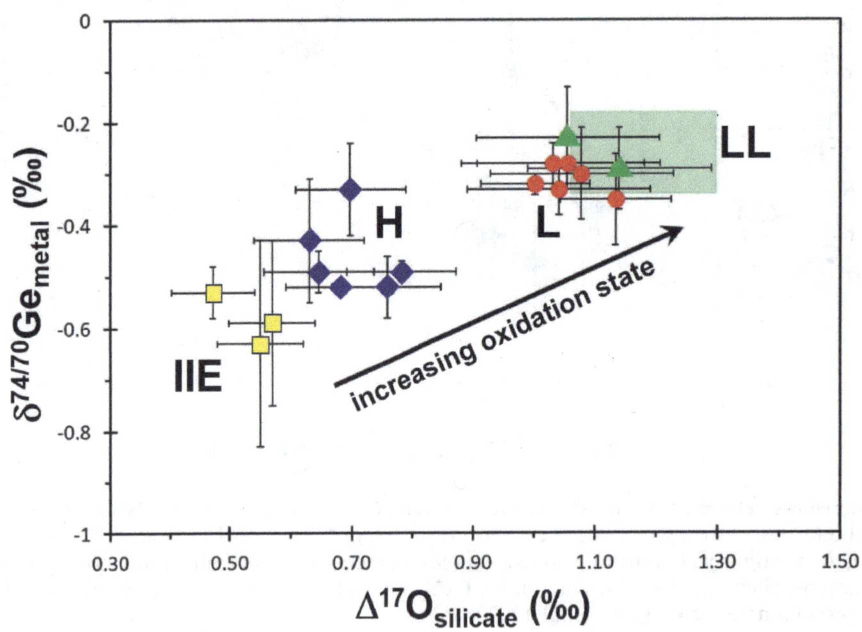

Fig. 10.13 $\delta^{74/70}Ge_{metal}$ vs. Δ^{17}O for H, L, and LL ordinary chondrites and the silicate fraction of IIE irons. These groups show the relationship between the Ge- and O-isotopic compositions. This is consistent with IIE iron meteorites being a fourth, more reduced OC group. Both parameters increase from IIE to H to L to LL. Points represent individual meteorites; the shaded green rectangle represents the range of values for LL chondrites. IIE irons (yellow squares), H chondrites (blue diamonds), L chondrites (red circles), and LL chondrites (green triangles) are represented. Diagram modified from Florin et al. (2020).

Semarkona material by pyrolysis has a δD value of ~3000‰ (e.g., McNaughton et al., 1981); the highest value recorded in bulk Semarkona is δD = 5300‰ (Yang and Epstein, 1983). As petrologic type increases from type 3.7 to 4, δD values flatten out and reach levels of roughly –60 to –150‰. This indicates

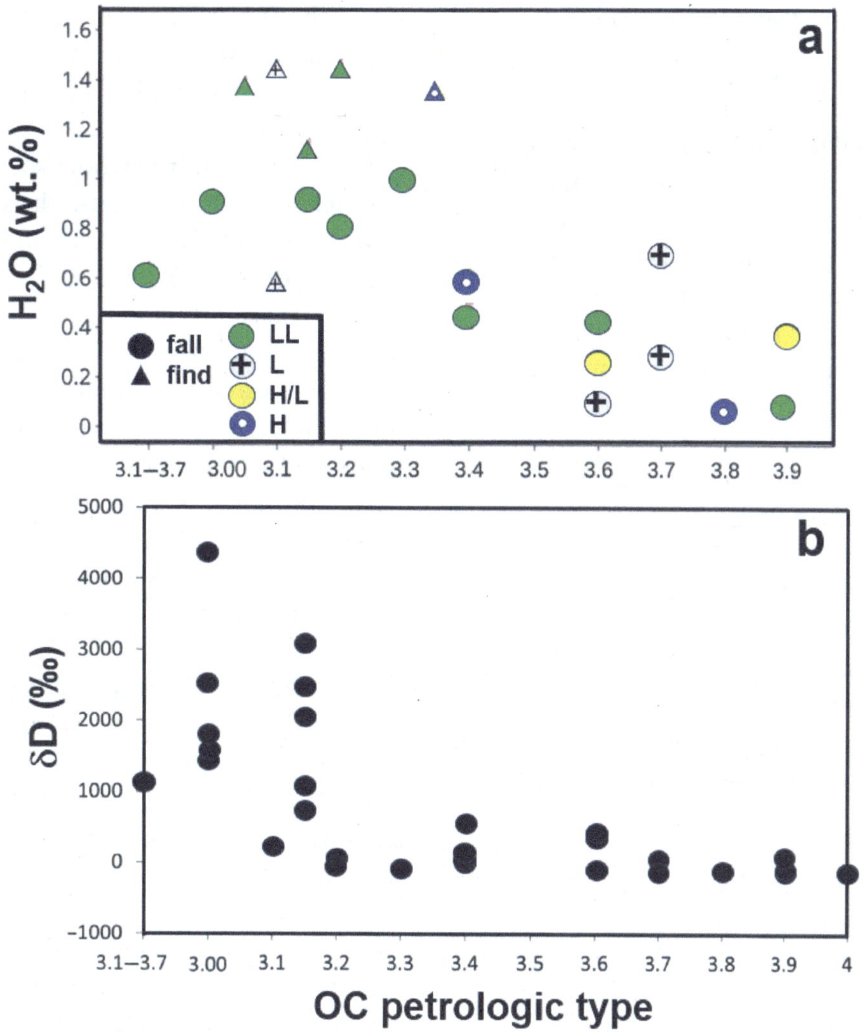

Fig. 10.14 Relationships between water, δD, and petrologic type of ordinary chondrites. (a) Bulk H_2O vs. subtype in type-3 ordinary chondrites. The point at the 3.1–3.7 subtype is Ngawi, a genomict LL3 breccia containing type 3.1–3.7 clasts (as well as LL4 and LL6 clasts; Scott and Taylor, 1982). The least equilibrated ordinary chondrites (those with the lowest subtypes) tend to contain more water than more equilibrated samples. Finds (triangles) tend to have more bulk water than falls (circles), presumably due to terrestrial contamination. H chondrites (blue with internal white circles), H/L chondrites (yellow), L chondrites (white with internal plus signs), and LL chondrites (green) are represented. (b) δD vs. OC petrologic type ranging from type 3.00 to 4. There is considerable variation in the δD values of the least equilibrated samples, but all measured portions of them have much higher δD values than those in more equilibrated OC. Data are from Grant et al. (2023) and the literature. Diagrams modified from Grant et al. (2023).

that thermal metamorphism destroys the δD-rich component. Because type-3 OC contain little insoluble organic matter (Alexander et al., 2007) it is apparent that the D-rich component is not organic material. Because water is driven out of OC during metamorphism, it seems likely that it is the water (bound in phyllosilicates) that has high D/H ratios.

Ordinary chondrites are noncarbonaceous materials that accreted in the Inner Solar System. Although many workers have suggested that, in the solar nebula, D/H ratios tended to increase from the warm regions of the Inner Solar System to the cooler zones farther from the Sun (e.g., Drouart et al., 1999; Alexander et al., 2012; Jacquet and Robert, 2013; Yang et al., 2013; Ito et al., 2022), the occurrence of high δD values in portions of the least equilibrated ordinary chondrites indicates that the situation is more complex. Grant et al. (2023) suggested that the high δD component of type-3 OC was derived either from the outer regions of the solar nebula or from "streamers" injected from the parental molecular cloud.

10.8.4 Nucleosynthetic Isotope Anomalies

Some meteorites contain anomalous isotopic abundances derived from solid materials originating outside the Solar System. These small grains, formed within the outflow of dying stars, were eventually incorporated into the molecular cloud from which the solar nebula and the Sun (and likely solar siblings; e.g., Ramírez et al., 2014) arose. Nucleosynthetic isotope anomalies in meteorites were reviewed by Qin and Carlson (2016). As has been pointed out by these authors, there is a decoupling of light and heavy nucleosynthetic elements. Meteorite whole-rocks tend to be depleted in s-process isotopes of Sr, Zr, Mo, Ru, Ba, and Nd relative to the Earth.

The average isotopic anomalies for 11 elements in ordinary chondrites are listed in Table 10.14. In many cases, the differences between the three OC groups (where comparable data are available) are smaller than the experimental uncertainties. (Iron meteorites tend to have slightly heavier $^{57}Fe/^{54}Fe$ ratios [by 0.1–0.2 per mil] than chondrites due to the enrichment of solid metal in heavier Fe isotopes relative to liquid metal during core formation [Ni et al., 2020].)

Table 10.14 *Average isotopic anomalies for selected elements in ordinary chondrites.*

Element	Isotope	Concentration
Ca	$\varepsilon^{48}Ca$	−0.20 ± 1.07
Ti	$\varepsilon^{46}Ti$	−0.09 ± 0.03
Ti	$\varepsilon^{50}Ti$	−0.64 ± 0.12
Cr	$\varepsilon^{54}Cr$	−0.42 ± 0.03
Ni	$\varepsilon^{62}Ni$	−0.07 ± 0.04
Ni	$\varepsilon^{64}Ni$	−0.17 ± 0.12
Sr	$\varepsilon^{84}Sr$	0.15 ± 0.04
Zr	$\varepsilon^{96}Zr$	0.51 ± 0.14
Mo	$\varepsilon^{92}Mo$	0.77 ± 0.28
Mo	$\varepsilon^{94}Mo$	0.69 ± 0.19
Mo	$\varepsilon^{95}Mo$	0.29 ± 0.10
Mo	$\varepsilon^{97}Mo$	0.14 ± 0.06
Mo	$\varepsilon^{100}Mo$	0.19 ± 0.15
Ru	$\varepsilon^{96}Ru$	0.35 ± 0.14
Ru	$\varepsilon^{98}Ru$	0.15 ± 0.21
Ru	$\varepsilon^{100}Ru$	−0.30 ± 0.03
Ru	$\varepsilon^{102}Ru$	−0.12 ± 0.04
Ru	$\varepsilon^{104}Ru$	0.13 ± 0.16
Ba	$\varepsilon^{135}Ba$	0.08 ± 0.04
Ba	$\varepsilon^{137}Ba$	−0.02 ± 0.06
Nd	$\varepsilon^{142}Nd$	−0.21 ± 0.03
Nd	$\varepsilon^{145}Nd$	0.03 ± 0.03
Nd	$\varepsilon^{148}Nd$	0.05 ± 0.14
Sm	$\varepsilon^{144}Sm$	0.02 ± 0.07

Data from Qin and Carlson (2016).

Ordinary-chondrite whole-rocks exhibit negative anomalies in the neutron-rich isotopes ^{48}Ti, ^{54}Cr, ^{62}Ni, ^{64}Ni, and ^{48}Ca. The abundances of many of these isotopes in OC are similar to those in other noncarbonaceous materials (e.g., enstatite chondrites, R chondrites, HEDs, main-group pallasites, IIIAB irons, angrites, ureilites, lunar rocks, martian rocks). In contrast, carbonaceous chondrite whole-rocks tend to show positive anomalies in ^{48}Ti, ^{54}Cr, ^{62}Ni, ^{64}Ni, and ^{48}Ca, suggesting the formation of these isotopes in distinct nucleosynthetic events. This would be consistent with there being separate formation locations for carbonaceous (CC) and noncarbonaceous (NC) rocks (Chapter 6).

Frossard et al. (2024) analyzed the Cr-isotopic compositions of leachates of L3.2 NWA 8007 and two enstatite chondrites (EH3 Sahara 97158 and EL3 PCA 91020). The leachate of the ordinary chondrite showed variable Cr-isotopic compositions whereas those of the enstatite chondrites did not. The Cr-isotopic homogeneity in the enstatite chondrites probably reflected the instability of the presolar Cr carriers under reducing conditions. The authors concluded that there was one major presolar ^{54}Cr-rich carrier in the solar nebula and that this carrier was depleted in the carbonaceous regions of the Solar System relative to the noncarbonaceous regions.

Isotopes that are neutron rich (e.g., ^{50}Ti, ^{54}Cr, ^{62}Ni, ^{64}Ni, ^{48}Ca) likely formed by related astrophysical processes in similar environments. In many cases, the abundances of these isotopes are correlated with one another (e.g., $\varepsilon^{54}Cr$ and $\varepsilon^{62}Ni$; $\varepsilon^{48}Ca$ and $\varepsilon^{50}Ti$; Fig. 2 of Qin and Carlson, 2016). Molybdenum, Ru, and Ba are remarkable in having seven isotopes produced by different nucleosynthetic means – the p (proton capture reaction), s (slow), and r (rapid) processes: ^{92}Mo (p), ^{94}Mo (p, s), ^{95}Mo (s, r), ^{96}Mo (s), ^{97}Mo (s, r), ^{98}Mo (s, r), ^{100}Mo (r); ^{96}Ru (p), ^{98}Ru (p), ^{99}Ru (s, r), ^{100}Ru (s), ^{101}Ru (s, r), ^{102}Ru (s, r), ^{104}Ru (r); ^{130}Ba (p), ^{132}Ba (p), ^{134}Ba (s), ^{135}Ba (s, r), ^{136}Ba (s), ^{137}Ba (s, r), ^{138}Ba (s, r). (Some extant ^{135}Ba could have formed from the extinct isotope ^{135}Cs; small amounts of ^{138}Ba can be produced from ^{138}La.)

S-process (slow) nucleosynthesis occurs in asymptotic giant branch (AGB) stars (see Section 12.1) wherein an atomic nucleus seed captures a neutron, creating an isotope with an atomic mass that is 1 unit higher. Stable isotopes can continue to increase in mass via neutron capture over time scales of $n \times 10^1$ to $n \times 10^3$ years. Unstable isotopes will undergo beta decay, creating a new element with an atomic number that is 1 unit higher. In beta minus (β^-) decay, a neutron is transformed into a proton, resulting in the emission of an electron plus an electron antineutrino. The heaviest elements produced by the s process are ^{209}Bi ($t_{1/2} = 2.01 \times 10^{19}$ a) and ^{210}Po ($t_{1/2} = 138.376$ d).

In r-process (rapid) nucleosynthesis, seed nuclei rapidly acquire nearby free neutrons, thereby creating massive nuclei (including Se, Br, Kr, Ag, Te, I, Xe, Os, Ir, Pt, Au, U, and possibly superheavy elements). Neutrons are acquired much faster than beta decay can occur, perhaps at a rate of 100 neutron captures per second. An environment conducive to this process can be found in the merging of inspiraled binary neutron stars that had previously orbited one another. The first such merger was discovered on August 17, 2017 through the detection of gravitational waves (GW170817) (e.g., Belczynski et al., 2018). The source of the gravitational waves had an electromagnetic counterpart (a distinctive thermal emission pattern dubbed a "kilonova") that revealed the presence of newly synthesized lanthanide isotopes created by the r process (Kasen et al., 2017). Beginning 1.7 s after the detection of the gravitational waves there was a short gamma-ray burst (GRB170817A) in the same region of the sky; it may have been caused by a giant flare produced by the merger (Salafia et al., 2018).

10.9 Prior's Rules and Systematic Changes in Oxidation State

George T. Prior, Keeper of Mineralogy at the British Museum of Natural History from 1909 to 1927, made two general observations about variations in meteorite mineralogy as a function of oxidation state (Prior, 1916): "The less the amount of Ni-Fe in chondritic stones, the richer it is in Ni, and the richer in Fe are the magnesium silicates." These observations have been recognized by researchers (e.g., Mueller and Olsen, 1967) and enshrined in meteorite textbooks since the mid twentieth century as "Prior's Rules" (e.g., Mason, 1962; McCall, 1973; Wasson, 1974; Sears, 1978; Dodd, 1981; Heide and Wlotzka, 1995; Norton, 2002; Hutchison, 2004; Sears, 2004; Rubin and Ma, 2021). However, as Keil (1969) pointed out, Prior (1916) had assumed that chondrites varied in redox state but not in bulk chemical composition. That assumption is belied by the differences among OC groups in their concentrations of bulk Fe (Tables 10.15a, 10.15b) and many other elements (Table 10.7).

Most of the mineralogic differences among the H, L, and LL chondrite groups (as well as IIE; Tables 10.3, 10.4) result from systematic differences in mean oxidation state (Tables 10.15a, 10.15b). For example:

Table 10.15a *Systematic variations in mean mineralogical, bulk chemical, and bulk physical properties of ordinary-chondrite groups. Most are attributable to changes in oxidation state.*

–	Olivine Fa (mol%)	Low-Ca pyroxene Fs (mol%)	Modal (vol.%/vol.%) olivine/low-Ca-pyroxene ratio	Chromite TiO_2 (wt.%)	Chromite FeO (wt.%)	Ilmenite FeO (wt.%)	Kamacite Co (mg/g)
Source of data	a	b	c	d	d	b	a
H	18.8	17.2	1.4	2.3	31.2	40.9	4.7
L	24.7	21.3	2.0	2.8	33.0	41.9	8.0
LL	29.4	24.1	3.6	3.2	34.5	44.2	26.8*

* Median values are reported for LL kamacite Co and Ni because the few very high values of Co (up to 370 mg/g) and low values of Ni (as low as 14.0 mg/g) in LL4 to LL6 chondrites skew the means.

Sources of data: a = Rubin (1990); b = Gomes and Keil (1980); c = Hutchison (2004); d = Snetsinger and Keil (1969).

Table 10.15b *Additional systematic variations in mean mineralogical, bulk chemical, and bulk physical properties of ordinary-chondrite groups, most of which are attributable to changes in oxidation state.*

–	Kamacite Ni (mg/g)	Ge in metal (μg/g)	Metallic Fe-Ni-Co (wt.%)	Kamacite/taenite modal ratio	Bulk Fe (wt.%)	Bulk Ni/Mg	Bulk density (g cm^{-3})
Source of data	a	e	f	g	f	h	i
H	69.4	64.6	17.8	9	27.45	0.115	3.53
L	65.3	135.9	8.33	3	21.93	0.083	3.42
LL	50.0*	102.5	3.56	1	19.63	0.067	3.28

* Median values are reported for LL kamacite Co and Ni because the few very high values of Co (up to 370 mg/g) and low values of Ni (as low as 14.0 mg/g) in LL4 to LL6 chondrites skew the means.
Sources of data: a = Rubin (1990); e = Florin et al. (2020); f = Jarosewich (1990); g = The kamacite/taenite modal ratio is calculated based on the mean Ni content of kamacite in each group (this table; data from Rubin, 1990), the mean Ni content in the metal of each group (Afiattalab and Wasson, 1980) and the assumption that taenite in each group contains 400 mg/g Ni (Afiattalab and Wasson, 1980); h = Kallemeyn et al. (1989); i = Consolmagno et al. (2006). Data for metallic Fe-Ni-Co, bulk Fe, and bulk density are for observed falls only.

(1) Mean metallic Fe-Ni-Co decreases from H (17.8 wt.%) to L (8.33 wt.%) to LL (3.56 wt.%) (Jarosewich, 1990) as increasing proportions of iron change from metallic (Fe^0) to ferrous (Fe^{2+}). The H→L→LL sequence shows (2) increases in the FeO contents of olivine, low-Ca pyroxene, chromite, and ilmenite, and increases in the TiO_2 content of chromite (e.g., Mueller and Olsen, 1967; Bunch et al., 1967; Snetsinger and Keil, 1969; Gomes and Keil, 1980; Rubin, 1990) as more Fe^{2+} becomes available and higher Ti^{4+}/Ti^{3+} ratios develop in the set of chondrule olivine grains (e.g., Simon et al., 2008) with increasing bulk oxidation state, (3) increases in the Co and Ni concentrations in kamacite (Sears and Axon, 1976; Rubin, 1990) as more metallic iron is oxidized, (4) increases in the Ge content of the metal fraction (Chou and Cohen, 1973; Chou et al., 1973; Florin et al., 2020), a consequence of the diminution of Fe, (5) decreases in the kamacite/taenite modal abundance ratios (Afiattalab and Wasson, 1980) (because less metallic Fe is available to produce kamacite – the low-Ni metal phase),[2] and (6) higher olivine/low-Ca-pyroxene ratios (Gastineau-Lyons et al., 2002; Hutchison, 2004) as more Fe^{2+} is produced from metallic Fe.

The oxidation state of a mineral assemblage is a function of the oxygen fugacity (f_{O2}) that pertained during the final equilibrium among the major phases and an associated gas phase. The f_{O2} value is equivalent to the partial pressure of oxygen in an ideal gas. As indicated by Rubin et al. (1988b), the relevant redox reactions among metallic Fe, olivine, low-Ca pyroxene, and oxide in ordinary chondrites are:

$$3Fe + 2O_2(g) = Fe_3O_4$$

$$2MgSiO_3 + 2Fe + O_2(g) = Mg_2SiO_4 + Fe_2SiO_4$$

$$Mg_2SiO_4 + 2FeSiO_3 = Fe_2SiO_4 + 2MgSiO_3$$

wherein Fe_3O_4 is magnetite, $MgSiO_3$ is enstatite, $FeSiO_3$ is ferrosilite, Mg_2SiO_4 is forsterite, and Fe_2SiO_4 is fayalite.

Brett and Sato (1984) measured the intrinsic f_{O2} as a function of temperature for several ordinary-chondrite falls (Fig. 10.15). The data for LL3.01 Semarkona showed appreciable scatter (±0.4 log f_{O2} units at a given T), consistent with this meteorite's containing an unequilibrated assemblage of phases formed over a range of f_{O2} conditions. These phases tended to equilibrate during measurement. The limited data show that ordinary chondrites are more reduced than the iron-wüstite (I-W) buffer and that H chondrites are more reduced than L or LL chondrites. Although LL6 Cherokee Springs appears a bit more reduced than L5 Farmington, there is a large standard error in the Cherokee Springs data, leaving the result uncertain (Brett and Sato, 1984).

[2] The kamacite/taenite modal ratio reported in Table 10.15b is a new calculation (different from that in Table 8.1 of Rubin and Ma, 2021, which was based on the number of analyzed kamacite and taenite grains in Afiattalab and Wasson, 1980). The new values in Table 10.15b are based on the mean Ni content of kamacite in each OC group, the mean Ni content in the metal of each group (Afiattalab and Wasson, 1980), and the assumption that taenite in each OC group contains 400 mg/g Ni (Afiattalab and Wasson, 1980). Because of uncertainties in this approach, the listed values have only one significant figure.

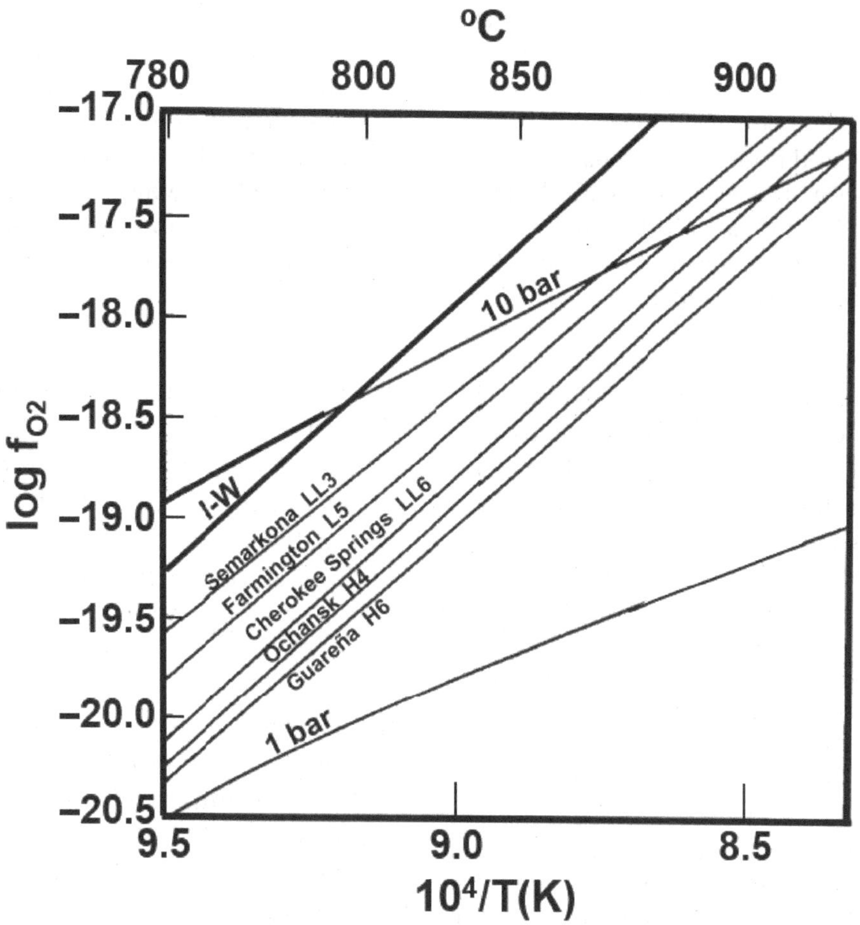

Fig. 10.15 Plot of linear regression lines for log f_{O2} vs. 1/T (where T is temperature in Kelvins) for several H, L, and LL ordinary chondrites. The graphite surface is shown for 1 bar and 10 bar pressures. I-W is the iron-wüstite buffer. If iron and wüstite are both present in an assemblage, the oxidizing conditions remain constant on the I-W curve. Ordinary chondrites lie below I-W; H chondrites (which are the most reduced OC group) lie farthest below the I-W line. Diagram modified from Rubin et al. (1988b).

As temperatures increase during thermal metamorphism, the oxygen fugacity increases (Fig. 10.15). This is consistent with the higher olivine Fa values and higher kamacite Co contents in type-6 OC relative to type-3 OC (Table 13.2b). Reduction by graphite was not an important process during OC metamorphism; this is because little graphite is present in most OCs, and at high temperatures graphite tends to dissolve in the metallic Fe-Ni.

10.10 The Metal-Silicate Fractionation

In going from H to L to LL, there are: (a) decreases in total Fe (Jarosewich, 1990), (b) decreases in siderophile/lithophile element ratios (e.g., Ni/Mg) (Kallemeyn et al., 1989), and (c) decreases in bulk density (Consolmagno et al., 2006), consonant with decreasing proportions of total Fe. The differences in total Fe and in the siderophile/lithophile element ratios are nebular effects attributable to the metal-silicate fractionation (e.g., Urey and Craig, 1953; Tandon and Wasson, 1968; Müller et al., 1971; Wasson, 1972).

A plausible mechanism that may account for this fractionation was offered by Rubin (2022a). He suggested it was caused, in part, by the loss of metal from chondrules during chondrule melting and the subsequent failure of chondrite whole-rocks to reacquire this metal quantitatively.

There is abundant evidence that metal was lost from chondrules:

(1) Metallic Fe-Ni melts are immiscible in silicate melts. A spinning molten or partly molten chondrule would tend to expel rounded metal-sulfide blebs along the plane of the chondrule equator. If the chondrule were solidifying at this time, depressions left by the expelled metal-rich blebs may remain as "chondrule craters" (e.g., Fig. 10.16a). Most of these craters are hemispherical depressions lined or partly filled with metallic Fe-Ni and/or sulfide (Grossman and Wasson, 1985); these are remnants of the expelled opaque blebs. About 10% of OC chondrules contain such depressions (Gooding and Keil, 1981).
(2) There are rare opaque assemblages occurring outside of chondrules that could represent a subset of the expelled metal-rich blebs (Fig. 10.16b). These objects have sometimes been called "metallic chondrules."
(3) Some OC chondrules are surrounded by rings of metal and sulfide blebs and ribbons (e.g., Allen et al., 1980). Evaporative loss of some of this material while chondrules were close to their peak temperatures could have contributed to the metal-silicate fractionation.
(4) Some individual Type-II chondrules in ordinary and carbonaceous chondrites have highly fractionated (non-flat) abundance patterns of siderophile elements (e.g., Wasson et al., 2000). Such chondrules contain high abundances of Fe relative to those of other siderophile and chalcophile elements (Fig. 10.17). Because much of the Fe occurs as FeO in the silicate portion of these ferroan chondrules, the Fe was largely retained when the metal blebs were lost (even though there were probably only small amounts of metal to begin with). In contrast, the low-FeO chondrules have rather flat abundance patterns for refractory and common siderophiles (Ir, Ni, Co, Fe), indicating that there was little fractionation during chondrule melting even if opaque blebs were expelled. This is because nearly all the Fe in the low-FeO chondrules was in the metal and was lost along with the other siderophiles.

Low-FeO chondrules' loss of metallic Fe-Ni-rich globules that were not later re-acquired by the chondrite whole-rock would cause a decrease in the bulk chondrite Fe/Si and Ni/Si ratios. This is consistent with the CI-normalized Ni/Si ratios of H, L, and LL chondrites being appreciably less than unity (H: 0.94; L: 0.66; LL: 0.53; Table 10.7; Lodders, 2021).

If the metal lost from chondrules during chondrule formation had been quantitatively re-acquired by the whole-rock during chondrite agglomeration, this process would not have contributed to the metal-silicate fractionation. However, it appears that much of the metal was lost: "metallic chondrules" constitute only ~0.5% of OC chondrules (Gooding and Keil, 1981).

Where did the lost metal go? One possibility is that some of the lost metal settled to the nebular midplane prior to parent-body accretion (Wasson, 1985). Another is that, during the several-million-year period when nebular gas was

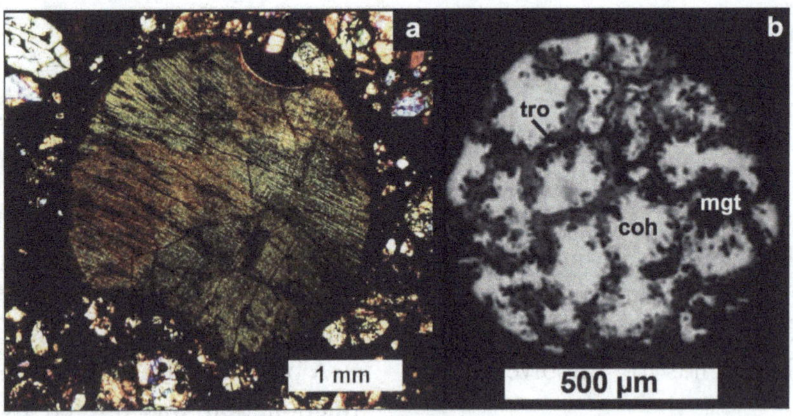

Fig. 10.16 Chondrules that have lost opaque blebs. (a) Fine-grained barred-pyroxene (BP) chondrule in LL3.7 NWA 3126 showing two depressions ("craters") at opposite ends, presumably located along the chondrule's equator. Such depressions may be the site of metallic droplets' expulsion from the spinning precursor semi-molten droplet via centrifugal force. Metal droplets and surrounding molten silicate would have been immiscible. Transmitted light, crossed polars. Image courtesy of John Kashuba. (b) An opaque nodule from LL3.01 Semarkona consisting of polycrystalline grains of cohenite (coh) mantled by magnetite (mgt) and a thin layer of troilite (tro). The nodule was probably initially composed mainly of metallic Fe-Ni with minor troilite prior to whole-rock aqueous alteration. Image modified from Grossman and Wasson (1985).

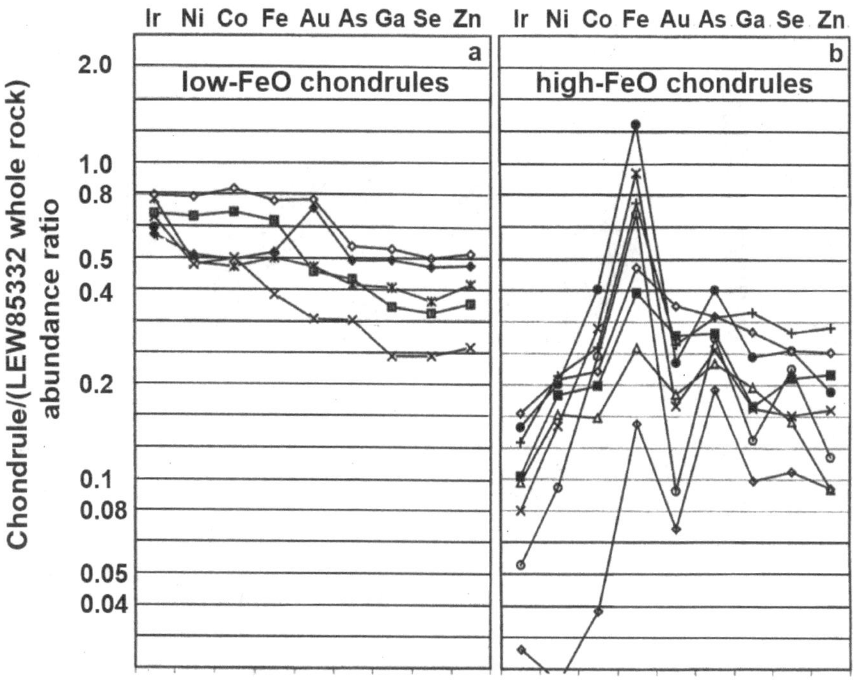

Fig. 10.17 Bulk siderophile and chalcophile elements in selected low-FeO (Type-I) and high-FeO (Type-II) chondrules in the ungrouped type 3.0–3.1 carbonaceous chondrite LEW 85332. Compositions were determined by INAA and normalized to the LEW 85332 whole-rock. (a) The low-FeO chondrules tend to have rather flat abundance patterns for refractory and common siderophiles (Ir, Ni, Co, Fe) and show increasing depletions in the more volatile siderophiles (Au, As, Ga) and chalcophiles (Se, Zn). (b) The high-FeO chondrules tend to have highly fractionated siderophile abundance patterns: Fe is much higher than any of the siderophile or chalcophile elements. Horizontal lines show 0.10 intervals in log abundance ratios. Diagram modified from Wasson et al. (2000).

still present (e.g., Ciesla and Charnley, 2006), metal globules drifted far away from the chondrules due to differences in their aerodynamic behavior.

Such behavior is a function of the product of a particle's size and density (e.g., Weidenschilling, 1977; Nakagawa et al., 1986). This agrees with the conclusions of Kuebler et al. (1999), who studied chondrules and metal-troilite grains in ordinary chondrites and found that aerodynamic stopping times in the nebula varied with particle size and density. An isolated 75-μm diameter opaque globule (composed of 80 wt.% metal and 20 wt.% troilite) with a density of ~7.2 g cm^{-3} would experience dramatically different radial drift rates in the nebula than a 500-μm-diameter chondrule with a density of ~3.3 g cm^{-3}. (The latter case has a diameter-density product three times larger.)

Figure 10.18 shows that loss of metal and changes in redox state are both partly responsible for the metal-silicate fractionation. If metal loss alone caused the metal-silicate fractionation, the different chondrite groups would exhibit little horizontal spread in Fig. 10.18. Furthermore, as shown in Fig. 10.17, fractionations among the set of siderophile elements are also functions of both metal loss and oxidation state.

10.11 Size Scales of Homogeneity in Oxidation State and O-Isotopic Composition

Vernazza et al. (2014) proposed that the occurrence of large numbers of compositionally similar OC asteroids was a natural consequence of planetesimal formation. They suggested that multiple H and LL parent asteroids formed through fast accretion on time scales of less than 0.3 Ma. It is clear, however, that in each of the three major OC groups, most members (or at least a plurality) hail from the same parent asteroid. There are several factors supporting this assertion: (1) Heymann (1967) found that about two-thirds of L chondrites suffered He losses during a major collisional event (that likely destroyed the asteroid) at a time now determined to be ~470 Ma ago (Korochantseva et al., 2007).

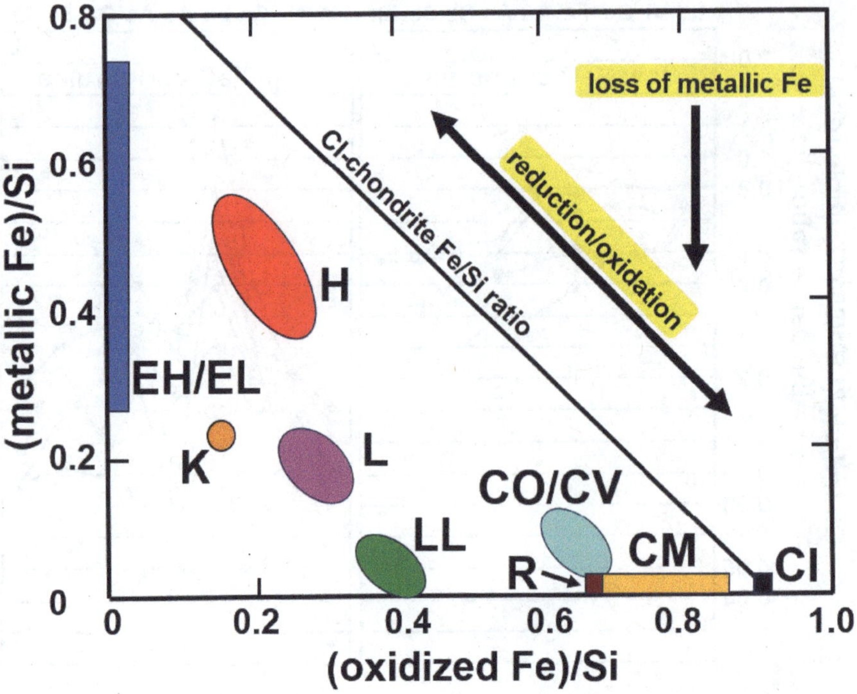

Fig. 10.18 Concentration of metallic Fe vs. concentration of oxidized Fe in chondrite groups. Both parameters are normalized to the concentration of Si. Loss of metal and changes in oxidation state are both partly responsible for the metal-silicate fractionation. The CI Fe/Si ratio represents the mean Solar System composition. Diagram adapted from McSween and Huss (2010).

Because this single collision affected most L chondrites, it is clear that most L chondrites were present on the same parent body at that time. (2) About 45% of H chondrites, mainly those of petrologic types 4 and 5, plot within a CRE-age cluster peaking at 7–8 Ma (Graf and Marti, 1995). Almost all these meteorites were launched from the same parent asteroid by an impact event ~7.5 Ma ago. (3) An analogous situation occurs in LL chondrites: about a third of these meteorites fall within a CRE-age cluster peaking at 15 Ma (Graf and Marti, 1994). (4) Among L chondrites, Marti and Graf (1992) found a large fraction of L5 chondrites and an even larger fraction of L6 chondrites cluster within a CRE peak at 40 Ma; this is a strong indication of a major collisional event on a large remnant of the original L parent body at that time.

Most estimates of the diameters of OC parent bodies are on the order of 200 km (e.g., Trieloff et al., 2003; Kleine et al., 2008; Henke et al., 2012). Blackburn et al. (2017) obtained $^{207}Pb/^{206}Pb$ data on phosphates from a suite of ordinary chondrites and estimated the minimum diameters of the H and L parent asteroids to be ~260–280 km. Gaffey and Gilbert (1998) suggested the 185-km-diameter asteroid 6 Hebe was the H-chondrite parent body; this body is ~30% smaller in diameter than the Blackburn et al. estimate. Nevertheless, if these estimates are approximately correct, the OC parent bodies would have been among the largest asteroids: today there are only 26 asteroids >200 km in diameter and only ~250 asteroids that are >100 km. Most extant asteroids that are $\gtrsim$120 km in diameter are probably primordial (Bottke et al., 2005).

Rubin et al. (2008) explored the size scales over which OC properties were relatively homogeneous:

Global size scale (diameters of ~200–300 km): OC are relatively homogeneous in mean chondrule size. Nelson and Rubin (2002) measured the sizes of 644 intact chondrules in four moderately to highly unequilibrated LL3 chondrites and found their mean apparent diameters to vary by ≤15%. OC are also homogeneous in bulk chemical composition. Kallemeyn et al. (1989) analyzed 22 H chondrites (21 falls and one unweathered find) by INAA. They found that the mean Mg- and CI-normalized abundances of lithophile elements (Al, Ca, Sc, La, Sm, Eu, Yb, Lu, V, Mg, Cr, Mn, Na, K) in these samples have a narrow range (0.90–0.95 × CI) and a low standard deviation (0.91 ± 0.01). Other chondrite groups have different (but similarly narrow) ranges in bulk chemical composition (e.g., Wasson and Kallemeyn, 1988).

Properties that are heterogeneous at global size scales include oxidation state (as represented by olivine Fa) (e.g., Table 10.1a; Figs. 10.19a, 10.20) and bulk O-isotopic composition (Fig. 10.19b).

Kilometer size scale: Anders (1978) estimated that the ~7.5-Ma collisional event on the H-chondrite parent body excavated material with a volume on the order of 1 km^3. If this is correct, H chondrite samples that are of the same petrologic type and have identical CRE ages were probably located within ~1 km of each other on the H-chondrite asteroid prior to the collision. The olivine compositional distributions of two H5 chondrites (ALHA79046 and TIL 82415) with identical CRE ages of 7.6 Ma are shown in Fig. 10.21. Both chondrites are of S2 shock stage and have similar modal abundances of metallic Fe-Ni (15.2 and 16.8 wt.%, respectively) and troilite (6.0 and 5.4 wt.%, respectively). Nevertheless, the two olivine distributions

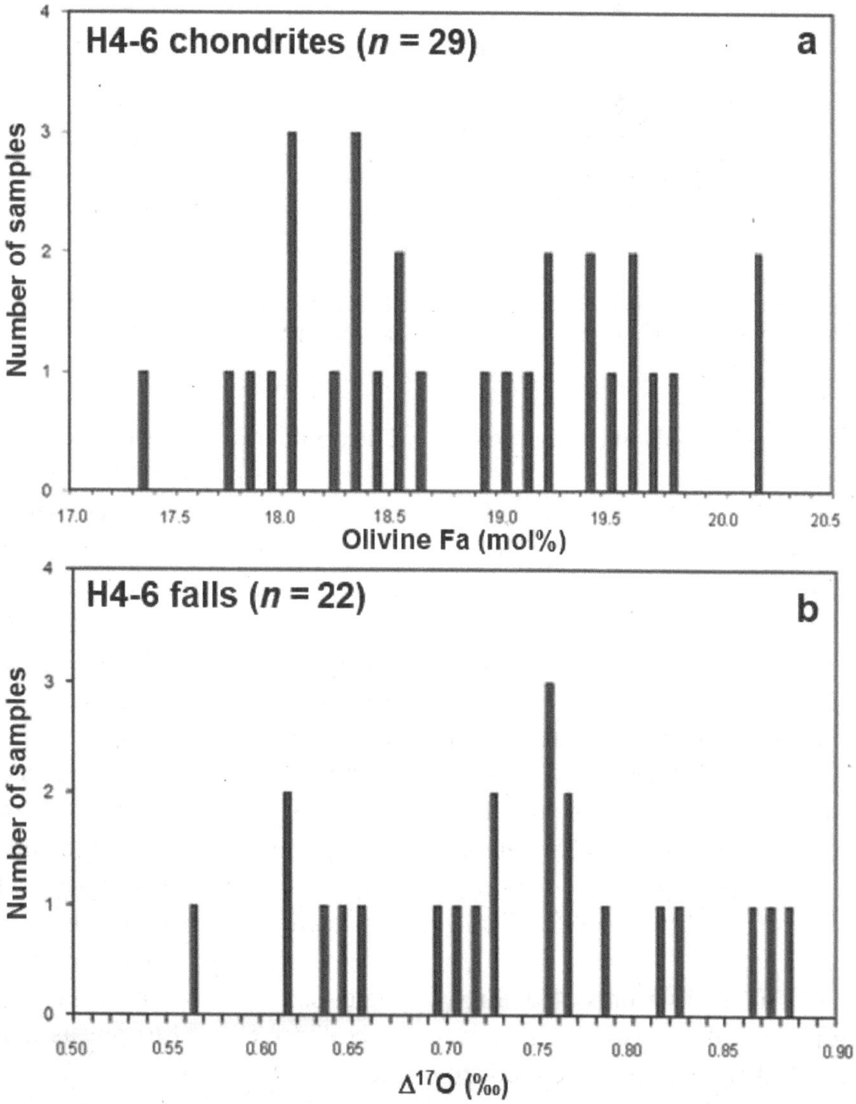

Fig. 10.19 Global compositional variations in oxidation state and O-isotopic composition among H chondrites. (a) Oxidation state as indicated by variations in the mean olivine composition among the 29 equilibrated H chondrites analyzed by Rubin (1990). (b) Oxygen-isotopic composition as represented by variations in the mean bulk $\Delta^{17}O$ compositions of the 22 equilibrated H-chondrite falls analyzed by Clayton et al. (1991). Figure modified from Rubin et al. (2008).

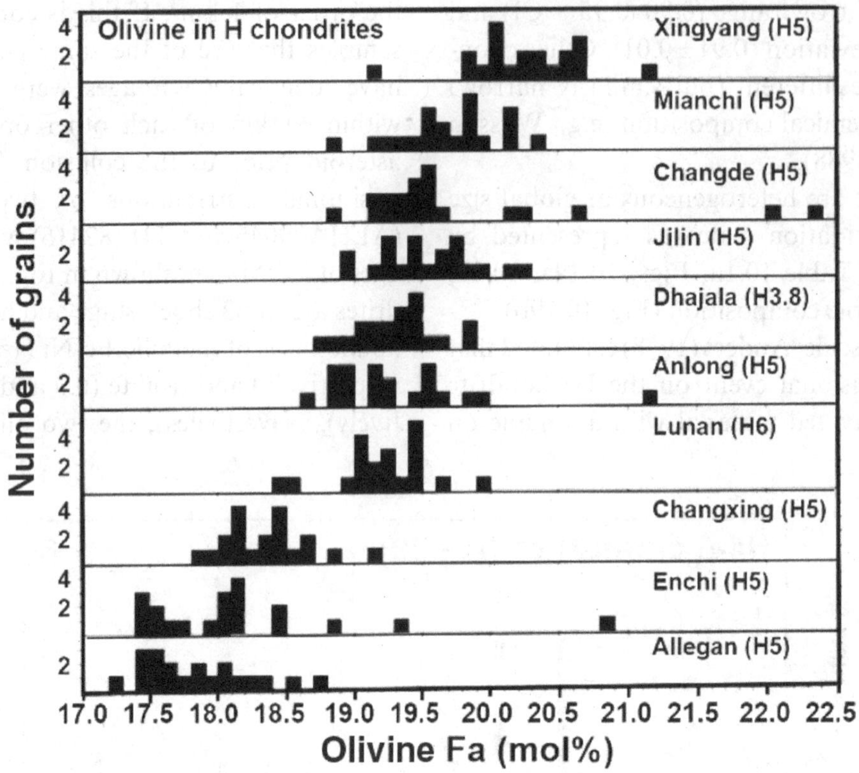

Fig. 10.20 Diverse olivine compositional distributions of equilibrated H chondrites illustrating the variability of oxidation states on the H-chondrite parent body. Although Dhajala is type 3.8, its olivines are moderately equilibrated. Diagram based on Rubin et al. (1988b).

are very different, overlapping only at the extremes: TIL 82415 (Fa17.2–18.1; mean Fa17.7 ± 0.3, $n = 26$) and ALHA79046 (Fa18.1–19.2; mean Fa18.5 ± 0.3, $n = 28$). It is apparent that at the ~1-km size scale, ordinary chondrites were heterogeneous with respect to oxidation state.

Although more data are required, it has been suggested that O-isotopic composition is homogeneous at the kilometer scale. Rubin et al. (2008) measured the O-isotopic compositions of several 1- to 2-mg-sized samples of ALHA79046 and TIL 82415. Three acid-washed samples of each meteorite were analyzed – the mean $\Delta^{17}O$ values of the two meteorites are nearly identical: 0.80 ± 0.01‰ and 0.81 ± 0.02‰, respectively.

Meter size scale: The S asteroids 433 Eros (34.4 × 11.2 × 11.2 km) and 25143 Itokawa (535 × 294 × 209 m) are likely of ordinary-chondrite composition. Spacecraft images have shown the presence of multi-meter-sized boulders on Eros (Thomas et al., 2001) and Itokawa (Fujiwara et al., 2006; Noguchi et al., 2010; Tancredi et al., 2015). The boulders are of two types (Noguchi et al., 2010): (1) rounded boulders that appear to have internal heterogeneities in hardness (they may be breccias); and (2) angular boulders that seem more resistant to impact-fragmentation and could be of a different lithology than the rounded boulders. The number densities of large (>20-m-sized) boulders average roughly 3–5 km^{-2} on Eros and ~15 km^{-2} on Itokawa. Most boulders on Eros (a relatively large body) probably are ejecta from impact craters; in contrast, the majority of boulders on Itokawa likely formed during collisional disruption and gravitational reassembly of the sub-kilometer body. Unfortunately, the relationship between these boulders and available samples of meter-sized OC or fragments of meter-sized OC meteoroids remains unclear.

Holbrook (L/LL6) fell as a shower of ~14,000 stones in Navajo County, Arizona in 1912. A total of 218 kg was recovered. Rubin et al. (2008) made the assumptions that 15% of the original mass of Holbrook was recovered and that, prior to atmospheric entry, the meteoroid was a sphere with a

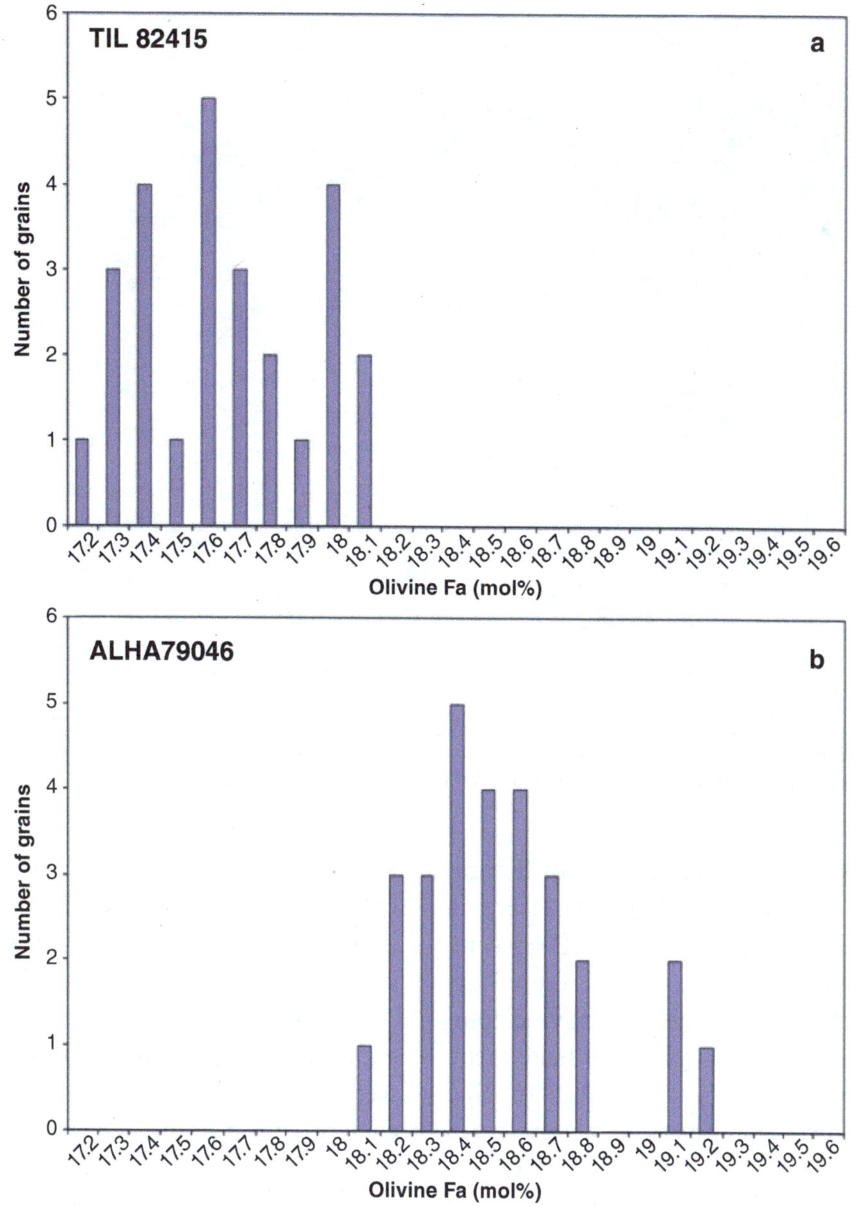

Fig. 10.21 Olivine compositional distributions of two Antarctic H5 chondrites: (a) TIL 82415 and (b) ALHA79046. The meteorites have identical CRE ages of 7.6 Ma. The distributions are distinct, overlapping only at their extreme values. Although these rocks were launched from their parent asteroid during the same collisional event (and were thus previously in close proximity), they have very different mean oxidation states. Data from Rubin (2005a). Figure modified from Rubin et al. (2008).

bulk density of 3.2 g cm^{-3}. They calculated that this model Holbrook meteoroid had a pre-atmospheric diameter of roughly 1 meter. Rubin et al. (2008) examined eight individual Holbrook stones with original masses of 53–408 g; it is not known how far apart these pieces resided on the parent meteoroid. The mean olivine Fa values of the individual stones are essentially identical, ranging from 25.6 ± 0.1 mol% to 25.8 ± 0.2 mol%. Oxygen isotope ratios were determined for four of the samples. They also indicate homogeneity: $\Delta^{17}O$ ranges from 1.26‰ to 1.31‰ with an average $\Delta^{17}O$ value of 1.30 ± 0.02‰. If Holbrook is representative of equilibrated ordinary chondrites, it would appear that, on the meter scale, OC are homogeneous in oxidation state and O-isotopic composition.

Decimeter size scale: Estacado H6 was found in semi-arid Hale County, Texas in 1883 as a single

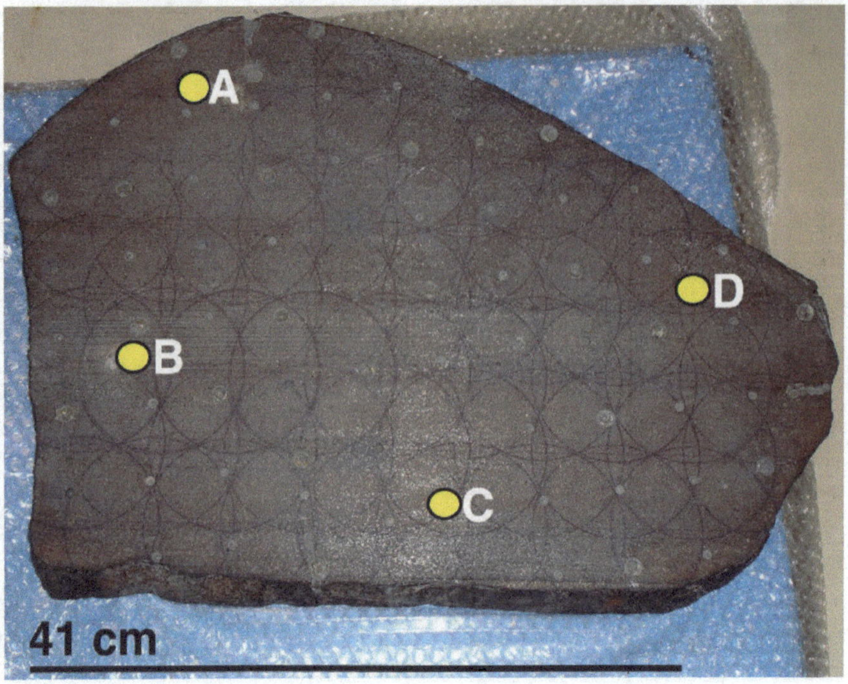

Fig. 10.22 Slab of the Estacado H6 chondrite (sample BM 1906, 259) from the Natural History Museum in London. The locations of the four drill cores (A, B, C, D) are shown with yellow circles. The circular patterns were drawn on the slab during a previous, unrelated study. Image modified from Rubin et al. (2008).

290 kg stone. The distances between the drill cores in the large slab of Estacado (Fig. 10.22) are in the decimeter range: AB = 17 cm, AD = 34 cm, BD = 36 cm. All core samples are more than 3 cm from the surface of the meteorite. The olivine compositions in the cores are essentially identical; mean core olivine Fa values range from 18.5 to 18.6 mol%. The O-isotopic compositions of the cores are also essentially identical; mean core $\Delta^{17}O$ values range from 0.75 to 0.76‰ with an average of 0.76 ± 0.01‰. On the decimeter scale, ordinary chondrites are homogeneous in oxidation state and O-isotopic composition.

Centimeter size scale: Typical meteorite thin sections are round glass disks 1 inch (2.54 cm) in diameter supporting thin meteorite slices commonly 1–2 cm^2 in surface area. Rubin (1990) analyzed 20 or more randomly distributed olivine grains in thin sections of 20 different equilibrated H and L chondrites. The average standard deviation of these 20 sets of analyses is 0.4 ± 0.2 mol% Fa. This can be compared to the recommended ranges for the olivine Fa compositions in each of the three OC groups (Table 10.1a): the H-group range is 4.4 mol% Fa wide, the L range is 4.0 mol%, and the LL range is 6.0 mol%. The much smaller ranges in olivine Fa in individual thin sections indicate relative homogeneity in oxidation state on the centimeter scale.

Clayton et al. (1991) analyzed replicates of 10 equilibrated OC falls; in most cases, the samples were probably located within a few centimeters of each other (or, at most, a few tens of centimeters) in the precursor meteoroid. The average spread of the replicate analyses in $\Delta^{17}O$ is 0.09 ± 0.08‰; 6 of the 10 analyses have a spread of ≤0.05‰. The observed ranges are about the same as the estimated analytical uncertainty in $\Delta^{17}O$ in the analyses (0.08‰; Clayton and Mayeda, 1983). It thus appears that ordinary chondrites are relatively homogeneous in O-isotopic composition at the centimeter size scale.

Summary: At the global size scale (~200–300 km), there is uniformity in mean chondrule size and bulk chemical composition. These homogeneities persist down to sub-centimeter size scales. Individual thin sections of type-3 OC have approximately the same mean chondrule size (e.g., Nelson and Rubin, 2002). INAA samples are 250–300 mg (Kallemeyn et al., 1989) and approximately 0.5–1.0 cm in diameter; in

Table 10.16 *Sizes of chondrules, CAIs, and AOAs in chondrites.*

–	Mean chondrule diameter (μm)	Mean CAI size (μm)	Maximum CAI size (μm)	Mean AOA size (μm)
Source of data	a	b,c	b,c	d
CM	270	36	169	225
CO	150	–	–	260
CV	**900**	**93**	**1088**	**3000**
CR	**700**	**55**	**497**	**600**
CK	**900**	**650**[e]	**20000**[e]	**900**
R	400	63	176	–
H	450	59	302	–
L	500	31	382	–
LL	550	68	218	200
EH	250	40	119	–
EL	500	64	64	–

a = Friedrich et al. (2015); b = Dunham et al. (2022); c = Dunham et al. (2023); d = Rubin (2013b); e = Metamorphosed CK CAIs from Chaumard et al. (2014). Chondrites with large chondrules are in bold.

most cases, replicate samples yield very similar bulk chemical compositions. In contrast, at this global scale, there is heterogeneity in oxidation state and O-isotopic composition. At the kilometer size scale, there is heterogeneity in oxidation state. The scant data available suggest that there may be homogeneity in O-isotopic composition. At the meter, decimeter and centimeter size scales, there is uniformity in oxidation state and O-isotopic composition.

There is relative uniformity in chondrule size and bulk chemical composition from centimeter-to-global size scales. This suggests that the major physical components of ordinary chondrites (chondrules, matrix dust clumps, grains of metallic Fe-Ni, and sulfide) were fairly well mixed in the agglomeration zones of individual OC parent bodies.

Aerodynamic size sorting is at least partly responsible for this mixing; this process occurred early in Solar System history when nebular gas was still present. Table 10.16 shows that chondrite groups with large average chondrule sizes (CV, CR, CK) tend to have the largest amoeboid olivine aggregates (AOAs) and the highest maximum sizes of CAIs. The chondrite groups with the smallest average chondrule sizes (CM, CO, EH) tend to have small CAIs and AOAs. This is consistent with size sorting in the nebula.

In contrast to these conclusions, Jacquet (2014) conjectured that the narrow size ranges of different particle types in individual chondrite groups are due to the growth of chondrule precursor aggregates by sticking. However, because CAIs formed from their precursors near the Sun prior to being distributed throughout the solar nebula (e.g., MacPherson, 2005; MacPherson et al., 2005) and chondrules formed locally (e.g., Rubin, 2011), it seems more likely that the similarities in sizes among chondrules, CAIs, and AOAs in individual chondrite groups (Table 10.16) result from local aerodynamic size sorting of compact, previously melted objects.

Ordinary chondrites have heterogeneous O-isotopic compositions and oxidation states at global size scales. At meter and smaller size scales, both parameters are homogeneous. It is clear that these properties correlate, as illustrated by the correlation between $\Delta^{17}O$ and olivine Fa (a proxy for oxidation state) in ordinary-chondrite groups (Fig. 10.23a).

There are two distinct structural/chemical types of porphyritic chondrules in unequilibrated chondrites: Type-I chondrules have low FeO and tend to be smaller, while Type-II chondrules have high FeO and tend to be somewhat larger. Although the groups overlap in their O-isotopic compositions, most datasets (e.g., Kita et al., 2010; Baeza, 2019) show that ^{16}O-rich samples (particularly those rich in olivine) tend to have lower FeO (Fig. 10.24). Many

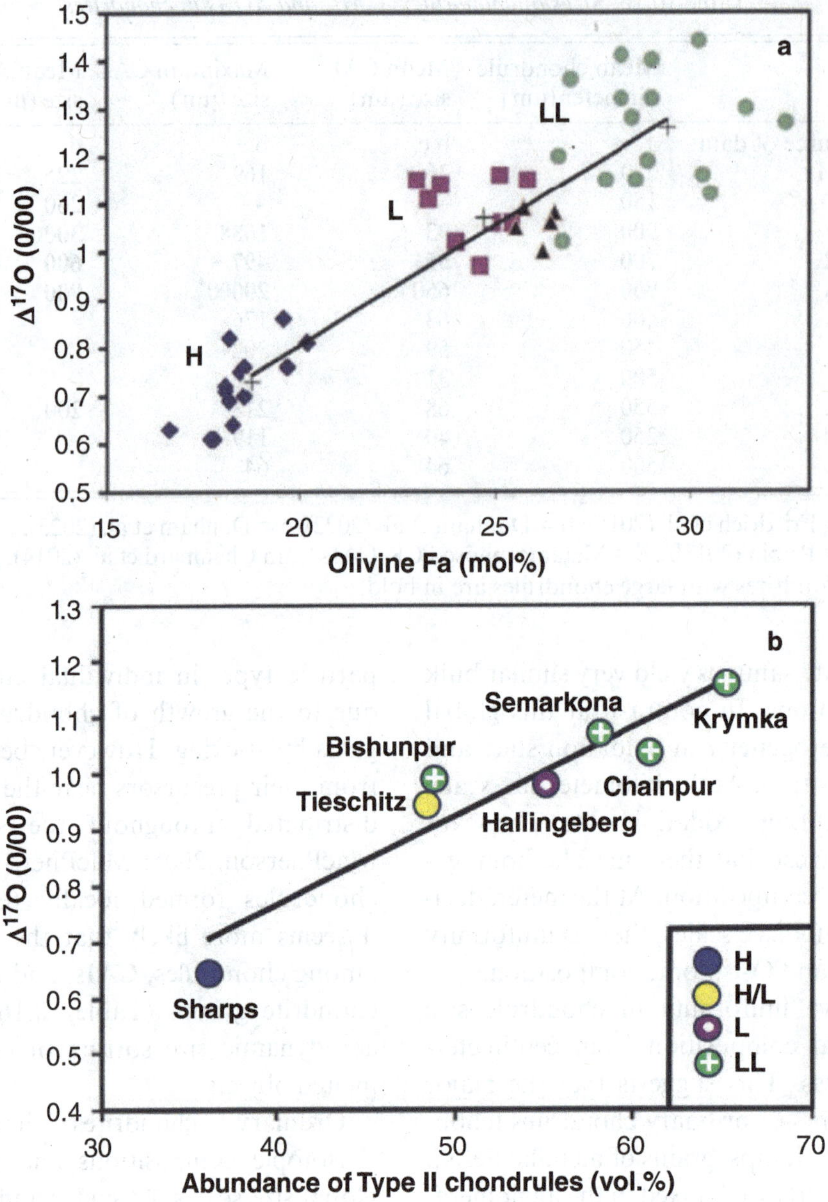

Fig. 10.23 Relationships among O-isotopic composition, oxidation state, and the modal abundance of Type-II chondrules in ordinary chondrites. (a) $\Delta^{17}O$ vs. olivine Fa for equilibrated H, L, and LL chondrites, illustrating a positive correlation between $\Delta^{17}O$ and oxidation state. The more oxidized OC groups have higher mean $\Delta^{17}O$ values. Diagram modified from Rubin (2005a). (b) Positive correlation between $\Delta^{17}O$ and the modal abundance of FeO-rich (Type-II) chondrules in type-3 OC. Image from Rubin (2024), modified from Zanda et al. (2006).

small chondrules and isolated olivine grains also tend to be ^{16}O-rich (Clayton et al., 1991; Marrocchi et al., 2024). Different chondrites have different proportions of Type-I and Type-II chondrules (Zanda et al., 2006). Figure 10.23b shows that the modal abundance of Type-II chondrules correlates with $\Delta^{17}O$. It also correlates with olivine Fa values. It seems plausible that the abundance of Type-II chondrules is rather uniform on meter and smaller size scales, but more heterogeneous on global size scales and possibly on kilometer size scales.

Type-I and Type-II chondrules in ordinary chondrites have different average chondrule sizes (Sears et al., 1996). Takagi (2005), for example, studied LL3.01 Semarkona and found the average size of Type-II chondrules (800 ± 550 μm, $n = 41$) is

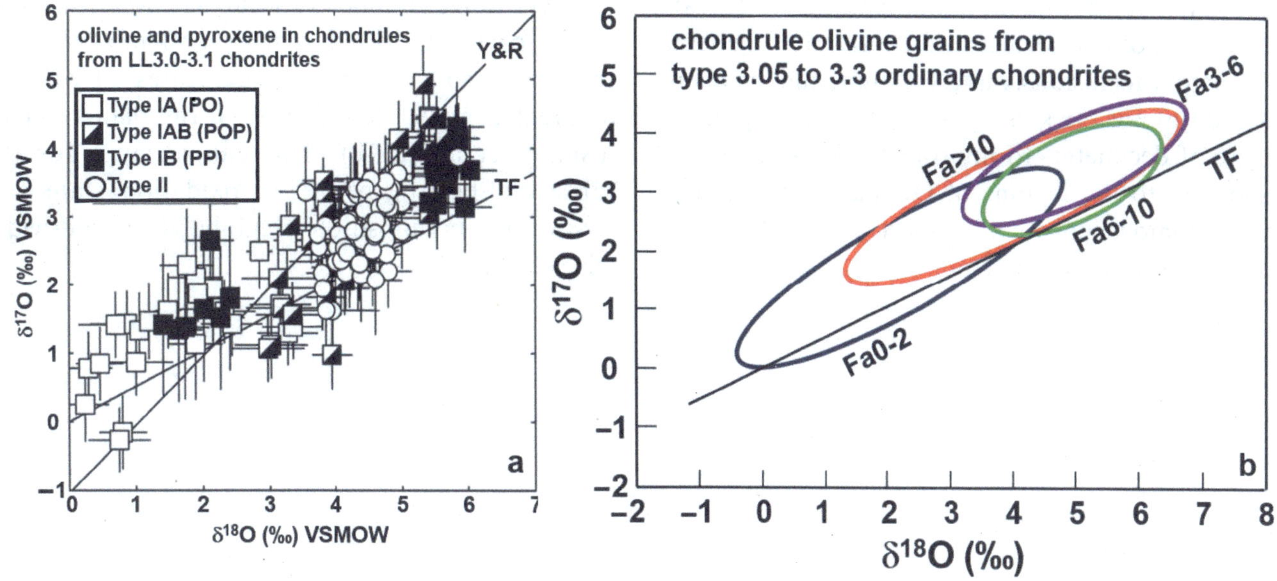

Fig. 10.24 O-isotopic compositions of olivine and pyroxene grains from Type-I and Type-II chondrules in type-3 ordinary chondrites. (a) LL3.0–3.1 chondrites. There is a tendency for the grains from Type-II chondrules to form a tight cluster. Olivine grains from Type-IA chondrules tend to be ^{16}O rich; pyroxene grains from Type-IB chondrules tend to be ^{16}O poor. Y&R = Young and Russell line (Young and Russell, 1998); TF = terrestrial fractionation line. VSMOW is the Vienna SMOW standard (see Section 10.8). Diagram modified from Kita et al. (2010). (b) Ranges in O-isotopic compositions of chondrule olivine grains from type-3 OC of low petrologic type: WSG 95300 (H3.3), GRO 06054 (L3.05), DOM 10556 (L3.1), and MIL 05050 (L3.1). (Although the Meteoritical Bulletin Database lists the petrologic types of these meteorites as H3.3, L3.6, L3.6, and L3.6 (probably), respectively, the study of Baeza [2019] suggested the lower subtypes for three of the samples.) There is significant overlap in O-isotopic composition for olivine grains of different Fa contents, but the most reduced olivine grains tend to be ^{16}O rich. TF is the terrestrial fractionation line. Diagram from Rubin (2024), modified from Baeza (2019).

appreciably greater than that of Type-I chondrules (620 ± 270 μm, $n = 20$). Rubin (2024) used the data of Gooding (1979) to show that Type-I and Type-II chondrules have different average bulk Fe contents (91 vs. 129 mg/g Fe) (due to loss of immiscible metal melts from Type-I chondrules, but not from Type-II chondrules) and, consequently, different average densities (2.91 vs. 3.20 g cm^{-3}).

As discussed earlier, the aerodynamic behavior of a particle is a function of the product of its size and density (e.g., Weidenschilling, 1977; Nakagawa et al., 1986). The average diameter and density values of Type-I and Type-II chondrules in Semarkona (Type-I: 620 μm, 2.91 g cm^{-3}; Type-II: 800 μm, 3.20 g cm^{-3}) yield diameter-density products that are 42% larger for average Type-II chondrules than for average Type-I chondrules. This could lead to different radial drift rates in the solar nebula and cause partial separation of Type-I and Type-II chondrules in the accretion zones of ordinary chondrites prior to the formation of planetesimals.

During accretion, LL chondrites acquired the most Type-II chondrules, L chondrites acquired an intermediate amount, and H chondrites acquired the fewest (Fig. 10.23b). This led to the observed differences among the H, L, and LL groups in oxidation state and O-isotopic composition (Rubin, 2024). It thus seems that variations in the proportions of Type-I and Type-II chondrules were significant at large (100–1000-km) size scales.

The observed homogeneities in oxidation state and O-isotopic composition at smaller size scales on OC parent bodies suggest that there were distinct sub-kilometer regions in the nebula within which the different types of chondrules were well mixed.

Simon et al. (2018) proposed that turbulent diffusion in the solar nebula caused individual chondrules to traverse diverse nebular regions for extended periods before the chondrules clumped together into aggregates. These nebular regions differed in their thermal environment and the amount of available dust. The aggregates presumably consisted mainly of a well-mixed batch of various types of chondrules and chondrule fragments, refractory inclusions, matrix clumps, and metal and sulfide grains. It seems likely that the aggregates were of decimeter- to multi-meter sizes and corresponded to the regions of OC parent bodies that were homogeneous with respect to oxidation state and O-isotopic composition.

Over longer periods of time, as chondrules and other particles drifted radially and mixed across broader nebular regions, aggregates of varying character mixed together to form planetesimals (Simon et al., 2018). This can account for the observed heterogeneities in oxidation state and O-isotopic composition at global (asteroid-wide) size scales.

11

Physical Components of Ordinary Chondrites

11.1 Chondrules

What is your substance, whereof are you made,
That millions of strange shadows on you tend?
William Shakespeare, Sonnet 53

What is a chondrule, Oh tell me wise man,
And how do they fit into Nature's big plan?
Edward Olsen (1989), "The Chondrule"

Variations on these poetic queries concerning the nature and origin of chondrules have been made countless times since chondrules were first described by Jacques-Louis, Comte de Bournon (1802) as "small bodies, some of which are perfectly globular, others rather elongated or elliptical" within four ordinary-chondrite falls (H5 Tabor, L6 Wold Cottage, LL5 Siena, and LL4 Benares (a)). Farrington (1915) pointed out that "The conditions which have brought about the formation of chondri are not well understood, though the question has been much discussed and various hypotheses have been suggested." The situation has remained unchanged for decades (e.g., Merrill, 1920, 1934; Foshag, 1941; Roy, 1957; Wood, 1963; McSween, 1977; Gooding and Keil, 1981; Ash, 1994; Rubin, 2000; Russell et al., 2018; Rubin et al., 2024).

11.1.1 Ferromagnesian Chondrules

Type-3 ordinary chondrites contain ~70–75 vol.% chondrules and chondrule fragments (Fig. 11.1). By far, the most abundant chondrules are ferromagnesian in composition and consist mainly of olivine and/or low-Ca pyroxene plus silicaceous mesostasis. The interference colors of the chondrule phenocrysts form striking displays when viewed microscopically in transmitted light using crossed polars (Fig. 11.2).[1]

[1] As summarized by Rubin and Ma (2021), isotropic materials have a single refractive index (n); light moving through these materials travels at the same velocity in all directions. In contrast, anisotropic minerals have more than one refractive index; the velocity of light varies with crystallographic direction. A beam of light passing through an anisotropic mineral is divided into two polarized rays that vibrate in planes perpendicular to each other. One of the two rays, the *fast ray*, has a lower index of refraction (n_f) and moves faster through the crystal; the other ray, the *slow ray*, has a higher index of refraction (n_s) and moves more slowly.

Every anisotropic crystal has one or two orientations, called *optic axes*, in which a beam of light is not divided into two rays. Thin sections of anisotropic minerals viewed with a petrographic microscope through crossed polarizers (and not along an optic axis) display *interference colors*, due to transmitted light being divided into two rays. In the time it takes for the slow ray to pass through a mineral, the fast ray passes through that same distance plus an additional distance; the difference between these distances is called *retardation* (Δ). Once light rays leave the mineral and reach open air, they travel at the same velocity (because air is isotropic) and the retardation value remains constant. The value of Δ depends on the thickness of the mineral (d) and the difference between the indices of refraction of the slow ray and the fast ray (i.e., n_s–n_f) so that: $\Delta = d(ns - nf)$. Because petrographic thin sections have a standard thickness of 30 μm, the differences in the Δ values among minerals in such sections will depend solely on (n_s–n_f), assuming the mineral grains are thicker than the thickness of the section. The unitless quantity (n_s–n_f) is known as birefringence (δ). Different crystal orientations have different values for δ, ranging from 0 for the optic axis to a maximum value characteristic of the particular mineral. Maximum birefringence values can be diagnostic.

For a specific orientation of a grain, the slow and fast rays may be of equal magnitude but vibrate in opposite directions. The waves will cancel each other out and the mineral will appear dark. This occurs when the retardation is a whole multiple of the wavelength: $\Delta = i\lambda$, where i is an integer and λ is wavelength. In contrast, in those mineral orientations where the retardation is equal to one-half wavelength ($\Delta = (i + ½)\,\lambda$), the waves are vibrating in the same direction and constructively interfere. Light will pass through the upper polarizer.

Interference colors are produced when polychromatic light (i.e., white light) passes through the mineral and the crossed polarizers. For a particular retardation, some wavelengths will cancel each other out and others will constructively interfere to different extents. When one wavelength is cancelled out, its complementary color becomes visible. First-order interference colors are produced when the path difference between the rays is 1λ; second-order colors are produced at 2λ, third-order colors at 3λ, etc. To identify a common mineral, all that might be needed is a chart of interference colors (a.k.a. a Michel-Lévy Chart) available on-line and in optical mineralogy textbooks.

Fig. 11.1 Slab of the NWA 2833 L3.8 chondrite showing abundant chondrules. Image courtesy of John Kashuba.

Fig. 11.2 The variety of chondrules in H/L3.6 Tieschitz. Although many chondrules are quasi-circular in two dimensions, some are irregular in shape, and others are fragmented. Textural types include porphyritic olivine, porphyritic olivine-pyroxene, porphyritic pyroxene, barred olivine, radial pyroxene, and cryptocrystalline. Most chondrules appear to be 500 ± 200 μm in diameter. Transmitted light crossed polars image courtesy of John Kashuba.

There are two principal divisions of ferromagnesian chondrules (porphyritic and nonporphyritic) and five common textural types (as formalized by Gooding and Keil, 1981).

Porphyritic chondrules include porphyritic olivine (PO), porphyritic pyroxene (PP), and porphyritic olivine-pyroxene (POP). Porphyritic textures are classic igneous features consisting of coarse phenocrysts embedded in a fine-grained groundmass or glass. There are two structural/chemical types of porphyritic chondrules in unequilibrated chondrites (McSween, 1977), that could be called "oxidation types": Type-I (low-FeO) chondrules and Type-II (high-FeO) chondrules (described below). Many porphyritic chondrules contain numerous small olivine and low-Ca pyroxene grains and grain fragments in addition to coarse phenocrysts (e.g., Fig. 11.3a,b,c). PP chondrules contain <10 vol.% olivine; PO chondrules contain <10 vol.% low-Ca pyroxene; POP chondrules contain ≥10 vol.% olivine and ≥10 vol.% low-Ca pyroxene.

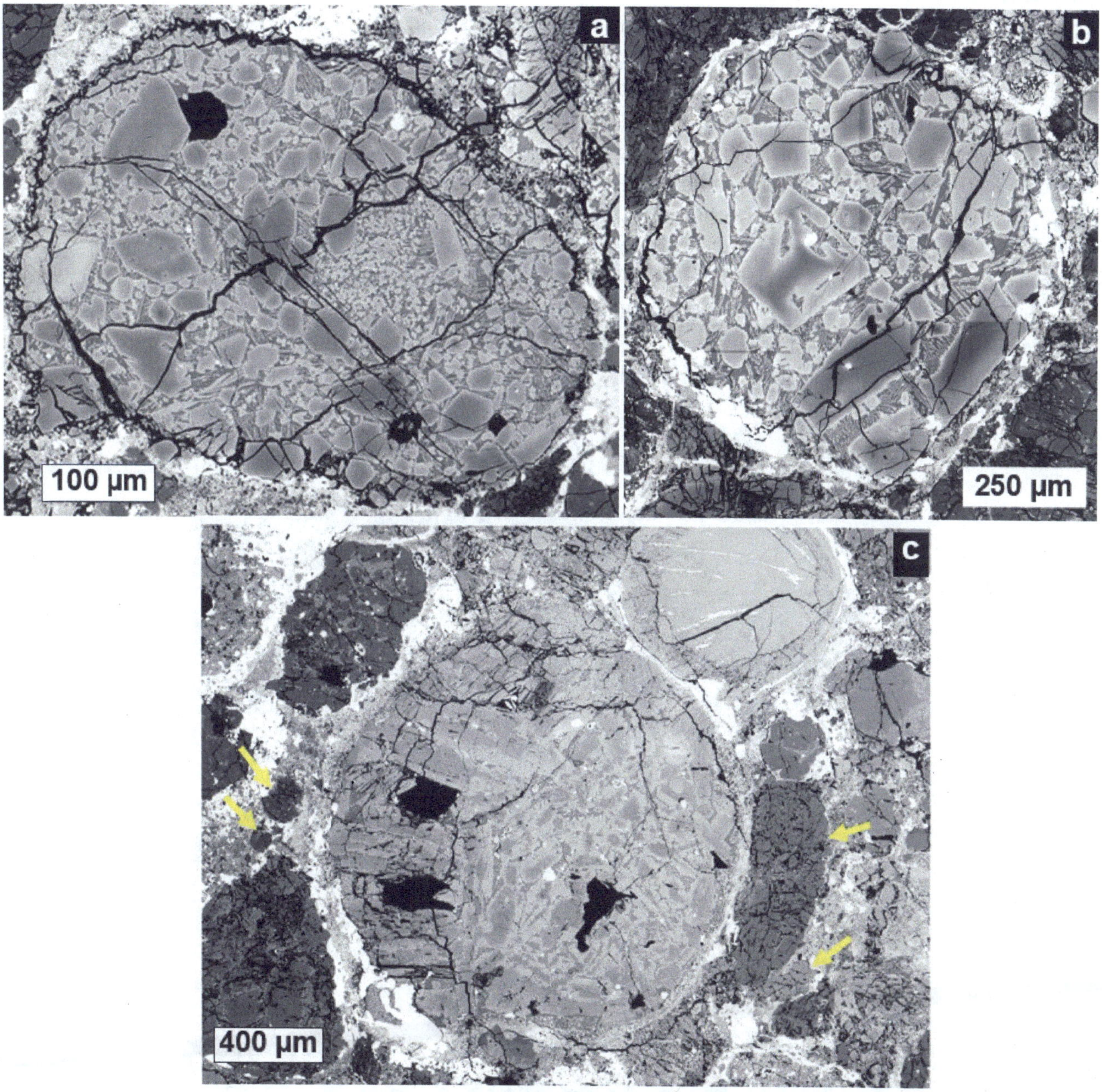

Fig. 11.3 Type-II chondrules from LL3.01 Semarkona. (a, b) Type-IIA (FeO-rich PO) chondrules which contain a few large olivine grains, several smaller olivine grains, and numerous small euhedral, subhedral, and irregular olivine grains. Also present in the chondrules are patches of mesostasis with numerous Ca-pyroxene crystallites and a few small blebs of metallic Fe-Ni (white). (c) Type-IIB (FeO-rich PP) chondrule with low-Ca and high-Ca pyroxene phenocrysts and small grains of metallic Fe-Ni. Arrows indicate isolated mafic-silicate grains. BSE images.

In many cases, the mafic silicate grains in porphyritic chondrules are roughly equant and display crystal faces; some grains have skeletal or hopper morphologies indicative of rapid cooling (Fig. 11.3b). In some finer-grained low-FeO porphyritic chondrules, the olivine ± low-Ca-pyroxene grains are rounded (Fig. 11.4). Small grains of Ca-pyroxene occur as overgrowths flanking low-Ca pyroxene grains in many PP and POP chondrules; crystallites of Ca pyroxene are also present as isolated or clumped, quasi-acicular (needle-like) grains in the mesostasis (Fig. 11.5). Opaque phases (mainly kamacite ± minor troilite) occur commonly as small blebs or droplets within the mesostasis. The occurrence of REE-bearing sulfide (with a flat REE pattern and a positive Eu anomaly) within a low-FeO chondrule from LL3.15 Bishunpur indicates the sulfide grains are igneous, crystallizing from the chondrule melt during cooling (Kong et al., 2000).

Also present in a minority of Type-I (low-FeO) and Type-II (high-FeO) porphyritic (and barred olivine) chondrules in LL3.01 Semarkona are small grains of primary plagioclase of diverse compositions (An_{2-99}, but with little K_2O ($Or_{<3}$)) (Lewis and Jones, 2019). Merrillite ($Ca_9MgNa(PO_4)_7$) occurs commonly in Type-II chondrules in Semarkona as small grains, typically a few micrometers in size (e.g., Grossman and Brearley, 2005; Rubin and Baecker, 2021).

In many POP chondrules, elongated grains of low-Ca pyroxene line the inner margins of the chondrules (e.g., Fig. 11.6) with their long axes subparallel to the chondrule surface. Low-Ca pyroxene commonly occurs in the interiors of these chondrules as well (e.g., Fig. 11.4; Rubin, 2018c).

The phenocrysts in Type-II chondrules commonly exceed 500 μm in maximum dimension; some exceed 1000 μm (e.g., Jones, 1990, 1996). Such large sizes have prompted the suggestion that porphyritic chondrules cooled relatively slowly (~50 K hr^{-1}) over the crystallization temperature range (e.g., Desch and Connolly, 2002), consistent with many dynamic crystallization experiments (e.g., Lofgren, 1996).

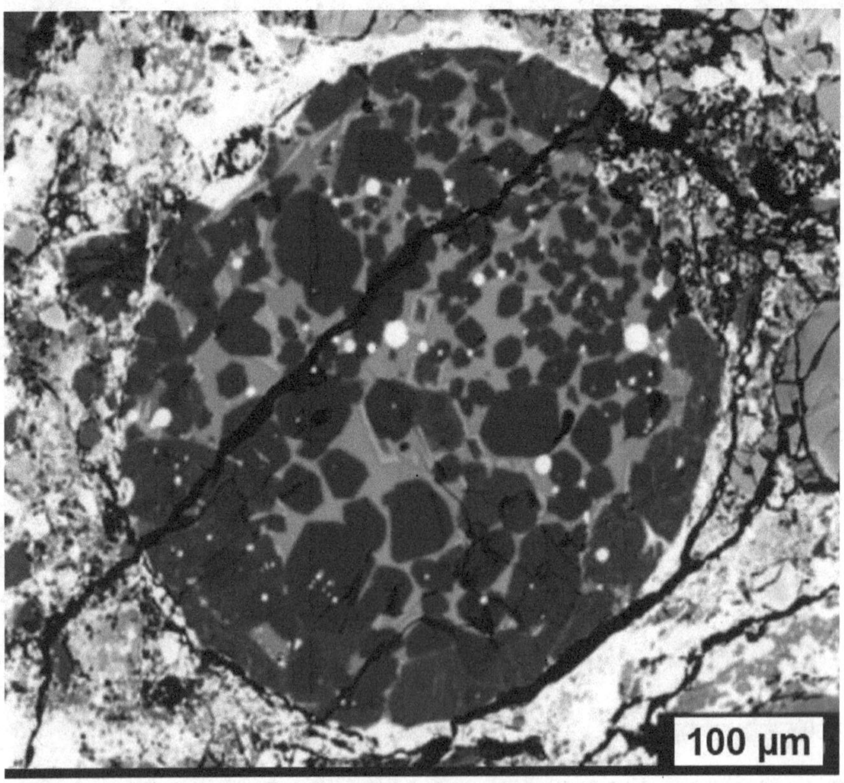

Fig. 11.4 Chondrule H4o (Type-IAB) from LL3.01 Semarkona containing numerous small phenocrysts, many of which are somewhat rounded. The chondrule is 380 × 440 μm in size and consists of ~40 vol.% olivine, ~25 vol.% clinoenstatite, ~30 vol.% mesostasis, ~3 vol.% Ca-pyroxene, and ~2 vol.% metallic Fe-Ni. Most phenocrysts at the chondrule margin are clinoenstatite. BSE image from Rubin (2018c).

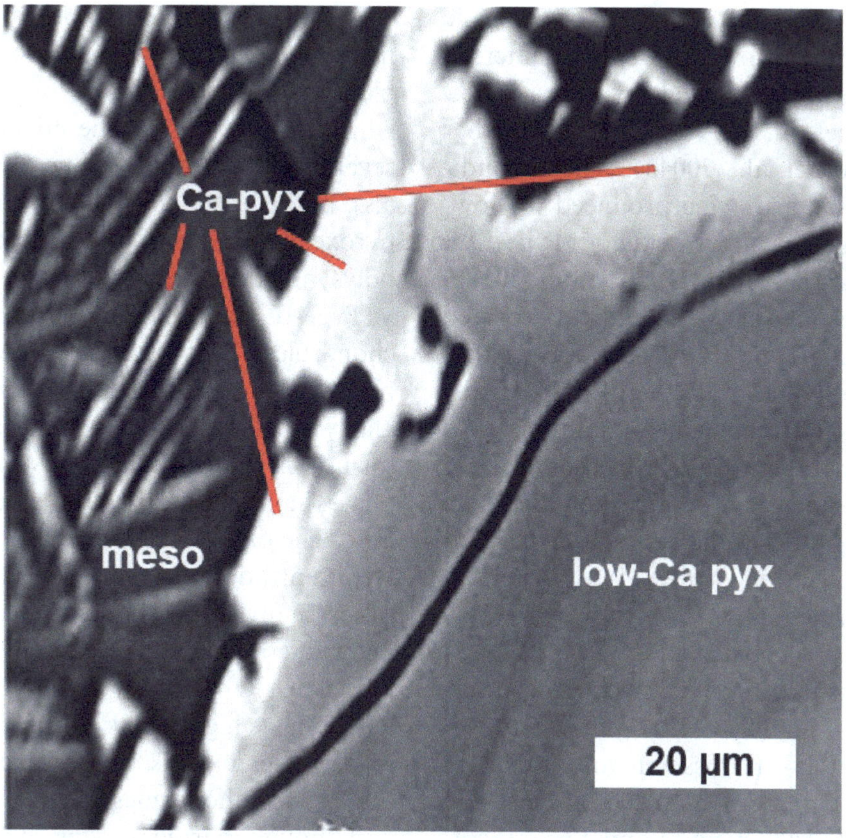

Fig. 11.5 Portion of Type-IIAB (FeO-rich POP) chondrule E4m from LL3.01 Semarkona showing Ca-pyroxene grains (white) flanking a coarse low-Ca pyroxene phenocryst as well as occurring as acicular crystallites in the mesostasis (meso). BSE image modified from Baecker et al. (2017).

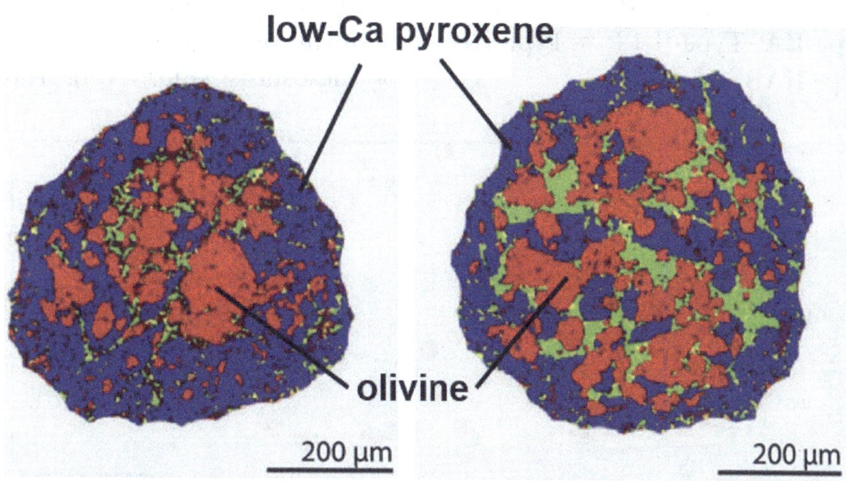

Fig. 11.6 False-color elemental maps of two porphyritic olivine-pyroxene (POP) chondrules in Moorabie (L3.8-an). Low-Ca pyroxene (blue); olivine (red); Ca-pyroxene (yellow); mesostasis (light green); metallic Fe-Ni (dark purple). Coarse phenocrysts of low-Ca pyroxene occur mainly at the chondrule perimeter where little mesostasis is present; olivine and mesostasis tend to be concentrated in the chondrule core. Image modified from Barosch et al. (2020).

To prevent evaporative loss of bulk volatiles from chondrule melts during such slow cooling, some workers have suggested that the ambient environment where chondrules formed had high total gas pressures or high partial pressures of lithophile elements (e.g., Ebel and Grossman,

1998; Hewins et al., 2005). Several studies have proposed that some chondrules were immersed in nebular gas with high partial pressures of SiO (e.g., Nagahara et al., 1999, 2008; Tissandier et al., 2002; Krot et al., 2004a). These conditions were proposed as having produced the low-Ca-pyroxene-rich margins of many Type-IAB (POP) chondrules (Figs. 11.4, 11.6). Rubin (2018c) suggested that these chondrules could have formed instead by remelting Type-IB (PP) chondrules that contained relict olivine grains derived from disrupted Type-IA (PO) chondrules. Another possibility is that high-temperature evaporation of FeO-bearing chondrule droplets increased the SiO_2/(FeO+MgO) ratio of the residual melt, leading to the crystallization of low-Ca pyroxene at quenching chondrule margins.

Type-I and Type-II chondrules differ in the FeO/(FeO+MgO) ratios in their mafic silicates. The cutoff between Type-I and Type-II is somewhat arbitrary: $Fa_{\lesssim 2}$ and $Fa_{\gtrsim 2}$ (Grossman and Brearley, 2005); $Fa_{<4}$ and $Fa_{\geq 4}$ (Takagi, 2005); $Fa_{\geq 8}$ and $Fa_{>8}$ (Zanda et al., 2006); $Fa_{\leq 10}$ and $Fa_{\geq 10}$ (Scott and Taylor, 1983; Krot and Wasson, 1995; Hewins, 1997; Brearley and Jones, 1998). The following nomenclature is often used: Type-I PO = Type IA; Type-I PP = Type IB; Type-I POP = Type IAB; Type-II PO = Type IIA; Type-II PP = Type IIB; Type II POP = Type IIAB.

Type-I chondrules tend to have smaller phenocrysts and clear, isotropic glassy mesostases with blebs of kamacite. Some have poikilitic textures (in which large grains of one mineral contain at least several smaller grains of one or more other minerals). In the least equilibrated chondrites, the coarse pyroxene grains tend to be clinoenstatite exhibiting polysynthetic twinning.

Type-II chondrules tend to be larger than Type-I chondrules. They generally have larger euhedral phenocrysts surrounded by a darker, more-turbid, glassy to microcrystalline mesostasis containing few (or no) blebs of metallic Fe-Ni. Additional opaque phases in the mesostasis can include troilite and chromite. The olivine grains are characterized by higher P concentrations as well as higher FeO/(FeO+MgO) ratios; many grains show strong compositional zoning with core-to-rim increases in the concentrations of FeO, Cr_2O_3, MnO, and CaO. Type-II chondrules tend to have higher bulk $\Delta^{17}O$ values; they also tend to be richer in bulk Fe and have higher densities than Type-I chondrules (Fig. 11.7).

Although barred olivine (BO) chondrules are occasionally considered a variety of porphyritic chondrule, most researchers view them as a separate textural type. BO chondrules are characterized by subparallel continuous or discontinuous plates of olivine of varying thickness separated by patches of mesostasis (Fig. 11.8). Rare BO chondrules

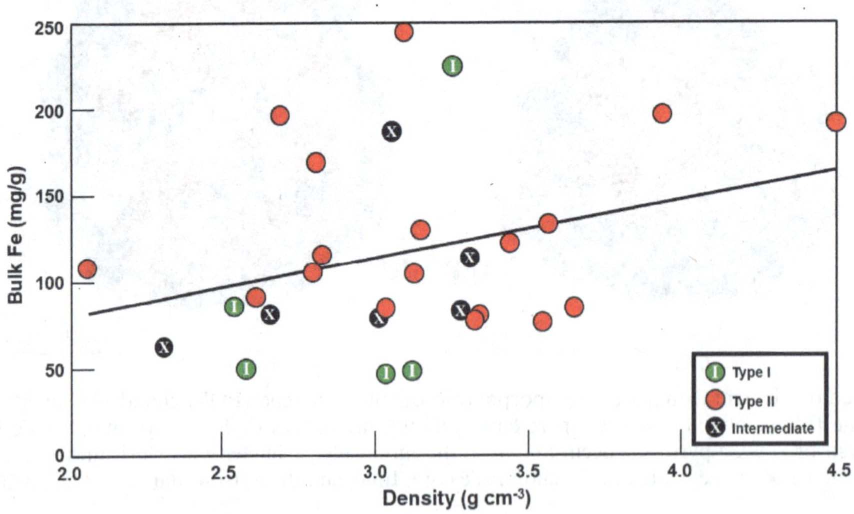

Fig. 11.7 Bulk Fe vs. calculated density for separated porphyritic chondrules in type 3.0–3.4 chondrites (Hallingeberg, L3.4; Chainpur, LL3.4; Semarkona, LL3.01) studied by Gooding (1979). The best-fit trend line is shown. For this diagram, the following criteria were used: Type-I chondrules ($Fa_{<5}$ or $Fs_{<3}$); Type II chondrules ($Fa_{>10}$ or $Fs_{>15}$). On average, Type-II chondrules have higher bulk Fe contents and higher densities than Type-I chondrules. Image from Rubin (2024).

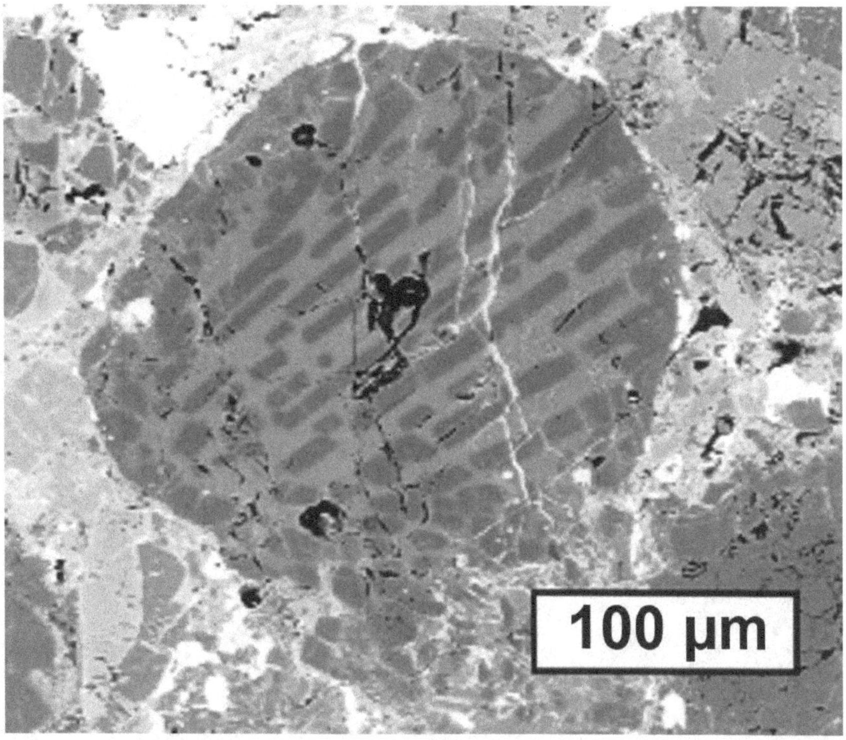

Fig. 11.8 Rimless barred olivine chondrule in H3.4 Sharps. Olivine bars (dark gray) are discontinuous and subparallel; they are surrounded by glassy mesostasis (medium gray). BSE image.

have bars that appear to radiate from a point on the surface; this type can be referred to as a radial olivine (RO) chondrule. Many BO chondrules are surrounded by a thin olivine spherical shell (a rim in two dimensions) (Fig. 11.9). In some cases, there are two or more sets of plates oriented in different directions (Fig. 11.9); some BO chondrules have pentagonal outlines. (In thin section, when viewed in plane polarized transmitted light, rimless BO chondrules have an uncanny textural resemblance to ridged potato chips, although on a much smaller scale [cf. Fig. 1 of Pedreschi et al., 2011].)

Among BO chondrules in unequilibrated ordinary chondrites, there are significant positive correlations among chondrule diameter, bar thickness, and rim thickness. In the nebula, smaller BO precursor droplets cooled faster than larger droplets (due to their higher surface-area/volume ratios) and grew thinner bars and rims (Rubin et al., 2024).

There is a bimodal distribution in the olivine FeO content in BO chondrules in type-3 OC, with a hiatus between 11 and 19 wt.% FeO (Fig. 11.10). The ratio of (FeO-rich)/(FeO-poor) BO chondrules decreases from 12.0 in H to 1.6 in L to 1.3 in LL (Rubin et al., 2024). This is the opposite of the case for porphyritic chondrules: the mean (FeO-rich)/(FeO-poor) modal ratio (i.e., the Type-II/Type-I ratio) increases from 0.8 in H to 1.8 in L to 2.8 in LL.

The limited data of Varela et al. (2025) show that FeO-poor BO chondrules are richer in bulk refractories (i.e., Al_2O_3, CaO, Ti, Hf, REE) and poorer in FeO and Cr_2O_3 than FeO-rich BO chondrules. The mesostases of FeO-poor BO chondrules appear to be poorer in SiO_2 and FeO than those in FeO-rich BO chondrules.

The other varieties of ferromagnesian chondrules are nonporphyritic. Radial pyroxene (RP) chondrules consist of thin laths of low-Ca pyroxene arranged in fan-like arrays typically radiating from a point at or near the chondrule surface (Fig. 11.11). In a minority of cases, there are two or more radiation points. A few RP chondrules contain non-radiating subparallel laths; they are texturally analogous to barred olivine chondrules

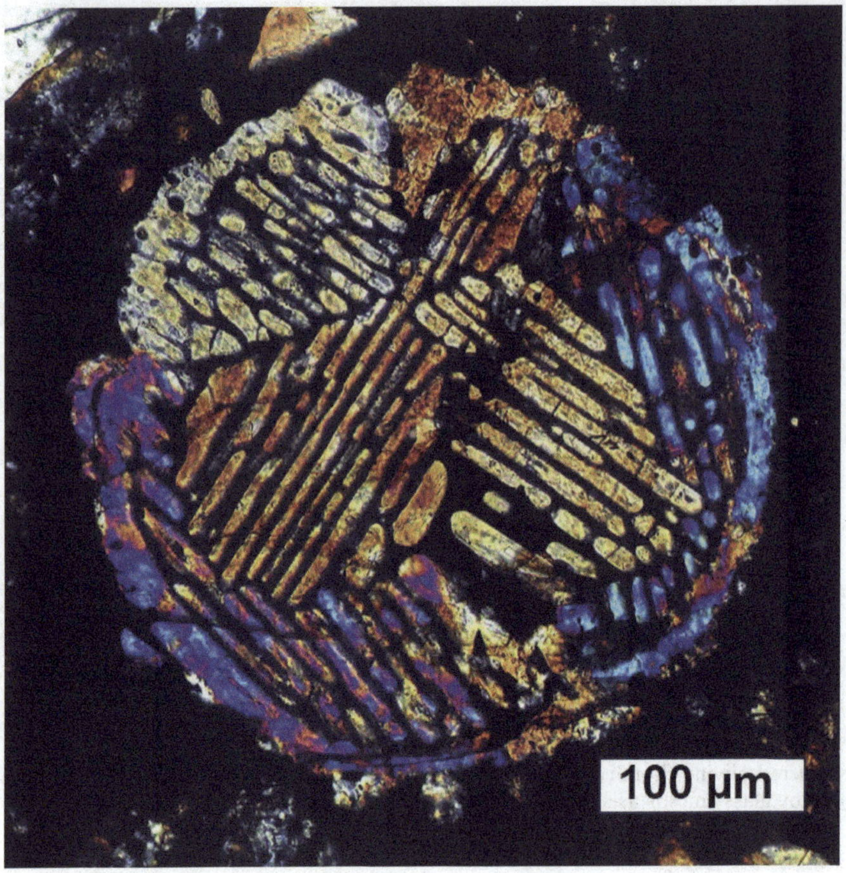

Fig. 11.9 Rimmed barred olivine chondrule with five distinct sets of olivine bars from H5 Faith. The texture likely developed from the presence of multiple seed nuclei during crystallization. Photomicrograph in transmitted light with crossed nicols. Image courtesy of John Kashuba.

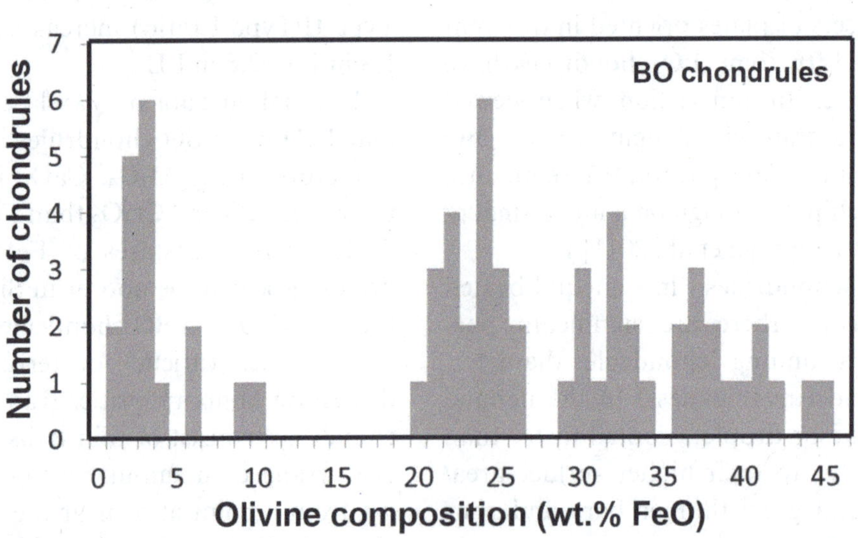

Fig. 11.10 Average composition of olivine (wt.% FeO) in 65 BO chondrules from type-3 ordinary chondrites studied by Rubin et al. (2024). Olivine compositions are bimodal, with a hiatus between 11 and 19 wt.% FeO. Diagram modified from Rubin et al. (2024).

Fig. 11.11 Radial pyroxene chondrules from type-3 ordinary chondrites. (a) Clovis (no.1), H3.6. (b) Sarir Quattusah 001 (SQ 001), L/LL3. (c) Zegdou, H3. Transmitted light, crossed nicols. Images courtesy of John Kashuba.

Fig. 11.12 Cryptocrystalline chondrule from the NWA 820 (L3-5) breccia. Many of the individual segments appear to exhibit a quasi-radiating texture. Image in transmitted light with crossed polars. Images courtesy of John Kashuba.

and are sometimes called barred pyroxene (BP) chondrules. Glassy mesostasis (commonly with included opaque phases) is present between the laths and bars of RP and BP chondrules.

Cryptocrystalline (C) chondrules are the finest-grained chondrules; they consist of numerous patchy domains containing ≤2-µm-sized low-Ca pyroxene grains and associated mesostasis (Fig. 11.12). Due to their fine grain size (far smaller than the thickness of a standard thin section), C chondrules typically appear darker in plane-polarized transmitted light than RP chondrules.

Fig. 11.13 Ellipsoidal granular olivine-pyroxene (GOP) chondrule with an aspect ratio of 1.55 in Barratta L4. This chondrule is probably an enveloping compound object with a coarser primary chondrule in the center mantled by a somewhat finer-grained secondary shell. Image in transmitted light with crossed nicols, courtesy of John Kashuba.

Granular olivine-pyroxene (GOP) chondrules consist typically of densely packed anhedral olivine and low-Ca pyroxene grains and small patches of associated mesostasis (with included opaque phases) (Fig. 11.13). Some granular pyroxene (GP) chondrules contain no olivine. Although these objects generally have microporphyritic textures, they are considered nonporphyritic chondrules.

Nelson and Rubin examined 719 intact chondrules in LL3 chondrites and determined their relative proportions: 47% POP, 21% PP, 16% PO, 6% BO, 3% RP+BP, 4% C, and 3% GOP+GP. The porphyritic/nonporphyritic ratio is 8.4. However, there is uncertainty in the proportions of the various types of porphyritic chondrules. It is a statistical certainty that most sectioning of spheroidal chondrules will be non-equatorial; this leads to apparent chondrule diameters being smaller, on average, than true three-dimensional diameters. Also, as shown in Fig. 11.14, random sectioning through a mineralogically zoned olivine- and low-Ca-pyroxene-bearing porphyritic chondrule could result in a two-dimensional misrepresentation of the actual olivine/low-Ca-pyroxene ratio; in particular, the abundance of PP chondrules may be overrepresented. Furthermore, the diagram suggests that, among porphyritic chondrules, PP chondrules would be the smallest, POP chondrules would be a little larger, and PO chondrules would be appreciably larger. This is in fact the case, as shown by the mean apparent diameters of LL3 porphyritic chondrules (Nelson and Rubin, 2002): PP (530 μm, $n=151$); POP (540 μm, $n=341$); PO (660 μm, $n=115$).

Numerous crystallization experiments aiming to reproduce natural chondrule textures have shown that porphyritic chondrules crystallized from incomplete melts that included nuclei and relict mafic silicate grains. Dodd (1978) and J. V. Smith (1979) considered these “lithic chondrules” to be mechanically abraded conglobated fragments of much larger igneous and metamorphic rocks instead of the modern view that they are partly molten millimeter-sized droplets that crystallized in the solar nebula.

In contrast, BO, RP, and C chondrules are sometimes called “droplet chondrules” because of their typical near-spherical shapes molded by surface tension; in most cases, they appear to have formed by quenching of totally molten silicate droplets.

The mesostases of ferromagnesian chondrules are feldspathic or silico-feldspathic, that is, they are rich

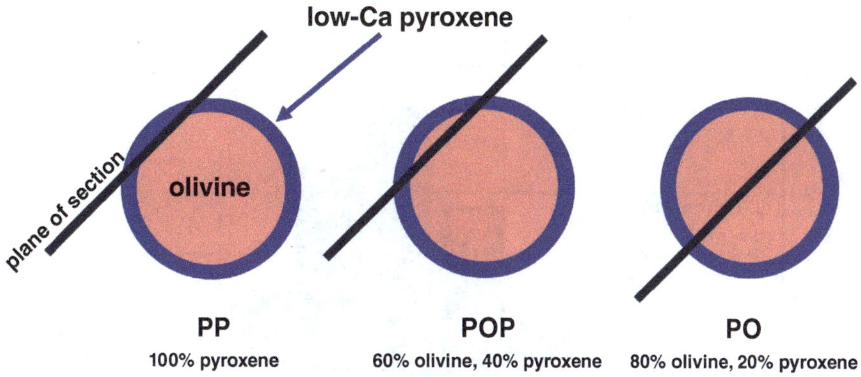

Fig. 11.14 Schematic diagram showing random sectioning of a mineralogically zoned POP chondrule with olivines (pink) concentrated toward the core and low-Ca pyroxene (blue) concentrated toward the margins. If the section slices through only the outer parts of the chondrule, the chondrule will appear to be PP in the plane of the thin section; if the section cuts through the outer part of the chondrule core, the chondrule will be identified as POP; if the section is cut through the center of the chondrule, it could be classified as PO with minor low-Ca pyroxene at the margin. Diagram based on Barosch et al. (2020).

Table 11.1 *Size ranges and mean diameters of the different textural types of intact chondrules in LL3 chondrites.*

Textural type	Number	Diameter range (μm)	Mean diameter (μm)
POP	341	110–2250	540
PP	151	170–2360	530
PO	115	140–2060	660
BO	40	120–2500	750
RP	18	270–2030	900
C	31	150–3110	560
GOP	23	140–1050	460

In every case, chondrule diameters were measured in thin section; these diameters are actually "apparent diameters" and are smaller than the true three-dimensional diameters. Data from Nelson and Rubin (2002).

in normative plagioclase or plagioclase plus silica. In Type-IIAB and -IIB chondrules, particularly those in mildly metamorphosed OC, there may be four pyroxene phases: low-Ca clinopyroxene, orthopyroxene, pigeonite, and Ca-pyroxene (typically augite).

Different types of ferromagnesian chondrules have fairly similar size ranges but different mean sizes (Table 11.1).

Table 11.1 shows that RP chondrules are larger than porphyritic chondrules, a result consistent with the findings of Ebel et al. (2024). Porphyritic chondrules are more friable than droplet chondrules, breaking readily along grain boundaries when mechanical stress is applied. The proportion of porphyritic chondrules decreases with increasing chondrule size interval (Nelson and Rubin, 2002), suggesting that large porphyritic chondrules are more easily fragmented by parent-body processes (presumably collisions) than droplet chondrules. This is consistent with the numerous porphyritic chondrule fragments in ordinary chondrites as well as the coarse isolated olivine and low-Ca pyroxene grains and polymineralic olivine-pyroxene assemblages in OC. Almost certainly, these isolated grains (Fig. 11.3) and assemblages were derived from mechanically disrupted porphyritic chondrules; it seems unlikely that any coarse mineral grains formed directly by condensation in the solar nebula as suggested by Steele (1986). Thus, the mean sizes of porphyritic chondrules in Table 11.1 should be considered lower limits on their initial (immediately post-formation) sizes.

Table 10.17 lists the mean apparent sizes of chondrules (all types) in ordinary chondrites: H 450 μm, L (500 μm), LL (550 μm). These chondrules have a log-normal size distribution (e.g., Fig. 11.15). About 60% of LL3 chondrules are within a factor of two of the mean size. Chondrule apparent diameters and chondrule-fragment apparent cross-sections are given in φ units which are defined by the relationship $\varphi = -\log_2 d$, where d is the average apparent diameter of the chondrule (or the average apparent cross-section of the fragment) in millimeters.

Chondrules are much larger than fragments; the chondrule size distribution is skewed toward the left, that is, toward larger sizes. The abrupt cutoff of the fragments at 3.5 φ units (88 μm) in an otherwise

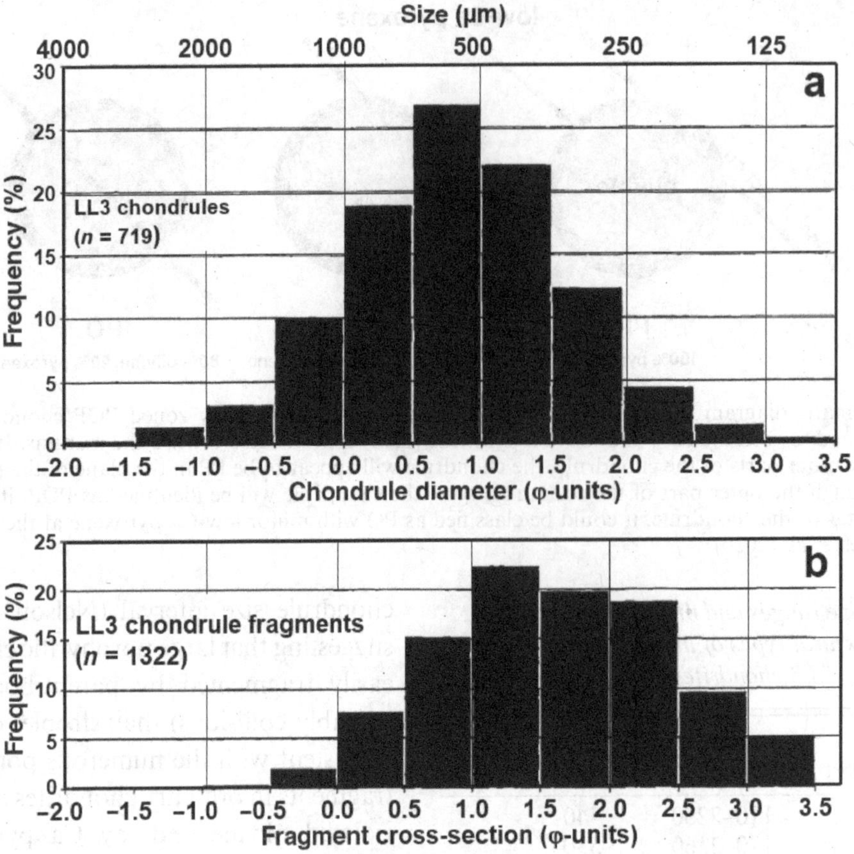

Fig. 11.15 Size-frequency distributions of intact chondrules and of chondrule fragments measured in thin section in five LL3 chondrites. (a) The 719 chondrules have a mean apparent diameter of 0.80 φ units (570 µm). (b) The 1322 chondrule fragments have a mean apparent cross-section of 1.60 φ units (330 µm). The fragments are significantly smaller than the intact chondrules. Objects smaller than 88 µm (3.5 φ units) were excluded. Diagram modified from Nelson and Rubin (2002).

broad log-normal distribution (Fig. 11.15b) is a consequence of the intentional omission of numerous fragments of smaller size in the study by Nelson and Rubin (2002). (At the other end of the size spectrum, Nelson and Rubin [2002] reported a 1340-µm-sized isolated olivine grain; it probably broke out of a very large Type-IIA or Type-IIAB chondrule.)

As is apparent from Fig. 11.14, random sectioning of a chondrule is unlikely to cut through the diameter, thus rendering apparent diameter (as measured in thin section) smaller than the true diameter. Several studies have proposed statistical corrections for the conversion of two-dimensional to three-dimensional particles: (Eisenhour, 1996; Sahagian and Proussevitch, 1998; Cuzzi and Olson, 2017; Metzler, 2018; Benito et al., 2019; Floyd et al., 2024; Hezel et al., 2025). None of the methods has yet achieved universal use.

As of this writing, there have been only a few studies (e.g., Friedrich et al., 2022; Kadlag et al., 2023) that obtained 3D images of OC chondrules through tomography. Friedrich et al. (2022) imaged 665 chondrules in one H chondrite (H3.4 Sharps) and found a log-normal mean of 260 µm; 550 chondrules in one L chondrite (L4 Saratov) and found a log-normal mean of 470 µm; and a total of 664 chondrules in three LL chondrites (LL3.3 ALH 83010; LL3.4 Chainpur; LL3.4 Ragland) and found a combined log-normal mean of 603 µm. These values, particularly that of Sharps, differ somewhat from the established mean chondrule diameter values (H, 450 µm; L, 500 µm; LL, 550 µm). Additional tomographic work is needed to refine the means.

Many particle size distributions can be well described by a continuous mathematical function. Hughes (1978a) and Martin and Hughes (1980) found that individual chondrules separated from several ordinary chondrites have mass distributions that follow a Weibull statistical distribution function (Weibull, 1951). Rubin et al. (1982a) found that OC chondrules separated by Gooding (1979) also conform to a Weibull distribution.

The Weibull function can be defined by the equation:

$$Y = 100\{1 - \exp[-b(d - \gamma)^n]\}$$

where Y is the mass percentage of different-sized objects passing through a sieve of mesh size d; b is the reciprocal of the average size of the measured particles; γ is the location parameter (which is directly related to the minimum d value), and n is the reciprocal of the spread of the size-frequency curve. The Rosin distribution function (Rosin and Rammler, 1933) is a special case of the Weibull function where $\gamma = 0$. Both the Rosin and Weibull functions approximate log-normal distributions (e.g., Fig. 11.15).

Crushing processes and random breakage of material produce particle distributions that follow "Rosin's law" (e.g., Krumbein and Tisdel, 1940). Various coals (Geer and Yancey, 1938), glacial tills (Krumbein and Tisdel, 1940), nuée ardente deposits (Krumbein and Tisdel, 1940), and the soil component (~1 mm fraction) of the lunar regolith (Martin and Mills, 1977) have particle mass-distributions that obey Rosin's law.

Martin and Hughes (1980) noted that the Weibull function can be produced from a Rosin function by the removal of an appreciable number of small particles. This seems to be the case for OC chondrules: it is common that when experimenters separate chondrules, they inadvertently undercount the smallest particles. This problem is illustrated by the large difference between the average size of separated LL chondrules (1280 µm; Gooding, 1979) and LL chondrules measured in thin section (570 µm; Nelson and Rubin, 2002). It therefore seems likely that OC chondrules (or their precursor dustballs) initially conformed to a Rosin size distribution, perhaps a result of size sorting in the nebula. The Weibull distribution of separated chondrules (Hughes, 1978a; Martin and Hughes, 1980) appears to be an experimental artifact.

11.1.2 Chondrule Bulk Chemical and Isotopic Compositions

A suite of chondrules separated from LL3.01 Semarkona, one of the least equilibrated ordinary chondrites (Grossman, 1985), was analyzed by INAA by Grossman and Wasson (1983, 1985). They found the refractory lithophile elements to be unfractionated relative to CI carbonaceous chondrites (Fig. 11.16), indicating that these

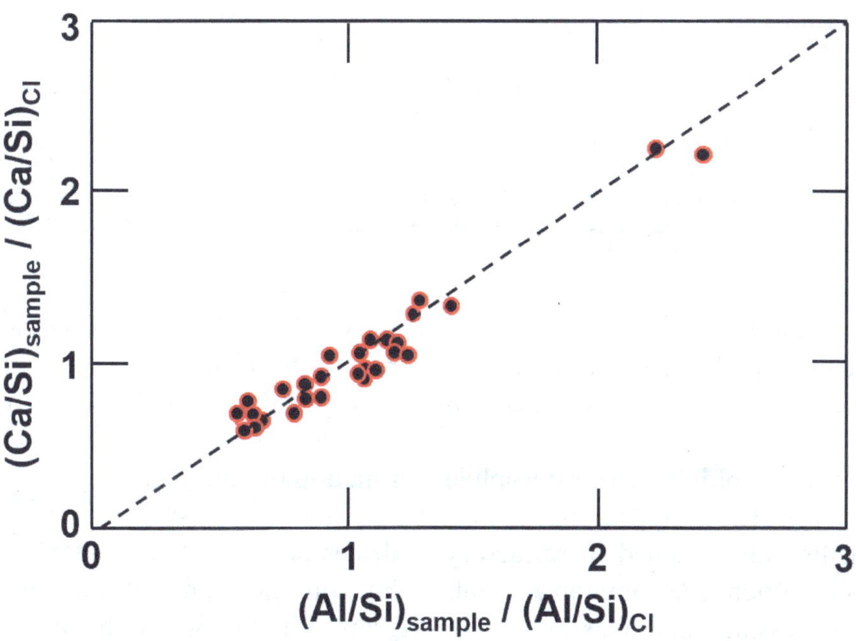

Fig. 11.16 The Si-normalized abundance ratio of Ca vs. that of Al for Semarkona chondrules. The trend line is the mean abundance ratio for all chondrite groups as determined by Kerridge (1979). Diagram modified from Grossman and Wasson (1983).

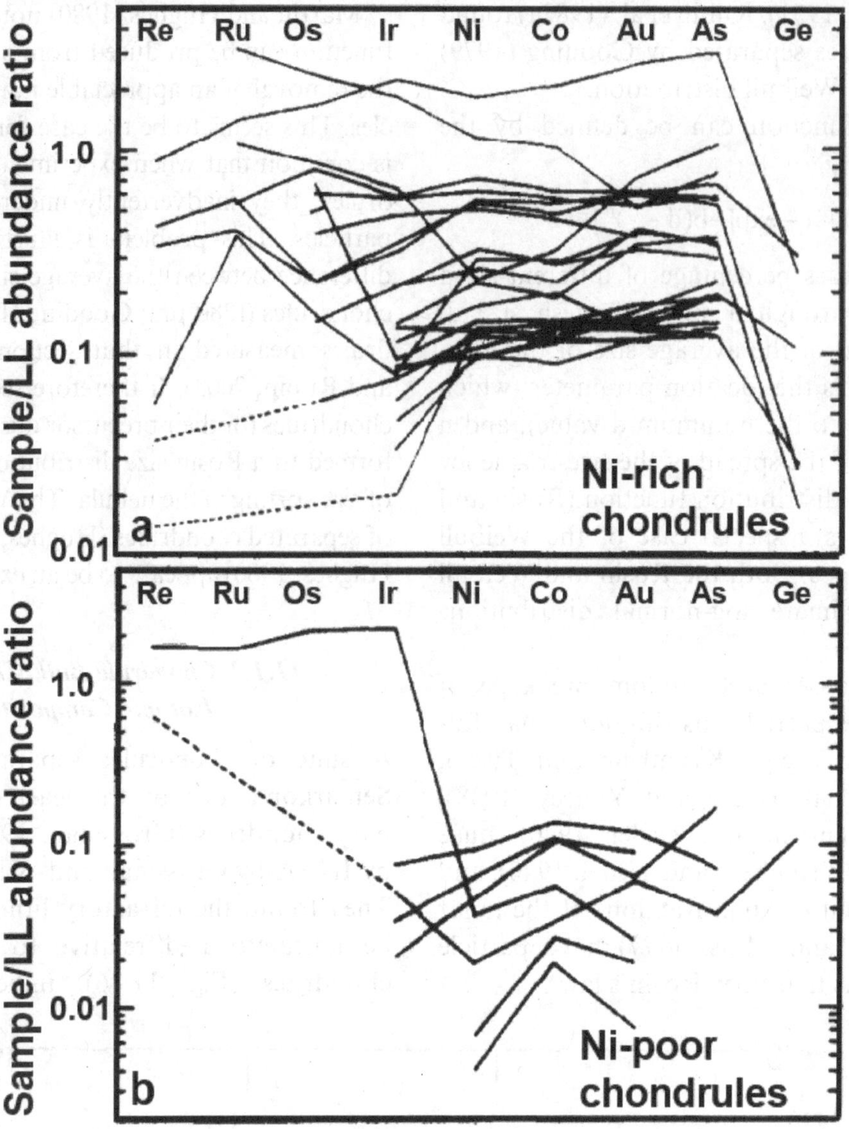

Fig. 11.17 LL-chondrite and Cr-normalized siderophile-element abundance patterns for Semarkona chondrules. Elements are plotted from left to right in order of increasing volatility. (a) Ni-rich chondrules (Ni > 1.0 mg/g). (b) Ni-poor chondrules (Ni < 1.0 mg/g). Refractory siderophiles (Re, Ru, Os, Ir) tend to correlate; in many chondrules their abundances are different than that of Ni. The volatile siderophiles (Au, As, Ge) tend to decrease in abundance as a function of volatility. Diagram modified from Grossman and Wasson (1985).

phases were likely in a single solid nebular component that was widely distributed in the OC region and acquired by the chondrule precursors prior to melting.

A different story is told by the siderophile elements (Fig. 11.17). There appear to be two metallic components: one enriched in refractory siderophiles and one enriched in common and volatile siderophiles (Grossman and Wasson, 1985). The refractory siderophiles (Re, Ru, Os, Ir) tend to correlate with one another and be fractionated relative to Ni and the volatile siderophiles. In contrast, the common (Ni, Co) and volatile (Au, As, Ge) siderophiles show increasing depletions as a function of volatility.

The bulk O-isotopic compositions of OC chondrules lie above the terrestrial fractionation (TF) line on the standard three-isotope oxygen diagram (Fig. 11.18a), in the same general field as OC whole-rocks (Fig. 10.6). This is because chondrules

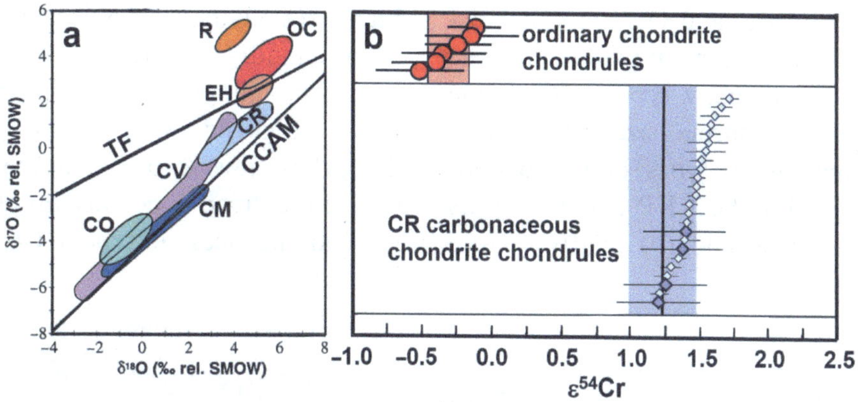

Fig. 11.18 Isotopic compositions of chondrules. (a) The O-isotopic composition of mechanically separated chondrules from most major chondrite groups plot in separate regions of the diagram. Outliers were excluded. TF = terrestrial fractionation line. CCAM = carbonaceous chondrite anhydrous minerals line. Diagram modified from Rubin (2000) with data compiled from numerous literature sources. (b) Values of ε^{54}Cr for individual chondrules from ordinary chondrites and CR carbonaceous chondrites. The mean compositions of the whole-rocks are shown as solid vertical lines surrounded by shaded regions denoting 2σ uncertainties. Diagram modified from Schneider et al. (2020).

constitute ~70–75 vol.% of type-3 OC whole-rocks (e.g., Figs. 11.1, 11.2). Al-rich chondrules extend the OC chondrule field to more-^{16}O-rich values (down to $\delta^{18}O = -15.7 \pm 1.8$‰; $\delta^{17}O = -13.5 \pm 2.6$‰) (Russell et al., 2000).

As discussed in Chapter 6, noncarbonaceous (NC) and carbonaceous (CC) whole-rocks plot in widely separated regions in diagrams of O-, Ti-, Cr-, and Ni-isotopic composition (Fig. 6.2), reflecting formation in the Inner and Outer Solar System, respectively (Warren, 2011a, 2011b). But what about the chondrules they contain? Analyses of O, Ti and Cr isotopes in individual chondrules (e.g., Gerber et al., 2017; Schneider et al., 2020) show that noncarbonaceous chondrules (i.e., those from ordinary and enstatite chondrites) contain relatively uniform Δ^{17}O, ε^{50}Ti, and ε^{54}Cr compositions, similar to those of their host chondrites. In contrast, carbonaceous-chondrite chondrules are more isotopically heterogeneous; some chondrules differ in isotopic composition from their host (e.g., Fig. 11.18b). There is no overlap in multi-isotope space between the carbonaceous and noncarbonaceous chondrules. The isotopic heterogeneity of carbonaceous-chondrite chondrules may result from mixing NC-like dust with a compositional component similar to CAIs and AOAs (Schneider et al., 2020). This mixing process may have been akin to the one responsible for producing the isotopic differences between the NC and CC whole-rocks (Fig. 6.2). The rapid accretion of Jupiter (Kruijer et al., 2017a, 2017b, 2017c) acted as a barrier, reducing the radial drift of small particles toward the Inner Solar System (e.g., Lambrechts et al., 2014), preserving the isotopic distinctions.

Neukampf et al. (2023) measured the Li isotopic compositions of olivine grains in Type-I and Type-II chondrules in two CR chondrites (Renazzo and NWA 852) and three ordinary chondrites (LL3.10 NWA 12581, LL3.15 NWA 12462, H3.10 NWA 13501). They found large variations in δ^{7}Li within individual grains; these isotopic variations likely were produced by kinetic fractionation during chondrule formation. Olivine in Type-II chondrules shows smaller Li-isotopic variations.

11.1.3 Microchondrules and Macrochondrules

In addition to the "normal-sized" ferromagnesian chondrules, there is a population of microchondrules, defined arbitrarily as being ≤40 μm in diameter. In ordinary chondrites, they occur as (1) abundant objects within FeO-rich fine-grained silicate-rich rims around normal-sized low-FeO (Type-I) chondrules (Figs. 11.19, 11.20), (2) the most common chondrules in rare clasts in type-3 ordinary chondrites and OC regolith breccias (Fig. 11.21), and (3) rare individuals in the fine-grained matrices of many type-3 OC. Some chondrule rims around normal-sized chondrules contain two compositional types of microchondrules: (a) abundant low-FeO RP and C objects rich in low-Ca pyroxene and (b) relatively rare high-FeO-olivine-rich objects (Figs. 11.19, 11.20). Microchondrule textures are moderately diverse: cryptocrystalline, microporphyritic, radial, and glassy (Bigolski et al., 2016). Most microchondrules have low porosity, but Dobrică and Brearley (2016) identified a porous variety within the fine-grained matrices in type-3 OC and in the rims

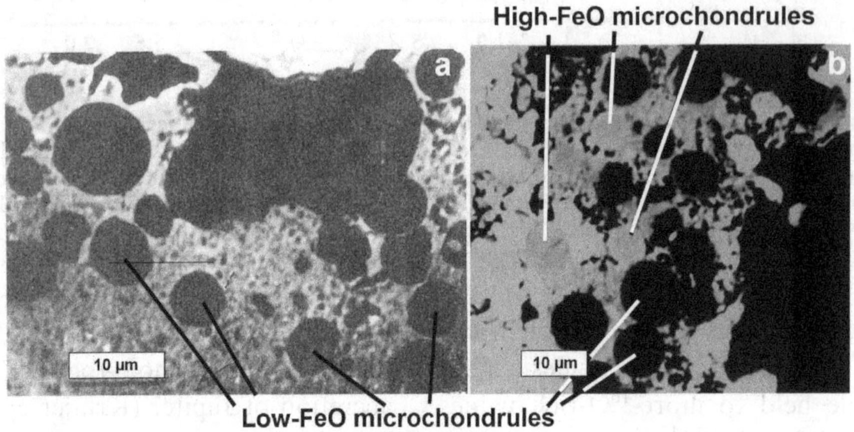

Fig. 11.19 Microchondrules within a fine-grained silicate rim around a normal-sized (850 × 1500 µm) Type-IA PO chondrule in L3.4 EET 90161. (a) Numerous low-FeO, low-Ca pyroxene microchondrules (black) and a large irregular pyroxene grain fragment (black; top). (b) Two varieties of microchondrules – abundant low-FeO, low-Ca pyroxene microchondrules (black) and less abundant high-FeO olivine microchondrules (light gray). A large, irregularly shaped pyroxene fragment (black) is at lower right. BSE images. Image modified from Krot et al. (1997a).

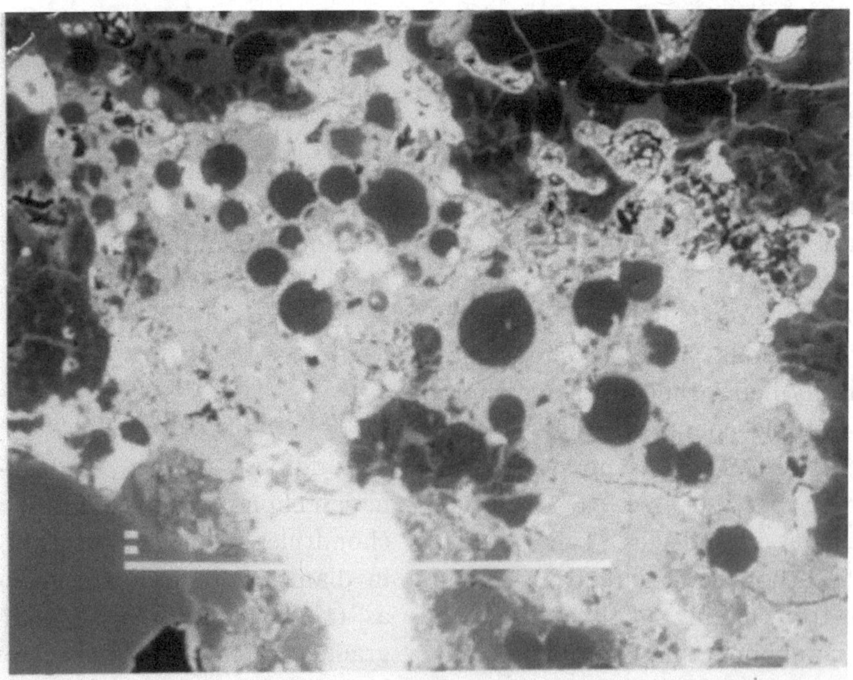

Fig. 11.20 Microchondrules within a fine-grained silicate rim around a normal-sized Type-I chondrule in the Piancaldoli LL3.10 chondrite. Low-FeO low-Ca pyroxene microchondrules (black) are abundant; also present are rare FeO-rich olivine microchondrules (light gray). Long white bar is 100 µm in length.

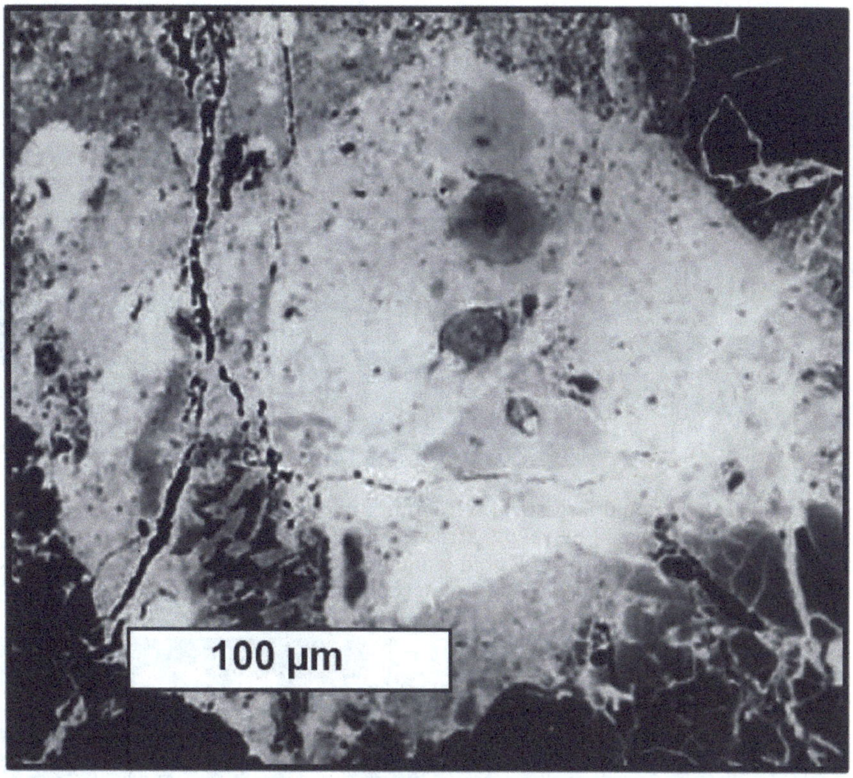

Fig. 11.21 Small clast (light gray) containing olivine microchondrules in LL3.2 Krymka. Two "large" microchondrules (top center) and many small ones are visible in the clast. The two largest microchondrules show "burn marks" from electron microprobe analysis. The large microchondrule at the top has a barred olivine texture; it is composed of subparallel vertical olivine bars and interjacent glass. The large microchondrule right below it has a mottled appearance; it appears to have a granular texture. Image modified from Rubin (1989).

around normal-sized Type-I chondrules. The porous microchondrules have irregular or rounded pores and contain glass with substantial FeO (18.4–23.9 wt.%) and moderate concentrations of Al_2O_3 (5.2–7.8 wt.%) and Na_2O (3.0–6.6 wt.%). Rubin et al. (1982a) estimated that microchondrules of all varieties constitute about 0.02 vol.% of H3, 0.006 vol.% of L3, and 0.0005 vol.% of LL3 chondrites.

The total range in diameters for OC chondrules spans nearly five orders of magnitude – from 0.8 μm (a microchondrule within a chondrule rim in LL3.10 Piancaldoli; Rubin et al., 1982a; Fig. 11.22a) to ~5 cm (a golfball-sized macrochondrule in L3.2 Gunlock; Prinz et al., 1988; Fig. 11.22b). It is possible that the paucity of very large chondrules is due to the fact that the sticking probability of particles to their precursor dustballs decreased with increasing dustball radius (Arakawa et al., 2023).

Weisberg et al. (1988) defined macrochondrules as chondrules >0.5 cm in maximum dimension. Centimeter-sized macrochondrules are quite rare: in a survey of 833 ordinary chondrites, Bridges and Hutchison (1997) reported only nine objects they identified as macrochondrules ≥1 cm in diameter. They all have porphyritic or barred olivine textures. The largest object in their study is a 1.3-cm-wide BO chondrule from L6 De Nova. Hellmann et al. (2020) reported a 2-cm-diameter BO chondrule from L5/6 NWA 8192. However, Gunlock still holds the record.

11.1.4 Compound Chondrules

About 3% of the chondrules in ordinary chondrites are compound objects consisting of two or more conjoined chondrules. There are three principal varieties (Wasson et al., 1995): *enveloping compound chondrules* (Fig. 11.23a,b), nested objects in which the inner primary chondrule is surrounded by a spherical shell (typically with a porphyritic

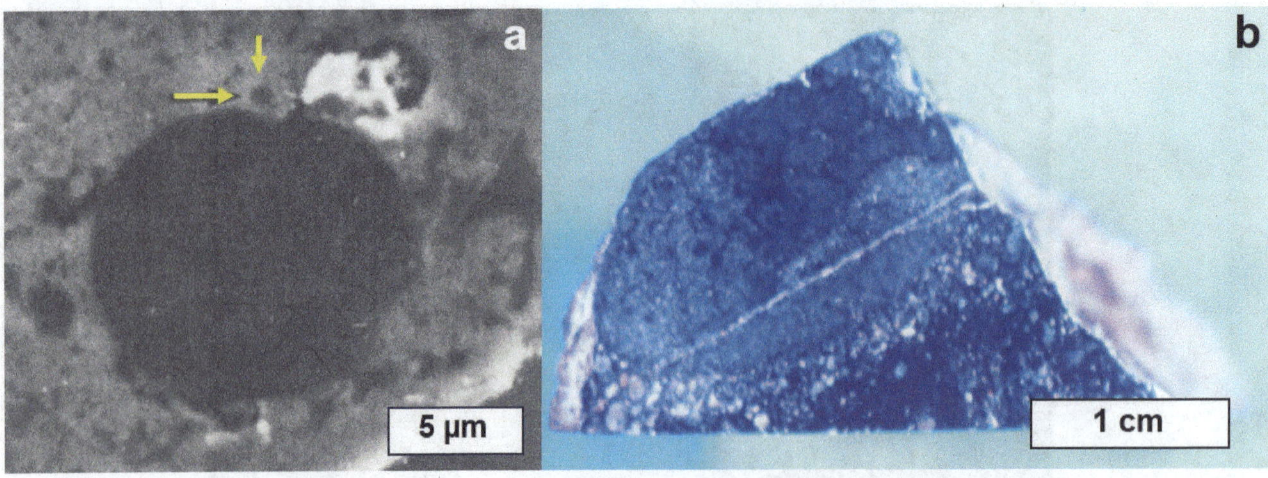

Fig. 11.22 Examples of the smallest and largest chondrules in ordinary chondrites. (a) The smallest reported micro-chondrule (0.8 μm in diameter), indicated by yellow arrows, within fine-grained silicate matrix in a chondrule rim in LL3.10 Piancaldoli. BSE image, modified from Rubin et al. (1982a). (b) Fragment of the separated ~5-cm-diameter macrochondrule in L3.2 Gunlock surrounded by host material. Image courtesy of Michael Weisberg.

Fig. 11.23 The variety of compound chondrules in ordinary chondrites. (a) Enveloping BO compound chondrule in L5 Mount Tazerzait, showing recrystallization and integration with the matrix due to parent-body thermal metamorphism. (b) Enveloping compound chondrule consisting of a small internal BO within a PO in H3 Zegdou. (c) Sibling BP compound chondrule in H3.7 Brownfield (1937). (d) Adhering compound chondrule consisting of a BO primary and several secondary BO adhesions in L3.3 Dhofar 008. Images in transmitted light with crossed polars, courtesy of John Kashuba.

or barred-olivine texture); *sibling compound chondrules* (Fig. 11.23c), bi-lobate objects consisting of two partly merged chondrules of similar size; and *adhering compound chondrules* (Fig. 11.23d) consisting of a large primary chondrule and one or more (or in rare cases, more than a dozen) small droplet chondrules that are attached to (and typically somewhat spread out along) the surface, more closely resembling mounds than spheroids. Many BO chondrules have two or more nested olivine rims, presumably representing chondrule surfaces that were later coated with fine-grained material that melted during subsequent chondrule-heating events.

11.1.5 Non-ferromagnesian Chondrules

A small proportion of OC chondrules have unusual mineralogy. These include silica-rich chondrules (some containing merrihueite or roedderite), chromite-rich chondrules, Al-rich chondrules, glass-rich chondrules (a variety of Al-rich chondrule), and metallic chondrules.

Silica-rich chondrules in ordinary chondrites consist mainly of ferroan low-Ca pyroxene and silica (probably cristobalite with >99 wt.% SiO_2) with minor to accessory amounts of Ca-pyroxene, glass, merrillite, metallic Fe-Ni, and troilite (Brigham et al., 1986). In many of these chondrules, silica occurs as laths or more-equant grains (Fig. 11.24), indicative of crystallization from the chondrule melt. Several ordinary chondrites also contain silica-pyroxene-fayalite clasts that may be silica-rich chondrule fragments.

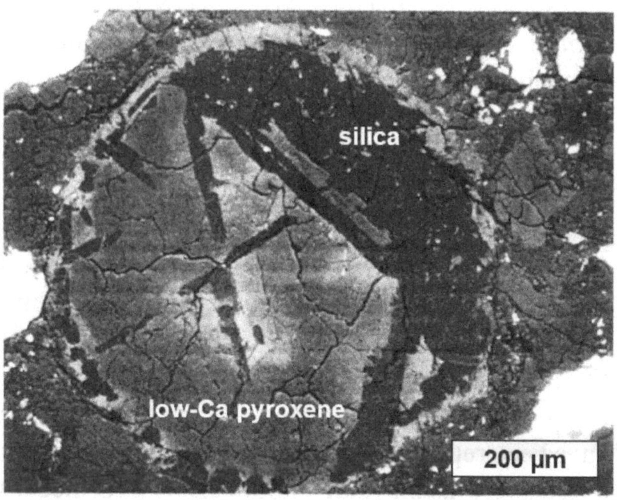

Fig. 11.24 A silica-rich chondrule consisting mainly of silica (very dark gray) and low-Ca pyroxene (medium gray) from H/L3.9 Bremervörde. The brighter regions of the low-Ca pyroxene are richer in FeO. BSE image modified from Brigham et al. (1986).

Some silica-bearing chondrules and chondrule fragments contain grains of merrihueite ($(K,Na)_2(Fe,Mg)_5Si_{12}O_{30}$) or roedderite ($(Na,K)_2Mg_5Si_{12}O_{30}$) (Krot and Wasson, 1994). These minerals are the Al-poor members of the osumilite group of cyclosilicates; they have a framework structure consisting of SiO_4 tetrahedra forming hexagonal double rings. Additional phases in these chondrules include ferroan low-Ca pyroxene, fayalitic olivine, troilite, and feldspathic mesostasis. Krot and Wasson (1994) suggested that these objects have a complicated history involving (a) fractional condensation in the nebula to produce Si-rich and Al-poor precursors, (b) melting of the precursors to create silica-pyroxene-rich chondrules, (c) reaction of silica with alkali-rich fluids on a parent body to form merrihueite and/or roedderite, and (d) reaction of silica with FeO on the same parent body to produce fayalitic olivine and ferroan pyroxene. There may also have been minor thermal metamorphism and one or more episodes of collisionally induced fragmentation.

Chromium-rich chondrules are rare objects, constituting <0.1% of ordinary-chondrite chondrules (Krot and Rubin, 1993). They are not found in highly unequilibrated OC; they have been reported only in OC of type 3.7 or 3.8 and higher. They are thus unlikely to be nebular products. Chromium-rich chondrules exhibit a wide variety of textures – porphyritic, granular, barred, radial, and cryptocrystalline. There are two types of Cr-rich chondrules: (1) *Chromian-spinel chondrules* consist of up to ~50 vol.% chromite or chromian-spinel ($(Mg,Fe^{2+})(Cr,Al,Fe^{3+})_2O_4$) grains embedded in plagioclase or feldspathic mesostasis of plagioclase composition (Fig. 11.25). Many of these chondrules also contain accessory ilmenite, Ca-pyroxene, merrillite, and olivine. (2) *Chromium-rich mafic silicate chondrules* consist of chromite, olivine, plagioclase and accessory merrillite, troilite, metallic Fe-Ni, Ca-pyroxene, and ilmenite. In most cases, the chromite grains are tiny (of micrometer size) and embedded in mesostasis; this renders the chondrules nearly opaque. Wlotzka (2005) suggested that Cr-rich chondrules formed by parent-body thermal metamorphism of spinel-bearing Al-rich chondrules. Krot and Rubin (1993) and Rubin

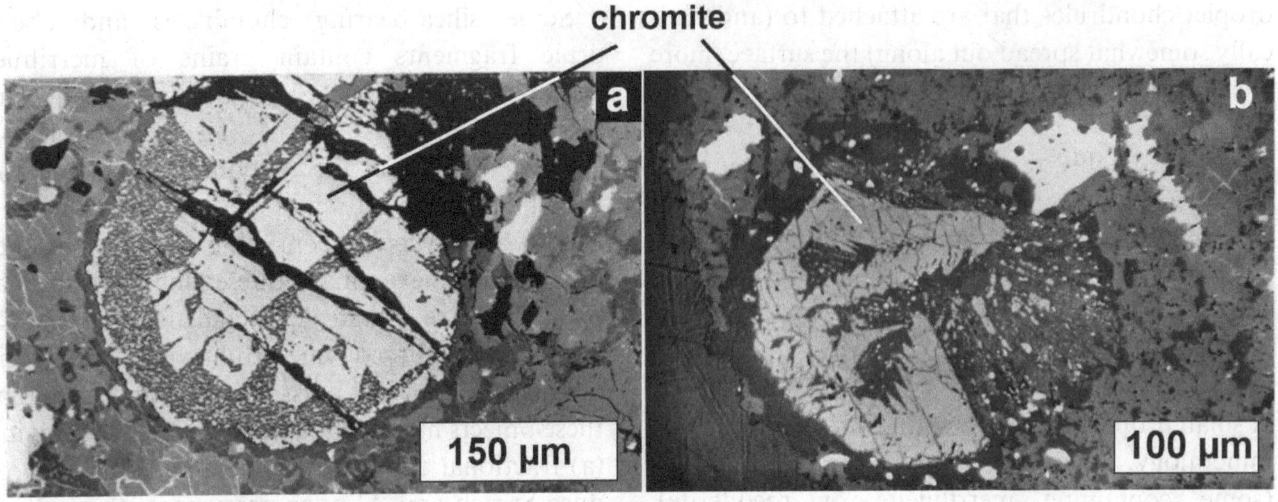

Fig. 11.25 Chromite-rich chondrules. (a) The chondrule has coarse chromite grains in the core and fine-grained chromite grains throughout the groundmass. The chondrule is rimmed by a thin rind of plagioclase. Chondrule from the Plainview (1917) H5 regolith breccia. (b) Coarse skeletal chromite surrounded by a feldspar-rich groundmass. Chondrule from LL6 Mangwendi. Both images in reflected light, modified from Ramdohr (1967b).

(2003) suggested formation by small-scale, localized impact-melting of chromite-rich assemblages in equilibrated or partially equilibrated OC.

Al-rich chondrules were defined by Bischoff and Keil (1984) as round objects with igneous textures that contain ≥10 wt.% bulk Al_2O_3; most of these chondrules are a few hundred micrometers in apparent diameter. There are four principal groups of Al-rich chondrules in ordinary chondrites, differing in bulk chemical composition: (1) 62% are Ca-Al-rich chondrules (≤5.0 wt.% Na_2O), (2) 16% are (Ca,Na)-Al-rich chondrules (>5.0 wt.% Na_2O) (e.g., Fig. 11.23), (3) 17% are Na-Al-rich chondrules (≤2.6 wt.% CaO, >4.1 wt.% Na_2O), and (4) 5% are Na-Al-Cr-rich chondrules (<4.8 wt.% CaO, >4.7 wt.% Na_2O, ≥2.4 wt.% Cr_2O_3). Also present in ordinary chondrites are Al-rich fragments, refractory mineral clasts, irregularly shaped CAIs, and a single 220-µm-diameter Al-Ti-rich inclusion from H3.8 Dhajala which may or may not be a chondrule.

The Ca-Al-rich chondrules contain spinel, Al-Ti diopside (fassaite), olivine, low-Ca pyroxene, Ca-pyroxene, plagioclase, and glass (e.g., Bischoff and Keil, 1984; M. Zhang et al., 2024). The (Ca,Na)-Al-rich chondrules are composed of spinel, Al-Ti diopside, low-Ca pyroxene, Ca-pyroxene, olivine, plagioclase, nepheline, and glass. The Na-Al-rich chondrules consist of skeletal olivine, minor Ca-pyroxene and spinel, and abundant sodic glass. The Na-Al-Cr-rich chondrules are fine-grained, nearly opaque objects containing abundant oxides (chromite, chromian-hercynian spinel, ilmenite), minor skeletal olivine, plagioclase, and low-Ca pyroxene.

A unique Ca-Al-rich chondrule (G7) in LL3.01 Semarkona (Fig. 11.26), studied by Rubin (2004a), contains 2 vol% low-Ca pyroxene with up to 16.3 wt.% Al_2O_3; it is among the most Al-rich low-Ca pyroxene grains known. Calculations suggest that 20% of its tetrahedral sites and 13% of its octahedral sites are occupied by Al. These pyroxene grains exhibit approximately parallel extinction, making it likely that the grains are orthorhombic. That would be consistent with the report of O'Hara and Schairer (1963) showing that Al_2O_3 stabilizes the orthoenstatite structure relative to protoenstatite in the diopside-pyrope system at atmospheric pressure. Skeletal olivine grains with the chain structure (Donaldson, 1976) are indicative of rapid cooling. The bulk composition of the chondrule is within the range of amoeboid olivine aggregates (AOAs) in carbonaceous chondrites. Because rare AOAs have been reported in Semarkona (Itoh et al., 2007), it seems plausible that chondrule G7 formed by remelting an AOA-bearing precursor assemblage.

Glass-rich chondrules are a variety of Al-rich chondrule containing either abundant Si-, Al-, and Na-rich glass or a Ca-Al-rich microcrystalline

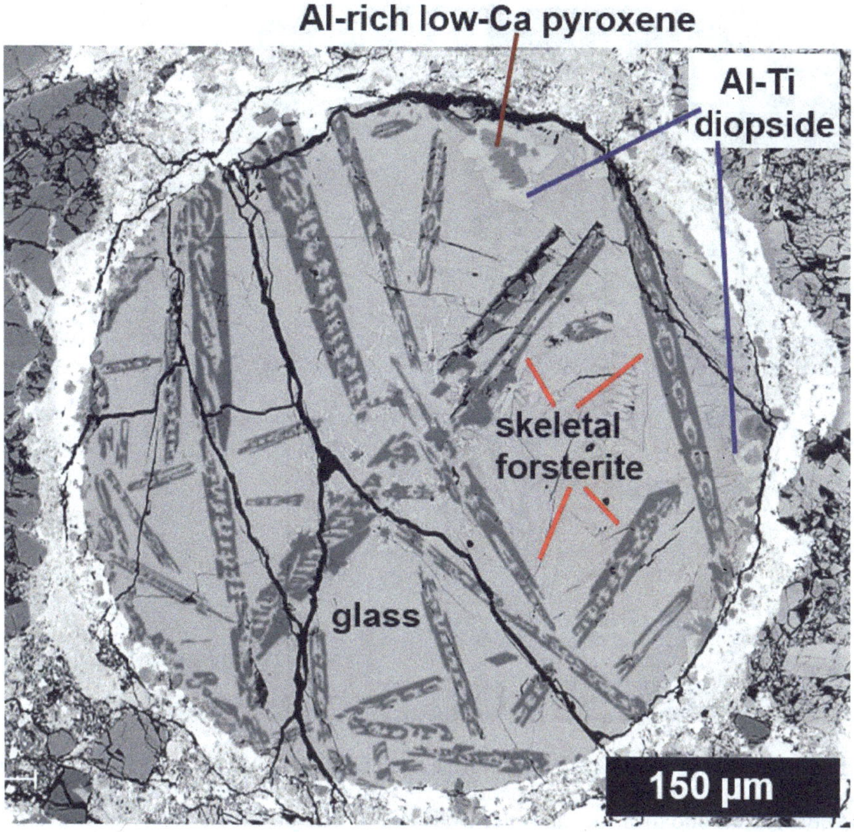

Fig. 11.26 Ca-Al-rich chondrule G7 from LL3.01 Semarkona. The 540-µm-diameter chondrule consists of 73 vol.% light-brown glassy mesostasis, 22 vol.% skeletal forsterite grains, 3 vol.% Al-Ti diopside (fassaite) and 2 vol.% Al-rich low-Ca pyroxene grains. Opaque phases are absent. Image modified from Rubin (2004a).

mesostasis. Many have dendritic and skeletal crystals of magnesian olivine; some also contain Al-rich low-Ca pyroxene and Al-Ti diopside (fassaite) (e.g., Fig. 11.26). Mg-Al spinel may also occur. There are two varieties: porphyritic glass-rich chondrules with 55–85 vol.% glass (Figs. 11.26, 11.27) and nonporphyritic chondrules with 90–99 vol.% glass (Krot and Rubin, 1994).

Nagahara et al. (2008) described a glass-rich chondrule (CH5) from LL3.01 Semarkona with 10–100-µm-sized forsterite grains, many with skeletal, dendritic, or tabular morphologies (Fig. 11.27). The glass is chemically zoned: it is richer in Al_2O_3 and poorer in MgO near the forsterite. Its mean composition (in wt.%) is about 51% SiO_2, 21% Al_2O_3, 17% CaO, 9% MgO, 1% TiO_2, and 1% total of FeO, MnO, and Na_2O.

Glass-rich droplet chondrules can devitrify during whole-rock thermal metamorphism as evident in H5 Burdett (Fodor et al., 1971).

Metallic chondrules constitute ~0.5% of OC chondrules (~9 out of 1674 examined in thin sections of type-3 and -4 ordinary chondrites; Gooding and Keil, 1981). An altered metallic chondrule from LL3.01 Semarkona (Fig. 10.16b) consists of polycrystalline grains of cohenite mantled by magnetite and a thin layer of troilite. Pristine metallic chondrules from H/L3.6 Tieschitz measure up to 500 µm across (e.g., Fig. 11.28a) and consist of intergrown metallic Fe-Ni and troilite. Their spheroidal shapes indicate their origin as liquid droplets molded by surface tension. Bevan and Axon (1980) estimated that metallic chondrules solidified at cooling rates between 1 and $10^6\,K\,s^{-1}$. The metal chondrules are plausibly related to rounded metal blebs within Type-I chondrules (e.g., Fig. 11.28b). Crater-like depressions at chondrule surfaces (Fig. 10.16a) are commonly lined with metal and troilite (Grossman and Wasson, 1985); they likely mark the spots where opaque

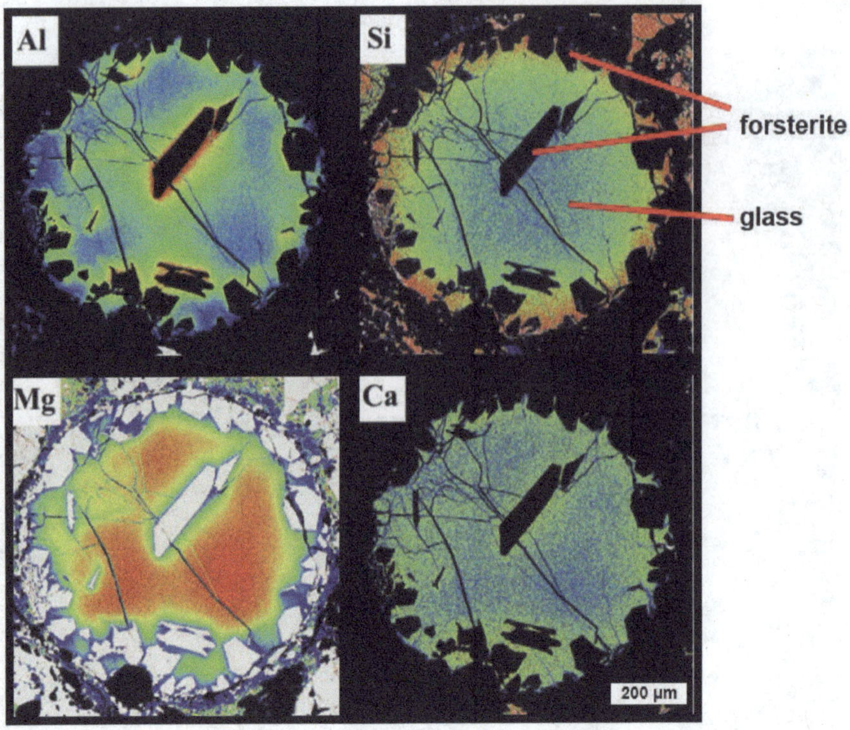

Fig. 11.27 Elemental X-ray maps of Ca-Al-rich chondrule CH5 from LL3.01 Semarkona. The 800-µm-diameter chondrule consists of 78 vol.% clear glassy mesostasis and 22 vol.% forsterite. The chondrule is almost completely free of metallic Fe-Ni. Colors correspond to elemental concentration increasing from black to blue to green to yellow to red to white. The maps are not normalized to a standard concentration – the glass contains much more Al_2O_3 (~21 wt.%) than MgO (~9 wt.%) even though the Al X-ray map appears blue and the Mg map appears red in much of the glass. Image modified from Nagahara et al. (2008).

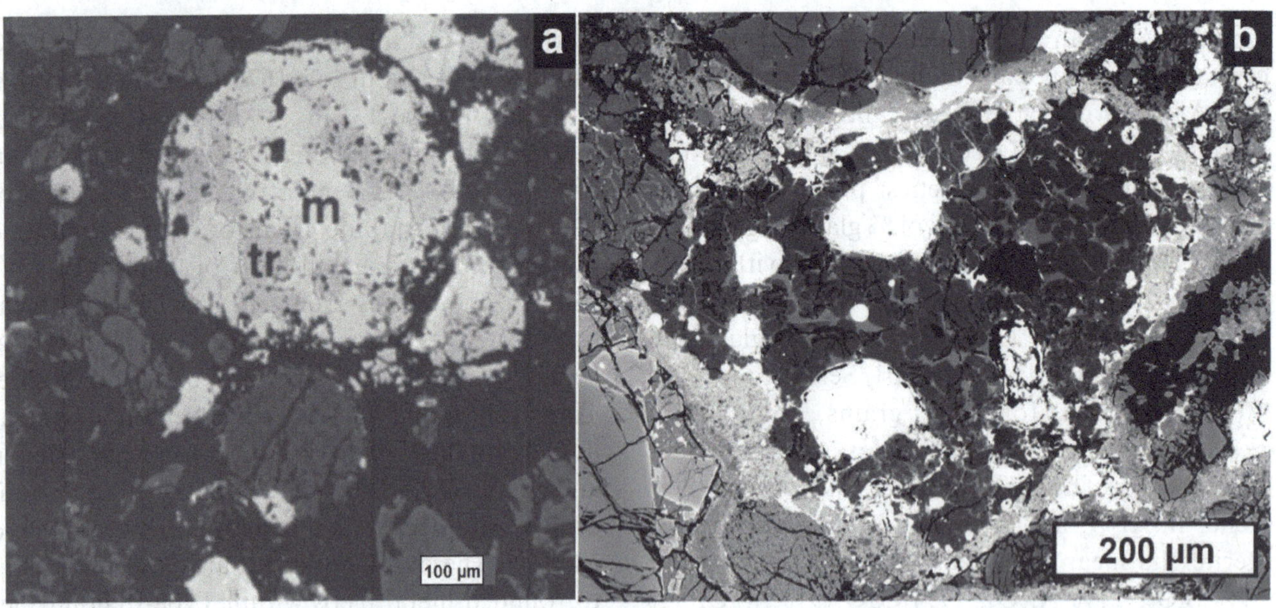

Fig. 11.28 Metallic-Fe-Ni-rich spheroids in ordinary chondrites. (a) Spheroidal metallic chondrule in H/L3.6 Tieschitz consisting of metallic Fe-Ni (m) (light gray) and troilite (tr) (medium gray) coexisting with ferromagnesian chondrules and chondrule fragments (dark gray) and grains of metallic Fe-Ni (light gray) and sulfide (medium gray). Reflected light image modified from Bevan and Axon (1980). (b) Type-I chondrule with metallic-Fe-Ni-rich spherules (up to ~100 µm in diameter) in LL3.01 Semarkona. BSE image.

droplets escaped from spinning chondrules via centrifugal force. Metallic chondrules may represent the minor fraction of escaped opaque blebs that reaccreted to the host chondrite.

11.2 Chondrule Rims

11.2.1 BO Chondrule Rims

Rims around barred olivine (BO) chondrules formed at the same time as the olivine bars, during the cooling of a molten precursor droplet. Typically, the rims are in optical continuity with the bars (or a prominent set of bars) indicating that these olivine grains all are part of a single crystal (Fig. 11.29a). Experimental evidence indicates that slower cooling results in thicker rims (Morita et al., 2025). In some cases, there is a second (or third, or fourth...) rim outside the primary rim (e.g., Fig. 11.29c,d,e). The primary BO chondrules either (1) encountered a molten or semi-molten droplet that adhered to its surface and spread across a portion of it (Fig. 11.29b) or (2) acquired a dusty rim (or became incorporated into a dustball) and experienced reheating (Fig. 11.29c,d,e). In the latter case, the heating was sufficient to melt the porous dust but not melt the primary BO chondrule. In some instances, such processes occurred several times.

11.2.2 Fine-Grained Rims

Although chondrules are modeled as isolated free-floating objects in the nebula, the existence of fine-grained rims and igneous rims around chondrules indicates that many chondrules encountered other solid particles that adhered to their surfaces. In some cases, the chondrules were reheated afterwards (perhaps multiple times, producing igneous rims and the spherical shells of enveloping compound chondrules); in other cases, they were not reheated (resulting either in rimless chondrules or in chondrules with fine-grained rims).

In the least metamorphosed type-3 ordinary chondrites, many chondrules are surrounded by 10–30-μm-wide, fine-grained rims; the rims are composed mainly of submicrometer-sized silicate grains (e.g., Christophe Michel-Lévy, 1976; Scott et al., 1984; Alexander et al., 1989b). The rims contain olivine (and, to a lesser extent, low-Ca pyroxene); amorphous material enriched in normative albite and partly altered to smectite occurs in chondrule rims in LL3.15 Bishunpur. Some rims contain tiny, scattered grains of metallic Fe-Ni and/or troilite. Because of their fine grain size (<0.1 μm; Ashworth, 1977), far smaller than the 30-μm thickness of typical thin sections, chondrule rims appear opaque when viewed in transmitted light in the microscope (Fig. 11.30a). Texturally and mineralogically, the rims resemble the silicate matrix material in type-3 chondrites (and are sometimes referred to as "matrix rims" or "matrix-like rims"), but they are finer-grained and somewhat less porous than typical matrix material (Ashworth, 1977). Some fine-grained rims contain abundant microchondrules (e.g., Fig. 11.20).

Individual rims are relatively homogeneous in composition, varying moderately in FeO but with a fairly constant MgO/SiO_2 ratio. A typical chondrule rim in H/L3.6 Tieschitz contains (in wt.%): ~33% SiO_2, 3% Al_2O_3, 32% FeO, 15% MgO, 1% CaO, 2% Na_2O, and 1% combined total of other components (TiO_2, Cr_2O_3, MnO, K_2O, P_2O_5, and S) (Scott et al., 1984). The low bulk total (~87 wt.%) is mainly an indication of rim porosity; C and H_2O are of low concentration and contribute little to the low total.

In somewhat more metamorphosed type-3 OC, a higher proportion of chondrule rims are recrystallized; they are up to 50 μm thick and composed of 3–5-μm-sized grains of olivine and low-Ca pyroxene. A few contain ribbons of troilite (Fig. 11.30b) and have been dubbed "lacy sulfide rims" by Allen et al. (1980). Some type-3 OC resemble type-4 samples in having chondrules with rims containing massive metal ± sulfide, giving the appearance of "armored chondrules" (Fig. 11.31). Many armored chondrules rimmed in metallic Fe-Ni occur in CR chondrites (e.g., Wasson and Rubin, 2010).

11.2.3 Igneous Rims

Originally dubbed "coarse-grained rims" by Rubin (1984a, 1984b), igneous rims have microporphyritic or granular textures and consist primarily of 5–40-μm-sized grains of low-Ca pyroxene and

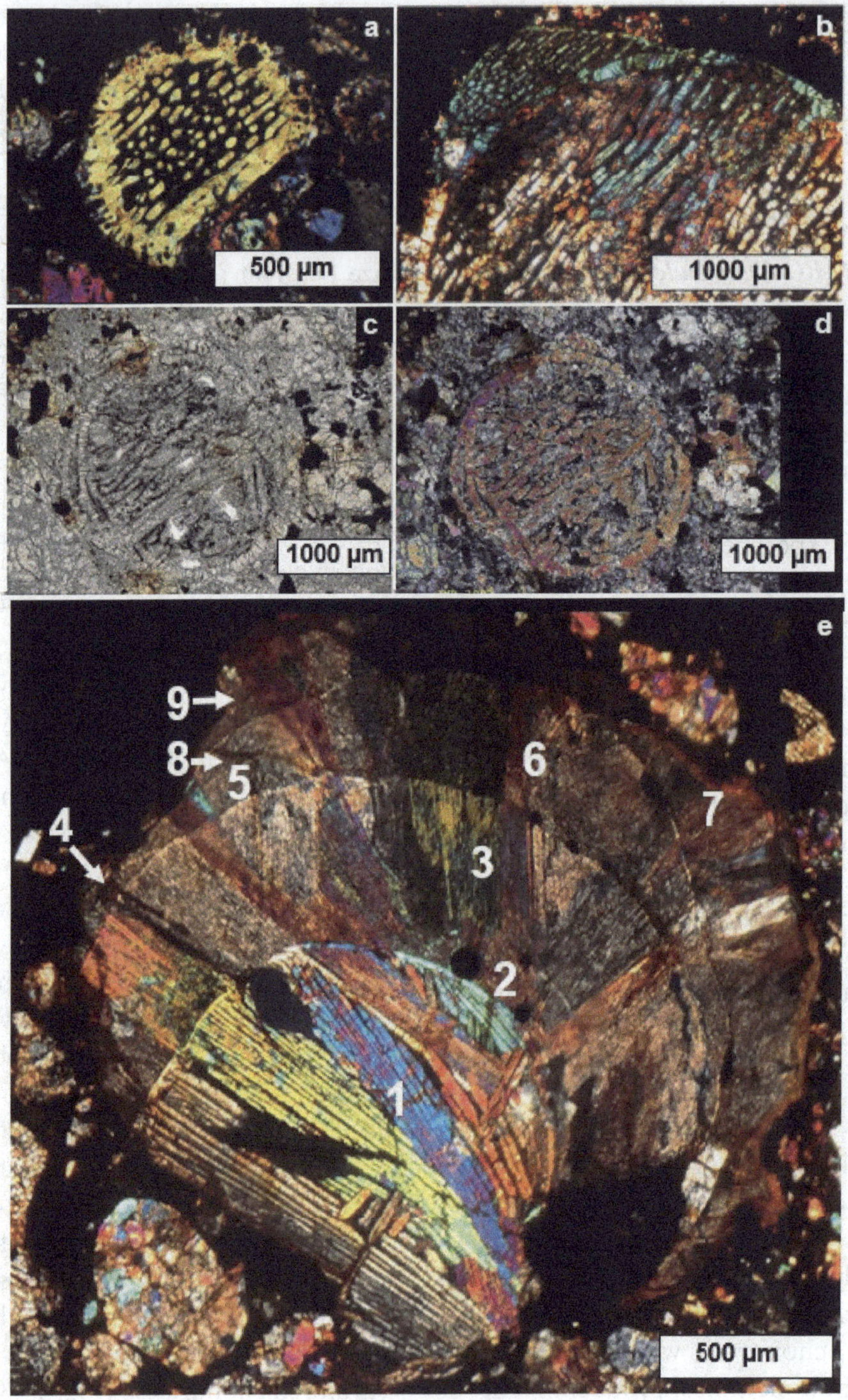

Fig. 11.29 Barred olivine chondrules. (a) Rimmed BO chondrule showing optical continuity in H5 Cole Creek. The BO is part of a compound chondrule that includes POP material at lower right. Transmitted light with crossed polars. (b) Detail of a BO chondrule from L3.3 Dhofar 008 (from Fig. 11.23d) showing an adhering BO mound above the olivine rim at upper left. Transmitted light with crossed polars. (c) Recrystallized BO chondrule with multiple rims from L6 Jalu. Transmitted light image. (d) The same chondrule as in (c) shown in transmitted light with crossed polars. (e) Compound BO chondrule from LL3 NWA 12009 with nine distinct portions that formed successively: 1 = primary chondrule; 2 = adhering chondrule; 3 = secondary shell; 4 = adhering chondrule; 5 = adhering chondrule; 6 = secondary shell; 7 = secondary shell; 8 = adhering chondrule; 9 = adhering chondrule. Transmitted light with crossed polars. Images (a) and (b) courtesy of John Kashuba; images (c) and (d) courtesy of J. M. Derochette; image (e) courtesy of Mila Zinkova.

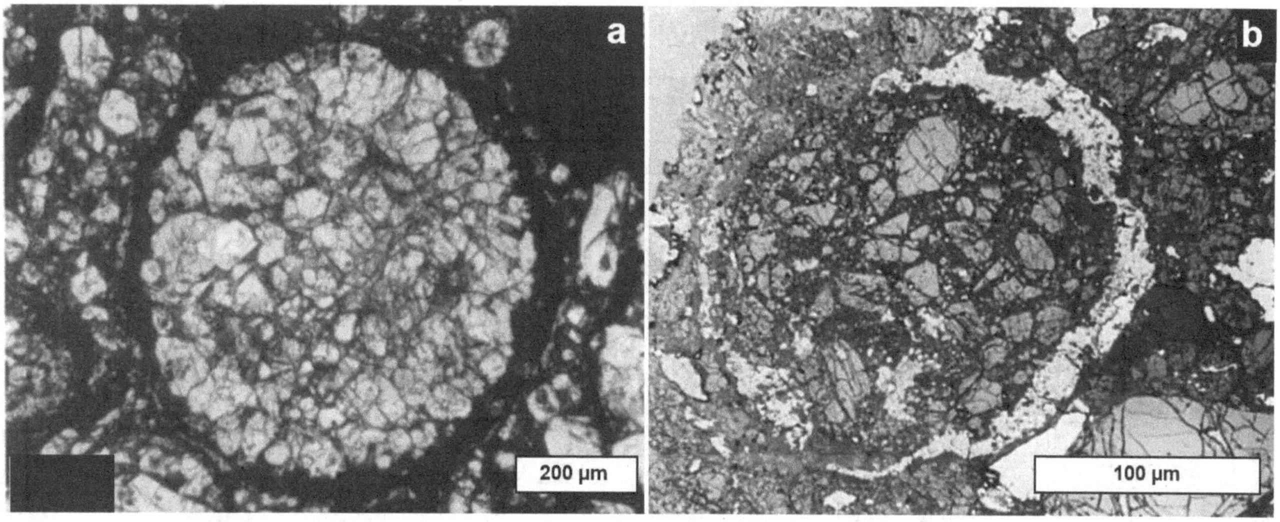

Fig. 11.30 Chondrule rims. (a) Transmitted-light photomicrograph of a 10–50-µm-wide chondrule rim around a PO chondrule in H/L3.6 Tieschitz. Transmitted light image from Scott et al. (1984). (b) A chondrule from L3.4 Hallingeberg with a lacy sulfide rim. Reflected light image modified from Allen et al. (1980).

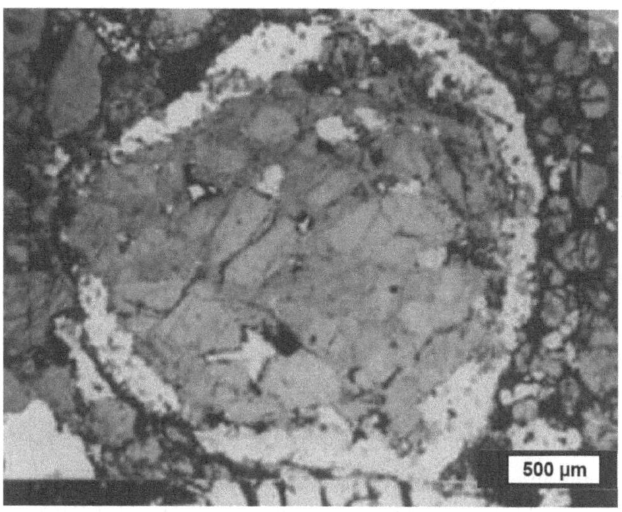

Fig. 11.31 Armored PO chondrule in H4 Conquista surrounded by a thick rim of metallic Fe-Ni. Reflected light image modified from Keil et al. (1978).

olivine (Fig. 11.32a). Low-Ca pyroxene is generally the more abundant phase. The rims form partial to complete mantles around ~10% of chondrules in type-3 OC; they surround all textural types of chondrules, both low FeO and high FeO. The rims are commonly of varying thickness but have a mean width of ~160 µm. Many of the rims also contain opaque phases (metallic Fe-Ni and/or sulfide; Fig. 11.32b) and appear relatively dark in transmitted light (Fig. 11.33). The boundaries between some rims and their host chondrules contain layers of metallic Fe-Ni and troilite. Many igneous rims are themselves surrounded by fine-grained rims; the boundaries between the igneous and fine-grained rims are sharp.

Krot and Wasson (1995) discerned evidence of melting in igneous rims (hence the name): (1) partly resorbed pyroxene grains at the chondrule margins and at the interface with the igneous rims, (2) round blebs of metallic Fe-Ni and/or troilite, (3) feldspathic mesostasis, and (4) euhedral olivine and low-Ca pyroxene grains of relatively uniform size.

There may be differences between igneous rims around low-FeO (Type-I) and high-FeO (Type-II) chondrules: (1) Those around low-FeO chondrules appear to have no relict grains, while some igneous rims around high-FeO chondrules contain relict grains. (A possible explanation: It is plausible that Type-II chondrules generally formed later than Type-I chondrules; thus, the dustball precursors of the igneous rims around Type-II chondrules may have spent more time as isolated bodies in the nebula, increasing the chances they would incorporate relict grains from shattered chondrules.) (2) Olivine and low-Ca pyroxene in igneous rims around low-FeO chondrules are similar in composition to the grains in the host chondrules; in contrast, olivine and low-Ca pyroxene in igneous rims around

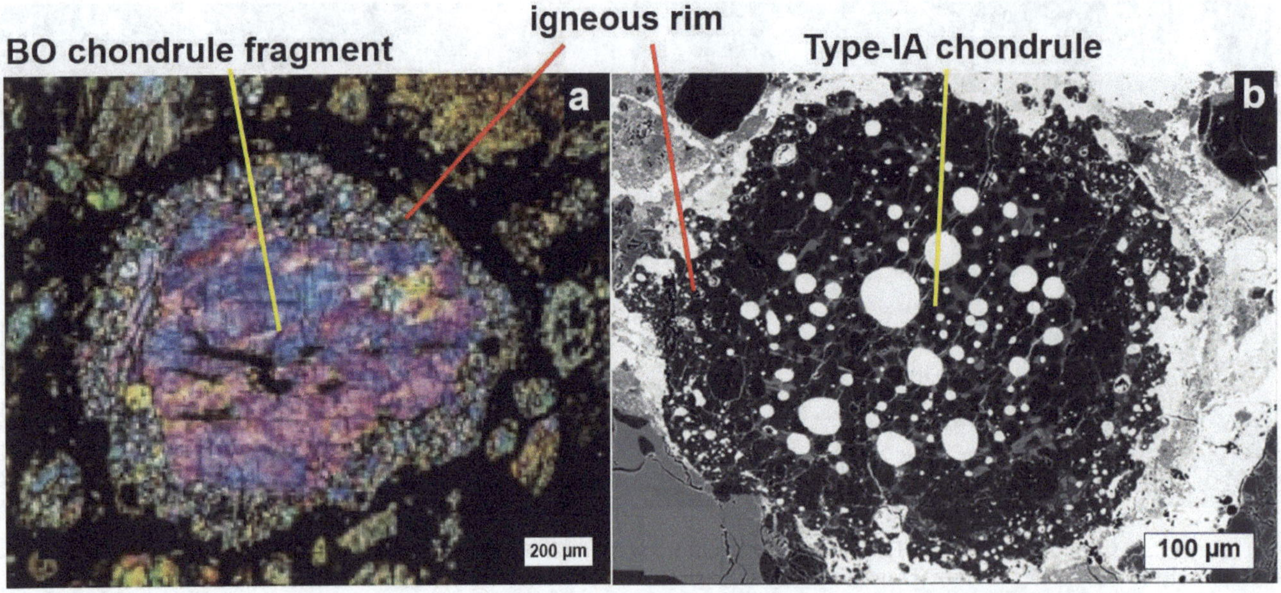

Fig. 11.32 Igneous rims around chondrules. (a) Rim around a BO chondrule fragment in L3.4 Catalina 001. Transmitted light with crossed polars; image modified from Morris and Garvie (2013). (b) Rim around a low-FeO porphyritic olivine chondrule (Type IA) in L3.05 QUE 97008. Metallic Fe-Ni blebs are white. BSE image, modified from Connolly and Jones (2016).

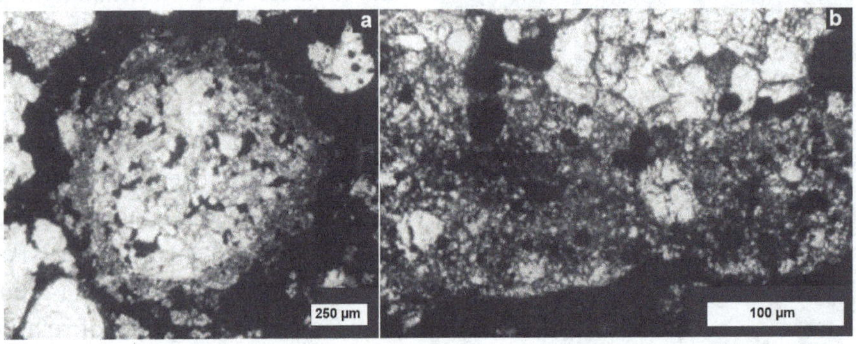

Fig. 11.33 Igneous rim around a POP chondrule with moderate amounts of troilite (black) in LL3.01 Semarkona. (a) Entire chondrule with igneous rim. (b) Higher-magnification view of top left of the igneous rim shown in (a). The dark grains in the rim are mainly troilite. Relict mafic silicate grains (white) are present. The host chondrule is at the top of the image. Transmitted light images from Rubin (1984b).

many high-FeO chondrules are more ferroan than the grains within the host chondrules. (Possible explanation: Olivine and pyroxene grains in igneous rims tend to be much smaller than chondrule phenocrysts. During the oxidation of metallic Fe to FeO, the smaller grains in the igneous rims would equilibrate faster and become more ferroan than chondrule phenocrysts.) These differences strongly suggest that low-FeO and high-FeO chondrules formed in different environments.

Enveloping compound chondrules form a continuum with igneous rims, but the outer spherical shell of an enveloping compound chondrule typically has a coarser grain size, a lower modal abundance of dispersed opaque phases, and a more uniform thickness than an igneous rim. Secondary shells around porphyritic chondrules (and some BO chondrules) typically have porphyritic textures; some shells around BO chondrules have BO textures and grade into normal BO chondrule rims (Fig. 11.9). Secondary layers around primary chondrules need not be spherical shells: some are large and quasi-spheroidal (Fig. 11.34a); others are ellipsoidal (Fig. 11.34b). Some primary

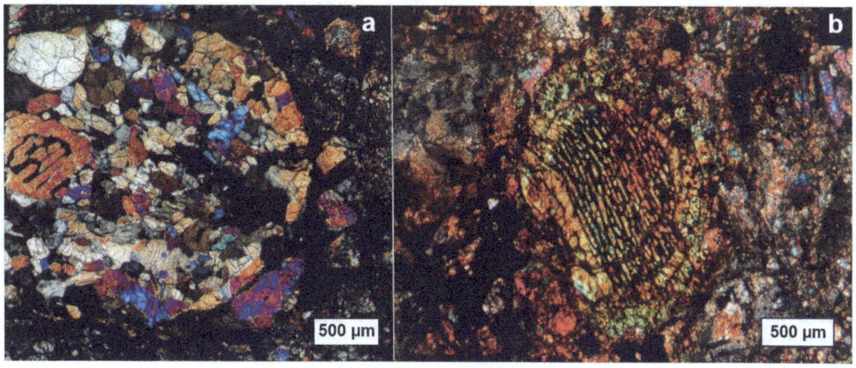

Fig. 11.34 Enveloping compound chondrules. (a) BO primary chondrule within a much larger secondary POP chondrule mantle in L3.3 Dhofar 008. Isolated olivine grains, probably derived from broken porphyritic chondrules, occur to the right of the chondrule. (b) BO primary chondrule surrounded by several olivine rims forming an ellipsoidal BO shell in L4 Dalgety Downs. Images courtesy of John Kashuba.

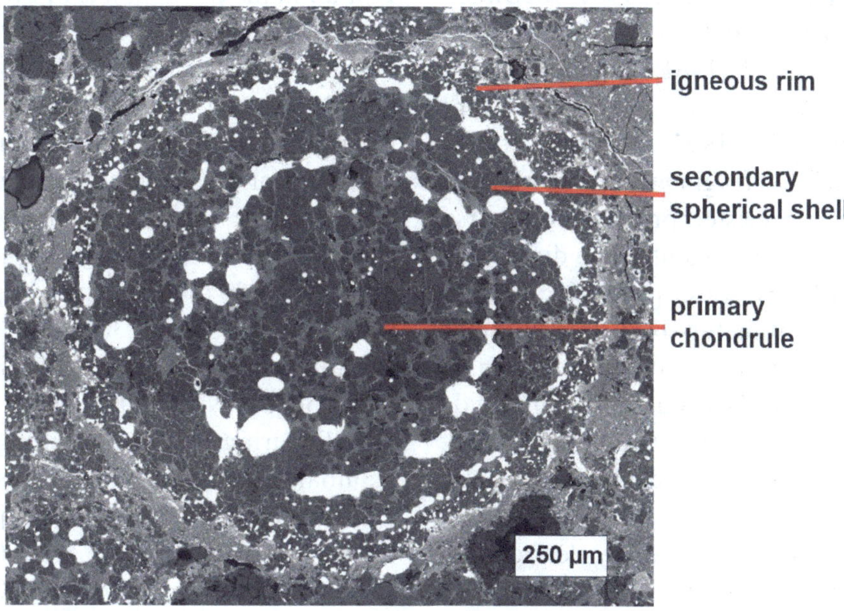

Fig. 11.35 Enveloping compound chondrule from the LAP 02342 CR2 carbonaceous chondrite. The primary object (center) is a Type-IA (low-FeO PO) chondrule. It is surrounded by a secondary shell of similar texture. The secondary shell is surrounded by an igneous rim. Both the primary chondrule and the secondary shell are rimmed by a discontinuous ring of metallic Fe–Ni blebs (white). This chondrule experienced at least three melting episodes separated by the acquisition of additional dusty material: the first melting event formed the primary chondrule, the second formed the secondary shell, and the third formed the igneous rim. BSE image modified from Rubin (2010a).

chondrules, particularly in CR and CV carbonaceous chondrites, are surrounded by spherical shells *and* igneous rims (Fig. 11.35).

11.2.4 Dustballs

Experimental and theoretical studies (Blum, 2004; Dominik et al., 2006) have shown that nebular fines tended to agglomerate into dustballs with ~90% porosity. Such dustballs would have been moderately resistant to disruption during low-energy collisions because collisional energy would have been distributed over entire dustballs (e.g., Paszun and Dominik, 2008; Ormel, 2008).

The igneous rims and secondary spherical shells around primary chondrules represent mantles of nebular dust that surrounded these chondrules before the chondrules were remelted. Igneous rims around chondrules in CV3 Allende are more similar in bulk composition to mean Allende chondrules than to the particular chondrules that they surround (Rubin and Wasson, 1987). This indicates that the igneous rims formed from moderately homogeneous dust clumps into which preexisting chondrules of various textural types and compositions were incorporated.

The dust clumps most likely were composed of several distinct components: interstellar dust containing presolar grains, Early-Solar-System condensates, and variable amounts of chondrule and CAI fragments. The mantles probably resembled matrix material in type-3 chondrites and consisted mainly of fine-grained olivine, pyroxene, plagioclase, amorphous materials, metallic Fe-Ni, and sulfide (e.g., Ashworth, 1977, 1981; McSween and Richardson, 1977; Huss et al., 1981; Ikeda et al., 1981; Fujimaki et al., 1981; Nagahara, 1984; Scott et al., 1982, 1984, 1988b; Alexander et al., 1989a, 1989b; Brearley, 1993; Brearley et al., 1995). Any chondrule fragments embedded in the dust would have consisted primarily of coarse chunks of these same components. The dust mantles around chondrules were probably derived from millimeter-to-centimeter-sized, free-floating, irregularly shaped, high-porosity dustballs. It is possible that the sizes of these dustballs in individual chondrule-forming regions were the main factor in determining ultimate chondrule size (Sekiya, 1997; Zsom et al., 2010).

These dustballs may have formed by electrostatic clumping of dust particles (Love and Pettit, 2004) or by gas-drag-induced concentration of dust in local high-pressure regions of the nebula (Haghighipour, 2004). Low-velocity collisions between chondrules and dustballs would, in some instances, cause the chondrules to become incorporated into the dustballs. If even small amounts of dust surrounded a particular chondrule, electrostatic forces could have caused the dust mantle to grow.

There were probably more chondrules surrounded by dust mantles in local nebular environments in which a lot of ambient dust was present. Chondrite groups (e.g., CV, CK, CR) that contain large chondrules, many chondrules with thick igneous rims, and many enveloping compound chondrules most likely were assembled in dusty regions of the nebula. Ordinary chondrites probably accreted in less dusty regions. This is consistent with the modal abundances of fine-grained silicate matrix material in these groups: for example, CV – 35 vol.%; CK – 50 vol.%; CR – 42 vol.%; OC – 12 vol.%; Table 6 of Rubin, 2010a). (Matrix material is presumed to be composed in part of reprocessed nebular dust [e.g., Ashworth, 1977; Huss et al., 1981; Nagahara, 1984; Scott and Krot, 2005; Wasson and Rubin, 2009; Rubin et al., 2009]. Chondrule and CAI fragments also make up a portion of the matrix.)

11.3 Relict Grains in Chondrules

Not all mafic phenocrysts in every ferromagnesian chondrule crystallized during a single cooling event from the chondrule's hot precursor droplet. At least 15% of the chondrules contain relict grains – relatively large solid precursors that survived the primary chondrule melting event (Jones, 1996). There are four principal varieties of relict grains (Fig. 11.36). (1) "Dusty" mafic silicate grains (mainly olivine) which appear dark in transmitted light (Fig. 11.36a) and occur in low-FeO chondrules. These grains contain numerous small blebs of low-Ni metallic Fe formed by reduction of FeO during chondrule melting (Nagahara, 1981; Rambaldi, 1981). In many cases, the metal grains are aligned along crystallographic directions in the relict olivine (Fig. 11.36b). (2) Magnesian relict grains occuring in high-FeO chondrules and appearing dark in BSE images (Fig. 11.36c). Many occur as partly resorbed masses within the cores of ferroan mafic phenocrysts. Some are ^{16}O-rich relative to their host chondrules (e.g., Kita et al., 2010), consistent with their likely derivation from Type-I chondrules. (3) Reduced relict olivine grains in Type-I chondrules which tend to have variable Δ^{17}O values and low Ca-Al-Ti contents compared to host olivine phenocrysts (Marrocchi et al., 2018). (4) Very coarse phenocrysts in both Type-I and Type-II chondrules, unlikely to have crystallized from the same melt at the same time as the much smaller mafic grains. The two largest olivine phenocrysts in the Type-IIA chondrule in Fig. 11.36d constitute 38 and 12 vol%, respectively, of the chondrule in the plane of the section.

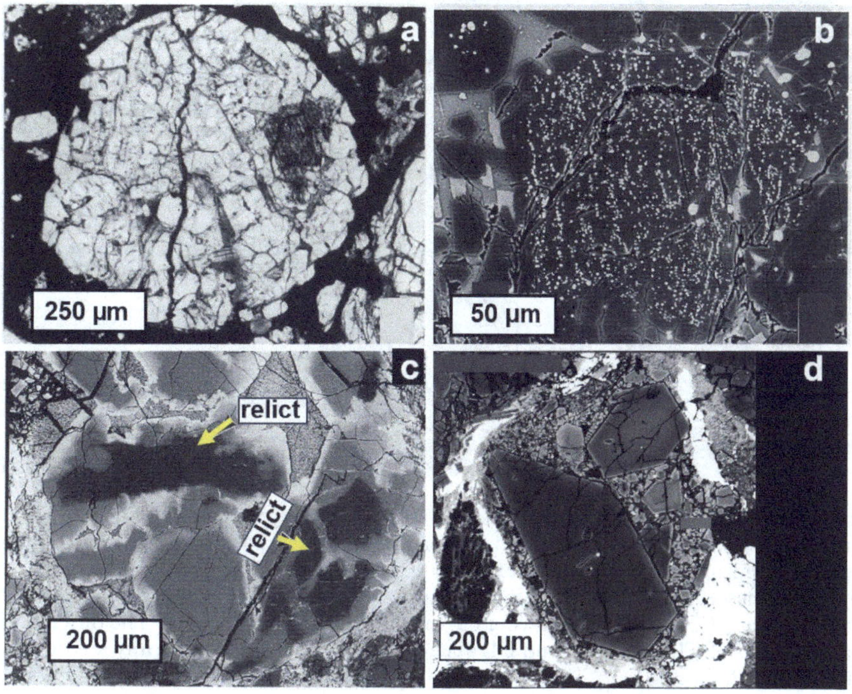

Fig. 11.36 Different varieties of relict grains in OC chondrules. (a) Porphyritic chondrule from LL3.01 Semarkona. The dark grain at right is an FeO-rich relict olivine grain. It is commonly called a "dusty olivine grain." Transmitted light image. (b) High-magnification view of the dusty olivine grain from (a). The grain appears dark because it contains numerous low-Ni metallic Fe grains aligned in crystallographic directions. BSE image. (c) Portion of a Type-IIB (FeO-rich PP) chondrule from Semarkona showing that the cores of some of the ferroan low-Ca pyroxene grains contain partially resorbed enstatite relicts (dark gray). BSE image. (d) Very coarse phenocrysts in Type-IIA (FeO-rich PO) chondrule B2i from Semarkona. BSE image. Images (a), (b), (c) modified from Jones (1996). Image (d) modified from Rubin and Baecker (2021).

Other relict grains may be less obvious. Rubin and Baecker (2021) showed that some olivine phenocrysts in Type-II chondrules appear unremarkable in the microscope or in BSE images but show complex overgrowth layers in P X-ray images (Figs. 11.37a, 11.38).

Interdiffusion in olivine of common divalent cations (e.g., Fe, Mg, Mn, Ca, Ni) is so rapid at olivine crystallization temperatures (e.g., Dohmen et al., 2007) that the initial compositional zoning of these cations can be smoothed out even during primary crystallization (Miyamoto et al. 1986; Fig. 3 of Jones, 1990). This process could mimic the compositional distributions produced during normal igneous zoning (e.g., Fig. 4a of Sano and Yamashita, 2004). In contrast, P is a moderately incompatible, pentavalent cation that diffuses slowly within olivine; the distribution of P in a crystal is more likely to preserve primary compositional zoning. In the olivine crystal structure, P^{5+} substitutes for Si^{4+} in the tetrahedral sites; charge balance is maintained by the presence of some vacant octahedral sites. Some Type-IIA chondrules in Semarkona exhibit correlations in their P- and Cr-zoning patterns (McCanta et al., 2008b); charge balance could be maintained in these cases by the substitution of both Cr^{3+} and P^{5+} for Si^{4+}.

Type-IA (low-FeO PO) chondrules have low P concentrations (e.g., Jones and Scott, 1989) and do not exhibit P zoning (McCanta et al., 2016). The average P_2O_5 concentrations in olivine grains in Type-I chondrules from LL3.01 Semarkona (<0.06 wt.%) are much lower than those in Type-II chondrules (up to 0.17 wt.%) (Jones and Scott, 1989, Jones, 1990). This difference is likely due to the near-quantitative occurrence of P from Type-I chondrules within non-silicate phases. Phosphorus zoning in olivine grains is apparent mainly in Type-IIA (high-FeO PO) chondrules. Additional examples of normally hidden complex zoning patterns in low-FeO ordinary- and carbonaceous-chondrite chondrules can be revealed by high-resolution

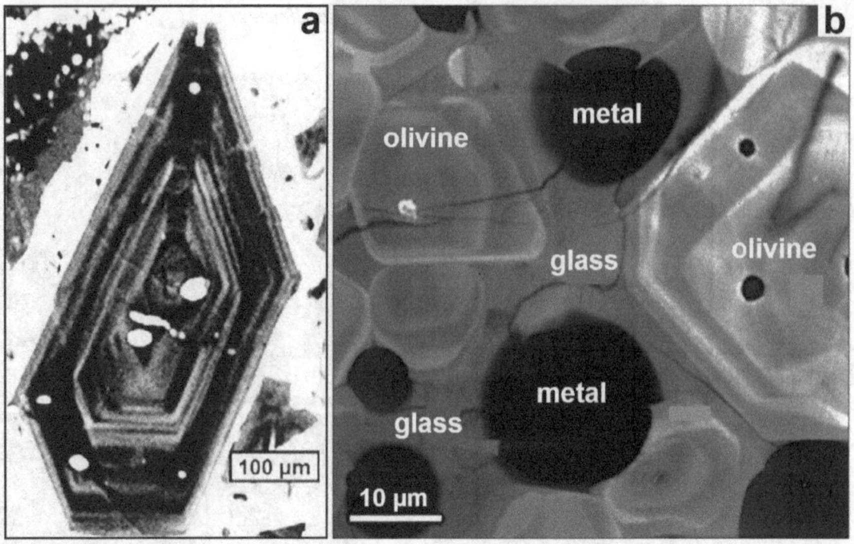

Fig. 11.37 Olivine grains with overgrowths. (a) P X-ray map of the large phenocryst from Fig. 11.36d of Type-IIA chondrule B2i from LL3.01 Semarkona. The grain shows a complicated pattern consisting of numerous dark/bright (P-poor/P-rich) nested layers indicative of a complex crystallization history. The two oval-shaped melt inclusions (white) contain relatively P-rich mesostasis. Image from Rubin and Baecker (2021). (b) Cathodoluminescence (CL) image of a metal-rich PO chondrule in CO3.0 Y 81020 showing olivine grains with overgrowths. The differences in CL brightness reflect different proportions of CL activators and CL quenchers. Darker regions generally are richer in Fe. Image modified from Libourel and Portail (2018).

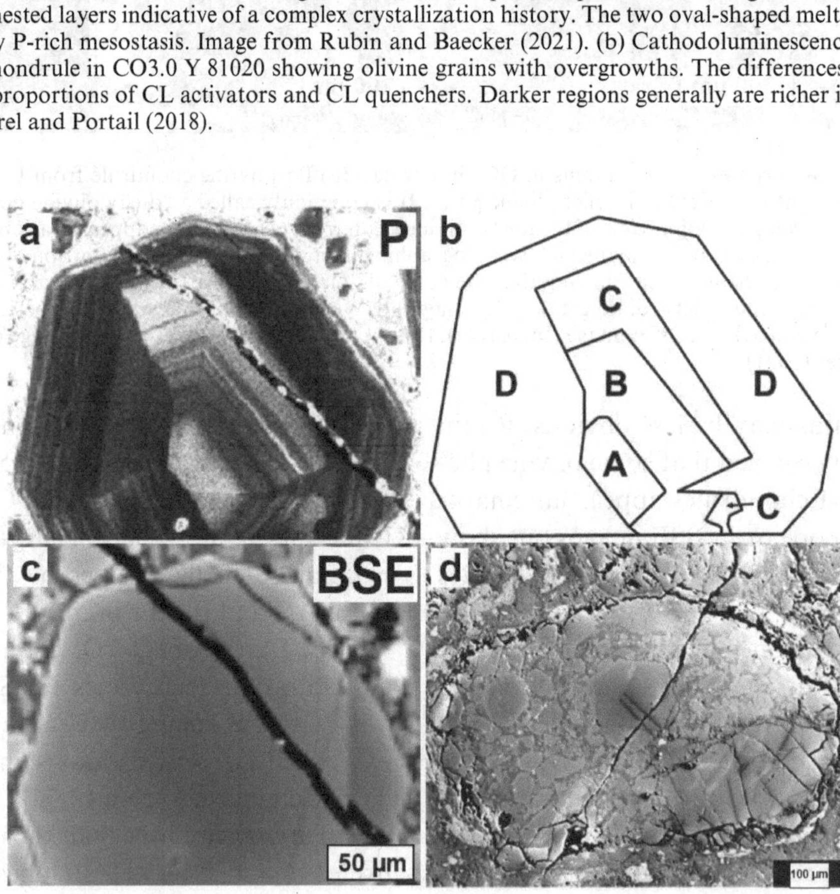

Fig. 11.38 Chondrule H5k from LL3.01 Semarkona. (a) P X-ray map of Grain H5k-2 showing complex pattern of interlayered P-rich and P-poor olivine. (b) Schematic diagram of Grain H5k-2 showing the principal parts of the grain. Area A is the P-rich core; this is flanked on the right side by Area B, a rhythmically layered region that is generally low in P. Atop Area B (and conformable to it) is Area C; this region is also rhythmically layered but contains appreciably higher P. A continuous wedge-shaped chunk of Areas B and C is missing at the lower right. Areas B and C also appear to be broken at their left, above the Area A core. The entire A-B-C assemblage is surrounded by Area D, a layered low-P region. Thin layers within Area D appear to be growing from the fractured B/C surface (NNW of Area A), but at a different angle than the layers in Areas B and C. (c) BSE image of Grain H5k-2 showing minor normal compositional zoning (with FeO increasing toward the grain edge) but no unusual features. (d) BSE image of the entire chondrule. Grain H5k-2 is in the center. The subparallel dark lines mark EPMA tracks. Images from Rubin and Baecker (2021).

cathodoluminescence (CL) images (e.g., Fig. 11.37b) (Libourel and Portail, 2018).

It seems likely that relict grains were derived from phenocrysts (or phenocryst fragments) of preexisting chondrules that had been collisionally disrupted in the nebula and subsequently incorporated into porous dustballs that experienced major melting (e.g., Rambaldi and Wasson, 1982; Jones and Danielson, 1997; Rubin, 2000). Other components trapped in these dustballs could include intact chondrules, chondrule fragments (polymineralic assemblages and coarse isolated mineral grains), CAIs, and small CAI fragments. Potential products include chondrules with relict grains, enveloping compound chondrules, chondrules with igneous rims, and compound chondrule–CAI objects.

It is possible that some relict grains in OC chondrules were derived from the Outer Solar System. Kita et al. (2010) found that four ferromagnesian chondrules from LL3 chondrites had ^{16}O-rich compositions that plot near the carbonaceous chondrites anhydrous minerals (CCAM) line. Piralla et al. (2021) reported two types of ^{16}O-rich relict olivine grains in OC chondrules: (1) grains with $\Delta^{17}O < -15‰$ that possess low abundances of minor elements and (2) grains with less-negative $\Delta^{17}O$ values (averaging −5.5‰) and varying abundances of minor elements. These findings could indicate a non-trivial mixing of materials from the Outer Solar System to the OC agglomeration zone prior to chondrule formation.

11.4 CAIs and AOAs

There are two principal varieties of refractory inclusions in chondritic meteorites: Ca-Al-rich inclusions (CAIs) and amoeboid olivine aggregates (AOAs), also known as amoeboid olivine inclusions (AOIs). Their mean modal abundances in type-3 ordinary chondrites are exceedingly low – CAIs: ~0.01 vol.% (Dunham et al., 2023); AOAs: 0.0005–0.001 vol.% (rough estimate based on data in Itoh et al., 2007). Nevertheless, Kimura et al. (2002) found that three unshocked (S1) primitive and paired H3.2–3.4 chondrites (Y-792947, Y-793408, Y-82038) have an average CAI abundance of 0.19 vol.%, roughly 20 times higher than in other type-3 OCs. The CAIs in the three Yamato meteorites tend to be spinel rich; they range in apparent size from 17 to 306 μm and average 80 μm.

CAIs: The most comprehensive study of CAIs in noncarbonaceous chondrites (including OCs) to date is by Dunham et al. (2023), who identified and analyzed 148 OC CAIs. The inclusions were found to be heterogeneously distributed in their host meteorites. With increasing OC petrologic subtype, CAIs tend to increase in average size and in modal abundance. In type 3.00–3.2 OC, the average CAI diameter is 17 μm and the mean modal abundance is 0.001 vol.%; in 3.3–3.9 OC, the average CAI diameter is 73 μm and the mean modal abundance is 0.02 vol.%. (The abundant CAIs in the primitive Yamato H3.2–3.4 chondrites do not fit this trend.) Selected examples of CAIs from Dunham et al. (2023) are shown in Fig. 11.39.

The 148 CAIs have a total diameter range of 4 to 382 μm; 76% are of irregular shape and 24% are round. Among all these CAIs, 15% are fragmented, 20% have diopside rims, and 20% have rims containing parent-body alteration phases such as sodalite and nepheline. Dunham et al. (2023) described nine different types of CAIs (Table 11.2).

The OC CAIs appear to be from the same broad population as those in carbonaceous chondrites: (1) McKeegan et al. (1998) found that the O-isotopic compositions of two CAIs from LL3.01 Semarkona ($\delta^{18}O = -41.0\%$, $\delta^{17}O = -44.0‰$,

Table 11.2 *Varieties of CAIs in type-3 ordinary chondrites.*

Variety of CAI	Percentage of CAI varieties	Mean diameter (μm)
Corundum-bearing	1%	9
Hibonite-rich	10%	41
Spinel-rich	36%	19
Hercynite-rich	26%	63
Melilite-bearing	6%	94
Spinel-pyroxene	10%	41
Spinel-anorthite	1%	12
Olivine-bearing	4%	79
Rich in alteration phases	5%	132
Total	99%	–

Data from Dunham et al. (2023).

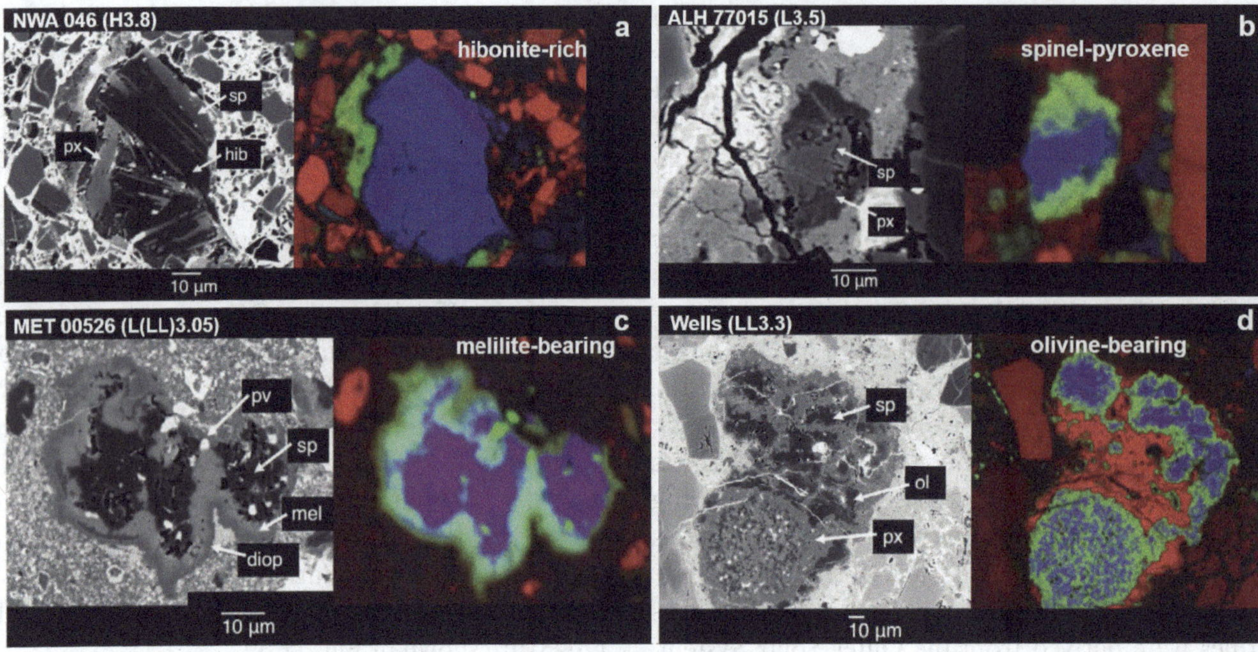

Fig. 11.39 CAIs in type-3 ordinary chondrites shown in BSE images and accompanying X-ray maps. (a) Hibonite-rich CAI in NWA 046. Hibonite, pyroxene and spinel occur as elongated coarse grains with hibonite predominating. (b) Spinel-pyroxene CAI in ALH 77015. The spinel core is flanked by pyroxene. (c) Melilite-bearing CAI in MET 00526. An irregular rim of monomineralic layers of diopside and melilite surround a spinel-rich core. (d) Olivine-bearing CAI in Wells. Olivine is mainly concentrated in the central region of the CAI. px = pyroxene; sp = spinel; hib = hibonite; pv = perovskite; mel = melilite; diop = diopside; ol = olivine. Images modified from Dunham et al. (2023).

$\Delta^{17}O = -22.7$‰, $n = 5$ spots) and LL3.6 Quinyambie ($\delta^{18}O = -19.1$‰, $\delta^{17}O = -26.2$‰, $\Delta^{17}O = -16.3$‰, $n = 6$ spots), are similar to those in carbonaceous-chondrite CAIs. (2) Huss et al. (2001) analyzed four OC CAIs, each a few hundred micrometers in size; two from Semarkona (a melilite-bearing CAI and a hibonite-rich CAI), one from Quinyambie (a melilite-bearing CAI), and one from L3.8-an Moorabie (a hibonite-rich CAI). All four CAIs contain radiogenic ^{26}Mg (i.e., ^{26}Mg*) produced by the decay of ^{26}Al. The inferred initial $^{26}Al/^{27}Al$ ratios – that is, $(^{26}Al/^{27}Al)_0$ – are $\sim 5 \times 10^{-5}$, indistinguishable from those in carbonaceous-chondrite CAIs.

AOAs: Rubin (2004a) showed the glassy Ca-Al-rich chondrule G7 from Semarkona (Fig. 11.26) has a bulk chemical composition similar to that of AOAs and suggested that the chondrule could have formed by remelting an AOA.

Itoh et al. (2007) later reported the first three known OC AOAs in Semarkona. These objects are irregularly shaped and heterogeneously distributed; they range in size from 50 to 500 µm. They contain micrometer-sized grains of magnesian olivine, fassaite (Al-Ti diopside), and anorthite. Also present are small (2–10-µm-size) grains of metallic Fe-Ni. Alteration phases (e.g., nepheline, sodalite, fayalite) are absent. The AOAs are rich in ^{16}O (with δ^{18}O and δ^{17}O values near –50‰, and $\Delta^{17}O = -25$ to -20‰). The inferred initial $^{26}Al/^{27}Al$ ratios (i.e., $(^{26}Al/^{27}Al)_0$) are $5.6 \pm 0.9 \times 10^{-5}$, similar to those of AOAs and fine-grained CAIs from carbonaceous chondrites. As with the OC CAIs, the OC AOAs appear to have been derived from the same nebular region as their counterparts in carbonaceous chondrites.

Rubin et al. (2017) found a 85×110 µm AOA in a clast that constitutes a new kind of matrix-rich ordinary chondrite in the NWA 10214 LL3-6 breccia. In addition, Dunham et al. (2023) reported that seven refractory inclusions (5% of the CAIs that they identified in type-3 OC) contain olivine and spinel and may be related to AOAs.

11.5 Compound Chondrule–CAI Objects

Chondrules and CAIs appear to have formed in different environments. Chondrules in different chondrite groups have distinctive sizes, different proportions of textural types, distinguishable O-isotopic compositions, different abundances and thicknesses of igneous rims, and different proportions of compound chondrules. Chondrules seem to have formed locally in different nebular regions where their whole-rocks later accreted (e.g., Rubin, 2010a, 2011; Ebel et al., 2024). Non-altered CAIs in carbonaceous chondrites (CB chondrites excepted; Krot et al., 2005b) and noncarbonaceous chondrites seem to constitute a single broad population with similar ^{16}O-rich bulk compositions, similar refractory minerals, and similar inferred initial $^{26}Al/^{27}Al$ ratios (i.e., $(^{26}Al/^{27}Al)_0$) of $\sim 5\times10^{-5}$. CAIs are modeled as having formed near the Sun and then dispersed throughout the solar nebula. Despite the different origins of chondrules and CAIs, there are numerous examples of conjoined inclusions – that is, compound chondrule–CAI objects. Most occur in carbonaceous chondrites (e.g., Misawa and Fujita, 1994; Wakaki et al., 2013; Rubin, 2015b; Krot et al., 1999, 2017) (e.g., Fig. 11.40).

Compound chondrule–CAI objects also occur in ordinary chondrites. Bischoff and Keil (1983a) identified one in H3.4 Sharps. The CAI fragment consists of anhedral hibonite grains flanked by spinel; it is embedded in the chondrule core (Fig. 11.41). The hibonite-spinel assemblage is surrounded by mesostasis enriched in nepheline and small grains of a different phase, probably sodalite. The chondrule host contains low-Ca pyroxene, Ca-pyroxene, olivine, calcic plagioclase, and small grains of metallic Fe-Ni.

Pack et al. (2004) identified two PO chondrules from Dar al Gani 369 (H/L3.5) and Dar al Gani 378 (H/L3) with unusual REE patterns (with negative chondrite-normalized Sm, Eu, and Yb anomalies). The authors proposed the chondrules' peculiar compositions to have resulted from previously incorporated relict grains composed of reduced ultrarefractory CAIs.

Additional evidence for a connection between chondrules and CAIs is the inferred refractory chemical components of normal ferromagnesian chondrules. Grossman and Wasson (1983, 1985) inferred that precursors of OC chondrules from LL3.01 Semarkona included a chemical component enriched in refractory lithophile elements (e.g., Ca, Al, Ti, REE) and another enriched in refractory siderophiles (e.g., Re, Ru, Os, Ir). Factor analyses of these data (e.g., Sears, 2004) sorted the elements and components into four distinct statistical clusters dubbed "factors" (Fig. 11.42): (Factor 1) a metal-sulfide-rich component consisting of metallic Fe-Ni, troilite (FeS), several siderophile elements (Fe, Co, Ni, Ge, As, Au, Ir), two chalcophile

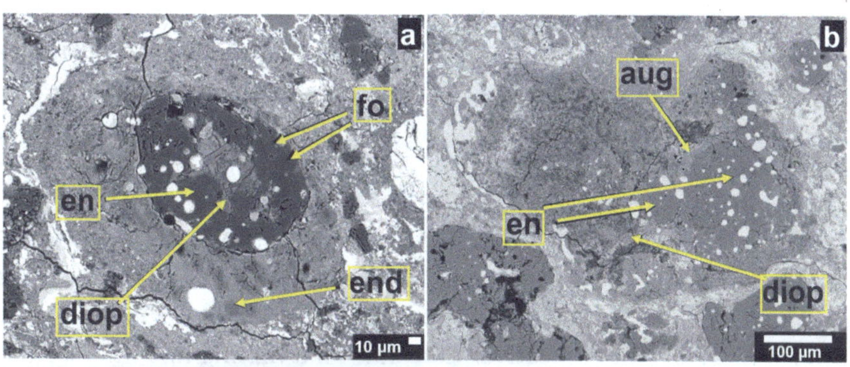

Fig. 11.40 Compound chondrule–CAI objects from the Paris CM carbonaceous chondrite. (a) Inclusion #45 consists of a Type-IAB (low-FeO POP) chondrule (center), containing phenocrysts of forsterite and enstatite (both dark gray), conjoined with a simple pyroxene-bearing CAI. The CAI forms a quasi-symmetric mantle around the chondrule. Although the CAI contains small grains of Ca-pyroxene (medium gray), most of it (and much of the chondrule interior) consists of matrix-like material (medium light gray). (b) Inclusion #26 consists of a Type-I chondrule (right side) conjoined with a simple distended pyroxene CAI. Parts of this CAI form a thin rim around the chondrule at right. Most of the original CAI has been transformed into matrix-like material (medium-to-dark gray). en=enstatite; fo=forsterite; diop=diopside; end = endiopside. BSE images from Rubin (2015b).

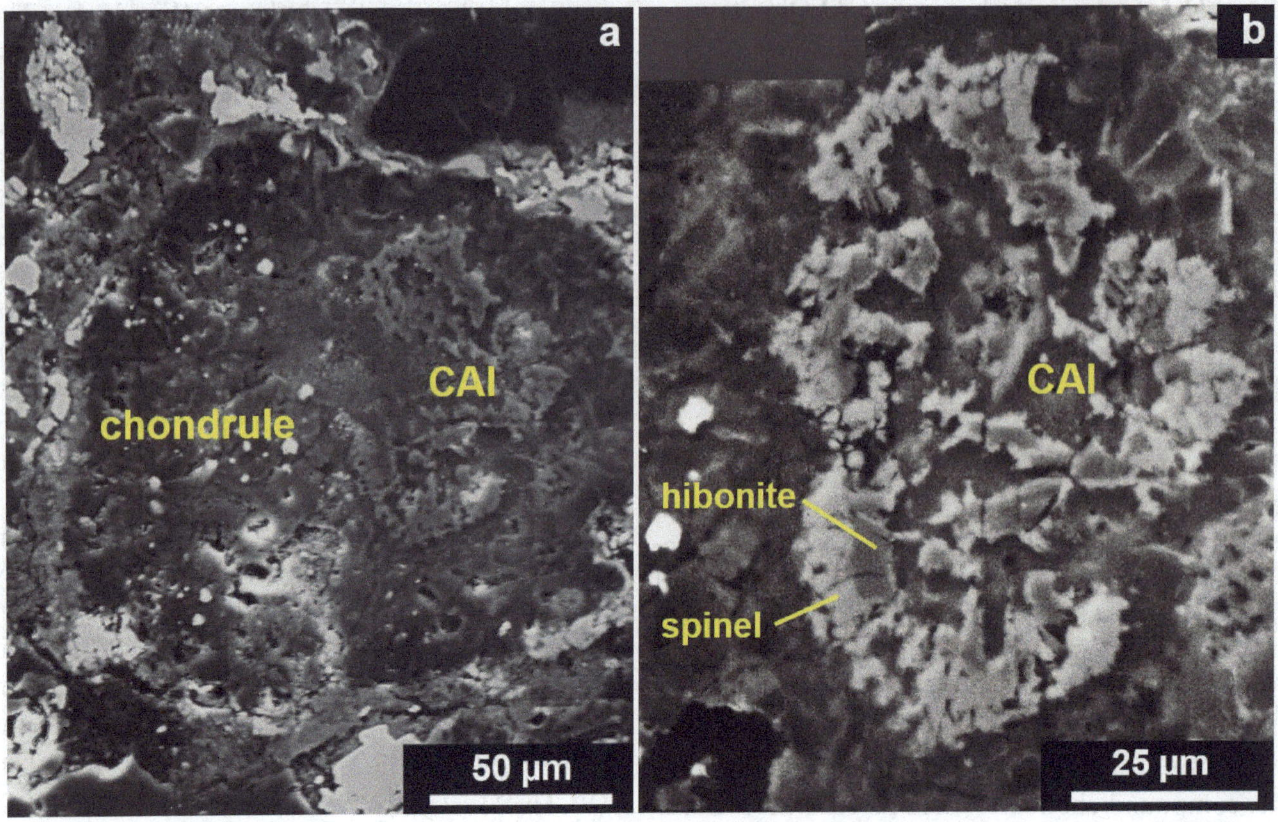

Fig. 11.41 A compound CAI–chondrule object from H3.4 Sharps. (a) Entire compound object. (b) High-magnification view of the enclosed CAI. Secondary electron images courtesy of Addi Bischoff, adapted from Bischoff and Keil (1983b).

elements (Se, Zn), and matrix material (which contains sulfide, Ni-rich metal, and carbide; Section 13.3); (Factor 2) a refractory-lithophile-rich component containing Al, Sm, Sc, V and Mg; (Factor 3) a volatile-lithophile-rich component containing Na, K, Ga, and glassy mesostasis (which is enriched in a feldspathic chemical component including Si, Al, Ca, and Na; e.g., Table 12.2); and (Factor 4) variable proportions of olivine and low-Ca pyroxene (i.e., chondrule phenocrysts). (Although Ga is a moderately volatile *metal*, it is enriched in the volatile *lithophile* component [Factor 3]. This is likely due to Ga being readily oxidized [e.g., Chung and Lee, 2013]. The ionic radii of Ga and Al are not that different [62 pm and 53.5 pm, respectively], allowing Ga_2O_3 to substitute for Al_2O_3 in a variety of silicate phases [e.g., Greenwood, 1963].)

In addition to the refractory chemical components inferred by Grossman and Wasson (1983, 1985) for OC chondrules, Rubin and Wasson (1987, 1988) inferred the existence of refractory-lithophile-rich components for CV3 and CO3 chondrules. Analogously, Grossman et al. (1985) inferred a refractory-lithophile component (enriched in Al, Sc, and Hf) in EH3 chondrules. All these refractory components were probably derived from CAI fragments that were incorporated within chondrule precursor dustballs and were later melted during chondrule formation. Petrographic evidence of the refractory components was lost; chemical evidence was retained.

However, M. Zhang et al. (2024) analyzed six Al-rich chondrules from type-3.05 OCs and found that the chondrules' bulk O-isotopic compositions fall on a mixing line intersecting that of CAIs and AOAs as well as some ferromagnesian Type-IA chondrules. The authors concluded that the OC Al-rich chondrules and some of the Type-IA chondrules did not form by melting a mixture of CAIs and normal ferromagnesian chondrule materials,

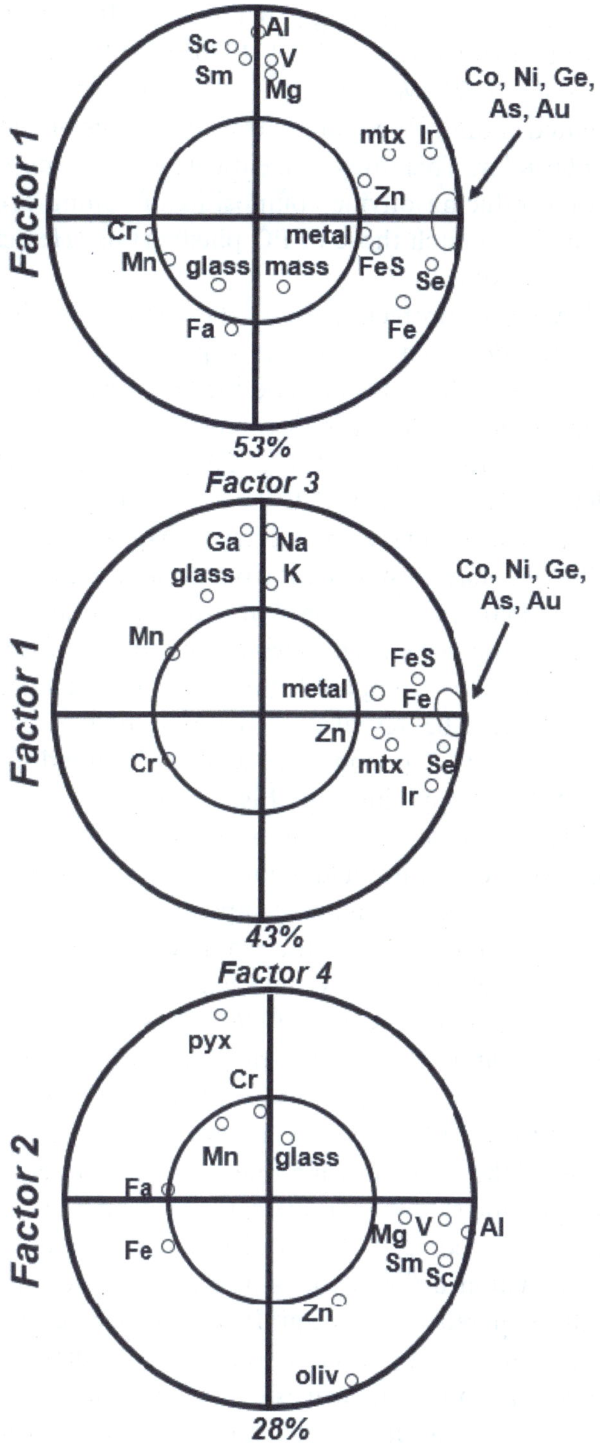

Fig. 11.42 Factor-loading plot for chondrules from LL3.01 Semarkona summarizing statistical relationships among the four most important clusters of chemical and petrologic components: Factor 1 (metallic Fe-Ni and sulfide), Factor 2 (refractory lithophiles), Factor 3 (volatile lithophiles), Factor 4 (olivine/low-Ca-pyroxene variations). mtx = matrix; pyx = pyroxene; oliv = olivine. Diagram modified from Sears (2004), courtesy of D. W. G. Sears.

but rather from the interaction of CAI-like melts and ^{16}O-poor nebular gas.

Another model (Ebel et al., 2025) holds that metal-olivine inclusions (MOIs) within amoeboid olivine aggregates (AOAs) derived from carbonaceous chondrites that drifted through the OC chondrule-forming regions. Different abundances of these MOIs in the H, L, and LL zones might account for the differences among these groups in chemical and isotopic composition.

11.6 Models of Chondrule Formation

Nineteenth-century models of chondrule formation included characterizing them as abraded fragments of preexisting meteorites (i.e., xenoliths; Reichenbach, 1860; Tschermak, 1895), as quenched magmatic segregations formed in place within the chondrites (i.e., melt pockets; Brezina, 1885), and as solidified igneous droplets, perhaps of a "fiery rain" (Sorby, 1864, 1877). Sorby's model chondrites could be characterized as tuffaceous agglomerates.

Below I offer a Shakespearean sonnet about chondrule formation (regrettably, iambic pentameter is not strictly followed).

A Chondrule Sonnet with a Chondrule Bonnet

In H II regions, the stars developed,
Replete with protoplanetary shrouds,
Chondrules (with ancient stardust enveloped),
Settled slowly through the gaseous clouds.

In the nebula, these small orbs were cast,
By lightning bolts around a nascent Sun,
Components fused by a thunderous blast,
In near-limitless numbers, they were spun.

Spheroids of olivine, pyroxene, glass,
Orbited at sub-Keplerian pace,
Products of several-milligram mass,
Wandered through interplanetary space.

To chondrites now, they have been assigned,
With fine-grained matrix they lie entwined.

11.6.1 Chondrule Cooling Rates

An important constraint on chondrule-formation models is the rapid rate of chondrule cooling. High D/H ratios in some LL3 chondrules (Deloule and Robert, 1996) and large variations in the $^{11}B/^{10}B$ ratio within individual chondrules from CV3 Allende and two type-3 OC (Chaussidon and Robert, 1995, 1996) were likely inherited from interstellar dust. The preservation of these isotopic anomalies implies incomplete melting and rapid chondrule cooling; otherwise, isotopic homogenization would have occurred. For example, Rubin (2000) reported that Marc Chaussidon (Institut de Physique du Globe de Paris) had found significant equilibration of B isotopes in glass held at 1615°C for 180 s and nearly complete homogenization in 900 s.

Yu et al. (1996) produced normally zoned olivines similar to those in Type-II chondrules from LL3.01 Semarkona by heating the charges for about 1 min and cooling them initially at ~5000 K/hr and, after about 20 min, cooling them at ~500 K/hr. Under these cooling conditions at an oxygen fugacity near the iron-wüstite (Fe-FeO) buffer, ~95% of the original Na was retained.

Experimental reproduction of chondrule textures is dependent not only on cooling rate, but also on peak temperature, duration at peak temperature, and the presence, size, and nature of relict nuclei. In addition to nuclei present in the melt, exogeneous grains could impact a molten droplet and initiate crystallization (Connolly and Hewins, 1995). Radial pyroxene textures were produced in charges by Hewins et al. (1981) at cooling rates of 3000 K/hr. Tsuchiyama et al. (2004) suggested that typical BO chondrules cooled at ~1000 K/hr. Skeletal, dendritic, and glassy chondrules (e.g., Fig. 11.26) must have been heated above the liquidus and quenched, probably at several thousand degrees per hour.

Several studies (e.g., Bell, 1986; Radomsky et al., 1986; Lofgren, 1989, 1996; Weinbruch and Müller, 1995; Weinbruch et al., 1998) have estimated chondrule cooling rates in the range of 10 to 1000 K/hr. This range is identical to that required for producing olivine compositional zoning profiles that match those of PO phenocrysts (Hewins et al., 2005).

A major problem with all these studies is their assumption that chondrules experienced a single episode of heating and cooling. However, many chondrules (and probably most) were reheated multiple times (e.g., Rubin, 2010a, 2013a, 2017; Rubin and Baecker, 2021). The normal zoning exhibited by olivine phenocryst Fe-Mg zoning profiles may have resulted not from a single episode of fractional crystallization, but rather from annealing during several episodes of chondrule heating and cooling. Complex histories of Type-I chondrules can be revealed by cathodoluminescence (e.g., Fig. 11.37b) and those of Type-II chondrules by P X-ray mapping (e.g., Fig. 11.38).

Kinetic melting models have shown that relict grains of enstatite could not have survived for more than a few seconds at high temperatures. For example, a 1-mm-sized enstatite grain heated to 1600°C would melt completely (and congruently) in 0.12 s (Greenwood and Hess, 1996). (This temperature is well within the inferred range of average chondrule melting temperatures: 1480–1880°C [Scott et al., 1996].) Greenwood and Hess proposed that, even if the actual melting rates were two orders of magnitude slower than the calculated rates, 10- to 100-μm-sized relict grains of enstatite would still melt within a few seconds at these temperatures. Although relict enstatite grains are much rarer than relict olivine grains in chondrules, these studies indicate that those chondrules containing relict enstatite grains must have cooled rapidly, consistent with flash heating mechanisms for chondrule formation.

11.6.2 Chondrule-Precursor Heating Mechanisms

There are numerous similarities among the populations of ferromagnesian chondrules in ordinary, carbonaceous, enstatite, and R chondrites (e.g., Scott and Taylor, 1983; Rubin, 2010a). The majority of chondrules in these different groups probably formed by the same limited set of heating processes. Numerous astrophysical and physico-chemical models have been proposed as the principal mechanism of chondrule formation. Some categories are not mutually exclusive: (1) condensation (Suess, 1949; Wood, 1962a; Blander and Katz, 1967; Bartholomay and Larimer, 1982; Cameron and Fegley, 1982; Larimer and Bartholomay, 1983; Wood and Hashimoto, 1993; Yoneda and Grossman, 1994; Ebel and Grossman, 1997, 1998; Petaev and Wood, 1998); (2) impact melting on planetary or asteroidal surfaces (Urey, 1952, 1967; Urey and Craig, 1953; Fredriksson, 1963; Ringwood, 1966; Fredriksson et al., 1970, 1973; Wlotzka, 1969; King et al., 1972a, 1972b; Kurat, 1982; Sears, 2004); (3) collisions between largely molten planetesimals (Zook, 1980, 1981; Wänke et al. 1981, 1984; Asphaug et al., 2011; Johnson et al., 2015; Hasegawa et al., 2016; Sanders and Scott, 2012); (4) exothermic chemical reactions (Clayton, 1980); (5) jetting during particle-to-particle collisions (Kieffer, 1975a; Kerridge and Kieffer, 1977); (6) meteor ablation in protoplanetary atmospheres (Podolak and Cameron, 1974; Podolak et al., 1993); (7) nebular lightning (Cameron, 1966; Whipple, 1966; Morfill et al., 1993; Wasson and Rasmussen, 1994; Horányi et al., 1995; Horányi and Robertson, 1996; Desch and Cuzzi, 2000; Muranushi, 2010; Johansen and Okuzumi, 2018; Kaneko et al., 2023); (8) supernova shock waves (Wark, 1979); (9) aerodynamic drag heating at the accretion shock front of the protoplanetary disk (Wood, 1984; Ruzmaikina and Ip, 1994); (10) magnetic reconnection flares (Sonett, 1979; Levy and Araki, 1989; Cameron, 1995); (11) gas dynamic shock waves (Hood and Horányi, 1991, 1993; Boss and Graham, 1993; Hood and Kring, 1996; Wood, 1996; Connolly and Love, 1998; Iida et al., 2001; Ciesla and Hood, 2002; Desch and Connolly, 2002; Boss and Durisen, 2005; Desch et al., 2012); (12) planetesimal bow shocks (Hood, 1998; Weidenschilling et al., 1998; Ciesla et al., 2004; Morris et al., 2012; Boley et al., 2013; Mann et al., 2016; Mai et al., 2018; Morris and Boley, 2018; Nagasawa et al., 2019); (13) FU Orionis outbursts (Huss, 1988); (14) bipolar outflows (Skinner, 1990; Liffman, 1992; Liffman and Brown, 1996; Shu et al., 1996, 1997); (15) intense gamma-ray bursts (McBreen and Hanlon, 1999); (16) radiative heating of dust clumps by electromagnetic radiation from an unspecified "localized region" (Eisenhour et al., 1994; Eisenhour and Buseck, 1995); (17) radiative heating of preexisting dust clumps by incandescent lava on the surfaces of nearby planetesimals (Herbst and Greenwood, 2016); and (18) instantaneous heating of primitive clumps of material injected into streams after low-speed collisions of large hot planetesimals (Herbst and Greenwood, 2024). In a hybrid model offered by Stewart et al. (2025), chondrules formed in nebular shock fronts powered by vapor plumes generated by planetesimal collisions. There is a welter of possibilities.

Libourel and Portail (2018) studied the cathodoluminescence (CL) properties of low-FeO chondrules and suggested chondrules formed in the nebula by gas-assisted near-equilibrium epitaxial growth under conditions of high partial pressures of Mg and SiO gas. However, some of their images (e.g., Fig. 11.37b) appear to show relict olivine grains surrounded by second-generation overgrowths. The internal structures of low-FeO olivine grains revealed by CL (Fig. 11.37b) resemble those of high-FeO olivine grains revealed in P X-ray maps (Figs. 11.37a, 11.38a). This suggests that both low-FeO and high-FeO chondrules underwent similar processing, presumably involving multiple heating events.

Taylor et al. (1983) provided evidence that most chondrules could not have formed by impact processes on asteroidal surfaces. An additional problem for these models is the narrow size range of chondrules, a characteristic inconsistent with the variability expected from impact-melting or planetesimal collisions (Jacquet, 2014). For example, the Abee enstatite chondrite breccia contains impact-produced clasts ranging in maximum dimension from 1 mm to 22 cm (Rubin and Scott, 1997; Kempton, 1996).

The varieties of bow shock and gas dynamic shock wave models have difficulties melting individual chondrules multiple times (cf. Rubin, 2017a, 2017b). They also cannot readily account for the significant melting of just the dusty rind around a chondrule (to form an igneous rim) while the primary chondrule is left unmelted (Figs. 11.32, 11.33). It would be even more difficult for such models to melt only those portions of low-Ca pyroxene grains at the peripheries of some normal-sized chondrules to form microchondrules (Fig. 11.19a).

Rubin (2000) showed that nebular flash-heating mechanisms (e.g., lightning) are the most consistent with the physical, chemical, and isotopic properties of typical chondrules. But as new models of chondrule formation are proposed and old ones reexamined and refined (e.g., Connolly and Jones, 2016), the problem of chondrule formation remains unsolved (cf. Farrington, 1915). Thomas Edison may as well have been speaking of chondrule-formation models (instead of filaments for the incandescent lightbulb) when he said "I have not failed. I've just found 10,000 ways that won't work."

Nevertheless, as Charles Kingsley (1855) quipped, "There are more ways of killing a cat than choking it with cream." In like manner, there are probably multiple ways to form chondrules. The presence of up to a few percent millimeter-sized chondrule-like igneous spherules in lunar regolith breccias (e.g., Keil et al., 1972; King et al., 1972a; Symes et al., 1998), howardites (Brownlee and Rajan, 1973), and suevite from the Ries Crater, Germany (Graup, 1981) suggests that a small fraction of the chondrules in OC regolith breccias may also have formed by impact processes (Krot and Rubin, 1993). However, the fact that the liquidus of basaltic material (as in howardites and many lunar rocks) (1278°C; Médard and Grove [2008]) is much lower than that of chondrites (1600°C; Elkins-Tanton et al. [2011]) indicates that impact-melted chondritic spherules require more energy to produce. In addition to their rarity, impact-derived chondrules may be difficult to recognize. At the Conference on Chondrules and their Origins held in Houston in 1982, John Wasson (UCLA) tried to clarify this issue by pointing out that real meteoritic chondrules (as distinguished from impact-melt spherules) resulted from the "chondrule-forming process."

Few issues in cosmochemistry are as contentious as chondrule formation. To achieve a deeper appreciation of the problem, it is important for serious researchers to understand the arguments of their opponents. As British philosopher John Stuart Mill (1859) wrote: "He who knows only his own side of the case, knows little of that. His reasons may be good, and no one may have been able to refute them. But if he is equally unable to refute the reasons on the opposite side; if he does not so much as know what they are, he has no ground for preferring either opinion."

11.6.3 Formation Locations

The isotopic differences between ordinary, carbonaceous, and enstatite chondrites (e.g., Fig. 10.6) show that different chondrite groups formed in separate nebular regions (although some chondrules appear to have been transported from the Inner to the Outer Solar System [Fukuda et al., 2024]). On the bases of variations in physical, chemical, and isotopic properties, Rubin (2011) proposed that the different chondrite groups formed at different heliocentric distances, ranging from enstatite chondrites (forming nearest to the Sun at roughly 1 AU [e.g., Baedecker and Wasson, 1975; Sears, 1980; Wasson, 1985, 1988; Nakashima et al., 2006; Scott, 2007]) to CI chondrites (forming at the greatest distances): EH,EL – H,L,LL – R – CR – CV,CK – CM,CO – CI. Groups listed together are members of the same clan and probably formed in close proximity. The division between carbonaceous and noncarbonaceous chondrites occurs between R and CR; this marks the distinction between the inner and outer regions of the Solar System (Section 14.1). With this arrangement of chondrite groups (Table 11.3), there are systematic variations in chondrite whole-rock properties: (1) the modal abundance of CAIs increases from enstatite and ordinary chondrites through R chondrites in the Inner Solar System, increases in the inner portions of the Outer Solar System from CR to CV,CK (where it peaks) and then decreases with greater heliocentric distances; (2) the mean abundance of refractory lithophile elements increases in the Inner Solar System from enstatite chondrites to R chondrites, increases in the inner portions of the Outer Solar System from CR to CV,CK (where it peaks) and then decreases with greater heliocentric distances; (3) the bulk O-isotopic composition (as measured by $\Delta^{17}O$) increases in the Inner Solar System from enstatite chondrites to R chondrites; in the Outer Solar System it decreases (becomes

Table 11.3 *Petrographic, chemical and isotopic properties of different chondrite groups.*

–	EH EL	H-L-LL	R	CR	CV CK	CM CO	CI
Mean modal abundance of CAIs (vol.%)	0.01 0.01	0.01	0.04	0.6	3.0 4.0	1.2 1.0	0.0
Mean abundance of refractory lithophiles (CI- and Mg-normalized)	0.88 0.87	0.89–0.90	0.97	1.02	1.35 1.24	1.13 1.11	1.00
$\Delta^{17}O$ (‰)	0.0 0.0	0.7–1.3	2.9	−1.0	−4.2 −4.6	−5.2 −4.7	−6.5
Mean modal abundance of matrix material (vol.%)	8 10	12	42	42	35 50	60 34	99

Data from Rubin (2011) except the modal abundance of CAIs in H-L-LL chondrites (Dunham et al., 2023) and the modal abundance of R-chondrite matrix material (Rubin and Ma, 2021). The enstatite, ordinary and R chondrites are from the Inner Solar System; the carbonaceous chondrites are from the Outer Solar System.

Table 11.4 *Physical properties of chondrules in different chondrite groups.*

–	EH EL	H-L-LL	R	CR	CV CK	CM CO
Mean chondrule diameter (μm)	230 500	450, 500, 550	400	700	900 900	270 150
Percentage of chondrules with igneous rims	0.2 0.3	10	0.8	26	50 32	0.1 0.1
Mean thickness of igneous rims (μm)	30 30	160	120	270	225 190	60 30
Percentage of enveloping compound chondrules	≤0.2 0.3	0.2	0.4	1.4	1.6 5	0.2 0.1
Percentage of RP+BP+C chondrules	18 13	7	1.6	0.7	0.3 0	1.6 3

All data from Rubin (2010a) except for chondrule diameters (Friedrich et al., 2015) and the thickness of igneous rims for CV chondrites (the median thickness is given here). CI chondrites are chondrule-free and are omitted from the table.

increasingly negative) systematically from CR to CV,CK to CM,CO to CI; (4) the average modal abundance of matrix material increases in the Inner Solar System from enstatite chondrites to R chondrites; in the Outer Solar System it is high among all carbonaceous chondrites (and very high in CI).

In Table 11.4 the chondrites are arranged in the same order and the properties of their constituent chondrules are listed. Similar systematic trends can be seen: (1) The mean chondrule diameter is low in the Inner Solar System, increases in the inner portions of the Outer Solar System from CR to CV,CK (where it peaks) and then decreases with greater heliocentric distances; (2) the percentage of chondrules with igneous rims is ≤10% in the Inner Solar System; it increases in the inner portions of the Outer Solar System from CR to CV,CK (where it peaks) and then decreases with greater heliocentric distances; (3) the mean thickness of igneous rims is ≤160 μm in the Inner Solar System; it increases in the inner portions of the Outer Solar System from CR to CV,CK (where it peaks) and then decreases with greater heliocentric distances; (4) the percentage of enveloping compound chondrules is ≤0.4 in the Inner Solar System; it increases in the inner portions of the Outer Solar System from CR to CV,CK (where it peaks) and then decreases with greater heliocentric distances; (5) the total percentage of RP, BP, and C chondrules shows a mirror-image trend – it decreases in the Inner Solar System from enstatite chondrites to R chondrites; it decreases in the inner portions of the Outer Solar System from CR to CV,CK (where there is a trough) and then increases with greater heliocentric distances. These systematic changes in chondrule properties among chondrite groups arranged by inferred heliocentric distance strongly indicate that chondrules were formed locally throughout the solar nebula.

11.6.4 Timing of Chondrule Formation

The absolute age of the Solar System is often equated with the formation of CAIs 4.56730 ± 0.00016 Ga before present (determined by Pb–Pb isotopic dating; Amelin et al., 2002; Bouvier and Wadhwa, 2010; Connelly et al., 2012). ^{26}Al–^{26}Mg chronometry indicates that anorthite-bearing chondrules from type 3.00–3.05 chondrites formed 1.8 ± 0.04 Ma to 2.16 ± (0.12 / 0.11) Ma after CAIs (Siron et al., 2021). ^{26}Al–^{26}Mg chronometry also shows that normal ferromagnesian chondrules in OC formed ~1.8 to 2.7 Ma after CAIs (Kita et al., 2000; Rudraswami and Goswami, 2007; Villeneuve et al., 2009; Kita and Ushikubo, 2012; Hellmann et al., 2019; Pape et al., 2019). These relatively late dates rest upon two assumptions: (1) ^{26}Al was more-or-less homogeneously distributed in the nebula throughout the chondrule-forming regions (consistent with the conclusions of Gregory et al., 2020) and (2) the oldest-dated chondrules in each chondrite group were not remelted late in Solar System history. Both assumptions can be challenged: (1) Connelly et al. (2012) and Bollard et al. (2017) reported some Pb–Pb ages for chondrules that are as old as CAIs. Such old Pb–Pb ages are consistent with heterogeneities in the distribution of ^{26}Al among chondrule-forming regions. (2) The high abundance of relict grains in chondrules (and the evidence from P X-ray maps that many Type-II chondrules were melted more than once; Section 11.3) raise the possibility that the oldest-dated chondrules may themselves have a complex prehistory.

Pape et al. (2019) combined their own results with literature data to show that there was only a limited time period in which OC chondrules, CO chondrules, and CV chondrules formed. Thus, chondrules appear to have been produced nearly contemporaneously in the Inner and Outer Solar System (in the noncarbonaceous and carbonaceous regions, respectively).

Several workers have shown from ^{26}Al–^{26}Mg chronometry that most Type-I chondrules tend to be somewhat older than most Type-II chondrules in primitive LL3 and CO3 chondrites (Mostefaoui et al., 2002; Tachibana et al., 2003; Kurahashi et al., 2008). This is consistent with well-established models showing that materials within individual nebular regions tended to become more oxidized with time as metallic Fe reacted with gaseous water molecules to produce iron oxide (e.g., Grossman, 1972; Lewis, 1972; Ebel and Grossman, 2000; Lodders, 2003; Ciesla and Charnley, 2006; Ebel, 2006).

Nevertheless, not all Type-I chondrules formed before all Type-II chondrules. The presence of FeO-rich dusty olivine grains (presumably derived from disrupted Type-II chondrules) within FeO-poor olivine phenocrysts in Type-I chondrules indicates that some Type-I chondrules formed after some Type-II chondrules (e.g., Jones and Danielson, 1997; Schrader et al., 2020). This is probably a consequence of chondrule recycling (e.g., Alexander, 1995; Jones, 1996; Ruzicka et al., 2008). In CR chondrites, ^{26}Al–^{26}Mg systematics indicate that Type-I and Type-II chondrules formed more-or-less contemporaneously (Schrader et al., 2017b).

11.7 Formation of CAIs

CAIs: The similarities between pristine CAIs in carbonaceous and noncarbonaceous chondrites in mineralogy, mineral chemistry, bulk chemical composition, O-isotopic composition, and inferred initial ^{26}Al/^{27}Al ratios (i.e., (^{26}Al/^{27}Al)$_0$) have led to a model in which CAIs (in carbonaceous and noncarbonaceous chondrites) all formed near the Sun before being dispersed throughout the solar nebula (e.g., MacPherson, 2005; MacPherson et al., 2005). This conclusion was bolstered by the discovery of ^{10}B excesses in a few CAIs from CV3 chondrites (McKeegan et al., 2000; MacPherson et al., 2003); these excesses were produced by the decay of ^{10}Be ($t_{½}$ = 1.387 ± 0.012 Ma; Korschinek et al., 2010). Because ^{10}Be is not a product of stellar nucleosynthesis but is likely created by spallation reactions in the nebula, it can be deduced that the CAIs in which ^{10}Be was incorporated formed near the proto-Sun.

The formation of CAIs occurred very early in Solar System history (4.56730 ± 0.00016 Ga; Amelin et al., 2002; Bouvier and Wadhwa, 2010; Connelly et al., 2012) and involved several processes including condensation, melt distillation, episodes of partial melting, and collisions (along with repetition of some of these processes in a different order). Different types of CAIs experienced varied formation histories. Heat was supplied by the high ambient temperatures near the proto-Sun and via irradiation by solar energetic particles (McKeegan et al., 2000; Scott et al., 2018).

Once CAIs were produced, disk winds could have propelled them to the periphery of the disk (Fig. 11.43). Alternatively, viscous expansion and turbulent diffusion of the protoplanetary disk could have transported CAIs far from the Sun (e.g., Ciesla, 2010; Yang and Ciesla, 2012; Desch et al., 2018). CAIs were even incorporated into comets (e.g., 81P/Wild 2) and chondritic porous interplanetary dust particles (probably derived from comets) tens of AU from the proto-Sun. These samples contain ~0.5–1 vol.% CAIs (Joswiak et al., 2017). The existence of disk winds has been inferred from observations of protostar TMC1A by the Atacama Large Millimeter/submillimeter Array (ALMA) (Bjerkeli et al., 2016). The early growth of proto-Jupiter beyond 10–20 Earth masses could have created a pressure bump – a high-pressure region in the disk outside its orbit (i.e., at greater distances from the proto-Sun). Pressure bumps can prevent most CAIs from spiraling back into the Inner Solar System. The gas in this region of the disk is accelerated and may reach super-Keplerian speeds; the effects of gas drag diminish near the pressure bumps, hindering the inward drift of millimeter-to-centimeter-sized particles (Weidenschilling, 1980; Bitsch et al., 2018).

This is consistent with the low modal abundances of CAIs in noncarbonaceous chondrites (0.01–0.04 vol.%) compared to those in non-CI carbonaceous chondrites (0.6–4.0 vol.%) (Table 11.4). Some bodies appear to be even richer in CAIs. The high Ni- and CI-normalized abundance ratios of highly siderophile elements in IVB and IID iron meteorites may be due to their

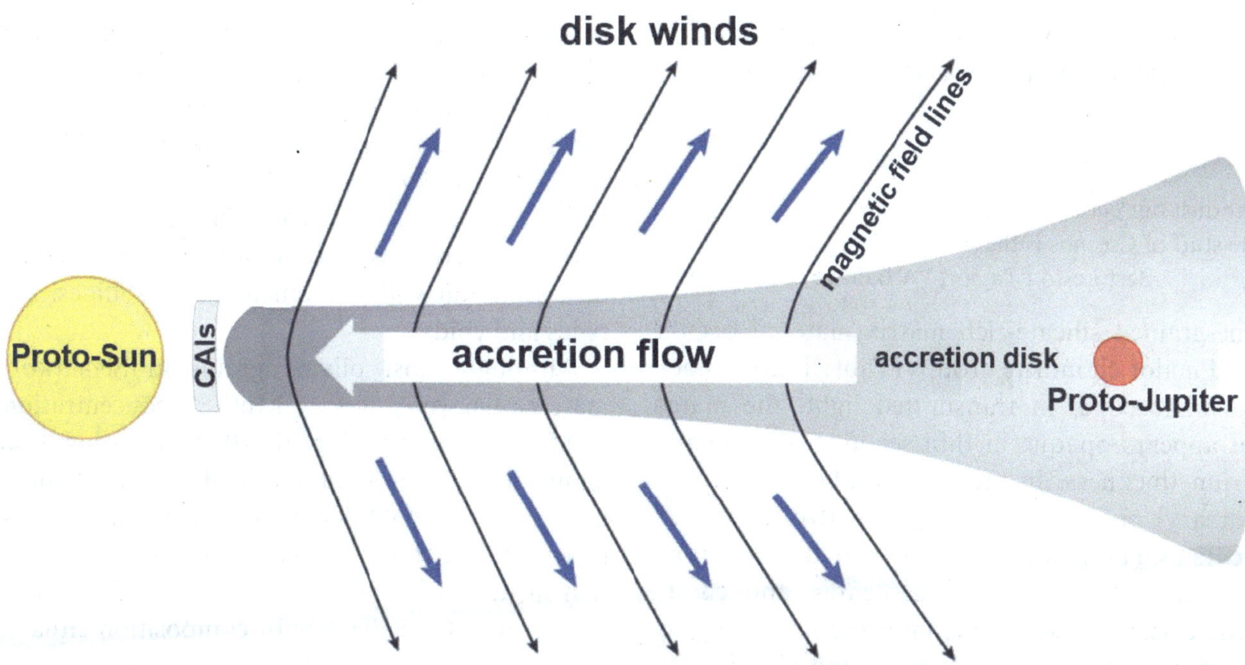

Fig. 11.43 Schematic diagram of the accretion disk surrounding the proto-Sun. Also present are disk winds (which ALMA observations suggest may be responsible for the shedding of angular momentum from disks). CAIs appear to have formed in a region close to the proto-Sun. The disk winds could have carried CAIs to the outer portions of the disk. Proto-Jupiter may have hindered the CAIs from spiraling into the Inner Solar System. Diagram modified from Scott et al. (2018); © AAS. Reproduced with permission.

carbonaceous-chondrite precursors having contained ~20 vol.% refractory metal nugget-bearing CAIs prior to melting (B. Zhang et al., 2022). L-type asteroids may contain up to 20–30 vol.% CAIs (Burbine et al., 1992; Sunshine et al., 2008; Devogèle et al., 2018), consistent with Jupiter having formed early in Solar-System history, about 0.2 Ma after CAIs (Aberham and Desch, 2024).

AOAs: Amoeboid olivine aggregates are the most abundant refractory objects in most carbonaceous-chondrite groups (e.g., Weber and Bischoff, 1997). They are very rare in ordinary chondrites. The ^{16}O-rich isotopic compositions of AOAs are similar to those of CAIs (e.g., Hiyagon and Hashimoto, 1999) and they probably formed near the proto-Sun in the same general vicinity as CAIs (Fig. 11.43). Krot et al. (2014b) proposed that AOAs formed as aggregates of condensed forsterite, metallic Fe-Ni, and small CAIs. In many cases, melilite and spinel in the CAI precursors reacted with SiO and Mg in ^{16}O-rich nebular gas to produce Al-rich diopside and anorthite. Low-Ca pyroxene formed in some AOAs after forsterite reacted with ^{16}O-rich SiO gas. After being mixed into chondrule-forming regions farther from the proto-Sun, some AOAs may have been reheated in the presence of ^{16}O-poor gas (Krot et al., 2004b).

11.8 Matrix Material

Star dust our end, from dust we came:
The stuff of Cosmos is the same.

Bert Leston Taylor, "A Ballade of Star Dust"

Fine-grained silicate-rich matrix material occurs in all major chondrite groups (Table 11.3). Under the microscope, in transmitted light, the material appears opaque in thin sections of standard 30-μm thickness. In the least equilibrated type-3 ordinary chondrites, opaque matrix material occurs as: (1) rims surrounding chondrules, chondrule fragments, opaque aggregates, and chondritic clasts, (2) isolated clumps (or lumps) up to 2 mm in size in the whole-rock, and (3) discrete 100–400-μm-sized clumps enclosed within a few chondrules (e.g., Scott et al., 1984) (Fig. 11.44). The mineral assemblage consists of ~0.1-μm-size grains of olivine ($Fa_{\sim 1-91}$), somewhat coarser grains of low-Ca pyroxene ($Fs_{\sim 1-30}$), Ca-pyroxene, albitic plagioclase, metallic Fe-Ni (kamacite, taenite, awaruite), troilite, magnetite, spinel, chromite, anorthitic plagioclase, pentlandite, carbide (cohenite, haxonite), merrillite, twinned calcite crystals, and rare examples of other phases. Also present are rare microchondrules and small angular fragments of normal-sized chondrules, chondritic clasts, and CAIs.

Several phases in the matrix are secondary products formed during alteration on the parent asteroid: phyllosilicates, Ni-rich sulfide, and smectite (Hutchison et al., 1987; Alexander et al., 1989a), fayalite–silica associations (Wasson and Krot, 1994), as well as carbide–magnetite assemblages and Ni-rich metallic Fe (Krot et al., 1997b).

As discussed in Rubin and Ma (2021), the fine-grained matrix in Semarkona (LL3.01) has experienced moderate aqueous alteration. In addition to fine-grained olivine, the matrix contains twinned monoclinic low-Ca pyroxene, and abundant Na-bearing Fe-rich smectite; other phases include maghemite, Ni-rich pyrrhotite and numerous grains of Ni-rich metal (taenite and awaruite), pentlandite, nickelphosphide, cohenite, edscottite, and wüstite (e.g., Nagahara, 1984; Keller, 1998). Some altered olivine and pyroxene grains contain voids. A few moderately altered sulfide grains have been partially replaced by Fe-rich phyllosilicate (Fe-rich chlorite). Nevertheless, Dobrică and Brearley (2020) reported pristine amorphous silicates in the Semarkona matrix occurring as a continuous groundmass and in association with nanometer-scale sulfides, carbides and voids.

Although most olivine grains in Semarkona matrix have low MnO concentrations (≤0.5 wt.%; e.g., Frank et al., 2014), a few low-iron, manganese-enriched grains (LIME olivines) contain up to 1.4 wt.% MnO (Klöck et al., 1989). The matrix in several OC contains amorphous material that is rich in normative albite (e.g., LL3.15 Bishunpur) or has an intermediate bulk composition situated between that of albite and ferroan olivine (e.g., LL3.3 ALHA76004 and LL3.3 ALHA77015).

Tables 11.5a and 11.6 provide the bulk chemical composition of opaque matrix material from several highly unequilibrated OC, as determined

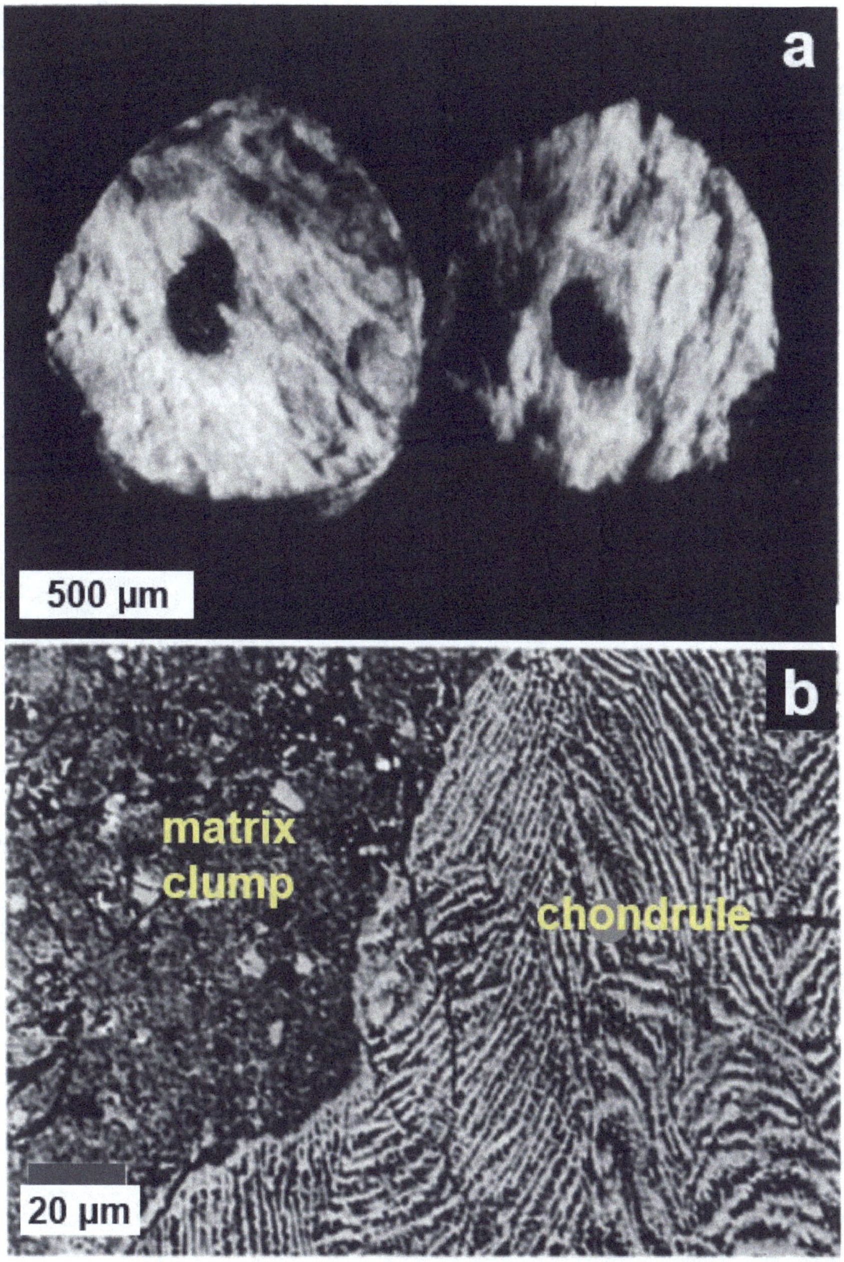

Fig. 11.44 Cryptocrystalline chondrule TI-21 (1.1 mm in diameter) from H/L 3.6 Tieschitz (Gooding, 1979) that contains an internal dark clump of matrix material. (a) Split chondrule with the dark quasi-ellipsoidal matrix clump in the center. (b) Interface between the matrix clump and the igneous portion of the chondrule (SEM image). The chondrule portion has a cryptocrystalline texture consisting of dendritic olivine (Fa_{42}), very similar to that in the matrix clump. The chondrule is very rich in bulk Fe (22.5 wt.%; Table B7 of Gooding, 1979). Images modified from Scott et al. (1984).

by electron microprobe analysis. The trace element composition of a 0.67-mg clump of matrix-rich material (consisting of 65 vol.% opaque matrix and 35 vol.% PP chondrule material) from H3.4 Sharps is listed in Table 11.7a (Rubin and Pernicka, 1989). The abundances of 10 out of 17 elements in this clump are within ±11% of those in the Sharps whole-rock (Fig. 11.45). This can be compared to the bulk composition of Sharps matrix determined by broad-beam electron microprobe analysis (Table 11.7b). Most occurrences of opaque matrix material within an individual type-3 OC differ mainly in their FeO/(FeO+MgO) content (Table 11.6; Fig. 11.46).

Table 11.5a *Bulk composition (wt.%) of opaque matrix material from the least-metamorphosed samples (type 3.01–3.2).*

–	Semarkona (LL3.01)	Bishunpur (LL3.15)	Krymka (LL3.2)	mean (n=3)
Number of analyses	61	34	53	–
SiO_2	42.1	46.7	37.5	42.1±4.6
TiO_2	0.08	0.10	0.07	0.08±0.02
Al_2O_3	4.7	4.1	3.8	4.2±0.5
Cr_2O_3	0.27	0.41	0.31	0.33±0.07
FeO	22.2	19.1	27.2	22.8±4.1
NiO	–	–	–	–
MnO	0.16	0.29	0.40	0.28±0.12
MgO	12.8	15.1	9.4	12.4±2.9
CaO	1.1	1.6	2.0	1.6±0.5
Na_2O	2.4	2.1	1.7	2.1±0.4
K_2O	0.68	0.71	0.42	0.60±0.16
P_2O_5	0.22	0.17	0.20	0.20±0.03
S	–	–	–	–
Total	86.7	90.4	83.0	86.7±3.7

Data from broad-beam electron microprobe analyses by Huss et al. (1981) normalized to their original totals.

Table 11.5b *Bulk composition (wt.%) of translucent matrix material in LL3.4 ALHA79003 determined by broad-beam electron microprobe analyses by Scott et al. (1982).*

–	ALHA79003 (LL3.4)
Number of analyses	12
SiO_2	39±2
TiO_2	0.14±0.07
Al_2O_3	4.2±2
Cr_2O_3	0.5±0.2
FeO	26±3
NiO	0.4±0.5
MnO	–
MgO	27±2
CaO	0.8±0.5
Na_2O	2.1±1
K_2O	0.3±0.2
P_2O_5	–
S	0.15±0.3
Total	100.6

Table 11.5c *Bulk composition (wt.%) of recrystallized matrix material in the more-metamorphosed type-3 ordinary chondrites*

–	Tieschitz (H/L3.6)	Khohar (L3.6)	Ngawi (LL3.6)	Parnallee (LL3.6)	Mezö Madaras (L3.7)	Dhajala (H3.8)	Bremervörde (H/L3.9)	mean (n=7)
Number of analyses	17	33	15	42	11	39	41	–
SiO_2	43.8	47.0	43.6	44.8	42.4	47.6	47.4	45.2±2.1
TiO_2	0.15	0.14	0.10	0.13	0.10	0.10	0.17	0.13±0.03
Al_2O_3	7.7	3.5	2.4	3.6	2.8	3.4	3.1	3.8±1.8
Cr_2O_3	0.34	0.64	0.38	0.42	0.47	0.48	0.48	0.46±0.10
FeO	17.8	18.5	18.0	17.4	19.3	11.3	13.1	16.5±3.0
MnO	0.39	0.29	0.34	0.35	0.39	0.25	0.29	0.33±0.05
MgO	17.6	24.6	26.4	25.4	29.4	27.2	28.2	25.5±3.9
CaO	3.3	2.7	2.4	1.9	2.0	2.2	2.3	2.4±0.5
Na_2O	3.1	2.0	0.86	2.1	1.2	1.6	1.6	1.8±0.7
K_2O	0.24	0.25	0.25	0.29	0.16	0.27	0.21	0.24±0.04
P_2O_5	0.29	0.41	0.25	0.22	0.20	0.20	0.31	0.27±0.08
Total	94.7	100.0	95.0	96.6	98.4	94.6	97.2	96.6±2.1

Data from broad-beam electron microprobe analyses by Huss et al. (1981) normalized to their original totals.

Relative to bulk CI chondrites or bulk ordinary chondrites (e.g., Wasson and Kallemeyn, 1988), mean OC opaque matrix (e.g., Table 11.5a) has low Mg/Si ratios and high Na/Si and K/Si ratios. In addition, matrix is enriched in chalcophile elements (Se, Zn, and by inference, S) relative to its bulk chondrite host (Fig. 11.45). If matrix and opaque assemblages are considered together, this set of non-chondrule bulk material would have high Fe/Si relative to the whole-rock. In contrast, the set of bulk chondrules in most carbonaceous and ordinary-chondrite groups has high Mg/Si, low Fe/Si, low chalcophile-element/

Table 11.6 *Representative analyses of different occurrences of opaque matrix material in type-3 ordinary chondrites.*

–	Tieschitz (H/L3.6) chondrule rim	Piancaldoli (LL3.4) lump	Quinyambie (LL3.6) clast-rich lump	Ngawi (LL3.6) lump inside chondrule
Number of analyses	6	6	5	5
SiO_2	32.8	46	44.7	30.4
TiO_2	0.10	0.09	0.10	0.08
Al_2O_3	3.2	6.7	4.1	1.9
Cr_2O_3	0.27	0.2	0.65	0.32
FeO	31.8	19	20.9	43.4
MnO	0.44	0.19	0.32	0.42
MgO	15.4	14.4	23.1	12.5
CaO	1.1	0.5	2.0	0.40
Na_2O	1.7	2.7	1.7	0.81
K_2O	0.31	1.7	0.08	0.15
P_2O_5	0.14	0.06	0.40	0.12
Ni	<0.11	0.7	0.9	0.9
S	0.03	0.4	2.0	0.13
Total	87.3	92.6	101.0	91.5

Data determined by broad-beam electron microprobe analyses by Scott et al. (1984).

Table 11.7a *Bulk composition of matrix material in Sharps (H3.4).*

–	Sample 1
Na (mg/g)	4.24
Mg (mg/g)	154
Al (mg/g)	11.1
Si (mg/g)	167
K (μg/g)	530
Sc (μg/g)	8.18
V (μg/g)	105
Cr (mg/g)	4.24
Mn (mg/g)	2.54
Fe (mg/g)	307
Co (μg/g)	916
Ni (mg/g)	17.6
Zn (μg/g)	139
Ga (μg/g)	8.6
Se (μg/g)	15.8
Sm (ng/g)	340
Ir (ng/g)	790
Au (ng/g)	256

The composition of sample 1 (0.67 mg) was determined by INAA (except for Si, determined by broad-beam electron microprobe analysis). Rubin and Pernicka (1989).

Table 11.7b *Bulk composition of matrix material in Sharps (H3.4).*

–	Sample 2 (wt.%)
SiO_2	35.3
TiO_2	0.11
Al_2O_3	1.9
Cr_2O_3	0.43
FeO	29.2
MnO	0.35
MgO	20.8
CaO	1.4
Na_2O	0.49
K_2O	0.22
P_2O_5	0.34
Total	90.5

Sample 2 data were obtained by broad-beam electron microprobe analysis (Huss et al., 1981)

Si, and low volatile-element/Si ratios relative to their whole-rock values (e.g., Bland et al., 2005; Braukmüller et al., 2018; Ebel et al., 2024). The specific compositional depletions and enrichments among chondrules must offset the enrichments and depletions in matrix-plus-opaque-assemblages for the whole-rocks to possess a chondritic bulk chemical composition. This correspondence (which is just a reflection of the equation: $(x + (1 - x) = 1)$) is often called "complementarity." The complementary compositions of chondrules and matrix are particularly evident in CR carbonaceous chondrites (e.g., Ebel et al., 2024) due in large part to the nearly CI-like Mg/Si and Fe/Si ratios of CR whole-rocks (e.g., Mason and Wiik, 1962; Kallemeyn et al., 1994).

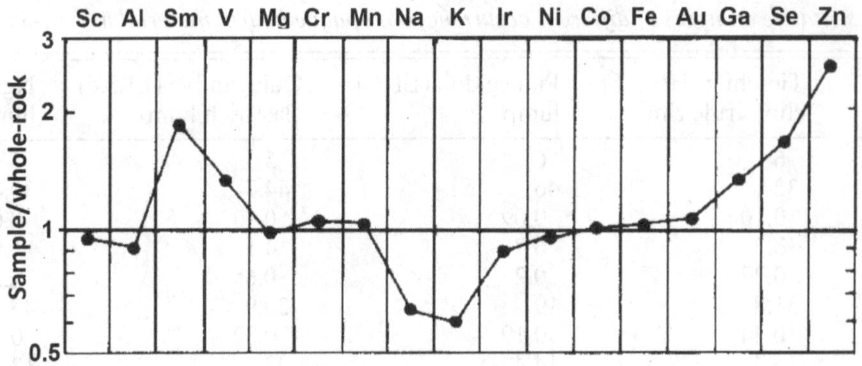

Fig. 11.45 Bulk compositional abundance of a 0.67-mg matrix-rich lump in H3.4 Sharps (consisting of 65 vol.% opaque matrix and 35 vol.% PP chondrule material) normalized to the Sharps whole-rock (Kallemeyn et al., 1989) and to Mg (Rubin and Pernicka, 1989). Ten out of 17 elements have abundances within ±11% of the whole-rock.

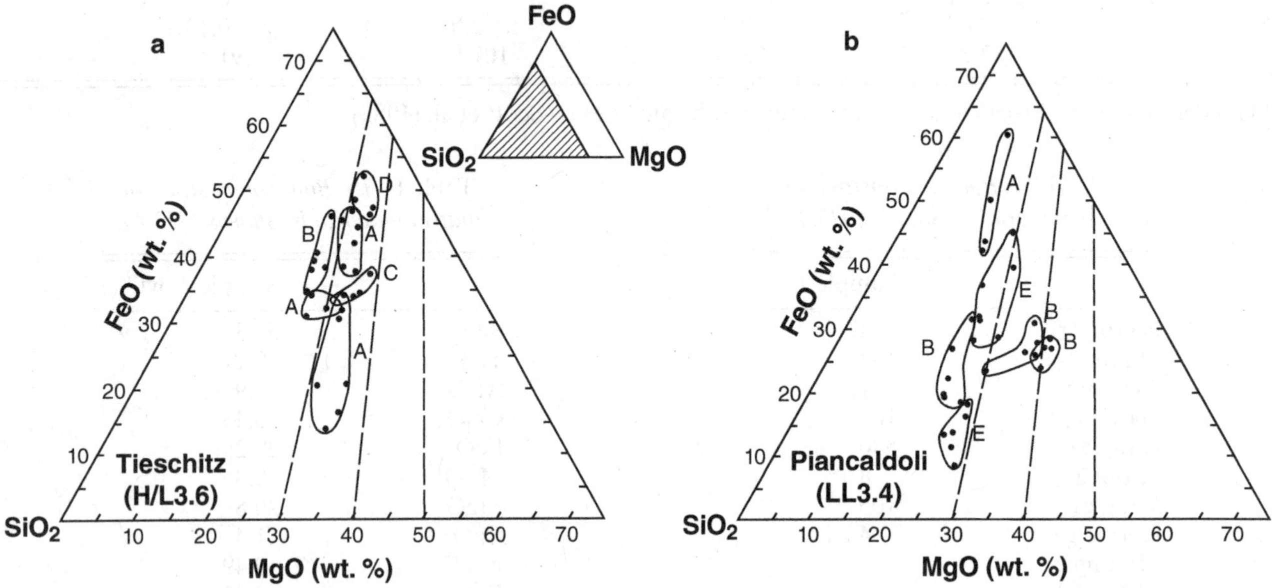

Fig. 11.46 Bulk compositions of selected occurrences of matrix material in (a) Tieschitz and (b) Piancaldoli projected onto an FeO-MgO-SiO_2 plot. Individual analyses are represented by points plotted within each enclosed region. The letters identify the type of occurrence: A = chondrule rim; B = matrix clump; C = clast-rich clump; D = matrix associated with troilite and metallic Fe-Ni; E = clumps enclosed within chondrules. Most analyses fall along lines of constant MgO/SiO_2 (dashed lines). Diagram modified from Scott et al. (1984).

The alleged paradoxical nature of complementarity arises from the assumption that "refractory inclusions, chondrules, and matrix originated in different nebular regions" (Ebel et al., 2024). Although refractory inclusions (CAIs) likely formed near the Sun and were transported to large heliocentric distances before starting to drift sunward (Section 11.7), chondrules seem to have formed in different nebular regions from local batches of available dust; this inference is supported by the distinct O-isotopic compositions of chondrules in different chondrite groups (Fig. 11.18a). It is not clear if each of the local dusty reservoirs had a CI-like bulk chemical composition prior to chondrule formation.

A significant fraction of the volatile elements lost from chondrules during the heating involved in chondrule formation would have condensed onto neighboring dust particles. Some of the metallic Fe-Ni and sulfide lost from spinning molten chondrule droplets would have been incorporated into nearby dustballs or otherwise accreted into the chondrite whole-rocks during agglomeration. Materials lost from chondrules and not subsequently accreted to the bulk chondrites can account, at least in part, for the depletions in volatile lithophiles, volatile siderophiles, and chalcophiles in chondrite whole-rocks relative to CI chondrites (e.g., Figs. 1 and 2 of Wasson and Kallemeyn, 1988). This scenario is a reasonable alternative to the class of single-reservoir models inferred from complementarity by Ebel et al. (2024). In fact, the scenario presented above is even mentioned by Ebel et al. as a potentially viable model: they noted that the coexistence of low-temperature matrix components and high-temperature chondrule components in a chondrite could indicate that chondrules experienced "a highly localized heating process while spatially coexisting with the low-temperature components..." As far as ordinary chondrites are concerned, Ebel et al. (2024) pointed out that LL chondrites (and other OC groups) do not formally meet the criteria for complementarity anyway because they have somewhat lower bulk Mg/Si ratios and much lower bulk Fe/Si ratios than CI chondrites.

Opaque matrix consists partly of silicate-rich primitive material derived from porous millimeter-to-centimeter-sized dustballs within the solar nebula. (Also present are chondrule fragments, metal and sulfide grains, and secondary phases formed on the parent body.) Nebular dustballs are the presumed progenitors of fine-grained chondrule rims, coarser-grained igneous rims, matrix clumps, and secondary spherical shells in enveloping compound chondrules. T Tauri and Herbig Ae star-disk systems appear to contain abundant millimeter-size particles near their midplanes (Chiang, 2004; Youdin and Chiang, 2004); these particles may be porous dust aggregates (e.g., Blum, 2004; Dominik et al., 2006).

The primitive nature of opaque matrix material in chondritic meteorites is reflected in their abundances of presolar grains. These are the oldest minerals within meteorites (e.g., Clayton and Nittler, 2004; Lodders and Amari, 2005); many predate the Solar System by hundreds of millions of years, some by more than a billion years (e.g., Gyngard et al., 2014; Heck et al., 2020). Presolar grains occur in carbonaceous, ordinary, and enstatite chondrites. Semarkona contains the highest concentrations of presolar grains among OC: 177 μg/g diamond, 3.13 μg/g SiC, and 0.034 μg/g graphite (Table 11.8; Huss and Lewis, 1995).

The presolar grains range from 2 nm to 25 μm (e.g., Gyngard et al., 2018); some of the larger grains form aggregates >200 μm in size. Presolar phases include C polymorphs (diamond and graphite); moissanite (SiC); cohenite ($(Fe,Ni)_3C$); Ti-, Mo-, Zr- and Ru-rich carbides; metallic iron; Ni-bearing kamacite; nierite (Si_3N_4); corundum (Al_2O_3); spinel ($MgAl_2O_4$); hibonite ($CaAl_{12}O_{19}$); TiO_2; magnesiochromite ($Mg(Cr,Al)_2O_4$); and silicates (predominantly olivine, low-Ca pyroxene, Al-rich silicate, and silicaceous glass). Although most presolar silicates are magnesian, a few contain more FeO than MgO. Organic compounds are also present.

Table 11.8 *Abundances of presolar grains in type-3 ordinary chondrites compared to those in CI Orgueil.*

–	Diamond (μg/g)	SiC (μg/g)	Graphite (μg/g)
Orgueil (CI)	1436	28.5	10.3
Semarkona (LL3.01)	177	3.13	0.034
Bishunpur (LL3.15)	125	1.85	0.022
Krymka (LL3.2)	137	0.48	<0.009
Ragland (LL3.4)	83	0.180	<0.00031
ALHA77214 (L3.4)	61	0.028	–
Julesburg (L3.6)	0.36	0.00096	–
Tieschitz (H/L3.6)	46	0.102	–
Mezö-Madaras (L3.7)	62	0.124	–
Dimmitt (H3.7)	3.3	0.00153	–
Grady (1937) (H3.7)	<0.09	<0.00108	–

Data from Huss and Lewis (1995) with updated OC subtypes.

Fig. 11.47 Spherulitic texture of white matrix (wmx) in H/L3.6 Tieschitz; it is adjacent to opaque matrix (a.k.a. black matrix). BSE image modified from Dobrică and Brearley (2014).

Table 11.9 *Bulk composition (wt.%) of white matrix and blocky mesostasis in porphyritic chondrules from H/L3.6 Tieschitz.*

–	White matrix	Chondrule blocky mesostasis
Number	2 Analyses	3 Samples
SiO_2	61.5	50.5 ± 2.6
TiO_2	<0.14	0.36 ± 0.25
Al_2O_3	19.3	20.1 ± 0.9
Cr_2O_3	<0.15	0.14 ± 0.06
FeO	3.3	3.8 ± 3.0
MnO	<0.15	<0.15
MgO	2.01	1.0 ± 0.6
CaO	3.27	0.88 ± 0.43
Na_2O	7.3	8.0 ± 1.8
K_2O	0.11	0.45 ± 0.16
S	<0.06	0.09 ± 0.08
Cl	<0.06	0.22 ± 0.09
Total	96.8	85.5

Data from Hutchison et al. (1998) determined by SEM and energy-dispersive analysis.

Huss and Lewis (1995) found the diamond/SiC ratios among presolar grains to be similar in different classes of chondrites. This suggests that there was a homogeneous reservoir of presolar grains sampled by the solar nebula inherited from the parent molecular cloud. Differences among the presolar grain populations in different chondrite groups are ascribable mainly to secondary processing on chondrite parent asteroids.

In addition to opaque matrix material (Huss et al., 1981), Tieschitz (H/L3.6) contains "white matrix" (Fig. 11.47; Table 11.9) with clastic grains of albite and nepheline (or an intergrowth of the two) as well as olivine, Ca-pyroxene, sulfide, and metallic Fe-Ni (Christophe Michel-Lévy, 1976; Hutchison et al., 1998). This material has a different origin from opaque matrix – it is not primitive. Hutchison et al. (1998) identified two types of mesostases within Tieschitz chondrules: "smooth" mesostasis (representing primary chondrule liquid) and "blocky" mesostasis (a parent-body alteration product of smooth mesostasis). White matrix resembles blocky mesostasis in composition (with the main exception being its higher silica content) (Table 11.9) and probably formed on the Tieschitz parent body during secondary alteration (Hutchison et al., 1998). This is consistent with the common presence in the Tieschitz matrix of geode-like voids and veins filled with the sodic-calcic amphiboles winchite (□$NaCa(Mg_4Al)Si_8O_{22}(OH)_2$) and barroisite (□$NaCa(Mg_3Al_2)(Si_7Al)O_{22}(OH)_2$) (Dobrică and Brearley, 2014). It is likely that white matrix formed at the same time as these amphiboles in the Tieschitz parent body at near-peak metamorphic temperatures.

Scott et al. (1982) reported that, in addition to opaque matrix, LL3.4 ALHA79003 contains translucent matrix (Table 11.5b). Both types of matrix occur as chondrule rims in this meteorite. As its name implies, this material is not opaque when viewed in transmitted light. It is compositionally more homogeneous than opaque matrix, is poorer in FeO, CaO, and S, and is richer in MgO. It seems likely that translucent matrix is just a variety of recrystallized matrix (see below).

Recrystallized matrix, composed mainly of somewhat coarser mafic silicate grains than opaque matrix, occurs principally in type 3.4–4 OC. It is translucent when viewed microscopically in transmitted light. It contains lower bulk S and FeO, and higher MgO than opaque matrix (e.g., Table 11.5c; Scott et al., 1984). The modal abundance ratio (vol.%/vol.%) of (recrystallized matrix)/(opaque matrix) tends to increase with petrologic subtype (Huss et al., 1981): type 3.00–3.2 ($n = 3$), 0.18 ± 0.12; type 3.4 ($n = 3$), 1.3 ± 1.9; type 3.6 ($n = 4$), 3.7 ± 3.4; type 3.7 – 3.9 ($n = 3$), 10 ± 15. The overall trend and the variability of this ratio are best illustrated by the extremes: Semarkona (LL3.01) and Bishunpur (LL3.15) have (recrystallized matrix)/(opaque matrix) ratios of 0.05 and 0.29, respectively; Dhajala (H3.8) and Bremervörde (H/L3.9) have ratios of 28 and 2.1, respectively.

Parent-body thermal metamorphism is responsible for transforming opaque matrix material into recrystallized matrix. Another effect of metamorphism is on the modal abundance of presolar grains: their abundance tends to decrease with increasing subtype (Fig. 11.48). Presolar grains are systematically destroyed at increasing sustained temperatures.

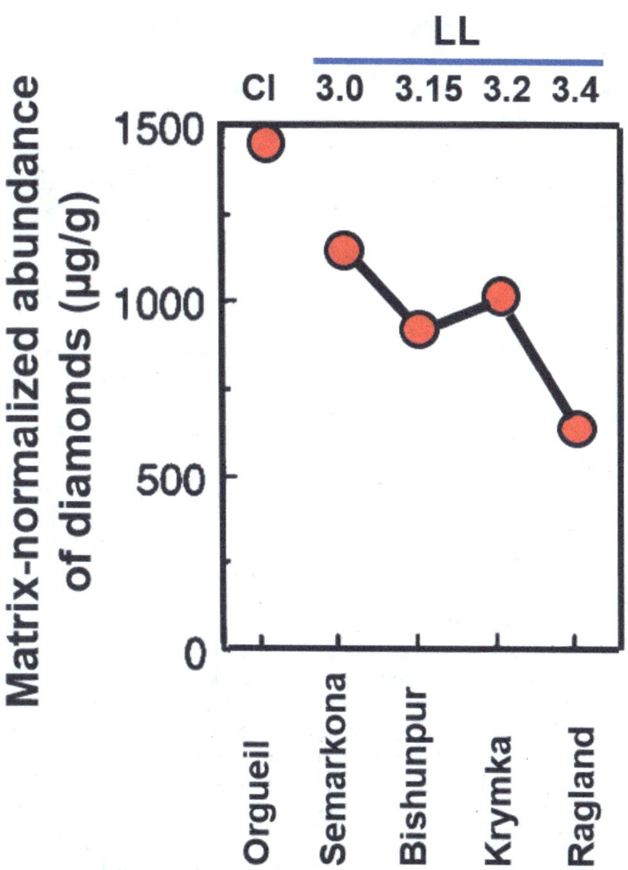

Fig. 11.48 Matrix-normalized abundances of diamonds (µg/g) are shown for CI Orgueil and a few LL3 chondrites. There is a tendency for diamond abundances based on etched residues to decrease with increasing petrologic subtype. Diagram modified from Huss and Lewis (1995).

11.9 Organic Matter

Organic matter is present in primitive chondrites of all classes – carbonaceous, ordinary, and enstatite chondrites. Because the inferred initial ratio of presolar nanodiamonds to insoluble C is similar in different chondrite groups (0.03–0.04), it seems likely they all sampled the same set of organic materials from the solar nebula during accretion (Alexander et al., 1998). Roughly 5–25% of organic matter occurs as low-molecular-weight compounds that are readily extractable with common organic solvents (e.g., Gilmour, 2005); this material includes amino acids (e.g., glycine, α-aminoisobutyric acid, D-alanine, L-alanine), aromatic hydrocarbons (e.g., benzene, toluene, naphthalene), and aliphatic hydrocarbons (e.g., *n*-alkanes). The remaining organic matter in chondrites is characterized as insoluble (e.g., Becker and Epstein, 1982); it consists of high-molecular-weight macromolecular material (e.g., $C_{100}H_{48}N_{1.8}O_{12}S_2$; Zinner, 1988) including abundant kerogen-like polycyclic aromatic hydrocarbons (PAH).[2]

Alexander et al. (1998) studied organic matter in ordinary chondrites and divided the meteorites into two sets: (1) Samples that release heavy N at low temperatures. For example, LL3.01 Semarkona has a peak isotopic composition of $\delta^{15}N = 196‰$ that is released below the peak yield temperature steps. (2) Samples that do not release heavy N at low temperatures. For example, in

[2] Amino acids are organic compounds consisting of a central carbon atom (Cα) – also known as an alpha (α) carbon – bonded to H_2N (an amino group), COOH (a carboxylic group), a hydrogen atom, and a side chain (i.e., an R group or residue, typically containing various combinations of C, H, O, S, and N). Aromatic hydrocarbons have ring-shaped structures with pi bonds (covalent bonds in which two lobes of an orbital of one atom overlap those of another). Polycyclic aromatic hydrocarbons (PAH) consist of multiple aromatic rings. Aliphatic hydrocarbons come in two varieties: saturated (in which all the C-C bonds are single, i.e., alkanes) and unsaturated (in which there are some C atoms with double bonds [alkenes] or triple bonds [alkynes]). Kerogen consists of insoluble organic matter containing a diverse set of organic materials.

LL3.15 Bishunpur and LL3.2 Krymka, the heaviest N (approaching $\delta^{15}N=100‰$) is released at temperatures higher than the peak yield steps.

Parent-body thermal metamorphism causes graphitization and oxidation of organic matter in primitive ordinary chondrites. These processes are reflected in two correlations (Fig. 11.49): (1) There is a positive correlation of the C/N ratio with N peak yield temperature. Graphitization increases the C/N ratio. Oxidation (perhaps due to the action of water) destroys the less-stable insoluble organic matter (which is equivalent to the low-temperature, isotopically heavy N component), thereby increasing the N peak yield temperature. (2) There is a negative correlation between the matrix-normalized C abundance and the N peak yield temperature as oxidation destroys nanodiamonds and some of the organic matter. This decreases the total abundance of C. Inman (L/LL3.4) is somewhat anomalous; it is more metamorphosed than Bishunpur (LL3.15) and Krymka (LL3.2) but has a lower N peak yield temperature (Fig. 11.49). Quirico et al. (2003) attributed the anomalous behavior of Inman (an OC find, ploughed up in Kansas) to terrestrial weathering.

The systematic destruction of organic matter in type-3 ordinary chondrites can be used as an alternative quantitative measure of metamorphic grade; it is particularly sensitive at the lowest subtypes. Quirico et al. (2003) found that with increasing thermal metamorphism, organic matter undergoes *maturation*: its structural order increases through the expulsion of non-carbon atoms, the linkage of structural units, and the orientation of aromatic (ring-shaped) layers. At greater degrees of maturation, the aromatic layers become condensed, oriented, and stacked. Raman spectroscopy can be used to evaluate the structural order of organic matter in meteorites. Several prominent bands are evident in the Raman spectrum of polyaromatic organic matter; these include the G-band peaking at ~1600 cm^{-1} and the D-band peaking at ~1350 cm^{-1} (Fig. 11.50). The width of the D band (actually, the full width at half maximum [FWHM]) correlates with the subtype of unequilibrated OC (Fig. 11.51). The

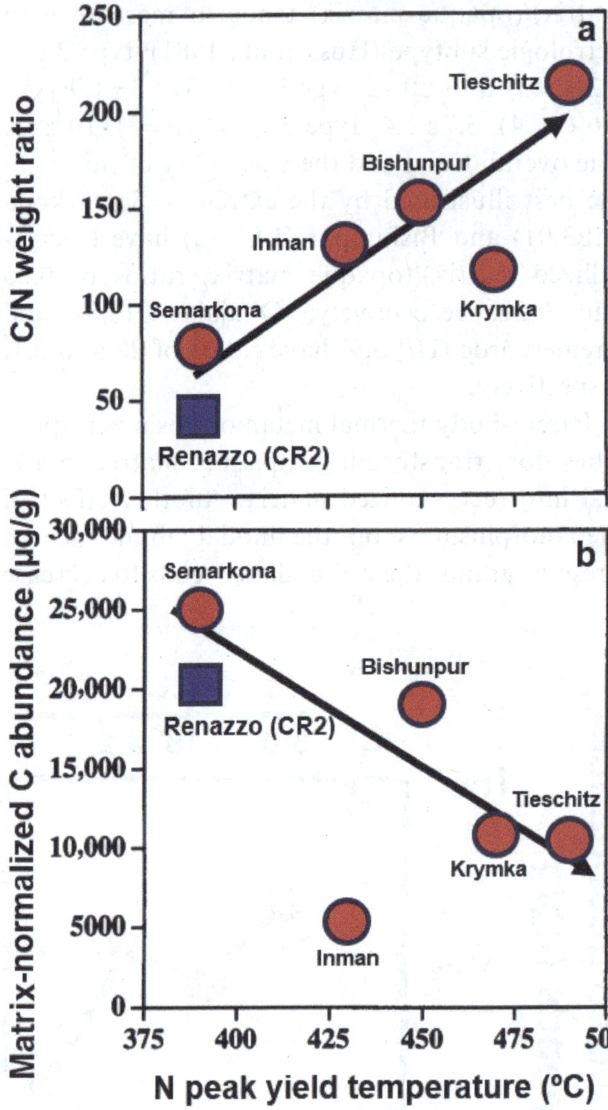

Fig. 11.49 The nature of organic matter in type-3 ordinary chondrites. (a) Bulk C/N ratio vs. N peak yield temperature for insoluble organic material in type-3 OC (red circles) compared to CR2 Renazzo (blue square). Heating of organic matter initially similar to that in Renazzo increases the C/N ratio and the N peak yield temperature. (b) Matrix-normalized C content vs. N peak yield temperature for insoluble organic material in type-3 OC compared to CR2 Renazzo. Oxidation of organic matter initially similar to that in Renazzo destroys nanodiamonds and some of the organic matter. The OC groups and subtypes are: Semarkona (LL3.01); Bishunpur (LL3.15); Krymka (LL3.2); Tieschitz (H/L3.6); Inman (L/LL3.4). Diagram modified from Alexander et al. (1998).

$FWHM_D$ parameter is inversely correlated with matrix-normalized concentrations of H, C, and N (Vacher et al., 2025).

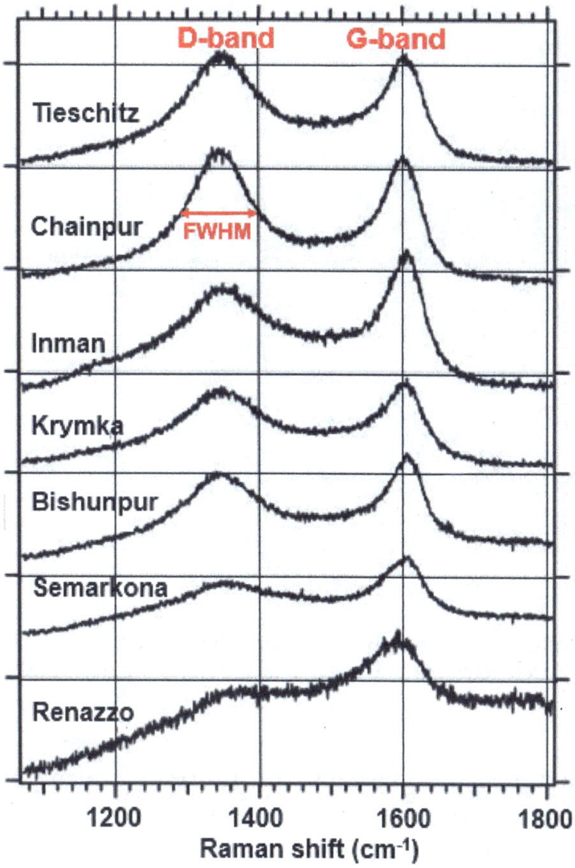

Fig. 11.50 Raman spectra of CR2 Renazzo and six type-3 ordinary chondrites: Semarkona (LL3.01), Bishunpur (LL3.15), Krymka (LL3.2), Inman (L/LL3.4), Chainpur (LL3.4), Tieschitz (H/L3.6). The prominent D and G bands are shown. The full width at half maximum (FWHM) of the D band is shown from Chainpur. Diagram modified from Quirico et al. (2003).

11.10 Carbon-Rich Aggregates and Clasts

Carbon-rich aggregates constitute 0.7–3.2 vol.% of a few type-3 OC (e.g., H3.4 Sharps, L3.5 ALHA77011, LL3.7 ALHA77278, H3.6 ALHA81024), occurring as isolated 5- to 1000-μm inclusions (Fig. 11.52). Although originally characterized as consisting mainly of graphite and magnetite (Scott et al., 1981a, 1981b; McKinley et al., 1981), subsequent petrographic, geochemical, isotopic, and TEM (transmission electron microscopy) studies (Scott et al., 1988a; Brearley, 1990) showed the aggregates to consist of irregularly shaped grains of poorly graphitized carbon along with amorphous carbon, small (<3 μm) grains of metallic Fe-Ni (up to >50 vol.% in some Sharps aggregates), and minor chromite.

Carbon-rich clasts are present in a few ordinary-chondrite regolith breccias; these clasts are 1–5 mm in size and were found in Plainview (1917) (clast PV1; Fig. 11.53), Dimmitt (DT1 and DT2; Fig. 11.54) and Weston (WN1). The clasts contain (in vol.%): 14–36% C-rich aggregates, 16–54% chondrules, 18–54% isolated silicate grains, 2–6% metallic Fe-Ni, and 1–16% troilite. Fine-grained opaque silicate matrix is not present in these clasts; the interchondrule space is occupied by C-rich aggregates instead.

The clasts have not been thermally metamorphosed; some of the chondrules contain glassy

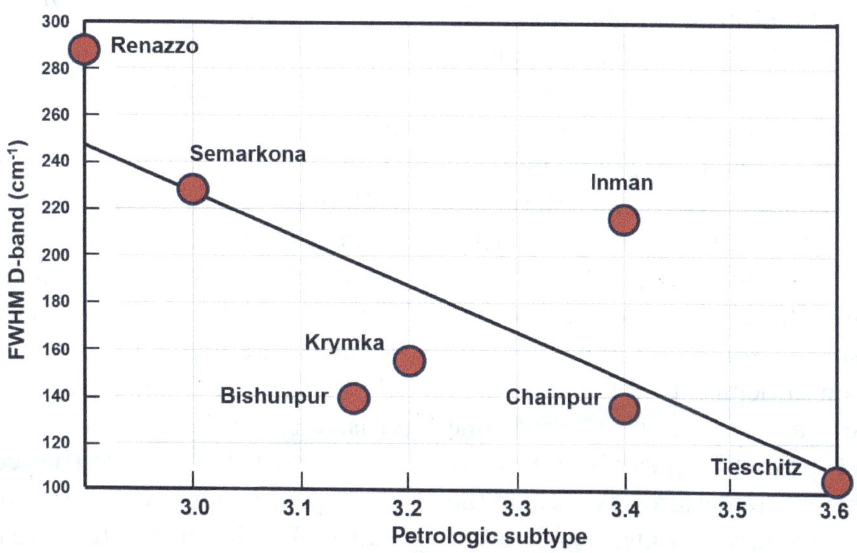

Fig. 11.51 Negative correlation between the full width at half maximum (FWHM) of the D band (determined by Raman spectroscopy) and the petrologic subtype of type-3 ordinary chondrites. CR2 Renazzo is also plotted at an arbitrary point to the left of Semarkona (at a lower degree of thermal metamorphism). OC subtypes are: Semarkona (LL3.01), Bishunpur (LL3.15), Krymka (LL3.2), Inman (L/LL3.4), Chainpur (LL3.4), Tieschitz (H/L3.6). Data from Quirico et al. (2003).

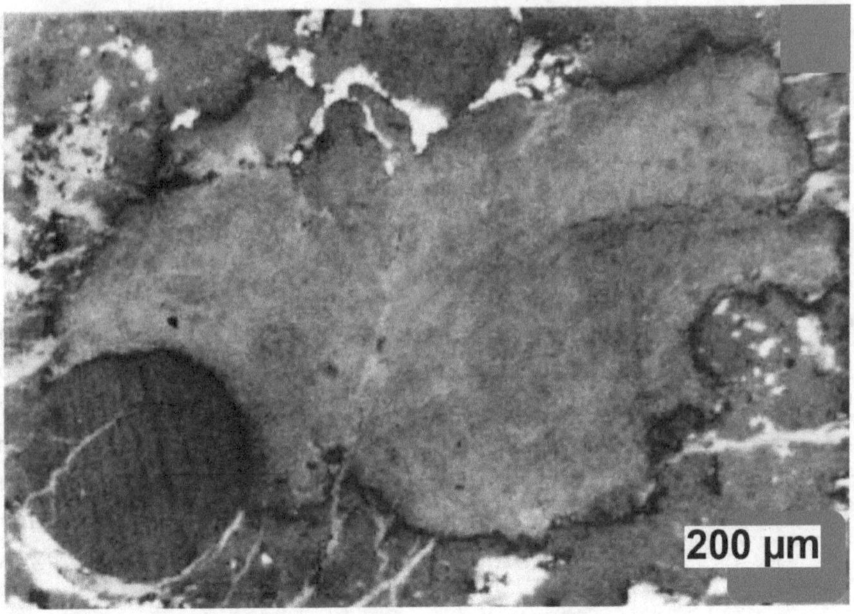

Fig. 11.52 An irregular isolated C-rich aggregate (medium gray) molded around a barred olivine chondrule (dark gray, lower left) in the Dimmitt H-chondrite regolith breccia. Light-colored veins are due to terrestrial weathering. Reflected light image from Scott et al. (1981a).

mesostases. Olivine and low-Ca pyroxene grains are heterogeneous in composition: e.g., olivine in PV1 ranges from Fa_{3-43} mol% ($n = 275$) with a mean of $Fa_{17\pm7.7}$; low-Ca pyroxene ranges from Fs_{1-38} mol% ($n = 241$) with a mean of $Fs_{10.6\pm6.3}$ (Scott et al., 1981b).

Although PV1 contains 13 wt.% C (compared to 0.1–0.6 wt.% C in most type-3 OC), it most closely resembles normal H chondrites (such as its Plainview host) in chondrule size (~300 μm), bulk Mg-normalized mean refractory lithophile abundance ratio (1.00 ± 0.09 × H) and bulk O-isotopic composition ($\Delta^{17}O = 0.66‰$) (Rubin et al., 2005). The clast likely formed from Plainview host material and experienced complex processing. All the chondrules in PV1 are interfused, indicating that temperatures must have briefly reached ~1100°C (the approximate solidus temperature of H-chondrite silicate). The most likely source of this heating was a collision. Carbon enrichment of the clast may have resulted from a second impact event, one involving a C-rich projectile (possibly a Jupiter-family comet). This event may also have heated the clast and provided a source of water. As the clast cooled after impact, it experienced hydrothermal alteration at low water/rock ratios as evidenced by the thick rims of ferroan olivine around low-FeO olivine cores. PV1 was later incorporated into the Plainview region of the H-chondrite parent-body regolith and was extensively fractured and faulted.

It is unclear if other C-rich clasts have a similar history.

11.11 Halite

Halite (NaCl), common table salt, has (as of this writing) been reported as a pre-terrestrial phase in three ordinary-chondrite breccias: Zag (H3-6 fall), Monahans (1998) (H5 fall), and Sidi El Habib 001 (H5-an fresh find, but possible fall) (Zolensky et al., 1999a; Rubin et al., 2002; Che et al., 2023a, 2023b). (A few days after a bright fireball was witnessed on December 20, 2020 by meteorite hunters in western Algeria, Sidi El Habib 001 was found as a group of fresh fusion-crusted stones near the Mali border.) (Halite has been reported as a terrestrial contaminant along with quartz sand within fractures of the Contis-Plage H5 chondrite; this weathered meteorite [W3] was found on a dune near a beach.)

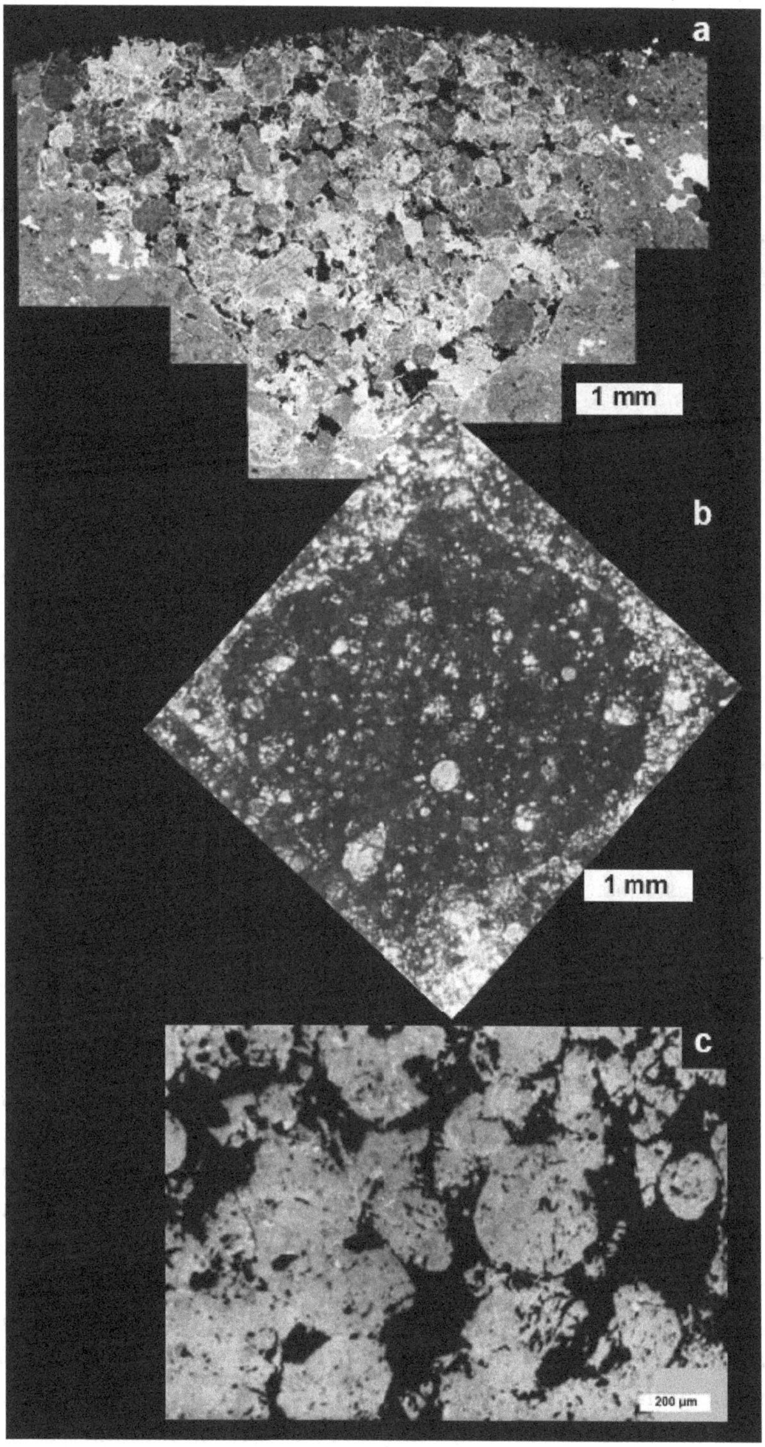

Fig. 11.53 Subrounded, 5-mm-diameter, C-rich clast PV1 from the Plainview (1917) H-chondrite regolith breccia. (a) PV1 and the surrounding Plainview host. BSE image from Rubin et al. (2005). (b) PV1 in approximately the same orientation and size scale as image (a). Transmitted light view showing most of the clast to be opaque. Metallic Fe-Ni is depleted in the clast relative to the host. Image from Scott et al. (1981b). (c) Portion of PV1 showing well-formed chondrules as well as chondrule fragments surrounded by C-rich aggregates. Reflected light image from Scott et al. (1981a).

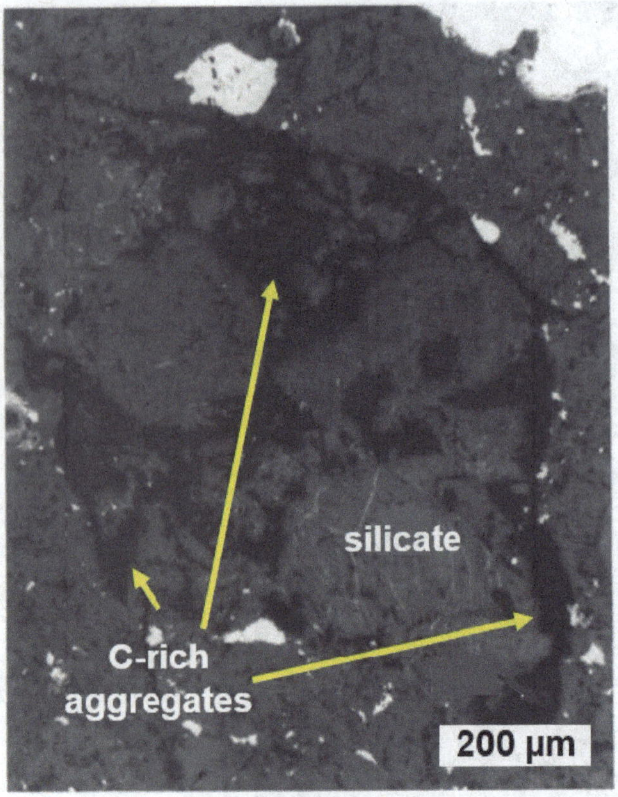

Fig. 11.54 Clast DT2 from the Dimmitt H-chondrite regolith breccia showing the C-rich aggregates (black) serving as a matrix into which chondrules and chondrule fragments (medium gray) are embedded. Reflected light image modified from Scott et al. (1981b).

The gray matrix of Monahans (1998) contains randomly distributed dark blue to purple, transparent to opaque, well-crystallized halite grains exhibiting sharp cubic morphology; the grains range up to 5 mm in size. Crystals of sylvite (KCl) are distributed heterogeneously as inclusions within the larger halite crystals. Halite contains minor K and Br; sylvite contains significant Na and minor Br. Aqueous fluid inclusions (up to ~15 μm) are present within the halite grains (Zolensky et al., 1999a). (Although fluid inclusions have occasionally been reported in other ordinary chondrites [and other meteorites] [e.g., Warner et al., 1983], many of these inclusions may be artifacts of thin-section preparation [Rudnick et al., 1985].)

The matrix of Zag contains scattered grains of dark-blue, transparent-to-opaque halite. Most grains are a few hundred micrometers in size, although aggregates can range up to 1 cm. Sylvite was not identified. The halite grains are well crystallized but do not exhibit sharp cubic morphology; many grains are irregular in shape. Zag halite grains contain less K and Br than those in Monahans (1998). Aqueous fluid inclusions, ranging up to ~10 μm, are much more abundant in Zag halite than Monahans (1998) halite. There are two varieties of aqueous fluid inclusions in Zag halite: (1) primary inclusions that were incorporated into growing halite crystals and (2) secondary inclusions that became trapped within preexisting halite grains when fluid-filled fractures healed. Zolensky et al. (2015, 2016) and Chan et al. (2018a) reported that the solids trapped within halite in Zag and Monahans (1998) include ^{15}N-rich amino acids, macromolecular carbon materials, aliphatic carbon compounds, aromatic, ketone, imine and/or imidazole compounds, Fa_{1-41} olivine, Ca-pyroxene, low-Ca pyroxene, feldspars, phyllosilicates, magnetite, sulfides, metallic Fe-Ni, lepidocrocite, carbonates, diamond, apatite, and zeolites.

(The discovery of diverse organic compounds in Zag and Monahans [1998] was covered by the media [e.g., Strickland, 2018], but this was not the first report of life or its components within ordinary chondrites. Hahn [1880] described fossil sponges, corals, and crinoids in several OC. Although these findings were verified by one zoologist [Weinland, 1881a, 1881b, 1882], they were quickly debunked by another [Vogt, 1882]. Referring to Hahn, Vogt wrote "One does not know at what to be most astonished, the author's complete ignorance of the laws of evolution or the audacity with which he presents his views in terms worthy of the oracle of Delphi.")

Sidi El Habib 001 contains recrystallized halite-free clasts and a less recrystallized main lithology (the host) containing numerous halite grains varying from submicrometer in size to >1 mm. The large halite grain in Fig. 11.55 contains abundant 2–10-μm-size rounded to cubic inclusions of sylvite. Halite in Sid El Habib 001 contains minor K and Br. Throughout the host, halite appears to be replacing chlorapatite (Che et al., 2023a). Also present in the host are grains of sphalerite ((Zn,Fe)S), suggesting that sphalerite and halite crystallized from the same hydrothermal fluids (Che et al., 2023b). A few sphalerite grains contain small amounts of Mn (up to 0.11 wt.%) or Cr (up to 0.03 wt.%).

I–Xe dating of Zag halite by Whitby et al. (2000) yielded an age of 4.25 ± 0.03 Ga; $^{39}Ar/^{40}Ar$ age-dating of Monahans (1998) halite by Garrison and Bogard (2000) yielded an age of >4.33 Ga. These grains formed by evaporation of brines early in Solar-System history.

Fig. 11.55 A large light-blue halite crystal (~1.5×2.5mm) in Sidi El Habib 001 with a quasi-rectangular shape and numerous fractures. The perimeter of the grain is moderately irregular with some minor embayments. Reflected light image modified from Che et al. (2023a).

The origin of OC halite is controversial. One possibility is that the halite grains are exogenous. Chan et al. (2018a) analyzed the fluid inclusions in Zag and Monahans (1998) and identified a variety of soluble and insoluble organic compounds including ketones and amino acids. They found the amino acid content within Zag halite to be different from that in the Zag matrix, suggesting derivation from different sources. Chan et al. (2018a) concluded that the halite grains formed via hydrovolcanism on a C-type asteroid; they suggested 1 Ceres, the largest asteroid (and a dwarf planet with a mean diameter of 939.4 km). According to this scenario, at the end of a hydrothermal cycle, the halite crystals precipitated from the fluid on the surface of Ceres, trapping brines and an assortment of organic compounds. If the halite crystals are in fact from Ceres, they might have been expelled from the asteroid by hydrothermal activity. (The equatorial escape velocity from Ceres is 516 m s^{-1}.) The Cerealia Facula dome in Occator Crater on Ceres (Fig. 11.56) contains bright deposits of a mixture of hydrated sodium chloride ($NaCl \cdot 2(H_2O)$; hydrohalite), ammonium chloride (NH_4Cl), and sodium carbonate (Na_2CO_3; natrite) (De Sanctis et al., 2020); these deposits formed relatively recently after deep-seated brines reached the asteroid surface. Similar deposits may have formed much earlier in Solar-System history. (Relatively high concentrations of ammonium [~13.6 μmol g^{-1}] were reported in a

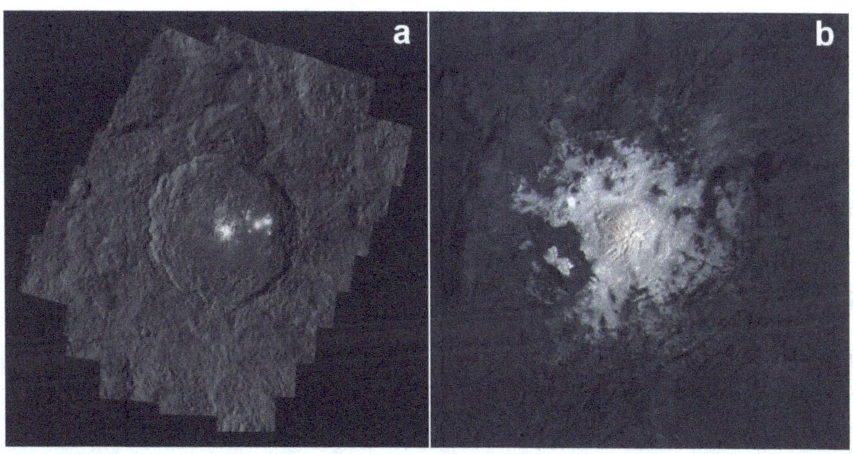

Fig. 11.56 Bright spot on the asteroid Ceres. (a) *Dawn* spacecraft image of Occator Crater (92 km in diameter and 4 km deep) on Ceres. (b) The prominent bright deposit known as Cerealia Facula (with regions of varying composition) within the Occator Crater. Images courtesy of NASA.

hot-water extract from a bulk sample of the Bennu CI-like asteroid [Glavin et al., 2025].)

If the OC halite crystals hail from Ceres, it is likely that exposure to ionizing radiation in interplanetary space was responsible for causing the halite grains to become blue-purple in color (due to capture of electrons in anion vacancies). Eventually, some of the expelled grains accreted to the regolith of the H-chondrite parent body.

An alternative origin for the OC halite grains is formation on the H-chondrite asteroid itself. If this were so, the body would have experienced aqueous alteration at an early epoch, not long after accretion. Jones et al. (2016) reported that chlorapatite in Zag is relatively abundant in the same matrix regions housing halite. Both phases are rich in Cl and might have formed from a Cl-rich fluid heated by impacts or by the decay of ^{26}Al. This seems consistent with the replacement of chlorapatite by halite in Sidi El Habib 001 and the association of halite and sphalerite in the same lithology of this meteorite (Che et al., 2023b). The discovery of F-bearing halite and sylvite among returned Bennu samples, a CI-like asteroid (McCoy et al., 2025), supports an indigenous origin of these salts in OC asteroids.

Meteorite finds tend to get rained on; some falls get rained on as well (particularly those recovered long after reaching the ground). Many meteorites, whether falls or finds, are cut on saws using water as a lubricant (e.g., Tobin, 2019). The paucity of ordinary chondrites containing coarse halite grains is probably due, at least in part, to the dissolution of halite in samples exposed to water. (Halite has a solubility of 6.13 mol/kg at 20°C.)

11.12 Opaque-Mineral Assemblages

The main opaque mineral groups in unweathered type-3 ordinary chondrites include metallic Fe-Ni (major kamacite and taenite, accessory tetrataenite, and, in a few very oxidized LL chondrites, awaruite), sulfide (primarily troilite, but in some cases minor-to-accessory pentlandite, pyrrhotite, and mackinawite), and oxides (predominantly chromite with minor-to-accessory ilmenite and rutile, and, in some aqueously altered chondrites, magnetite) (e.g., Rubin and Ma, 2021).

In going from H to L to LL, there is an overall oxidation state increase. For the metal fraction, this is illustrated by: (1) decreases in the modal abundance of metallic Fe-Ni (Jarosewich, 1990), (2) increases in the taenite/kamacite ratio (Afiattalab and Wasson, 1980), (3) decreases in the average grain size of kamacite (Rubin, 1990), (4) increases in the kamacite Co content (Sears and Axon, 1976; Rubin, 1990), (5) decreases in the kamacite Ni content (Rubin, 1990), and (5) increases in the Ge content of metallic Fe-Ni (Chou and Cohen, 1973; Chou et al., 1973; Florin et al., 2020). A few very oxidized LL chondrites (e.g., LL6 Appley Bridge and LL5 Parambu) contain rare metallic Fe grains with 37 wt.% Co and 2.5 wt.% Ni (Afiattalab and Wasson, 1980; Rubin, 1990). (These changes in the metal fraction are accompanied by systematic variations in the silicate and oxide phases: increases in FeO/(FeO+MgO) in olivine, low-Ca pyroxene, chromite, and ilmenite [Bunch et al., 1967; Mueller and Olsen, 1967; Snetsinger and Keil, 1969; Gomes and Keil, 1980; Rubin, 1990] and increases in the olivine/low-Ca-pyroxene modal ratio [Hutchison, 2004].)

Florin et al. (2023) analyzed siderophile elements in individual metal grains in a suite of H chondrites. They found significant variations in the concentrations of refractory siderophiles (Re, Os, W, Ir, Ru, Mo, Pt) and moderately volatile siderophiles (Pd, Au, Ga, Ge) in H3-chondrite metal. It appears that the range of Mo, W, and Ga concentrations was established by localized melting of the metal grains under different redox conditions. Metal in H4 to H6 chondrites is more compositionally homogeneous, a result of equilibration during parent-body metamorphic heating. The similar abundances of refractory siderophiles in H3 metal and L3, LL3, and L/LL3 metal (Okabayashi et al., 2019) led Florin et al. (2023) to suggest that there had been a single chemically homogeneous metal component that was heterogeneously distributed among the ordinary-chondrite parent bodies.

Shocked ordinary chondrites commonly contain small grains of metallic Cu (Rubin, 1994a); the modal abundance in many shocked type 4–6 OC is ~$1.2-1.5\times10^{-4}$ vol.%. Copper grains appear pale pink in freshly polished sections and darker reddish

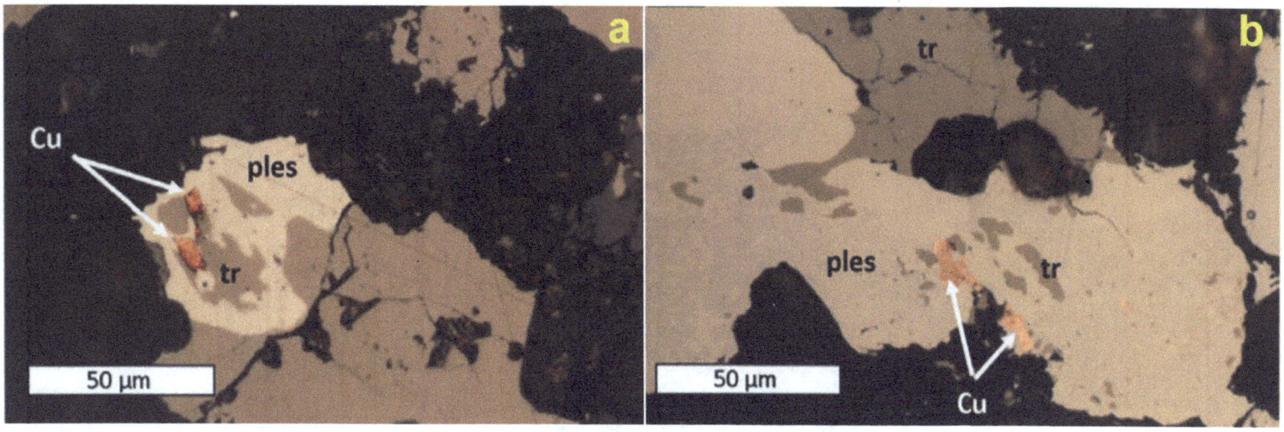

Fig. 11.57 Occurrences of metallic Cu in H5 Pultusk. (a,b) Plessite-troilite aggregates containing small irregular grains of metallic Cu at the interface between metallic Fe-Ni (plessite; white, ples) and small irregular troilite (tan, tr) grains. Dark phases are silicate. Reflected light images modified from Łuszczek and Krzesińska (2020).

copper in older sections with somewhat tarnished or oxidized surfaces. In most cases, copper grains are present within much larger metallic Fe-Ni grains at the boundaries with small, irregularly shaped troilite grains (Fig. 11.57). In rare cases, metallic Cu occurs between metallic Fe-Ni and troilite in elongated shock veins (e.g., L/LL6 Trysil) and as rare 0.5-µm -thick rinds at the outside edges of small horseshoe-shaped metallic Fe-Ni cells within coarse grains of shock-melted troilite (e.g., L6 Sixiangkou) (Fig. 11.58). The Cu is not pure: it contains ~1.5–2.0 wt.% Ni .

Martensite (α_2, a distorted bcc structure) formed from high-temperature taenite during quenching; it is present in some shocked ordinary chondrites (e.g., Rubin, 1990). A few OC melt-rock clasts (e.g., in the Dimmitt H-chondrite regolith breccia) contain opaque blebs consisting mainly of schreibersite intergrown with metallic Fe-Ni (Rubin et al., 1983a) (Fig. 11.59). The matrices of aqueously altered type-3 OC contain magnetite, carbides (cohenite and haxonite), Ni-rich metal (including tetrataenite and awaruite), and pentlandite (Taylor et al., 1981; Krot et al., 1997b).

Many opaque assemblages in OC are composed of metallic Fe-Ni and associated troilite (e.g., Fujita et al., 1999). Experiments by Lauretta et al. (1997) showed that S is mobilized from preexisting sulfide grains during metamorphic heating and forms rinds around extant metal grains. Adjacent silicate can be incorporated into the sulfide rinds. Although many large metallic Fe-Ni grains observed in polished sections appear free of sulfide, most of these grains may be attached to troilite in three dimensions (e.g., Kuebler et al., 1999). In contrast, many chromite grains appear monomineralic, and others are intergrown with kamacite and/or troilite.

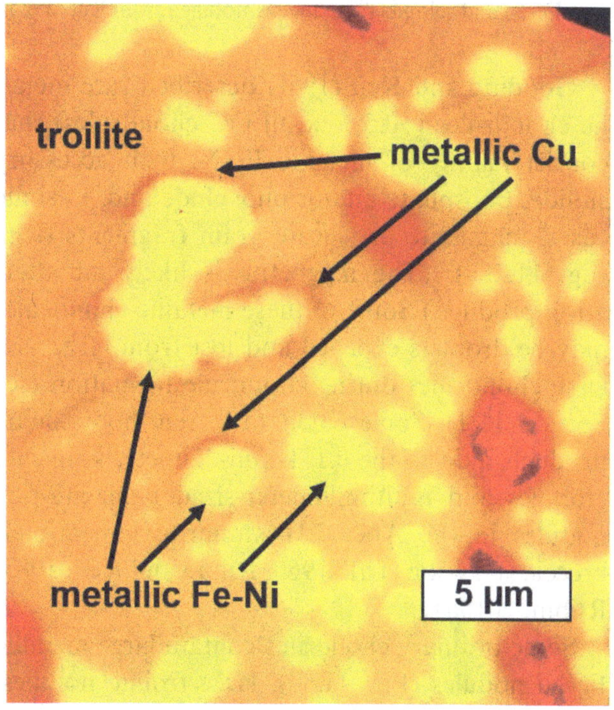

Fig. 11.58 A rapidly solidified metallic-Fe-Ni-troilite assemblage in L6 Sixiangkou (shock-stage S5) featuring rounded and horseshoe shaped blebs of metallic Fe-Ni embedded in troilite. A few of the metallic Fe-Ni blebs are partially rimmed by thin rinds of metallic Cu. The entire assemblage formed by shock melting.

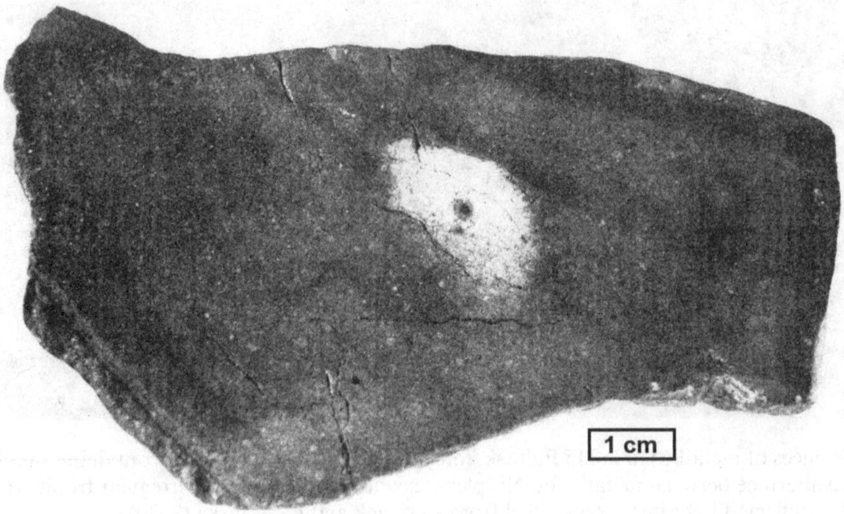

Fig. 11.59 A slab (UNM c10.9) of the Dimmitt H-chondrite regolith breccia containing a quasi-ellipsoidal light-colored poikilitic impact-melt-rock clast (DT4). The round dark assemblage in the center of the clast is composed of metallic Fe-Ni and schreibersite. Image modified from Rubin et al. (1983a).

Gooding and Keil (1981) described rare metallic chondrules (~0.5% of all OC chondrules) that consist mainly of metallic Fe-Ni with accessory sulfide, phosphate and/or phosphide and, in some cases, magnetite or silicate grain fragments (e.g., Fig. 11.28a). (The magnetite is likely an alteration product.) Some of these metallic chondrules may be droplets expelled and lost from larger silicate chondrules during chondrule formation (see Section 11.1). Some could be parent-body shock products akin to the 0.1–1-mm-diameter kamacite globules and multimillimeter-sized kamacite-rich nodules in the Abee EH-chondrite impact-melt breccia (Dawson et al., 1960; Rubin and Keil, 1983; Rubin, 2015a).

Some ordinary chondrites contain large metallic Fe-Ni nodules (Fig. 11.60), large troilite nodules, or large metallic-Fe-Ni-troilite nodules that range up to ~3 cm in maximum dimension (Rubin, 1985; Xie and Wang, 1992). The metallic Fe-Ni nodules have an unusual chemical composition; typically, they are depleted in refractory siderophile elements (e.g., Ir, Pt, Re, Os, Ru, Rh) and Cu by large factors (up to 240) relative to fine-grained metallic Fe-Ni in the same meteorite (Widom et al., 1986; Rubin, 1995b, 1999). Nevertheless, the nodules typically are not depleted in W or Mo (both normally refractory), nor in common (Fe, Ni, Co) or volatile (Au, As, Ga, Sb) siderophile elements (Widom et al., 1986; Kong et al., 1998; Rubin, 1999). Thin metallic veins (actually thin sheets in three dimensions) in some equilibrated ordinary chondrites (e.g., H6 Densmore (1950)) are of similar composition.

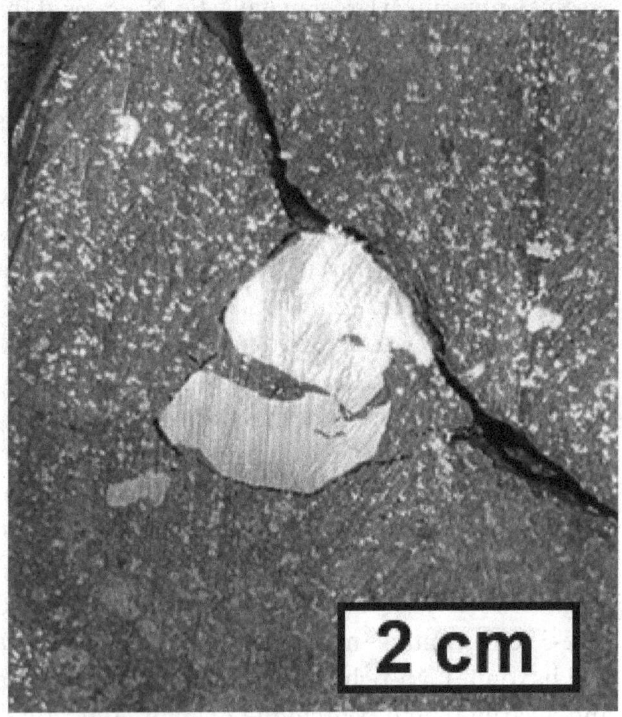

Fig. 11.60 Large metallic Fe-Ni nodule (light gray) in a slab of H5 Faucett. Also present are smaller nodules as well as normal fine-grained metallic Fe-Ni throughout the meteorite. Silicates appear dark gray. Image courtesy of eBay seller maccers531.

The metallic Fe-Ni nodules are modeled as having formed by a complicated shock process involving impact melting and vaporization, oxidation of W and Mo to form volatile oxides, concomitant partial oxidation of Re and Os, rapid fractional condensation of refractory siderophiles, transport of the residual vapor, condensation of Cu, S, and Se into vugs and fractures to form sulfide, and condensation of the remaining siderophiles to form metal nodules and veins. During vapor cooling and condensation (or during subsequent parent body annealing) W, Mo, Re, and Os oxides were reduced to their elemental states (e.g., Widom et al., 1986; Rubin, 1995b, 1999).

Rubin (1995b) analyzed a ~4×9×12-mm irregular metallic Fe-Ni nodule (dubbed RC1) situated between silicate melt and a relict chondritic clast in the Rose City H5 impact-melt breccia. The nodule is compositionally heterogeneous. Approximately 65 wt.% of RC1 is refractory rich; it is enriched in Os and Ir by 30–40% relative to bulk H-chondrite metal. About 20 wt.% of the nodule is of intermediate composition; it is depleted in Os and Ir by 31–35% relative to H-chondrite metal. About 15 wt.% of RC1 is refractory poor; it is depleted in Os and Ir by 75%. A cross section of the nodule and the corresponding compositions are shown in Fig. 11.61. Common and volatile siderophile elements are essentially unfractionated in all three regions; W is fractionated to only a moderate degree.

Metal-sulfide melts and silicate melts are mutually immiscible. When metal-bearing chondrites are melted (either via collisions or radiogenic heating), the fine metal and sulfide grains tend to coagulate and separate from the silicate melt. The density of molten metal with a composition of 90 wt.% Fe and 10 wt.% Ni is ~7 g cm^{-3}; the density of a liquid of half metallic Fe-Ni and half FeS is ~5.5 g cm^{-3}; the density of molten chondritic silicate is ~3 g cm^{-3}. Thus, a metallic liquid or a metal-sulfide liquid would tend to sink toward the gravitational center of the body and the molten silicate would tend to rise toward the surface. If these liquids encounter unmelted materials, they may flow around them, flow through fractures within them, or begin to melt them (via resorption). A variety of features can be produced, including the following:

1. Molten metal can form channels through unmelted or partially melted silicate clasts as in the H-chondrite impact-melt breccias Portales Valley (e.g., Rubin et al., 2001; Ruzicka et al., 2005) (Fig. 11.62a), Rose City (e.g., Mason and Wiik, 1966; Rubin, 1995b) (Fig. 11.63),

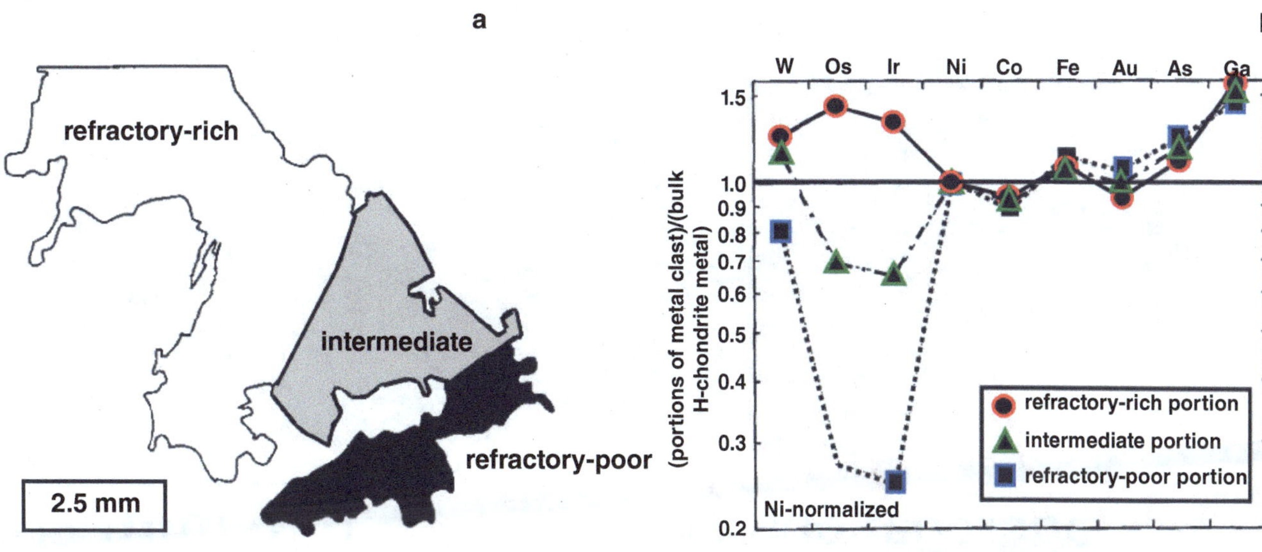

Fig. 11.61 An irregular, compositionally heterogeneous metallic Fe-Ni nodule from the Rose City H-chondrite impact-melt breccia. (a) A diagram of the central cross-section of the nodule showing the compositionally distinct regions. (b) Siderophile-element/Ni ratios of the three nodule regions normalized to the mean bulk composition of metallic Fe-Ni in nine H-chondrite falls. Elements are plotted from left to right in order of increasing volatility. In all three regions, W is less fractionated than Os and Ir, suggesting that it may have formed volatile oxides at relatively low oxygen fugacities and behaved more like the volatile siderophiles (Au, As, and Ga). Images modified from Rubin (1995b).

Fig. 11.62 Nital-etched slabs of two H-chondrite impact-melt breccias sporting Widmanstätten structures. (a) Portales Valley (1998 fall from New Mexico) contains massive regions of metallic Fe-Ni with Widmanstätten patterns (white to light gray) separating irregular silicate clasts (dark gray) of different sizes. Clasts at left and at top appear to be fractured blocks invaded and separated by metal veins. There are two distinct orientations of the Widmanstätten structure: the dividing line between them is the large elongated, quasi-rectangular silicate clast extending NNE from bottom center to just beyond the center of the slab. The uniformly distributed white dots in the silicate clasts are metallic Fe-Ni and troilite. Image from Rubin et al. (2001). (b) Punggur (2021 fall from Lampung, Indonesia) shows several metal-rich regions (medium gray) with differently oriented Widmanstätten patterns. Coarse silicate clasts (darker shade) are also present. Image courtesy of Roberto Vargas. Portales Valley and Punggur are among the very few chondrite impact-melt breccias with metallic Fe-Ni that cooled slowly enough at subsolidus temperatures to produce Widmanstätten patterns.

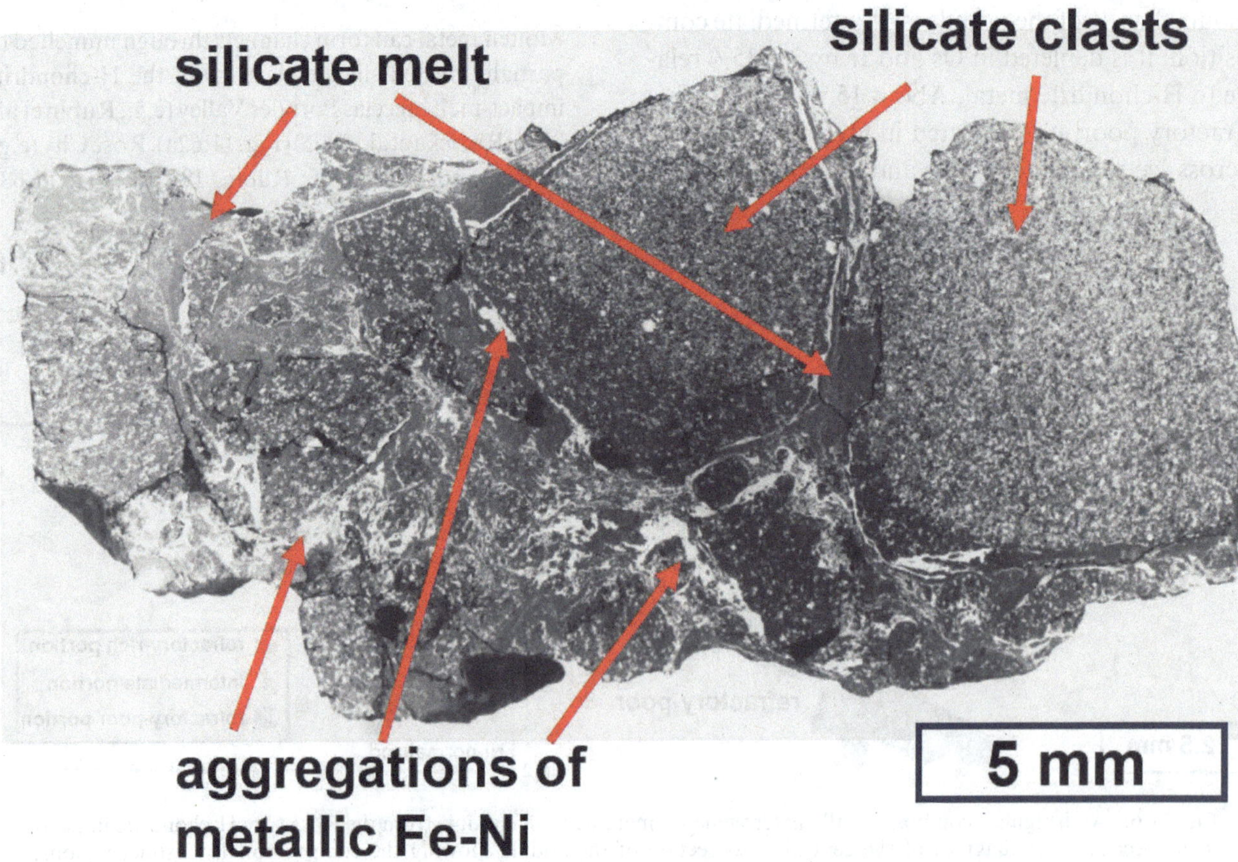

Fig. 11.63 Slab of the Rose City H-chondrite impact-melt breccia showing rounded and irregular silicate clasts separated by veins of silicate melt. Some aggregations of metallic Fe-Ni occur within the silicate melt; in some places, the metal forms an incomplete rind around the adjacent clast.

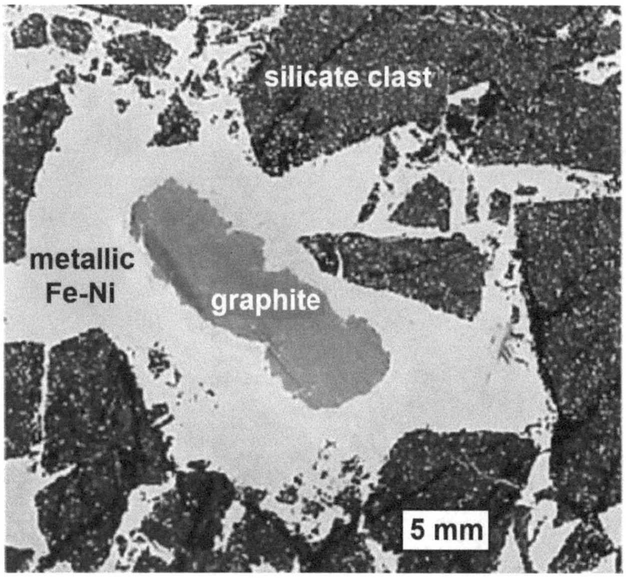

Fig. 11.64 Elongated graphite nodule with an irregular outline within metallic Fe-Ni in the Portales Valley H6 chondrite breccia. Reflected light image modified from Ruzicka et al. (2005). Image courtesy of Alex Ruzicka and the Cascadia Meteorite Laboratory.

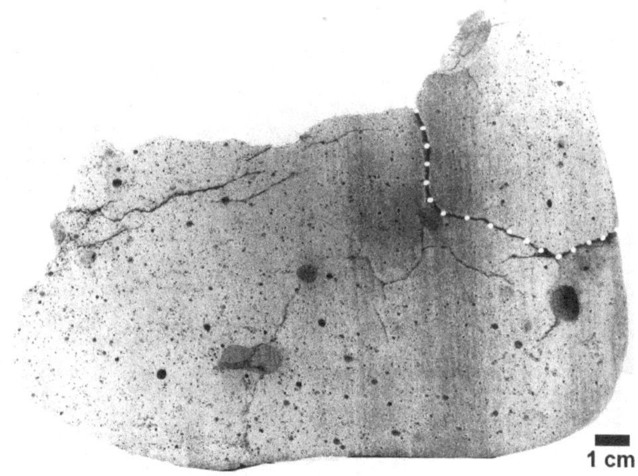

Fig. 11.65 A sawn face of the PAT 91501 L-chondrite impact-melt breccia. The rock has a uniform texture and contains vesicles up to ~3 mm in diameter. Also present are rounded to elongated, centimeter-sized nodules of metallic Fe-Ni and troilite (dark gray). The dark near-vertical streaks are an artifact of sawing. The string of white dots at upper right were placed prior to removal of a piece for sampling. Image from Mittlefehldt and Lindstrom (2001) and courtesy of the NASA Johnson Space Center.

and Punggur (Fig. 11.62b). Metal-rich rinds can solidify around unmelted silicate masses (Fig. 11.63). Also present within Portales Valley metallic Fe-Ni are rare graphite nodules (Fig. 11.64).

2. Metal and sulfide can solidify into scattered centimeter-sized rounded nodules as in the PAT 91501 L-chondrite impact-melt rock (e.g., Mittlefehldt and Lindstrom, 2001) (Fig. 11.65).
3. Within the Jelica LL6 breccia there is an impact melt-rock clast that contains a 65×107 μm grain of euhedral tetrataenite (ordered FeNi) attached to a similarly sized grain of troilite (Rubin, 1994b) (Fig. 11.66).
4. In the unusual case of the rapidly cooled LL-chondrite impact-melt rock LAR 06299 (Rubin and Moore, 2011), metal and sulfide droplets nucleated on gas bubbles in the melt and aligned themselves (Fig. 11.67) with the gravitational field of the asteroid according to density, with bubbles at the top end, troilite in the middle, and metallic Fe-Ni at the bottom:

gas bubble ($\rho < 0.001$ g cm^{-3}) ↑ TOP
troilite ($\rho \approx 4.7$ g cm^{-3})
metallic-Fe-Ni ($\rho \approx 8.0$ g cm^{-3}) ↓ BOTTOM

Similar structures occur within the L-chondrite impact-melt rock, PAT 91501 (Benedix et al., 2008). It is plausible the metal-sulfide-bubble assemblages were close to neutrally buoyant in these impact melts.

Shock veins consisting of metallic Fe-Ni and/or troilite occur in many shocked ordinary chondrites; these veins vary in thickness from <1 to >100 μm (Fig. 11.68; Fig. 3(1) of Lonsdale, 1937; Fig. 10 of Brandstätter et al., 2013). Some opaque veins consist mainly of small blebs of metal and troilite that form curvilinear trails within silicate grains. This feature produces "shock darkening" or "silicate darkening" (Rubin, 1992), which can significantly increase the opacity of silicates and produce "black chondrites" (e.g., Britt and Pieters, 1991) such as L5 Farmington (Fig. 11.69), LL4 Albareto, LL5 Guidder, and LL6 Dongtai. There are also shock-darkened centimeter-sized chondritic clasts in a few OC regolith breccias, such as H6 Nulles (e.g., Williams et al., 1985) and Plainview (1917) (Fodor and Keil, 1976a).

Several distinct petrographic varieties of chromite ($FeCr_2O_4$) occur in ordinary chondrites. There are intact unfractured chromite grains in unshocked OC. Some shocked ordinary chondrites contain fractured chromite grains (Fig. 11.70a), some with numerous troilite veins (Fig. 11.70b). Also present are thin veinlets of chromite (typically 0.5–2 μm thick and 10–300 μm long) (Fig. 11.70c,d) including:

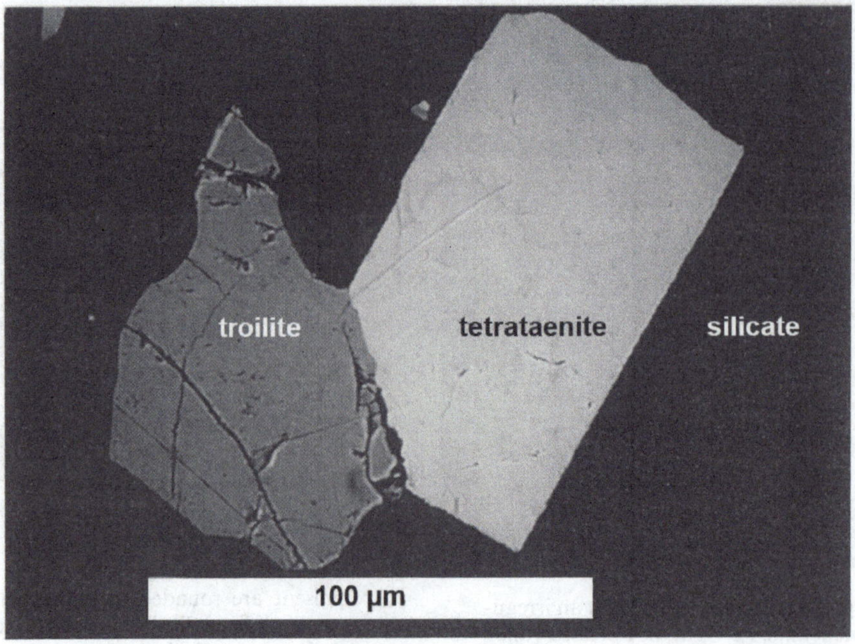

Fig. 11.66 A euhedral tetrataenite grain (white) and attached troilite (medium gray) surrounded by silicate (black) within an impact melt-rock clast in LL6 Jelica. The tetrataenite grain has the morphology of a truncated rectangle in plan view; the troilite nucleated and grew at the truncated side of the tetrataenite grain. SEM image modified from Rubin (1994b).

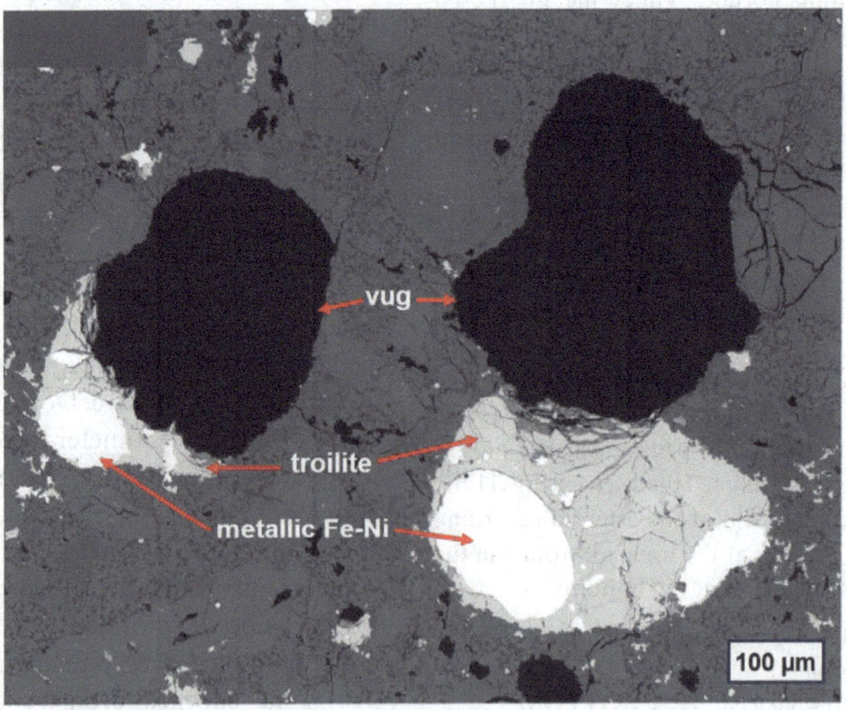

Fig. 11.67 LAR 06299. Neighboring metal-sulfide nodules attached to large vesicles or vugs (formerly gas bubbles in the precursor melt; black). The vesicles occur on the same side of the nodules (the top). The oblate metal blebs (white) occur on the opposite sides of the nodules (the bottom). Between the metal blebs and the vesicles are oblate masses of troilite (medium gray). The orientations of these assemblages were fixed after the surrounding melt solidified; they reflect the direction of gravity in the parent asteroid. Image modified from Rubin and Moore (2011).

Fig. 11.68 Metallic Fe-Ni and troilite shock veins in ordinary chondrites. (a, b) Whole thin section views of LAP 04581 (LL5, S2). (a) An elongated troilite vein (black) extends across the thin section. A few chondrules and chondrule fragments are discernable. Chondrules (chd) 1 and 2 are flattened and oriented subparallel to the long axis of the troilite vein. The inter-chondrule matrix is recrystallized. Transmitted light image from Rubin and Swindle (2011). (b) The same elongated troilite vein (pale yellow) extends across the thin section. It consists of two massive ends and a thin central region. The dark lines running roughly NW-SE at the right and left sides of the section are cracks in the broken slide. Reflected light image from Rubin and Swindle (2011). (c) Metallic Fe-Ni shock vein (white) in Kernouvé (H6, S1) transecting silicate grains (dark gray). (d) Portion of vein showing that silicate peninsulas within the vein are preferentially oriented toward the left. This likely resulted from flow of the metallic liquid from right to left. The existence of this shock vein demonstrates that Kernouvé was appreciably shocked before it was annealed to S1 levels. Images (c) and (d) are in reflected light, modified from Rubin (2003).

Fig. 11.69 Ordinary chondrites that experienced different degrees of shock. Black chondrite L5 Farmington is at left and normal ordinary chondrite H4 Farmville (H4) is at right. Image modified from Britt and Pieters (1991).

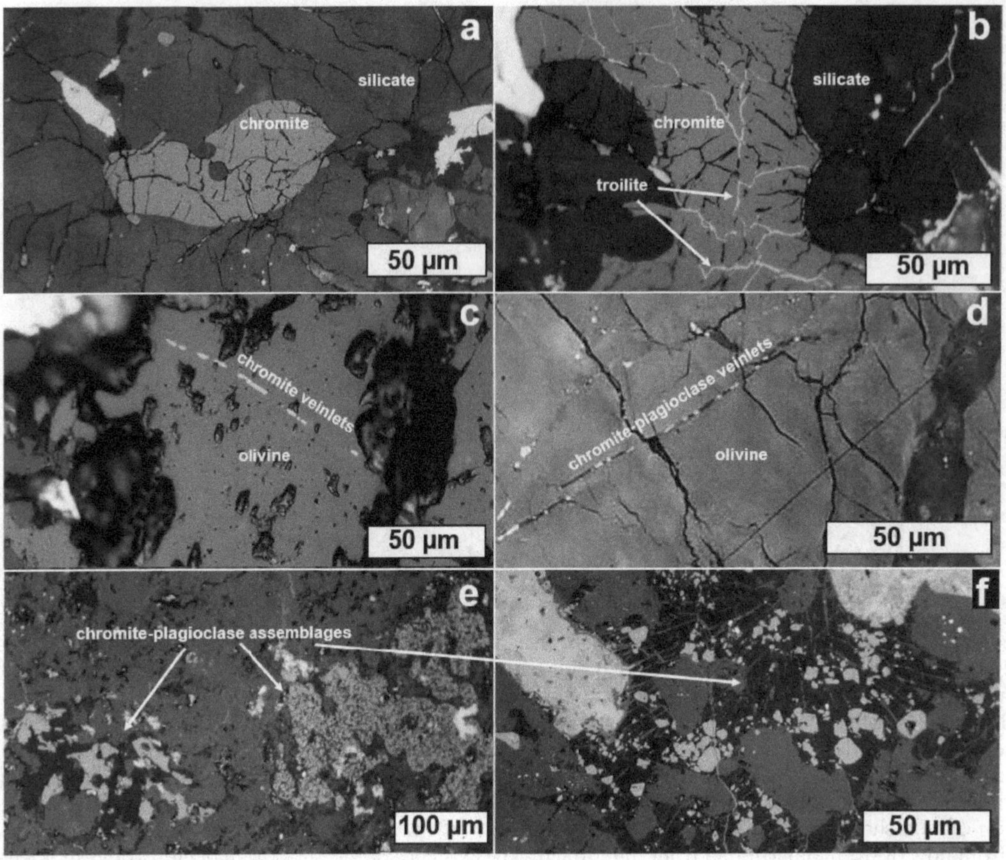

Fig. 11.70 Varieties of chromite occurrences in shocked ordinary chondrites. (a) Unmelted but fractured chromite grain (medium gray) in Jartai (L6, S4). Similar fractures (narrow black lines) occur in the surrounding silicate grains. (b) Unmelted chromite grain (medium gray) containing very thin veins of troilite (white) in Jartai (L6, S4). A few troilite veins also occur in the surrounding silicate (dark gray). (c) Chromite needles (light gray) forming a veinlet that transects adjacent silicate grains (medium gray) in Portales Valley (H6, S1). (d) Chromite veinlets (light gray), some with interstitial glass of plagioclase composition (dark gray), within silicate grains (medium gray) in Jartai (L6, S4). At right is a melt pocket (dark gray) consisting of plagioclase glass. (e) Fine-grained chromite-plagioclase assemblage (right) and coarse-grained chromite-plagioclase assemblage (left) in Hualapai Wash (L6, S4). The assemblages have very different plagioclase/chromite modal abundance ratios (i.e., 0.3 and 2, respectively). The coarse chromite grains in the assemblage at left may be unmelted grain fragments. (f) Network of interconnected chromite-plagioclase assemblages containing much more plagioclase than chromite in the Chico L-chondrite impact-melt breccia. Large light-gray masses at upper left and upper right are limonite patches formed from terrestrially weathered metallic Fe-Ni grains. Light gray lines transecting the plagioclase are veins of limonite. Reflected light images modified from Rubin (2003).

(1) thin, continuous elongated chromite bands, (2) discontinuous rounded and oblong tiny chromite grains, and (3) discontinuous rounded and oblong tiny chromite grains with interspersed plagioclase or glass of plagioclase composition. The veinlets formed from chromite-plagioclase melts that were injected into fractures in neighboring silicate grains; chromite crystallized in the fractures and the residual plagioclase-rich melt continued to flow, in some cases pooling to form plagioclase-rich melt pockets.

Rubin (2003) described chromite-plagioclase assemblages in ordinary chondrites (Fig. 11.70e,f). They occur in nearly every shock-stage S3-S6 OC; the assemblages range in size from 20–300 µm and consist of 0.2–20-µm-size euhedral, subhedral, anhedral, and rounded chromite grains surrounded by plagioclase or glass of plagioclase composition. Plagioclase has a low impedance to shock compression. Heat from shock-melted plagioclase caused adjacent chromite grains to melt; chromite grains crystallized from this melt. Those chromite grains in assemblages that are completely surrounded by plagioclase are generally richer in Al_2O_3 than unmelted, matrix chromite grains in the same meteorite. Chromium-rich chondrules (e.g., Fig. 11.25) may also be shock products, formed by small-scale impact-melting of chromite-rich assemblages (Krot and Rubin, 1993; Rubin, 2003).

Some OC impact-melt breccias (e.g., the Smyer H chondrite) contain troilite-rich regions (Figs. 11.71, 11.72) consisting of unmelted, crushed silicate grains surrounded by troilite and transected by thin troilite veins (some as thin as 0.1 μm) (e.g., Rubin, 2002a). Few grains of metallic Fe-Ni occur in these zones, leading to FeS/Fe modal ratios ranging from 25:1 up to ~500:1, far higher than the eutectic weight ratio of 7.5:1 (e.g., Brandes and Brook, 1998). The texture of these troilite-rich regions suggests that the sulfide formed from a fluid of very low viscosity. The moderately high viscosity (0.2 poise) and large surface tension of liquid FeS, its inability to wet silicate grain surfaces at low oxygen fugacities, and the super-eutectic FeS/Fe ratios in the troilite-rich regions indicate that the fluid was a vapor. It seems likely that, upon release of shock pressure, some troilite evaporated and dissociated. Molecules of S_2 were transported and condensed into fractures and around tiny silicate grains; there, they combined with Fe from small adjacent metallic Fe-Ni grains to form troilite. The Ni content at the edges of some of these metal grains increased significantly, and Co from these Ni-rich grains diffused into nearby kamacite.

Two portions of the surface of asteroid 433 Eros (Fig. 11.73) that were excited by energetic solar flares (Trombka et al., 2000) have S/Si weight ratios (0.036 ± 0.007 and 0.052 ± 0.006) that are much lower than those of mean H (0.116), L (0.113), and LL (0.111) chondrite falls (Jarosewich, 1990). Eros is a near-Earth S-type asteroid, likely similar in composition to ordinary chondrites (Binzel et al., 2002; Chapman, 2004b). The low S at the surface of Eros could have resulted from volatilization and loss of S during impact events akin to those documented for Smyer.

Additional opaque phases in some primitive type-3 ordinary chondrites (e.g., H3.8 WIS 91627; L3.05 EET 90161; LL3.01 Semarkona) are

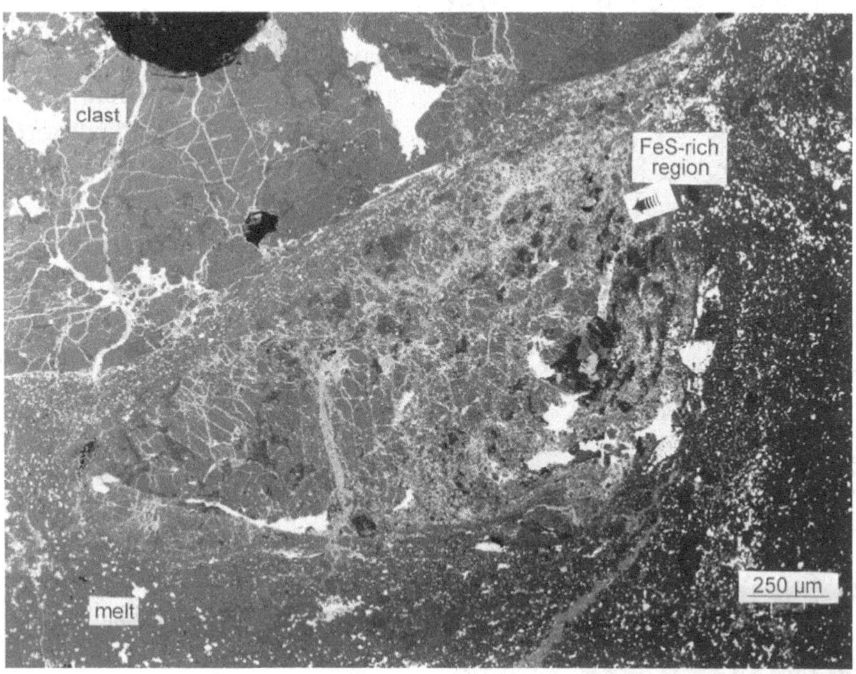

Fig. 11.71 A portion of the Smyer H-chondrite impact-melt breccia showing an unmelted chondritic clast (medium gray; upper left) transected by numerous thin veins of troilite (white). Also present in the clast are a few large metallic Fe-Ni grains (white). The black semicircle at upper left is part of a bubble in the thin-section epoxy. A clamshell-shaped troilite-rich region occurs in the center of the image, separated from the clast by a thin channel of silicate-rich melt containing numerous tiny metal blebs. The troilite-rich region contains shattered silicate grains transected and surrounded by troilite. Silicate-rich melt also occurs below and at the sides of the troilite-rich region. BSE image from Rubin (2002a).

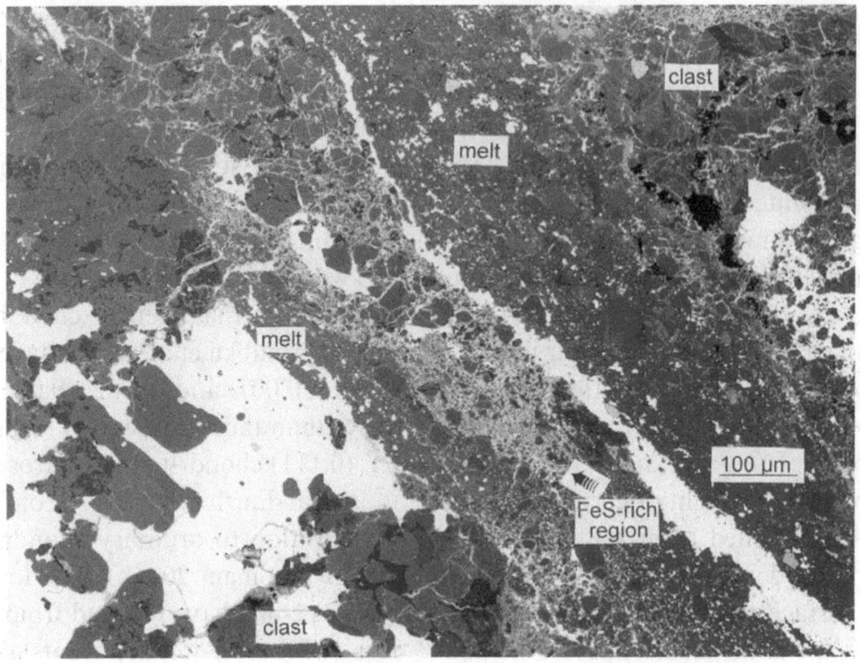

Fig. 11.72 The edge of an unmelted chondritic clast (lower left) containing large metal–troilite assemblage (white) and included silicate grains (gray). Immediately above and to the right of the clast is a layer of silicate-rich melt, followed by a small troilite-rich region, a thin metallic Fe-Ni vein (white), a thick channel of silicate-rich melt, and, at upper right, another unmelted chondritic clast containing a coarse metal grain. BSE image from Rubin (2002a).

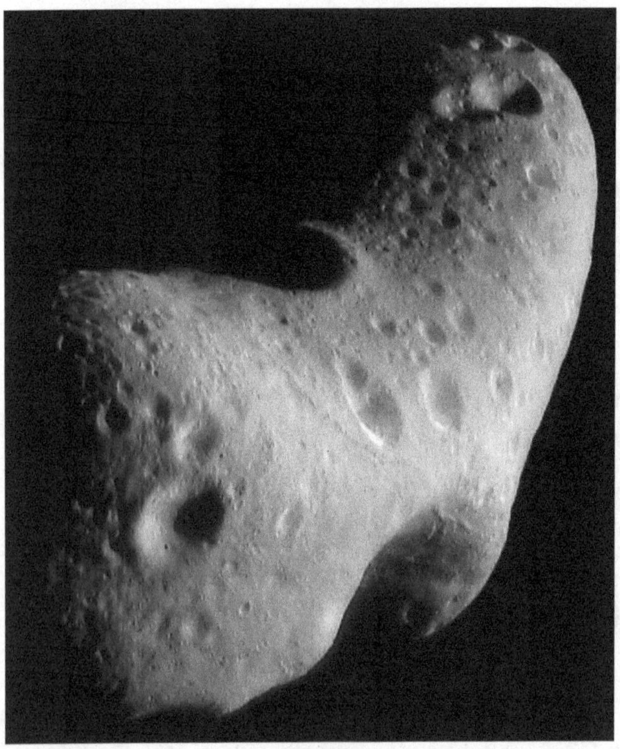

Fig. 11.73 Northern hemisphere of asteroid 433 Eros. Image constructed from a mosaic of six separate images draped over a computer model of the asteroid's shape. The crater at the top is 5.3 km in diameter. Image courtesy of NASA/JPL/JHUAPL.

present in carbide-magnetite assemblages (e.g., Fig. 11.74). These phases include Fe-oxide (magnetite), Ni-rich metallic Fe (primarily tetrataenite and awaruite), Co-rich kamacite, Fe-carbides (cohenite, haxonite, edscottite), and Ni-rich sulfide (pentlandite) (Taylor et al., 1981; Krot et al., 1997b). Not all phases are present in every carbide-magnetite assemblage. The assemblages formed from preexisting metal-sulfide nodules by parent-body hydrothermal alteration involving C-O-H-bearing fluids (including gaseous CO and H_2O).

Alteration proceeded as a front that not only replaced fine-grained particles in the matrix, but also penetrated some chondrules and dissolved glassy mesostasis. Rubin (2013a) illustrated this process in a multi-layered BO chondrule in LL3.01 Semarkona (Fig. 11.75). The local alteration process halted before affecting the entire chondrule.

Type-I chondrules in type-3 OC contain blebs of kamacite and metal of plessitic composition as well as blebs of troilite (e.g., Fig. 11.32b). The precursors of these chondrules contained fine-grained silicate particles as well as grains of

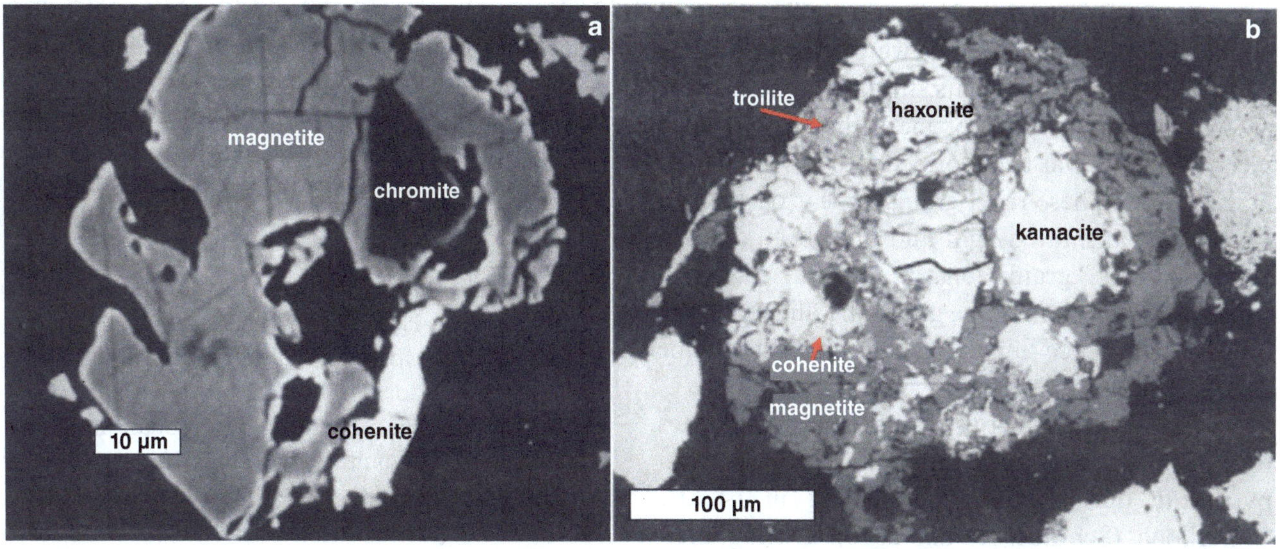

Fig. 11.74 Carbide-magnetite assemblages in type-3 ordinary chondrites. (a) Intergrown cohenite (white) and magnetite (medium gray) containing a grain of chromite (black) in H3.8 WIS 91627. (b) An assemblage containing magnetite (dark gray) that is replacing kamacite, haxonite, cohenite, Ni-rich taenite (all white or very light gray) and troilite (medium gray) in H3.8 RKPA80205. SEM images modified from Krot et al. (1997b).

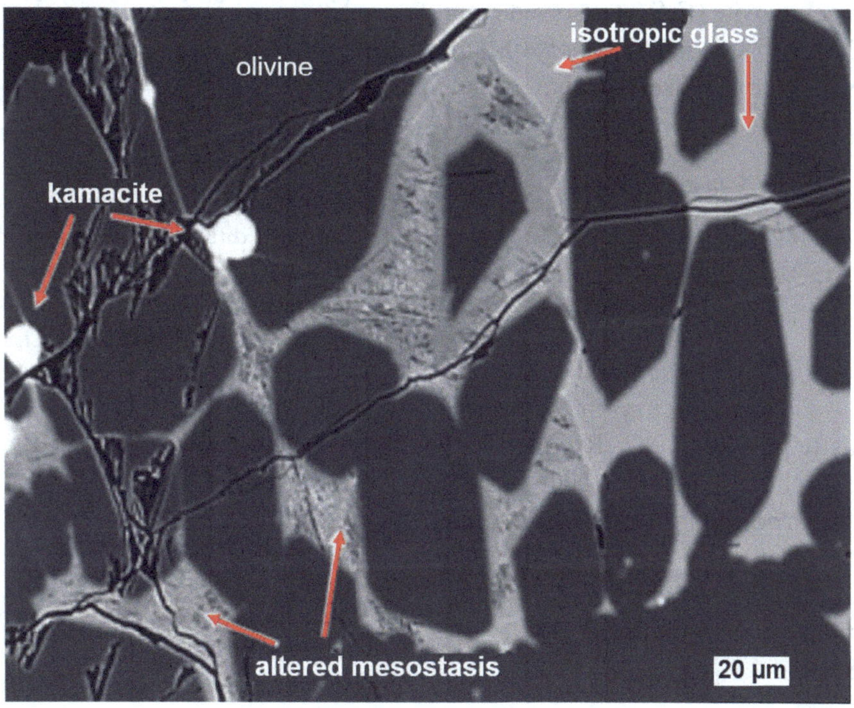

Fig. 11.75 Boundary between altered mesostasis and isotropic glass within a multi-layered barred olivine chondrule in LL3.01 Semarkona. The altered mesostasis consists of elongated holes and blades of more resistant material. The olivine grain at top center is surrounded by a sheath of less mottled mesostasis that is compositionally indistinguishable from the altered mesostasis adjacent to it. The boundary between the altered mesostasis and the isotropic glass is sharp and somewhat lobate; it probably marks the farthest extent (locally) of glass dissolution during parent-body alteration. BSE image modified from Rubin (2013a).

metallic Fe-Ni and troilite. In some cases, coarser phenocryst grain fragments from earlier generations of chondrules were also present.

Ferroan silicate grains (commonly olivine) derived from chondrules that experienced collisions and fragmentation in the nebula were occasionally incorporated into low-FeO dustballs (Type-I chondrule precursors) that were melted during chondrule-formation events. In many cases, where the ferroan grains did not melt, they experienced reduction: FeO was transformed into low-Ni metallic Fe grains. These relict grains appear dark when viewed in transmitted light and have been dubbed "dusty olivine" grains (e.g., Rambaldi, 1981). In some cases, the metallic Fe grains are aligned along preferential lattice directions in the host olivine crystal (e.g., Fig. 11.36b).

Some opaque minerals in ordinary chondrites formed by exsolution. Pentlandite exsolved from pyrrhotite during chondrule cooling in LL3.01 Semarkona (Schrader et al., 2016). During metamorphic heating of the parent body, chromite formed by exsolution in olivine (Ramdohr, 1973; Ashworth, 1979; Grossman and Brearley, 2005), chromite and rutile formed within ilmenite (Buseck and Keil, 1966), mackinawite (FeS_{1-x}) exsolved from pentlandite ($(Fe,Ni)_9S_8$) (Koeberl et al., 1990), and horomanite $(Fe,Ni,Co,Cu)_9S_8$ exsolved from troilite.

11.13 Clasts

I am large, I contain multitudes.

Walt Whitman, "Song of Myself"

My name is Legion: for we are many.

Mark 5:9; KJV

A large variety of clasts occurs in ordinary-chondrite breccias (e.g., Keil, 1982; Rubin, 1985; Scott et al., 1989; Bischoff et al., 1993, 2006, 2018). Some clasts were incorporated prior to whole-rock thermal metamorphism (and have equilibrated mineral compositions), some clasts after metamorphism (and contain unequilibrated minerals) (Rubin et al., 1983b).

(a) Genomict OC breccias (Wasson, 1974; Bischoff et al., 2006) contain angular clasts of different petrologic types but of the same OC group (e.g., Dar al Gani 300; Bischoff et al., 2018). If a genomict breccia is rich in solar-wind gases, it would also be a regolith breccia (e.g., H4-6 Noblesville; Lipschutz et al., 1993).

(b) Regolith breccias (Fig. 11.76) are characterized by a dark porous clastic matrix (rich in solar-wind-implanted noble gases) surrounding light-colored low-porosity angular chondritic clasts (typically of petrologic type 4 to 6). Some rocks contain silicate grains marred by solar-flare particle tracks, indicating that they were present at the surface of their parent bodies (e.g., Bhandari et al., 1996). These breccias are more common among H chondrites than L or LL chondrites (Bischoff et al., 2018). Examples of OC regolith breccias include H5 Pantar (Fredriksson and Keil, 1963), H5 Cangas de Onis (Williams et al., 1985), H6 Nulles (Williams et al., 1985), H4-6 Noblesville (Lipschutz et al., 1993), H4 Tysnes Island (Keil and Fodor, 1980), H5 Plainview (1917) (Fodor and Keil, 1976a), H4 Weston (Noonan and Nelen, 1976), H3.7 Dimmitt (Rubin et al., 1983a), L5 Assam (Schultz et al., 1971), L3-4 Mafra (Lange et al., 1979), L4 Rio Negro (Fodor et al., 1977), L5 Innisfree (D. Smith, 1980), LL Holman Island (Black, 1972), and LL6 St. Mesmin (Dodd, 1974). Crabb and Schultz (1981) found that the abundances of regolith breccias in the main OC groups are H (14%), L (3%), and LL (8%). Bischoff et al. (2006) revised these percentages: H (15.3%), L (3%), LL (5.5%).

(c) In addition to resembling regolith breccias in their containing some equilibrated OC clasts, carbonaceous-chondrite clasts, and impact-melt-rock clasts, primitive breccias (Scott and Taylor, 1982) may include type-3 OC clasts of the same chondrite group but with different modal abundances of chondrules and fine-grained matrix (e.g., LL3.10 Piancaldoli; Rubin et al., 1982a), type-3 clasts of a different OC group (e.g., an H3 clast in LL3.2 Krymka; Bischoff et al., 2018), and ungrouped type-3 OC clasts (e.g., LL3 NWA 10214; Rubin et al., 2017). Some clasts may have been acquired during parent-body accretion (Kracher et al., 1985), others from later-stage low-velocity collisions.

(d) Impact-melt rock clasts occur in regolith and fragmental breccias; they tend to be lighter-colored than their hosts and relatively depleted in metallic Fe-Ni and sulfide. They were described in detail by Rubin (1985). Clasts range in size from <50 μm (Rubin et al., 1983b) to ~4 cm (Binns, 1967a; Keil, 1982), they vary in shape from spheroidal to angular and blocky (Fig. 11.77), and display textures ranging from poikilitic to porphyritic to skeletal. In most cases, the clasts' bulk and mineral compositions are similar to the silicate portions of their host chondrites; the clasts are modeled as having formed from their hosts by impact-melting (e.g., Fodor et al., 1972, 1976; Fodor and Keil, 1973, 1975, 1976a, 1976b, 1978; Keil

Fig. 11.76 The Cangas de Onis H-chondrite regolith breccia showing light-colored H6 clasts amidst a dark-colored equilibrated clastic matrix. Image modified from Rubin (1982).

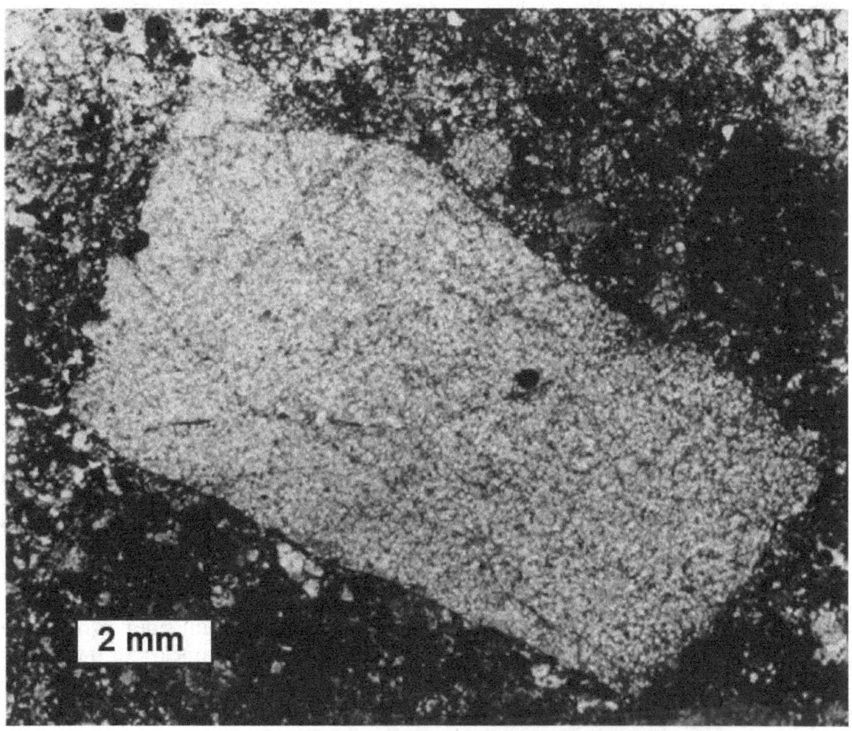

Fig. 11.77 Light-colored impact melt-rock clast (PV4) in the Plainview (1917) H-chondrite regolith breccia. The clast is composed of equigranular olivine, orthopyroxene, K-rich interstitial glass, and minor Ca-rich pyroxene (all light gray), and accessory metallic Fe-Ni and troilite (both black). Transmitted light image modified from Fodor and Keil (1976a).

and Fodor, 1973, 1980; Prinz et al., 1977; Wilkening, 1978; Keil et al., 1980; Rubin et al., 1983a, 1983b). Some clasts were melted completely (e.g., JC1 in L6 Johnson City; Rubin et al., 1983b; Clast 1 in L3.7 NWA 092; Goodrich et al., 2021); others contain relict, unmelted material (including many such clasts

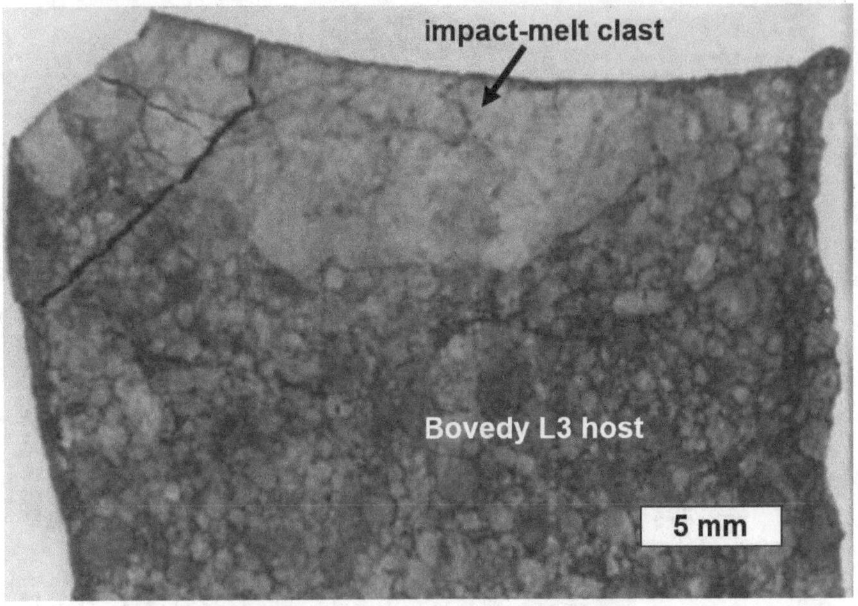

Fig. 11.78 Photograph of a large light-colored compositionally zoned impact-melt rock clast in L3.7 Bovedy. Image modified from Rubin et al. (1981b).

Fig. 11.79 Skeletal-olivine-bearing 0.6-mm melt-rock clast (DT14) in the Dimmitt H chondrite regolith breccia (Rubin et al., 1983a). The clast contains elongated olivine grains at one end (right side); progressively smaller olivine grains grade into a microcrystalline mesostasis toward the other end, which appears to be a quenched margin (left side). Transmitted light.

within the Dimmitt H-chondrite regolith breccia; Rubin et al., 1983a). A few clasts are compositionally zoned – one such example is the 1 × 2-cm-sized, metal- and sulfide-poor, poikilitic fragment in L3.7 Bovedy (Fig. 11.78). It exhibits a progressive increase in the bulk MgO/FeO ratio and in the mean MgO content of olivine and low-Ca pyroxene across its length (Rubin et al., 1981b). Other OC clasts are texturally zoned. For instance, the olivines in clast DT14 from Dimmitt vary from elongated 100-μm-long grains at one end to microcrystals within mesostasis at the other (Fig. 11.79; Rubin et al., 1983a). Some OC clasts suffered reduction (Williams et al., 1985). Also occurring within brecciated ordinary chondrites are alkali-rich impact-melt clasts (Kempe and Müller, 1969; Fodor et al., 1974; Fodor and Keil, 1978; Ikeda and Takeda, 1979; Wlotzka et al., 1983). Although an impact origin for a moderately large ($>4.5 \times 7 \times 4$ mm) silica-rich orthopyroxenite clast (Bo-1) in L3.7 Bovedy was dismissed by Ruzicka et al. (1995) in favor of heating by radionuclide decay or electromagnetic induction, it seems possible Bo-1 formed instead from impact melting of an OC target enriched in silica-rich chondrules.

(e) Shock-darkened clasts are permeated with thin curvilinear veinlets of metallic Fe-Ni and troilite (Williams et al., 1985); they are akin to black chondrites.

(f) Xenolithic or foreign clasts occur in a few OC breccias. These xenoliths include materials from the Inner Solar System (e.g., a eucritic clast in L3-6 NWA 869; Hyde et al., 2020) and the Outer Solar System (e.g., carbonaceous-chondrite clasts; e.g., Fodor et al., 1976). Shen et al. (2025) described a suite of clasts in L3.4 NWA 14763 with unusual O- and Cr-isotopic compositions; their spectral signatures resemble those of B-type asteroids. Because many xenoliths were metamorphosed, Goodrich et al. (2021) argued that most of the xenoliths in OC breccias were acquired after the epoch of thermal metamorphism, that is, ~60 Ma after CAI formation (or not too long after that).

Materials from the Inner Solar System are represented in rare cases by intact OC clasts of a different OC group than the host. These include: H-chondrite clasts in the LL6 St. Mesmin regolith breccia (Dodd, 1974), an LL5 clast in the Dimmitt H3.7 regolith breccia (Rubin et al., 1983a), an L4 clast in the LL6 breccia NWA 5764 (Gattacceca et al., 2017), L4/5 clasts in the LL4-6 breccia Glanerbrug (Niemeier and Bischoff, 2006), an ungrouped OC clast in LL3 NWA 10214 (Rubin et al., 2017), an H3 clast in LL3.2 Krymka (Bischoff et al., 2018), an LL clast in L3 Tanezrouft 039 (Funk et al., 2011), H-chondrite fragments in LL3.6 Ngawi (Fodor and Keil, 1975), L-chondrite components in LL5 Paragould (Fodor and Keil, 1978), unusual BO-chondrule-like clasts with H-chondrite O-isotopic compositions in L6 Y-793241 (Prinz et al., 1984), an L-chondrite-like inclusion in the Fayetteville H4 regolith breccia (Wieler et al., 1989b), and abundant L or H/L clasts in the Elmshorn H3-6 breccia (Bischoff et al., 2024b).

There are also several examples of OC breccias containing xenoliths of a different OC group, as identified principally by their O-isotopic compositions: (1) an H-chondrite melt-rock clast in LL5 MET 01004 (Corrigan et al., 2015), (2) an H-chondrite melt clast in L6 Chantonnay (Yolcubal et al., 1997), (3) lithic clasts of H chondrites in the L6 chondrites Y-75097 and Y-793241 (Nakamura et al., 1994), (4) an LL impact-melted clast in L3.7 NWA 092 (Goodrich et al., 2021), (5) an LL melt-rock clast in L6 Peace River (Herd et al., 2013), and (6) an H-chondrite melt-rock clast in L5 Barwell (Hutchison et al., 1988). A few additional examples are mentioned by Bischoff et al. (2006) and listed in Table 1 of Lipschutz et al. (1989).

Materials from the Outer Solar System come mainly in the form of carbonaceous chondrites (primarily CM; e.g., Fodor and Keil, 1976a; Fodor et al., 1976; Rubin and Bottke, 2009) (e.g., Fig. 11.80). Zolensky et al. (2018b) pointed out that many carbonaceous-chondrite xenoliths are chondrule- and CAI-free and consist mainly of phyllosilicate, magnetite, carbonate, and pyrrhotite. They appear to contain a large variety of organic materials. Some of these xenoliths also contain minor olivine and low-Ca pyroxene. Well-studied carbonaceous clasts in OC occur in Zag (H3-6 breccia), Tsukuba (H5 breccia with H6 clasts), Plainview (1917) (H5 regolith breccia), NWA 8369 (H5 breccia), and Carancas (H4-5 breccia) (Chan et al., 2018b; Zolensky et al., 2018c).

(g) Hydrated matrix-rich clasts containing phyllosilicates and the amphiboles tremolite ($Ca_2Mg_5Si_8O_{22}(OH)_2$) and cummingtonite ($(Mg,Fe^{2+})_2(Mg,Fe^{2+})_5Si_8O_{22}(OH)_2$) have been identified in LL3.05 NWA 5206 and L3.15 NWA 6925 (J. M. Johnson et al., 2016; Kerraouch et al., 2025).

(h) Microchondrule-bearing clasts occur in LL3.2 Krymka (Fig. 11.21; Rubin, 1989) and L4 Rio Negro (Fodor et al., 1977; Rubin et al., 1982a).

(i) Carbon-rich clasts in Krymka were described by Semenenko and Girich (1995) and Semenenko et al. (2004).

(j) A 6.1×7.0 mm OC clast in the LL3-6 breccia NWA 10214 contains abundant fine-grained matrix material and is more reduced than H chondrites and IIE chondritic clasts (Rubin et al., 2017).

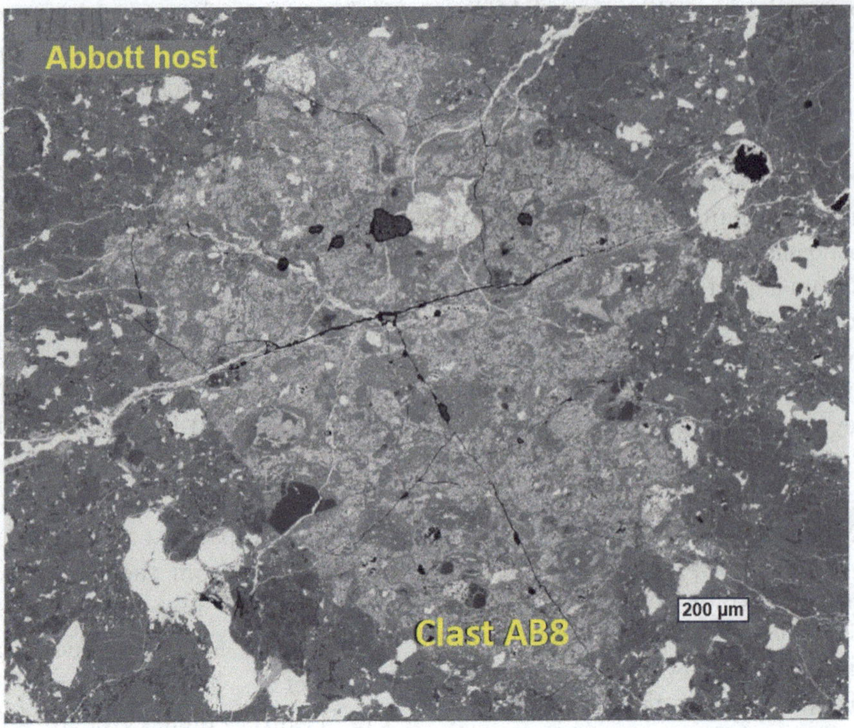

Fig. 11.80 Clast AB8 from the Abbott H-chondrite regolith breccia. This clast is a highly altered CM chondrite that forms a sharp boundary with the Abbott host. Light-colored phases in the clast consist of magnetite and sulfide. Fractures (black) and veins of limonite (FeO(OH)) (white) that formed during terrestrial weathering penetrate the clast from the host. White phases in the host are mainly metallic Fe-Ni and troilite. BSE image modified from Rubin and Bottke (2009).

(k) Calcic-plagioclase-bearing, differentiated, igneous clasts and chondrule-like objects are found in several type-3 and type-4 OC (e.g., Hutchison and Graham, 1975; Gibson et al., 1977; Fodor and Keil, 1978; Hutchison, 1982; Grossman and Rubin, 1986; Kennedy et al., 1992; Misawa et al., 1992; Bridges et al., 1995). Some are of basaltic or noritic composition; one consists mainly of feldspar and nepheline. These clasts could be xenolithic achondrite projectile fragments, impact-melt-rock clasts of host chondrite material, devolatilized chondrules, or fragments of melt sheets derived from large impact craters on chondritic asteroids. Different clasts may have different origins.

11.14 Agglutinates

Agglutinates are irregularly shaped, vesicular, glass-bonded aggregates of rock, glass, and mineral fragments. Differently sized fractions of lunar soils contain 9–55 vol.% agglutinates (Heiken, 1975), and lunar regolith breccias average 22 vol.% agglutinate glass (Drozd et al., 1976). In contrast, agglutinates are extremely rare in ordinary chondrites (probably ≤0.01 vol.%) – verified agglutinates occur in the Fayetteville H-chondrite regolith breccia (Kerridge and Kieffer, 1977; Rubin, 1982; Taylor, 1982) (Fig. 11.81).

Micrometeorite impact into regolith causes the formation of agglutinates (McKay et al., 1972a); thus, glass in lunar agglutinates tends to have the bulk chemical composition of the soil in which the agglutinates were formed (Marvin et al., 1971; Taylor et al., 1972, 1978; Gibbons et al., 1976; Hu and Taylor, 1977). The bulk composition of brown glass in the Fayetteville agglutinate in Fig. 11.81 is very similar to that of the silicate-normalized portion of H-group chondrites (such as the Fayetteville host), except for depletions in Na_2O, K_2O, and P_2O_5 and an enrichment in FeO (Rubin, 1982). These differences are probably due to the selective volatilization of Na, K, and P during impact melting and the inadvertent analysis of

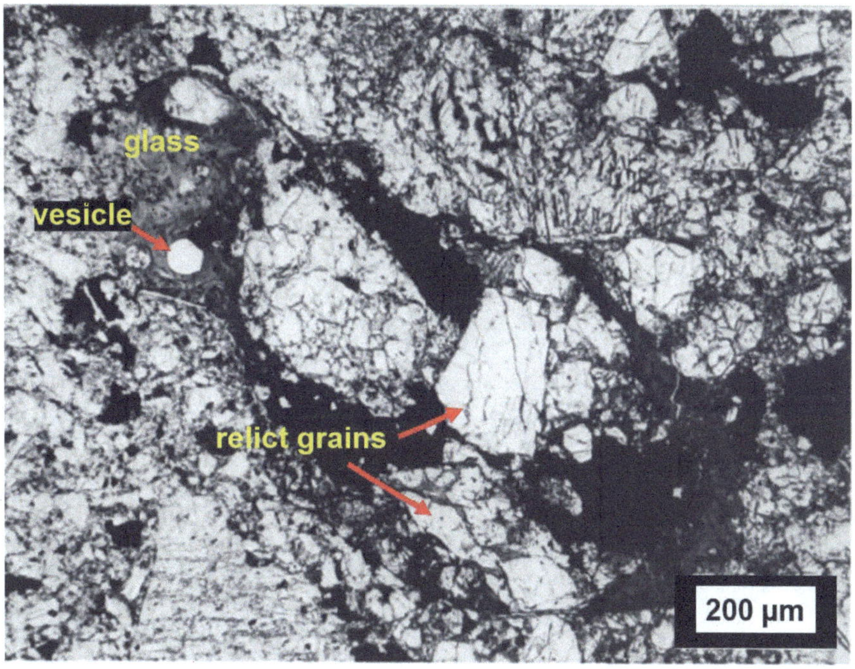

Fig. 11.81 A rare 0.4×1.3-mm agglutinate in the Fayetteville H-chondrite regolith breccia consisting of flow-banded, vesicular brown glass, grains of olivine and low-Ca pyroxene with resorbed margins, and droplets of metallic Fe-Ni and troilite (black). Object discovered by G. J. Taylor. Transmitted light image modified from Rubin (1982, 1985).

tiny metallic Fe-Ni and Fe-sulfide droplets within the agglutinate glass.

The modal abundance of agglutinates in a regolith is a function of the total micrometeorite bombardment and is a measure of regolith maturity. The low abundance of agglutinates in OC regolith breccias relative to lunar regolith breccias indicates that OC regoliths were less mature when the breccias were consolidated than the present-day lunar regolith (e.g., Kerridge and Kieffer, 1977; Taylor, 1982; McKay and Basu, 1983). This is consistent with other indicators of regolith maturity including the lower abundances in OCs (relative to lunar regolith) of impact glass, melt-rock clasts, glass spherules, particles with micrometeorite craters, and grains rich in solar-flare particle tracks, as well as the lower amounts of solar-wind-implanted rare gases.

Factors that may account for the low apparent abundance of agglutinates in ordinary-chondrite regolith breccias include the following:

(1) There is a factor-of-three lower micrometeorite flux in the vicinity of the asteroid belt relative to that at 1 AU (e.g., Langevin and Maurette, 1981), diminishing agglutinate production.

(2) The average relative velocity of micrometeorite impacts into asteroidal regoliths is lower than for the Moon (~5 km s-1 vs. 15–20 km s-1), causing less melting (Langevin and Maurette, 1981).

(3) There is greater dilution of agglutinates in asteroidal regoliths due to fresh ejecta from large craters (because ejecta travel farther on low-surface-gravity asteroids) (Housen et al., 1979).

(4) Asteroid surfaces are dominated by spallation processes that cause little melting of surficial materials (Hörz and Schaal, 1981).

(5) The ultramafic composition of ordinary chondrites (and their correspondingly higher melting temperatures) compared to basaltic lunar mare materials inhibit agglutinate production in OC.

(6) Regolith breccias (lunar and chondritic) are less mature than present-day surface soils (McKay et al., 1986, 1989) because breccias record an earlier epoch of regolith maturity than continuously maturing soils. Thus, chondritic meteorite breccias are presumably less mature than present-day chondritic regoliths (and present-day lunar soils).

(7) Lunar-meteorite regolith breccias are less porous than Apollo regolith breccias (1–11% vs. 25 ± 7%; e.g., Warren, 2001) probably because a high degree of compaction is necessary for a lunar regolith breccia

to survive launch from the Moon, a sojourn in interplanetary space, and traversal of Earth's atmosphere. There appear to be lower abundances of agglutinates in lunar-meteorite regolith breccias than in Apollo regolith breccias (McCay, 1991), perhaps due to compaction and squashing of agglutinates within meteorites into unrecognizable fragments (Warren, 2001). Nevertheless, Warren et al. (1983) observed abundant squashed agglutinate material in the ALH 81005 anorthositic lunar breccia. This suggests that an additional potential reason for the low abundance of agglutinates in meteorite regolith breccias (including OC regolith breccias) is that any agglutinates that may have been present initially were crushed into unrecognizability.

12

Mineralogy of Ordinary Chondrites

The mineralogy of meteorites, including ordinary chondrites, was most recently reviewed by Rubin and Ma (2021). In that work, comparisons are made between minerals in OC and those in other meteorite groups.

12.1 Presolar Grains

Presolar grains occur in primitive ordinary, carbonaceous, and enstatite chondrites (as well as in interplanetary dust particles [IDPs] and micrometeorites). Nittler et al. (1994) found that presolar oxide grains in H/L3.6 Tieschitz exhibit a wide range of O-isotopic compositions, indicating derivation from multiple stellar sources of differing mass and composition. Seifert et al. (2023) concluded that the four presolar olivine and pyroxene grains they analyzed from L3.05 MET 00526 and L3.05 QUE 97008 were derived from a type II supernova and from the circumstellar envelope of a red giant branch (RGB) star or an asymptotic giant branch (AGB) star.[1]

The broad similarities of presolar grains in different meteorite groups in mineralogy, stoichiometry, structure, crystallinity, chemical composition, and isotopic composition suggest there was a roughly homogeneous distribution of these grains throughout the solar nebula, presumably inherited from the parental molecular cloud (e.g., Barosch et al., 2022; Seifert et al., 2023). These are the oldest minerals within meteorites (e.g., Clayton and Nittler, 2004; Lodders and Amari, 2005); many predate the Solar System by hundreds of millions of years, some by more than a billion years (e.g., Gyngard et al., 2014; Heck et al., 2020).

The phases identified within presolar grains in primitive (mainly carbonaceous) chondrites (e.g., Zinner, 2005, 2014) include C polymorphs (diamond, graphite), moissanite (SiC), cohenite ($(Fe,Ni)_3C$), Ti-, Mo-, Zr- and Ru-rich carbides, metallic iron, Ni-bearing kamacite, nierite (Si_3N_4), corundum (Al_2O_3), spinel ($MgAl_2O_4$), hibonite ($CaAl_{12}O_{19}$), TiO_2 (probably rutile), magnesiochromite ($Mg(Cr,Al)_2O_4$), silicates (predominantly olivine, low-Ca pyroxene, Al-rich silicate, and silicaceous glass), and organic compounds. Although most presolar silicates are magnesian, some have FeO/MgO ratios >1. Diamonds are typically euhedral and ~2 nm in size; graphite grains range from 1–20 μm.

Although a minority of the micrometer-sized alumina (Al_2O_3) grains in LL3.01 Semarkona, LL3.15 Bishunpur, and H3.10 Roosevelt County 075 examined by Takigawa et al. (2014) have smooth surfaces, most have irregular rough surfaces. The proportion of amorphous or poorly crystalline alumina grains is higher among presolar Al_2O_3 grains than those formed in the Solar System. This suggests that the initial morphology of presolar corundum grains was smooth and that irradiation in the

[1] Type II supernovae (SNII) are massive stars (8 to ~50 solar masses), that, at the end of their lives, undergo rapid collapse and cataclysmic explosion. They exhibit spectral lines of hydrogen. Red giant branch (RGB) stars are of low-to-intermediate mass and have recently evolved off the main-sequence (central swath) of the Hertzsprung–Russell (H–R) diagram. They have a He core surrounded by a mantle of H that is fusing into He. Asymptotic giant branch (AGB) stars are of low or intermediate mass (~0.5 to 8 solar masses) and have evolved into bright red giant stars, commonly with luminosities several thousand times that of the Sun. They occupy a region of the H–R diagram just above (i.e., at somewhat higher visual magnitudes than) the RGB stars. AGB stars have a layered onion-like structure with a C- and O-rich core surrounded by successive shells of He fusing into C, H fusing into He, and a thick outer shell of similar composition to main-sequence stars.

interstellar medium or the protosolar disk damaged the surfaces of the grains and tended to degrade their crystallinity.

Barosch et al. (2022) identified three types of presolar grains in the fine-grained matrices of LL3.01 Semarkona, L/LL3.05 MET 00526, and L3.00 NWA 8276 by their anomalous O- and C-isotopic compositions. The average matrix-normalized modal abundances for Semarkona, MET 00526, and NWA 8276, respectively are: 151, 55, and 12 µg/g O-anomalous silicate and oxide grains; 53, 22, and 15 µg/g SiC; and 56, 3, and 1 µg/g carbonaceous grains (graphite and organics) (1 µg/g = 1 ppm).

An unusual composite presolar grain (dubbed F2-8) from Semarkona was studied with transmission electron microscopy (TEM) by Singerling et al. (2023). The micrometer-sized, amoeboid-shaped grain likely formed in an M-type AGB star envelope. F2-8 contains forsterite, Mg-Al spinel, Ca-pyroxene (a non-equilibrium condensation product), alabandite-oldhamite (an equilibrium condensation product), and magnetite (produced by parent-body aqueous alteration).

The presolar olivine and pyroxene grains studied by Seifert et al. (2023) are FeO-rich. This ferroan composition probably resulted from parent-body alteration of the fine-grained matrices of these type-3 OCs. Singerling et al. (2022) analyzed presolar grains within the matrix of Semarkona by TEM. They found the SiC and spinel grains to be stoichiometric and well crystallized, but most silicate grains to be non-stoichiometric and poorly crystalline. The silicates have a large range (i.e., 15–98) in Mg number (defined as ((Mg/(Mg + Fe)) × 100)) and have variable amounts of MgO, SiO_2, Al_2O_3, and Cr_2O_3. Although Singerling et al. (2022) attributed these compositional variations and low crystallinity to condensation under non-equilibrium conditions, parent-body aqueous alteration could be partly responsible.

12.2 CAIs and AOAs

The two principal varieties of refractory inclusions in chondritic meteorites, Ca-Al-rich inclusions (CAIs) and amoeboid olivine aggregates (AOAs), are very rare in type-3 ordinary chondrites: ~0.01 vol.% (Dunham et al., 2023) and 0.0005 – 0.001 vol.% (inferred from Itoh et al., 2007), respectively. Primary phases in OC CAIs include spinel ($MgAl_2O_4$), hercynite ($FeAl_2O_4$), hibonite ($CaAl_{12}O_{19}$), diopside ($CaMgSi_2O_6$), olivine ($(Mg,Fe)_2SiO_4$)), anorthite ($CaAl_2Si_2O_8$), corundum (Al_2O_3), and melilite ($(Ca,Na)_2(Al,Mg)(Si,Al)_2O_7$) (e.g., Table 11.2); secondary phases formed during parent-body aqueous alteration include nepheline ($(Na,K)AlSiO_4$) and sodalite ($Na_4(Si_3Al_3)O_{12}Cl$).

Because ordinary-chondrite CAIs are probably from the same broad population as those in carbonaceous chondrites (e.g., McKeegan et al., 1998; Huss et al., 2001), the primary mineral phases in carbonaceous-chondrite CAIs are likely to occur (if rarely) in unaltered CAIs in OCs. In addition to the phases listed above, potential OC CAI primary phases include grossite ($CaAl_4O_7$) and perovskite ($CaTiO_3$). Ultrarefractory phases identified in carbonaceous-chondrite CAIs that could be present in extremely rare OC CAIs include addibischoffite ($Ca_2Al_6Al_6O_{20}$), allendeite ($Sc_4Zr_3O_{12}$), davisite ($CaScAlSiO_6$), grossmanite ($CaTi^{3+}AlSiO_6$), hexamolybdenum (Mo,Ru,Fe), kangite ($(Sc,Ti,Al,Zr,Mg,Ca,\square)_2O_3$), krotite ($CaAl_2O_4$), kaitianite ($Ti^{3+}{}_2Ti^{4+}O_5$), kushiroite ($CaAlAlSiO_6$), machiite ($Al_2Ti_3O_9$), panguite ($(Ti^{4+},Sc,Al,Mg,Zr,Ca)_{1.8}O_3$), paqueite ($Ca_3TiSi_2(Al,Ti,Si)_3O_{14}$), rubinite ($Ca_3Ti^{3+}{}_2Si_3O_{12}$), tistarite ($Ti_2O_3$), and warkite ($Ca_2Sc_6Al_6O_{20}$).

Primary phases in the three known OC AOAs from LL3.01 Semarkona include magnesian olivine, fassaite (Al-Ti diopside), anorthite, and metallic Fe-Ni. No alteration phases were identified. Dunham et al. (2023) suggested that seven of the OC refractory inclusions in their study could be related to AOAs; these inclusions contain olivine and spinel. Although it has not yet been reported, perovskite is probably present in some OC AOAs.

12.3 Unaltered Type-3 Ordinary Chondrites

The principal primary phases in unaltered portions of the least equilibrated ordinary chondrites are olivine and low-Ca pyroxene (major clinopyroxene and rare orthopyroxene) with highly variable FeO/(FeO+MgO) ratios, metallic Fe-Ni (kamacite and taenite), troilite; and (nearly exclusively within chondrules) crystallites of Ca-pyroxene, rare grains of pigeonite, and rare tiny grains of merrillite. In some Type-II chondrules there are small, rare grains of primary chromite. Many chondrules also contain isotropic glassy silicaceous mesostasis; in some chondrules, only small portions of the mesostasis are unaltered. Rare CAIs, AOAs, and tiny presolar grains all have their characteristic allotments of primary minerals.

Jarosewich (1990) reported the bulk chemical composition of LL3.01 Semarkona, the least equilibrated ordinary chondrite. That analysis, combined with a (CIPW) norm calculation, yields its normative mineralogy (Table 12.1). The analysis has <45% SiO_2 and the normative minerals are dominated by olivine (~53 wt.%), pyroxene (~25 wt.%), and plagioclase (~10 wt.%). These characteristics match those of peridotite – an ultramafic rock that dominates the Earth's upper mantle (e.g., Palme and O'Neill, 2014). However, unlike terrestrial peridotites, type-3 OCs contain abundant chondrules, metallic Fe-Ni, and troilite.

The actual minerals in Semarkona differ from the normative minerals: Semarkona lacks coarse grains of plagioclase, ilmenite, and apatite; the chemical components of these phases are mostly within glassy chondrule mesostases. Coarse graphite is also absent from Semarkona; C occurs within organic matter in the fine-grained matrix (e.g., Alexander et al., 2007) and, in some type-3 OC, within C-rich aggregates (e.g., Scott et al., 1981a, 1988a; McKinley et al., 1981; Brearley, 1990) and/or C-rich clasts (Semenenko and Girich, 1995; Semenenko et al., 2004).

Table 12.1 *Bulk chemical composition of LL3.01 Semarkona and its normative mineralogy. and its normative mineralogy.*

Chemical component	wt.%	CIPW normative minerals	wt.%
SiO_2	40.16	Apatite	0.58
TiO_2	0.10	Ilmenite	0.19
Al_2O_3	2.23	Orthoclase	0.59
Cr_2O_3	0.49	Albite	8.12
FeO	17.57	Anorthite	1.48
MnO	0.35	Diopside	5.70
MgO	24.86	Hypersthene	19.27
CaO	2.04	Olivine	52.69
Na_2O	0.96	Chromite	0.72
K_2O	0.10	Metallic Fe-Ni	3.26
P_2O_5	0.24	Troilite	5.32
H_2O^+	1.22	Graphite	0.57
H_2O^-	0.40	–	–
Fe metal	2.05	–	–
Ni	1.16	–	–
Co	0.05	–	–
FeS	5.32	–	–
C	0.57	–	–
Total	99.87	Total	98.49

Wet-chemical analysis of Semarkona from Jarosewich (1990). Chromite, metallic Fe-Ni, troilite, and graphite were added to the CIPW normative minerals. The low total in the normative analysis is due to H_2O in the chemical analysis.

Olivine: Olivine is the dominant mineral in ordinary chondrites. It occurs mainly as phenocrysts in most porphyritic chondrules (e.g., Figs. 11.3, 11.4), as microphenocrysts in GOP chondrules (Fig. 11.13), as plates in BO chondrules (Figs. 11.8, 11.9), and as isolated grains in the matrix (e.g., Fig. 11.3) that probably were derived predominantly from fragmented PO or POP chondrules. In the least equilibrated OC, olivine and low-Ca pyroxene grains have heterogeneous compositions (Fig. 12.1). Grains of different composition occur adjacent to each other.

Low-Ca pyroxene: Type-IB (low-FeO porphyritic pyroxene) and IAB (low-FeO porphyritic olivine-pyroxene) chondrules in unequilibrated type-3 OC (e.g., type 3.0–3.5) contain low-Ca clinopyroxene phenocrysts exhibiting polysynthetic twinning on (100), inclined extinction, and shrinkage cracks perpendicular to the twinning planes (Fig. 12.2). During chondrule quenching from high temperatures, these low-Ca clinopyroxene phenocrysts formed as a metastable phase by inversion from magnesian protoenstatite (the stable polymorph between 1000°C and 1557°C) (Smyth, 1974) with an accompanying change in volume. As is the case for olivine, low-Ca pyroxene in the least equilibrated OC have heterogeneous compositions (Fig. 12.1). In most cases, low-Ca pyroxene contains higher abundances of incompatible elements than coexisting olivine (Xiang and Jin, 2024).

Ca-pyroxene: In type 3.0–3.5 OC, Ca-pyroxene is present mainly in chondrules (both Type-I and Type-II; e.g., Jones, 1994, 1996), where it occurs

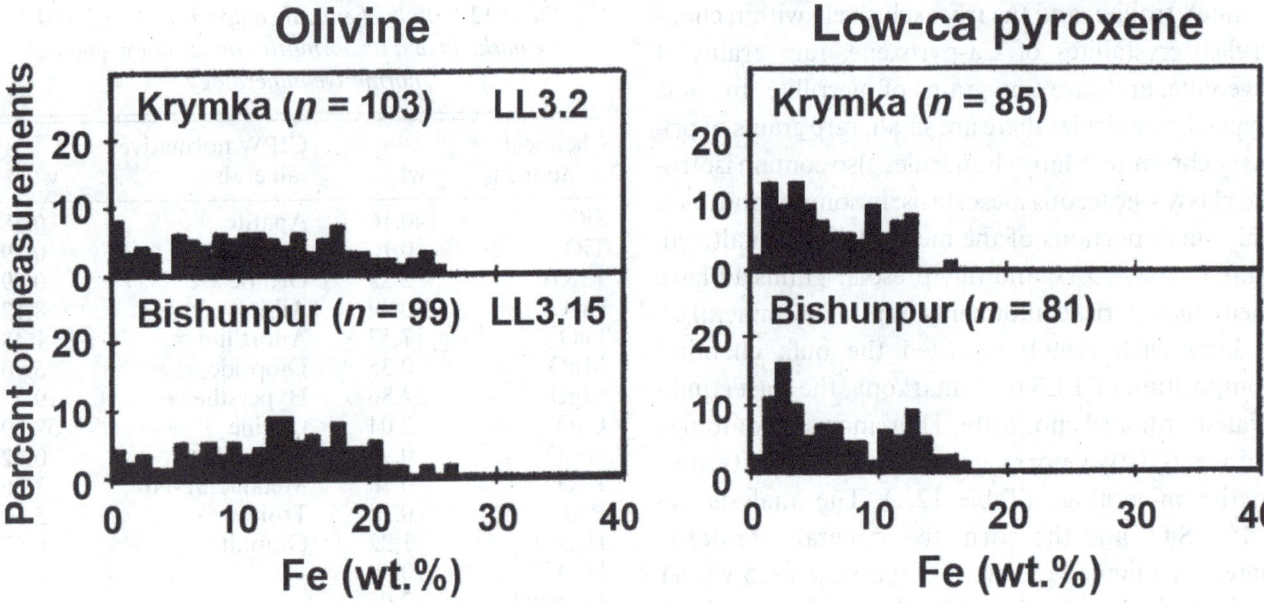

Fig. 12.1 Compositional distributions (wt.% Fe) of olivine and low-Ca pyroxene grains in two highly unequilibrated LL3 chondrites. The olivine distribution is fairly flat and extends from 0 to ~25 wt.% Fe; the low-Ca pyroxene distribution is much narrower with a broad peak between 1 and ~15 wt.% Fe. Diagram modified from Dodd et al. (1967).

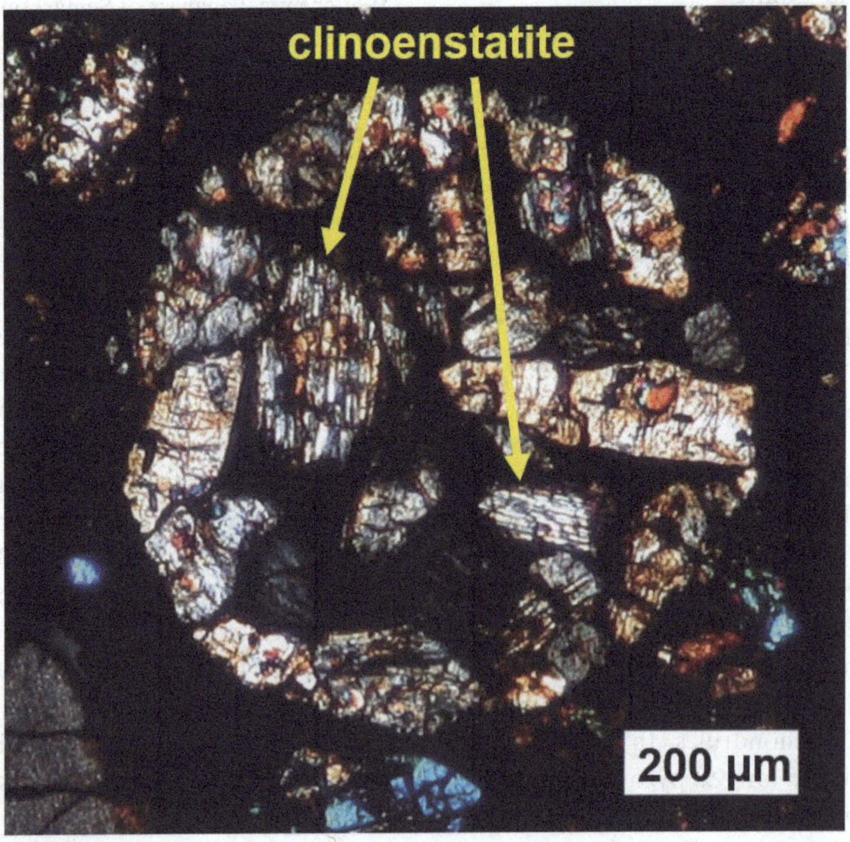

Fig. 12.2 Type-IB porphyritic pyroxene chondrule in LL3.10 NWA 3127 showing low-Ca clinoenstatite phenocrysts exhibiting polysynthetic twinning. Many of the phenocrysts along the chondrule margin have their long axes parallel to the margin. X-nicols image courtesy of John Kashuba.

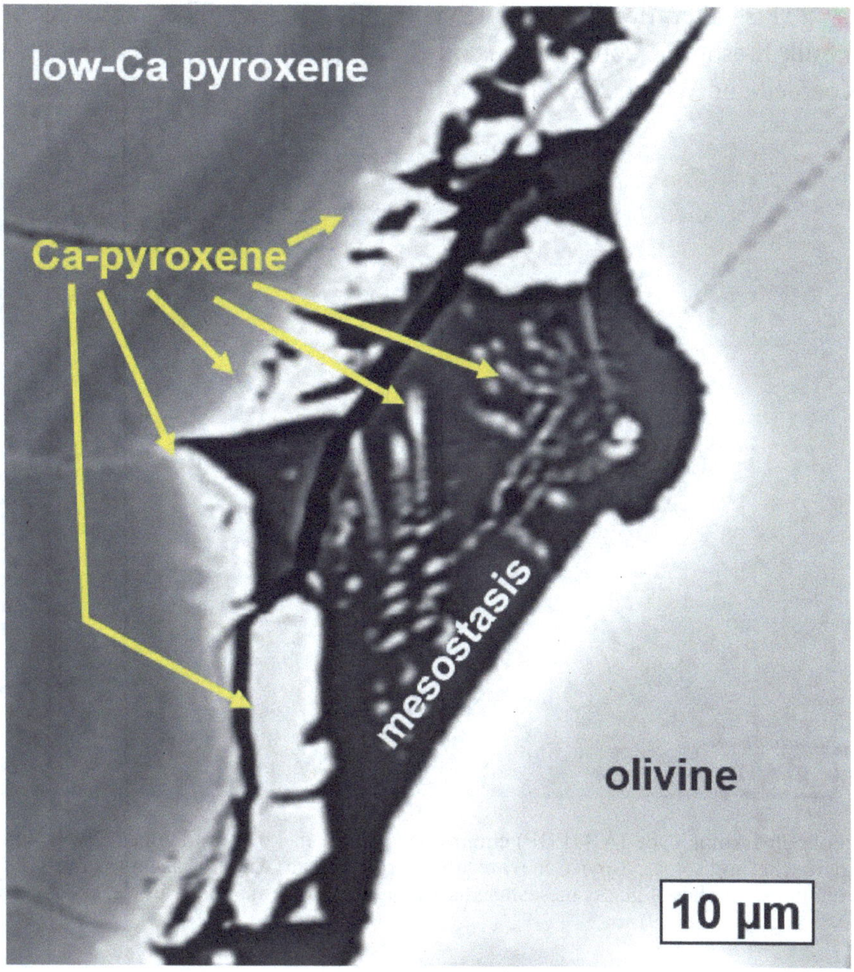

Fig. 12.3 Portion of Type-IIAB (FeO-rich POP) chondrule E4m from LL3.01 Semarkona showing Ca-pyroxene (white) overgrowths on coarse low-Ca pyroxene phenocrysts. Acicular Ca-pyroxene crystallites also occur in the mesostasis (very dark gray). Backscattered electron (BSE) image modified from Baecker et al. (2017).

as small crystallites within the mesostasis and as overgrowths on low-Ca pyroxene phenocrysts (Fig. 12.3). Ca-pyroxene also occurs in rare OC CAIs.

Plagioclase: Relatively coarse grains of plagioclase are very rare in type 3.0–3.5 OC. However, plagioclase does occur as anorthite phenocrysts in rare Al-rich chondrules (Russell et al., 2000; Huss et al., 2001), small oligoclase grains within some igneous rims around OC chondrules (Rubin, 1984b), and anorthite grains within very rare CAIs and AOAs.

Lewis and Jones (2019) observed plagioclase grains within the mesostases of 18% of Semarkona chondrules. The grains are generally lath-shaped, 1–8 µm wide and 30–60 µm long (Fig. 12.4). They have highly variable CaO/Na_2O ratios ($Ab_{0.3–95.7}$) and little K_2O ($Or_{0.2–2.4}$). Lewis and Jones (2019) concluded that these plagioclase grains are primary igneous phases based on grain morphology and composition. Plagioclase grains in Type-I (FeO-poor) chondrules tend to be more calcic; plagioclase grains in Type-II (FeO-rich) chondrules tend to be more sodic.

However, the plagioclase grains are located within altered mesostasis, adjacent to porous phyllosilicates and voids (Fig. 12.4). Grossman and Brearley (2005) attributed the juxtaposition of albite and K-rich glassy mesostasis to parent-body metasomatism and concluded that the crystallization of albite was controlled by the availability of nucleation sites. Because the fraction of chondrules containing albite increases in type-3 OC with metamorphic subtype (Grossman and Brearley, 2005), it seems more likely that most of the plagioclase in type-3 OC (including LL3.01 Semarkona) are secondary phases. (This alteration also partly masked the initial bulk K-isotopic compositions of OC chondrules [Koefoed and Wang, 2025].)

Phosphate: The mesostases in Type-II (FeO-rich porphyritic) chondrules in LL3.01 Semarkona average ~1.3 wt.% P_2O_5 and contain accessory

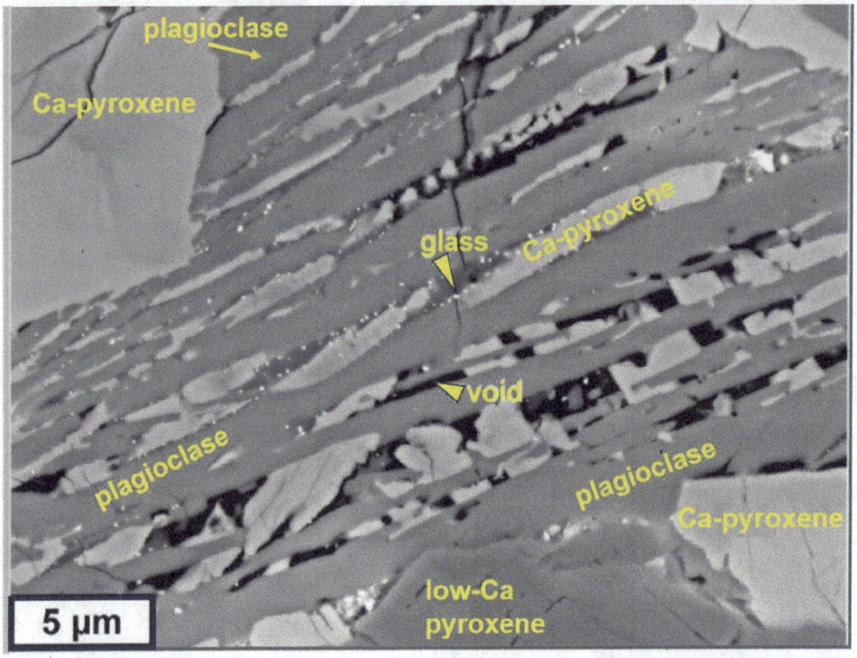

Fig. 12.4 Portion of Semarkona Type-IAB (POP) chondrule Ch20 from Lewis and Jones (2019). Interstitial to coarser grains of low-Ca pyroxene (and some Ca-pyroxene) are laths of plagioclase (An_{66}) adjacent to grains of Ca-pyroxene. Also present are some unaltered patches of glassy mesostasis as well as voids within altered and leached mesostasis. BSE image modified from Lewis and Jones (2019).

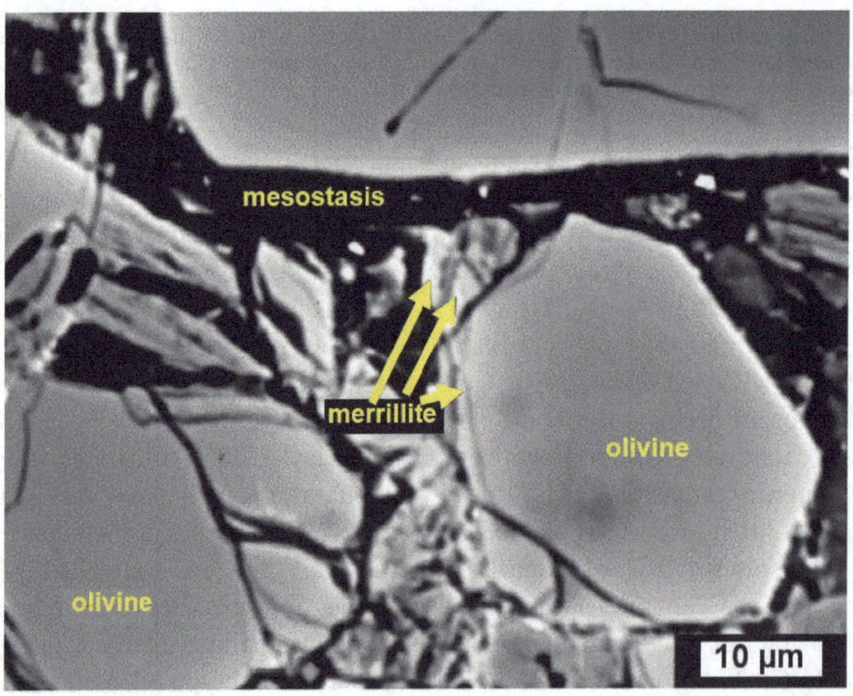

Fig. 12.5 Tiny merrillite grains (light gray) flanking a small olivine phenocryst in Semarkona chondrule H5k. Mesostasis appears black. BSE image modified from Rubin and Baecker (2021).

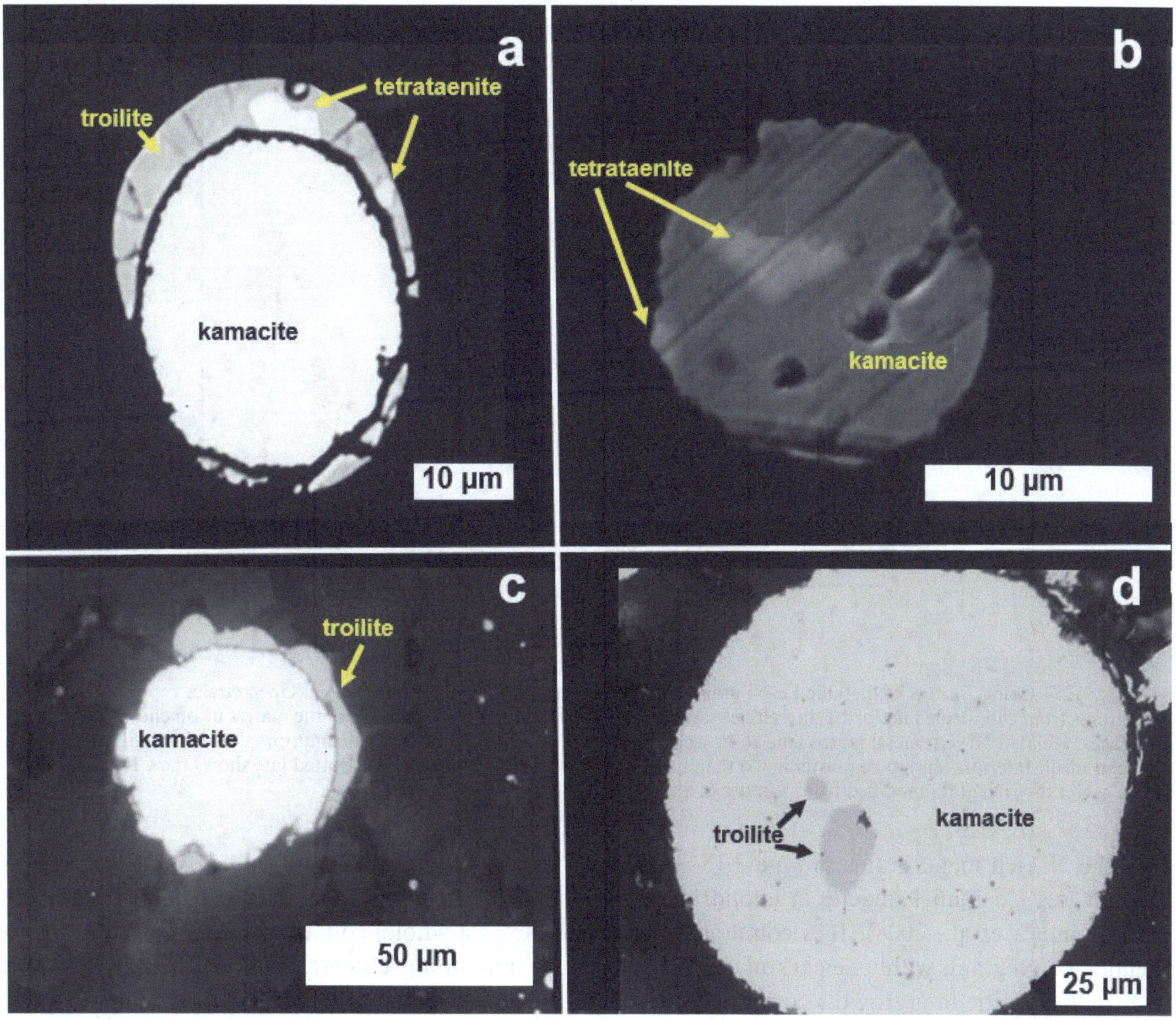

Fig. 12.6 Metal-troilite assemblages in type-3 ordinary chondrites. (a) Several small tetrataenite grains (white) within the troilite rim (light gray) around kamacite in an opaque nodule in a Type-I chondrule in LL3.01 Semarkona. BSE image. (b) An opaque spherule in a Type-I chondrule in LL3.10 Piancaldoli with a thin troilite rind (not visible). Tetrataenite (light gray) occurs in a thin patch at the left center edge and as a triangular patch near the center; the remaining portion of the spherule is kamacite (dark gray). BSE image. (c) Typical opaque spherule inside Type-IA chondrule 7 (olivine, Fa2.6) in Semarkona. The troilite (light gray) forms a relatively thin partial rind around the kamacite spherule (white). Reflected light image. (d) Polygonal troilite grain (light gray) within a kamacite-rich spherule (white) in Type-IA chondrule 16 (olivine, Fa1.4) in LL3.15 Bishunpur. Reflected light image. Images modified from Rubin et al. (1999).

2–10 µm grains of merrillite ($Ca_9MgNa(PO_4)_7$) (Grossman and Brearley, 2005; Rubin and Baecker, 2021). These small phosphate grains appear to have nucleated at the surface of olivine grains (Fig. 12.5) and crystallized within the mesostasis during chondrule cooling.

Metallic Fe-Ni: The two principal metallic Fe-Ni minerals in type-3 OC are kamacite (low-Ni metallic iron) and taenite; rare tetrataenite is also present (e.g., Fig. 12.6). Metallic Fe-Ni grains are present in chondrule interiors as well as at chondrule surfaces and as isolated grains in the matrix. Isolated Ni-rich metal grains in the matrix and at chondrule surfaces tend to be richer in Co than kamacite in chondrule interiors; in contrast, Ni-rich metal grains in chondrule interiors tend to be poorer in Co than the kamacite in chondrule interiors (Fig. 12.7). This relationship holds only for the least metamorphosed

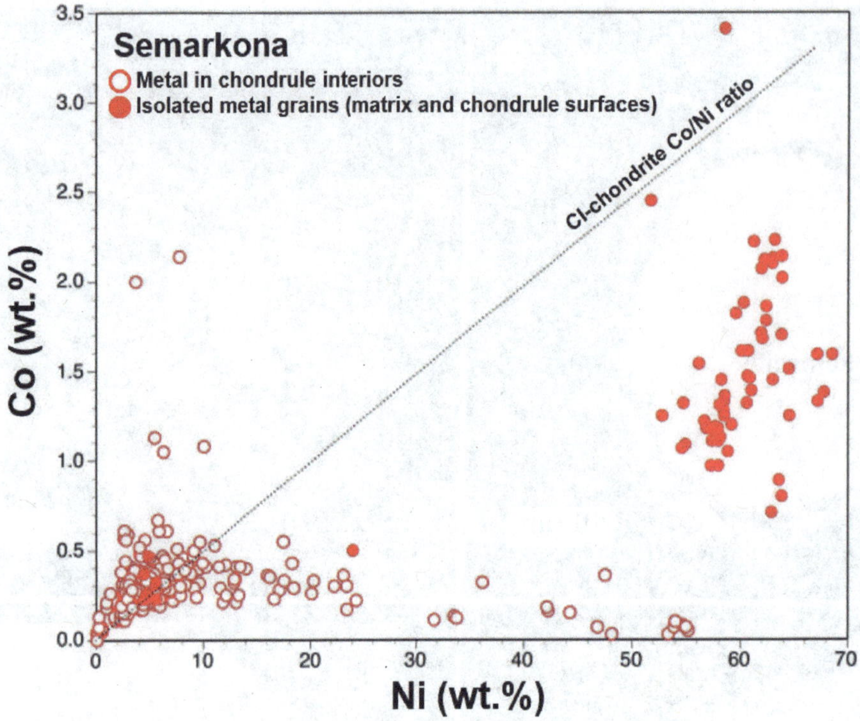

Fig. 12.7 Compositions of metallic Fe-Ni grains in LL3.01 Semarkona: Co vs. Ni (wt.%). Open circles represent metal grains from chondrule interiors. Filled circles represent isolated metal grains occurring in the matrix or on chondrule surfaces. Isolated Ni-rich metal grains tend to be richer in Co than kamacite in chondrule interiors; Ni-rich metal grains in chondrule interiors tend to be poorer in Co than kamacite in chondrule interiors. The dotted line shows the CI-chondrite Co/Ni ratio. Diagram modified from Kimura et al. (2008).

OC; the Ni-rich metal grains in type 3.15–3.9 OC contain less Co than kamacite in chondrule interiors (Kimura et al., 2008). It is common for the metallic Fe-Ni grains within Semarkona chondrules to exhibit plessitic intergrowths comprised of submicrometer patches of kamacite and Ni-rich metal (Fig. 12.8). In OC of higher subtype, the metallic Fe-Ni grains consist of coarser aggregates of kamacite and Ni-rich metal (Kimura et al., 2008).

Sulfide: The main sulfide phase in ordinary chondrites is troilite (FeS). Rubin et al. (1999) examined 226 intact chondrules of all textural types from eight unshocked to weakly shocked OC falls of low petrologic type; 68 chondrules were from LL3.01 Semarkona. They proposed that there was a high probability that specific troilite grains within a chondrule were primary (i.e., present among chondrule precursor components) if they were completely embedded within a mafic silicate phenocryst, were located within one-half radius of the apparent chondrule center, and were part of an opaque assemblage with an igneous texture (Fig. 12.6a,c,d). Based on these criteria, 13% of the chondrules in Semarkona (and in the suite as a whole) contain primary troilite. Most of the remaining chondrules contain troilite that is probably primary (rather than metamorphic) but does not meet all three criteria.

Troilite occurs next to tetrataenite in some opaque spherules within Type-I chondrules in Semarkona, implying that the Ni required to form the tetrataenite came from the troilite. Troilite can accommodate 5 mg/g Ni at high temperatures (>1170 K) but much less Ni at lower temperatures (Kullerud, 1963; C. Ma et al., 1998); because this is far higher than the metamorphic temperature inferred for Semarkona (~670 K; McSween et al., 1988), the troilite must be primary. Primary troilite fitting the three criteria occurs in a smaller fraction of Type-I than Type-II chondrules in Semarkona (9% vs. 33%).

Some oxidized L3 and LL3 chondrites contain pentlandite ($(Fe,Ni)_9S_8$): for example, ALHA77011 (L3.5), ALHA77278 (LL3.7), Gandom Beryan 011 (LL3), Ngawi (LL3.2/3.7), NWA 6722 (L3.5), NWA

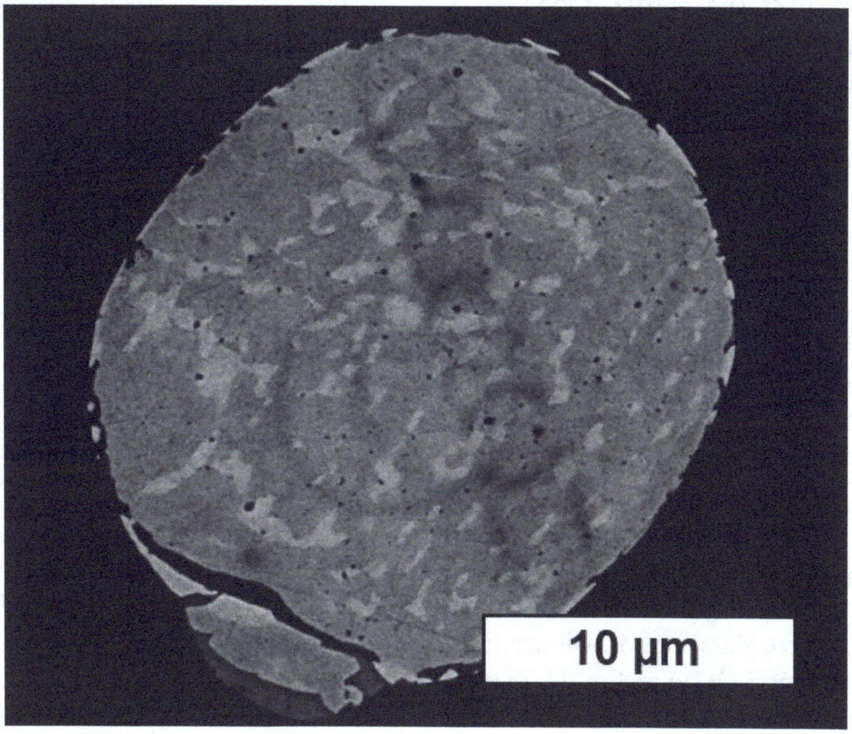

Fig. 12.8 A metallic Fe-Ni spherule with a plessitic intergrowth texture from a chondrule in LL3.01 Semarkona. The Ni-rich metal (bright-shaded phase) is homogeneously distributed. Field emission scanning electron microscope (FE-SEM) image modified from Kimura et al. (2008).

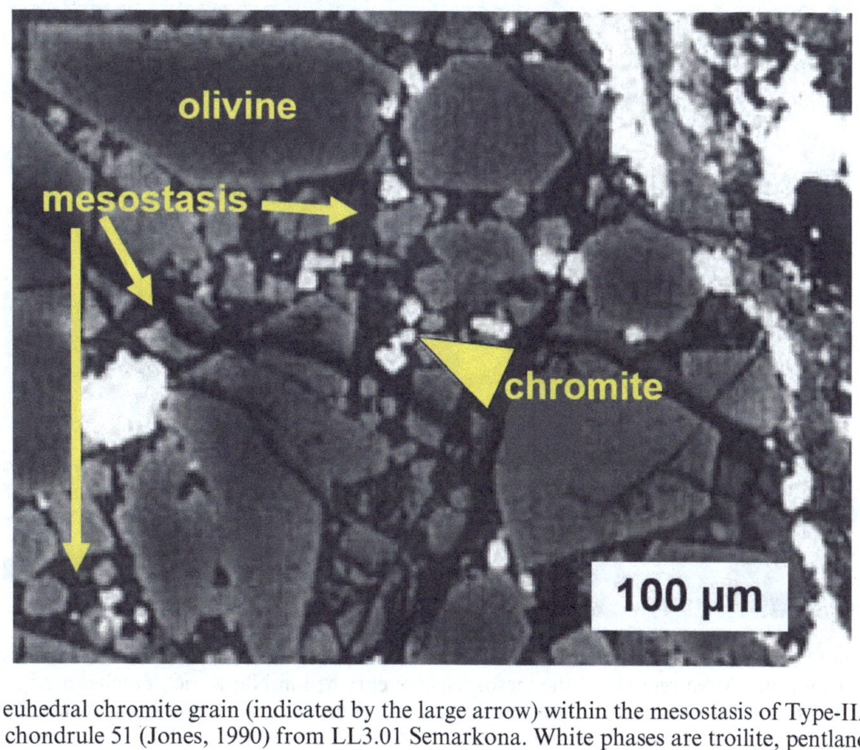

Fig. 12.9 Small euhedral chromite grain (indicated by the large arrow) within the mesostasis of Type-IIA (FeO-rich porphyritic olivine) chondrule 51 (Jones, 1990) from LL3.01 Semarkona. White phases are troilite, pentlandite, and metallic Fe-Ni. BSE image modified from Jones (1990).

7477 (L3.9), NWA 8258 (L3), NWA 13920 (LL3-6), NWA 14454 (LL3), NWA 13182 (LL3), NWA 14745 (LL3), NWA 15869 (L3), and Semarkona (LL3.01). However, many of these OC also contain associated magnetite and carbide (Taylor et al., 1981; Krot et al., 1997b), suggesting that most of the pentlandite grains are probably products of secondary aqueous alteration. Primary pentlandite in OC may occur exclusively in the most oxidized LL chondrites, such as those with $Fa_{>31}$ mol% (e.g., LL5 Parambu, Fa_{32}; Fig. 5 of Rubin, 1990).

Chromite: FeO-rich porphyritic chondrules in type-3 OC contain some small euhedral chromite grains up to ~10 μm in size in the mesostasis (Fig. 12.9; Jones, 1990). These grains crystallized during chondrule cooling, consistent with a decrease in the Cr_2O_3 content of the olivine grains near their edges.

Mesostasis: Unaltered glassy mesostasis occurs as patches in primitive type-3 OC (e.g., Fig. 11.75 from LL3.01 Semarkona). The CIPW norm of unaltered glass in a BO chondrule from Semarkona (Rubin, 2013a) shows the mesostasis to be silico-feldspathic with a minor mafic component (Table 12.2).

Heterogeneous concentrations in Na_2O and K_2O in the mesostases of many type-3 OC result from parent-body alteration (Grossman and Brearley, 2005). The

Table 12.2 *Average composition of unaltered mesostasis (wt.%) within a BO chondrule from LL3.01 Semarkona and the corresponding CIPW normative mineral phases.*

Chemical component	wt.%	CIPW normative minerals*	wt.%
SiO_2	59.1	Ilmenite	0.68
TiO_2	0.70	Orthoclase	0.12
Al_2O_3	18.5	Albite	5.2
Cr_2O_3	0.30	Anorthite	47.7
FeO	0.25	Diopside	21.4
MnO	0.07	Hypersthene	2.3
MgO	4.9	Quartz	21.4
CaO	15.4	Titanite	0.84
Na_2O	0.61	Chromite	0.44
K_2O	<0.04	–	–
Total	99.8	Total	100.0
Number of analysis	82	–	–

* Chromite was added to the normative analysis; the calculation assumes the K_2O value is half the given upper limit.
Data from Rubin (2013a).

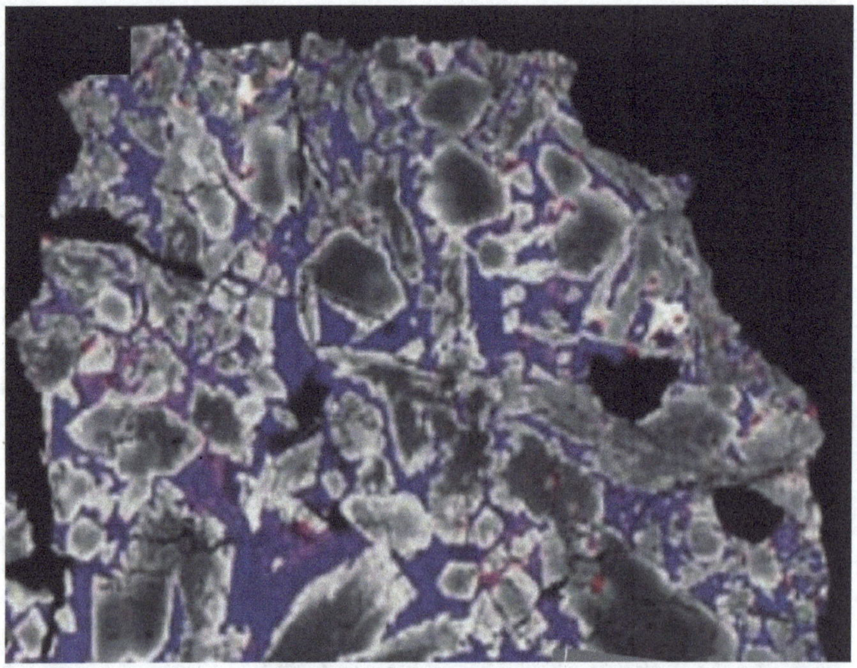

Fig. 12.10 Portion of Semarkona Type-II chondrule 7-4 from Grossman and Brearley (2005) showing inhomogeneous alkali-element distributions. Most regions of the mesostasis are enriched in Na; K-rich patches are heterogeneously distributed. Olivine (dark gray) is present mainly as euhedral and subhedral grains. Black-and-white BSE image overlain with x-ray maps of Na (blue) and K (red) in the mesostasis. Na-rich mesostasis patches predominate. Color-enhanced image modified from Grossman and Brearley (2005).

altered glass appears mottled and filled with voids (Figs. 11.75, 12.4). A portion of a Type-II chondrule from Semarkona illustrates the heterogeneous distribution of Na and K (Fig. 12.10); most of the mesostasis is sodic, but there are numerous potassic patches.

12.4 Olivine and Pyroxene of Unusual Composition

LIME olivine and low-Ca pyroxene: There is a population of olivine and low-Ca pyroxene grains with relatively low FeO and high MnO that occur mainly in IDPs, AOAs, and the matrices of type-3 ordinary chondrites (e.g., Klöck et al., 1989; Brearley, 1993; Krot et al., 2004b). These **l**ow-**i**ron, **m**anganese-**e**nriched (LIME) grains contain <5 wt.% FeO and have MnO contents ranging from ~0.70 up to ~5 wt.% (e.g., Table 12.3). LIME olivine and low-Ca pyroxene grains have much higher MnO/FeO ratios than common olivine or low-Ca pyroxene grains. In terrestrial rocks, orthopyroxene and olivine with high MnO are generally also rich in FeO; terrestrial rocks with low FeO and high MnO occur primarily in metamorphosed evaporites and calc-silicate deposits.

Two populations of olivine grains were identified by Klöck et al. (1989) in the matrix of LL3.01 Semarkona (Fig. 12.11): (1) FeO-rich olivines ($Fa_{2–30}$), commonly with FeO/MnO ratios of 20 to 50. These grains are similar in composition to chondrule olivine grains. (2) MgO-rich olivines ($Fa_{0.3–0.5}$) with FeO/MnO ratios of 0.5–1.0. Klöck et al. (1989) reported olivine of similar composition in the matrices of some carbonaceous chondrites.

Klöck et al. (1989) suggested that LIME grains formed in the nebula after forsterite (Mg_2SiO_4) condensed from nebular gas, followed by tephroite (Mn_2SiO_4) at ~1100 K (which formed a solid solution with forsterite). Condensation of olivine (which is Mg-rich) increased the Si/Mg ratio of residual nebula gas, facilitating the condensation of enstatite. Enstatite would have reacted with the gas and incorporated Mn.

Some LIME olivine and low-Ca pyroxene grains may have formed by a different mechanism. Rubin (1984a) reported numerous olivine and orthopyroxene grains with relatively low FeO/MnO ratios (7.7 and 0.33, respectively) within an igneous rim around a large porphyritic chondrule in CV3 Allende. Because the bulk enrichments of Cr, Na, and Mn in the igneous rim approximately mirror the depletions of these elements in bulk CV chondrites, Rubin (1984a) suggested that during flash-heating and cooling of the fine-grained chondrule-rim precursor

Table 12.3 *Compositions of selected olivine and low-Ca pyroxene grains from LL3.01 Semarkona (wt.%).*

–	Low-FeO olivine	High-FeO olivine	LIME olivine	Low-FeO low-Ca pyroxene	High-FeO low-Ca pyroxene	LIME low-Ca pyroxene
Source	1	2	3	1	4	5
SiO_2	42.2	39.6	41.72	58.7	56.9	58.27
TiO_2	–	–	–	0.07	0.02	–
Al_2O_3	–	–	–	0.45	0.08	1.21
Cr_2O_3	0.91	0.43	0.23	0.68	0.54	1.34
FeO	2.6	13.3	0.33	1.8	11.4	1.20
MnO	0.16	0.32	0.98	0.30	0.40	2.66
MgO	55.0	46.0	56.66	37.9	30.3	33.74
CaO	0.09	0.10	0.08	0.26	0.11	1.58
Total	100.5	99.75	100.0	100.2	99.8	100.0
Fa	2.6	13.9	0.33	–	–	–
Fs	–	–	–	2.6	17.3	1.9
Wo	–	–	–	0.5	0.2	3.2

1 = Jones (1994), Type-IAB chondrule 20; 2 = Jones (1990), Type-II chondrule A91; 3 = Klöck et al. (1989), Grain 3; 4 = Jones (1996) Type-II chondrule 70; 5 = Klöck et al. (1989), Grain 6.

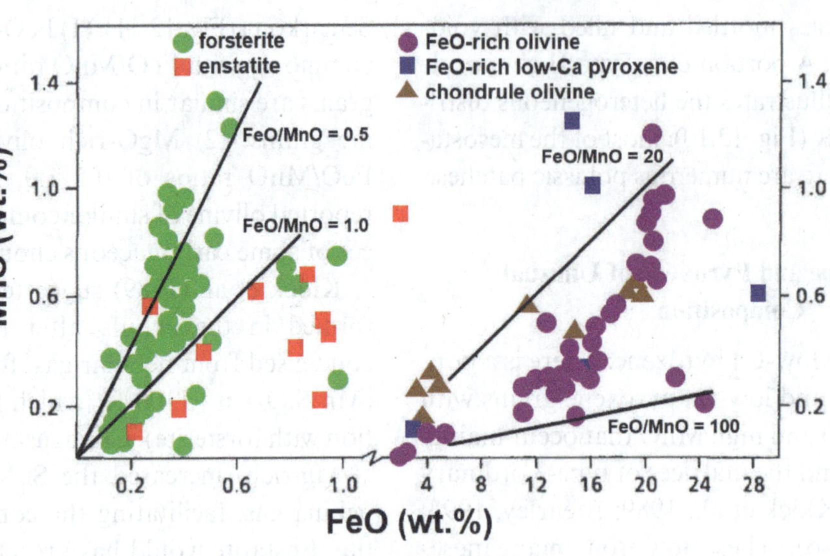

Fig. 12.11 Compositions of matrix olivine grains in LL3.01 Semarkona. There are two distinct populations: (1) ferroan olivine grains with high FeO/MnO ratios that are similar in composition to chondrule olivines and (2) magnesian olivine grains with low FeO/MnO ratios, similar in composition to some olivine grains in the matrices of CM2 Murchison and ungrouped carbonaceous chondrite EET 83226. Diagram modified from Klöck et al. (1989).

material, the rim acquired by condensation volatilized Cr, Na, and Mn that had been lost from chondrule precursors during chondrule-formation heating episodes. If this scenario is correct, the MnO-rich olivine and low-Ca pyroxene grains grew at elevated temperatures, probably ≥1000°C.

Chromian-manganoan augite: In his study of H/L3.6 Tieschitz, Hutchison (1987) discovered an angular fragment of a twinned clinoenstatite grain flanked by chromian-manganoan augite and surrounded by fine-grained opaque matrix material. The assemblage occurs within a large patch of white matrix. Because white matrix appears to be a product of parent-body thermal metamorphism (Section 11.8), the chromian-manganoan augite also likely formed at metamorphic temperatures.

Na_2O-bearing augite: Some low-FeO chondrules in type-3 OC contain Na_2O-bearing augite (Jones and Scott, 1989; Jones, 1994; Rubin et al., 1999). A typical grain from LL3.15 Bishunpur contains (in wt.%): 57.3% SiO_2, 2.4% Al_2O_3, 0.86% FeO, 0.81% MnO, 25.3% MgO, 13.6% CaO, and 0.12% Na_2O (Rubin et al., 1999). The Na_2O-bearing augite grains are associated with chondrule mesostases that themselves contain Na_2O and K_2O. Because the compositions of these mesostases were probably affected by parent-body aqueous alteration (e.g., Grossman and Brearley, 2005), it seems plausible that, contrary to the suggestion of Rubin et al. (1999), the Na_2O-bearing augite grains are not primary, but rather are products of parent-body alteration.

Aluminian low-Ca pyroxene: Rubin (2004a) studied a Ca-Al-rich chondrule from LL3.01 Semarkona (Fig. 12.12) and reported the occurrence of an Al_2O_3-rich low-Ca pyroxene grain associated with Al-Ti-diopside (fassaite) and Ca-Al-rich mesostasis. The low-Ca pyroxene grain exhibits near-parallel extinction, consistent with the orthorhombic structure. The entire chondrule (dubbed G7) is shown in Fig. 11.26. All phases (even forsterite) are enriched in CaO and Al_2O_3 (Table 12.4).

As chondrule G7 cooled after melting, the phases crystallized in the order of forsterite, Al-rich low-Ca pyroxene, and fassaite; the residual liquid vitrified during chondrule quenching. Phase relationships indicate that, at equilibrium, spinel and anorthite would have crystallized early and orthopyroxene would not have crystallized at all. Quenching prevented anorthite from crystallizing, facilitating the crystallization of orthopyroxene. Aluminum partitioned into the relatively large B tetrahedra of the orthopyroxene grain; because fassaite contains only one tetrahedral site, this phase contains higher concentrations of Al_2O_3 than orthopyroxene.

Table 12.4 *Compositions (wt.%) of the phases in Ca-Al-rich chondrule G7 from LL3.01 Semarkona.*

–	Forsterite (mean)	Fassaite (mean)	Al_2O_3-rich low-Ca pyroxene (mean)	Low-Ca pyroxene analysis with highest Al_2O_3	Mesostasis (mean)
Number of analyses	6	12	23	1	6
SiO_2	41.9	43.1	48.8	47.2	44.9
TiO_2	0.37	3.8	2.3	3.7	2.2
Al_2O_3	0.41	17.8	14.7	16.3	28.1
Cr_2O_3	0.04	0.20	0.13	0.14	0.08
FeO	0.14	0.18	0.13	0.20	0.12
MnO	0.03	0.06	0.03	0.04	0.05
MgO	56.1	13.7	31.2	29.8	5.6
CaO	0.28	20.9	2.4	2.5	17.4
Na_2O	<0.03	0.04	<0.03	<0.03	0.26
K_2O	<0.03	<0.03	<0.03	<0.03	<0.03
Total	99.3	99.8	99.7	99.9	98.7
Fa	0.14	–	–	–	–
Fs	–	0.35	0.22	0.35	–
Wo	–	52.1	5.2	5.7	–

Data from Rubin (2004a).

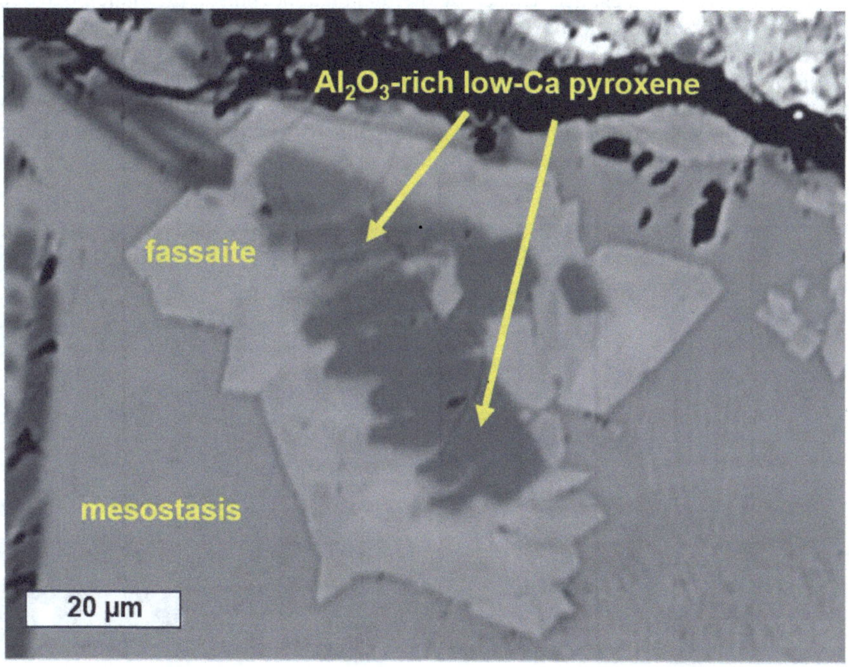

Fig. 12.12 A large grain of Al_2O_3-rich low-Ca pyroxene (dark gray) rimmed by fassaite (Al-Ti diopside). The low-Ca pyroxene appears to be composed of several elongated cigar-stub-shaped grains stacked perpendicular to their c-axes. The faceted fassaite grains also flank the pyramidal terminus of a skeletal olivine (upper left). The assemblage is surrounded by mesostasis in a Ca-Al-rich chondrule (G7) in LL3.01 Semarkona. BSE image modified from Rubin (2004a).

12.5 Secondary Phases Formed During Thermal Metamorphism

There are significant textural, mineralogical, physical, and bulk chemical changes in ordinary chondrites caused by parent-body thermal metamorphism (e.g., Fig. 12.13). These include devitrification of chondrule glass, textural integration between chondrules and matrix, mineral recrystallization, grain coarsening, erasure of compositional zoning in individual mineral grains, reduction in grain-to-grain

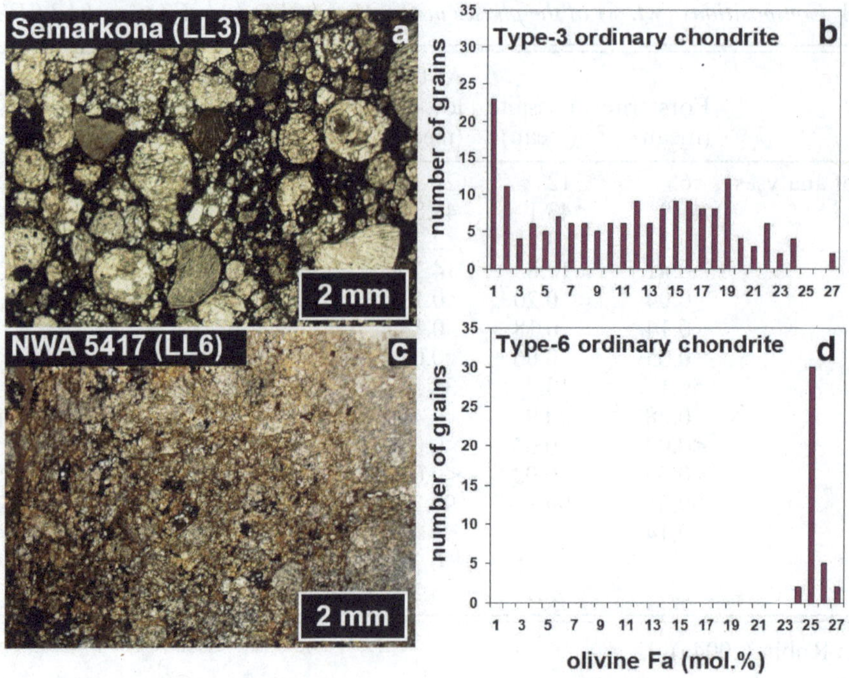

Fig. 12.13 Textural and compositional changes caused by thermal metamorphism. (a) Plane-polarized light microscopic image of chondrules in LL3.01 Semarkona showing sharp chondrule boundaries. Chondrules are separated by narrow patches of fine-grained silicate-rich opaque matrix (black). (b) Histogram showing the compositional distribution of olivine in a type-3 OC of low subtype. Olivine grains exhibit extreme grain-to-grain compositional heterogeneity. (c) Plane-polarized light microscopic image of LL6 NWA 5417. The chondrules are indistinct, having been texturally integrated with the matrix. Image is to the same scale as image (a) of Semarkona. (d) Histogram showing the compositional distribution of olivine in a type-6 OC. Olivine grains are of nearly uniform composition.

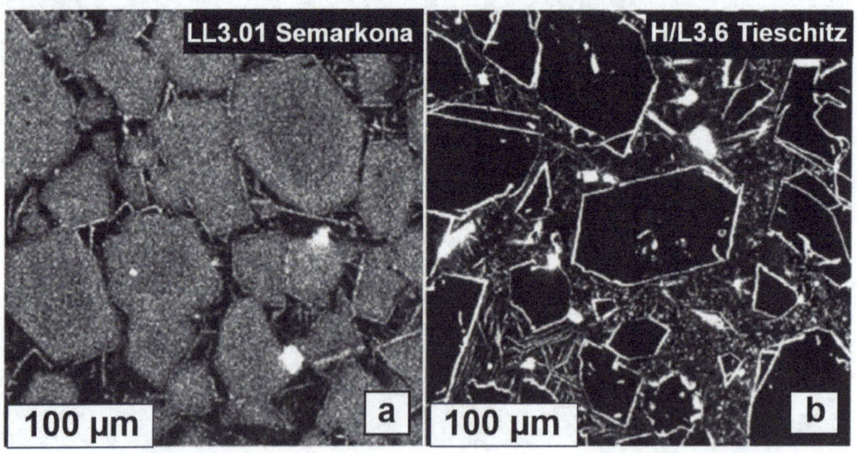

Fig. 12.14 Changes in the Cr distributions in Type-II chondrules in ordinary chondrites. (a) Olivine phenocrysts (gray) in LL3.01 Semarkona exhibit normal igneous zoning: Cr (white) is enriched near the chondrule margins. There are a few small chromite grains within olivine and in the mesostasis (black). Ca-pyroxene appears as thin white linear features in the mesostasis and near olivine margins. (b) H/L3.6 Tieschitz has experienced significant metamorphic changes. Olivine interiors appear black (with uniformly low Cr contents) except for some small chromite grains within them. All the Tieschitz olivine grains are flanked by Cr-rich rims. Cr-Kα X-ray maps from Grossman and Brearley (2005).

compositional variation among mineral species, loss of some primary mineral phases, destruction of presolar grains, growth of secondary minerals, transformation of low-Ca clinopyroxene to orthopyroxene, changes in thermoluminescence (TL) and cathodoluminescence (CL) properties, and partial loss of bulk C, H_2O, noble gases, and volatile siderophile elements.

Olivine grains in Type-II (FeO-rich) chondrules in type-3.00 chondrites exhibit normal zoning in Cr (Grossman and Brearley, 2005), with higher Cr concentrations near the grain edge (Fig. 12.14). This is consistent with growth of the grains by fractional crystallization. At the earliest stages of thermal metamorphism, in type-3.1–3.2 OC, thin streaks of a Cr-rich phase (probably chromite) form by exsolution in the olivine along crystallographic axes. In OC that reach type-3.3–3.6 levels, the interiors of olivine grains have developed uniformly low Cr concentrations with thin Cr-rich rims. Some small chromite grains have grown within the olivine phenocrysts.

Metallic Fe-Ni grains within low-FeO chondrules in primitive ordinary (and carbonaceous) chondrites contain P, Si, and Cr alloyed with metallic Fe in solid solution (e.g., Rambaldi et al., 1980; Rambaldi and Wasson, 1982; Zanda et al., 1994). These elements were incorporated into metallic Fe-Ni by condensation in the solar nebula (Wai and Wasson, 1977). Some metal grains contain small inclusions of phosphide and silicide (e.g., Huss et al., 2006; Yaroshevsky and Ivanov, 2010). The concentrations of P, Si, and Cr in metal grains are inversely correlated with bulk oxidation state (as represented by olivine Fa content) (Fig. 12.15). At higher oxidation states, higher proportions of these elements partition into silicate and oxide phases. With increasing metamorphic subtype, reduced P, Si, and Cr react with O, Ca, and Cl from other components in the meteorite (mainly chondrule mesostases) to produce phosphate (merrillite and chlorapatite), silica (quartz, cristobalite, and silica glass) and chromite (e.g., Zanda et al., 1994; Grossman and Brearley, 2005). In ordinary chondrites, many of these newly formed grains occur in opaque assemblages consisting of (1) troilite and merrillite with or without associated metallic Fe-Ni, (2) troilite and chlorapatite, and (3) metallic Fe-Ni and chlorapatite (Rubin and Grossman, 1985). Chlorapatite appears to replace some of the merrillite during metasomatic reactions.

With increasing metamorphism to type-4-6 levels, chromite and phosphate grains (merrillite and chlorapatite) coarsen (Fig. 12.16). For example, chromite in primitive type-3 chondrites (subtype $\lesssim$3.5) is present only as small grains within chondrules (in the mesostasis and/or within olivine phenocrysts). Some type-3.8 OC have moderate-sized chromite grains in the matrix. Coarse chromite grains are present in type 4–6 OC in the matrix as 10–200-µm-sized isolated blocky, equant, subhedral, anhedral, and rounded grains found in association

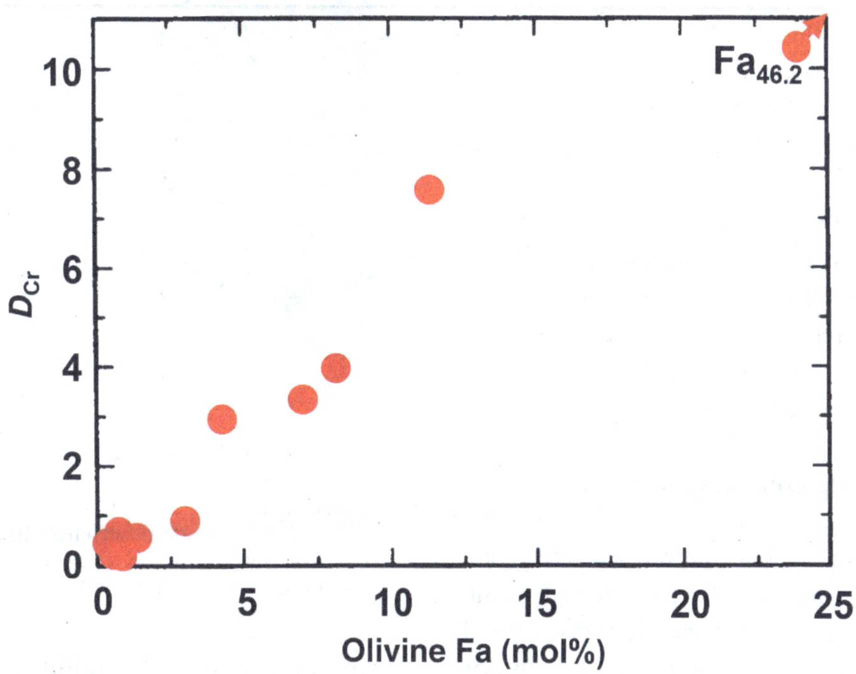

Fig. 12.15 The olivine-metal partition coefficient for Cr (D_{Cr}) increases with the average olivine Fa content in Semarkona chondrules. Olivine Fa is a proxy for bulk oxidation state. With increasing oxidation state, Cr partitions preferentially into silicate and oxide phases. Image modified from Zanda et al. (1994).

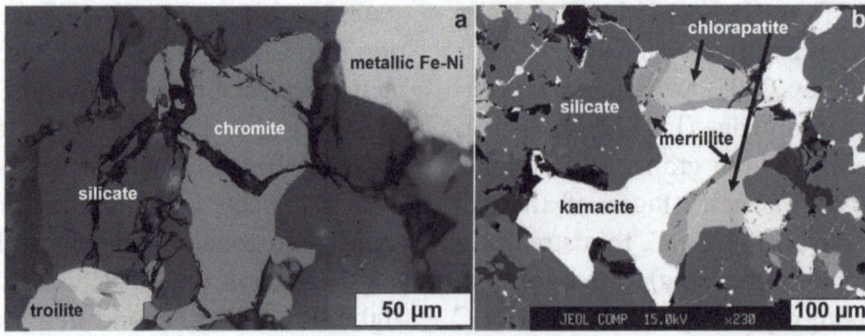

Fig. 12.16 Coarse secondary grains in ordinary chondrites. (a) Intact, relatively unfractured chromite grain adjacent to metallic Fe-Ni and troilite and surrounded by silicate phases in H6 Kernouvé. Black areas are plucked regions produced during thin-section preparation. Reflected light image modified from Rubin (2003). (b) Kamacite grain (white) rimmed by two phosphate phases in H6 Spade. The innermost rim (medium gray) is merrillite; the outer rim (light gray) is chlorapatite. The entire assemblage is surrounded by silicate (dark gray). BSE image modified from Rubin and Jones (2003).

with metallic Fe-Ni and/or troilite. With increasing petrologic type, chromite tends to become richer in FeO, MnO, and Al_2O_3 and poorer in Fe_2O_3 (Bunch et al., 1967). Chromite in H5 Pultusk is unusual among ordinary chondrites in containing ~1.1 wt.% ZnO (although some chromite grains within silicate inclusions in iron meteorites also contain ZnO; Bunch et al., 1967).

Metallic Fe-Ni and sulfide grains also coarsen and become more circular; small metal and sulfide grains tend to disappear.

At temperatures ≥1440°C, troilite can accommodate no more than ~5 mg/g Ni in systems with ~100 mg/g Ni (Kullerud, 1963; L. Ma et al., 1998). As local rock temperatures decline on OC parent bodies, Fe diffuses into troilite, while Ni diffuses out of troilite and into adjacent taenite. Reuter et al. (1988) showed that taenite (γ) transforms into both FeNi (γ') and kamacite just below 936°C; at 866°C, γ' undergoes an ordering reaction to form tetrataenite. During thermal metamorphism, tetrataenite continues to grow, producing large patches at the kamacite-troilite interface in metal-troilite spherules in chondrules within slightly metamorphosed OC such as H/L3.6 Tieschitz (Rubin et al., 1999).

During metamorphism, plessitic metal grains outside chondrules became surrounded by thin rims of tetrataenite (Fig. 12.17); the tetrataenite tended to grow thicker in regions where it abutted troilite. Some of these opaque assemblages developed cloudy taenite (Fig. 12.17), a type of plessite (generally containing 25–40 wt.% Ni) that has a cloudy brown appearance after etching with Nital (Scott, 1973).

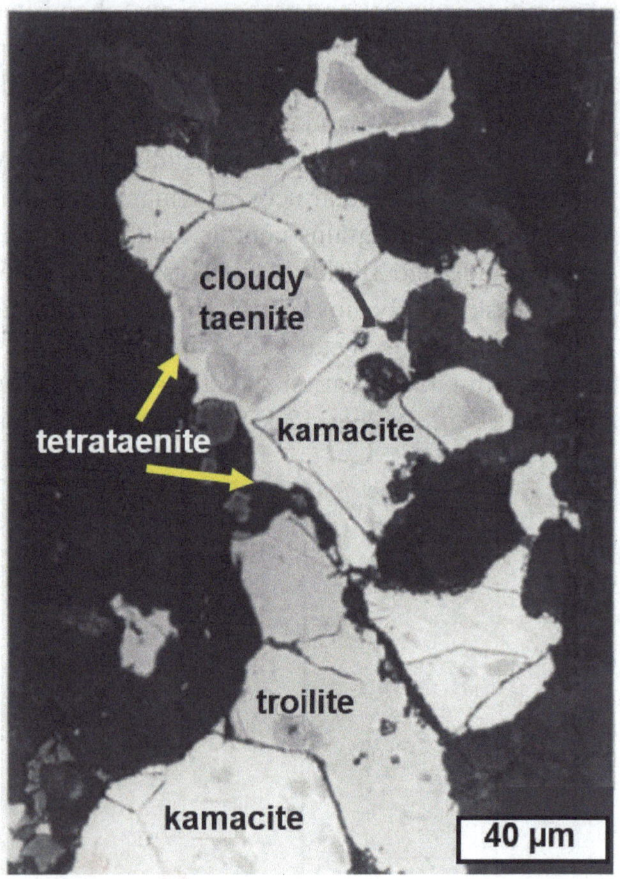

Fig. 12.17 An opaque assemblage consisting of kamacite, cloudy taenite, and troilite within an L4 clast in the Mezö-Madaras L3.7 chondrite breccia. The cloudy taenite occurrences have well-developed rims of tetrataenite. Reflected light image modified from Scott and Rajan (1981).

In their study of equilibrated ordinary chondrites, Reisener and Goldstein (2003) distinguished two types of metallic Fe-Ni particles: (1) zoned

Fig. 12.18 Optical micrograph of a bright-etched coarse-grained zoneless plessite in H6 Guareña. Image modified from Reisener and Goldstein (2003).

taenite + kamacite and (2) zoneless plessite (metal assemblages composed of tetrataenite within kamacite that do not exhibit systematic zoning in Ni; Fig. 12.18). Both types of metal particles formed during whole-rock cooling from peak metamorphic temperatures. Zoned taenite + kamacite formed from polycrystalline taenite by the nucleation of kamacite at taenite/taenite grain boundaries. Zoneless plessite formed in type-5 and -6 OC from monocrystalline taenite via the formation of martensite and its subsequent decomposition into tetrataenite and kamacite.

As oxidation state increases and more metallic Fe is transformed into ferrous iron (Fe^{2+}), the Ni and Co contents of residual metal increase. In going from H to L to LL, there are systematic decreases in the kamacite/taenite ratio (9, 3, 1) and increases in the mean kamacite Co content (4.7, 8.0, 26.8 mg/g) (Table 10.15). Because olivine Fa is a proxy for whole-rock oxidation state, there is a positive correlation between olivine Fa and kamacite Co in ordinary chondrites (Fig. 12.19).

Pentlandite ($(Fe,Ni)_9S_8$) forms at high oxygen fugacities in R and CK chondrites (Schrader et al., 2022). In those OC where it did not form during aqueous alteration, pentlandite occurs along with troilite primarily in highly oxidized LL chondrites (e.g., those containing olivine with high Fa) (Fig. 12.20). One exception is the H5-chondrite fall Ksar El Goraane, which contains four iron sulfides: troilite, pentlandite, pyrite (FeS_2), and pyrrhotite (Arif et al., 2024).

Most metallic Fe-Ni grains in highly oxidized LL chondrites are either awaruite (Ni_3Fe) or Co-rich kamacite. In the case of LL5 Parambu (which has olivine of Fa_{32}; Rubin, 1990), the kamacite contains 370 mg/g Co (Fig. 12.20).

With the rare exceptions of anorthite-bearing CAIs and Al-rich chondrules, coarse grains of plagioclase are absent in type-3 OC. Nevertheless, the chemical components of plagioclase (i.e., SiO_2, Al_2O_3, CaO, Na_2O, K_2O) reside within the feldspathic mesostases of abundant ferromagnesian chondrules (e.g., Mason, 1965; Dodd, 1969). In low-FeO chondrules, albite crystallizes within the mesostases by type ~3.4 (e.g., Grossman and Brearley, 2005); in high-FeO chondrules, the abundance of albite increases progressively from type 3.00 to higher petrologic types. The grain sizes of albite (or oligoclase) increase from < 1 μm in type-3 to ~ 2 μm in type-4 to 2 – 10 μm in type-5 to ≥ 50 μm in type-6 OC. In higher petrologic-type OCs, many plagioclase grains are located within recrystallized chondrules (Fig. 12.21).

Many chondrule feldspathic mesostases also have a normative pyroxene component (low-to-moderate concentrations of SiO_2, MgO, FeO, MnO, Cr_2O_3). The concentration of this component in residual mesostasis increases in metamorphosed OC after plagioclase crystallizes, causing the growth of optically continuous coarse pyroxene grains between olivine bars in a few BO chondrules.

The most abundant pyroxene phase in type-3 and some type-4 OC is low-Ca clinopyroxene exhibiting polysynthetic twinning and inclined extinction (Fig. 12.2); it occurs as phenocrysts in pyroxene-rich chondrules, within chondrule fragments, and as isolated grains in the matrix. This phase transforms into orthopyroxene at ~630°C (Boyd and England, 1965); this is above the maximum metamorphic temperature of some type-4 OC (~600–700°C; e.g., Wlotzka, 1985, 1987) and below the minimum metamorphic temperatures of type-5 and type-6 OC (~700–750°C and ~820–930°C, respectively; e.g., Dodd, 1981; Olsen and Bunch, 1984). This could account for the absence of low-Ca clinopyroxene in

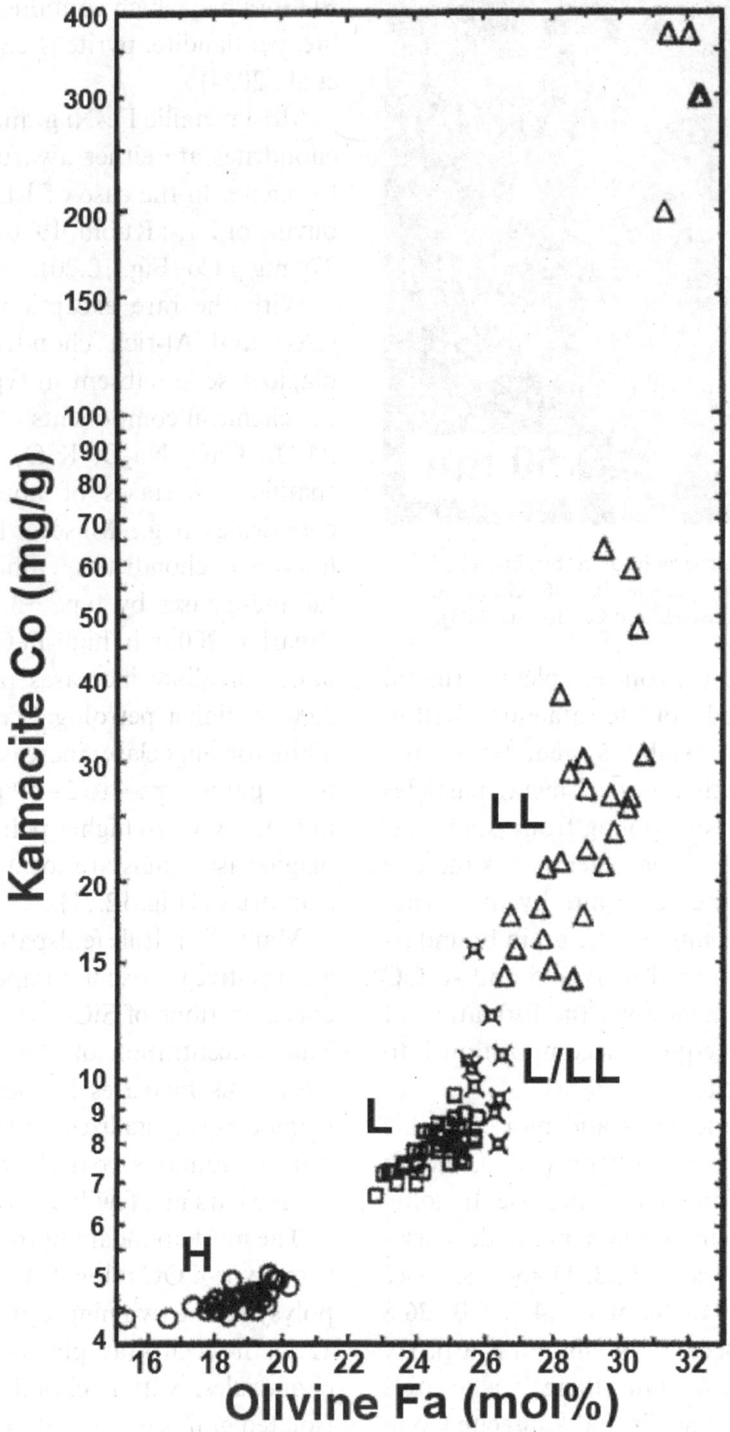

Fig. 12.19 Kamacite Co concentration (mg/g) vs. olivine Fa content (mol%) for 113 type-3–6 ordinary chondrites. There is an exponential relationship between these parameters. Diagram modified from Rubin (1990).

unshocked type-5 and -6 chondrites. (Nevertheless, low-Ca clinopyroxene exhibiting polysynthetic twinning can form by shock heating and quenching of low-Ca pyroxene grains in any low-Ca-pyroxene-bearing rock, irrespective of metamorphic grade, at shock stage≥S3 [Rubin et al., 1997]. It also could be subsequently transformed back into orthopyroxene by annealing.)

Coarse grains of diopside are absent in type-3 OC outside of rare CAIs; this phase occurs nearly

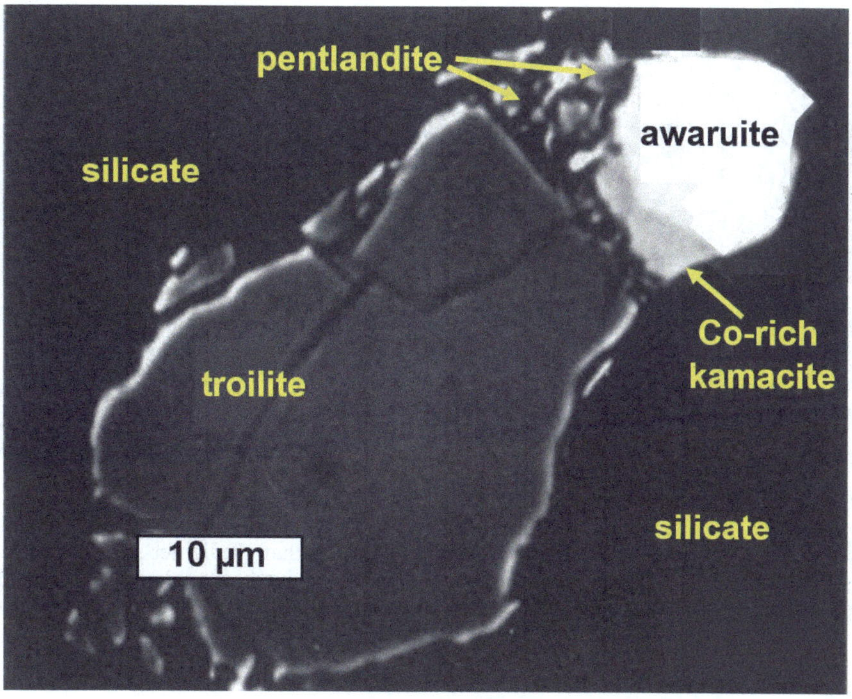

Fig. 12.20 An oxidized opaque assemblage in LL5 Parambu consisting of troilite (dark gray), pentlandite (fractured medium gray), awaruite (white), and kamacite with 370 mg/g Co (light gray). BSE image modified from Rubin (1990).

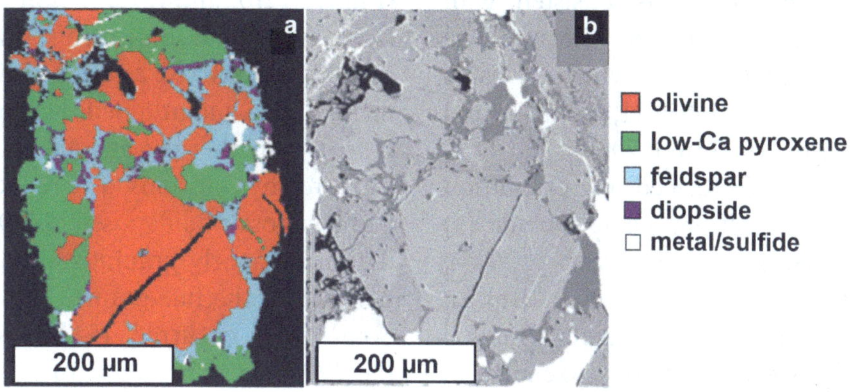

Fig. 12.21 Mineral phases in POP chondrule CH-3 from H5 Richardton (Kovach and Jones, 2010). (a) Mineral phase distributions. The chondrule contains (in vol.%): 51% olivine, 28% low-Ca pyroxene, 16% feldspar ($Ab_{84.7}Or_{4.2}$), 2% diopside, and 2% metallic Fe-Ni ± sulfide. (b) BSE image. The two images have slightly different orientations. Data and modified images from Kovach and Jones (2010).

exclusively in chondrules as crystallites in the mesostasis and as overgrowths on mafic phenocrysts (Fig. 11.5). During thermal metamorphism, diopside nucleates in the fine-grained matrix; it is present as submicrometer grains in type-4, 2–5-μm-sized grains in type-5, and as coarse grains (up to tens of micrometers in size) in type-6 OC (Huss et al., 2006). Nevertheless, the modal abundance of diopside appears to increase to only a minor extent from type-4 to type-6 OC (Gastineau-Lyons et al., 2002; Dunn et al., 2010a, 2010b).

Type-3 OC have heterogeneous olivine and low-Ca pyroxene compositional distributions (Fig. 12.1). During thermal metamorphism, diffusion causes the mineral-grain chemistry to approach equilibrium, decreasing grain-to-grain compositional variations (Fig. 12.13d). As more metallic Fe is transformed into $Fe^{2+}O$ during oxidation, the

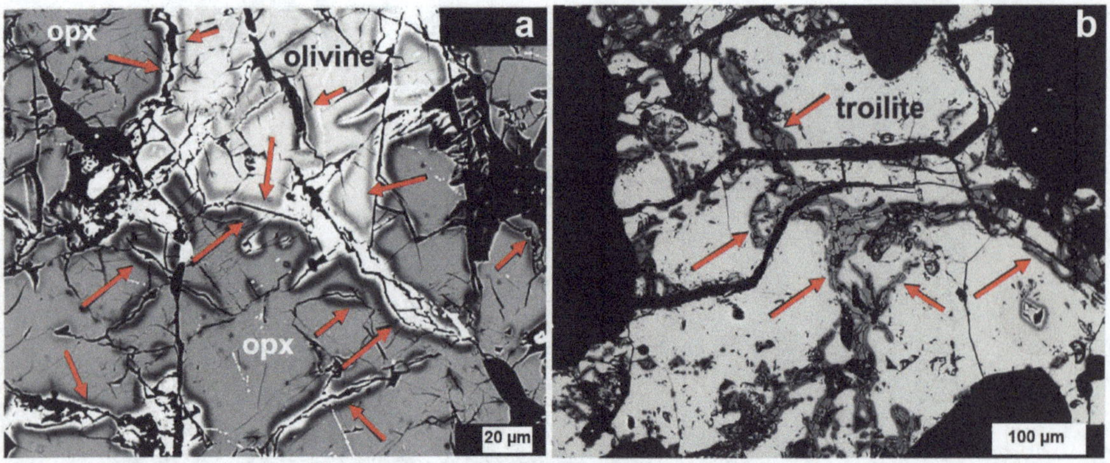

Fig. 12.22 The Osceola L6-chondrite fall exhibiting major reduction features. (a) Nearly every grain margin and every internal fracture in olivine (light gray) and orthopyroxene (opx; medium gray) in this thin section have 4–5-μm-thick BSE-dark bands (arrows). The bands are ~0.4 mol% lower in Fa and Fs than the lighter-shaded grain centers. (b) Troilite grain showing 4–12-μm-thick dark bands (probably poorly crystalline pyrrhotite) along grain boundaries and internal fractures (arrows). These dark bands average 0.4 wt.% less Fe than grain centers. BSE images modified from Rubin and Turrin (2024).

divalent-cation/Si ratio increases, facilitating the formation of $(Mg,Fe)_2SiO_4$ olivine (which has a 2:1 ratio) at the expense of $(Mg,Fe)SiO_3$ low-Ca pyroxene (which has a 1:1 ratio). This is consistent with the differences in the mean olivine/low-Ca-pyroxene modal ratio of the three OC groups, which increases along with oxidation state: 1.4 in H, 2.0 in L, and 3.6 in LL (Table 5.1 of Hutchison, 2004).

Some minerals in OC formed by exsolution during thermal metamorphism. These include (a) chromite in olivine (Grossman and Brearley, 2005), (b) chromite and rutile in ilmenite (e.g., Buseck and Keil, 1966), (c) ilmenite in Cr-spinel (Wlotzka, 2005), (d) calcic pyroxene in orthopyroxene (e.g., Keil and Fredriksson, 1964; Heyse, 1978), (e) alkali feldspar in albite (Lewis and Jones, 2022), (f) pentlandite ($(Fe,Ni)_9S_8$) in troilite (e.g., Bukovanská et al., 1983), and (g) mackinawite (FeS_{1-x}), cubanite ($CuFe_2S_3$), and chalcopyrite ($CuFeS_2$) in pentlandite (Bukovanská et al., 1983).

Some occurrences of common mineral phases in ordinary chondrites formed via reduction during thermal metamorphism. About 17% of L6 chondrites show significant reduction features in backscattered electron (BSE) images (Rubin and Turrin, 2024). These features include: (1) 4–5-μm-thick BSE-dark reduction rims on olivine and orthopyroxene grains and along fracture boundaries in these grains (Fig. 12.22a), (2) 4–12-μm-thick dark bands (probably poorly crystalline pyrrhotite) at the margins and along fractures in troilite grains (Fig. 12.22b), and (3) 2–5-μm-thick dark rinds of kamacite around some taenite grains. It seems likely reduction occurred after the silicate, sulfide, and metal grains had attained their present sizes during parent-body thermal metamorphism and had been fractured by collisions. The reductant was probably a low-viscosity fluid, possibly CO, derived from oxidation of poorly graphitized and amorphous carbon within the fine-grained matrix. The reduction event likely occurred 4543 ± 3 Ma ago, based on the $^{40}Ar/^{39}Ar$ plateau age of L6 Thamaniyat Ajras (Rubin and Turrin, 2024).

12.6 Assorted Accessory and Trace Phases

Silicates: Free silica (likely cristobalite) occurs as laths or more equant grains in rare silica-bearing chondrules and chondrule fragments in ordinary chondrites (Brigham et al., 1986). As discussed in Section 11.1.5, some of these chondrules contain merrihueite ($(K,Na)_2(Fe,Mg)_5Si_{12}O_{30}$) or roedderite ($(Na,K)_2Mg_5Si_{12}O_{30}$) (Krot and Wasson, 1994).

Oxides: The central region of a 220 × 430 μm Mg-Al-chromite fragment in H3.8 Raguli contains a subhedral olivine grain attached to radiating euhedral ilmenite crystals (Krot et al., 1993). One ilmenite crystal encloses a $0.5 \times 5.8\,\mu m$ grain of baddeleyite (ZrO_2). Two end-member pyrophanite grains ($MnTiO_3$) are present in the outer portion of the Mg-Al-chromite fragment. This was the first report of pyrophanite and baddeleyite in an ordinary chondrite. Krot et al. (1993) suggested that the fragment formed on a metamorphosed parent body from an impact melt of a chromite-rich assemblage containing Mn-rich ilmenite. Pyrophanite was later also reported within the strongly shocked Suizhou L6 chondrite (Xie and Gu, 2023 and references therein). In their study of H4 Ste. Marguerite, Christophe Michel-Lévy et al. (1995) described a $200 \times 300\,\mu m$ chromite-rich inclusion that contained 2–3-μm-wide patches of baddeleyite. Also present were a few blebs of Ru-bearing metallic Fe-Ni.

Whereas meteoritic ilmenite typically contains 0.6–9.8 wt.% MnO (Snetsinger and Keil, 1969; Bunch and Keil, 1971), the most MnO-rich ilmenite reported in an ordinary chondrite (Fig. 12.23) is present in H4/5 Tomiya (Shima et al., 1986). The grain contains 19.6 wt.% MnO and occurs as a $\sim 0.01 \times 0.02\,mm^2$ rectangular crystal within troilite. This Mn-rich ilmenite is whiter in color and exhibits stronger anisotropy in reflected light than normal OC ilmenite grains. Tomiya has been shocked: shock features include silicates with undulose extinction, some olivine grains with mosaic extinction and kinking, low-Ca pyroxene grains exhibiting fine deformation lamellae, cryptocrystalline melt pockets, abundant metallic Cu (adjacent to troilite and metallic Fe-Ni), and troilite grains with deformation lamellae (Shima et al., 1986). Shock processes may be indirectly responsible for the formation of this MnO-rich ilmenite.

Chromite and rutile formed by exsolution from ilmenite in ordinary chondrites during thermal metamorphism (Buseck and Keil, 1966). The Calama 451 type-3 OC (probably LL3) contains several oxide phases: rutile, ilmenite, armalcolite ($(Mg,Fe)Ti_2O_5$), spinel, and chromite.

Halides: As discussed in Section 11.11, halite (NaCl) was reported in three OC breccias: Zag (H3-6 fall), Monahans (1998) (H5 fall), and Sidi El Habib 001 (H5-an fresh find, but possible fall) (Zolensky et al., 1999a; Rubin et al., 2002; Che et al., 2023a, 2023b). Within some halite crystals in Monahans (1998), there are small, scattered inclusions of sylvite (KCl).

Sulfides: Section 11.11 discusses the presence of sphalerite ((Zn,Fe)S) in Sidi El Habib 001 (H5-an). It occurs along with halite, consistent with the suggestion that both sphalerite and halite crystallized

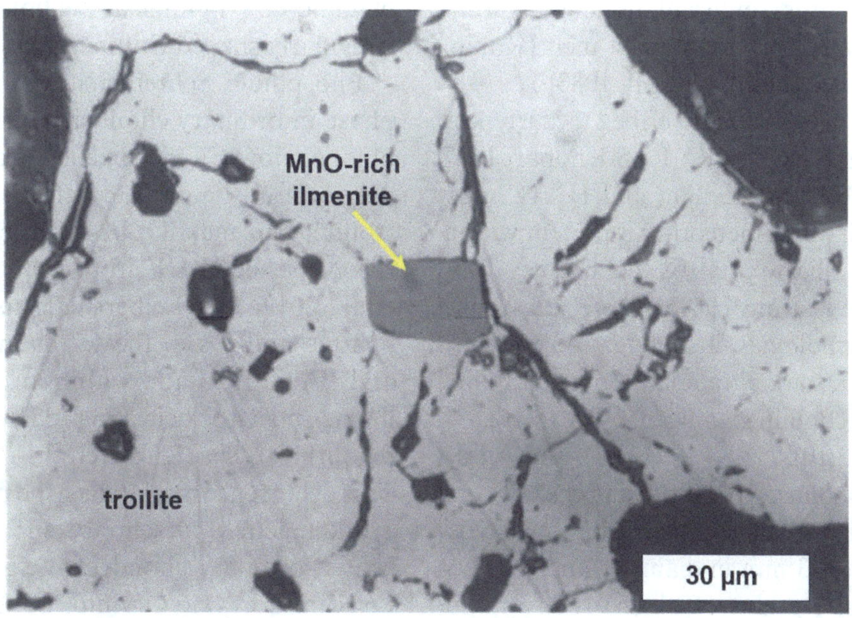

Fig. 12.23 Rectangular MnO-rich ilmenite grain within troilite in H4/5 Tomiya. Reflected light image modified from Shima et al. (1986).

in the meteorite from the same hydrothermal fluids (Che et al., 2023b). Pentlandite ($(Fe,Ni)_9S_8$) in ordinary chondrites occurs mainly in highly oxidized LL samples (e.g., LL5 Parambu; Fig. 12.20), although it is also present in L6 Ustí Nad Orlici (Bukovanská et al., 1983) and in samples subjected to aqueous alteration (e.g., Taylor et al., 1981). There are also occurrences in Ustí Nad Orlici of pyrrhotite ($Fe_{1-x}S$), mackinawite (FeS_{1-x}), cubanite ($CuFe_2S_3$), and chalcopyrite ($CuFeS_2$) (Bukovanská et al., 1983). Poorly crystalline pyrrhotite is probably present as thin dark bands in troilite at grain margins and along fractures in reduced L6 chondrites (Rubin and Turrin, 2024).

Graphite: All primitive chondrites contain presolar graphite grains, but coarse graphite is rare in ordinary chondrites. One exception is Portales Valley, an annealed H6 impact-melt breccia that contains interconnected coarse metal patches (with Widmanstätten patterns) interstitial to angular and subrounded silicate clasts (Kring et al., 1999; Rubin et al., 2001; Ruzicka et al., 2005). Ruzicka et al. (2000, 2005) described graphite nodules (up to ~17 × 20 – 25 mm in size) encased in coarse grains of metallic Fe-Ni (Fig. 11.64). They proposed that, during impact melting, C was scavenged from a massive region of the parent body and concentrated in the metallic melt. Graphite crystallized from this melt. (In some enstatite-chondrite impact-melt breccias, graphite laths (some with pyramidal terminations) crystallized from the impact melt (Rubin, 1983, 1997b, 2015a; Rubin and Keil, 1983].)

Mostefaoui et al. (2005) described occurrences of graphite in some type-3 OC: L3.6 Khohar, L3.7 Mezö-Madaras, L/LL3.4 Inman, and H3.7 Grady (1937). Graphite in these meteorites is present within the fine-grained silicate matrix, chondrule rims, and rare graphite-bearing clasts. There is a range of graphite morphologies including book, veins, fibrous, spherulitic, and granular. Some graphite grains in Khohar contain significant ^{15}N excesses.

NWA 3358 (H3.10), NWA 5697 (L3), and Adrar 003 (L/LL3.10) contain graphite-bearing clasts (Storz et al., 2021). The largest clast in the Storz et al. study is in Adrar 003; it is 3 mm in diameter and consists of 100 – 300 - μm-size grains of olivine (Fa_{6-25}) and low-Ca pyroxene ($Fs_{7-21}Wo_{3-5}$), and minor mesostasis and chromite. Graphite, which constitutes 1.6 vol.% of the clast, ranges in size from 10 – 100 μm and occurs mainly between silicate grains.

Kaeter et al. (2017) used Raman spectroscopy to detect micrometer-sized inclusions of graphite in olivine within LL5 Chelyabinsk.

Although not composed of crystalline graphite, C-rich aggregates (Section 11.10) are present in some type-3 OC and OC regolith breccias. These aggregates consist of irregular grains of poorly graphitized carbon along with amorphous carbon, small grains of metallic Fe-Ni, and minor chromite (Scott et al., 1988a; Brearley, 1990).

Nitride: The trigonal polymorph of Si_3N_4 (nierite) was identified within perchloric acid-resistant residues of L/LL3.10 Adrar 003, L/LL3.4 Inman, H/L3.6 Tieschitz, and EH4 Indarch (Lee et al., 1995). The phase occurs as small laths, ~0.4 × 2 μm in size, in some cases intergrown with whiskers of the hexagonal Si_3N_4 polymorph. Most nierite crystals in Indarch appear to have formed by exsolution of Si and N from kamacite, perryite, and schreibersite during thermal metamorphism. However, it is unclear how the nierite crystals in the ordinary chondrites formed. Because the hexagonal phase is enriched in ^{15}N (Russell et al., 1995), it may be presolar.

Metallic Cu: As discussed in Section 11.12, metallic Cu occurs in many shocked ordinary chondrites, commonly as small grains located within larger metallic Fe-Ni grains at the boundaries with small, irregularly shaped troilite grains (Rubin, 1994a).

Phosphide: Schreibersite ($(Fe,Ni)_3P$) is a rare phase in ordinary chondrites. It occurs along with olivine, low-Ca pyroxene, Ca-pyroxene, sodic plagioclase, merrillite, chlorapatite, chromite, troilite, and metallic Fe-Ni in chondritic clasts in the Netschaëvo IIE iron (Rubin, 1990; Schrader et al., 2017a). Schreibersite is also present in small opaque nodules in the L7 chondrites PAT 91516 and PAT 91528 and within an opaque bleb in an impact-melt-rock clast in the Dimmitt H-chondrite regolith breccia (Rubin et al., 1983a) (Fig. 11.59). This occurrence is related to the presence of secondary kamacite and schreibersite in severely shock-heated ordinary chondrites (e.g., L5 Farmington, L6 Ramsdorf, H6 Orvinio, L5 Lubbock, H5 Rose City) (Smith and Goldstein, 1977). (These latter schreibersite grains are ≤2 μm in size.) Associated

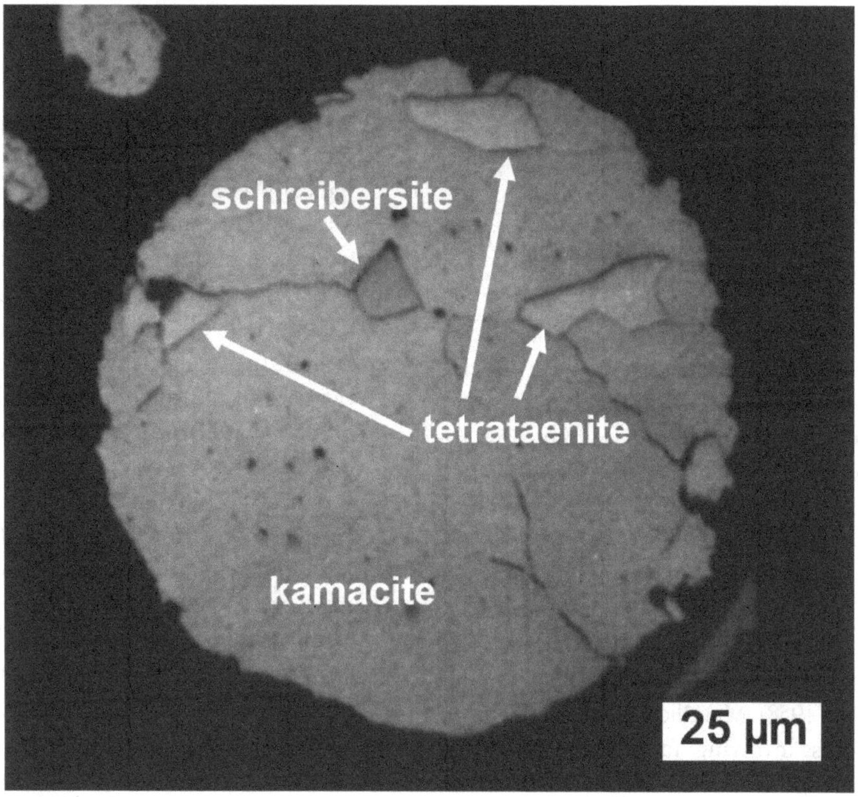

Fig. 12.24 Spheroidal kamacite bleb from a Type-I chondrule in LL3.15 Bishunpur. The respective Co and Ni contents (mg/g) of the constituent phases are: kamacite (6.4, 37), tetrataenite (1.7, 500), schreibersite (18, 518). Image modified from Rambaldi and Wasson (1981).

metallic Fe-Ni grains have >0.1 wt.% P in solid solution. The P enrichment of the metal was likely caused by reduction of phosphates during impact-heating accompanied by partial melting of metal-troilite assemblages (e.g., Begemann and Wlotzka, 1969; Taylor and Heymann, 1971). Schreibersite formed during subsequent slow cooling after nucleating on metal-troilite grain boundaries (Smith and Goldstein, 1977).

Some metallic Fe-Ni grains in primitive OC contain small inclusions of phosphide and silicide (e.g., Huss et al., 2006; Yaroshevsky and Ivanov, 2010). Gooding and Keil (1981) described rare metallic chondrules consisting primarily of metallic Fe-Ni with accessory sulfide, phosphate and/or phosphide and, in some cases, magnetite or silicate grain fragments (e.g., Fig. 11.28a). Many of these metallic chondrules were affected by parent-body aqueous alteration.

Rambaldi and Wasson (1981) described ~100-μm metallic blebs in Type-I chondrules in LL3.15 Bishunpur that contain small grains of schreibersite (along with tetrataenite) (Fig. 12.24). These opaque assemblages probably equilibrated down to low temperatures, perhaps ~400°C (Romig and Goldstein, 1980). The smaller schreibersite grains tend to be richer in Ni.

12.7 Minerals Formed During Shock Metamorphism

Different ordinary chondrites were subjected to different shock pressures. Some samples were shocked multiple times (e.g., Lambert et al., 1984; Kunz et al., 1997). Many samples were annealed after being shocked; some OC were shocked again after being annealed (Rubin, 2004b). The present petrographic features of OC reflect their shock, thermal, alteration, and weathering histories. Stöffler et al. (1991) divided the ordinary chondrites into distinct shock stages (Table 2.2), with each stage corresponding to a different range of maximum shock pressure: S1-S2 (<4–5 GPa), S2-S3 (5–10 GPa), S3-S4 (10–20 GPa), S4-S5 (20–35 GPa), S5-S6 (35–60 GPa), S6-S7 (>45 GPa). However, Sharp and DeCarli (2006) found these pressure values to be unreliable (Section 13.2.1). In any case, the shock pressures required to produce the petrographic indicators of shock stage are lower for target samples that were hot prior to collision.

Many phases were produced directly or indirectly by impact processes in OC at different shock stages (e.g., Rubin, 1985; Bischoff and Stöffler, 1992; Sharp and DeCarli, 2006):

(1) Chromite and plagioclase (which happened to occur in proximity prior to the impact event) were shock-melted, later crystallizing in chromite veinlets and chromite-plagioclase assemblages (Rubin, 2003). Temperatures probably reached ≥1100°C.
(2) Metallic Cu grains in shocked equilibrated ordinary chondrites most commonly occur within Ni-rich metal adjacent to small, irregular grains of troilite (Rubin, 1994a; Y. Li et al., 2021). Copper occurs in solid solution primarily in Ni-rich metal because fcc taenite can accommodate ~16 times more Cu than bcc kamacite (Hansen, 1958). Metallic Cu in OC may have formed by a process involving: (a) collisional heating of troilite, (b) breakdown of FeS into metallic Fe and S_2, (c) migration of S_2 into metal along concentration gradients to produce new small grains of troilite and Ni-enriched metal, (d) eventual supersaturation of Cu in Ni-rich metal grains, and (e) nucleation of metallic Cu at (metallic Fe-Ni)/troilite grain boundaries (which are high-surface-energy sites) (Tomkins, 2009; Rubin and Ma, 2021).
(3) Small irregular grains of troilite (e.g., Fig. 11.57) crystallized within larger masses of metallic Fe-Ni after metal-sulfide assemblages were melted and quenched (Rubin, 1994a).
(4) Rapidly crystallized metal-sulfide melts crystallized finely-intergrown globular assemblages of metal and sulfide (sometimes called "fizzed troilite" after their superficial resemblance to aerated drinks; Scott, 1982) (Fig. 12.25).
(5) Quenched metallic Fe-Ni melts produced martensite that annealed to form patches of plessite in some cases (e.g., J. Zhang et al., 1994). Martensite typically contains ~8–15 wt.% Ni and has a distorted body-centered cubic structure (α_2). It forms from taenite during rapid cooling (i.e., $\gamma \rightarrow \alpha_2$) and is usually associated with a shock-heating event. Plessite is a mixture of kamacite and taenite and/or tetrataenite. Buchwald (1975) described several varieties of plessite (e.g., spheroidized, acicular, net, cellular) with distinct ranges in size and shape determined by the Ni content of the original martensite. Wood (1964) proposed that plessite forms through the decomposition of martensite (i.e., $\alpha_2 \rightarrow \alpha + \gamma$), but Goldstein and Michael (2006) reported that parent taenite crystals were retained at the boundaries of martensite laths during cooling after martensite formation. They suggested a more-complex reaction: $\gamma \rightarrow \alpha_2 + \gamma \rightarrow \alpha + \gamma$. (Neither martensite nor plessite is an IMA-approved mineral, but both terms are entrenched in the meteorite literature. Martensite is also a common term in metallurgy.)
(6) Shock-melted metal and sulfide was injected into fractures within silicate grains to form opaque veins upon quenching (e.g., Bischoff and Stöffler, 1992).
(7) Curvilinear trails of small (metallic Fe-Ni)±troilite blebs entered narrow fractures in the silicate grains, causing the whole rocks to become darker (Rubin, 1992). This phenomenon, known as "shock darkening" or "silicate darkening," can form so-called "black chondrites." It results from the shock melting of metal and sulfide (which melt at a lower temperature than silicates; ~940° vs. ~1100°C). These features were reproduced experimentally by Hörz et al. (2005) and Moreau et al. (2024).

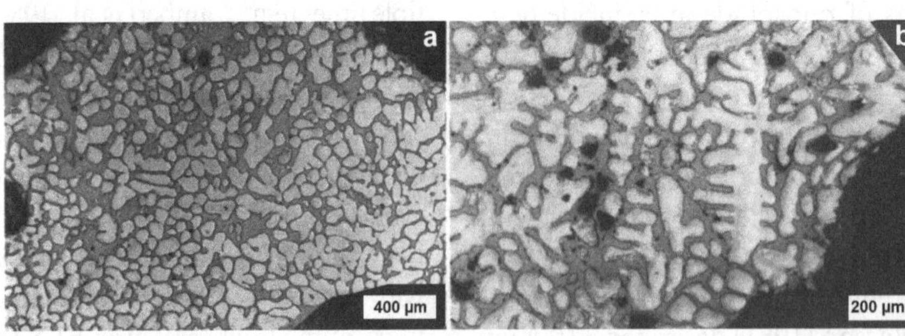

Fig. 12.25 Rapidly solidified metal-troilite inclusions in ordinary chondrites. (a) An inclusion in the Weston H3-6 regolith breccia. Metallic Fe-Ni (white) occurs as small particles of varied shape within troilite (dark gray). (b) Inclusion in H4 San Emigdio showing metallic Fe-Ni dendrites (single tree-shaped crystals) and rounded cells (white) within troilite (medium gray). Reflected light images from Scott (1982).

(8) Molten silicate that formed at ≥1100°C (Jurewicz et al., 1995) was injected under high pressure into fractures within unmelted silicate grains. Some stringers of melt quenched into dark-colored glassy (pseudotachylite-like) veins (Melosh, 2005); in other fractures, fine-grained mafic silicate phases crystallized. (Pseudotachylites derive their name from a textural resemblance to the basaltic volcanic glassy rock known as tachylite [cf. Figs. 2, 3 of Bae et al., 2012)].) Numerous nanophase iron particles (np-Fe^0) were reported within pyroxene glass in a shock melt vein from the shock-stage S5 GRV 022115 L6 chondrite (Guo et al., 2020).

(9) Melt pockets containing silicate glass and small grains of chromite, olivine, pyroxene, and plagioclase (Dodd and Jarosewich, 1979, 1982) formed by localized shock melting at temperatures ≥1100°C. The melt pocket glasses are enriched in normative plagioclase, a likely consequence of preferential impact melting of chondritic minerals with low impedances to shock compression. Experimental studies of shock-loaded basalt (Schaal et al., 1979) showed that the most compressible phase in the sample was plagioclase. The melt pockets were augmented by less extensive melting of phases such as pyroxene with lower compressibilities.

(10) Orthopyroxene grains were impact heated to temperatures between 1000°C and 1557°C (Smyth, 1974) and entered the protoenstatite field; upon quenching, they transformed into clinoenstatite with polysynthetic twinning parallel to (100).

(11) Clinopyroxene grains in some so-called type-7 OC (many of which are probably impact-melt breccias or impact-melt rocks) contain very fine exsolution lamellae most easily detected by backscattered electron imaging (e.g., Y. Li et al., 2020). The lamellae formed during rapid cooling from high temperatures.

(12) Impact-melt rock clasts formed in many ordinary-chondrite breccias by melting target material (Rubin, 1985); these clasts typically contain major olivine and low-Ca clinopyroxene and minor diopside with interstitial K-rich feldspathic glass (e.g., Fodor and Keil, 1975, 1976a, 1976b; Fodor et al., 1974).

(13) Very rare agglutinates (vesicular, glassy objects containing chunks of rock, glass, and mineral fragments) formed in the Fayetteville H4 regolith breccia (e.g., Kerridge and Kieffer, 1977; Rubin, 1982, 1985). In addition, Noonan and Nelen (1976) described two agglutinates in Weston (H4 regolith breccia), and Lorin and Pellas (1979) reported two, or possibly three, in Djermaia (H3-6 regolith breccia).

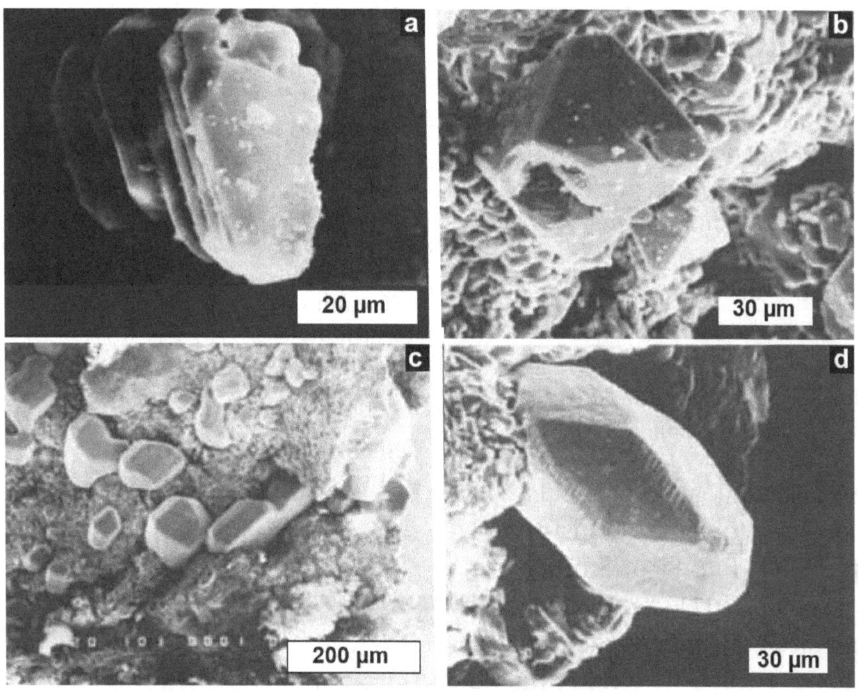

Fig. 12.26 Vapor-deposited mineral phases within vugs in L5 Farmington, a shock-blackened OC. (a) Stacks of euhedral low-Ca pyroxene grains attached to the vug wall. Small white patches atop the pyroxene grains are fayalitic olivine. (b) Octahedral crystals of chromite on top of low-Ca pyroxene. (c) Subparallel euhedral crystals of troilite protruding from vug wall. (d) Euhedral metallic Fe-Ni grain with numerous V-shaped markings. At left is a plagioclase grain with an irregular surface. Scanning electron microscope (SEM) images modified from Olsen (1981).

(14) Some regions in a few OC breccias contain crushed silicate grains surrounded by very-fine-grained troilite; the troilite/(metallic Fe-Ni) modal ratios range up to ~500:1 (Rubin, 2002a). This is much higher than the eutectic weight ratio of 7.5:1 (Brandes and Brook, 1998), suggesting that the troilite-rich regions formed by deposition from a shock-generated S-rich vapor.
(15) Millimeter-to-centimeter-sized vugs containing euhedral mineral crystals (metallic Fe-Ni, troilite, low-Ca pyroxene, albite, fayalitic olivine, merrillite, and chromite) are present in many ordinary chondrites (Olsen, 1981) (Fig. 12.26). These phases appear to have been deposited by impact-generated vapor; they resemble euhedral, vapor-deposited crystals (metallic Fe-Ni, troilite, ilmenite, pyroxene, plagioclase, merrillite, and apatite) within vugs in lunar breccias (e.g., McKay et al., 1972b).
(16) A 65×107 μm euhedral tetrataenite grain occurs within an impact-melt-rock clast in LL6 Jelica (Fig. 11.66). The metal grain crystallized as taenite from the melt, and after cooling to 320°C, underwent an ordering reaction to form tetrataenite (Rubin, 1994b).
(17) Pyrophanite ($MnTiO_3$) in impact-melted assemblages in H3.8 Raguli (Krot et al., 1993) and L6 Suizhou (Xie and Gu, 2023) crystallized from shock-induced melts.
(18) Phosphates were also affected by shock processes, causing mobilization and redistribution of Cl and F, accounting for rare apatite grains with F/Cl ratios $\gtrsim 2$ (e.g., Dobrică et al., 2024).
(19) Plagioclase transforms (either via shock disordering or fusion) into a dense glass (maskelynite) at high shock pressures (e.g., Milton and DeCarli, 1963; Chen and El Goresy, 2000; Jaret et al., 2015; Dobrică et al., 2024). Because maskelynite is not crystalline, it is considered a mineraloid.

The higher the calcic content of plagioclase, the lower the shock pressure required to transform plagioclase into maskelynite (Fig. 11 of Stöffler et al., 1986; Fig. 3 of Fritz et al., 2011). As pointed out by Fritz et al. (2011), more-anorthitic plagioclase contains more Al–O bonds (which are weaker than the common Si–O bonds in silicates) and can withstand less confining pressure; thus, lower shock pressures are required for maskelynitization. For example, An_{50} (as in martian basalts) maskelynitizes at 29 GPa, An_{75-95} (as in lunar basalts) at 20–24 GPa. Ordinary chondrites contain plagioclase of An_{10-15}, which would undergo maskelynitization at pressures on the order of 36–40 GPa.

Maskelynite is present in ~1% of H, ~11% of L, and ~4% of LL chondrites (Rubin, 2015c). Most of the maskelynite in H and LL chondrites was probably formed by hypervelocity impacts on these OC parent bodies. In contrast, much of the maskelynite in L chondrites likely resulted from high shock pressures generated by the catastrophic collisional disruption of the L asteroid ~470 Ma ago (e.g., Heymann, 1967; Bogard et al., 1976; Turner, 1988; Britt and Pieters, 1994; Bogard, 1995; Korochantseva et al., 2007; Swindle et al., 2014).

Table 12.5 *High-pressure phases in ordinary chondrites.*

Phase	Chemical formula	Source
Akimotoite	$(Mg,Fe)SiO_3$	1
Albitic jadeite	$(Na,Ca,\square_{1/4})(Al,Si)Si_2O_6$	1
Ahrensite	Fe_2SiO_4	1
Asimowite	Fe_2SiO_4	2
Bridgmanite	$MgSiO_3$	1
Chenmingite	$FeCr_2O_4$	2
Coesite	SiO_2	3
Elgoresyite	$(Mg,Fe)_5Si_2O_9$	2
Hemleyite	$FeSiO_3$	1
Hiroseite	$FeSiO_3$	2
Lingunite	$NaAlSi_3O_8$	1
Majorite	$Mg_3(MgSi)Si_3O_{12}$	1
Majorite-pyrope$_{ss}$	$Mg_3(MgSi)Si_3O_{12}$–$Mg_3Al_2(SiO_4)_3$ solid solution	1
Maskelynite	$(Na,Ca)(Si,Al)_4O_8$	4
Ohtaniite	$Mg_3(Si_{0.5}\square_{0.5})Si_2O_8$	5
Poirierite	Mg_2SiO_4	2
Ringwoodite	Mg_2SiO_4	1
Ringwoodite-Q (tetragonal)	$(Mg,Fe,Si)_2(Si,\square)O_4$	6
Shenzhuangite	$NiFeS_2$	7
Stishovite	SiO_2	1
TiO_2-II (orthorhombic)	TiO_2	2
Tuite	γ-$Ca_3(PO_4)_2$	1
Wadsleyite	Mg_2SiO_4	1
Wangdaodeite	$FeTiO_3$	2
Xieite	$FeCr_2O_4$	2

Sources: 1 = Tomioka and Miyahara (2017) and references therein; 2 = Xie and Gu (2023) and references therein; 3 = Miyahara et al. (2016); 4 = Van Schmus and Ribbe (1968); 5 = Bindi et al. (2024); 6 = C. Ma et al. (2019); 7 = Bindi and Xie (2017).

(20) High-pressure phases (Table 12.5) formed in the vicinity of silicate melt veins in many highly shocked OC (e.g., M. Chen et al., 1996; Gillet et al., 2000). As discussed by Tomioka and Miyahara (2017), these phases can form in two ways: via crystallization of phases from a shock melt under high pressure or by solid-state transformation of minerals in the host rock under high pressure (Fig. 12.27). Three conditions must be met before high-pressure phases can be produced and survive: (a) portions of the sample must be shock-heated to melting temperatures at high pressure to facilitate formation of the high-pressure phases prior to pressure release, (b) the high-pressure phases must be quenched before ambient pressure conditions reemerge and cause retrograde transformation to low-pressure polymorphs (Hu and Sharp, 2022), and (c) the sample has to avoid post-shock annealing (e.g., Rubin 2002b, 2004b) that also can lead to back transformation to low-pressure polymorphs (Hu and Sharp, 2017). (Although not necessarily a high-pressure phase, shenzhuangite ($NiFeS_2$), the Ni-analog of chalcopyrite, probably formed after shock-vaporization of troilite raised the ambient pS_2/pO_2 ratio and preexisting taenite underwent sulfidation [Bindi and Xie, 2017].)

As has been pointed out by Bindi et al. (2024), the shock veins in L6 Suizhou contain numerous high-pressure phases including akimotoite, asimowite, elgoresyite, hemleyite, hiroseite, lingunite, magnesiowüstite, majorite, majorite-pyrope garnet, ohtaniite, ringwoodite, tuite, wadsleyite, wangdaodeite, xieite, and (Mg,Fe)SiO_3-glass. The latter phase could be vitrified perovskite (Chen et al., 2004). Many high-pressure phases are polymorphs of common phases in unshocked OC: olivine (ahrensite, asimowite, poirierite, ringwoodite, wadsleyite), orthopyroxene (akimotoite, bridgmanite, hemleyite, hiroseite, majorite, ohtaniite), plagioclase (lingunite, maskelynite), chromite (chenmingite, xieite), merrillite (tuite), ilmenite (wangdaodeite), and rutile (TiO_2-II). In addition, L6 Château-Renard contains sodic pyroxene (with the composition and structure of omphacite) in association with ahrensite and wadsleyite (Baziotis et al., 2018). Although free silica is a rare phase in ordinary chondrites, there are two high-pressure polymorphs of SiO_2 in shocked OC: coesite and stishovite (Table 12.5). (Natural occurrences of both phases were discovered at Meteor Crater, Arizona in quartz sandstones of

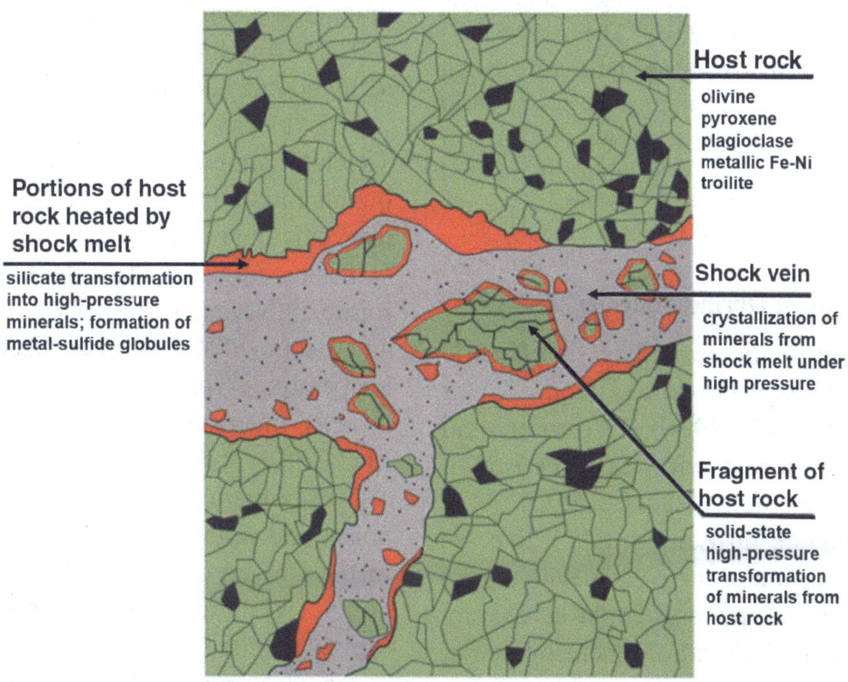

Fig. 12.27 Cartoon of a shock vein in an ordinary chondrite. High-pressure minerals can form by two distinct mechanisms: (1) crystallization of phases from the shock melt under high pressure and (2) solid-state transformation of minerals in the host rock under high pressure. Diagram modified from Tomioka and Miyahara (2017), courtesy of Naotaka Tomioka.

the Coconino Formation [Chao et al., 1960, 1962]. Vesicular silica glass [lechatelierite] is also present at Meteor Crater.)

Albitic jadeite likely formed at 11.5 GPa (Baziotis et al., 2025), wadsleyite at 14–18 GPa (Tomioka and Miyahara, 2017), tuite at 20–23 GPa at 1900–2000°C (Murayama et al., 1986; Xie et al., 2003, 2013; Zhai et al., 2014), xieite at 18–23 GPa at 1800–1950°C (Chen et al., 2008), and majorite over a range of pressures (~17–23 GPa) depending on composition (Tomioka et al., 2016; Ohtani et al., 1991; Kato and Kumazawa, 1985; Tomioka et al., 2016; Bazhan et al., 2017).

(21) As discussed in Section 11.12, some ordinary chondrites contain large (centimeter-sized) metal nodules and elongated metal veins and sheets that formed by a complicated process involving impact vaporization of bulk metal-sulfide assemblages, oxidation, translational motion, and fractional condensation of siderophile elements (Widom et al., 1986; Rubin, 1995b, 1999).

(22) Some OC were completely (or almost completely) melted by collisions. These rocks include PAT 91501 (Mittlefehldt and Lindstrom, 2001; Benedix et al., 2008), Y-74160 (Takeda et al., 1984), Ramsdorf (Yamaguchi et al., 1999), and LAR 06299 (Rubin and Moore, 2011). The main minerals (olivine, low-Ca pyroxene, Ca-pyroxene, Cr-rich spinels, metallic Fe-Ni, troilite, sodic plagioclase) crystallized largely or entirely from the impact-generated melt. The silico-feldspathic interstitial melt quenched into glass.

12.8 Minerals Formed During Aqueous Alteration

Parent-body aqueous alteration has significantly affected type-3 ordinary chondrites (e.g., Taylor et al., 1981; Krot et al., 1997b; Grossman and Brearley, 2005), consistent with the presence of chemically bound water (e.g., 1.2 wt.% H_2O^+ in the LL3.01 Semarkona fall; Jarosewich, 1990). In cases where aqueous alteration occurred at elevated temperatures, modification processes could be characterized as hydrothermal alteration, metasomatism, or fluid-assisted metamorphism.

Petrographic evidence for aqueous alteration in type-3 OC includes Fe-rich phyllosilicate (Fe-rich chlorite), Na-bearing Fe-rich smectite, sulfides (Ni-rich pyrrhotite, pentlandite), Ni-rich metal (taenite, tetrataenite, awaruite), nickelphosphide, carbides (cohenite, haxonite, edscottite), oxides (wüstite, magnetite, maghemite), and trails of small calcite crystals (Taylor et al., 1981; Nagahara, 1984; Hutchison et al., 1987; Alexander et al., 1989a; Keller, 1998; Grossman et al., 2000; Brearley, 2006). Also present are fayalite-silica associations (Wasson and Krot, 1994), carbide-magnetite assemblages (Krot et al., 1997b), and bleached radial pyroxene and cryptocrystalline chondrules containing porous outer regions in which mesostasis was broken down (Grossman et al., 2000).

The composition of the aqueous fluid likely varied with time as additional primary (and early secondary) phases were altered. Chlorapatite (Fig. 12.16b) formed from a fluid that was dry, acidic, and halogen-rich (with a high Cl/F ratio) (e.g., Jones et al., 2016). The fluid appears to have varied among the different OC bodies – there is a tendency for the Cl/F ratio in apatite to decrease from H to L to LL (Jones et al., 2014, 2016; Lewis and Jones, 2016). In LL-chondrite apatite, other anions besides OH^- (e.g., S^{2-} and possibly CO_3^{2-}) may have substituted for Cl^- and F^- (e.g., Jones et al., 2014). Carbide-magnetite assemblages (Fig. 11.74) formed from a CO_2–bearing fluid. Smectite (which contains chemically bound water) formed from an H_2O-rich fluid.

The "blocky" mesostasis of chondrules in H/L3.6 Tieschitz formed from "smooth" mesostasis by aqueous alteration. In the Tieschitz matrix there are geode-like voids and veins filled with sodic-calcic amphiboles (winchite ($\square NaCa(Mg_4Al)Si_8O_{22}(OH)_2$) and barroisite ($\square NaCa(Mg_3Al_2)(Si_7Al)O_{22}(OH)_2$)) (Dobrică and Brearley, 2014); they were likely formed at elevated temperatures from the same fluid that produced the "blocky" mesostasis and "white matrix" (Christophe Michel-Lévy, 1976; Hutchison et al., 1998).

New minerals can form by thermal metamorphism and sintering of aqueously altered materials. In the CM-an chondrite Dhofar 225, small grains of the oxysulfide cafeosite ($Ca_4Fe^{2+}{}_3Fe^{3+}{}_2\square O_6S_4$) formed via the heating of an assemblage of sulfides, hydroxides, phyllosilicates, Ca carbonates, and iron oxides (Ivanova et al., 2025). Similar phases could form in thermally metamorphosed aqueously altered ordinary chondrites.

13

Parent-Body Processes

13.1 Thermal Metamorphism

13.1.1 Ordering into Petrologic Types

Since the nineteenth century, many workers have observed that some ordinary chondrites appear more recrystallized than others (Sorby, 1877; Wahl, 1910; Merrill, 1921; Zavaritzky, 1948, 1950; Wood, 1962b; Dodd and Van Schmus, 1965; Dodd et al., 1967; Van Schmus and Wood, 1967; Binns, 1967b). The evidence for thermal metamorphism of ordinary chondrites was reviewed by Dodd (1969).

Van Schmus and Wood (1967) divided the ordinary chondrites into a series of "petrologic types" ranging from unmetamorphosed type-3 samples to highly metamorphosed type-6 samples. Beautiful images of OC thin sections of all petrologic types can be found in Lauretta and Killgore (2005) and Grady et al. (2014). Type-3 OC have very well defined chondrules; with increasing petrologic type, chondrules become progressively more texturally integrated with matrix material. Chondrules in type-6 chondrites are highly recrystallized and often difficult to discern (Fig. 13.1), although the remnants of BO chondrules (i.e., parallel olivine bars intercalated with fine silicate crystals) can commonly be identified (e.g., Fig. 13.2). Along with textural integration (Fig. 12.13a,c), other properties undergo systematic changes with increasing metamorphic grade (Table 13.1). (1) Individual mineral grains become more uniform in composition (i.e., internal compositional zoning is reduced); (2) grain-to-grain mineral compositional heterogeneity is diminished (Fig. 12.13b,d); (3) bulk concentrations of C, H_2O, and mobile volatile elements decrease; (4) there are systematic changes in thermoluminescence (TL) and cathodoluminescence (CL); (5) low-Ca clinopyroxene transforms into orthopyroxene; (6) secondary phases grow (e.g., plagioclase, chromite, ilmenite, phosphate, diopside); (7) chondrule glassy mesostasis devitrifies and recrystallizes; and (8) metallic Fe-Ni and troilite grains coarsen.

After studying the Shaw L chondrite, Dodd et al. (1975) noted its coarse texture and calcic orthopyroxene and suggested it be classified as type 7. Other meteorites were soon likewise designated. Currently, about 200 ordinary chondrites are classified in the Meteoritical Bulletin Database as petrologic type 7 (e.g., NWA 15762, H7; NWA 12961, L7; Los Vientos 231, LL7). Several OC are fragmental breccias with clasts ranging in petrologic type from 3, 4, 5, or 6 up to type 7 (e.g., NWA 14828, LL3-7). However, some meteorites classified as type 7 (e.g., PAT 91501, L7) are actually impact-melt rocks or impact-melt breccias (e.g., Rubin, 1995a; Mittlefehldt and Lindstrom, 2001). These impact-melted samples include Shaw itself (Berkeley et al., 1976; Taylor et al., 1979; Scott and Rajan, 1979). Such samples have been shock-heated over short time scales, not thermally metamorphosed over a long period. This difference invalidates their type-7 designation.

Tait et al. (2014) suggested that type-7 chondrites are those that underwent partial melting deep within an asteroid. They proposed several criteria for identifying type-7 OC: (1) uniform olivine compositions, (2) orthorhombic low-Ca pyroxene grains containing >1 wt.% CaO and having Ca-depleted rims, (3) exsolution lamellae of low-Ca pyroxene within Ca-pyroxene, (4) compositionally zoned plagioclase grains >>100 μm, with Ca-rich cores and K-rich

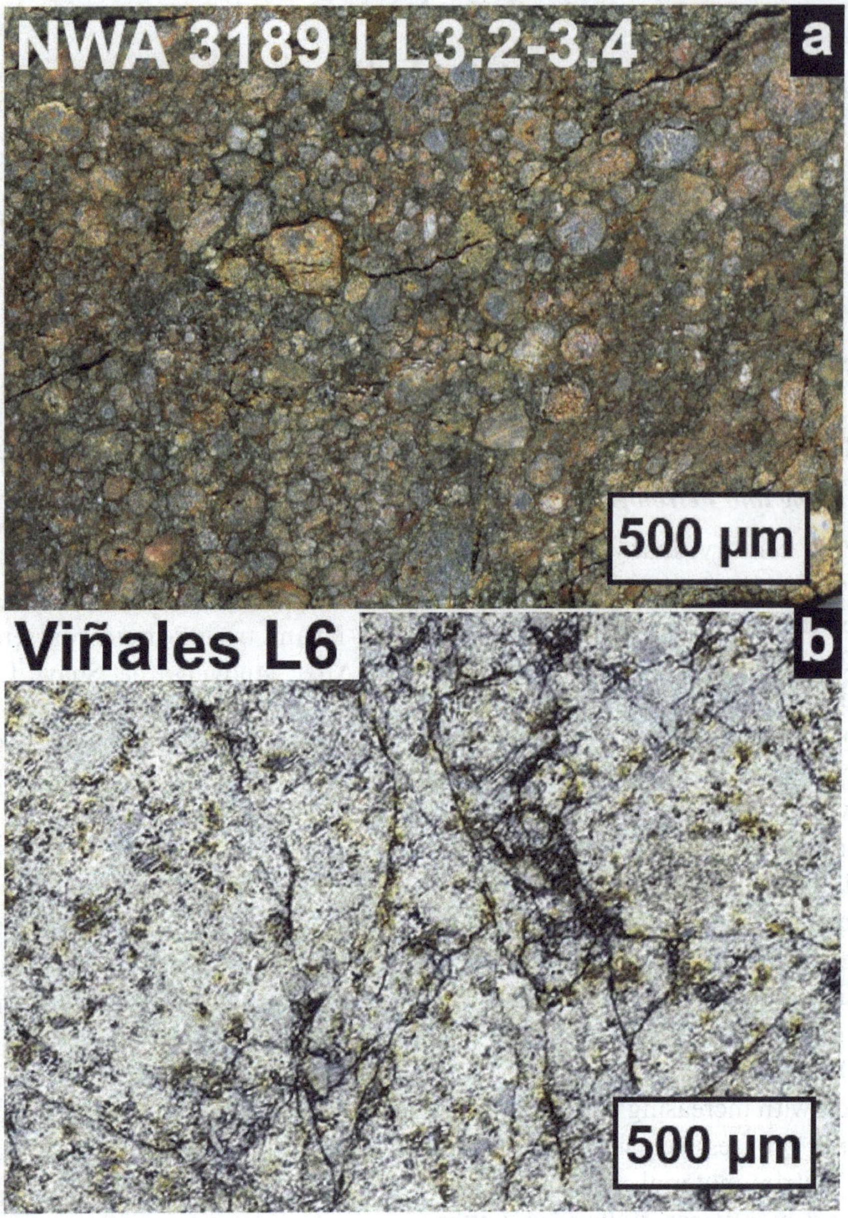

Fig. 13.1 Textural differences between type-3 and type-6 ordinary chondrites. Chondrules are readily discernible in (a) NWA 3189, LL3.15, but completely recrystallized in (b) Viñales, L6. Numerous dark shock veins traverse Viñales. Both images courtesy of James St. John.

rims, forming an interconnected skeletal network, (5) scarce relict chondrules (≤1 cm^{-2} in thin section) that lack glassy mesostases, (6) kamacite and taenite with exsolution relationships, (7) no fine-grained silicate matrix material, (8) bulk chemical fractionations in Al, Si, Ca, Na, K, siderophiles, and chalcophiles, and (9) bulk Eu anomalies. The most straightforward distinction between type 6 and type 7, easily determinable with the electron microprobe, is the >1.0 wt.% CaO in type-7 OC orthopyroxene. Brearley and Jones (1998) also reported that the Wo value of Ca-pyroxene decreases from 45 mol% in type-6 OC to 36–45 mol% in type-7 OC.

There is no unanimity in the meteorite-research community concerning the use of the type-7 designation. There are two principal reasons for reluctance about its adoption: First, as discussed above, at least some chondrites classified as type 7 are impact-melted rocks; and second, as pointed out by Huss et al. (2006), type-7 chondrites "fit comfortably

Fig. 13.2 Recrystallized BO chondrules in H6 Estacado. (a) BO chondrule fragment at lower left showing discontinuous subparallel olivine bars separated by grains of pyroxene, olivine and plagioclase. The subparallel olivine laths in the center of the image likely belong to another recrystallized BO chondrule fragment. (b) Enveloping compound chondrule consisting of a primary BO chondrule surrounded by a thick spherical shell with a PO texture. Crossed polar transmitted light images courtesy of John Kashuba.

within the Van Schmus and Wood (1967) definition of type 6." Because some meteorite classifiers use type 7 and some do not, the statistics of the overall abundance of type-7 OC are unreliable.

There are many more LL chondrites classified as type 7 than H or L chondrites (ignoring possible pairings). According to the February 20, 2024 Meteoritical Bulletin Database, 0.12% of H chondrites, 0.16% of L chondrites, and 1.30% of LL chondrites are classified as "type 6/7," "type 7," or "type 7 melt breccia." If only meteorites designated as "type 7" are considered, then the percentages are:

Table 13.1 *Variations of properties with petrologic type in ordinary chondrites.*

<table>
<tr><th rowspan="2">Category</th><th rowspan="2">Parameter</th><th colspan="13">Petrologic Type</th></tr>
<tr><th>3.0</th><th>3.1</th><th>3.2</th><th>3.3</th><th>3.4</th><th>3.5</th><th>3.6</th><th>3.7</th><th>3.8</th><th>3.9</th><th>4</th><th>5</th><th>6</th></tr>
<tr><td rowspan="4">Bulk Composition</td><td>C (wt.%)</td><td colspan="5">0.3–0.6</td><td colspan="3">0.2–0.5</td><td colspan="5"><0.3</td></tr>
<tr><td>H_2O (wt.%)</td><td colspan="7">1.0–2.5</td><td colspan="6"><1.0</td></tr>
<tr><td>Indium (ng/g)</td><td colspan="10">8–100</td><td colspan="3">0.1–8</td></tr>
<tr><td>$^{36}Ar_p$ (10^{-8} g^{-1} STP)</td><td colspan="3">45–60</td><td colspan="4">15–50</td><td colspan="3">5–15</td><td colspan="2">0.5–5</td><td>0.2–2</td></tr>
<tr><td rowspan="2">Texture</td><td>Chondrule textural integration</td><td colspan="6">None</td><td colspan="2">Incipient</td><td colspan="2">Discernable</td><td>Minor</td><td>Moderate</td><td>Extensive</td></tr>
<tr><td>Chondrule appearance</td><td colspan="10">Very sharply defined</td><td>Well-defined</td><td>Readily delineated</td><td>Poorly defined</td></tr>
<tr><td rowspan="3">Thermoluminescence (TL)</td><td>Sensitivity (×1000, relative to the Dhajala H3.8 chondrite)</td><td colspan="2"><1</td><td>1–2.2</td><td>2.2–4.6</td><td>4.6–10</td><td>10–22</td><td>22–46</td><td>46–100</td><td>100–220</td><td>220–460</td><td>300–600</td><td colspan="2">600–2000</td></tr>
<tr><td>Glow curve shape</td><td colspan="2">Irregular</td><td colspan="11">Sharp peak</td></tr>
<tr><td>Peak temperature (°C)</td><td colspan="2">170–200</td><td colspan="3"><140</td><td colspan="8">>140</td></tr>
<tr><td rowspan="4">Cathodoluminescence (CL)</td><td>Yellow CL in chondrule meso</td><td colspan="2">Common</td><td colspan="3">Rare</td><td colspan="8">Absent</td></tr>
<tr><td>Blue CL in chondrule meso</td><td colspan="2">Present</td><td colspan="3">Common</td><td colspan="8">Dominant</td></tr>
<tr><td>Red CL in chondrule olivine</td><td colspan="5">Present</td><td colspan="5">Rare to absent</td><td colspan="3">Absent</td></tr>
<tr><td>Matrix CL</td><td>Red</td><td>Some red areas</td><td colspan="3">Low CL</td><td colspan="6">Increasingly blue →</td><td colspan="2">Matrix absent</td></tr>
<tr><td rowspan="7">Matrix</td><td>Hydrated phases</td><td>Abundant</td><td colspan="2">Present</td><td colspan="8">Absent</td><td colspan="2">Matrix absent</td></tr>
<tr><td>Presolar graphite</td><td colspan="3">Present</td><td colspan="8">Absent</td><td colspan="2">Matrix absent</td></tr>
<tr><td>Presolar diamonds</td><td colspan="8">Decreasing →</td><td colspan="3">Absent</td><td colspan="2">Matrix absent</td></tr>
<tr><td>FeO/(FeO+MgO) in matrix normalized to whole-rock</td><td colspan="5">>1.4</td><td colspan="3">1.1–1.4</td><td colspan="3"><1.1</td><td colspan="2">Matrix absent</td></tr>
<tr><td>% matrix recrystallization</td><td colspan="4"><20</td><td colspan="6">>20</td><td>100</td><td colspan="2">Matrix absent</td></tr>
<tr><td>Sulfur (wt.%)</td><td>>1.0</td><td colspan="2">0.5–1.0</td><td colspan="8"><0.5</td><td colspan="2">Matrix absent</td></tr>
</table>

<table>
<tr><td rowspan="6">Feldspathic material</td><td>Texture of chondrule meso</td><td colspan="9">Increasing number density of crystallites →</td><td colspan="3">Increasingly recrystallized →</td></tr>
<tr><td>Isotropic glass</td><td colspan="4">Common</td><td colspan="2">Less common</td><td colspan="3">Rare</td><td colspan="3">Absent</td></tr>
<tr><td rowspan="2">Type-I chondrule meso</td><td colspan="4">Modal albite uncommon</td><td colspan="5">Modal albite present</td><td colspan="3">Mesostasis absent</td></tr>
<tr><td colspan="9">Decreasing normative anorthite →</td><td colspan="3">Mesostasis absent</td></tr>
<tr><td>Type-II chondrule meso</td><td>Albite rare</td><td colspan="3">Increasing modal albite →</td><td colspan="5">Modal albite present</td><td colspan="3">Mesostasis absent</td></tr>
<tr><td>Secondary feldspar (µm)</td><td colspan="9">Submicrometer size</td><td>2</td><td>2–25</td><td>50–150</td></tr>
<tr><td rowspan="4">Olivine</td><td>PMD of FeO</td><td colspan="5">>33</td><td colspan="3">15–33</td><td>5–15</td><td colspan="3"><5</td></tr>
<tr><td>Type-I chondrules</td><td colspan="3">FeO<2 wt.%
CaO>0.3 wt.% common</td><td colspan="5">FeO increases to equilibrium value
CaO decreases to equilibrium value</td><td>FeO near equilibrium
CaO near equilibrium</td><td colspan="3">FeO uniform
CaO uniform</td></tr>
<tr><td rowspan="2">Type-II chondrules</td><td colspan="8">FeO converges to equilibrium value</td><td>FeO near equilibrium</td><td colspan="3">FeO uniform</td></tr>
<tr><td>Cr_2O_3>0.3 wt.%</td><td>Cr_2O_3 heterogeneous</td><td colspan="10">Cr_2O_3<0.1 wt.%</td></tr>
<tr><td rowspan="3">Pyroxene</td><td>PMD of FeO low-Ca pyx</td><td colspan="7">>30</td><td colspan="2">20–30</td><td>5–20</td><td colspan="2"><5</td></tr>
<tr><td>Structure of low-Ca pyx</td><td colspan="9">Mainly monoclinic</td><td>Mixed</td><td colspan="2">Orthorhombic</td></tr>
<tr><td>Size of Ca-pyx grains (µm)</td><td colspan="9">Mainly submicrometer grains in chondrule mesostasis</td><td><1</td><td>2–5</td><td>5–30</td></tr>
<tr><td rowspan="8">Non-silicates</td><td>Phosphates</td><td colspan="2">Some Na- and Fe-rich</td><td colspan="7">Ca-rich, increasing modal abundance and grain size →</td><td colspan="3">Ca-rich, large grains</td></tr>
<tr><td>Chromite</td><td colspan="8">Tiny grains in chondrule mesostasis and within olivine phenocrysts</td><td>Moderate-size grains</td><td colspan="3">Coarse grains</td></tr>
<tr><td rowspan="2">Kamacite composition</td><td colspan="3">Co heterogeneity high</td><td colspan="3">Co heterogeneity moderate</td><td colspan="6">Co heterogeneity low</td></tr>
<tr><td>P-, Cr-, Si-bearing</td><td colspan="11">P-, Cr-, Si-free</td></tr>
<tr><td>Mean kamacite grain size (µm)</td><td colspan="9">~20–80</td><td>~50–100</td><td>~100–300</td><td>~100–300</td></tr>
<tr><td>Mean troilite grain size (µm)</td><td colspan="9">~20–80</td><td>~20–100</td><td>~20–150</td><td>~20–150</td></tr>
<tr><td>Taenite</td><td colspan="3">Unzoned</td><td colspan="9">Zoned</td></tr>
<tr><td>Carbides and magnetite</td><td colspan="3">Common</td><td colspan="2">Present</td><td colspan="7">Rare to absent</td></tr>
</table>

PMD is determined in the following way: Obtain the wt.% FeO in each grain of a particular mineral, calculate the mean of these values, determine the absolute value of the deviation of each analysis from that mean, obtain the average of these deviations and divide that value by the average FeO content, then multiply the result by 100. Table reproduced from Rubin and Ma (2021). $^{36}Ar_p$ = primordial ^{36}Ar; meso = mesostasis; pyx = pyroxene; and PMD = percent mean deviation, modified from Dodd et al. (1967).

H (0.11%), L (0.12%), LL (1.23%). Many recrystallized LL chondrites are designated type 7 instead of type 6 solely on the basis of the high CaO in their orthopyroxene. Caution must be exercised here, because orthopyroxene in equilibrated LL chondrites tends to contain more CaO than in equilibrated H or L chondrites. For example, among recent OC falls, orthopyroxene Wo values are typically 1.3–1.8 mol% in H6 and L6 chondrites, but average ~2.1 mol% in LL6 chondrites. Thus, some LL chondrites that are just as metamorphosed as some H and L chondrites could be classified as type 7 while their H and L chondrite counterparts are classified as type 6.

The three main OC groups are not identical in their distributions of petrologic types (Fig. 13.3). As of February 20, 2024, among the set of observed falls in the Meteoritical Bulletin Database, H chondrites comprise: H3 (6.3%), H4 (17.3%), H5 (51.3%), H6 (24.9%), H7 (0.2%); L chondrites comprise: L3 (3.3%), L4 (6.5%), L5 (22.7%), L6 (67.5%), L7 (0%); LL chondrites comprise: LL3 (15.8%), LL4 (8.4%), LL5 (26.3%), LL6 (47.4%), LL7 (2.1%). A slim majority of H-chondrite falls are type 5, a significant majority of L-chondrite falls are type 6, and a large plurality of LL-chondrite falls are type 6. LL-chondrite falls also include a much higher proportion of type-3 samples than the other two groups: 15.8% versus 3.3–6.3%.

The type-3 to -6 OC range was reexamined by Sears et al. (1980) who found that the thermoluminescence (TL) sensitivity of ordinary chondrites increases by a factor of 10^5 from type-3 to type-6. (TL sensitivity is an intrinsic property defined as the amount of TL induced in a sample by a standard dose of radiation. Prior to irradiation, any previously accumulated TL is erased from the sample by annealing it to ≥500°C. TL results for unbrecciated samples are reproducible to within ±5%.) The increase in TL sensitivity during progressive metamorphism is caused almost entirely by the devitrification of chondrule glass and the growth of secondary feldspar crystals (Guimon et al., 1986). The primary feldspathic glass in unrecrystallized chondrules is a poorer phosphor than crystalline feldspar because TL depends on defects and impurities within a crystalline lattice (Garlick, 1970). The TL technique is viable for all three OC groups because these groups are similar in feldspar modal abundance and mineral composition.

In practice, the TL sensitivity of the Dhajala H3.8 fall is set at unity. The overall range in TL sensitivity (normalized to that of Dhajala) from type-4 to type-6 OC is about an order of magnitude (Sears et al., 1980); in contrast, the range among all type-3 OC is about three orders of magnitude. This allows the division of type-3 ordinary chondrites into ten subtypes, from type 3.0 to 3.9 (e.g., Sears et al., 1980, 1982, 1991).

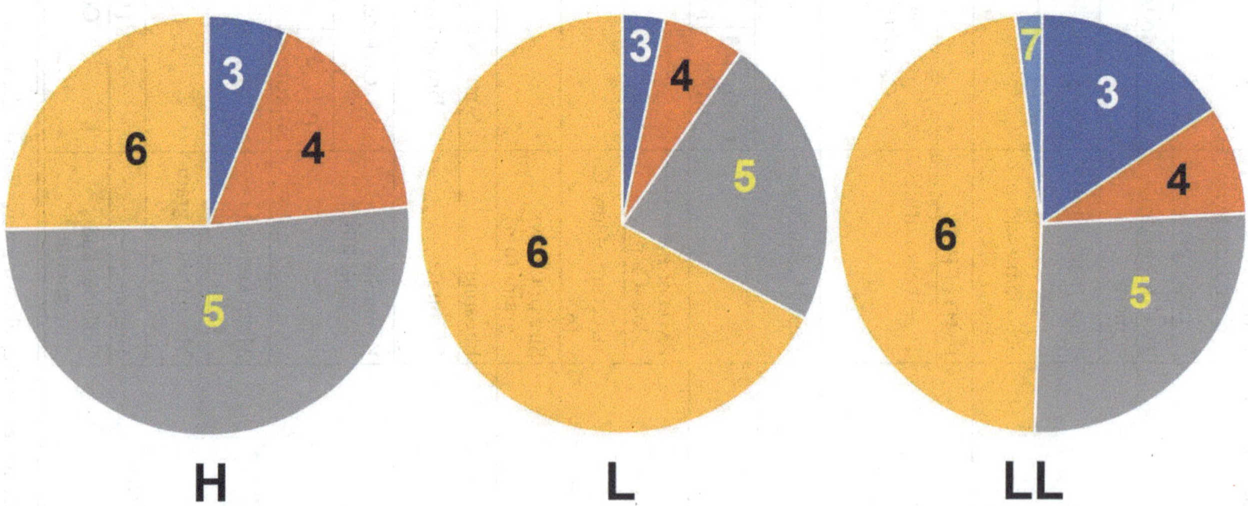

Fig. 13.3 Pie diagrams showing the proportions of petrologic types among observed falls for the three major ordinary-chondrite groups. Although the H-chondrite group contains 0.2% type-7 members, the strip is too narrow to be discerned on the pie chart. There are no L-chondrite falls classified as type 7.

The shape of the TL-sensitivity glow curve also varies among OC subtypes – from irregular below type 3.3 (attributable to the dominance of non-feldspar phosphors; Guimon et al., 1988) to having a well-defined peak for types 3.3–3.6. With increasing subtype (3.7–3.9), the peak gets wider and shifts toward higher temperatures (attributable to the order/disorder transition of feldspar at 500–600°C; Guimon et al., 1985; Sears et al., 1991).

Grossman and Brearley (2005) further subdivided the least metamorphosed type-3 ordinary and CO chondrites. In practice, olivine grains with >2 wt.% FeO can be analyzed with the electron microprobe; their Cr_2O_3 compositions can be plotted in Fig. 13.4 to distinguish the lowest OC subtypes: 3.00 to 3.01 (e.g., Semarkona, LL), 3.05 (e.g., NWA 3302, H), 3.10 (e.g., MET 96503, L), 3.15 (e.g., Bishunpur, LL), and ≥3.2 (e.g., Dhofar 008, L3.3). Kimura et al. (2008) later suggested that Semarkona should be classified as LL3.01 based on the texture and composition of metallic Fe-Ni. (CO chondrites have a different curve [e.g., Grossman and Brearley, 2005; Fig. 1 of Rubin and Li, 2019].)

Kimura et al. (2024) summarized the salient characteristics of type-3.0 chondrites in different chondrite groups. Those pertinent to ordinary chondrites include: (1) ferroan olivine grains with ≥0.25 wt.% Cr_2O_3 and ≥0.15 wt.% CaO, (2) largely unaltered chondrule mesostases, (3) chromite grains with $\lesssim$0.1 wt.% Al_2O_3 and MgO, (4) rare hydrous phases (perhaps <0.05 vol.%; Howard et al., 2015), (5) the presence of amorphous phases in the fine-grained matrix, (6) $\gtrsim$1 wt.% bulk S in the matrix, (7) some metal grains with plessitic textures, (8) positive correlation of Ni and Co in metallic Fe-Ni grains, (9) kamacite with $\gtrsim$0.15 wt.%P, $\gtrsim$0.12 wt.%Cr, and $\gtrsim$0.4 wt.%Si dissolved in kamacite, (10) $\gtrsim$100 μg/g O-bearing presolar grains in the matrix, (11) undisturbed Al-Mg systematics, (12) undisturbed O-isotopic compositions of chondrule phases, (13) TL sensitivity <1 (×1000 relative to H3.8 Dhajala), (14) yellow CL in chondrules, and (14) high average peak widths of the D band in Raman spectra.

Each type-3 OC is unique. With apologies to Tolstoy, every unequilibrated ordinary chondrite is unequilibrated in its own way. Different ordinary chondrites of the same group and petrologic subtype may vary in their olivine and low-Ca-pyroxene compositional distributions, degree of aqueous alteration, shock stage, extent of brecciation, varieties of organic matter, amount of metallic Fe-Ni, and abundances and types of clasts, aberrant grains, compound chondrules, CAIs, fine-grained silicate

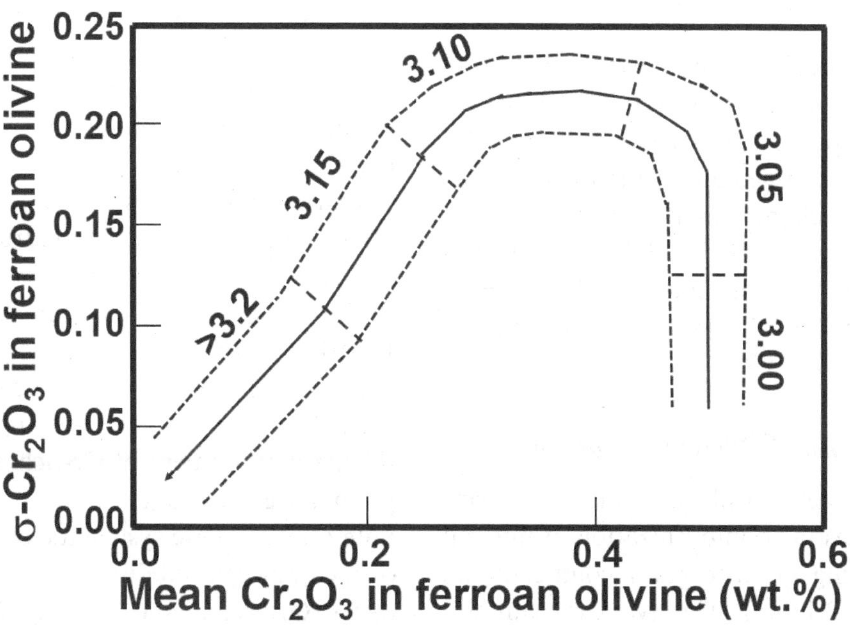

Fig. 13.4 Diagram of $\sigma-Cr_2O_3$ vs. mean Cr_2O_3 in ferroan olivine grains of type-3 ordinary chondrites. Figure is reproduced from Fig. 2.2 and adapted from Grossman and Brearley (2005).

matrix material, presolar grains, and extent of terrestrial weathering. For example, there are distinct differences among such LL3.1 chondrites as Los Vientos 171, NWA 2040, NWA 12751, Watonga, and Y-793384.

Jacquet and Doisneau (2024) proposed a new secondary classification scheme for unequilibrated chondrites of all carbonaceous and noncarbonaceous groups. It was based on the ratio of the mean olivine Fa contents of Type-I chondrules to those of Type-II chondrules ($m = Fa_{I}/Fa_{II}$). They defined a metamorphic scale ranging from M0.0 to M1, wherein M0.0 corresponds to subtype 3.0 and M0.9 corresponds to 3.9; M1 indicates petrologic type 4. Time will tell if the scheme catches on.

There are significant variations among the three OC groups in their proportions of the least equilibrated members – the petrologic 3 subtypes, as shown in Table 5.2.

H3.0–3.2: 6%
H3.3–3.6: 25%
H3.7–3.9: 69%

L3.0–3.2: 13%
L3.3–3.6: 51%
L3.7–3.9: 36%

LL3.0–3.2: 31%
LL3.3–3.6: 36%
LL3.7–3.9: 33%

The H3 chondrites are dominated by the more-metamorphosed samples (type 3.7–3.9) with very few members of type 3.0–3.2. A slight majority of L3 chondrites are of intermediate metamorphic grade (type 3.3–3.6). LL3 chondrites have the highest proportion of members of type 3.0–3.2 (31%) (including Semarkona, Bishunpur, and Krymka) and the lowest proportion of the more-metamorphosed samples from type 3.7–3.9.

13.1.2 Changes in Phase Composition

During thermal metamorphism, grain boundary diffusion facilitates the homogenization of mineral phases, diminishing grain-to-grain compositional variations (Fig. 12.13b,d). The average grain size increases: kamacite increases in mean grain size from ~20–80 μm in type-3 OC to ~50–100 μm in type-4 to ~100–300 μm in type-5 and -6; troilite increases in mean grain size from ~20–80 μm in type-3 OC to ~20–100 μm in type-4 to ~20–150 μm in type-5 and -6 (Table 13.1). Small metallic Fe-Ni and troilite grains in the matrix tend to disappear as the mean grain size increases. Textural integration occurs between chondrules and matrix (Fig. 12.13a,c).

Secondary phases that formed in OC during thermal metamorphism are discussed in Section 12.5. These include orthopyroxene, Ca-pyroxene, plagioclase, phosphate (merrillite and chlorapatite), chromite, ilmenite, and rutile.

At the early stages of thermal metamorphism, there are systematic changes in the compositions of mafic silicate phenocrysts in chondrules. Some olivine phenocrysts in porphyritic chondrules in type 3.3–3.6 OC exhibit significant compositional zoning (e.g., Fa_{3-20}; Scott et al., 1994). Figure 13.5 shows the compositions of olivine phenocrysts in olivine-rich porphyritic chondrules. Although Type-IA chondrules in type-3.0 chondrites contain low-FeO olivine grains, with increasing metamorphism, diffusive exchange between chondrules and matrix causes the olivine grains to become richer in FeO and poorer in CaO. Nevertheless, the overall chondrule texture remains relatively unchanged.

Enrichment of FeO and depletion in CaO in olivine phenocrysts occur in both Type-I and Type-II chondrules as thermal metamorphism progresses from type 3.0 to 3.3 to 3.6 to 4 (Fig. 13.5). In LL4-5 chondrites, olivine grains in Type-IA and Type-IIA chondrules are essentially identical in composition (McCoy et al., 1991).

During the earliest stages of metamorphism, olivine grains in ferroan chondrules tend to become depleted in Cr_2O_3 while the variability in Cr_2O_3 (i.e., $\sigma\, Cr_2O_3$) tends to decrease (Grossman and Brearley, 2005) (Fig. 13.4). Accompanying these changes are (a) the replacement of igneous zoning profiles in olivine grains with complex profiles, (b) the development of Cr-rich coatings on olivine grain margins, and (c) the exsolution of small chromite grains. These systematic changes are the basis of the fine-scale subdivision of type 3.00–3.2 chondrites (Grossman and Brearley, 2005).

Percent mean deviation (PMD) is a measure of compositional variability (defined at the bottom of

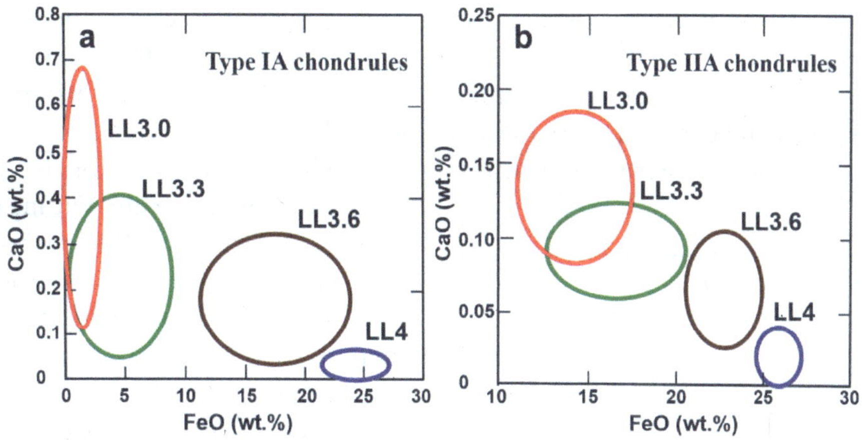

Fig. 13.5 Mean concentrations of FeO and CaO in olivine grains within LL3 olivine-rich porphyritic chondrules. (a) Type-IA chondrules. (b) Type-IIA chondrules. With increasing metamorphism, the chondrules retain their Type-I and Type-II textures even though their olivine grains have become more ferroan and less calcic. Diagrams based on literature data, modified from Scott et al. (1994).

Table 13.1). The PMD of FeO in olivine changes among OC from >33 in type 3.0–3.4 samples to 15–33 in type 3.5–3.7 to 5–15 in type 3.8–3.9 to <5 in type 4–6. Diffusive exchange of FeO and MgO is more sluggish in low-Ca pyroxene than olivine (e.g., Freer, 1981; Chakraborty, 1997), allowing low-Ca pyroxene to be more resistant to metamorphic equilibration. The PMD of FeO in low-Ca pyroxene changes among OC from >30 in type 3.0–3.6 samples to 20–30 in type 3.7–3.9 to 5–20 in type 4 to <5 in type 5–6.

Chondrule mesostasis undergoes structural and compositional changes during metamorphism. Whereas colorless, transparent, isotropic glass (containing small crystallites of Ca-pyroxene) is common in Type-I chondrules in type-3.00 OC (e.g., Fig. 11.5), it undergoes devitrification during heating and is mostly absent by type 3.6 (e.g., McCoy et al., 1991). The abundance of Ca-pyroxene crystallites also increases during low-level metamorphism (DeHart et al., 1992). In going from type 3.0 to 3.6, the mesostases in Type-I chondrules tend to become richer in FeO, Na_2O, and K_2O and poorer in TiO_2 and CaO (McCoy et al., 1991; Sears et al., 1990; Alexander et al., 2000). Albite begins crystallizing within mesostasis in type-3.4 OC (Grossman and Brearley, 2005). In going from type 3.0 to 3.6, mesostases in Type-II chondrules tend to become poorer in FeO and MnO (McCoy et al., 1991). The abundance of crystalline albite increases progressively with petrologic subtype. In both Type-I and Type-II chondrules, albite crystallization causes the K/Na ratio in residual mesostasis to increase to several times the CI-chondrite ratio (Grossman and Brearley, 2005). It also causes an increase in blue CL in these chondrules.

Kamacite likewise undergoes compositional change during metamorphism: the variability in Co and Ni concentration tends to decrease, as is most evident in the difference between type-3 and type-6 OC (with some exceptions) (Table 13.2a). Although P, Si, and Cr occur in solid solution in some metallic Fe-Ni grains in type-3.0 OC (Zanda et al., 1994), these elements diffuse out of the metal by subtype 3.1, reacting with O, Cl, Ca, and Fe outside the metal to form phosphides and chromite.

The Fe-Ni phase diagram (Fig. 13.6) shows that at elevated metamorphic temperatures, metallic Fe-Ni exists as a single phase – taenite (γ-iron, face-centered cubic). As the system cools, metallic Fe-Ni reaches the two-phase Fe-Ni field (the solvus); the temperature of this boundary is between 900°C and 500°C depending on the bulk Ni content of taenite. At the solvus, during cooling, kamacite (α-iron, body-centered cubic) nucleates and begins to crystallize along the {111} planes of the taenite. (Actual kamacite nucleation temperatures are dependent upon the bulk Ni and P concentrations of the parent taenite grain.) With further cooling, both kamacite and taenite simultaneously become richer in Ni as kamacite forms at the expense of taenite.

The intergrowth of kamacite lamellae and Ni-rich taenite lamellae forms the Widmanstätten pattern in which kamacite lamellae are oriented along the planes of an octahedron. This pattern is evident in most iron meteorites (the octahedrites) and pallasites, as well as portions of the Portales Valley and Punggur H-chondrite impact-melt breccias (e.g., Rubin et al., 2001; Ruzicka et al., 2005; Fig. 11.62).

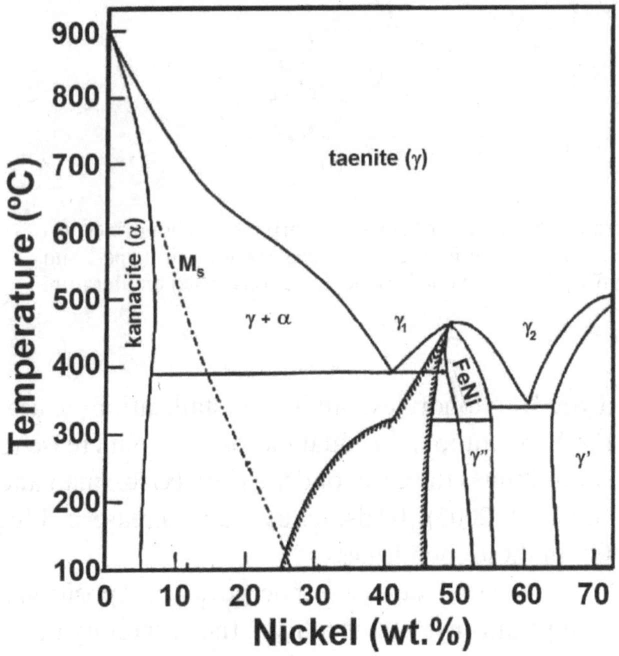

Fig. 13.6 Simplified Fe-Ni phase diagram modified from Reuter et al. (1987) by Rubin and Ma (2021). Kamacite (α) is the bcc (body-centered cubic) phase; taenite (γ) is the fcc (face-centered cubic) phase; M_s is the martensite start temperature; γ_1 is the paramagnetic fcc phase; γ_2 is the ferromagnetic fcc phase; γ' is ordered Ni_3Fe; γ'' is ordered FeNi. The field labeled FeNi above γ'' is the disordered high-temperature form of γ''.

Solid-state diffusion of Ni is more rapid in kamacite than in taenite. During cooling within the two-phase field, as diffusion becomes more sluggish, compositional traces across taenite lamellae (Fig. 13.7), as determined by electron-microprobe analysis, show a buildup of Ni in taenite at the α/γ boundaries and a dip in Ni concentration at the centers of the taenite grains. This "M-shaped" profile can be used to measure metallographic cooling rates: the Ni concentration at the trough near the centers of taenite lamellae and the distance from these troughs to the nearest α/γ boundary are matched against theoretically calculated cooling rate curves (e.g., Wood, 1964, 1967; Willis and Goldstein, 1981; Yang and Goldstein, 2005). The increases and decreases in Ni in Fig. 13.7 are mirrored by corresponding decreases and increases in Fe and Co (Fig. IV-4 of Wasson, 1985).

The increase in Ni concentration in taenite at the α/γ boundary is counterbalanced by a decrease in Ni in adjacent kamacite at the same boundary (Fig. 13.7). This latter feature, known as the "Agrell dip," is named for Stuart Agrell, the Cambridge University mineralogist who discovered the phenomenon (Agrell et al., 1963). Although most evident in iron meteorites, it also occurs in ordinary chondrites (e.g., Skirak et al., 2025).

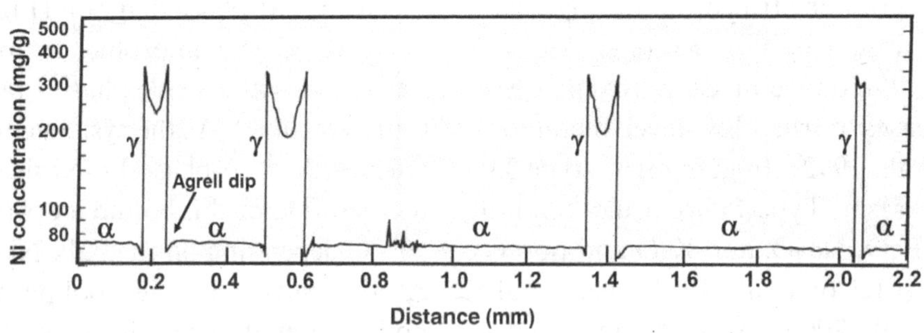

Fig. 13.7 Electron microprobe trace across a slab of the Buenaventura IIIAB iron meteorite showing variations in Ni concentration. In slowly cooled ordinary chondrites, traces across kamacite/taenite grain boundaries would show a similar pattern. Kamacite (α) grains have low Ni contents. The Ni content of an individual taenite (γ) grain is a function of grain width – thin taenite grains have high central Ni contents; wide taenite grains have lower central Ni contents. During whole-rock cooling, taenite grows at the expense of kamacite. Due to the sluggish pace of diffusion, the Ni concentration increases at the edges of taenite grains and is relatively depleted in taenite grain centers. This leads to the so-called M-shaped Ni profile. As Ni diffuses into taenite from kamacite, Ni is depleted in kamacite at the boundary with taenite. This depletion is the Agrell dip. Diagram is from Rubin and Ma (2021).

When the central-trough Ni content of taenite grains is plotted against the distance to the nearest edge of the grain on diagrams with theoretically calculated cooling rate curves, the metallographic cooling rate for a metal-bearing meteorite can be established (e.g., Fig. 13.8). Data fields with points that form a relatively smooth bottom edge subparallel to the cooling rate curves (e.g., H4 Conquista and H5 Beaver Creek) are said to plot coherently – the indicated cooling rates are deemed reliable. If the bottom edge of the data field is not smooth (e.g., L5 Ausson) the data are not coherent, and the inferred metallographic cooling rate is of lower reliability. Such incoherence may result from the inclusion of compositionally aberrant metal grains in the sample (a result of post-metamorphic brecciation), from heterogeneous shock, or from analytical error.

Fine-grained silicate matrix material (Section 11.8) also undergoes textural and compositional changes during thermal metamorphism. Matrix appears opaque in transmitted light in thin sections of standard (30 μm) thickness in type 3.0–3.2 OC but becomes more translucent as amorphous material disappears and recrystallization proceeds. As shown in Fig. 13.9, the set of small matrix olivine grains have heterogeneous compositions in OC of type 3.00; they become somewhat more ferroan and more heterogeneous in type 3.1 (as equilibration with FeO-rich grains begins) and become

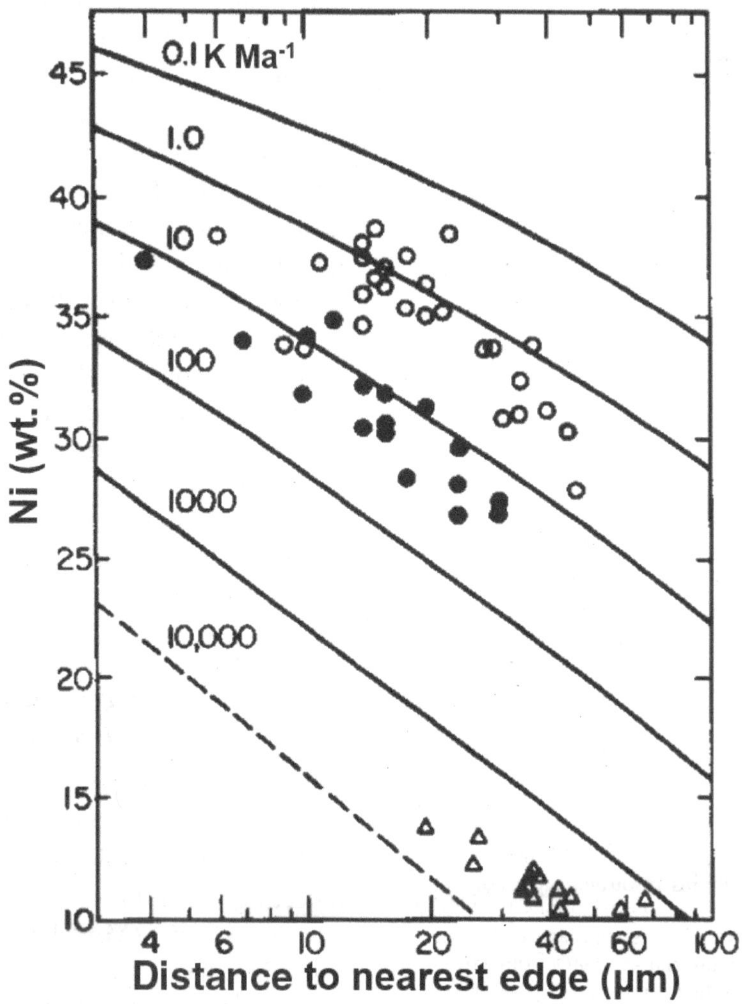

Fig. 13.8 Diagram showing the Ni content in the centers of taenite grains versus the distance to the nearest edge of the grain for three apparently unbrecciated ordinary chondrites. Cooling rate curves are from Willis and Goldstein (1981). Data for H4 Conquista (●) and H5 Beaver Creek (△) plot coherently; those for L5 Ausson (O) are somewhat incoherent and yield a less accurate cooling rate. Diagram from Taylor et al. (1987).

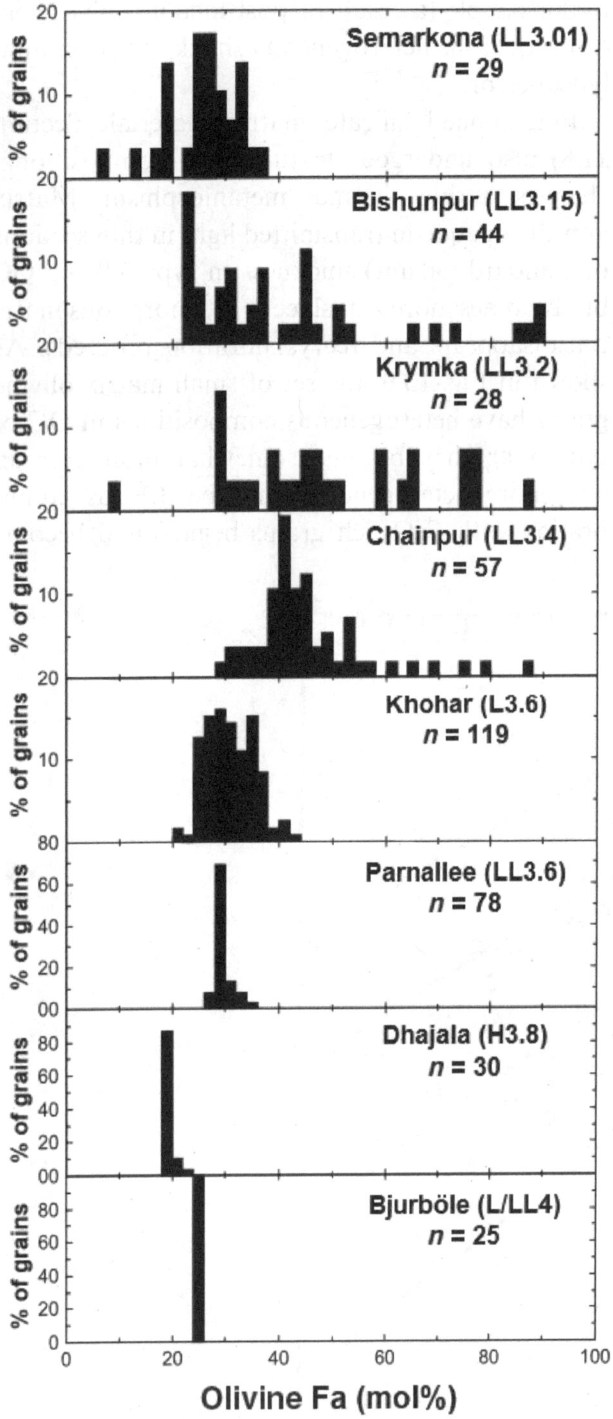

Fig. 13.9 Histograms of the compositional distributions of small matrix olivine grains in ordinary chondrites with increasing petrologic type. Distributions tend to become narrower as equilibrium is approached during parent-body thermal metamorphism. Data from Huss et al. (1991); diagram modified from Huss et al. (2006).

more homogeneous by type 3.6–3.8 as equilibrium is closely approached and effectively reached by about type 4 (Huss et al., 1981).

Matrix in OC of type 3.00 contains numerous small forsterite grains, leading to red CL. During metamorphism, the smallest of these grains equilibrate with FeO (and become more ferroan) by type ~3.2; the coarser grains equilibrate by type 3.5–3.6 (Sears et al., 1990). At that point, the CL of matrix exhibits a strong blue color.

Other changes in matrix material caused by thermal metamorphism include loss of bulk C (Huss et al., 1981); depletion in small metallic Fe-Ni and sulfide grains and growth of coarser grains (e.g., Dodd et al., 1967; Huss et al., 1981; Grossman and Brearley, 2005); nucleation and growth of plagioclase, chlorapatite, and chromite; depletion in bulk S, Na, and Ni (reflecting alkali enrichment in chondrules and recrystallization and coarsening of metallic Fe-Ni and sulfide grains; Wood, 1967; Grossman and Brearley, 2005); and the progressive destruction of presolar grains (e.g., Table 11.8; Fig. 11.48).

There are systematic increases in oxidation state from type 3 to type 6 in all three OC groups (Rubin, 1990, 2005a). This is represented by increases in mean olivine Fa, kamacite Co content, and kamacite Ni content (Table 13.2b; Fig. 10.8). Type-3 and type-4 OC have lower mean olivine Fa and kamacite Co values than type-5 and type-6 OC of the same group. This confirms the results of Scott et al. (1986), who found that the mean FeO contents of ordinary-chondrite olivine and low-Ca pyroxene increase by 3%–5% from type 4 to type 6. Although kamacite Ni contents are variable, it is clear that, in each OC group, the mean kamacite Ni contents in type-3 samples are lower than those in type-6 samples.

13.1.3 Changes in Ordinary-Chondrite Whole-Rocks

The average model porosities of type-6 OC (8.1%) are a bit lower than those of lower petrologic types (9.2–9.7%) (Consolmagno et al., 2008), likely reflecting mineral growth into voids during thermal metamorphism. (Model porosity is calculated by assuming that initial grain densities are uniform in each meteorite class, that all samples once had identical pristine mineral compositions, and that bulk densities were unaffected by terrestrial weathering [Consolmagno et al., 1998].)

Table 13.2a *Mean values and standard deviations (1σ) of olivine Fa for different petrologic types of ordinary chondrites.*

Group/type	Mean Fa (mol%)	σ Fa
H3	17.6	1.6
H4	18.3	0.7
H5	18.6	0.7
H6	19.4	0.6
L3	23.5	1.0
L4	24.0	0.7
L5	24.7	0.8
L6	25.1	0.5
LL3	27.7	0.2
LL4	28.5	2.1
LL5	29.5	1.2
LL6	29.7	1.7

Data from Rubin (1990).

Table 13.2b *Mean values and standard deviations (1σ) of kamacite Co and Ni for different petrologic types of ordinary chondrites.*

Group/type	Mean Co (mg/g)	σ Co	Mean Ni (mg/g)	σ Ni
H3	4.6	0.2	66.7	4.8
H4	4.6	0.1	70.7	2.9
H5	4.7	0.2	68.8	3.0
H6	4.8	0.2	68.7	1.8
L3	7.8	1.0	57.1	5.4
L4	7.8	0.6	64.8	4.8
L5	7.9	0.6	65.5	6.6
L6	8.2	0.5	65.7	4.3
LL3	17.4	19.7	53.5	10.0
LL4	18.9	6.3	58.0	9.7
LL5	31.7	14.3	51.7	6.3
LL6	27.4	14.1	54.9	11.6

Data from Rubin (1990). The high-Co, low-Ni metallic Fe phases were excluded from the LL mean values.

Thermal metamorphism also causes a loss of volatile species from ordinary chondrites. For example, as discussed in Section 10.8.1, bulk H_2O^+ (chemically bound water) decreases in OC falls from 1.06±0.54 wt.% H_2O^+ ($n = 10$) in type 3 to 0.65±0.72 wt.% ($n = 17$) in type 4 to 0.32±0.36 wt.% ($n = 26$) in type 5 to 0.27±0.33 wt.% ($n = 43$) in type 6 (Jarosewich, 1990). As shown in Table 13.1, bulk C decreases in OC from 0.3–0.6 wt.% in type 3.0–3.4 to 0.2–0.5 wt.% in type 3.5–3.7 to <0.3 wt.% in type 3.8–6. Among L-chondrite falls, median bulk C decreases systematically from L3 (0.32 wt.%, $n = 3$) to L4 (0.21 wt.%, $n = 4$) to L5 (0.16 wt.%, $n = 13$) to L6 (0.06 wt.%, $n = 34$) (Jarosewich, 1990). Primordial noble gases are also lost during metamorphism (e.g., Heymann and Mazor, 1968; Alaerts et al., 1979); $^{36}Ar_p$ (10^{-8} g^{-1} STP), for example, decreases from 45–60 in type 3.0–3.2 to 15–50 in type 3.3–3.6 to 5–15 in type 3.7–3.9 to 0.5–5 in type 4–5 to 0.2–2 in type 6 (Table 13.1).

Huss et al. (1996) studied P1, the principal component of "normal planetary" noble gases in primitive chondrites. They found that, with increasing thermal metamorphism, the abundance of the P1 component decreases and the median release temperature increases. A viable model for the origin of the P1 carrier is that it consists of gases trapped within mantles around carbonaceous grains that had formed by ion-molecule reactions on the surfaces of dust grains in the parental molecular cloud (Huss and Alexander, 1987). (The materials within the molecular cloud had been influenced by complex interactions of turbulence, gravity, magnetic fields, and radiation.) Organic materials are readily oxidized to CO and CO_2 during metamorphism, accounting for the correlation in OC between bulk C and the concentrations of noble gases (Huss et al., 2006).

In Section 10.7, it was shown that the concentrations of volatile mobile trace elements in OC tend to decrease with increasing degree of metamorphism. Thus, in going from H4 to H5 or from H5 to H6, there is a tendency for the concentrations of these volatile elements to decrease (Table 10.10). The concentrations of Cs, In, Cd, Tl, and Bi in type 4–6 OC are lower by one or two orders of magnitude than those in type-3 OC (e.g., Tandon and Wasson, 1968; Dodd, 1969; Lipschutz and Woolum, 1988).

The effects of parent-body thermal metamorphism on organic matter in ordinary chondrites were discussed in Section 11.9. As shown in Fig. 11.49, the N peak yield temperature is positively correlated with the C/N ratio and negatively correlated with the matrix-normalized C abundance. These correlations are caused by graphitization (which increases the C/N ratio) and oxidation (which destroys nanodiamonds and some insoluble organic matter) during thermal metamorphism.

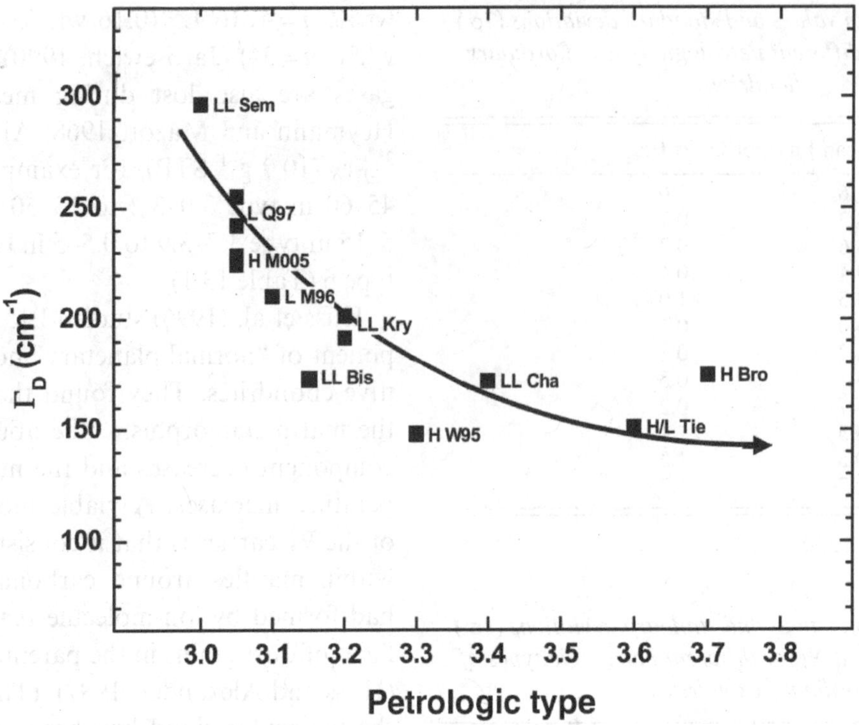

Fig. 13.10 The width of the D band (actually the full width at half maximum, Γ) determined by Raman spectroscopy vs. the petrologic subtype of 10 type-3 ordinary chondrites. The Γ_D value decreases steadily with increasing petrologic type. Sem = Semarkona (LL3.01); Q97 = Q 97008 (L3.05); M005 = MET 00526 (H3.05); M96 = MET 96515 (L3.10); Kry = Krymka (LL3.2); Bis = Bishunpur (LL3.15); W95 = WSG 95300 (H3.3); Cha = Chainpur (LL3.4); Tie = Tieschitz (H/L3.6); Bro = Brownfield (1937) (H3.7). There are replicate values for Q97 and M005. Diagram modified from Busemann et al. (2007).

Busemann et al. (2007) used confocal imaging Raman spectroscopy to analyze insoluble organic matter (IOM) in type-3 ordinary chondrites as well as unequilibrated members of other chondrite groups. This technique is based on the detection of inelastically scattered photons to determine the vibrational modes of molecules. Raman band positions are labeled D (for "disordered") and G (for "graphite") and are presented as shifts relative to the wavelength generated by the laser beam used to excite the molecules. The D band in Busemann et al. (2007) is situated at ~1355 cm^{-1} and is caused by "defects, crystal boundary effects, polycrystallinity and small domain sizes." The Greek letter Γ (uppercase gamma) represents the full width of the band at half the maximum peak height (FWHM). Figure 13.10 shows that the value of Γ for the D band decreases systematically among type-3 OC with increasing subtype. The narrowing of this band during the early stages of thermal metamorphism reflects kinetically controlled oxidation, destruction, and graphitization of primitive organic materials. These materials were initially ubiquitous and fairly homogeneous among unprocessed chondrites throughout the solar nebula (e.g., Alexander, 2005).

13.1.4 Metamorphic Temperatures

As discussed by Huss et al. (2006), peak metamorphic temperatures in type-3 ordinary chondrites can be estimated by phase changes or mineral reactions that have been determined experimentally to occur at low pressure (≤1 kb; Dodd, 1981; McSween and Patchen, 1989) and at specific temperatures. LL3.01 Semarkona reached a maximum temperature between ~200°C (based on the abundance of noble gases in presolar diamonds; Huss and Lewis, 1994) and 260°C (based on clay mineralogy; Alexander et al., 1989a). LL3.15 Bishunpur reached a maximum temperature of 300°C to 350°C (based on the Ni content of matrix metal; Rambaldi and Wasson, 1981). Although Brearley (1990) deduced that type 3.4–3.6 OC reached maximum temperatures of 350°C to 400°C (based on the degree of ordering of poorly graphitized carbon; Reitmeijer and MacKinnon, 1985), Sears et al. (1991) estimated that OC of type 3.5 reached higher maximum temperatures – between 500°C and 600°C (based on the order/disorder transition of feldspar). The Sears et al. value is similar to that inferred by Huss and Lewis (1994) for LL3.6 Parnallee (~600°C).

Equilibration temperatures commonly differ from peak metamorphic temperatures in that equilibration temperatures are based on diffusive equilibration between coexisting minerals. For example, several H3.7–3.8 chondrites have equilibration temperatures between 600°C and 700°C (based on the Cr-spinel/olivine thermometer; Wlotzka, 1985, 1987).

The maximum metamorphic temperatures of equilibrated ordinary chondrites have been estimated: type 4 (~600°C to 700°C; e.g., Wlotzka, 1985, 1987); type 5 (700°C to 750°C; Dodd, 1981); type 6 (~820°C to 930°C; Dodd, 1981; Olsen and Bunch, 1984; Slater-Reynolds and McSween, 2005). One could also define the maximum temperature for a type-6 OC to be that which causes incipient whole-rock melting, between ~940–950°C (e.g., Dodd, 1981) and ~1100°C (Jurewicz et al., 1995). Alternatively, if the type-7 designation is used, it could plausibly be assigned a peak temperature between ~930°C and ~1000°C or ~1100°C.

There are significant uncertainties with respect to these values. Nakamuta and Motomura (1999) determined that type-6 ordinary chondrites experienced metamorphic temperatures of 725–820°C based on their studies of Si-Al disorder in plagioclase. However, because plagioclase may be completely crystallized prior to peak metamorphic temperatures (Slater-Reynolds and McSween, 2005), metamorphic temperatures based on plagioclase are likely to be underestimated. Metasomatic temperatures based on plagioclase alteration also tend to give lower temperatures for type-6 OC (~600°C; Lewis et al., 2022).

A different approach is to use olivine-spinel thermometry: Kessel et al. (2007) found that the average equilibration temperatures for type-4 to type-6 chondrites span only a relatively small range (586–777°C). Although they found that maximum metamorphic temperatures for individual chondrites increase with petrologic type, the increases are small: H4 (733–739°C) to H5 (739°C) to H6 (754°C); LL4 (670°C) to LL5 (727–741°C) to LL6 (777°C). There is no consensus on what the correct average metamorphic temperatures are for different OC types; there may also be differences in the ranges for the three OC groups.

Although it has been suggested that type 4–6 OC may have all reached similar peak temperatures but cooled at different rates, with lower petrologic types cooling faster (Harvey et al., 1993; Kessel et al., 2007), this scenario seems inconsistent with the absence of a correlation between metallographic cooling rate and petrologic type in ordinary chondrites (Taylor et al., 1987; Scott et al., 2014). This method should accurately record cooling rates below ~500°C.

Different mineral systems provide varying equilibration temperatures due to differences in closure temperature. Huss et al. (2006) concluded that Ca-Mg partitioning between Ca-pyroxenes yields the most accurate peak metamorphic temperatures: 800°C to 950°C for H6; 850°C to 950°C for L6; and 900°C to 960°C for LL6 (McSween and Patchen, 1989; Harvey et al., 1993).

The three major ordinary-chondrite groups differ in silicate/metallic Fe-Ni ratio, mean chondrule size, and in the proportions of Type-I and Type-II chondrules. Their parent asteroids may have been of different sizes. They may not have accreted at exactly the same time and thus may have acquired somewhat different abundances of ^{26}Al, plausibly accounting for differences in the proportions of petrologic types (e.g., Fig. 13.3). The three ordinary-chondrite bodies certainly experienced distinct collisional histories. There is no reason to assume that the H, L, and LL parent asteroids had identical thermal histories.

13.1.5 Metamorphic Timescales

Radiometric dating provides the best estimates for the timescales of metamorphism among ordinary-chondrite parent asteroids.

If it is assumed that (1) ^{26}Al was distributed more-or-less homogeneously in the solar nebula and (2) OC parent asteroids were heated mainly by the decay of ^{26}Al (because they accreted early enough and grew large enough), then the relative abundance of ^{26}Al in chondrites and in their components can provide constraints on the early metamorphic history of OC parent bodies.

The inferred initial ^{26}Al/^{27}Al ratio in chondrules from LL3.01 Semarkona and LL3.15 Bishunpur are consistent with OC chondrule formation having mostly occurred 1–2 Ma after CAIs (Kita et al., 2000, 2005). Sugiura and Fujiya (2014) found correlations between meteorite accretion ages and ε^{54}Cr; they concluded that OC accreted 2.14 ± 0.1 Ma after the formation of CAIs. This is consistent with Hf-W isotope data (Hellmann et al., 2019) indicating that the metal-silicate fractionation among OC (attributed by Rubin [2022a], in part, to loss of metal globules from spinning molten chondrules) occurred between ~2 and ~2.7 Ma after CAIs. Rubin (2022a) suggested that IIE iron meteorites accreted earlier than other ordinary-chondrite groups. Kruijer and Kleine (2019) concluded from Hf-W model ages that the IIE irons underwent metal-silicate separation in their hot, partially molten parent asteroid at a time less than 4–5 Ma after CAIs. As discussed by Huss et al. (2006), the inferred initial ^{26}Al/^{27}Al ratios for a clast from L3.7 Bovedy and for plagioclase from H4 Ste. Marguerite are consistent with these two meteorites' having cooled below the blocking temperature of the Al-Mg system by ~6 Ma after CAIs. A basaltic clast from L5 Barwell had cooled down by ≥7 Ma after CAIs (Huss et al., 2001).

In summary:

(1) OC chondrules formed mainly 1–2 Ma after CAIs.
(2) OC parent bodies accreted 2.14 ± 0.1 Ma after CAIs.
(3) Type 3.0–3.6 OCs underwent thermal metamorphism between ~2.1 and ~6 Ma after CAIs.
(4) The IIE iron parent asteroid partially differentiated <4–5 Ma after CAIs.
(5) Type-3.7 to type-4 OCs underwent thermal metamorphism ~6 Ma after CAIs.
(6) Type-5 OCs underwent thermal metamorphism ≥7 Ma after CAIs.

The duration of OC metamorphism was estimated by Humayun and Campbell (2002). They measured W in OC metal by laser ablation inductively coupled plasma mass spectrometry (ICP-MS) and found that metal grains in the matrices of type-3 OC are depleted in W relative to metal grains in equilibrated OC. They concluded that W is incorporated into metal during parent-body metamorphism. The differences between ages determined by ^{182}Hf–^{182}W and ^{207}Pb–^{206}Pb (or ^{129}I–^{129}Xe) dating measure the time interval between reduction of W (which was initially lithophile) and incorporation into metal grains at the early stages of metamorphism (when OC reached petrologic type 3.0–4) and the closure temperatures of Pb (or Xe) during parent-body cooling. The interval for H chondrites was 2–12 Ma. (However, there are uncertainties in the Hf-W dates. Archer et al. [2019] studied seven equilibrated H chondrites and found that W in silicates and oxides was only partially equilibrated with metal by the time diffusive transport was halted during post-metamorphic cooling.)

Trieloff et al. (2003) performed ^{244}Pu-fission-track and Pb–Pb dating on seven to nine H chondrites. The Pb–Pb ages indicate the time it took ordinary chondrites to cool below ~480°C: H4 (4–6 Ma), H5 (10–16 Ma), H6 (45–63 Ma) after CAI formation. In contrast, ^{244}Pu fission tracks are retained at lower temperatures (~280°C for pyroxene; ~120°C for merrillite) and thus record later epochs in the cooling history of metamorphosed asteroids from their peak temperatures: H4 (~10 Ma), H5 (50–65 Ma), H6 (55–65 Ma) after CAIs.

13.1.6 Potential Heat Sources

Ordinary chondrites can be arranged in a sequence of increasing degrees of recrystallization and mineral equilibration (Van Schmus and Wood, 1967). One model to account for this sequence is autometamorphism, also known as hot accretion, in which batches of hot chondritic material (with internal temperatures of ~700°C to 900°C) accreted to an asteroid and then began to cool. In this model, type-3 chondrites cooled quickly, while type-4, -5, and -6 chondrites were buried at greater depths and cooled increasingly more slowly (e.g., Anders, 1964; Larimer and Anders, 1967; Hutchison et al., 1980; Hutchison, 1996). Various arguments have been cited that appear to support hot accretion, including the composition and fine structure of pyroxene grains (Ashworth, 1980; Ashworth et al., 1984; Watanabe et al., 1985), the composition and texture of metallic Fe-Ni (Molin et al., 1994; Artoli and Davoli, 1994; Hutchison et al., 1980; Smith and Launspach, 1991), the occurrence of misshapen chondrules (Hutchison et al., 1979), the texture of fine-grained silicate-rich matrix material (Hutchison and Bevan, 1983), the depletion in chondrite whole-rocks of volatile trace elements (Larimer and Anders, 1967; Binz et al., 1976), and the random orientations in chondrites at ~1-mm^3 scales of stable natural remanent magnetization (NRM) (Morden and Collinson, 1992). Some of these arguments are discussed in detail in Section 14.2. Rubin and Brearley (1996) evaluated all the arguments offered in support of hot accretion and showed that the data are more consistent with prograde metamorphism of initially cold material. They found the hot accretion model to be untenable.

According to the prograde metamorphism model, cold, unequilibrated, type-3 material was progressively heated within a parent body (Sorby, 1877; Merrill, 1921; Wood, 1962a; Van Schmus and Koffman, 1967; Dodd, 1969; Lux et al., 1980) – type-3 chondrites were heated the least, type-6 (or type-7) chondrites were heated the most. Among type-3 chondrites, samples of type 3.0 were heated the least and samples of type 3.9 the most. Mechanisms proposed to account for asteroidal heating include: (a) heating by decay of short-lived radionuclides, particularly ^{26}Al (e.g., Reeves and Audouze, 1968; Herndon and Herndon, 1977; Miyamoto et al., 1981; Grimm and McSween, 1993; Bennett and McSween, 1996), (b) collisional heating (e.g., Wasson et al., 1987; Cameron et al., 1990, 1991; Rubin, 1995a, 2002b, 2004b; Dixon et al., 2004), (c) electromagnetic induction in the protosolar wind during an early T-Tauri phase of the Sun (Sonett et al., 1970; Herbert and Sonett, 1979; Herbert, 1989), (d) FU-Orionis-type solar events (Wasson, 1992), and (e) exothermic chemical reactions provided by the internal energy of interstellar particles that had accreted into larger aggregates (Clayton, 1980). These mechanisms are not mutually exclusive; for example, some ordinary chondrites may have already been hot (from internal heating of their parent body due to ^{26}Al decay) when they were heated locally by collisions (e.g., Hugo et al., 2020; Rubin, 2025).

Although the decay of ^{60}Fe ($t_{½} = 2.26$ Ma) had been considered a possible source of asteroidal heating (e.g., Shukolyukov and Lugmair, 1993; Tachibana and Huss, 2003; Tachibana et al., 2005; Mostefaoui et al., 2003), more-recent measurements of a chondrule in LL3.01 Semarkona, troilite grains from several type-3 ordinary and carbonaceous chondrites, and mineral separates from the ungrouped achondrite Erg Check 002 provided evidence that the ^{60}Fe abundance in the early Solar System was very low: $^{60}Fe/^{56}Fe_{initial} = 7.71 \pm 0.47 \times 10^{-9}$ (Trappitsch et al., 2018; Kodolányi et al., 2022; Fang et al., 2025).

The same processes responsible for heating chondritic parent asteroids also likely caused some asteroids to melt, resulting in the creation of differentiated asteroidal meteorites: namely, CC and NC magmatic irons, HEDs, pallasites, mesosiderites, angrites, aubrites, and scores of ungrouped achondrites. One OC-related fully differentiated asteroid is the parent body of IVA irons.

Electromagnetic induction, FU-Orionis-type events, and exothermic chemical reactions have fallen into disfavor (e.g., Wood and Pellas, 1991; Bell et al., 2000; Rubin, 2000, 2002c) and are now seldom mentioned as viable asteroidal heating mechanisms. Most researchers currently favor heating by the decay of ^{26}Al, but collisional heating is still actively discussed. Both mechanisms undoubtedly heated asteroids; the question is which one was dominant early in Solar System history.

It is clear live ^{26}Al was present in the early Solar System; its decay product (excess ^{26}Mg) has been identified in diverse chondritic components (e.g., Lee et al., 1976; MacPherson et al., 1995). (There are three stable isotopes of Mg: ^{24}Mg [78.99%], ^{25}Mg [10.00%], ^{26}Mg [11.01%].) Numerous workers have inferred that ^{26}Al was distributed more-or-less homogeneously in the early Solar System and that the initial $^{26}Al/^{27}Al$ ratio was roughly 5×10^{-5}. Models of asteroid thermal evolution (e.g., Fig. 13.11) show that the cores of 100-km-diameter bodies could have reached peak temperatures >1000°C and maintained these temperatures for at least 10 Ma (e.g., Miyamoto et al., 1981; Bennett and McSween,

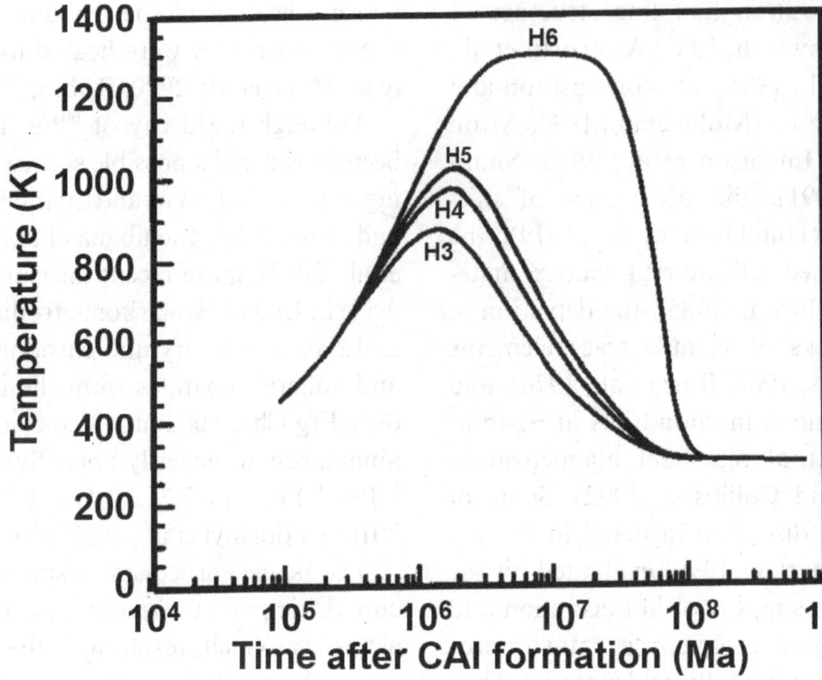

Fig. 13.11 Time–temperature curves for a model H-chondrite asteroid heated solely by the decay of ^{26}Al. The body is assumed to be 88 km in radius, with a low porosity and an initial $^{26}Al/^{27}Al$ ratio equivalent to that present in a chondritic body that accreted 2 Ma after CAI formation. It is further assumed that the ambient nebular temperature at the time of accretion was 300 K. Under these constraints, the chondritic asteroid develops an onion-shell structure wherein the most deeply buried materials reach the highest metamorphic temperatures; they reach their maximum temperatures later than the less metamorphosed materials situated closer to the asteroid surface. Diagram based on Bennett and McSween (1996) and modified from McSween et al. (2002).

1996; McSween et al., 2002). Based on inferred initial $^{26}Al/^{27}Al$ ratios in chondrules, Kunihiro et al. (2004) calculated that the maximum temperatures in OC asteroids could have reached ~1100°C due solely to the decay of ^{26}Al; this is appreciably higher than the commonly estimated metamorphic temperatures of type-6 OC (~820°C to 930°C; Dodd, 1981; Olsen and Bunch, 1984).

If these models represent reality, they could readily account for ordinary-chondrite metamorphism – the estimated model peak temperatures exceed those inferred for type-6 chondrites. The current consensus in the cosmochemistry community is that the dominant heating mechanism responsible for OC metamorphism is ^{26}Al decay.

Internal heating of chondrite parent bodies through the decay of ^{26}Al ($t_{1/2} = 717$ ka) is likely to produce an asteroid with an onion-shell structure (Fig. 5.12a), as discussed in Section 5.4. Such asteroids accreted early enough to acquire substantial concentrations of ^{26}Al; they grew large enough (and hence developed sufficiently low surface-area/volume ratios) to avoid cooling quickly by radiation into interplanetary space. In these bodies, the more highly metamorphosed samples resided in the center and less-metamorphosed materials were closer to the surface (e.g., Dodd, 1969; Anders, 1978; Minster and Allègre, 1979; Pellas and Storzer, 1981; Miyamoto et al., 1981; Trieloff et al., 2003).

Whereas collisions and ^{26}Al decay could both have operated 4.5–4.6 Ga ago, only collisions could have caused substantial heating of asteroids hundreds of millions of years later. A prime example is MIL 99301, an LL6 chondrite sporting numerous shock features (e.g., chromite-plagioclase assemblages, opaque trails within olivine grains, metallic Cu, polycrystalline troilite, myrmekitic plessite, coarse low-Ca clinopyroxene with polysynthetic twinning). However, olivine and plagioclase grains in this rock exhibit sharp optical extinction, giving MIL 99301 an apparent shock stage of S1 (i.e., unshocked) (Rubin, 2002b).

The impact responsible for the prominent shock effects in MIL 99301 must have occurred after the epoch of thermal metamorphism to type-6 levels; otherwise, the polycrystallinity of the troilite would have disappeared and the low-Ca clinopyroxene would have transformed into orthopyroxene. The occurrence of olivine with sharp optical extinction (instead of undulatory extinction or mosaicism) and the occurrence of crystalline plagioclase with sharp optical extinction (instead of undulatory extinction or partial maskelynitization) indicate that MIL 99301 experienced post-shock annealing. As pointed out in Section 5.4, during such annealing, the shock-induced damage of the olivine and plagioclase crystal lattices must have healed (Bauer, 1979). Because elemental diffusion in olivine is faster than in low-Ca pyroxene (Freer, 1981; Chakraborty, 1997), olivine grains can be healed at lower temperatures and/or shorter heating durations than those required to transform low-Ca clinopyroxene into orthopyroxene.

Dixon et al. (2004) showed that the annealing of MIL 99301 took place 4.26 Ga ago (i.e., 300 Ma after accretion). By this time – equivalent to more than 400 half-lives of ^{26}Al – there may not have been a single ^{26}Al atom left over from the era of accretion. Impacts must have been responsible for the annealing. MIL 99301 was probably annealed to metamorphic levels approximating petrologic type 4; more-significant heating would have transformed the low-Ca clinopyroxene into orthopyroxene.

Although MIL 99301 was annealed long after accretion, there is evidence that impact-induced heating also occurred *early* in Solar System history. Annealed impact-melt-rock clasts occur in many ordinary-chondrite fragmental breccias; the clasts were recrystallized along with their hosts. This indicates that impact events preceded or accompanied whole-rock thermal metamorphism (Rubin et al., 1983b).

Impact heating on asteroids cannot be an efficient global process (Keil et al., 1997; McSween et al., 2002). Nevertheless, advanced simulations imply that impact-heating was likely a significant local heat source (Davison et al., 2010, 2012, 2013), particularly in porous bodies. It seems plausible that the same impact event that shocked MIL 99301 also buried the rock beneath the crater floor in the vicinity of impact melts or deposited the rock within a hot, thick ejecta blanket where annealing occurred.

A possible scenario is that impact-induced metamorphism within slow-cooling masses in the proximity of a crater (Fig. 13.12) occurred on porous rubble-pile asteroids (e.g., Davison et al., 2012, 2013). According to this model, most of the asteroid's surface may have consisted of unrecrystallized type-3 OC material until collisions heated portions of the asteroid to different extents in the proximity of impact craters. This scenario would not account for the positive correlation between shock stage and petrologic types among OCs (Figs. 5.19, 5.21).

These models are not mutually exclusive. Collisions can cause heating and annealing of any target material irrespective of its initial temperature at any point in an asteroid's history.

In impact-crater formation, a residual component of the kinetic energy of the projectile is converted into thermal energy after the shock wave dissipates. Thermal energy is released via non-adiabatic decay of the shock wave (Raikes and Ahrens, 1979) and can induce static metamorphism in rocks beneath the crater floor. For example, maximum impact-induced temperatures in local regions within a few kilometers of the center of the Vredefort Dome in South Africa exceeded 1350°C (Gibson, 2002). (However, care must be taken in making analogies between impact craters on Earth [equatorial diameter =12,756 km] and on 200-km-diameter asteroids.)

During impacts, kinetic energy is converted into heat largely by $P \cdot \Delta V$ effects, where P is pressure and ΔV is the change in volume. In rubble piles, kinetic energy derived from the projectile is attenuated through relatively small volumes of material and efficiently converted into heat in the crater vicinity. The greater the porosity of the body, the larger the ΔV, and the more efficient the conversion of kinetic energy into heat (Melosh, 1989; Housen and Holsapple, 1999). Impacts into asteroidal rubble piles might deposit thick insulating layers of finely crushed debris above strongly heated target rocks, allowing deeply buried rocks to cool slowly and anneal (Rubin, 1995a).

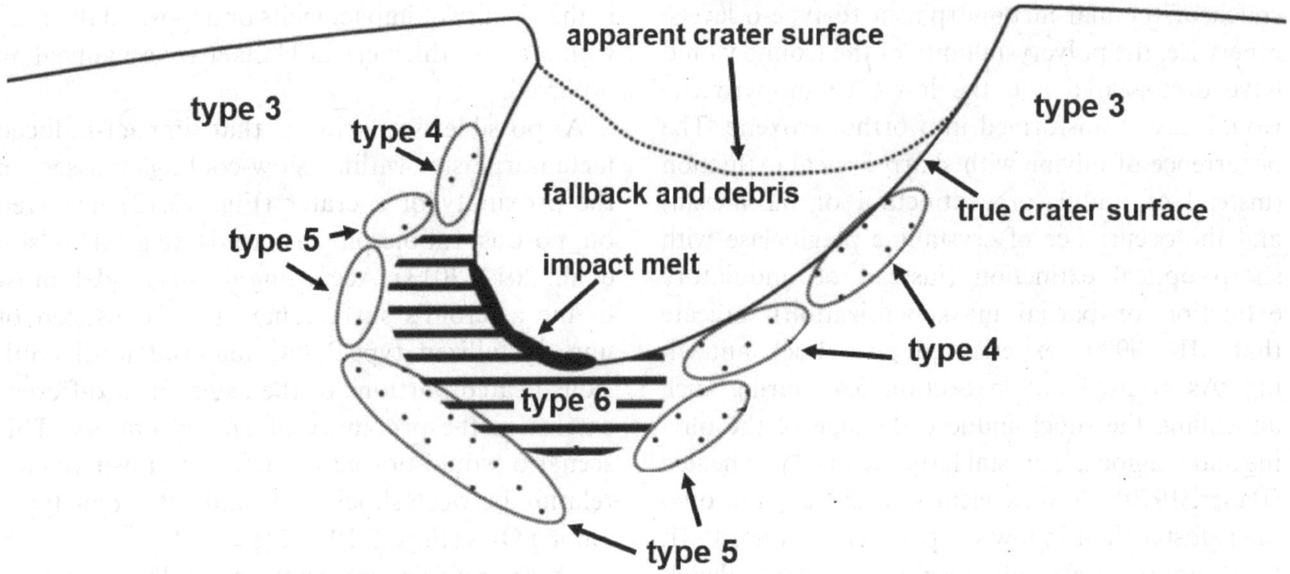

Fig. 13.12 Diagrammatic model of an impact crater formed on an initially highly porous ordinary-chondrite parent body with a surface composed mainly of type-3 material. The depth/diameter ratio of the true crater is close to unity. Much of the material at the collision site is compressed and heated rather than ejected from the crater. Materials adjacent to the impact melt at the bottom of the crater are heated significantly by contact metamorphism. Subsequent collisions affecting the hotter, type-6 materials would lead to higher apparent shock stages than collisions of similar kinetic energy affecting cooler materials of lower petrologic type. Diagram modified from Friedrich et al. (2014a) (based on a diagram from Rubin [2005b]).

13.2 Shock Metamorphism

13.2.1 Ordering into Shock Stages

Petrographic features in meteorites, now recognized as having been produced by shock processes, have long been noted (e.g., Tschermak, 1872, 1883; Brezina, 1904; Farrington, 1915). Several schemes were developed in the late 1960s to rank meteorites by their degree of shock (e.g., Heymann, 1967; Carter et al., 1968; Van Schmus and Ribbe, 1968; Taylor and Heymann, 1969). The most widely used classification scheme in the 1980s was one proposed by Dodd and Jarosewich (1979, 1982) who examined a suite of equilibrated L chondrites and described melt pockets within them. These pockets are equant to sinuous in shape and commonly a few hundred micrometers in size. They contain alkali-rich silico-feldspathic glass. The modal abundance of melt pockets correlates positively with the degree of deformation of olivine and plagioclase and negatively with the whole-rock abundance of ^{40}Ar.

Dodd and Jarosewich (1979) divided L chondrites into "shock facies" ranging from *a* to *f* (where "facies" is a geologic term referring to the particular character of a rock expressed by its formation, composition, and physical components). In *shock facies a*, melt pockets are absent, olivine is fractured, and plagioclase is at least 50% undeformed; in *facies e*, the melt pocket abundance is several volume-percent, olivine is fractured and exhibits mosaic extinction with marginal granulation, and plagioclase is at least 50% maskelynitized. In the highest shock category, *facies f*, melt pockets are present, olivine is recrystallized, and plagioclase is wholly maskelynitized. The scheme was later applied by researchers to all three OC groups. Many research papers from this period use this scheme.

The shock classification system now in wide use, introduced by Stöffler et al. (1991) for ordinary chondrites, was a refinement of previous attempts. It is based mainly on easily observable shock effects in olivine and plagioclase. The system was extended to carbonaceous chondrites (Scott et al., 1992) and (with the inclusion of shock effects in low-Ca pyroxene) to enstatite chondrites (Rubin et al., 1997; Izawa et al., 2011) (Table 2.2). Langenhorst (2002) pointed out that, in some cases, planar fractures develop in olivine at shock stage S2. Stöffler et al. (2018) expanded the classification scheme to shocked rocks on diverse planetary surfaces, encompassing crystalline, sedimentary, and chondritic rocks, as well as unconsolidated sediments and regoliths. Even though the shock stage value represents the whole-rock, many individual OCs (brecciated and unbrecciated) exhibit inhomogeneous shock effects over centimeter size scales (e.g., Huss, 1980; Rubin and Read, 1984; Gyollai et al., 2023).

In Table 2.2, chondrite whole-rocks are ranked by shock stage, increasing from stage S1 through S5. Shock-stage S6 applies only to localized regions of shocked rocks, in the vicinity of melt zones (e.g., Bischoff et al., 2019). S7 represents whole-rock melting. The whole-rock shock stage of an unbrecciated sample corresponds to the highest stage exhibited by at least 25% of the grains (Stöffler et al., 1991). An alternative to whole-rock shock-classification as an integer (e.g., S1, S2, S3, etc.) is to determine the mean shock stage of all measured grains and express it as a decimal (e.g., S2.3, S4.17) (Jamsja and Ruzicka, 2010; Hugo et al., 2020). In brecciated samples, the shock stages of host and clasts can be evaluated separately, but the whole-rock value is equivalent to the lowest shock stage of any major component. To make accurate assessments of shock stage, Stöffler et al. (1991) recommended examining at least 10–20 randomly distributed grains of moderately large size (>50–100 μm).

Olivine, low-Ca pyroxene, and plagioclase undergo systematic changes with increasing shock stage:

At S1 (unshocked), all three phases display irregular fractures, but exhibit sharp optical extinction (angle of rotation <2°) in the microscope under crossed polars.

At S2 (very weakly shocked), all three phases exhibit undulose extinction (wherein the grains undergo an angle of rotation ≥2°); low-Ca pyroxene (and in some cases olivine) develop some planar fractures.

At S3 (weakly shocked), olivine develops planar fractures and low-Ca pyroxene undergoes mechanical twinning, forming clinoenstatite lamellae on (100).

At S4 (moderately shocked), olivine and low-Ca pyroxene develop weak mosaic extinction. Individual grains develop numerous separate internal domains, a few micrometers across, that differ in their extinction positions by angles of rotation exceeding 3–5°. Plagioclase becomes partly isotropic.

At S5 (strongly shocked), olivine and low-Ca pyroxene exhibit strong mosaic extinction; plagioclase transforms into maskelynite. Olivine and orthopyroxene grains may develop kink bands (e.g., Carter et al., 1968; Carapezza et al., 1976; Gyollai et al., 2017).

At S6 (very strongly shocked), olivine undergoes solid-state recrystallization and may exhibit yellow-brown staining. Some grains may transform into ringwoodite (a high-pressure polymorph of olivine; Fig. 13.13); other olivine grains may melt. Some

Fig. 13.13 Ringwoodite (blue grains) in the shock-stage S6 chondrite Taiban (L5). The grains formed from olivine during shock metamorphism. Photomicrograph in plane-polarized light. Image from E. R. D. Scott, published in Taylor (2004).

low-Ca pyroxene grains may also melt, and some may be transformed into majorite (a high-pressure polymorph of orthopyroxene). Plagioclase may melt and transform into normal glass.

The augmented shock-stage classification of chondritic rocks proposed by Stöffler et al. (2018) ranges from C-S1 to C-S7. A few entries in the Meteoritical Bulletin Database use these terms. The C-S1 to C-S7 range is equivalent to the normal shock-stage range in chondrites of S1 to S7.

Other minerals also exhibiting systematic shock effects including brittle and plastic deformation are chromite (Rubin, 2003) and phosphate (merrillite and apatite; Dobrică et al., 2023, 2024).

Shock stages are not evenly distributed among the OC groups (Fig. 13.14). Bischoff et al. (2019) determined the shock stages of 2280 ordinary-chondrite whole-rocks represented by thin sections in the collection of the Institut für Planetologie in Münster, Germany. H chondrites ($n = 1193$) comprise: S1 (7.2%), S2 (41.1%), S3 (43.5%), S4 (7.5%), S5 (0.1%), impact-melt breccias (0.6%); L chondrites ($n = 947$) comprise: S1 (2.0%), S2 (16.1%), S3 (34.7%), S4 (40.3%), S5 (5.9%), impact-melt breccias (1.0%); and LL chondrites (LL = 140) comprise: S1 (2.9%), S2 (40.0%), S3 (40.7%), S4 (14.3%), S5 (1.4%), impact-melt breccias (0.7%). The mean shock stages are: H (2.52), L (3.32), LL (2.71). It is clear that L chondrites have the highest proportions of more-highly shocked members (presumably a result of the catastrophic disruption of the L parent body).

Rubin (2002b, 2004b) introduced the term "apparent shock stage" for ordinary chondrites that were once shocked to a higher shock stage and then annealed. During the annealing process, deformed olivine crystal lattices healed, causing the grains to acquire the sharp optical extinction characteristic of S1. This is their *apparent shock stage*; they may have had a much higher shock stage prior to annealing. Some shock features that were retained during annealing include chromite-plagioclase assemblages, chromite veinlets, metallic Cu, polycrystalline troilite, and coarse low-Ca clinopyroxene with polysynthetic twinning; these features provide strong evidence of an earlier high degree of shock.

Stöffler et al. (1991) relied on shock-recovery experiments to assign distinct ranges in shock pressure to the boundaries between each shock stage: S1/S2 (<5 GPa), S2/S3 (5–10 GPa), S3/S4 (15–20 GPa), S4/S5 (30–35 GPa), and S5/S6 (45–55 GPa).

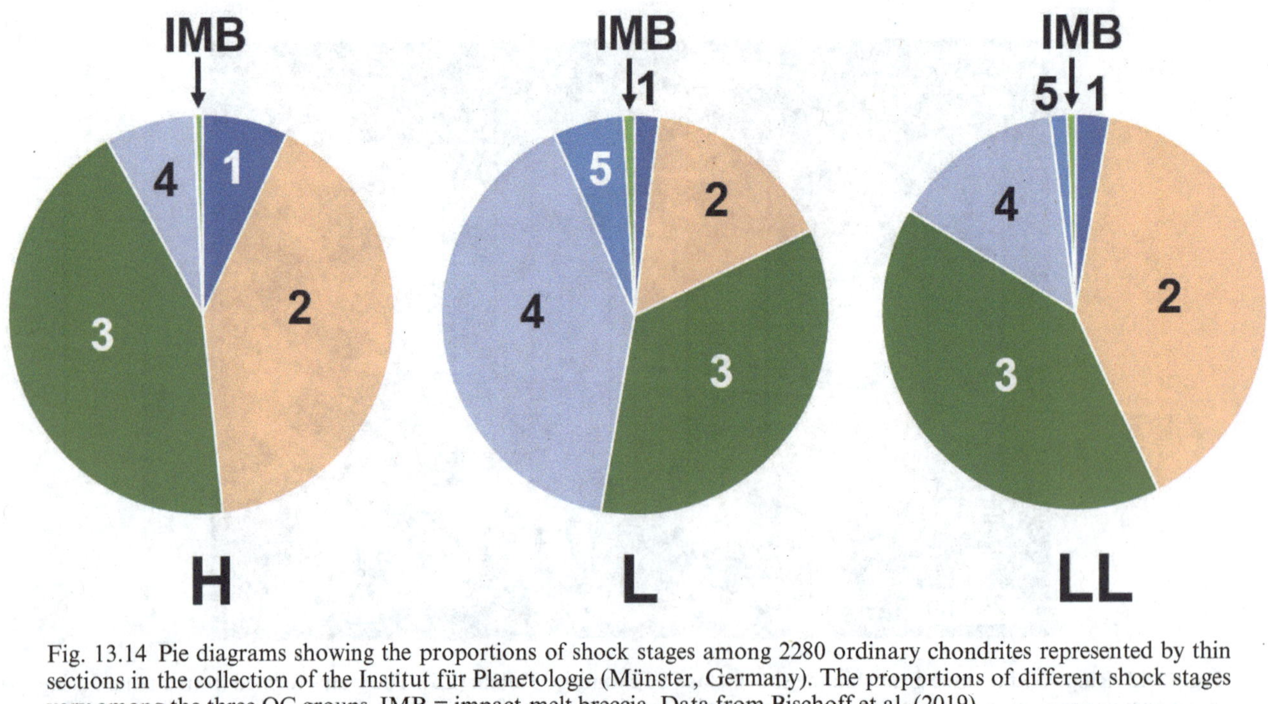

Fig. 13.14 Pie diagrams showing the proportions of shock stages among 2280 ordinary chondrites represented by thin sections in the collection of the Institut für Planetologie (Münster, Germany). The proportions of different shock stages vary among the three OC groups. IMB = impact-melt breccia. Data from Bischoff et al. (2019).

However, Sharp and DeCarli (2006) argued that these values are unreliable, citing several reasons: (1) The shock-recovery experiments were conducted mostly on non-porous samples, whereas chondrites tend to have significant porosity (Schmitt, 2000). (2) Shock-recovery experiments are of microsecond duration, far shorter than the timescales of natural shock events. (3) Many shock-recovery experiments use sample containers of high shock impedance, leading to artificially high shock pressures and relatively low shock temperatures in the enclosed samples (Bowden, 2002). (4) The production of specific shock effects depends on temperature as well as pressure. Shock effects can develop in heated samples at lower shock pressures than in cold samples, underscoring the role of kinetics in natural events. Moreau et al. (2019) emphasized that shock stage is a function of target porosity: high porosity improves the efficiency of shock melting. Shock features produced at a certain pressure in an OC with 5% porosity can be reproduced at a pressure ~10 GPa lower in an OC with 15% porosity.

13.2.2 Microstructural Mineral Dislocations

Natural crystals are imperfectly ordered and inhomogeneously constructed. They contain point defects, line defects, planar defects, and bulk defects, reflecting imperfections in zero, one, two, and three dimensions, respectively.

There are four major varieties of point defects: (1) vacant lattice sites (occurring where atoms are missing from a lattice site that is normally occupied), (2) interstitial defects (arising where atoms occupy positions in the lattice that are normally vacant), (3) antisite defects (occurring where atoms of different elements – substitutional atoms – exchange sites in the lattice), and (4) topological defects (present where atoms are arranged in different patterns than in the bulk crystal). The Burgers vector (b) represents the magnitude and direction of the lattice distortion.

There are two major types of line defects: (1) edge dislocations (in which a plane of atoms terminates within the crystal and the Burgers vector is perpendicular to the dislocation line) and (2) screw dislocations (wherein the planes of atoms are arranged in a helix around a dislocation line and the Burgers vector is parallel to the dislocation line).

There are three principal planar defects: (1) those at grain boundaries, commonly causing changes in crystal orientation, (2) stacking faults representing changes in the normal sequence of adjacent layers, and (3) planar defects located at twin boundaries, representing domains related by a mirror plane, a rotation axis, or an inversion axis.

Bulk defects occur in three dimensions. They include fractures and cracks, clusters of vacancies (voids), pores, and clusters of impurities forming foreign inclusions (commonly called precipitates).

When subject to shock-pressure pulses and their associated thermal effects, crystal lattices undergo deformation. The type and extent of deformation are functions of the intensity of shock and the temperature of the sample. Annealing of a deformed crystal also affects its deformation features. Deformed crystals are frequently studied by optical microscopy and transmission electron microscopy (TEM). The rock samples in which they occur can be examined for additional shock effects (Section 12.7) and checked for evidence of degassing.

As discussed by Ruzicka et al. (2015), the temperature at which crystal deformation occurred can be inferred from the Burgers vector, the screw/edge character, and the extent of any curvature. All dislocations can be characterized as either bound within subgrain boundaries or free (i.e., not bound). A useful parameter is the FFB ratio – the ratio of free dislocations to the total number of dislocations (i.e., free plus bound). The FFB ratio is a function of annealing (Goetze and Kohlstedt, 1973).

Deformation occurring at low temperatures or high strain rates produces long, straight $b=[001]$ screw dislocations in olivine. Deformation occurring at high temperatures or low strain rates produces less-ordered $b=[001]$ dislocations – that is, those with significant numbers of edge or mixed components, dislocation loops, and dislocations with $b=[001]$ (Ruzicka et al., 2015).

Because crystal deformations are due mainly to impact events and because impacts produce high strain rates in target materials (Spray, 2010), the principal variable affecting the resultant crystal deformation is sample temperature.

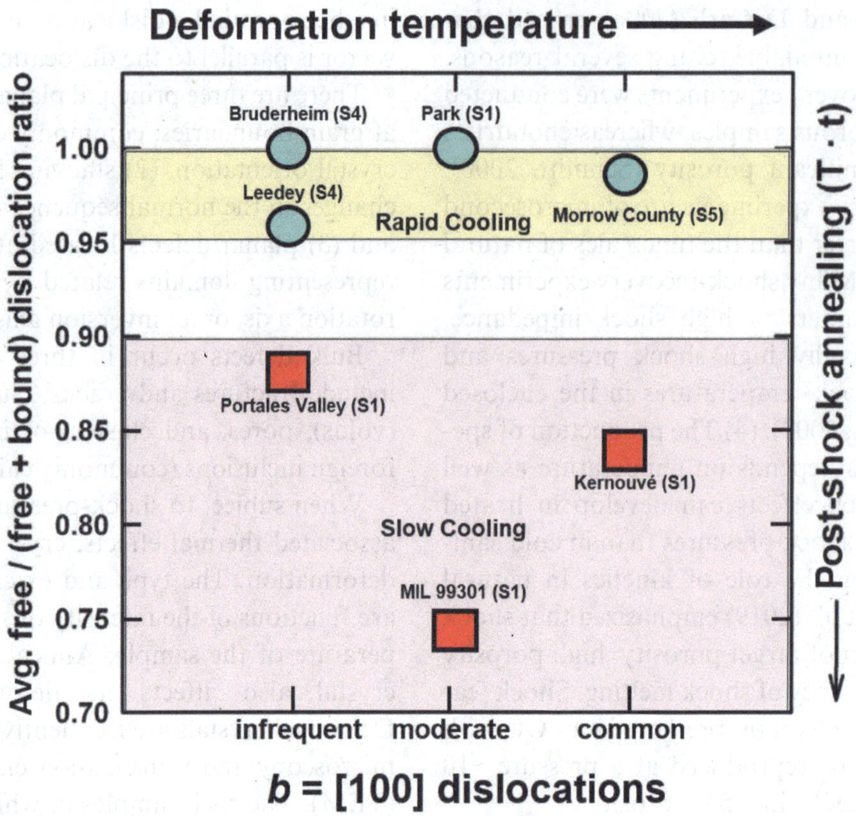

Fig. 13.15 Diagram illustrating the deformation and thermal histories of seven metamorphosed ordinary chondrites based on interpretations by Ruzicka et al. (2015) of TEM data on olivine grains. The four L6 chondrites (circles) were shocked and rapidly cooled; the other chondrites (squares) were shocked and then annealed at depth. Diagram modified from Ruzicka et al. (2015), courtesy of Alex Ruzicka, Cascadia Meteorite Laboratory.

Ruzicka et al. (2015) analyzed olivine grains in seven metamorphosed ordinary chondrites from different groups and with different shock stages to determine their shock and thermal histories (Fig. 13.15). They found no significant correlation between post-shock annealing and deformation temperature. The four L6 chondrites in their study (Leedey, Bruderheim, Park, Morrow County) have low FFB ratios, consistent with these rocks' having been shocked and quenched, plausibly as crater ejecta. Park may have already been hot when it was shocked and excavated. The varying ^{39}Ar–^{40}Ar ages of the samples indicate that they were shocked at different times.

In contrast, H6 Kernouvé (S1) had a different history. It has relatively abundant $b = [100]$ dislocations and low FFB ratios, consistent with its being shocked while hot and then slowly cooled (perhaps beneath the crater floor or within a hot, thick ejecta blanket near the crater rim). Shock features in Kernouvé include coarse veins of metallic Fe-Ni (Hutson, 1989; Rubin, 2002b; Friedrich et al., 2013) that likely were formed during this event. The old ^{39}Ar–^{40}Ar age of Kernouvé (~4.46 Ga; Turner et al., 1978) indicates that this shock event occurred early in Solar System history.

Ruzicka et al. (2015) also analyzed MIL 99301 (LL6, S1), which was described in Section 13.1.6. They concluded that this sample, like Kernouvé, was shocked while hot and slowly cooled at depth. During annealing, strain was obliterated within the olivine grains in MIL 99301. Dixon et al. (2004) interpreted the ^{39}Ar–^{40}Ar data as indicative of two shock events at 4.52 ± 0.08 Ga and 4.23 ± 0.03 Ga. Ruzicka et al. (2015) suggested MIL 99301 was degassed and slowly cooled during the first event while the asteroid was undergoing thermal metamorphism; the second event caused admixing of heterogeneously shocked material. This model differs from that of Dixon et al. (2004) who had proposed that the pervasiveness of shock effects in MIL 99301 (including the presence of unstrained S1 olivine grains with numerous cross-cutting opaque-bearing shock veins; Rubin, 2002b) were caused by the 4.23-Ga shock event.

Portales Valley (H6, S1) (Fig. 11.62a) has relatively few $b = [100]$ dislocations in its olivine, consistent with either a low deformation temperature or with annealing having destroyed many $b = [100]$ dislocations (Ruzicka et al., 2015). The meteorite may have experienced dislocation climb (the movement of a dislocation relative to its slip plane) in which $b = [101]$ dislocations in well-formed subgrain boundaries were formed, possibly by the combination of $b = [001]$ and $b = [100]$ dislocations. Portales Valley was probably shocked while hot early in Solar System history; separate samples yield ^{39}Ar–^{40}Ar ages of 4477 Ma (Garrison and Bogard, 2001) and 4458 Ma (Bogard and Garrison, 2009). Kruijer and Kleine (2019) analyzed the W-isotopic composition of a metal vein in Portales Valley and found it formed 11 ± 2 Ma after CAIs.

13.2.3 A Potpourri of Shock Effects

Bischoff et al. (2018) surveyed 2280 ordinary chondrites and found that 23% of H, 23% of L, and 79% of LL chondrites are brecciated. It is therefore not surprising that (as discussed in Section 12.7) there are myriad shock effects evident in ordinary chondrites (e.g., Rubin, 1985; Bischoff and Stöffler, 1992; Sharp and DeCarli, 2006). They range from low-level mechanical effects such as fracturing of grains, crushing, and rock fragmentation to those involving degassing, partial melting, whole-rock melting, and vaporization. During impact events, some chondrites develop shatter-cone structures (Section 9.3.3).

Mechanical effects are produced at low shock intensities. They include grain fracturing (e.g., Fig. 11.70a) and grain deformation (Section 13.2.2). Slickensides (smooth striated surfaces on some grains caused by shear deformation) were produced on phyllosilicates in the regolith of the 162173 Ryugu carbonaceous asteroid at low shock pressures ($\lesssim$4.5 GPa) and low temperatures ($\lesssim$830°C) (Miyahara et al., 2024). At low temperatures, shock waves can induce kamacite (body-centered cubic α iron) to form mechanical twin lamellae along (112) planes; these are known as Neumann bands (or Neumann lines) and, in three dimensions, are of lenticular shape. They can form in as many as 12 different directions within a single crystal (e.g., Uhlig, 1955).

Meteoritic martensite is metallic Fe-Ni, typically with ~8–15 wt.% Ni and a distorted body-centered cubic structure (α_2). It commonly forms from taenite (face-centered cubic) after a shock-heating event during rapid cooling (i.e., $\gamma \rightarrow \alpha_2$) when the martensitic transformation starting temperature (M_s) is reached (Fig. 13.6). When etched with Nital and viewed microscopically in reflected light, martensite grains generally appear festooned with dark brown to black needles or thin plates.

Collisions commonly fracture whole-rocks, crush target materials, and form poorly sorted debris on the parent body; minor melting along grain boundaries of these materials (Bischoff et al., 1983; Hörz et al., 2005) can cement the broken chunks into coherent fragmental breccias. Many OC regolith breccias consist of angular clasts (formed from fractured metamorphosed OC rocks) embedded within a fine-grained clastic matrix composed of crushed chondritic debris (e.g., H6 Nulles; Rubin, 1982; Williams et al., 1985).

Shock processes can also reduce the average model porosities of ordinary chondrites, with no additional reductions occurring after stage S4 is reached: S1, 13.0%; S2, 10.0%; S3, 8.3%; S4, 6.5%; S5, 6.8%; S6, 6.9% (Consolmagno et al., 2008).

Impact events on OC parent bodies preferentially shatter large porphyritic chondrules, accounting for the decreasing proportions of porphyritic chondrules relative to droplet chondrules at large size intervals (Nelson and Rubin, 2002). This process is also responsible for producing the numerous porphyritic chondrule fragments and large isolated olivine grains within type-3 and -4 OC.

Post-metamorphic brecciation can introduce compositionally aberrant grains into previously equilibrated samples (Scott et al., 1985; Rubin, 1990). After the aberrant grains are mixed with fragments of metamorphosed material, the rocks are recemented by minor impact-melting along grain boundaries. Aberrant olivine, low-Ca pyroxene, and/or kamacite grains with compositions significantly different from the majority in the whole-rock (i.e., more than three standard deviations from the mean) occur in at least half (and probably virtually all) equilibrated ordinary chondrites. Aberrant kamacite grains are not uniformly distributed among the OCs; they are more common in L and LL than in H chondrites. Scott et al. (1985) reported that L6 Roosevelt County 006 (mean Fa = 25.6 ± 0.8, $n = 37$; mean Fs = 21.0 ± 0.4, $n = 14$) contains aberrant olivine (Fa29.0) and low-Ca pyroxene (Fs16.8) grains. An extreme example of an aberrant kamacite grain is one with 106 mg/g Co in LL6 Dongtai, a meteorite that otherwise has a mean kamacite Co content of 17.6 ± 1.0 mg/g ($n = 10$) (Rubin, 1990).

Collisions can cause shearing of target materials. Microfaults and, in rare cases, microfault networks are associated with some sheared ordinary chondrites (Walton et al., 2025). Rubin and Swindle (2011) described the LAP 04581 LL5, S2 chondrite that contains flattened chondrules. The shapes of the chondrules were inferred to have formed after a rare oblique impact produced a petrofabric in unrecrystallized LL3 material (Fig. 11.68a,b). Chondrule flattening was facilitated by impact-induced collapse of pores in the surrounding matrix. Ar–Ar age-dating shows that this event occurred 4175 Ma ago, long after ^{26}Al had decayed away. Impact-induced annealing likely metamorphosed the sample to petrologic type 5 and annealed the olivine crystals to shock-stage S1 from a post-shock stage of S4-S5. After annealing, the rock was weakly shocked again, raising the shock stage to S2.

In general, chondrule flattening is associated with whole-rock petrofabrics (e.g., Cain et al., 1986) formed during collisions (e.g., Gattacceca et al., 2005). For example, many type-3 to type-6 ordinary chondrites exhibit a petrofabric (e.g., Dodd, 1965; Sneyd et al., 1988), typically a strong foliation and a weaker lineation. (A foliation in a rock is a two-dimensional feature – the parallel grouping of minerals in a plane perpendicular to the direction of maximum principal stress. A lineation is a one-dimensional feature – a linear alignment of phases within this plane.) In ordinary chondrites exhibiting a petrofabric, there are positive correlations among strain ellipsoids, magnetic susceptibility ellipsoids, and petrographic shock indicators (e.g., Sneyd et al., 1988; Gattacceca et al., 2005).

Silicate minerals, typified by olivine, become increasingly strained with increasing shock stage (Table 2.2). They develop irregular fractures (S1), undulose extinction (S2), planar fractures (S3), and mosaic extinction of increasing prominence (S4 to S5), and undergo solid-state recrystallization (S6).

Another shock effect in olivine is the development of yellow-brown staining at shock-stage S6. Takenouchi et al. (2015) reported that the NWA 2737 chassignite contains darkened olivine with metallic Fe nanoparticles formed by the reduction of FeO during shock. Brown olivine was also observed in several shergottites by optical microscopy; other shergottites contain clear olivine grains with darkened regions adjacent to shock-melt veins. These features are not restricted to martian meteorites: the NWA 4719 L6 chondrite, which has abundant shock veins and ringwoodite, contains olivine with darkened regions near shock-melt veins or patches where the olivine had been melted. It seems likely the brown staining of olivine is caused by thermal effects resulting from a shock pulse (Takenouchi et al., 2015).

Clinoenstatite containing polysynthetic twinning parallel to (100) was produced during shock from orthopyroxene grains that were heated to temperatures between 1000°C and 1557°C (Smyth, 1974); the grains entered the protoenstatite field, and then were quenched.

Keil (1982) reviewed the different varieties of breccias:

Monomict breccias consist of angular rock fragments of a single lithology; dimict breccias contain two distinct lithologies; and polymict breccias contain several distinct lithologies. Some clasts in these breccias are flanked by narrow melt veins (van der Bogert et al., 2003).

Fragmental breccias are generally unmelted rocks composed of cemented rock fragments. Genomict breccias are fragmental breccias composed of clasts of different petrologic types from the same chondrite group.

Regolith breccias are fragmental breccias containing solar-wind-implanted rare gases; many OC regolith breccias (Section 11.13) have a light/dark structure composed of light-colored metamorphosed chondritic clasts of relatively low porosity set amidst a dark-colored, fine-grained, moderately porous, clastic matrix in which the solar-wind gas resides. Regolith breccias commonly contain exotic (xenolithic) fragments of other chondrite groups that collided with the parent asteroid surface at low relative velocity. Many xenoliths are CM carbonaceous chondrites. Cases of intact OC clasts of one group occurring within another group in a regolith breccia include H clasts in LL6 St. Mesmin (Dodd, 1974) and an LL5 clast in H3.7 Dimmitt (Rubin et al., 1983a). (Although LL6 NWA 5764 contains an L4 clast and LL4-6 Glanerbrug contains L4/5 clasts, the hosts of these meteorites are not enriched in solar-wind gas and the meteorites are not regolith breccias [Gattacceca et al., 2017; Niemeier and Bischoff, 2006].) Also present within regolith breccias are clasts of impact-melt rocks (Section 11.13), in most (but not all) cases formed from target material by an impact. Impact-melt products of chondritic material were reviewed by Rubin (1985).

At somewhat higher shock pressures, portions of ordinary chondrites begin to melt. Objects produced by localized melting include chromite-plagioclase assemblages and chromite veinlets (Rubin, 2003); metallic Cu grains, commonly adjacent to small irregular troilite grains within Ni-rich metal (Rubin, 1994a); rapidly solidified metal-troilite melts ("fizzed troilite") (Scott, 1982); metal- and troilite-rich veins (e.g., Bischoff and Stöffler, 1992); curvilinear trails of small metal-troilite blebs causing whole-rock silicate darkening (Rubin, 1992), forming shock-darkened clasts within breccias (e.g., Fodor and Keil, 1976a; Williams et al., 1985; Grier et al., 2004), and producing black chondrites (Britt and Pieters, 1994); dark-colored glassy (pseudotachylite-like) veins (e.g., Melosh, 2005); glassy melt pockets (Dodd and Jarosewich, 1979, 1982); agglutinates (e.g., Rubin, 1982, 1985); and impact-melt rock clasts (e.g., Fodor and Keil, 1975, 1976a, 1976b; Fodor et al., 1974; Rubin, 1985). Among the melt-rock clasts are rare objects that contain euhedral tetrataenite (Rubin, 1994b), pyrophanite ($MnTiO_3$) (Krot et al., 1993; Xie and Gu, 2023), or baddeleyite (ZrO_2) (Krot et al., 1993; Christophe Michel-Lévy et al., 1995).

Whereas shock processes appear to have little or no effect on the Ni-isotopic composition of shocked OCs (Z. Li et al., 2025), Xu et al. (2023) studied K and Zn in shock melt pockets and unmelted portions of shocked OC and found the shock melts to be enriched in K and depleted in heavy K isotopes relative to adjacent unmelted material. In contrast, the shock melts were depleted in Zn and enriched in heavy Zn isotopes. The authors suggested that (1) Zn was likely evaporated from the shock melts, causing enrichment in heavy isotopes, and (2) the K isotope fractionation was probably caused by solid-melt diffusion influenced by the element's incompatibility.

At high shock pressures, greater amounts of melt are produced. There are ancillary shock effects. Some clinopyroxene grains in a few type-7 OC (rocks that are probably impact-melt breccias or impact-melt rocks) contain very fine exsolution lamellae (e.g., Y. Li et al., 2020) formed during quenching from high temperatures. Impact-melt breccias such as the Rose City H chondrite (Rubin, 1995b) (Fig. 11.63) and the Cat Mountain and NWA 8709 L chondrites (Kring et al., 1996; Ruzicka et al., 2020) and impact-melt rocks such as the PAT 91501 L chondrite (Mittlefehldt and Lindstrom, 2001; Benedix et al., 2008) (Fig. 11.65) reflect pervasive heating of target rocks during collisions. In rare cases (e.g., LAR 06299, LL chondrite; Fig. 11.67), the gravitational field of the asteroid causes different phases to align according to density within the melt (Rubin and Moore, 2011).

Mineral transformations also occur during shock. These include the formation of maskelynite from plagioclase (e.g., Milton and DeCarli, 1963; Chen and El Goresy, 2000; Jaret et al., 2015) and the production of high-pressure silicate, oxide, sulfide, and phosphate phases (Table 12.5) formed in proximity to silicate-rich shock-melt veins.

Impact-induced vaporization and condensation was involved in producing small vugs containing euhedral mineral grains formed by condensation (Olsen, 1981), large metal and troilite nodules (Widom et al., 1986; Rubin, 1995b, 1999), and regions of crushed silicate grains surrounded by extremely high abundances of fine-grained troilite (Rubin, 2002a).

The fine fractions of some ordinary chondrites contain condensates with fractionated compositions. One component, dubbed "mysterite," comes in two varieties: one enriched in many volatile elements (e.g., Bi, Br, Cd, Ge, In, Se, Zn, Au, Ag, Bi, Te, Tl), the other enriched only in volatile siderophiles (e.g., Ag, Bi, Tl) (Laul et al., 1973; Higuchi et al., 1977; Davis et al., 1977). Enrichments range up to 100× typical H6-chondrite abundances. Mysterite was identified in H6 Supuhee, LL3.2 Krymka, and L3.7 Mezö-Madaras. Although it was initially suggested that mysterite is a nebular condensate, it seems more likely to be a condensate of impact-generated vapor on the OC parent bodies. Such vapors can undergo fractional condensation (e.g., Widom et al., 1986; Rubin, 1995b), leading to compositional variations on different substrates. Another fractionated condensate, called "Holy Smoke," was found in ultrafine matrix material sieved from different OCs. It is enriched in both volatiles (e.g., Na, K, Zn, Sb, Pb) and refractories (e.g., W, REE) (Rambaldi et al., 1981). Holy Smoke was also probably formed from condensed impact-generated vapors on the parent bodies.

Analogous processes operate on the Moon: elements mobilized by impact vapor and derived from indigenous lunar materials occur in the regolith (e.g., Morgan et al., 1972). Laul et al. (1971) reported that Apollo 12 soil sample 12028,55 is highly enriched in the volatiles Bi, Cd, and Ag.

(The designation "mysterite" comes from a tradition of assigning names denoting inscrutability or unholiness to puzzling objects. Some examples: Devils Tower in Wyoming [a laccolithic butte with columnar jointing, made famous by a 1977 science-fiction movie]; Canyon Diablo [an arroyo in Arizona, 19 km from Meteor Crater]; Devil's Arse [also known as Peak Cavern] in Derbyshire, England – a cave renowned for the eerie flatulence sounds made by draining flood water; the velvet-black mineral aenigmatite – $Na_4(Fe^{2+}{}_{10}Ti_2)O_4(Si_{12}O_{36})$ – whose composition proved difficult to determine in the mid nineteenth century; the perplexing scrubwren [*Sericornis virgatus*], a bird species from Indonesia and Papua New Guinea that breeds inconsistently with other scrubwrens; and two peculiar fossils from the Middle-Cambrian Burgess Shale – *Anomalocaris* and *Hallucigenia* [made famous by Gould, 1989].)

Three-dimensional imaging of shocked ordinary chondrites by X-ray microtomography reveal

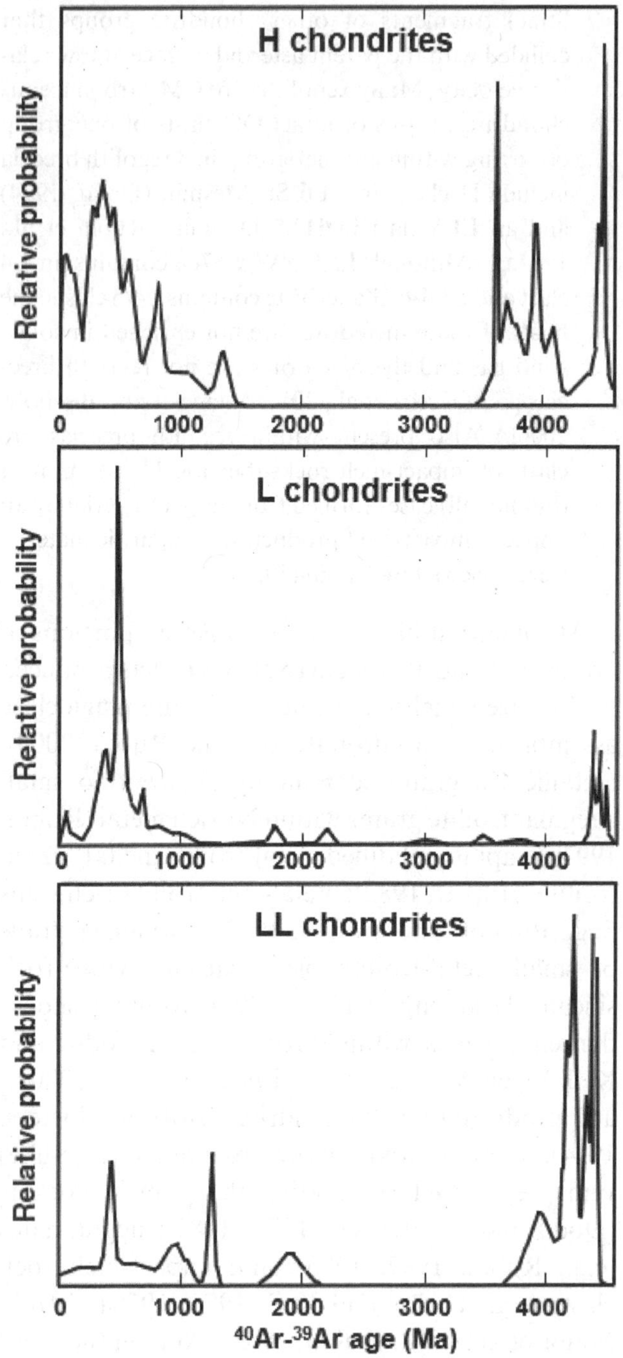

Fig. 13.16 Age probability distribution for $^{40}Ar/^{39}Ar$ analyses of ordinary chondrites. H chondrites cluster near 500 and 4000 Ma, L chondrites mainly near 500 Ma, and LL chondrites mainly at 4200–4300 Ma. The lines represent overall trends. Diagram after Swindle et al. (2015).

discordant petrofabrics indicative of multiple shock events in the H3.8 NWA 7298 breccia (Friedrich et al., 2014b). These features are consistent with evidence for multiple generations of whole-rock fractures in L6 Abernathy (Lambert et al., 1984).

Deposition of shocked OC within thick ejecta blankets or their burial beneath the floors of large impact craters could cause annealing of rocks that had been shocked to S4-S5 levels. Such emplacement could cause healing of the crystal lattices of constituent minerals and produce a low (S1) apparent shock stage in these rocks (Rubin 2002b, 2004b; Rubin and Jones, 2003). Annealing caused by later-stage shock events could also cause crystal-lattice healing in previously shocked samples.

There were multiple collisional events affecting individual ordinary-chondrite parent asteroids long after the era of accretion (Fig. 13.16). The H chondrites show a few distinct impact events that occurred less than 1000 Ma ago. Many L chondrites have an $^{40}Ar/^{39}Ar$ age of ~500 Ma, reflecting the catastrophic disruption of the L parent body 470 Ma ago (Korochantseva et al., 2007). Several LL chondrites have K–Ar degassing ages of approximately 4200–4300 Ma; some are younger than 4000 Ma. LL5 Chelyabinsk has an $^{40}Ar/^{39}Ar$ age of ~25 Ma, the lowest well-determined age of an ordinary chondrite yet measured.

13.3 Aqueous and Hydrothermal Alteration

Ordinary-chondrite falls contain chemically bound preterrestrial water (e.g., 2.03 wt.% H_2O^+ in the Sharps H3.4 fall; Jarosewich, 1990), consistent with the detection of the 3-μm water spectral absorption feature on the asteroid 6 Hebe (Rivkin et al., 2001). This asteroid had been suggested as a possible parent body for H chondrites (Farinella et al., 1993a; Migliorini et al., 1997; Gaffey and Gilbert, 1998).

Indigenous water was also discovered among returned samples of 25143 Itokawa, an S(IV) asteroid of LL-chondrite composition (Jin and Bose, 2019). The original bulk silicate portion of Itokawa is inferred to have had 160–510 μg/g water. The presence of water on an LL asteroid is consistent with the occurrence of 1.22 wt.% H_2O^+ in the Semarkona LL3.01 fall (Jarosewich, 1990).

Indigenous water occurred as water ice and less likely as chemically bound hydroxyl within phyllosilicates such as smectite. (Smectite has been identified in Semarkona [Hutchison et al., 1987].) Heat generated by the kinetic energy of collisions could cause melting of ice and dehydration of phyllosilicates (if any phyllosilicates in fact existed on the parent body prior to aqueous alteration; cf. Ciesla et al., 2003; Wasson, 2008). In addition, because even type-3 OC were heated during thermal metamorphism on their parent bodies (probably via ^{26}Al decay) to at least ~200°C (Huss and Lewis, 1994) – that is, well above the melting temperature of water ice (0°C) and higher than the dehydration reaction in smectite (<200°C; Derkowski and Kuligiewicz, 2023) – it seems probable that water was mobilized in OC asteroids, facilitating aqueous alteration. Alteration seems to have started about 3–5 Ma after CAIs (Krot et al., 2015). The alteration front within chondrule glassy mesostasis in Semarkona can be seen in Fig. 11.75 (Rubin, 2013a): unaltered isotropic glass is flanked by altered, mottled mesostasis riddled with elongated holes. Aqueous alteration of chondrule mesostasis is also evident among Al-rich chondrules; Ebert et al. (2022) analyzed the O-isotopic compositions of a suite of these chondrules in type ≤3.2 OC and found the mesostases to be similar in $\Delta^{17}O$ to fayalite and magnetite grains within OC matrix. (Alteration fronts have also been reported in the matrices of CM chondrites [e.g., Musolino et al., 2024].)

As shown by Grossman and Brearley (2005), the glassy mesostases of chondrules became increasingly rich in alkalis during fluid-assisted metamorphism (i.e., aqueous alteration during the early stages of thermal metamorphism). As albite nucleated and crystallized in the mesostases of Type-II chondrules, the residual glass developed a higher K/Al ratio. Alkalis also entered glass in Type-I chondrules. The chondrules behaved as open systems with alkalis transferring among chondrules and fine-grained silicate matrix material.

Chondrules from LL3.4 Chainpur and LL3.6 Parnallee are enriched in Na and Cl and contain tiny grains of sodalite, nepheline, and scapolite (Bridges et al., 1997). I-Xe ages of Parnallee chondrules range from 1.75 ± 0.16 Ma to 5.0 ± 0.70 Ma after Bjurböle chondrule closure (Ash et al., 1995); these ages probably reflect parent-body alteration or regolith processing (Swindle et al., 1991a).

The significant depletion of alkalis in the outer portions of RP and C chondrules in ordinary chondrites during alteration (Fig. 13.17) caused them to develop a "bleached" appearance (Kurat, 1969; Christophe Michel-Lévy, 1976; Skinner et al., 1989; Grossman et al., 2000). The small patches of glassy mesostases in these regions were dissolved and partly replaced by phyllosilicates. Troilite grains were replaced by pentlandite. As shown by Grossman et al. (2000), many of these altered chondrules can be divided into an outer zone, a bleached zone, and a (less altered) core. The outer zone is a porous region where mesostasis has been depleted; this zone has low concentrations of Na and Al and is compositionally intermediate between the core and the bleached zone. Within individual chondrules, the bleached zones are fairly constant in thickness; these zones tend to be porous and more transparent in transmitted light than the fine-grained chondrule cores.

Bleaching does not affect all RP or C chondrules equally in a given meteorite. Small chondrules have higher surface-area/volume ratios and undergo greater amounts of alteration per unit mass (Fig. 13.18). Very small RP and C chondrules can be transformed into completely bleached objects.

Choi et al. (1998) identified ~30–180-μm-sized magnetite grains in LL3.01 Semarkona and LL3.6 Ngawi. Point-counting of a ~2 cm^2 area of Semarkona yielded a magnetite modal abundance of 40 mg g^{-1}. The magnetite grains formed by aqueous alteration of metallic Fe on the LL parent body. The O-isotopic compositions of magnetite in Semarkona and Ngawi are similar; both show indigenous water on the asteroid was enriched in heavy O isotopes ($\Delta^{17}O \approx 5‰$).

TEM examination of Semarkona shows that fine-grained chondrule rims and fine-grained silicate matrix material are enriched in smectite. This phase is of moderately uniform composition, has some Fe in tetrahedral coordination, and is similar to nontronite ($Na_{0.3}Fe_2((Si,Al)_4O_{10})(OH)_2 \cdot nH_2O$) (Hutchison et al., 1987; Alexander et al., 1989a). The process of smectite formation also produced associated maghemite (γ-Fe_2O_3) and calcite ($CaCO_3$). Adjacent grains of low-Ca pyroxene contain numerous voids. Fayalitic olivine also occurs; it is accompanied by elongated subparallel voids and fibrous material compositionally similar to Fe-rich chlorite $(Fe,Mg)_3(Si,Al)_4O_{10}(OH)_2 \cdot (Fe,Mg)_3(OH)_6$ or septachlorite.

Kimura et al. (2024) proposed that chondrites of type 3.0 can be distinguished from those of type ≤2.9 by the presence in type-3.0 chondrites of few

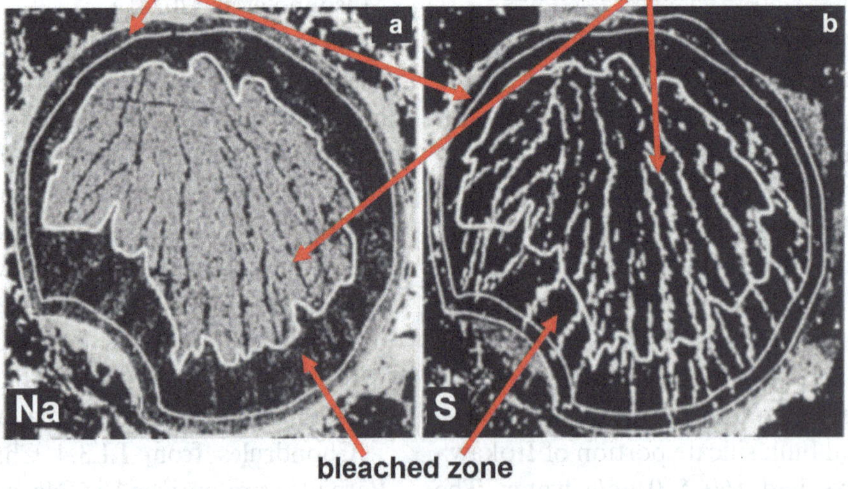

Fig. 13.17 X-ray maps of a bleached cryptocrystalline (C) chondrule from LL3.01 Semarkona, with superimposed white outlines on the boundaries between zones. (a) Na X-ray map showing the depletion of Na in the outer zone and bleached zone. (b) S X-ray map showing the presence of thin sulfide veins throughout the chondrule. Diagram modified from Grossman et al. (2000).

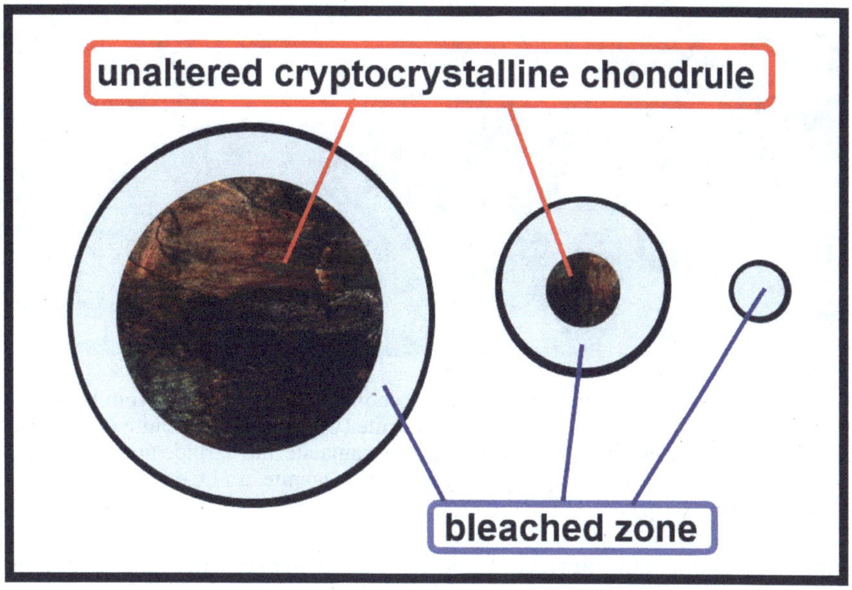

Fig. 13.18 Alteration of cryptocrystalline chondrules of different sizes to produce bleached chondrules. Smaller chondrules have higher surface-area/volume ratios and acquire higher proportions of bleached zones. Very small cryptocrystalline chondrules become completely bleached. Diagram modified from Rubin (2023).

and mild aqueous alteration features. However, these characterizations pertain mainly to carbonaceous chondrites; no ordinary chondrites are currently classified as petrologic type 2. Furthermore, type-3.01 Semarkona exhibits numerous aqueous-alteration features.

Alexander et al. (1989a) pointed out that smectite in LL3.15 Bishunpur occurs principally in the fine-grained matrix. It is relatively rich in CaO and contains up to 0.3 wt.% Cl. Other Cl-bearing phases in Bishunpur include chlorapatite ($Ca_5(PO_4)_3Cl$) and scapolite (mainly of marialite composition; $Na_4(Si,Al)_{12}O_{24}Cl$)) (Alexander et al., 1987), consistent with the presence of Cl-bearing aqueous fluids within LL3 chondrites. Jones et al. (2016) inferred that chlorapatite formed from a fluid with a high Cl/F ratio.

As discussed in Section 12.8, fluid-assisted metamorphism produced voids and veins filled with sodic-calcic amphiboles in the fine-grained silicate matrix of H/L3.6 Tieschitz (Dobrică and Brearley, 2014). The same fluid may have formed the "blocky" chondrule mesostases and "white matrix" in this meteorite (Christophe Michel-Lévy, 1976; Hutchison et al., 1998). I-Xe ages determined for Tieschitz chondrules (Nichols et al., 1991) define closure ages ranging from 1.3 to 4.9 Ma after Bjurböle chondrule closure. These ages are consistent with model-dependent cooling rates ranging from a few to a few hundred degrees per Ma and may reflect regolith processing.

Lewis et al. (2022) found that small grains of primary calcic plagioclase within OC chondrules are altered during thermal-metamorphism/metasomatism. In type-3.2 to type-3.9 OC, it alters to sodalite, scapolite, and nepheline; in type-3.6 to type-5 OC it alters to albite. In addition, plagioclase grains develop alteration features (e.g., zoning, micropores, alteration lamellae) in type-3 to type-4 OC.

A hibonite-spinel-rich CAI from the NWA 046 H3.8 chondrite is partly surrounded by secondary phases including Fe,Zn,Cr-spinel, magnetite, nepheline, and sodalite (Han and Dunham, 2025). Alteration appears to have occurred during Fe-alkali-halogen metasomatism (driven by hydrothermal fluids) on the H-chondrite parent asteroid.

Aqueous alteration of metallic Fe-Ni in Semarkona produced magnetite, carbide, Ni-rich metallic Fe, and Ni-rich sulfide (Taylor et al., 1981). These phases occur within carbide-magnetite assemblages (Fig. 13.19) as observed by Krot et al. (1997b) in 20 type-3 H, L, and LL chondrites including H3.8 WIS 91627, L3.05 EET 90161, and LL3.7 ALHA77278. In some cases,

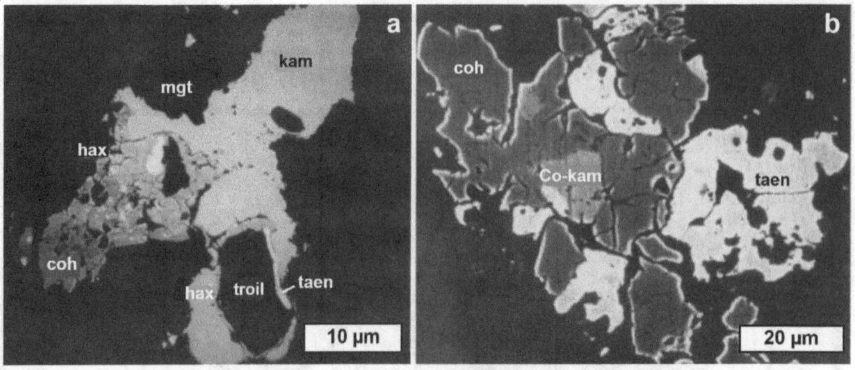

Fig. 13.19 Aqueously altered opaque assemblages in ordinary chondrites. (a) Inclusion from H3-6 Study Butte consisting of kamacite (light gray; kam), troilite (black; troil) Ni-rich taenite (white; taen), haxonite (medium gray; hax), cohenite (dark gray; coh), and magnetite (black; mgt). Magnetite replaces kamacite and carbide in the outer parts of the inclusion. (b) Inclusion from L3.05 EET 90161 consisting of cohenite, Ni-rich taenite, and Co-rich kamacite (medium-dark gray; Co-kam). Images modified from Krot et al. (1997b).

carbide-magnetite-sulfide veins cut across fine-grained chondrule rims, attesting to their late origin.

Present in these carbide-magnetite assemblages are magnetite, Ni-rich metallic Fe (mainly tetrataenite and awaruite), Co-rich kamacite, Fe-carbides (cohenite, haxonite, edscottite), and Ni-rich sulfide (pentlandite). The carbides contain variable contents of Ni (1–14 wt.%) and Co (0–1.3 wt.%). The kamacite contains up to 35 wt.% Co, and awaruite contains up to 70 wt.% Ni (Krot et al., 1997b). Not every phase occurs in every assemblage.

The assemblages formed from C-O-H-bearing fluids (including gaseous CO and H_2O) that altered preexisting metallic Fe-Ni and troilite. The high D/H ratios in the bleached chondrules and matrix of Semarkona imply that the aqueous fluid had a high D/H ratio (Grossman et al., 2000).

Swindle et al. (1991b) analyzed 17 Semarkona chondrules and determined I-Xe ages ranging over a period of ~10 Ma (from -4.4 ± 2.9 Ma to 5.4 ± 0.5 Ma relative to the Shallowater aubrite). Krot et al. (2006) suggested that LL-chondrite alteration took place throughout this period.

A detailed list of minerals produced in type-3 OC during aqueous alteration appears in Section 12.8.

Other products of fluid alteration on OC parent bodies include fayalite-silica associations (Wasson and Krot, 1994). These contain high-FeO olivine (Fa_{56-99} mol%) associated with silica and, in some cases, ferroan low-Ca pyroxene ($Fs_{\leq 66}$ mol%). The free silica probably formed in the nebula (Section 11.1.5) and was incorporated into the parent body during accretion. This is consistent with the presence of rare silica-bearing chondrules in some OC (Brigham et al., 1986). To form fayalite, the silica later reacted with oxidized iron that plausibly was itself derived from metallic Fe that had interacted with water vapor. A reformulation of the reaction favored by Wasson and Krot (1994) is:

$$SiO_2 + 2Fe(s) + 2H_2O(g) = Fe_2SiO_4 + 2H_2(g)$$

silica + metallic Fe + water vapor = fayalite + molecular hydrogen

During this process, Ni was not oxidized and thus did not enter fayalite grains. Instead, the residual metal grains became enriched in Ni; that is, the taenite/kamacite modal ratio increased.

Even equilibrated ordinary chondrites show some evidence of fluid alteration. Type-6 OC whole-rock falls average 0.30 ± 0.39 wt.% H_2O^+ ($n = 43$) (Jarosewich, 1990). And I once found a large bleached chondrule in LL6 EET 96080 (reported in Grossman et al., 2000).

The Kakowa L6 chondrite contains shock-melt veins replete with high-pressure phases (e.g., ringwoodite, wadsleyite, majorite, and albitic jadeite) as well as subsequently produced fractures containing detrital materials: corundum+fayalite+forsterite+albite+margarite+silica+troilite+metallic Fe-Ni embedded within a matrix enriched in Fe and Pb (Baziotis et al., 2022). (Margarite is a hydrous phase with an ideal formula of $CaAl_2(Al_2Si_2)O_{10}(OH)_2$.) It seems likely the detritus was derived from OC material that underwent low-grade thermal metamorphism (equivalent to petrologic type ≤4) and aqueous alteration (Baziotis et al., 2022).

As discussed in Section 11.11, halite (NaCl) was identified in three ordinary-chondrite breccias: Zag (H3-6 fall), Monahans (1998) (H5 fall), and Sidi El Habib 001 (H5-an fresh find, possibly a fall) (Zolensky et al., 1999a; Rubin et al., 2002; Che et al., 2023a, 2023b). Crystals of sylvite (KCl) are heterogeneously distributed within halite crystals in Monahans (1998).

Also present in Monahans (1998) halite grains are aqueous fluid inclusions (Zolensky et al., 1999a). Zag halite contains two varieties of aqueous fluid inclusions: primary inclusions incorporated into growing halite crystals and secondary inclusions trapped within preexisting halite grains when fluid-filled fractures healed.

Throughout Sidi El Habib 001, halite appears to be replacing chlorapatite (Che et al., 2023a). The presence of sphalerite ((Zn,Fe)S) in this meteorite implies that sphalerite and halite likely crystallized from the same hydrothermal fluids (Che et al., 2023b).

The halite grains in Zag and Monahans (1998) formed by evaporation of brines early in Solar System history. Zag halite formed 4.25 ± 0.03 Ga ago (Whitby et al., 2000); Monahans (1998) halite formed >4.33 Ga ago (Garrison and Bogard, 2000). Busfield et al. (2004) performed I-Xe analysis of halite from Monahans (1998) and compared the data to those from Zag. Halite in both samples appears to have formed ~4559 Ma ago. At this early date, ^{26}Al decay could have been responsible for melting extant ice and mobilizing brines on the H-chondrite parent body. This is about the same time frame as carbonate formation in CI chondrites (Brearley, 2006).

Chan et al. (2018a) suggested the H-chondrite halite grains are actually foreign clasts formed on a C-type asteroid, perhaps 1 Ceres. Alternatively, the halite grains could be indigenous to the H-chondrite parent asteroid, products of early aqueous/hydrothermal alteration. (However, Rubin et al. [2000] found no direct petrographic evidence that halite in Zag and Monahans (1998) formed in situ. It is of course possible they formed elsewhere on the parent body and are analogous to aberrant olivine and kamacite grains [Section 13.2.3].)

Zag contains both merrillite ($Ca_9NaMg(PO_4)_7$) and apatite (Fig. 13.20). Because the porous H4

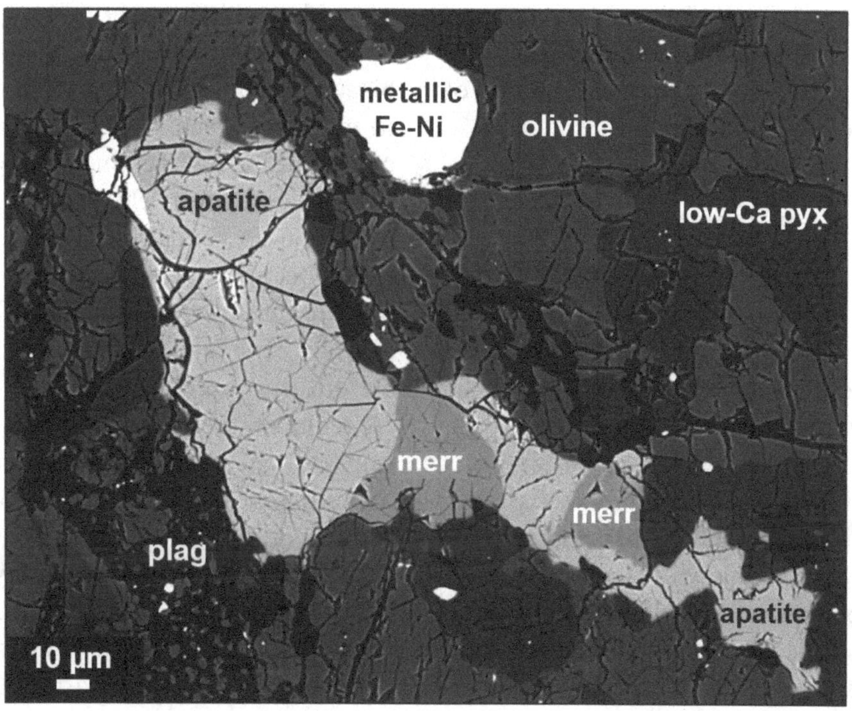

Fig. 13.20 Intergrown apatite (light gray) and merrillite (merr; medium gray) surrounded by olivine (medium dark gray), low-Ca pyroxene (pyx; dark gray), and plagioclase feldspar (plag; black) in an H6 lithology of Zag. Also present are grains of metallic Fe-Ni (white). BSE image modified from Jones et al. (2016).

matrix regions of Zag where halite occurs have high chlorapatite/merrillite ratios (Jones et al., 2016), both halite and chlorapatite could have formed from a Cl-rich fluid. Jones et al. (2016) found that the Cl/F ratio of apatite in H chondrites of all petrologic types exhibits a similar amount of heterogeneity. They concluded that apatite compositions in different meteorites are dependent on local conditions, perhaps due to low fluid/rock ratios.

The Zag breccia contains apatite with moderately heterogeneous halogen concentrations (Cl: 2.9–6.2 wt.%; F: 0.03–1.9 wt.%) but uniformly low water contents (in the range of a few hundred μg/g) (Jones et al., 2016). Although this compositional heterogeneity could have been produced from a fluid of evolving composition, it is also possible that the heterogeneity reflects stochastic impacts (e.g., Baldwin, 1964) into the asteroid regolith. One scenario suggested by Jones et al. (2016) involved impact melts causing devolatilization of adjacent chondritic material, after which the halogen-bearing vapor suffused through nearby apatite grains, changing their Cl/F ratios. Potential sources of the halogens were probably chondrule mesostasis and/or fine-grained matrix material (Jones et al., 2014). (The low inferred abundances of Cl and F in most analyses of chondrule mesostases may indicate these elements were leached from chondrule glass at very early stages of metamorphism or aqueous alteration.)

13.4 Parent-Body Differentiation

Magnetization studies have led to suggestions that chondritic bodies were partly differentiated and had molten cores (e.g., Elkins-Tanton et al., 2011; Bryson et al., 2019). Alternatively, it is plausible that the magnetic fabrics in these meteorites could have been produced in matrix grains by low-velocity impacts (Muxworthy et al., 2017). The common occurrence of coarse type-6 clasts in OC regolith and genomict breccias (e.g., H4-6 Noblesville; Lipschutz et al., 1993) and the paucity of Ca-plagioclase-bearing achondritic clasts in these rocks (cf. Hutchison and Graham, 1975; Bridges et al., 1995) seem inconsistent with OC asteroids' having molten cores.

Nevertheless, some OC parent bodies melted and fully differentiated. For example, IVA irons are from the metallic core of a differentiated OC asteroid; the chondritic precursor of this asteroid was related to L or LL ordinary chondrites (Clayton et al., 1983). A fragment of the Steinbach IVA-an iron contains metal and sulfide with trace-element compositions very similar to those in LL chondrites (Andronikov et al., 2025). The heat source that melted the IVA parent asteroid was likely ^{26}Al decay. If bodies accreted early enough, they may have acquired large numbers of ^{26}Al-bearing CAIs; and if the bodies were of sufficient size, they may have retained their heat long enough to reach melting temperatures.

The initial melt from slowly heated OC material appears at ~940–960°C (Usselman, 1975; McCoy et al., 1996, 1997) and contains ~85 wt.% FeS (Kullerud, 1963). Silicate partial melting of an ordinary chondrite begins between ~1323 and 1380 K (1050–1107°C) (Kushiro and Mysen, 1979; Takahashi, 1983; Jurewicz et al., 1995; T. E. Johnson et al., 2016). This first melt is broadly basaltic in composition, dominated by pyroxene and plagioclase. Tomkins et al. (2020) pointed out that a few highly recrystallized type-6 OC (e.g., H6 O'Malley 003) contain fine-grained plagioclase+augite±orthopyroxene assemblages that might have crystallized from a basaltic melt. Heating an equilibrated OC assemblage to 1200°C at an oxygen fugacity of IW-1 leads to melting of metallic Fe-Ni + sulfide, plagioclase, phosphate, and a fraction of the olivine, orthopyroxene, and chromite (Feldstein et al., 2001).

Type-6 OC and impact-melted type-7 OC generally do not contain augite grains with low-Ca pyroxene exsolution lamellae. The type-6 samples did not experience sufficiently high temperatures for diffusive exchange to enhance the Fe and Mg contents of their augite grains (Lindsley, 1983). Impact-melted OC experienced rapid heating, inhibiting the incorporation of Fe and Mg into augite. In contrast, many primitive achondrites possess augite grains with low-Ca pyroxene exsolution lamellae (Tomkins et al., 2020). These meteorites reached sufficiently high temperatures and cooled slowly enough for such features to form.

Although McCoy et al. (2006) concluded that a sulfide-rich melt within an asteroid could not migrate to form a core under static conditions, the ambient

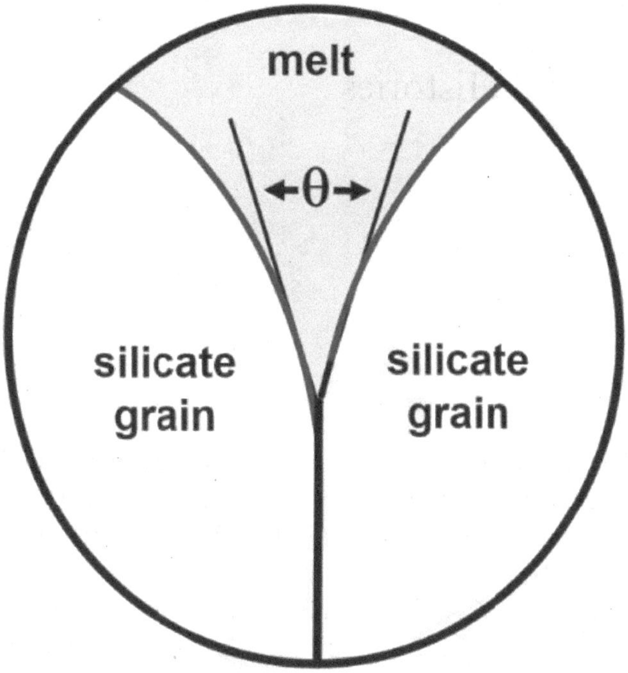

Fig. 13.21 The dihedral angle (θ) of a melt situated between two silicate grains. At low melt fractions, small dihedral angles will facilitate the formation of interconnected melt networks and permit melt migration. Diagram based on McCoy et al. (2006).

conditions were probably not static. Asphaug et al. (2015) determined that, for a collisionally evolved population of asteroids, bodies with diameters exceeding a few hundred meters would likely become rubble piles. Large fractures and pores in an asteroidal rubble pile would facilitate migration of metal-sulfide melts to the gravitational center of the body. However, before small amounts of melt can undergo grain-scale percolation, coalesce, and flow into fractures, they must avoid being trapped as small pockets among silicate grain boundaries. The dihedral angle at triple junctions between the metal-sulfide melt and two solid silicate grains (e.g., Fig. 13.21) controls the connectivity of different small batches of liquid (e.g., Cerantola et al., 2015).

The dihedral angle is defined as $\theta = 2\cos^{-1}(\gamma_{ss}/2\gamma_{sl})$, wherein γ_{ss} is the solid-solid surface energy and γ_{sl} is the solid-liquid surface energy. At low melt fractions, the dihedral angle needs to be <60° to allow efficient interconnection of liquids. This angle is a function of oxidation state, static pressure, and extent of active deformation (e.g., Cerantola et al., 2015; Ghanbarzadeh et al., 2017). For asteroid-sized bodies that are experiencing little or no active strain, the dihedral angle is governed principally by oxidation state. Under moderately reducing conditions (as may be expected for IIE or H chondrites), the dihedral angle is 70°–110°; the percolation threshold in a medium consisting of silicate grains of different sizes and shapes must exceed 10–19% melt to produce an interconnected melt network (Ghanbarzadeh et al., 2017). With increasing melt fraction, adjacent melt pockets would tend to merge to form this network.

After the percolation threshold is reached, as the melt fraction in the mantle decreases during drainage, the network becomes disconnected and the remaining melt is trapped. If the percolation threshold is not reached, some melt may still drain toward the core, but much will be left within the silicate mantle.

Yoshino et al. (2003) pointed out that asteroids with diameters ≥60 km will retain most of their internal heat for periods far exceeding several ^{26}Al half-lives. The mantles of these asteroids may have experienced only ~40–75% melting (McCoy et al., 2006), with unmelted material becoming highly metamorphosed or partly melted. As the asteroid slowly cools, the molten mantle would undergo fractional crystallization; initially, olivine would crystallize from the melt, settle through the silicate liquid, and come to rest atop the molten metal core.

The cooling core itself would likely experience fractional crystallization as taenite crystals settled toward the center. Many of these cores may have developed an internal dynamo, plausibly powered by thermal convection caused by core crystallization.

IIE irons may have been derived from a partly differentiated body. Rubin (2022a) proposed that IIE irons agglomerated earlier than H-L-LL chondrites, acquired higher bulk concentrations of ^{26}Al, and were heated to the temperature of the Fe,Ni-FeS eutectic. This facilitated the separation of an FeS-rich metallic melt from unmelted silicate. Some of the melt drained into fractures and reached the gravitational center of the IIE asteroid – a dynamo was generated by convection currents caused by crystallization in the small core (Maurel et al., 2020). Most IIE metal remained as restites within the crust/mantle region alongside chondritic wall rock that had been recrystallized to petrologic type 6.

14

Parent-Body Collisional Histories

14.1 Accretion Ages and Models

To every thing there is a season ... A time to gather stones together

Ecclesiastes 3:1, 3:5; KJV

The totality is not, as it were, a mere heap, but the whole is something besides the parts

Aristotle, *Metaphysics*

Time zero (t=0) in Solar System history is commonly defined as the epoch 4.56730 ± 0.00016 Ga ago (Amelin et al., 2002; Bouvier and Wadhwa, 2010; Connelly et al., 2012) when the first calcium-aluminum-rich inclusions (CAIs) formed. Most CAIs have an inferred initial $^{26}Al/^{27}Al$ ratio of $5.23 \pm 0.13 \times 10^{-5}$ (Jacobsen et al., 2008); it is commonly assumed (e.g., Desch et al., 2018) that this ratio pertained among newly minted CAIs at t=0. As reviewed by Bogard (2011), this date is in accord with most meteorite whole-rock radiometric ages based on systems involving long-lived parent nuclides (e.g., U-Th-Pb; Sm-Nd; Rb-Sr) as well as short-lived extinct nuclides (e.g., ^{53}Mn-^{53}Cr; ^{26}Al-^{26}Mg; ^{107}Pd-^{107}Ag; ^{182}Hf-^{182}W; ^{146}Sm-^{142}Nd; ^{129}I-^{129}Xe): they all yield formation ages of ~4.565 Ga.

^{26}Al–^{26}Mg chronometry indicates that normal ferromagnesian chondrules formed ~2–2.7 Ma after CAIs (Kita et al., 2000; Rudraswami and Goswami, 2007; Villeneuve et al., 2009; Kita and Ushikubo, 2012; Pape et al., 2019; Hellmann et al., 2019). Chondrules are the most abundant millimeter-sized particles in the Solar System.

Sugiura and Fujiya (2014) found correlations between meteorite accretion ages and $\varepsilon^{54}Cr$. They concluded that the earliest accretion ages (1 ± 0.5 Ma after CAIs) are for angrites, magmatic iron meteorites, main-group pallasites, mesosiderites, HED samples (presumably from Vesta), and ureilites. Pedersen et al. (2019) concluded that the H-chondrite parent body accreted 1.1 or 1.8 Ma after CAIs (depending on whether the initial $^{26}Al/^{27}Al$ ratio in Inner Solar System solids is based on angrites or CV CAIs, respectively). The ^{182}Hf-^{182}W data of Kleine et al. (2005) support early differentiation of the parent asteroids of magmatic iron meteorites. Ordinary chondrites accreted later: 2.14 ± 0.1 Ma after the formation of CAIs (Sugiura and Fujiya, 2014). This is consistent with Hf-W isotope data (Hellmann et al. 2019) indicating that the metal-silicate fractionation among OC occurred between ~2 and ~2.7 Ma after CAIs. (The metal-silicate fractionation involved the separation of siderophile from lithophile elements in the solar nebula [e.g., Urey and Craig, 1953].)

Rubin (2022a) suggested that the order of OC asteroid accretion, from earliest to latest, was IIE, H, L, LL. Because mean chondrule size increases systematically from IIE to H to L to LL (Table 10.4), this hypothesis is consistent with the model of Rubin (2005a), who proposed that chondrules tended to grow larger with time through repeated episodes of dust-accretion and chondrule remelting. In addition, as more metallic Fe was oxidized in the OC region of the nebula, the dust that accreted to chondrules tended to become richer in bulk FeO/(FeO+MgO) as time passed. Early agglomeration of IIE chondrites and later agglomeration of H, L, and LL chondrites could account for the general trends in the OC groups among mean chondrule size, whole-rock oxidation state (as reflected in the mean values of olivine Fa, low-Ca-pyroxene Fs, kamacite Co, and bulk modal abundance of metallic Fe-Ni), and whole-rock $\Delta^{17}O$ content (Tables 10.3, 10.4).

The principal nebular constituents of ordinary chondrites are chondrules (and chondrule fragments), grains of metallic Fe-Ni and sulfide, and fine-grained silicate-rich matrix material (compacted and altered nebular dust). Also present in type-3 OCs are rare CAIs and extremely rare amoeboid olivine aggregates (AOAs).

In nearly all young stars, micrometer-sized dust grains grow to pebble sizes within the protoplanetary disk (Testi et al., 2003, Wilner et al., 2005). These pebbles are accreted rapidly in a process known as pebble accretion (Johansen and Lambrechts, 2017). There are several processes that potentially could have been responsible for planetesimal formation: (1) The accumulation of solid particles in quiescent nebular regions between turbulent vortices may have produced porous "sandpiles" 10–100 km in diameter (Cuzzi et al., 2001; Hartlep and Cuzzi, 2020). (2) Magnetorotational instabilities producing turbulence in the disk could have formed large-scale pressure bumps. Pebbles became highly concentrated in these zones, triggering gravitational collapse into bodies a few thousand kilometers in size (Johansen et al., 2009; Simon et al., 2012). (3) There may have been a "streaming instability" caused by interpenetrating streams of gas and solids. Particles and gas moved at different velocities; filaments were produced that contained slightly higher concentrations of particles. Aerodynamic drag concentrated the particles into zones of high density (perhaps 10^4 times the local gas density) that acted as seed nuclei for planetesimal growth (Youdin and Goodman, 2005; Johansen and Youdin, 2007). This process could have led to the formation of 100-km-sized bodies in the main asteroid belt within just a few orbital periods (e.g., Johansen et al., 2015; Drążkowska et al., 2023).

Once 1–10 km-sized planetesimals had formed, these bodies would have been too large to have been affected by gas drag. They could have accreted into 100-km-sized objects by colliding with each other at low relative velocities. The largest of the bodies in a given region had their collisional cross sections enhanced by gravitational focusing (Weidenschilling and Cuzzi, 2006); this would have led to a positive feedback loop, eventually causing these bodies to dominate their accretion zone.

The giant planets accreted beyond the water snow line, in a region where much of the H_2O was present as ice crystals. Interior to this region, much of the H_2O was present as water vapor within nebular gas; there was thus a lower abundance of solid particles available for accretion.

For sailing vessels navigating the seas, tacking involves a change of course achieved by turning the ship's bow toward and through the wind. The "Grand Tack" model of the early Solar System involves the migration of giant planets as they ploughed through the nebular gas that remained in the protoplanetary disk (Walsh et al., 2011, 2012; Raymond and Morbidelli, 2014). Jupiter first migrated toward the Sun; Jupiter and Saturn later migrated outward. These movements caused severe gravitational scattering of asteroids. Many bodies were perturbed into resonances with the giant planets; many developed high eccentricities and had close encounters with Jupiter; many asteroids collided with the terrestrial planets, the Moon, and the Sun. The asteroid belt was depleted in mass by a factor of 1000 or more. No large bodies remained in the belt beyond 2 AU.

Over the course of tens of millions to hundreds of millions of years, gravitational perturbations repopulated the asteroid belt (Fig. 14.1); this new crop of objects developed the present complex distribution of eccentricities and inclinations. S types dominate the asteroids in the Inner Solar System; C types dominate the Outer Solar System.

A few extant asteroids have been proposed as the parent bodies of established meteorite groups: (a) 6 Hebe (195 km, 2.43 AU) with H chondrites (e.g., Gaffey and Gilbert, 1998; Bottke et al., 2010); (b) 8 Flora (146 km, 2.20 AU) with LL chondrites (Vernazza et al., 2008; Dunn et al., 2013) (contradicting Gaffey [1984] who had maintained earlier that Flora was likely a differentiated body); (c) the Gefion asteroid family (2.74 AU) (named for 1272 Gefion [~7 km] the family member with the lowest assigned number) or the Massalia asteroid family (2.3088 AU) (named for 20 Massalia [~145 km]) with L chondrites (Nesvorný et al.,

Table 14.1 *Systematic changes in bulk parameters for chondrite groups.*

–	EC	OC	R	CR	CV, CK	CM, CO	CI
CAIs (vol.%)	0.01	0.01	0.04	0.6	3.0 to 4.0	1.0 to 1.2	0.0
Refractory lithophile mean abundances (CI- and Mg-normalized)	0.87 to 0.88	0.89 to 0.90	0.97	1.02	1.24 to 1.35	1.11 to 1.13	1.00
Δ^{17}O (‰)	0.0	0.7 to 1.3	2.9	−1.0	−4.2 to −4.6	−4.7 to −5.2	−6.5

EC = EH and EL enstatite chondrites; OC = H, L, and LL ordinary chondrites.
All data from Rubin (2011), except for CAI abundances in OC (Dunham et al., 2023).

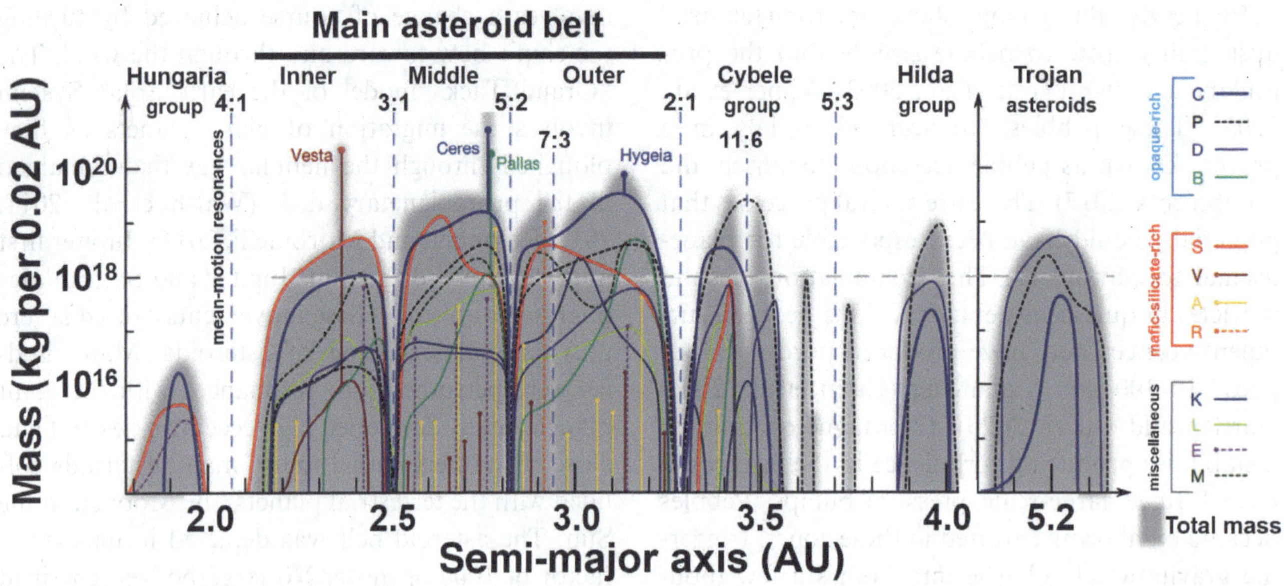

Fig. 14.1 The compositional mass distribution in the asteroid belt and beyond. The gray zones represent the total mass within each 0.02-AU bin. Individual asteroid spectral classes are represented by different colors. Diagram modified from DeMeo and Carry (2014).

2009; Marsset et al., 2024) (although the L parent asteroid was destroyed by collisions ~470 Ma ago [Korochantseva et al., 2007]); and (d) 4 Vesta (525 km, 2.36 AU) with HED achondrites (e.g., McCord et al., 1970; Consolmagno and Drake, 1977; Binzel and Xu, 1993; Binzel, 2001; Burbine et al., 2001; McSween et al., 2013).

A model for the formation locations of the parent bodies of different meteorite groups is shown in Fig. 14.2. Jupiter opened a gap in the solar nebula starting about 0.6 Ma after CAIs formed (e.g., Desch et al., 2018), thus separating carbonaceous materials beyond Jupiter from noncarbonaceous (NC) materials interior to Jupiter's orbit. Except for R chondrites, the precursors of noncarbonaceous materials formed with only minor amounts of nebular dust. Most CAIs in this region spiraled into the Sun, leaving the NC planetesimals with few CAIs. Jupiter created a pressure bump beyond the gap which acted to concentrate nebular dust and CAIs.

As shown in Table 14.1, there are systematic changes in CAI abundance from EC (enstatite chondrites) to OC to R to CR, peaking at CV and CK just past the pressure bump and then decreasing from CM to CO to CI. Mean CI- and Mg-normalized abundances of refractory lithophile elements follow the same trend. The bulk Δ^{17}O content (see Fig. 3.4) follows a pattern of increasingly positive values from EC to OC to R and with increasingly negative values from CR to CV and CK to CM and CO to CI.

Although Fig. 14.2 places the H-, L-, and LL-chondrite accretion locations in order of increasing heliocentric distance, Vernazza et al. (2014)

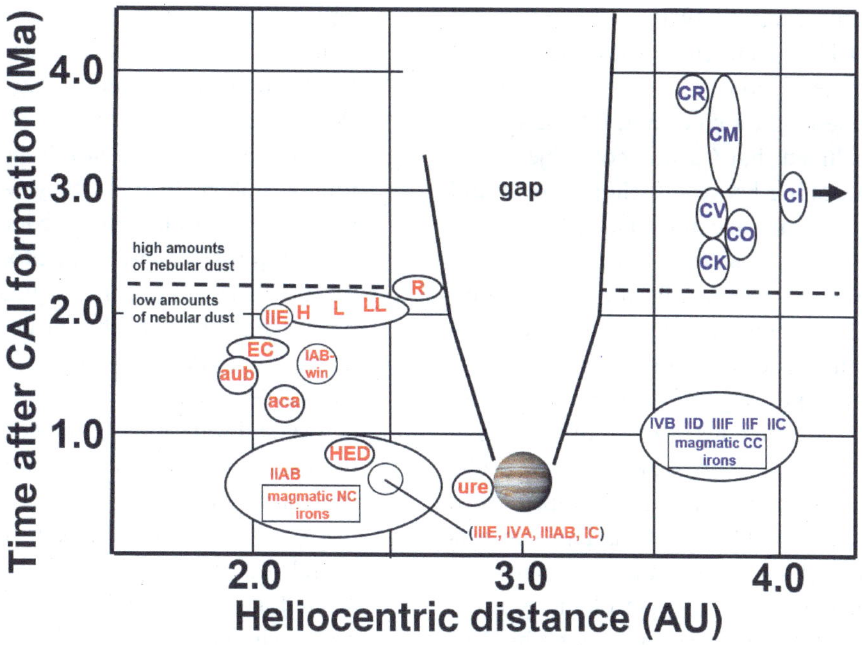

Fig. 14.2 Model for approximate formation locations, formation ages, and initial dust-allotments of the precursors of many meteorite groups. EC = enstatite chondrites; aub = aubrites (enstatite achondrites); aca = acapulcoites and lodranites; ure = ureilites; IAB-win = IAB iron meteorites (nonmagmatic) and winonaites; IIE iron meteorites are also nonmagmatic. Noncarbonaceous (NC) materials are labeled in red; carbonaceous (CC) materials are labeled in blue. The gap opened by Jupiter at 3 AU beginning at ~0.6 Ma after CAI formation contains little nebular gas. CI chondrites accreted at appreciably greater heliocentric distances, perhaps >15 AU. Diagram modified from Desch et al. (2018) with additions and changes taken from Rubin (2011, 2022a) and B. Zhang et al. (2024). © AAS. Reproduced with permission.

suggested that each OC group formed from multiple parent bodies and that, on average, LL-chondrite asteroids formed closer to the Sun than H asteroids. Desch et al. (2018) likewise placed LL chondrites closest to the Sun, followed by H and L. It is safe to say that no consensus has been reached on the accretion distances of the ordinary chondrites.

14.2 Accretion: Hot or Cold?

A question: When ordinary chondrites accreted, 2.14 ± 0.1 Ma after the earliest set of CAIs had formed, was the agglomerating OC material hot or cold? The model known as autometamorphism or hot accretion (outlined in Section 13.1.6) posits that chondritic material had internal temperatures of ~700°C to 900°C (or thereabouts) when they accreted to form their parent asteroids. Material near the surface cooled quickly by radiating heat into space; these materials became type-3 chondrites. Deeply buried material cooled more slowly, forming type-4, type-5, and type-6 chondrites at increasing depths due to greater amounts of insulating overburden (e.g., Anders, 1964; Larimer and Anders, 1967; Hutchison et al., 1980; Hutchison, 1996).

Haack et al. (1992) demonstrated the physical implausibility of models of hot accretion in a nebular environment. The slow chondrite cooling rates mandated by these models are inconsistent with the rapid cooling rates of chondrules (Section 11.6.1). Furthermore, to achieve sufficient heat retention on chondritic asteroids residing in a much cooler nebular environment, hot-accretion scenarios require accretion rates four orders of magnitude greater than those inferred theoretically for planetary bodies.

Rubin and Brearley (1996) addressed the numerous arguments that had been advanced in support of hot accretion and showed that each of them is based on inadequate data, flawed experiments, and/or improbable interpretations. A few examples, updated from that paper, are discussed here:

(1) *Apparent lack of correlation between OC petrologic type and Ca in low-Ca pyroxene.* Models of prograde metamorphism assume that, with increasing temperature, orthopyroxene acquires Ca at the expense

of augite. Hutchison et al. (1980) analyzed low-Ca pyroxene in 10 H3 to H6 chondrites and did not find a positive correlation between the Ca content of low-Ca pyroxene grain cores and OC petrologic type. They concluded that Ca was not subjected to solid-state partitioning between orthopyroxene and diopside during thermal metamorphism. They proposed a model involving igneous partitioning of Ca in hot H6 chondrites via crystallization of substantial amounts of very-low-Ca clinopyroxene or protopyroxene from chondrule liquids; orthopyroxene formed by inversion during slow cooling. However, the Ca-abundance histograms in Fig. 2 of Hutchison et al. (1980) show that the pyroxene in type-3 OC peaks at lower Ca contents than that in type-5 or -6 OC (i.e., ~0.1 vs. ~0.4–0.6 wt.% Ca). This suggests that there *is* a correlation between the Ca content of low-Ca pyroxene cores and OC petrologic type. In fact, the type-7 petrologic type was defined by Tait et al. (2014) based in part on orthopyroxene grains containing >1 wt.% CaO (a higher value than those in OCs of lower petrologic type).

(2) *Misshapen chondrules.* Although chondrules are often referred to as spherical or spheroidal (e.g., Hughes, 1978b), some chondrules that do not appear to be compound objects (e.g., Fig. 11.23c) or to have undergone shock-induced shear (Fig. 11.68a) are clearly ellipsoidal (e.g., Fig. 11.3a). Some of their aspect ratios exceed 1.5 (e.g., Fig. 11.13). Hutchison et al. (1979) found that a few chondrules in H/L3.6 Tieschitz are dumbbell-shaped and appear to have molded themselves around one another while at least partly molten. Ruzicka et al. (2024) reported apparent in situ impingement deformation among adjacent chondrules in Tieschitz, LL3.3 NWA 5781, LL3.7 NWA 5421, and one lithology within LL3.2 NWA 5205. These observations are consistent with those of Rubin and Wasson (2005) who reported that ~60% of the Type-I chondrules in LL3.01 Semarkona are non-spherical. Hutchison et al. (1979) concluded that such misshapen chondrules were partly molten at the time they accreted together with cooler chondrules and other chondritic components to the surface of a hot (~830°C) OC asteroid. Ruzicka et al. (2024) postulated still-warm chondrules accreting at ~900–950°C. Similar arguments were made by Metzler (2012), who reported chondritic clasts up to 10 cm in size within type-3 OC that contain chondrules with deformed shapes conforming around their neighbors (Fig. 14.3). He termed these clasts "cluster chondrites" and suggested them to be products of hot accretion. However, there are several alternative models for forming non-spherical chondrules. Rubin and Wasson (2005) calculated that if non-spherical Type-I chondrules in CO3 chondrites had formed from totally molten droplets, they would have collapsed into spheres within a millisecond, far too quickly for their 20-μm-sized olivine phenocrysts to have grown from the melt. These olivine grains must therefore be relicts from an earlier chondrule generation. The final heating episode experienced by the non-spherical chondrules involved only minor amounts of melting and crystallization. The immediate precursors of the individual non-spherical chondrules may have been irregularly shaped chondrule fragments whose fractured surfaces were rounded during melting. Some chondrules that appear to have molded themselves around another chondrule are likely adhering compound chondrules (e.g., Fig. 11.23d); Jacquet (2021) modeled most lobate chondrules as compound objects. These objects formed in the nebula after a molten droplet bumped into a preexisting, previously solidified primary chondrule and spread partway across its surface (Wasson et al., 1995). Adjacent to the chondrules in type-3 chondrites are patches of fine-grained opaque silicate-rich matrix material. This material contains presolar grains (Section 11.8) and has a basically unfractionated bulk chemical composition (Fig. 11.45). Opaque matrix material has never been melted and was not extruded from chondrules or separately melted during chondrule formation. Presolar grains would have been destroyed. Thus, any complementarity in shape between adjacent chondrules separated by matrix material (as reported in H/L3.6 Tieschitz and LL3.2 Krymka by Holmèn and Wood, 1986) is due to coincidence, jostling, and wedging during chondrite compaction, or to uniaxial deformation during shock events. Chondrules that are deformed around their neighbors without intervening matrix material are likely just compound chondrules that remelted in the solar nebula. It is also possible that small chondritic clasts were melted in the nebula by the chondrule-formation heating mechanism. Finally, it is unlikely that non-spherical chondrule shapes were caused by natural oscillations of liquid droplets alternately elongating and contracting along their principal axes. Fully molten chondrule droplets with plausible viscosities (1–5 poise) would collapse into spheres without oscillating (Rieber et al., 2000). Incompletely molten chondrules would

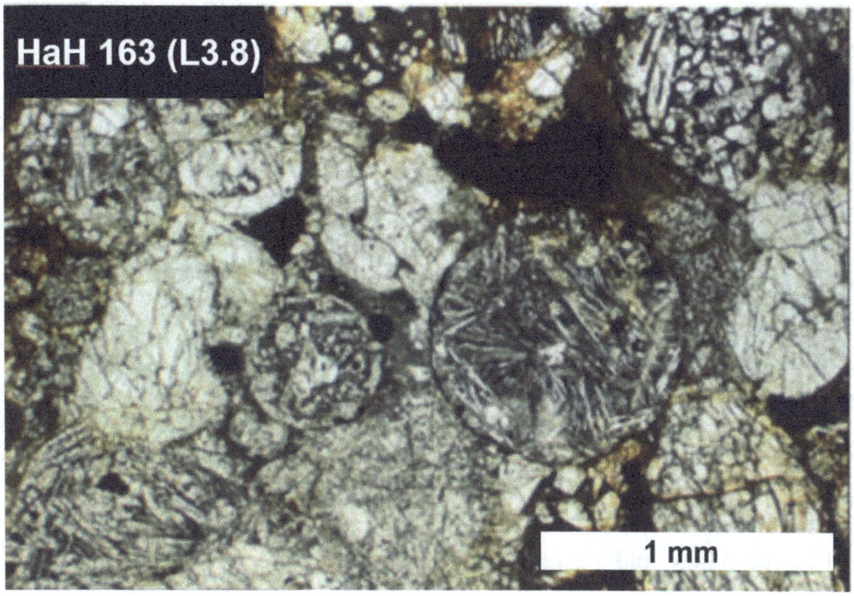

Fig. 14.3 A "cluster-chondrite clast" in L3.8 Hammadah al Hamra 163. Some chondrules are misshapen and conform around their neighbors. Image modified from Metzler (2012).

have higher effective viscosities and would not oscillate. Thus, non-spherical chondrules were not caused by the quenching of oscillation-induced deformations. Nevertheless, Miura et al. (2008) carried out three-dimensional hydrodynamic simulations of rotating molten droplets encountering a gas flow. They found the droplets become prolate – elongated along the rotation axis. If these simulated droplets quench while in that state, ellipsoidal chondrules will result.

(3) *Negative correlation between volatile trace elements and petrologic type.* The concentrations of volatile trace elements in OC become increasingly depleted with increasing petrologic type (e.g., Larimer and Anders, 1967; Binz et al., 1976). These data were interpreted as indicating that type-3 OC formed at lower nebular temperatures than more equilibrated OC and were therefore able to acquire by condensation higher amounts of volatile trace elements. However, Ikramuddin et al. (1977) annealed samples of H/L3.6 Tieschitz at low pressure and showed that volatile elements were lost with increasing temperature, consistent with prograde thermal metamorphism.

(4) *Natural Remanent Magnetization (NRM).* Morden and Collinson (1992) found that millimeter-sized samples of 11 crushed ordinary chondrites exhibited random orientations of the stable NRM component. They also determined that the samples have a magnetic fabric (analogous to a structural fabric in rocks deformed by tectonic activity or shock); in most samples the magnetic fabric is continuous. Morden and Collinson ruled out prograde metamorphism because heating above ~580°C (the Curie point of tetrataenite), as would be expected in type-4 to type-6 OC (Section 13.1.4), would erase the random orientations of the NRM. However, most ordinary chondrites seem to be post-metamorphic fragmental breccias containing compositionally aberrant silicate and metal grains introduced after the rocks had cooled from peak metamorphic temperatures (Section 13.2.3). One example is from H5 Anlong – out of 20 olivine grains in this meteorite analyzed by Rubin et al. (1988b), 19 are equilibrated with a mean composition of $Fa_{19.2\pm0.4}$; one grain has a composition of $Fa_{21.1}$, more than four standard deviations from the mean. Another example is from L4 Dalgety Downs – out of 12 kamacite grains analyzed by Rubin (1990), 11 are equilibrated with a mean kamacite Co content of 8.6 ± 0.3 mg/g Co; one grain contains 13.1 mg/g Co, fully 15 standard deviations from the mean. If these grains had been present in their host chondrites prior to metamorphism, they too would have equilibrated. Thus, the random orientations of the stable NRM component among millimeter-size OC samples reported by Morden and Collinson (1992) could have resulted from post-metamorphic brecciation. The continuous magnetic fabric in many equilibrated OC is probably due to final lithification

caused by minor melting along grain boundaries during an impact event (e.g., Bischoff et al., 1983) after the epoch of thermal metamorphism.

Rubin and Brearley (1996) concluded that the "textural, compositional, isotopic, and magnetic properties of type-3 to -6 ordinary chondrites and their constituent minerals are more readily understood in terms of prograde metamorphism of initially cold, unequilibrated type-3 material and episodes of brecciation, shock heating, and, in some cases, aqueous alteration." The model of hot accretion is unsupported.

14.3 Ordinary-Chondrite Contributions to the Accretion of Earth, Mars, and Vesta

The terrestrial planets, the Moon, and Vesta are not geochemically identical to any individual group of meteorites. These large bodies accreted from compositionally diverse planetesimals that may have included a substantial contribution from ordinary chondrites. This is not surprising – like ordinary chondrites, Inner Solar System bodies (e.g., Earth, Mars, Vesta) are composed mainly of noncarbonaceous materials (see Chapter 6).

Mixing models involving the bulk chemical and isotopic compositions of meteorites can be constructed to yield plausible parental materials for terrestrial planets and protoplanets.

Earth: Mass-balance calculations using O-isotopic data indicate that the Earth accreted from a mixture of approximately 70% EH chondrites, 21% H chondrites, 5% CV chondrites, and 4% CI chondrites (Lodders, 2000). These calculations yield an FeO content for the bulk silicate Earth that is relatively close to the accepted value (~8 wt.%; e.g., McDonough and Sun, 1995). It is also possible that noncarbonaceous achondrites such as aubrites contributed to the mix of materials that accreted to form the Earth (and other terrestrial planets). Young et al. (2023) proposed that Earth was formed predominantly of enstatite chondrites. This suggestion was similar to that of Javoy et al. (2010), who noted that bulk Earth has a chemical composition similar to that of the Abee EH impact-melt breccia and modeled the Earth as having accreted solely from enstatite-chondrite material. A refinement: Zn-, Cr-, and Ti-isotopic anomalies imply that the bulk Earth is composed of 95% enstatite chondrites and 5% CI chondrites (Steller et al., 2022). Although Garai et al. (2025) proposed that Earth's building blocks consisted mainly of pebbles of chondritic components (metal grains, chondrules, CAIs, AOAs) derived from carbonaceous chondrites, this model is inconsistent with Steller et al. (2022), who found that Earth derived ~70% of its Zn, H, and N from Inner Solar System bodies.

Mars: The Oxygen-Isotope Mixing (OIM) model of Lodders and Fegley (1997) indicates that Mars accreted from a mixture of approximately 85% H chondrites, 11% CV chondrites, and 4% CI chondrites. Delaney (1994) also used O isotopes to suggest a mixture of ~90% H chondrites and ~10% CM chondrites. Decades earlier, without the use of O-isotopic data, Anderson (1972) had calculated that Mars formed from ~75% ordinary chondrites and ~25% carbonaceous chondrites.

Vesta: Lodders (2000) calculated that Vesta accreted from a mixture of ~78% H chondrites, ~22% CV chondrites, and a small amount (0.3%) of CI chondrites. Other Vesta estimates include those of Boesenberg and Delaney (1997) (30% CM and 70% H) and Righter and Drake (1997) (30% CV and 70% L).

Results differ among researchers in part because of uncertainties in the bulk compositions of these large bodies. Their interiors are inaccessible to direct sampling and there have yet to be sample return missions from Mars or Vesta. Martian meteorites and HED samples are not representative of the entire geological complexity of these differentiated worlds. The data supplied by telescopes, orbiting spacecraft, rovers, and helicopters are valuable but limited.

14.4 History of OC Parent-Body Collisions

To every thing there is a season… A time to cast away stones…

Ecclesiastes 3:1, 3.5; KJV

Asteroids formed by collisions. During accretion, planetesimals tended to collide at low relative velocities and grow into larger bodies. 4 Vesta is an example of this process (e.g., Davis et al., 1985). Bland et al. (2014) showed that low-velocity collisions can heat the fine-grained matrix in unconsolidated, porous, unequilibrated materials to high temperatures for brief periods, leading to sample compaction.

Later in Solar System history, many asteroids were destroyed by major collisions (such as the L-chondrite parent-body breakup event ~470 Ma ago; Korochantseva et al., 2007). Some rubble-pile asteroids (e.g., 25143 Itokawa) were reassembled after the collisional disruption of a larger body (e.g., Jourdan et al., 2017). That collisional events have occurred throughout the history of the Solar System is attested to by the numerous impact craters scarring the surfaces of most visited asteroids (e.g., Figs. 11.56, 11.73).

From the present size distribution of the asteroid belt, Beitz et al. (2016) inferred that 200-km-diameter asteroids were subjected to 10^{14} collisions by projectiles >20 cm in diameter throughout the course of Solar System history. They suggested that the ordinary chondrites in our collections were not derived directly from 200-km-diameter regolith-covered bodies, but rather from smaller regolith-free asteroidal fragments. However, there are numerous OC regolith breccias; they constitute 15.3% of H chondrites, 3% of L chondrites, and 5.5% of LL chondrites (Bischoff et al., 2006). The common occurrence of these regolith breccias casts doubt on the general applicability of the Beitz et al. result. Either some of the smaller asteroidal fragments retained regoliths, or the extant OC regolith breccias were solidified prior to the breakup of the primary parent body.

14.4.1 Cosmic-Ray Exposure Ages

One way to gauge the collisional history of an ordinary-chondrite parent body is to determine the cosmic-ray exposure (CRE) ages of many individual members of a particular group. Cosmic rays consist predominantly of protons with energies >5 MeV; these particles have sufficient energy to penetrate an object in interplanetary space to depths of ~1 meter. Cosmogenic nuclides are produced by this encounter. The CRE age of a sample marks the length of time it took from its asteroid/meteoroid progenitor being collisionally reduced to a roughly meter-sized object until that object traversed the Earth's atmosphere (after which it was protected from cosmic-ray bombardment).

The CRE-age distributions of meteorite groups (including ordinary chondrites) were reviewed by Eugster (2003) and Herzog (2005). Most OC have CRE ages between 3 and 80 Ma; only a few samples have ages <1 Ma (Fig. 14.4). Tall peaks in the distribution represent major collisions on the parent body that produced large numbers of meter-sized fragments. Although it is possible that there was more than one parent-body source for individual chondrite groups, it is assumed that the vast majority of samples from the same group that share a CRE peak were in close proximity on the same parent asteroid at the time of the collision.

The paucity of OC samples with short (<1 Ma) CRE ages reflects either (1) low numbers of recent collisions responsible for producing numerous meter-size fragments or (2) the several-million-year time period it takes typical meter-sized bodies from that region of the Solar System to reach the Earth. The absence of OC samples with long (≥100 Ma) CRE ages reflects the friability of meter-sized bodies with OC composition; such bodies are destroyed by collisions (and turned into small fragments and dust grains) in interplanetary space over long time scales. In contrast, iron meteorites are far more durable – most CRE ages of irons are within the range of 200–1000 Ma (Herzog, 2005).

The background distribution results from numerous smaller collisions on the parent asteroid, each one yielding only a few samples that made it into our collections.

About 45% of H chondrites (mainly H4 and H5 samples) populate a CRE-age peak of ~7.5 ± 0.5 Ma (Fig. 14.5). All petrologic types of H chondrites have a peak at ~33 Ma. The H6 chondrites have a prominent peak at 24 Ma that is absent from the CRE-age distributions of lower petrologic types. Eugster (2003) interpreted these data to indicate either that the H-chondrite parent body had an onion-shell structure (Fig. 5.12a), allowing impacts of different energies to excavate materials from different depths, or that there were multiple H-chondrite bodies.

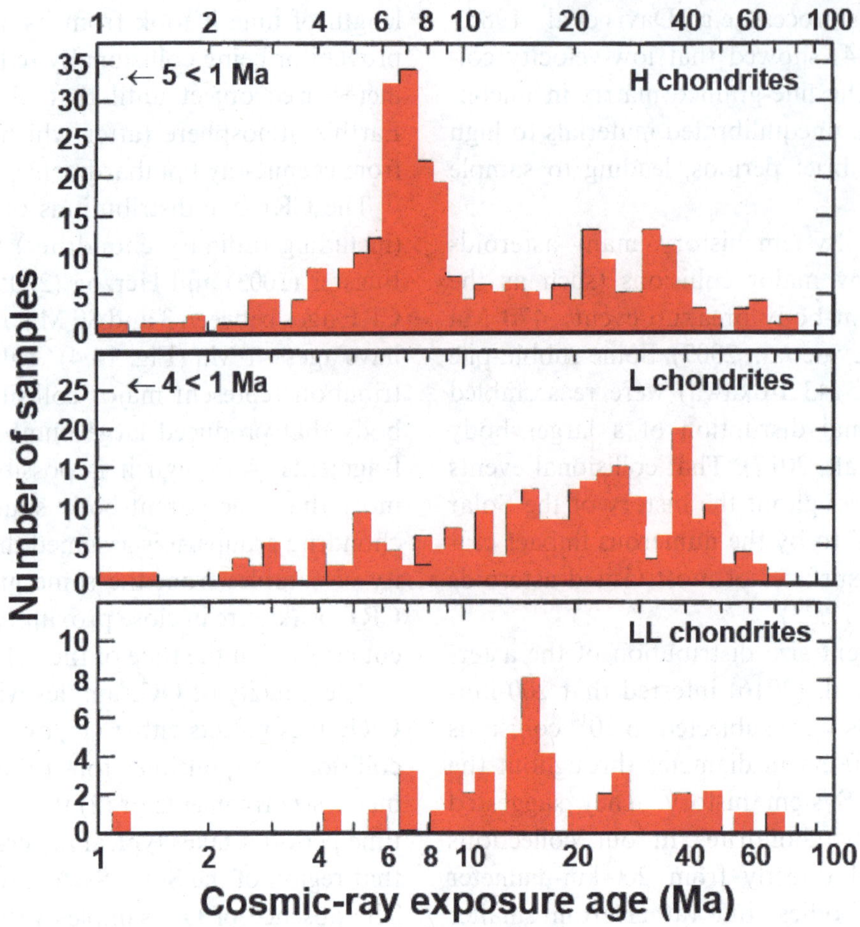

Fig. 14.4 Cosmic-ray exposure (CRE) age distributions for ordinary chondrites. The numbers of H and L chondrites with low (off-scale) CRE ages are indicated by arrows. Diagram modified from Eugster (2003).

An alternative is the metamorphosed-planetesimal model (Fig. 5.12b), in which separate impacts involved different regions of moderately uniform petrologic type of a single H-chondrite parent body. The fact that the ~33-Ma peak includes all petrologic types strongly indicates that all types were present on the same parent body at that time. If the body had an onion-shell structure, the impact would have had to have been massive enough to excavate the deeply buried H6 chondrites. It is also possible that the body had earlier been catastrophically disrupted and gravitationally reassembled (Taylor et al., 1987), causing H-chondrite materials of different petrologic types to be mixed together at the surface (and throughout the body). That scenario would be consistent with the juxtaposition in the matrices of H-chondrite regolith breccias of taenite grains that appear to exhibit quite diverse metallographic cooling rates (Fig. 5.14). If these grains were derived from a body with an onion-shell structure, they were likely derived from very different depths.

The most prominent peak in the CRE-age distribution of L chondrites is at ~40 Ma (Fig. 14.4). There is also a small peak at ~5 Ma, a likely peak at ~15 Ma, and possibly a broad peak centered around 28 Ma (Eugster, 2003; Herzog, 2005). Samples with CRE ages ≥50 Ma are exclusively (or nearly exclusively) L5 and L6 chondrites. Because the L-chondrite parent asteroid was disrupted 470 ± 6 Ma ago (e.g., Bogard, 1995; Bogard et al., 1976, 1995; McConville et al., 1988; Turner, 1988; Schmitz et al., 2003; Korochantseva et al., 2007; Nesvorný et al., 2009; Swindle et al., 2014), all the CRE ages shown in Fig. 14.4 (which are <80 Ma) reflect fragmentation of large pieces of the disrupted original body.

About 20% of LL chondrites have a CRE peak at ~15 Ma (Fig. 14.4). Herzog (2005) suggests that there is also a broad peak stretching from 27 to 33 Ma.

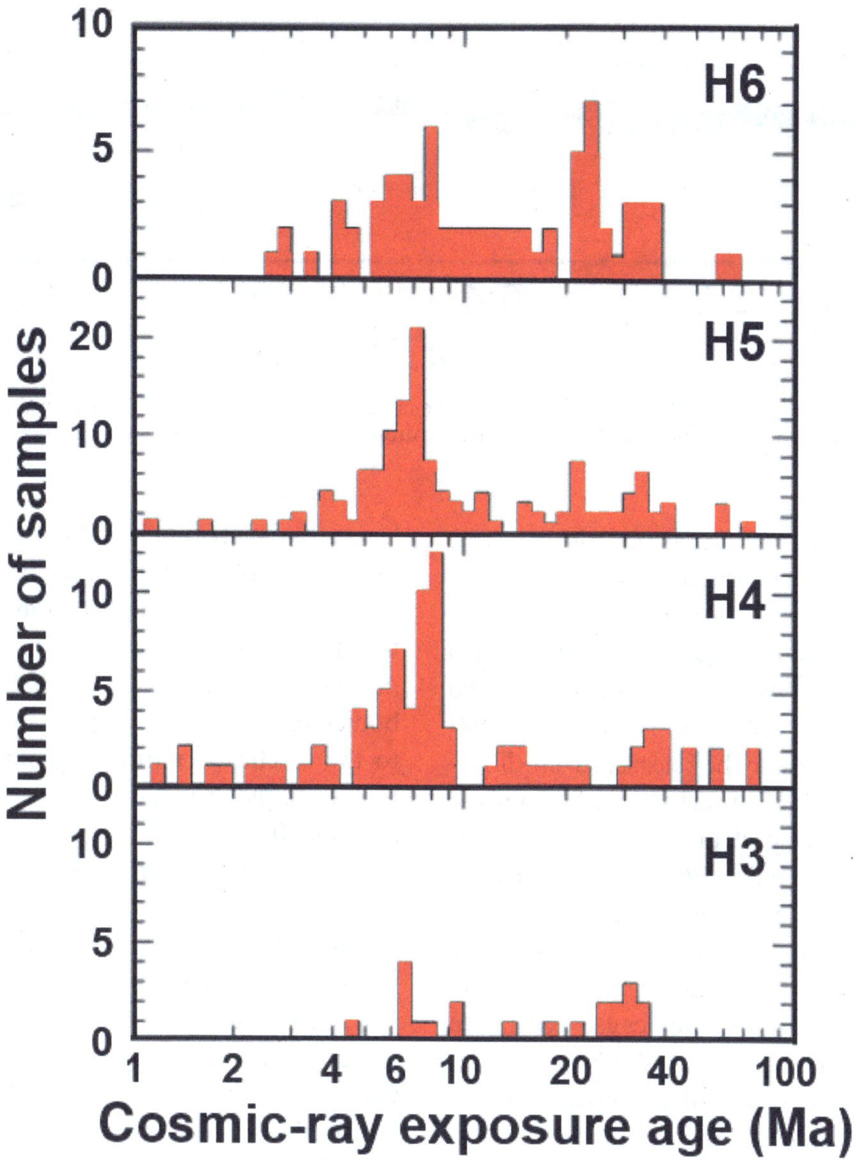

Fig. 14.5 CRE-age distributions for H chondrites of different petrologic types. Diagram modified from Herzog (2005).

There are few data on the CRE ages of IIE irons. Average $T_{38(Ar\text{-}Ar)}$ ages (provided by Greg Herzog, Rutgers University) include Netschaëvo (9 Ma), Watson (13 Ma), Kodaikanal (13 Ma), and Weekeroo Station (542 Ma). There is a possibility that these meteorites were released by two impacts: an early impact >500 Ma ago and a more-recent one 11±2 Ma ago. Much more data are needed to resolve the collisional history of the IIE parent body.

As pointed out by Swindle et al. (2015), there are no $^{40}Ar/^{39}Ar$ ages that match the CRE ages of any meteorite group. This indicates that the collisions that ejected the masses that constitute the meteorites in our collections were not sufficiently energetic to cause extensive degassing of radiogenic ^{40}Ar.

14.4.2 Argon–Argon Ages

The ^{39}Ar-^{40}Ar dating method marks the time a cooling mineral reached its closure temperature after which there was no significant diffusion of the relevant isotopes out of the mineral. As discussed by Bogard (2011), the method involves irradiation of the sample by fast neutrons, thereby converting some ^{39}K (a stable isotope) into unstable

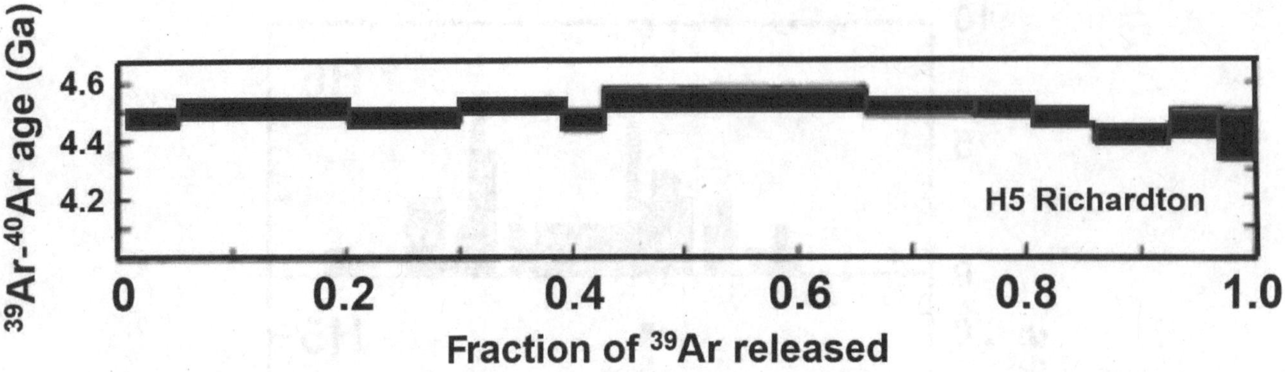

Fig. 14.6 Idealized graph of ^{39}Ar-^{40}Ar age as a function of cumulative ^{39}Ar release for H5 Richardton. The diagram has a well-defined plateau. Based on the total Ar released, Richardton has an Ar–Ar age of 4.50 ± 0.03 Ga. Diagram based on Turner et al. (1978).

^{39}Ar ($t_{½}$ = 269 a). The ^{39}Ar atoms occupy the same K lattice sites as ^{40}Ar produced by ^{40}K decay ($t_{½}$ = 1.25 Ga). After irradiation, the meteorite sample is heated at increasingly higher temperatures in a stepwise fashion; the ^{40}Ar/^{39}Ar ratio of the released Ar gas is measured at each step. At different temperature steps, the Ar may be released from different lattice sites or from different mineral phases. A typical Ar–Ar age-spectrum diagram plots the ^{39}Ar-^{40}Ar age and the bulk K/Ca ratio against the fraction of the cumulative released ^{39}Ar at each step. In well-behaved systems (e.g., Fig. 14.6), the diagram features a plateau defining the Ar–Ar age. In contrast, impact heating can cause partial to complete ^{40}Ar degassing; highly shocked samples generally do not produce well-defined plateaus.

LL chondrites vary in Ar–Ar age from 27–4530 Ma (Righter et al., 2015; Swindle et al., 2014). Similarly, the Ar–Ar ages of LL-chondrite samples recovered from asteroid 25143 Itokawa vary from 1300–4219 Ma (Park et al., 2015; Jourdan et al., 2017, 2023).

All in all, ordinary chondrites form three clusters of Ar–Ar age: (1) a young cluster with ages <1.3 Ga, (2) a broad cluster with ages ranging from ~3.5–4.1 Ga (mainly in H chondrites; Fig. 13.16), and (3) an old cluster with ages mainly in the range ~4.43–4.53 Ga (e.g., Bogard, 1995; Swindle and Kring, 2008; Dixon et al., 2004; Benedix et al., 2008; Weirich et al., 2011). The two younger age clusters (as compiled by Bogard, 1995) are shown in Fig. 14.7.

Bogard (1995, 2011) inferred that samples from the youngest age cluster experienced impact heating and Ar–Ar-age-resetting while they were buried at least a few meters below the asteroid surface. Four (and possibly many more) separate impact events were inferred for samples from the youngest cluster at 0.3 Ga, 0.5 Ga, 0.9 Ga, and 1.2 Ga before present. It can be assumed that the H, L, and LL bodies experienced different impact histories due to the stochastic nature of collisions (Fig. 13.16). For example, the cluster of L-chondrite Ar–Ar ages at ~0.5 Ga (Fig. 14.7) reflects the collisional disruption of the L parent asteroid ~470 Ma ago (Korochantseva et al., 2007). This event apparently did not affect the H or LL parent bodies.

Samples from the oldest OC cluster (~4.43–4.53 Ga) had their Ar–Ar ages reset during or shortly after the epoch of parent-body thermal metamorphism. One possibility is that these ages were not reset by impacts but instead reflect closure of the K–Ar chronometer as OC parent bodies cooled from their peak metamorphic temperatures (Bogard, 2011).

However, some highly shocked ordinary chondrites, including impact-melt breccias and impact-melt rocks, have old Ar–Ar ages. For example, the PAT 91501 L-chondrite impact-melt rock (Mittlefehldt and Lindstrom, 2001) (Fig. 11.65) has an Ar–Ar age of 4.461 ± 0.008 Ga (Benedix et al., 2008). The Portales Valley H6 impact-melt breccia (Figs. 11.62a, 11.64) has Ar–Ar ages ranging from 4458–4477 Ma and an I-Xe age of 4559.9 ± 0.5 Ma (Bogard and Garrison, 2009). Shaw, an L6/7 impact-melt breccia (Taylor et al., 1979; Scott and Rajan, 1979), has an Ar–Ar age of 4.43 Ga (Turner et al., 1978).

MIL 99301 is an LL6 chondrite that appears to have been strongly shocked and subsequently

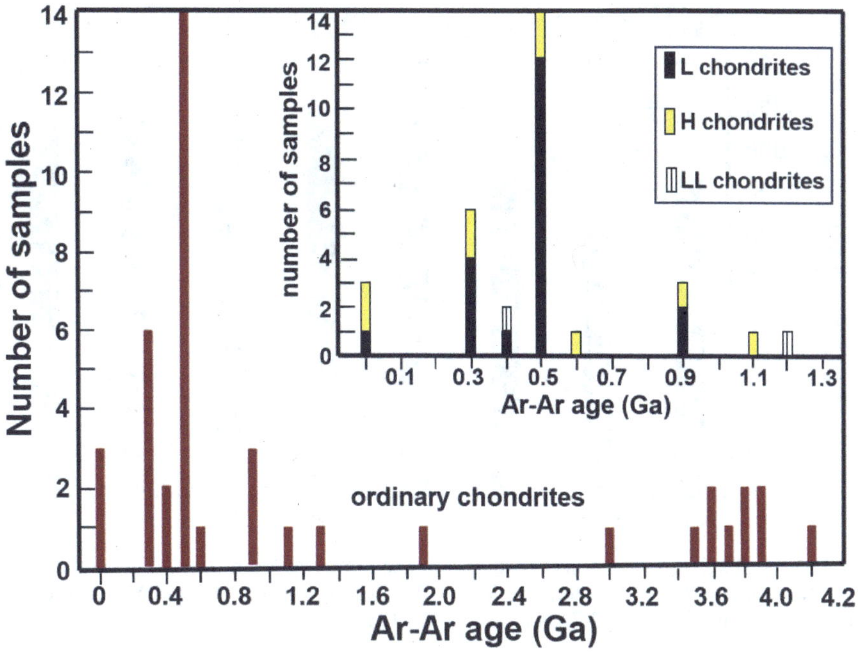

Fig. 14.7 Histogram of impact-reset ^{39}Ar-^{40}Ar ages of ordinary chondrites (brown). The large number of L chondrites with an Ar–Ar age of 0.5 Ga reflects the disruption of the L parent body around this time. The inset figure shows H, L, and LL chondrites with ^{39}Ar-^{40}Ar ages between 0 and 1.3 Ga. Diagram modified from Bogard (1995).

annealed (Rubin, 2002b). The ^{39}Ar-^{40}Ar data suggest two events (Dixon et al., 2004), one at 4.52 ± 0.08 Ga, the other at 4.23 ± 0.03 Ga. The first age could indicate shock while the sample was hot (Ruzicka et al., 2015) during the epoch of thermal metamorphism, the second event, a later-stage impact event after cooling.

The occurrence of strongly shocked samples with old Ar–Ar ages indicates that major collisions to OC parent asteroids were frequent early in Solar System history (e.g., Ciesla et al., 2013; Davison et al., 2013). One example is Shaw. Taylor et al. (1979) deduced that the impact event that affected Shaw created a crater at least several kilometers in diameter. The Shaw sample was modeled as having cooled at ~3.5°C ka^{-1} through the temperature range 660–500°C; it was probably buried at a depth of a few hundred meters.

As described in Section 10.3, there are two types of silicate inclusions in IIE irons: (1) extensively recrystallized, chondritic, chondrule-bearing inclusions in Netschaëvo, Techado, Garhi Yasin, and Mont Dieu (e.g., Olsen and Jarosewich, 1971; Bild and Wasson, 1977; Rubin, 1990; Schrader et al., 2017a; Van Roosbroek et al., 2015; McDermott et al., 2016) and (2) alkali-rich silicate inclusions with nonchondritic bulk compositions (e.g., Bunch and Olsen, 1968; Wasserburg et al., 1968; Bunch et al., 1970; Olsen and Jarosewich, 1970; Ruzicka et al., 1999; Ruzicka and Hutson, 2010; Ruzicka, 2014).

The ages of these IIE silicate inclusions were compiled by Bogard (2011); Ar–Ar ages are shown in Fig. 14.8. The Ar–Ar and other radiometric data show two age clusters: one at ~3.7 Ga and one at ~4.3–4.6 Ga. Each age cluster includes both chondritic and non-chondritic silicate inclusions. The silicates with ages >4.5 Ga indicate that silicate-metal mixing occurred very early in the history of the IIE parent body. As suggested by Bogard (2011), the slightly younger Ar–Ar ages (4.47–4.49 Ga) of silicate inclusions in Colomera, Weekeroo Station, Techado, and Tarahumara may reflect parent-body K-Ar closure during cooling after metal-silicate mixing.

14.4.3 U/Th-He Age Distributions

Based on the compilation of rare-gas data from Schultz and Kruse (1989), Wasson and Wang (1991) calculated U/Th-He gas-retention ages for ordinary chondrites. They found that most H chondrites and most LL chondrites have high ages: ~81% exceed 2.2 Ga. There appears to be a roughly random

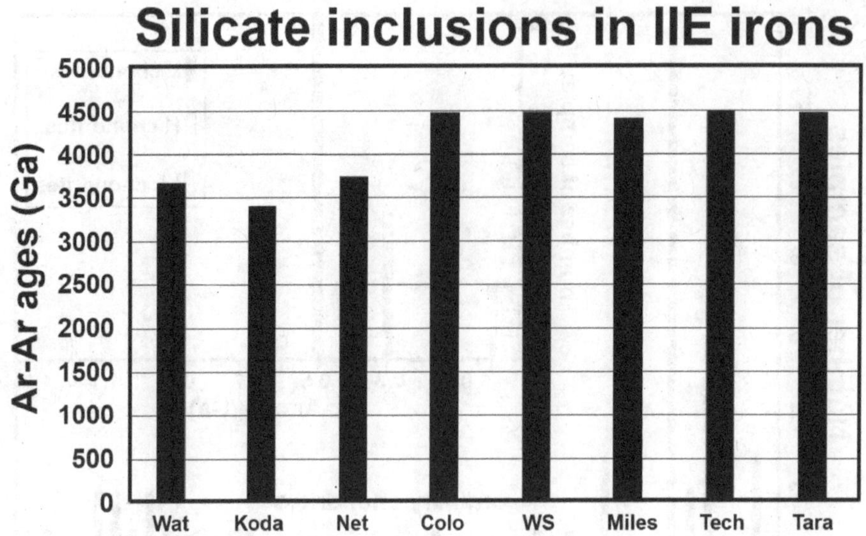

Fig. 14.8 The Ar–Ar ages of silicate inclusions in IIE irons based on data from Bogard (2011). Wat = Watson; Koda = Kodaikanal; Net = Netschaëvo; Colo = Colomera; WS = Weekeroo Station; Miles = Miles; Tech = Techado; Tara = Tarahumara. There are two age clusters: ~3.7 Ga and ~4.3–4.6 Ga.

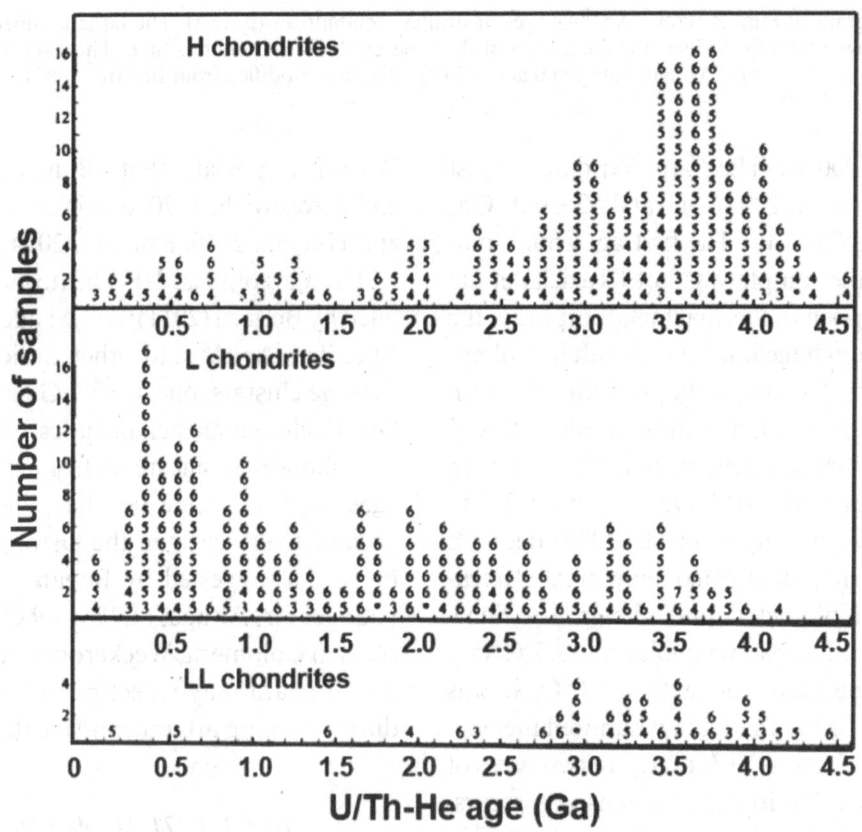

Fig. 14.9 Histograms of the U/Th-He gas-retention age distributions for ordinary chondrites. Each point of the diagram is expressed by a number (3, 4, 5, 6) representing the petrologic type of the analyzed sample. The prominent H- and L-chondrite peaks are due mainly to type-5 and -6 samples. Diagram modified from Wasson and Wang (1991).

distribution of U/Th-He gas-retention ages among the different petrologic types of H and LL chondrites (Fig. 14.9).

L chondrites have appreciably lower ages than the other two OC groups: ~69% of L chondrites have ages ≤2.2 Ga and ~35% have ages ≤0.9 Ga.

There are only a few L chondrites with U/Th-He ages exceeding 3.5 Ga. These data show that the L chondrites experienced more extensive outgassing than the other OC groups; this is presumably due to a significant extent to the breakup of the L parent asteroid ~470 Ma ago. There is also a roughly random distribution of petrologic types with age among the L chondrites except for the apparent absence of L3 chondrites in the 1.9–4.6-Ga interval.

Ages were also determined for 12 L/LL chondrites; seven of them have ages ≥3.5 Ga. This is statistically inconsistent with derivation from the L parent body but suggests that the samples designated L/LL are either anomalous LL chondrites (from the LL parent body) or are from an independent, otherwise-unsampled L/LL parent asteroid.

14.5 Regolith Formation

Asteroidal regolith constitutes the uppermost layer of an asteroid, extending from the surface down to depths of several meters. It comprises fine dust particles, millimeter-to-centimeter-sized rock and mineral fragments, cobbles, and boulders up to several meters across. Many of the rocks and rock fragments are breccias; they include monomict breccias (consisting of a single lithology), polymict breccias (containing more than one rock type), and impact-melt breccias (unmelted rock fragments flanked by solidified shock melt). CI chondrites exhibit fractal-like brecciation wherein clasts are observable at a large range of size scales (Friedrich et al., 2025).

Active regolith processes involve bombardment at all size scales. There are massive projectiles that create large impact craters, and secondary cratering events produced by ejecta from the large craters. Regoliths are subjected to a continuous rain of micrometeorites, cosmic-ray particles, and solar-flare particles. Because asteroids lack the protective blanket of an atmosphere, regoliths are exposed to the solar wind and become saturated in solar-wind gases. Exposure to these gases causes ion damage in the outermost 100 nm of mineral grains (e.g., Berger and Keller, 2015).

Large projectiles impact large main-belt asteroids at a range of relative velocities but with a mean of ~5.3 km s^{-1} (Bottke et al., 1994a, 1994b) (Fig. 14.10).

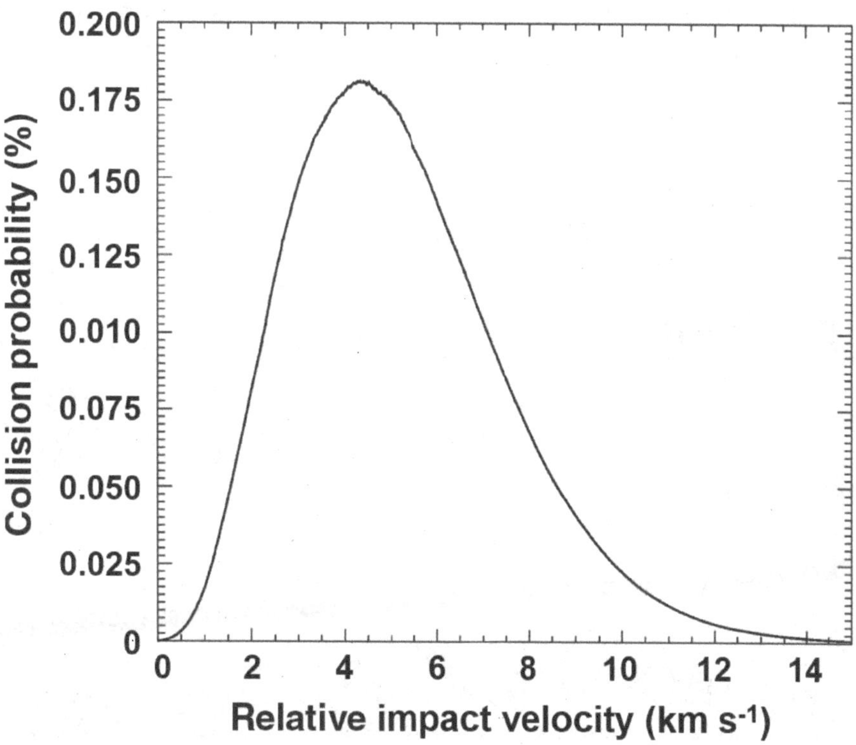

Fig. 14.10 Smoothed relative velocity distribution for mutual collisions among large (diameter > 50 km) main-belt asteroids. The mean and median collision velocities are 5.29 and 5.03 km s^{-1}, respectively (Bottke et al., 1994a). Diagram courtesy of William Bottke (with modifications).

Small asteroids probably have a roughly similar relative collisional velocity distribution (Bottke et al., 2020). In most cases, the pressure generated within a projectile upon impact is far greater than the projectile's strength, leading to severe deformation, melting, or even vaporization of the projectile. However, there is a selection effect: meteoroids belonging to the small percentage of projectiles with unusually low relative velocities are preferentially preserved (e.g., Wilkening, 1977; Anders, 1978; Wasson and Wetherill, 1979). Another cohort that is preserved is the set of surviving xenoliths derived from the small amounts of material spalled off the posterior surfaces of larger projectiles. These dislodged materials had low velocities relative to the surface of the target asteroid (Melosh, 1989; Bland et al., 2008). Fragments of all these foreign meteoroids were incorporated into the regolith as xenoliths.

Regoliths on two S asteroids of ordinary-chondrite composition have been examined up close: 433 Eros and 25143 Itokawa.

Eros has an extensive regolith of variable thickness, ranging up to several tens of meters deep (Barnouin-Jha et al., 2001; Cheng, 2002; Robinson et al., 2002) – the variability is likely due to the asymmetric distribution of ejecta (Murdoch et al., 2015). There are smooth terrains on Eros, dubbed "ponds," that consist mainly of submillimeter-sized grains (Fig. 14.11a). The ponds occur preferentially near the asteroid's equator, where the surface gravity is minimal. Ponds exhibit evidence of slumping, indicative of at least moderate cohesiveness (possibly a result of impact heating). Metallic Fe-Ni grains in the ponds may have been depleted by settling or size sorting (Cheng et al., 2002). The ponds may have formed by impact-induced seismic activity causing flow and deposition in low-lying regions of fine-grained regolithic debris.

It is possible that chondritic meteorites could contain fragments of indurated pond material. These may include clasts with bedding features such as those identified in CV3 Vigarano (Zolensky et al., 2022), CV3 Allende, CM2 Cold Bokkeveld, H3.4 Sharps and eight Antarctic OC: Y-790149 (H), Y-790199 (H), Y-790448 (LL3.2), Y-791863 (H6), Y-793374 (L3.6), Y-793472 (H6), Y-793535 (H6), and Y-82055 (L3.6) (Zolensky et al., 2004). The bedding features consist of numerous parallel arcuate bands. In the Vigarano clasts, cross beds are apparent; in the Sharps clast (Fig. 14.11b), the bands typically form closed loops and are ~10–30 μm thick and spaced ~7–40 μm apart (Zolensky et al., 2004). Bands appear bright in backscattered electron (BSE)

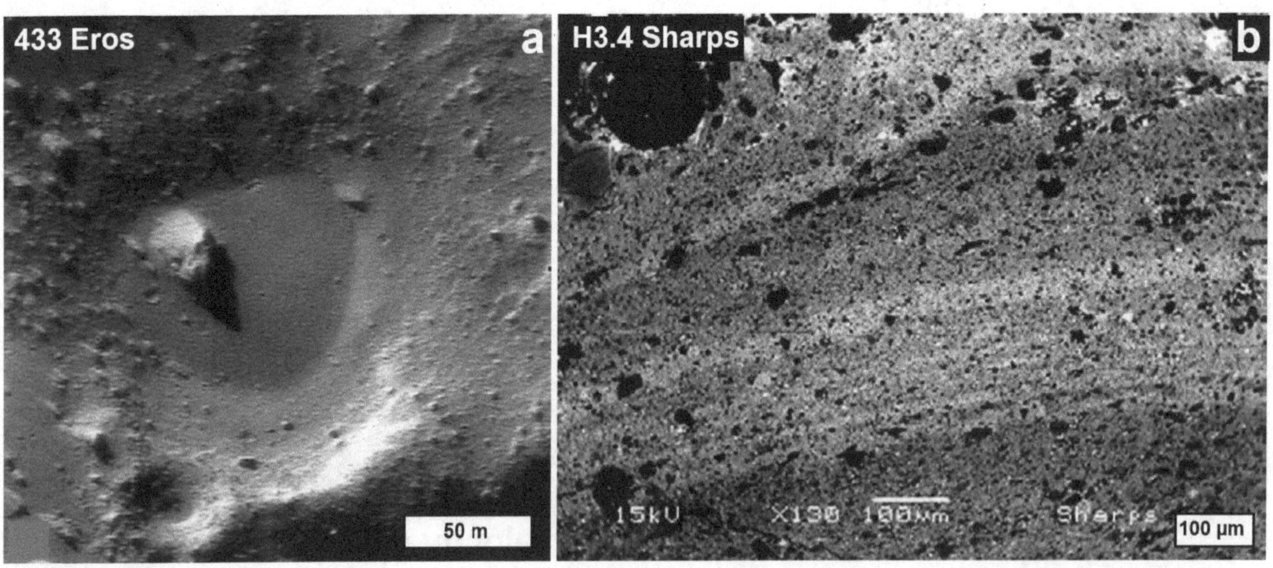

Fig. 14.11 Pond materials. (a) A typical pond deposited at the bottom of an impact crater on S asteroid 433 Eros. (NASA/JHU, image MET 155888598, resolution 0.55 m/pixel). (b) Subparallel arcuate beds in a clast from H3.4 Sharps that might represent layered pond material. Backscattered electron (BSE) image modified from Zolensky et al. (2004).

images because they contain high proportions of fine-grained ferroan olivine.

Itokawa's surface has some smooth patches with regolith depths of ~2.5 m (Cheng et al., 2007; Barnouin-Jha et al., 2001). The small regolith particles on Itokawa range from ~10 to 100 μm; however, the regolith overall is dominated by particles >1 mm in size (Yano et al., 2006; Miyamoto et al., 2007). The paucity of fine particles on Itokawa is a function of the small size of the asteroid ($0.535 \times 0.294 \times 0.209$ km); Itokawa has an escape velocity of only ~20 cm s^{-1} (e.g., Tsuchiyama, 2014). Micrometeorite impacts that fragment surface materials cause many fine regolithic particles to escape into space. Larger regolith particles (those >1 mm) may also escape the asteroid surface, possibly due to electrostatic levitation induced by solar radiation pressure (Lee, 1996; Scheeres, 2005).

Freshly exposed bedrock on airless bodies immediately begins to degrade. An *immature* regolith (characterized by coarse grains and small amounts of glass) forms in a relatively short period. As time passes and bombardment by micrometeorites, galactic and solar cosmic rays, and solar-flare particles continues, the regolith undergoes maturation. On the Moon, *mature* regoliths consist mainly of small particles; they contain abundant glass, numerous agglutinates, and fragments of lithified regolith breccias (e.g., McKay et al., 1991).

As discussed in Section 11.14, asteroidal regoliths are less mature than lunar regolith due to: (1) the factor-of-three lower micrometeorite flux in the vicinity of the asteroid belt relative to that at 1 AU where the Moon resides (e.g., Langevin and Maurette, 1981), (2) the average relative velocity of micrometeorite impacts into asteroidal regoliths being lower than for the Moon (~5 km s^{-1} vs. 15–20 km s^{-1}), causing less melting (Langevin and Maurette, 1981), (3) the domination of asteroid surfaces by spallation processes, which cause little melting of surficial materials (Hörz and Schaal, 1981), (4) the ultramafic composition of ordinary chondrites (and their correspondingly higher melting temperatures) compared to basaltic lunar mare materials (Keil, 1982), and (5) the earlier epoch of regolith maturity incorporated into OC regolith breccias compared to that of continuously maturing lunar soils (McKay et al., 1986, 1989).

Agglutinates, glass shards, and glass spherules, which are abundant on the Moon, are extremely rare in OC regolith breccias (<0.01 vol.%). As discussed in Section 11.14, among the few verified OC agglutinates are those that occur in the Fayetteville H-chondrite regolith breccia (Kerridge and Kieffer, 1977; Rubin, 1982; Taylor, 1982) (Fig. 11.81).

Although unconsolidated asteroidal regolith cannot make it to Earth without hitching a ride on a spacecraft (Tsuchiyama, 2014), our meteorite collections house ordinary-chondrite regolith breccias (enriched in solar-wind-implanted gases) that can be studied in the lab.

Many of the ordinary-chondrite regolith breccias have a light/dark structure (Fig. 11.76) as first described by Fredriksson and Keil (1963). The light-colored clasts are typically metamorphosed (type-4, -5, or -6) chondritic material of relatively low porosity. The dark material is a fine-grained clastic matrix representing consolidated regolith from the parent asteroid. The solar-wind-implanted light rare gases occur exclusively within this dark matrix (Konig et al., 1961, 1962). The abundance of ^{4}He ranges up to $\sim 2.0 \times 10^{-2}$ cm^{3} g^{-1} STP (e.g., Schultz and Kruse, 1978). The matrix is also appreciably richer in other volatile elements than the adjacent light-colored coarse clasts (Müller and Zähringer, 1966; Bart and Lipschutz, 1979; McSween and Lipschutz, 1980).

Solar flares emit energetic particles (predominantly Fe-group nuclei) that can penetrate mineral grains at the surfaces of airless bodies to depths of a few millimeters. The trail of ionization damage can be imaged with the TEM (e.g., Berger and Keller, 2015). The density of solar-flare particle tracks can be used to determine surface exposure age; the track density is a function of the length of time a particle resided within a few millimeters of the surface (e.g., Lal and Rajan, 1969; Lorin et al., 1970; Wilkening, 1971; Pellas, 1972; Macdougall et al., 1973, 1974; Rajan, 1974; Lal, 1980). Surface exposure is not continuous; impact gardening stirs the regolith, sporadically burying and excavating near-surface particles. The occurrence of mineral grains with solar-flare particle damage in OC regolith breccias indicates that components of these breccias were at the very surface of their parent asteroid.

On the Moon, regolith chemical composition varies among sampled sites, indicating that, in any particular region, the regolith is dominated by local bedrock but also contains components from farther afield (e.g., McKay et al., 1991). The same appears to be true of OC asteroids. The H-chondrite regolith serves as an example. The dark clastic matrix in some ordinary-chondrite regolith breccias (e.g., H5 Plainview (1917); H4 Tysnes Island; H4 Weston; H3.7 Dimmitt; H5 Cangas de Onis) is less equilibrated than the adjacent light-colored clasts (Fodor and Keil, 1978; Keil and Fodor, 1980; Noonan and Nelen, 1976; Rubin et al., 1983a; Williams et al., 1985). In contrast, the dark matrix in H6 Nulles is nearly identical to its light-colored clasts and must be composed nearly entirely of comminuted H6 material (Williams et al., 1985). The regolith on the H-chondrite parent body at the time and place of Nulles' consolidation consisted mainly of pulverized H6 material and coherent H6 pebbles, cobbles, and boulders. This appears consistent with the global spread of ejecta from the 2022 impact of the DART spacecraft on the small ($116 \times 174 \times 177$ m) asteroidal rubble-pile moon Dimorphos (Raducan et al., 2024).

The low surface gravity of asteroids relative to that of the Moon would cause crater ejecta to spread over larger regions, contaminating local regoliths with distal materials.

Regolith breccias form from porous unconsolidated regolith (and its load of pebbles and cobbles) by shock-lithification engendered by energetic impacts (e.g., Kieffer, 1975b; Ashworth and Barber, 1976). Shock experiments show that pores within OC powders (representative of asteroidal regoliths) collapse completely at moderate shock pressures (≤14.5 GPa); grain-boundary melting begins at <27 GPa (Hörz et al., 2005). The cement holding particles in the regolith breccias together is shock-melted feldspathic material, generally occurring around grain boundaries. The more cohesive the OC regolith breccia, the lower the content of solar-wind-implanted ^{4}He and ^{20}Ne (Bischoff et al., 1983), presumably reflecting more-extensive degassing during more-intensive shock heating. Typical asteroidal impacts of ~5 km s^{-1} will generate stress in the targets corresponding to about 67 GPa and produce ~50% melting of the affected target (Hörz et al., 2005).

Fragments of foreign projectiles in the asteroidal regoliths are preserved as xenoliths in OC regolith breccias. As discussed in Section 11.13, most foreign clasts fall into two populations: (1) rare samples of a different OC group (Dodd, 1974; Rubin et al., 1981b, 2017; Gattacceca et al., 2017; Niemeier and Bischoff, 2006; Bischoff et al., 2018; Funk et al., 2011) and (2) carbonaceous chondrites (predominantly CM) (e.g., Fodor and Keil, 1976a; Fodor et al., 1976; Rubin and Bottke, 2009) (e.g., Fig. 11.80). Some xenoliths are impact-melt rocks identified by the fact that their bulk O-isotopic compositions differ from those of their hosts (e.g., Table 1 of Lipschutz et al., 1989)

Ordinary-chondrite regolith breccias contain some shock-darkened clasts, riddled with micrometer-sized grains of metallic Fe-Ni and troilite. A prominent example is a 2.0×2.0 cm very dark clast in Nulles described by Williams et al. (1985).

Also present in OC regolith breccias is an assortment of impact-melt-rock clasts (Rubin, 1985). They range in size from <50 μm (Rubin et al., 1983a, 1983b) to ~4 cm (Binns, 1967a; Keil, 1982); they range in shape from spheroidal to angular, and in texture from poikilitic to porphyritic to skeletal (Figs. 11.77, 11.78, 11.79). Most melt-rock clasts have the same bulk O-isotopic compositions as their hosts, indicating derivation from host material.

14.6 Ordinary Chondrites Associated with Abundant Non-OC Material

As discussed in Section 11.13, there are several examples of ordinary chondrites of one group occurring as clasts within a brecciated sample of a different OC group. They include H-chondrite clasts in LL6 St. Mesmin (Dodd, 1974), an LL5 clast in H3.7 Dimmitt (Rubin et al., 1983a), an L4 clast in LL6 NWA 5764 (Gattacceca et al., 2017), L4/5 clasts in LL4-6 Glanerbrug (Niemeier and Bischoff, 2006), an ungrouped OC clast in LL3 NWA 10214 (Rubin et al., 2017), an H3 clast in LL3.2 Krymka (Bischoff et al., 2018), an LL clast in L3 Tanezrouft 039

Fig. 14.12 Jagged-edged ordinary-chondrite clast (right) and coarse enstatite grains from aubrite host (white, left). The clast contains numerous PP, RP, BO, PO, and POP chondrules ranging from 600 to 2000 μm in apparent diameter. The clast exhibits extensive silicate darkening due to shock dispersion of melted opaque phases. Plane-polarized transmitted light. Diagram modified from Rubin (2010b).

(Funk et al., 2011), H-chondrite fragments in LL3.6 Ngawi (Fodor and Keil, 1975), L-chondrite components in LL5 Paragould (Fodor and Keil, 1978), and L or H/L clasts in H3-6 Elmshorn (Bischoff et al., 2024b).

In addition to colliding with OC asteroids (and, of course, the Earth), ordinary-chondrite projectiles have pelted other bodies. These include Vesta (e.g., NWA 3197; NWA 10674, as will be discussed) and the Moon (there is a possible LL-chondrite fragment identified among Luna 16 samples; Demidova et al., 2020). However, no ordinary-chondrite meteorites have yet been identified on Mars.

Cumberland Falls: Rubin (2010b) described six chondritic clasts from the Cumberland Falls aubrite polymict breccia. Four of the clasts texturally resemble ordinary chondrites (e.g., Fig. 14.12); two clasts are impact-melt breccias containing shocked OC clasts adjacent to shock melt. These chondritic clasts, although unmelted, have also been shocked. They exhibit silicate darkening and contain olivine grains with undulose extinction and planar fractures, and a few olivine grains exhibit mosaic extinction. The inferred OC parentage of the chondritic clasts is supported by: (1) an LL-chondrite-like bulk chemical composition, (2) O-isotopic compositions ($\delta^{18}O$ = +5.53 to +5.8‰; $\delta^{17}O$ = +3.5 to +4.00‰; $\delta^{17}O$ = +0.48 to +0.99‰) similar to those of some LL3 chondrites, (3) chondrule sizes within the LL-chondrite range, (4) the presence of BO chondrules (very rare in enstatite chondrites) and RP chondrules (rare in carbonaceous chondrites), (5) an OC-like modal abundance of chondrules (~70 vol%), and (6) core compositions of low-Ca pyroxene grains (Fs_{20-21}) that are within or close to the ranges of those in equilibrated L and LL chondrites.

The chondritic clasts differ from ordinary chondrites in being appreciably reduced. They contain forsteritic olivine ($Fa_{0.4-0.7}$), 30–100-μm-sized grains of daubréelite ($FeCr_2S_4$), and opaque intergrowths of troilite, metallic Fe-Ni, and schreibersite ($(Fe,Ni)_3P$). (Daubréelite and schreibersite are nearly absent in ordinary chondrites.) All six clasts may be remnants of a single ordinary-chondrite projectile that was heterogeneously shocked when it collided with the Cumberland Falls host. Heat generated during the collision caused partial equilibration between the highly reduced aubrite host and the OC clasts. Olivine in the clasts developed forsteritic compositions; phosphates may have been reduced to phosphide. Because diffusion is more sluggish in low-Ca pyroxene than in olivine (e.g., Freer, 1981; Chakraborty, 1997), the pyroxene grains did not completely equilibrate; grain cores retained their initial more-ferroan compositions.

Another possible OC clast within an aubrite occurs in Pesyanoe. The 230 × 420-μm-sized clast studied by Rosenshein et al. (2006) is FeO rich and contains olivine ($Fa_{14.4}$) and orthopyroxene ($Fs_{13.2}Wo_{3.0}$) grains a little more magnesian than those in H chondrites. The clast contains ~18 vol.% glass, plausibly quenched impact melt formed during the collision with the aubrite parent body. Heat generated by the impact could also have caused partial reduction of the mafic silicates in the clast (with some FeO transforming into Fe^0), thereby lowering the olivine's FeO/(FeO+MgO) ratio.

Galim: The Galim meteorite is a polymict breccia that fell as a shower of stones in Cameroon in 1952. Seven pebbles were recovered. At least three pebbles (and probably five) are fragments of a heavily shocked (S5-6) LL6 chondrite called Galim (a). Two pebbles (one of which was lost) constitute Galim (b), an impact-melted EH chondrite. Rubin (1997c) noted that many of the reduced phases typical of EH chondrites (such as Si-bearing metallic Fe-Ni and Ti-bearing troilite) are absent from Galim (b). It seems likely that Galim (b) was shock-melted while in contact with a more oxidized source (i.e., Galim (a)), During this event, metal from Galim (b) was oxidized and lost Si; troilite was oxidized and lost Ti.

The absence of evidence for reduction in the Galim (a) LL6 chondrite may indicate that the amount of LL material greatly exceeded that of EH material. This suggests that shock metamorphism took place on the LL parent asteroid. The Galim breccia probably formed by a two-stage process: (1) accretion to the LL parent body of an EH projectile at low relative velocities (causing oxidation in Galim (b) and strong shock effects in Galim (a)), and (2) shock metamorphism of the heterogeneous assemblage by the subsequent impact of another projectile at significantly higher relative velocities. It is not clear whether the second impact launched the assemblage into interplanetary space.

Almahata Sitta: Asteroid 2008 TC_3 fell in the Nubian Desert in Sudan on October 7, 2008 about 19 hours after its discovery. The thousand-or-so collected fragments constitute the Almahata Sitta polymict breccia, originally classified as an anomalous polymict ureilite (Jenniskens et al., 2009) but which may instead have been a polymict C1 meteoroid containing abundant ureilitic material (Bischoff et al., 2022). After extensive examination of the diverse components of this meteorite, Bischoff et al. (2022) summarized their modal abundances: ureilites (~67%), EH3 to EH6 and EL3 to EL6 enstatite chondrites (~26%), ordinary chondrites (~4%), and miscellaneous specimens (~3%) including carbonaceous chondrites, a unique R-like chondrite, and enstatite achondrites. The OC clasts include H4, H4/5, H5, H5/6, L3, LL3, and LL4 chondrites (e.g., Warren and Rubin, 2010; Horstmann and Bischoff, 2014; Goodrich et al., 2018; Bischoff et al., 2022).

The abundance of metamorphosed (type 4–6) enstatite and ordinary chondrites among the collected samples of Almahata Sitta indicates that the parent body of asteroid 2008 TC_3 must have accreted after the period of thermal metamorphism of the EH, EL, and OC parent bodies. Almahata Sitta is not the only polymict meteorite breccia to contain ordinary-, carbonaceous- or enstatite-chondrite clasts. Such clasts are present in both Inner Solar System rocks (such as polymict ureilites) and Outer Solar System rocks such as the Kaidun carbonaceous chondrite (to be discussed). This observation leads to the inference that meteoroids composed of these materials once flooded the Solar System. Only a small fraction of these meteoroids was likely to have had sufficiently low relative velocities to survive intact after colliding with the regoliths of their asteroidal targets.

Goodrich (2023) estimated that polymict ureilites constitute ~5% of all ureilites; these meteorites are modeled as representing regolith that had formed on secondary ureilitic bodies after the original ureilite parent asteroid was collisionally disrupted ~5 Ma after CAI formation. A few volume-percent of polymict ureilites consist of diverse xenoliths of chondritic (ordinary, enstatite, carbonaceous, R) and achondritic parentage (e.g., Downes et al., 2008; Goodrich et al., 2021).

Among the numerous OC xenoliths in polymict ureilites are: (a) mineral fragments derived from H, L, and LL chondrites, (b) intact L and LL chondrites of petrologic type 3 to 6, and (c) L/LL breccias (Goodrich et al., 2021). The OC mineral fragments include olivine-clinobronzite assemblages in Nilpena that resemble phases in H3 chondrites (Jaques and Fitzgerald, 1982). In addition, Dar al Gani 319, North Haig, and Nilpena contain BO, PO, POP, and RP chondrules, isolated mineral grains, and small clasts similar to those in type-3 OC; also present in Dar al Gani 319 are equilibrated chondrite fragments of OC O-isotopic composition (although their olivine is more-ferroan [Fa_{32-41}] than in ordinary chondrites) (Prinz et al., 1986; Ikeda et al., 2000, 2003; Ross et al., 2010).

NWA 10674 and NWA 3197: According to the Meteoritical Bulletin Database (MBD), ~20 vol.% of the NWA 10674 polymict eucrite consists of dark H6 clasts (up to 3 cm in diameter) with abundant metallic Fe-Ni and rare recrystallized chondrules. Olivine averages $Fa_{18.6}$ in composition. Other clasts in NWA 10674 include small CM2-like carbonaceous-chondrite clasts, rare centimeter-sized felsic anorthite-rich clasts, minor vitric breccia clasts, and abundant eucrite clasts and clast fragments.

The NWA 3197 howardite contains large clasts of shocked and recrystallized H-chondrite material ($Fa_{15.3-18.9}$) with relict chondrules and abundant altered grains of kamacite. Also present in NWA 3197 are clasts and mineral fragments of gabbroic eucrites and diogenites. Both meteorites were described by Tony Irving and Scott Kuehner (University of Washington). The descriptions can be found on the Meteoritical Bulletin Database.

Ordinary-chondrite material occurs in other HED meteorites. As described in the MBD, within the NWA 16719 polymict eucrite there is a clast containing an ovoid ~200×250-μm cellular intergrowth of kamacite and troilite texturally resembling shock-melted assemblages in ordinary chondrites. Olivine ($Fa_{15.0\pm1.9}$, $n = 5$) and orthopyroxene ($Fs_{20.2\pm2.5}$, $n = 5$) in the clast are close to the ranges in H chondrites. The Jiddat al Harasis 556 howardite contains recrystallized chondrule fragments with central olivine compositions of Fa_{20-23}. Matrix O-isotopic data are consistent with ~10–15% admixed OC (likely H-chondrite) material.

Bencubbin: The Bencubbin CBa chondrite contains an LL4±1 clast (Lovering, 1962; McCall and De Laeter, 1965; Kallemeyn et al., 1978). Also present in Bencubbin are an enstatite-chondrite clast (McCall, 2018), a carbonaceous-chondrite clast (Kallemeyn et al., 1978) (later characterized as a "dark xenolith" by Weisberg et al., 1990), and an R4 clast (Isa et al., 2014). The assortment of foreign objects within a single meteorite led one researcher to characterize Bencubbin not as a meteorite, but as a meteorite collection.

Kaidun: Although Kaidun is classified in the Meteoritical Bulletin as a CR2 chondrite, it is a complex breccia containing a stunningly diverse variety of components including clasts of EH3-4, EH5, EL3, CV3, CM1, CM2, and R3 chondrites, ungrouped C1 and C2 lithologies, alkali-enriched materials, a phosphide-enriched object, impact-melt rocks, and Ca-rich achondrites (Zolensky and Ivanov, 2003). There is also a porphyritic olivine-pyroxene clast that texturally resembles POP chondrules in type-3 ordinary chondrites (Mikouchi et al., 2005). Zolensky and Ivanov (2003) pointed out that Kaidun accumulated materials from throughout the main asteroid belt. The high petrologic types of some of the xenoliths indicate that radial transport must have occurred after the period of parent-body metamorphism ended; this epoch is tens of millions of years after CAI formation, long after nebular gas had dissipated.

The small (submillimeter-sized) xenoliths in Kaidun could be fragmented remnants of larger (meter-to-hundred-meter-sized) interplanetary meteoroids. Bodies of this size would have been subjected to Yarkovsky forces. The *diurnal Yarkovsky effect* is a thermal radiation force acting on a rotating body that causes that body's semimajor axis to drift (e.g., Öpik, 1951; Peterson, 1976; Rubincam, 1995, 1998; Farinella et al., 1998; Vokrouhlický et al., 2000; Bottke et al., 2001, 2002, 2006; Vokrouhlický and Bottke, 2001; Chesley et al., 2003; Morbidelli and Vokrouhlický, 2003). Asteroids are warmed by the Sun but reach their highest temperatures in local afternoon after the warmed surface rotates away from the subsolar point. As the asteroid emits the absorbed radiation in the infrared, momentum is carried away from the body. There is a mismatch between the directions of radiation absorption and emission; that is, more momentum leaves the asteroid from hot regions than from cooler regions, producing a net force on the object. If the asteroid rotates in a prograde manner (counterclockwise when viewed from above the Sun's north pole), the net force is in the same direction as the orbital motion, causing an increase in its orbital velocity and a consequent increase in its semimajor axis. If the asteroid rotates retrograde, the net force is in the opposite direction, causing a decrease in its orbital velocity and a decrease in its semimajor axis. Asteroids rotating in different directions could therefore scatter ejecta both toward and away from the Sun. It seems possible that such a mechanism is responsible for the great diversity of xenoliths in the Kaidun regolith.

In any case, after incorporation into the Kaidun regolith, xenoliths were subjected to varying degrees of shock heating and aqueous alteration.

Comet 81P/Wild 2: The Stardust mission was launched in February 1999 (Brownlee et al., 2003) and returned to Earth carrying material captured from active Jupiter-family comet 81P/Wild 2 in January 2006. This comet resided in the Kuiper Belt for billions of years before its orbit was modified by Jupiter. The spacecraft's collector tray consisted of 90 blocks of silica aerogel with a total surface area exceeding 1000 cm^2. Most of the captured cometary particles impacted the aerogel at a velocity of 6.1 km s^{-1}.

The diameters of the best-preserved particles are in the range of roughly 2 to 30 μm; these grains tended to be strong enough to remain intact during capture (Brownlee et al., 2012). Frank et al. (2014) analyzed a suite of olivine grains from the set of coarse Stardust particles and found their compositional range ($Fa_{0–97}$) to be similar to those collectively encompassed by 5–30-μm-size olivine grains from unequilibrated carbonaceous, ordinary, and enstatite chondrites. Although the majority of the low-Ca pyroxene grains in Wild 2 most closely resembles grains from carbonaceous chondrites in chemical and O-isotopic composition (Joswiak et al., 2024), a few grains are similar to those in type-3 ordinary chondrites.

Frank (2023) showed that chondritic components of Wild 2 samples include Type-I and Type-II chondrules, Al-rich chondrules, CAIs, and AOAs. Although M. Zhang et al. (2023) suggested that these chondrules may have formed in the icy Outer Solar System, Frank (2023) identified pyroxene grains from CR, CH-CB, LL, and aubritic sources. One of these particles (80-1) is a low-Ca pyroxene grain ($Fs_{26}Wo_{2.0}$) with FeO (17.0 wt.%), MnO (0.68 wt.%), Cr_2O_3 (0.39 wt.%), CaO (1.0 wt.%), and TiO_2 (<0.07 wt.%) in the range of LL chondrites of petrologic type >3.5. No other chondrite group commonly contains pyroxene grains of this composition.

There must have been widespread radial transport of rocky materials from the Inner Solar System to the Kuiper Belt during the epoch of cometary accretion (Zolensky et al., 2006; Westphal et al., 2009). Disk winds (Fig. 11.43) or viscous expansion and turbulent diffusion of the protoplanetary disk could have been responsible for transporting these grains.

15

Magnetic Fields in the Solar Nebula

Many young stellar objects are surrounded by a protoplanetary disk that persists for just a few million years. During this astronomically brief period, dust grains aggregate into ~100-km-sized planetesimals and ~1000–100,000-km-sized planets (e.g., Chambers, 2004; Johansen et al., 2014). It seems likely that magnetic fields are involved in causing significant transfer of mass and angular momentum through these disks (e.g., Alfvén, 1981; Bai and Stone, 2013; Weiss et al., 2021); they may also be partly responsible for planetesimal accretion (e.g., Cuzzi and Hogan, 2003; Johansen et al., 2007). The magnetic fields in protoplanetary disks were probably inherited from their parental molecular clouds. In these clouds, star formation is influenced by gravity, magnetic fields, and gas turbulence (Li, 2021).

Although the inherited magnetic fields of protoplanetary disks may have had intensities of only ~10 μT (Desch and Mouschovias, 2001), processes operating within the disks (such as nebular shocks) could have enhanced magnetic field strengths by close to an order of magnitude (Desch and Connolly, 2002).

Contemporary astronomical instruments are incapable of directly measuring magnetic fields in molecular clouds or protoplanetary disks (Crutcher, 2012; Stephens et al., 2014). Nevertheless, paleomagnetic measurements of primitive components in chondrites could shed light on the strength of the magnetic field of the solar nebula. Studies by Fu et al. (2014a) of five relict dusty-olivine-bearing chondrules from LL3.01 Semarkona (e.g., Fig. 11.36a,b) revealed that the chondrules had been embedded in a nebular magnetic field of 54 ± 21 μT. This is of the same order of intensity as the present geomagnetic field at the surface of the Earth (22–67 μT, depending on location).

Unlike CV3 Allende chondrules (Fu et al., 2014b), chondrules from Semarkona avoided parent-body remagnetization. The small size of the metal particles within relict dusty olivine grains from Semarkona chondrules suggests that the particles may consist of a single magnetic domain. These metal grains are expected to have retained stable magnetization since their host chondrules formed ~4.6 Ga ago and cooled through the 765–350°C blocking temperature of the HC/HT component (e.g., Uehara and Nakamura, 2006; Lappe et al., 2011). (HC is magnetic coercivity, the ability of a ferromagnetic material to resist an external magnetic field without demagnetizing. HT is the tangential magnetic field at a material discontinuity.)

Wang et al. (2017) found that volcanic angrites formed in a very low, near-zero magnetic field (<0.6 μT) at 4563.5 ± 0.1 Ma; this is about 3.8 Ma after CAIs. The implication is that nebular gas had dissipated by that time.

Weiss et al. (2021) summarized the inferences drawn from paleomagnetic studies of various kinds of meteorites and from theoretical models. They concluded that the magnetic field strengths in the solar nebula were: (a) 54 ± 21 μT between ~1 and 3 AU from the Sun until >1.22 Ma after CAI formation, (b) $\gtrsim$6 μT between 3 and 7 AU >2.51 Ma after CAIs, (c) <0.6 μT in the Inner Solar System by 3.94 Ma after CAIs, and (d) <0.3 μT in the Outer Solar System by 4.89 Ma after CAIs. The much lower nebular magnetic field strengths by ~4 Ma after

CAI formation suggest that the gas had dissipated by that time. The inferred lifetime of nebular gas is consistent with the observed lifetimes of other protoplanetary disks (Weiss et al., 2021).

After making assumptions about the heliocentric distance at which Semarkona accreted and the vertical distribution of nebular dust, Fu et al. (2014a) inferred that the measured nebular magnetic intensity of the nebula was most consistent with chondrule formation by nebular shocks or planetesimal collisions. Arguments against such models are presented in Section 11.6. It is fair to say the problem of chondrule formation remains unsolved. It has remained unsolved since 1802 (Section 11.1).

16

Ordinary-Chondrite Formation

A Summing Up

The Apollo program was a seminal event, marking great advances in our understanding of Solar System history. Before the return of the first lunar samples, it was known that some carbonaceous and ordinary chondrites had quite short cosmic-ray exposure ages (e.g., Anders, 1962); this finding had led some researchers (e.g., Urey, 1965) to suggest that the Moon was the source of CI-chondrites. Ringwood (1965) proposed that ordinary and enstatite chondrites formed from planetesimals of CI-chondrite composition that had impacted the Moon. According to Ringwood's model, the projectiles were shock melted, creating reducing conditions that transformed FeO in chondritic minerals into metallic Fe. Ringwood surmised that an explosion accompanying the reduction process produced chondrules that littered the lunar surface, joining the newly formed metallic Fe. None of these processes has proven to be correct.

A decade after Apollo 11, J. V. Smith (1979) offered a model of ordinary-chondrite formation involving six steps, with H, L, and LL chondrites each experiencing a broadly similar history:

(1) formation of a debris cloud of diverse components during a major collision of planetesimals,
(2) mechanical sorting of the cloud under the influence of a central gravitational field,
(3) sequential accretion with change of composition, especially of volatile elements,
(4) thermal metamorphism varying with depth,
(5) formation of a regolith, yielding mainly monomict breccias with varying content of adsorbed solar gases, but producing some polymict breccias incorporating late-accreting material, and
(6) disruption of the regolith and underlying metamorphosed material during the collision that started fragments on their paths toward the Earth.

Scenarios of the formation and history of ordinary chondrites were offered by Wood (1962b) (pre-Apollo) and Wasson (1974) (post-Apollo) that are similar to the current perspective (presented here) and differ only modestly from that of J. V. Smith (1979):

1. Prior to the accretion of chondritic planetesimals, there were diverse solid components embedded within nebular gas; these included chondrules, CAIs, AOAs, grains of metallic Fe-Ni and sulfide, fragments of all these objects, porous aggregates of fine-grained organic-matter-bearing dust, and tiny presolar grains. These materials did not form by planetesimal collisions. An assortment of presolar grains was inherited from the parental molecular cloud. Other primitive materials formed in the solar nebula by such processes as condensation, localized flash melting, partial evaporation, agglomeration, and particle–particle collisions. Some materials may have experienced more than one process and/or the same process(es) operating multiple times.
2. Size-sorting occurred in the nebula; chondrite groups with large chondrules tend to have large CAIs and large AOAs. These particles (or their precursor dust aggregates) were separated in the nebula by aerodynamic processes based on their sizes and densities. Gravity did not play a major role at this stage (other than the fact that all the particles and the surrounding gas were in orbit around the proto-Sun).
3. It is assumed that individual planetesimals did not grow by heterogeneous accretion (i.e., with different lithologies accreting sequentially). Instead, the planetesimals likely agglomerated as fairly homogeneous porous bodies from locally available material. Initially, they were probably not significantly stratified by bulk chemical or isotopic composition, although lithostatic pressure may have reduced the porosity of deeply buried materials.

4. The ordinary chondrites experienced thermal metamorphism likely due mainly to the decay of ^{26}Al; early in Solar System history, their parent asteroids probably developed onion-shell structures, with petrologic type increasing with depth. The H, L, and LL bodies probably did not have molten cores. The OC asteroids may have been collisionally disrupted and gravitationally reassembled while still hot. Partial loss of volatiles occurred during metamorphism.
5. Each of the three major OC bodies was bombarded by meteoroids and micrometeoroids and formed a regolith. Collisions fused regolithic material into cohesive regolith breccias enriched in solar-wind-implanted rare gases. Solar-flare particles penetrated mineral grains at the surfaces of these asteroids, reaching depths of a few millimeters. Impacts occurred at low and high relative velocities, creating monomict, genomict, polymict, and impact-melt breccias, as well as impact-melt rocks. Fragments of some projectiles (predominantly carbonaceous chondrites) collided with OC bodies at low relative velocities and survived. They were incorporated into the regoliths as xenocrysts that have been identified in many OC breccias.
6. All OC parent bodies suffered major collisions. The L body was destroyed ~470 Ma ago. Impacts into all OC parent asteroids (and into massive fragments of the original bodies) launched small OC asteroids and meteoroids into interplanetary space. Some reached resonances, developed Earth-crossing orbits, and crashed to Earth. There they began to weather.

Epilogue

In 1844, Edgar Allan Poe published one of the first modern detective stories: "The Purloined Letter." The plot revolves around an unscrupulous government minister who had stolen a compromising letter penned by a royal lady; he used the threat of releasing the letter to blackmail her into granting him political favors. When the prefect of police was alerted, he had his men search the minister's apartments. They failed to find the letter. They looked in every nook and cranny, checked the walls and carpets, probed the furniture, opened every shelved book and scrutinized its binding. The minister himself was waylaid on the street, but the letter was not on his person.

Weeks later, the dispirited prefect told famed detective C. Auguste Dupin of the case. Dupin visited the minister (against whom he held a grudge) and espied the letter hidden in plain sight. It had been folded inside out, intentionally soiled, readdressed, and stuffed into a cheap card-rack dangling from the mantelpiece. The very obviousness of the letter had caused it to evade the notice of the police; their attention had been focused on the likelihood of the letter's clever concealment. On Dupin's next visit to the minister, there was a disturbance in the street (prearranged by Dupin); while the minister peered out the window, Dupin filched the letter and substituted a facsimile he had crafted himself. As a reward for handing over the twice-purloined letter, the prefect wrote Dupin a personal check for 50,000 francs.

Vital clues to the nature of the Solar System held by ordinary chondrites can also be hidden in plain sight. Even highly weathered equilibrated samples, seemingly unworthy of a second look, may bear the markings of thermal metamorphism, shock metamorphism, and post-shock annealing (Rubin, 2004b). Ferroan chondrule phenocrysts appearing to have formed by crystallization during a single cooling event, may conceal a complex history involving multiple episodes of crystallization, fragmentation, and remelting (e.g., Rubin and Baecker, 2021). Some olivine phenocrysts in OC chondrules may be enriched in ^{16}O and hail from the Outer Solar System (Kita et al., 2010; Piralla et al., 2021). Statistical analyses of large numbers of ordinary chondrites reveal correlations between petrologic type and the FeO/(FeO+MgO) contents of mafic silicates (e.g., Rubin, 1990), between petrologic type and shock stage (Stöffler et al., 1991; Rubin, 1995a, 2004b, 2025), and between CaO and FeO concentrations in chondrule-olivine phenocrysts (Scott et al., 1994).

Ordinary chondrites (e.g., Fig. E.1), the most abundant of all meteorites, lie strewn across the planet. They have pelted every continent and many remote islands; most have fallen into the sea. The Earth itself may have been made in part from ordinary-chondrite progenitors (e.g., Lodders, 2000). As pointed out by C. S. Lewis (1946): "All that is Earth has once been sky …" To study the heavens, we just need to keep our eyes open and dig. The rocks beneath our feet may conceal the secrets of the cosmos. *Sic itur ad astra.*

Fig. E.1 A 119-g remnant of the NWA 5590 L3 ordinary-chondrite find showing abundant chondrules of varying size and type. Image courtesy of John Kashuba.

References

Abell G. O. (1964) *Exploration of the Universe*, Holt, Rinehart and Winston, New York, 646 pp.

Aberham V. and Desch S. J. (2024) How CAI-rich can asteroids be? *86th Annu. Meet. Meteorit. Soc.,* abstract#3036.

Aboulahris M., Chennaoui Aoudjehane H., Rochette P., et al. (2019) Characteristics of the Sahara as a meteorite recovery surface. *Meteorit. Planet. Sci.* 54, 2908–2926.

Afiattalab F. and Wasson J. T. (1980) Composition of the metal phases in ordinary chondrites: Implications regarding classification and metamorphism. *Geochim. Cosmochim. Acta* 44, 431–446.

Agrell S. O., Long J. V. P., and Ogilvie R. E. (1963) Nickel content of kamacite near the interface with taenite in iron meteorites. *Nature* 198, 749–750.

Akai J. (1992) T-T-T diagram of serpentine and saponite and estimation of metamorphic heating degree of Antarctic carbonaceous chondrites. *Proc. NIPR Symp. Antarct. Met.* 5, 120–135.

Alaerts L., Lewis R. S. and Anders E. (1979) Isotopic anomalies of noble gases in meteorites and their origins-III. LL-chondrites. *Geochim. Cosmochim. Acta* 43, 1399–1415.

Alekseev V. A. (1988) Meteor shower statistics. *Meteoritika* 47, 24–31.

Alexander C. M. O'D. (1995) Formation of chondrules by recycling and volatile loss (abstract). *Lunar Planet. Sci.* 26, 15–16.

Alexander C. M. O'D. (2005) From supernovae to planets: The view from meteorites and interplanetary dust particles. In *Chondrites and the Protoplanetary Disk* (edited by Krot A. N., Scott E. R. D. and Reipurth B.), pp. 972–1002. ASP Conference Series vol. 341.

Alexander C. M. O.'D. and Grossman J. N. (2005) Alkali elemental and potassium isotopic compositions of Semarkona chondrules. *Meteorit. Planet. Sci.* 40, 541–556.

Alexander C. M. O'D., Hutchison R., Graham A. L., and Yabuki H. (1987) Discovery of scapolite in the Bishunpur (LL3) chondritic meteorite. *Mineral Mag.* 51, 733–735.

Alexander C. M. O'D., Barber D. J., and Hutchison R. (1989a) The microstructure of Semarkona and Bishunpur. *Geochim. Cosmochim. Acta* 53, 3045–3057.

Alexander C. M. O., Hutchison R., and Barber D. J. (1989b) Origin of chondrule rims and interchondrule matrices in unequilibrated ordinary chondrites. *Earth Planet. Sci. Lett.* 95, 187–207.

Alexander C. M. O'D., Russell S. S., Arden J. W., et al. (1998) The origin of chondritic macromolecular organic matter: A carbon and nitrogen isotopic study. *Meteorit. Planet. Sci.* 33, 603–622.

Alexander C. M. O'D., Grossman J. N., Wang J., et al. (2000) The lack of potassium-isotopic fractionation in Bishunpur chondrules. *Meteorit. Planet. Sci.* 35, 859–868.

Alexander C. M. O'D., Fogel M., Yabuta H., and Cody G. D. (2007) The origin and evolution of chondrites recorded in the elemental and isotopic compositions of their macromolecular organic matter. *Geochim. Cosmochim. Acta* 71, 4380–4403. https://doi.org/10.1016/j.gca.2007.06.052.

Alexander C. M. O'D., Bowden R., Fogel M. L., et al. (2012) The provenances of asteroids, and their contributions to the volatile inventories of the terrestrial planets. *Science* 337, 721–723.

Alexeev V. A., Gorin V. D., Ivliev A. I., Kashkarov L. L., and Ustinova G. K. (2001) Combined study of thermoluminescence, tracks, and radionuclides in the recently fallen Kunya-Urgench chondrite. *Geochem. Int.* 39, 1043–1055.

Alfvén H. (1981) Origin of solar system. *Adv. Space Res.* 1, 5–20.

Allen J. S., Nozette S., and Wilkening L. L. (1980) A study of chondrule rims and chondrule irradiation records in unequilibrated ordinary chondrites. *Geochim. Cosmochim. Acta* 44, 1161–1175.

Almas K. S., Ash R. D., and Walker R. J. (2024) Effects of hot desert weathering on highly siderophile elements in ordinary chondrites. *Meteorit. Planet. Sci.* 59, 3072–3086.

Amelin Y., Krot A. N., Hutcheon I. D., and Ulyanov A. A. (2002) Lead isotopic ages of chondrules and calcium-aluminum-rich inclusions. *Science* 297, 1678–1683.

Amin K., Huang J. M., Hu K. J., Zhang J., and Ristroph L. (2019) The role of shape-dependent flight stability in the origin of oriented meteorites. *PNAS* 116, 16180–16185.

Anders E. (1962) Two meteorites of unusually short cosmic-ray exposure age. *Science* 138, 431–433.
Anders E. (1964) Origin, age and composition of meteorites. *Space Sci. Rev.* 3, 583–714.
Anders E. (1978) Most stony meteorites come from the asteroid belt. In *Asteroids: An Exploration Assessment*, Vol. NASA CP 2053 (edited by Morrison D. and Wells W. C.). U.S. Government Printing Office, Washington, DC, 57–75.
Anderson D. L. (1972) Internal constitution of Mars. *J. Geophys. Res.* 77, 789–795.
Anderson S. L., Benedix G. K., Godel B., et al. (2024) The Arpu Kuilpu meteorite: In-depth characterization of an H5 chondrite delivered from a Jupiter Family Comet orbit. *Meteorit. Planet. Sci.* 59, 3087–3110.
Andronicos C. L., Bindi L., Distler V. V., et al. (2018) Comment on "Composition and origin of holotype Al-Cu-Zn minerals in relation to quasicrystals in the Khatyrka meteorite" by M. Ivanova et al. (2017). *Meteorit. Planet. Sci.* 53, 2430–2440.
Andronikov A. V., Andronikova I. E., Pour O., and Bohdalek P. (2025) A Potůčky fragment of the Steinbach IVA-an meteorite: Implications from the trace-element composition of the constituent phases. *Meteorit. Planet. Sci.* 60, 1166–1193.
Anonymous (1874) Queer Rains. *Sci. Amer.* 31, 193.
Antier K. (2023) Pieces of 2023 CX1 recovered in Normandy! *International Meteor Organization (IMO)*, posted February 15, 2023. www.imo.net/pieces-of-2023-cx1-recovered-in-normandy/.
Antoniadi E. M. (1939) On ancient meteorites, and on the origin of the crescent and star emblem. *J. Roy. Astron. Soc. Canada.* 33, 177–184.
Arai T., Kobayashi M., Yamada M., et al. (2019) METEOR: Two-year observation from the International Space Station. *82nd Annu. Meet. Meteorit. Soc.*, abstract#6493.
Arakawa S., Okuzumi S., Tatsuuma M., et al. (2023) Size dependence of the bouncing barrier in protoplanetary dust growth. *Astrophys. J. Lett.* 951, L16. https://doi.org/10.3847/2041-8213/acdb5f.
Archer G. J., Walker R. J., Tino J., et al. (2019) Siderophile element constraints on the thermal history of the H chondrite parent body. *Geochim. Cosmochim. Acta* 245, 556–576.
Arif S., Chennaoui Aoudjehane H. and Bouvier A. (2024) Shedding light on the beginnings of the solar system: The presence of sulfides in ordinary chondrites – Insights into the Ksar El Goraane meteorite and other Moroccan falls. *86th Annu. Meet. Meteorit. Soc.*, abstract#3036.
Armstrong J. C., Wells L. E., and Gonzalez G. (2002) Rummaging through Earth's attic for remains of ancient life. *Icarus* 160, 183–196.
Artoli G., and Davoli G. (1994) Low-Ca pyroxenes from LL-group chondritic meteorites: Crystal structural studies and implications for thermal histories. *Earth Planet. Sci. Lett.* 128, 469–478.
Ash R. (1994) Chondrules: Small spheres of influence. *Nature* 372, 219–220.
Ash R. D. and Pillinger C. T. (1995) Carbon, nitrogen, and hydrogen in Saharan chondrites: The importance of weathering. *Meteoritics* 30, 85–92.
Ash R. D., Grady M. M., Morse A. D., and Pillinger C. T. (1991) Renazzo-like chondrites: A light clement stable isotope study (abstract). *Meteoritics* 26, 314.
Ash R. D., Gilmour J. D., Whitby J. A., et al. (1995) The history of the Parnallee meteorite as revealed by iodine-xenon dating (abstract). *Meteoritics* 30, 483–484.
Ashworth J. R. (1977) Matrix textures in unequilibrated ordinary chondrites. *Earth Planet. Sci. Lett.* 35, 25–34.
Ashworth J. R. (1979) Two kinds of exsolution in chondritic olivine. *Mineral. Mag.* 43, 535–538.
Ashworth J. R. (1980) Chondrite thermal histories: Clues from electron microscopy of orthopyroxene. *Earth Planet. Sci. Lett.* 46, 167–177.
Ashworth J. (1981) Fine structure in H-group chondrites. *Proc. Royal Soc. London* A374, 179–194.
Ashworth J. R. and Barber D. J. (1976) Lithification of gas-rich meteorites. *Earth Planet. Sci. Lett.* 30, 222–233.
Ashworth J. R. and Mallinson L. G. (1985) Transmission electron microscopy of L-group chondrites, 2. Experimentally annealed Kyushu. *Earth Planet. Sci. Lett.* 73, 33–40.
Ashworth J. R., Mallinson L. G., Hutchison R., and Biggar G. M. (1984) Chondrite thermal histories constrained by experimental annealing of Quenggouk orthopyroxene. *Nature* 308, 259–261.
Asphaug E., Jutzi M., and Movshovitz M. (2011) Chondrule formation during planetesimal accretion. *Earth Planet. Sci. Lett.* 308, 369–379.
Asphaug E., Collins G., and Jutzi M. (2015) Global-scale impacts. In *Asteroids IV* (edited by Michel P., Demeo F. E. and Bottke W. F.), pp. 661–677, University of Arizona Press, Tucson.
Azoulay P., Fons-Rosen C., and Graff Zivin J. S. (2019) Does science advance one funeral at a time? *Amer. Econ. Rev.* 109, 2889–2920.
Badyukov D. D., Raitala J., Kostama P., and Ignatiev A. V. (2015) Chelyabinsk meteorite: Shock metamorphism, black veins and impact melt dikes, and the Hugoniot. *Petrology* 23, 103–115.
Bae S. G., Choo C. O., and Jang Y. D. (2012) Mineralogical characteristics of tachylite occurring in Basic Dike, Basaltic Agglomerate Formation, Ulleung Island and its implications of volcanic activity. *J. Miner. Soc. Korea* 25, 63–76.
Baecker B., Rubin A. E., and Wasson J. T. (2017) Secondary melting events in Semarkona chondrules revealed by compositional zoning in low-Ca pyroxene. *Geochim. Cosmochim. Acta* 211, 256–279.
Baedecker P. A. and Wasson J. T. (1975) Elemental fractionation among enstatite chondrites. *Geochim. Cosmochim. Acta* 39, 735–765.

Baeza L. (2019) Oxygen isotope systematics of ordinary chondrite chondrules: Insights into the inner solar system planetary reservoir. Master's thesis, Australian National University, Canberra, Australia, 174 pp.

Bagnall P. M. (1991) *The Meteorite & Collector's Handbook*, Willmann-Bell, Richmond, Virgina, 160 pp.

Bai X.-N. and Stone J. M. (2013) Wind-driven accretion in protoplanetary disks. I. Suppression of the Magnetorotational instability and launching of the magnetocentrifugal wind. *Astrophys. J.* 769:76, 21 pp.

Baker R. H. (1959) *Astronomy – A Textbook for University and College Students, 7th edition,* D. Van Nostrand Company, Princeton, New Jersey.

Baldwin R. B. (1942) The meteoritic origin of lunar craters. *Popular Astronomy* 50, 365–369.

Baldwin R. B. (1949) *The Face of the Moon,* University of Chicago Press, Chicago.

Baldwin R. B. (1964) Lunar crater counts. *Astronom. J.* 69, 377–392.

Barnes V. E. (1958) Properties of tektites pertinent to their origin. *Geochim. Cosmochim. Acta* 14, 267–278.

Barnes S. J., Naldet A. J., and Gorton M. P. (1985) The origin of the fractionation of platinum-group elements in terrestrial magmas. *Chem. Geol.* 53, 303–323.

Barnouin-Jha O. S., Garvin J. B., Cheng A. F., et al. (2001) Preliminary impact crater dimensions on 433 Eros from the NEAR laser range finder and imager. *Lunar Planet. Sci.* XXXII, abstract#1786.

Barosch J., Hezel D. C., Sawatzki L., Halbauer L., and Marrocchi Y. (2020) Sectioning effects of porphyritic chondrules: Implications for the PP/POP/PO classification and correcting modal abundances of mineralogically zoned chondrules. *Meteorit. Planet. Sci.* 55, 993–999.

Barosch J., Nittler L. R., Wang J., et al. (2022) Presolar O- and C-anomalous grains in unequilibrated ordinary chondrite matrices. *Geochim. Cosmochim. Acta* 335, 169–182.

Bart G. and Lipschutz M. E. (1979) On volatile element trends in gas-rich meteorites. *Geochim. Cosmochim. Acta* 43, 1499–1504.

Bartholomay H. A. and Larimer J. W. (1982) Dust–gas fractionation in the early solar system (abstract). *Meteoritics* 17, 180–181.

Bartoschewitz R., Appel P., Barrat J.-A., et al. (2017) The Braunschweig meteorite – A recent L6 chondrite fall in Germany. *Chem. Erde.* 77, 207–224.

Bauer J. F. (1979) Experimental shock metamorphism of mono- and polycrystalline olivine: A comparative study. *Proc. Lunar Planet. Sci. Conf. 10th*, 2573–2596.

Baxter P. J. and Gresham A. (1997) Deaths and injuries in the eruption of Galeras Volcano, Columbia, 14 January 1993. *J. Volcanol. Geothermal Res.* 77, 325–338.

Bazhan I. S., Litasov K. D., Ohtani E., and Ozawa S. (2017) Majorite-olivine–high-Ca pyroxene assemblage in the shock-melt veins of Pervomaisky L6 chondrite. *Amer. Mineral.* 102, 1279–1286.

Baziotis I., Asimow P. D., Hu J., et al. (2018) High pressure minerals in the Château-Renard (L6) ordinary chondrite: Implications for collisions on its parent body. *Sci. Rep.* 8, 9851.

Baziotis I. P., Ma C., Guan Y., et al. (2022) Unique evidence of fluid alteration in the Kakowa (L6) ordinary chondrite. *Sci. Rep.* 12, 5520. https://doi.org/10.1038/s41598-022-09465-6.

Baziotis I., Ma C., Klemme S., et al. (2025) The high-pressure, vacancy-stabilized component in clinopyroxenes. *Amer. Mineral.* https://doi.org/10.2138/am-2024-9609.

Becker R. H. and Epstein S. (1982) Carbon, hydrogen, and nitrogen isotopes in solvent-extractable organic matter from carbonaceous chondrites. *Geochim. Cosmochim. Acta* 46, 97–103.

Beckerling W. and Bischoff A. (1995) Occurrence and composition of relict minerals in micrometeorites from Greenland and Antarctica – Implications for their origins. *Planet. Space Sci.* 43, 435–449.

Beech M., Brown P., Hawkes R. L., et al. (1995) The fall of the Peekskill meteorite: Video observations, atmospheric path, fragmentation record and orbit. *Earth Moon Planets* 68, 189–197.

Begemann F. and Wlotzka F. (1969) Shock induced thermal metamorphism and mechanical deformations in the Ramsdorf chondrite. *Geochim. Cosmochim. Acta* 33, 1351–1370.

Beitz E., Blum J., Parisi M. G., and Trigo-Rodriguez J. (2016) The collisional evolution of undifferentiated asteroids and the formation of chondritic meteoroids. *Astrophys. J.* 824, 18 pp.

Belczynski K., Askar A., Arca-Sedda M., et al. (2018) The origin of the first neutron-star – neutron star merger. *Astron. Astrophys.* 615. https://doi.org/10.1051/004-8361/201732428.

Bell D. J. (1986) Experimental crystallization of an olivine-normative chondrule composition. BS dissertation, Rutgers University.

Bell J. F., Davis D. R., Hartmann W. K., and Gaffey M. J. (1989) Asteroids: The big picture. In *Asteroids II* (edited by Binzel R. P., Gehrels T. and Matthews M. S.), pp. 921–945, University of Arizona Press, Tucson.

Bell K. R., Cassen P. M., Wasson J. T., and Woolum D. S. (2000) The FU Orionis phenomenon and solar nebula material. In *Protostars and Planets IV* (edited by Mannings V. et al.,), pp. 897–926. University of Arizona, Tucson.

Bellemare P. M. (1996) Meteorite sparks a cult. *J. Royal Astron. Soc. Canada* 90, 287–291.

Bellesi M., Manzetti F., Pratesi G., and Giuli G. (2020) Chromites in ordinary chondrite fusion crusts. 14th Europlanet Science Congress 2020. www.epsc2020.eu/,id.EPSC2020-886.

Bellucci J. J., Nemchin A. A., Grange M., et al. (2019) Terrestrial-like zircon in a clast from an Apollo 14 breccia. *Earth Planet. Sci. Lett.* 510, 173–185.

Benedix G. K., Ketcham R. A., Wilson L., et al. (2008) The formation and chronology of the PAT 91501 impact-melt L chondrite with vesicle-metal-sulfide assemblages. *Geochim. Cosmochim. Acta* 72, 2417–2428.

Benedix G. K., McCoy T. J., and Lauretta D. S. (2003) Is NWA 1463 the most primitive winonaite? (abstract) *Meteorit. Planet. Sci.* 30, A70.

Benito S., Cuervo C., Pöhl F., and Theisen W. (2019) Improvements on the recovery of 3D particle size distributions from 2D sections. *Mater. Character.* 156, 109872.

Benner L. A. M., Busch M. W., Giorgini J. D., Taylor P. A., and Margot J.-L. (2015) Radar observations of near-Earth and main-belt asteroids. In *Asteroids IV* (edited by Michel P. et al.), pp. 165–182, University of Arizona Press, Tucson.

Bennett M. E. and McSween H. Y. (1996) Revised model calculations for the thermal histories of ordinary chondrite parent bodies. *Meteorit. Planet. Sci.* 31, 783–792.

Benoit P. H. and Sears D. W. G. (1993) Breakup and structure of an H-chondrite parent body: The H-chondrite flux over the last million years. *Icarus* 101, 188–200.

Berger E. L. and Keller L. P. (2015) Transmission electron microscope measurements. *Lunar Planet Sci.* 46, abstract#1543.

Berkeley J. L., Fodor R. V., Taylor G. J., Keil K., and Fruland R. M. (1976) The Shaw meteorite: Evidence for an igneous origin. *Meteoritics* 11, 249–250.

Bevan A. W. R. and Axon H. J. (1980) Metallography and thermal history of the Tieschitz unequilibrated meteorite – Metallic chondrules and the origin of polycrystalline taenite. *Earth Planet. Sci. Lett.* 47, 353–360.

Bevan A. W. R. and Binns R. A. (1989) Meteorites from the Nullarbor Region, Western Australia: II. Recovery and classification of 34 new meteorite finds from the Mundrabilla, Forrest, Reid and Deakin areas. *Meteoritics* 24, 135–141.

Bevan A. W. R., Downes P. J., Henry D. A., Verrall M., and Haines P. W. (2019) The Gove relict iron meteorite from Arnhem Land, Northern Territory, Australia. *Meteorit. Planet. Sci.* 54, 1710–1719.

Bhandari N., Castagnoli G. C., Bonino G., and Suthar K. M. (1996) Solar flare tracks in Ro Negro (L4) chondrite. In *Proceedings of the 24th International Cosmic Ray Conference* (edited by Iucci N. and Lamanna E.), pp. 1181–1183, International Union of Pure and Applied Physics, Trieste, Italy.

Bigolski J. N., Weisberg M. K., Connolly H. C., and Ebel D. S. (2016) Microchondrules in three unequilibrated ordinary chondrites. *Meteorit. Planet. Sci.* 51, 235–260.

Bild R. W. and Wasson J. T. (1977) Netschaëvo: A new class of chondritic meteorite. *Science* 197, 58–62.

Bindi L. and Xie X. (2017) Shenzhuangite, $NiFeS_2$, the Ni-analogue of chalcopyrite from the Suizhou L6 chondrite. *Eur. J. Mineral.* 30, 165–169.

Bindi L., Steinhardt P. J., Yao N., and Lu P. J. (2011) Icosahedrite, $Al_{63}Cu_{24}Fe_{13}$, the first natural quasicrystal. *Amer. Mineral.* 96, 928–931.

Bindi L.,Yao N., Lin C., et al. (2015) Decagonite, $Al_{71}Ni_{24}Fe_5$, a quasicrystal with decagonal symmetry from the Khatyrka CV3 carbonaceous chondrite. *Amer. Mineral.* 100, 2340–2343.

Bindi L., Lin C., Ma C., and Steinhardt P. J. (2016) Collisions in outer space produced an icosahedral phase in the Khatyrka meteorite never observed previously in the laboratory. *Sci. Rep.* 6, 38117.

Bindi L., Xie Z., Sharp T. G., and Xie X. (2024) Unveiling deep Earth's hidden potential: Insights from a new high-pressure phase discovered in a shocked meteorite. *Discov. Miner.* 1, 1–6.

Binns R. A. (1967a) An exceptionally large chondrule in the Parnallee meteorite. *Mineral. Mag.* 36, 319–324.

Binns R. A. (1967b) Structure and evolution of non-carbonaceous chondritic meteorites. *Earth Planet. Sci. Lett.* 2, 23–28.

Binz C. M., Ikramuddin M., Rey P., and Lipschutz M. E. (1976) Trace elements and interelement relationships in unequilibrated ordinary chondrites. *Geochim. Cosmochim. Acta* 40, 59–71.

Binzel R. P. (2001) Forging the fourth link between planetary worlds: Vesta and the HEDs. *Meteorit. Planet. Sci.* 36, 479480.

Binzel, R. P. and Xu, S. (1993) Chips off of asteroid 4 Vesta: Evidence for the parent body of basaltic achondrite meteorites. *Science* 260, 186–191.

Binzel R. P., Lupishko D. F., Di Martino M., Whiteley R. J., and Hahn G. J. (2002) Physical properties of near-Earth objects. In *Asteroids III* (edited by Bottke W. F., Cellino A., Paolicchi P. and Binzel R. P.), pp. 255–272, University of Arizona Press, Tucson.

Binzel R. P., Rivkin A. S., Stuart J. S., et al. (2004) Observed spectral properties of near-Earth objects: Results for population distribution, source regions, and space weathering processes. *Icarus* 170, 259–294.

Binzel R. P., DeMeo F. E., Turtelboom E. V., et al. (2019) Compositional distributions and evolutionary processes for the near-Earth object population: Results from the MIT-Hawaii Near-Earth Object Spectroscopic Survey (MITHNEOS). *Icarus* 324, 41–76.

Biot J.-B. (1803) *Relation d'un voyage fait dans le departernent de l'Orne pour constater la réalité d'un méteoré observé à l'Aigle le 26 floréal an 11*, Baudouin, Imprimeur de I'Institut National de France. 47 pp.

Birch S. P. D. and Umurhan O. M. (2024) Retention of CO ice and gas within 486958 Arrokoth. *Icarus* 413, 116027.

Bischoff A. (2001) Meteorite classification and the definition of new chondrite classes as a result of successful meteorite search in hot and cold deserts. *Planet. Space Sci.* 49, 769–776.

Bischoff A. and Geiger T. (1995) Meteorite classification and the definition of new chondrite classes as a result of successful meteorite search in hot and cold deserts. *Meteoritics* 30, 113–122.

Bischoff A. and Keil K. (1983a) Ca-Al-rich chondrules and inclusions in ordinary chondrites. *Nature* 303, 588–592.

Bischoff A. and Keil K. (1983b) Catalog of Al-rich Chondrules, Inclusions and Fragments in Ordinary Chondrites. In *UNM Institute of Meteoritics Special Publication 22*, University of New Mexico, Albuquerque, 35 pp.

Bischoff A. and Keil K. (1984) Al-rich objects in ordinary chondrites: Related origin of carbonaceous and ordinary chondrites and their constituents. *Geochim. Cosmochim. Acta* 48, 693–709.

Bischoff A. and Stöffler D. (1992) Shock metamorphism as a fundamental process in the evolution of planetary bodies: Information from meteorites. *Eur. J. Mineral.* 4, 707–755.

Bischoff A., Rubin A. E., Keil K., and Stöffler D. (1983) Lithification of gas-rich chondrite regolith breccias by grain boundary and localized shock melting. *Earth Planet. Sci. Lett.* 66, 1–10.

Bischoff A., Geiger T., Palme H., et al. (1993a) Mineralogy, chemistry, and noble gas contents of Adzhi-Bogdo – An LL3-6 chondritic breccia with L-chondritic and granitoidal clasts. *Meteoritics* 28, 570–578.

Bischoff A., Palme H., Ash R. D., et al. (1993b) Paired Renazzo-type (CR) carbonaceous chondrites from the Sahara. *Geochim. Cosmochim. Acta* 57, 1587–1603.

Bischoff A., Scott E. R. D., Metzler K., and Goodrich C. A. (2006) Nature and origins of meteoritic breccias. In *Meteorites and the Early Solar System II* (edited by Lauretta D. S. and McSween H. Y.), pp. 679–712, University of Arizona Press, Tucson.

Bischoff A., Horstmann M., Pack A., Laubenstein M., and Haberer S. (2010) Asteroid 2008 TC_3 – Almahata Sitta: A spectacular breccia containing many different ureilitic and chondritic lithologies. *Meteorit. Planet. Sci.* 45, 1638–1656.

Bischoff A., Vogel N., and Roszjar J. (2011) The Rumuruti chondrite group. *Chem. Erde.* 71, 101–133.

Bischoff A., Schleiting M., Wieler R., and Patzek M. (2018) Brecciation among 2280 ordinary chondrites – Constraints on the evolution of their parent bodies. *Geochim. Cosmochim. Acta* 238, 516–541.

Bischoff A., Schleiting M., and Patzek M. (2019) Shock stage distribution of 2280 ordinary chondrites – Can bulk chondrites with a shock stage S6 exist as individual rocks? *Meteorit. Planet. Sci.* 54, 2189–2202.

Bischoff A., Alexander C. M. O'D., Barrat J.-A., et al. (2021) The old, unique C1 chondrite Flensburg – Insight into the first processes of aqueous alteration, brecciation, and the diversity of water-bearing parent bodies and lithologies. *Geochim. Cosmochim. Acta* 293, 142–186.

Bischoff A., Bannemann L., Decker S., et al. (2022) Asteroid 2008 TC_3, not a polymict ureilitic but a polymict C1 chondrite parent body? Survey of 249 Almahata Sitta fragments. *Meteorit. Planet. Sci.* 57, 1339–1364.

Bischoff A., Patzek M., Di Rocco T., et al. (2023) Saint-Pierre-le-Viger (L5-6) from asteroid 2023 CX_1 recovered in the Normandy, France – 220 years after the historic fall of L'Aigle (L6 breccia) in the neighborhood. *Meteorit. Planet. Sci.* 58, 1385–1398.

Bischoff A., Patzek M., Barrat J. A., et al. (2024a) Cosmic pears from the Havelland (Germany): Ribbeck, the twelfth recorded aubrite fall in history. *Meteorit. Planet. Sci.* 59, 1–35.

Bischoff A., Patzek M., Alosius R. M. L., et al. (2024b) The anomalous polymict ordinary chondrite breccia of Elmshorn (H3-6) – Late reaccretion after collision between two ordinary chondrite parent bodies, complete disruption, and mixing possibly about 2.8 Gyr ago. *Meteorit. Planet. Sci.* 59, 2321–2356.

Bissinger H. G. (1990) *Friday Night Lights*, Addison-Wesley, Boston, 419 pp.

Bitsch B., Morbidelli A., Johansen A., et al. (2018) Pebble-isolation mass: Scaling law and implications for the formation of super-Earths and gas giants. *Astron. Astrophys.* 612, 15 pp.

Bjerkeli P., van der Wiel M. H. D., Harsono D., Ramsey J. P., and Jørgenson J. S. (2016) Resolved images of a protostellar outflow driven by an extended disk wind. *Nature* 540, 406–409.

Black D. C. (1972) On the origins of trapped helium, neon and argon isotopic variations in meteorites – I. Gas-rich meteorites, lunar soil and breccia. *Geochim. Cosmochim. Acta* 36, 347–375.

Blackburn T., Alexander C. M. O'D., Carlson R., and Elkins-Tanton L. T. (2017) The accretion and impact history of the ordinary chondrite parent bodies. *Geochim. Cosmochim. Acta* 200, 201–217.

Blagdon C. (1784) An account of some fiery meteors; with observations. *Philos. Trans. R. Soc. London* 74, 201–232.

Bland P. A., Berry F. J., and Pillinger C. T. (1995) Variation in weathering between the Great Bend meteorite and LEW 85319 (abstract). *Lunar Planet. Sci.* 26, 131–132.

Bland P. A., Smith T. B., Jull A. J. T., et al. (1996) The flux of meteorites to the Earth over the last 50,000 years. *Monthly Notices R. Astronom. Soc.* 283, 551–565.

Bland P. A., Sexton A. S., Jull A. J. T., et al. (1998a) Climate and rock weathering: A study of terrestrial age dated ordinary chondritic meteorites from hot desert regions. *Geochim. Cosmochim. Acta* 62, 3169–3184.

Bland P. A., Berry F. J., and Pillinger C. T. (1998b) Rapid weathering in Holbrook: An iron-57 Mössbauer spectroscopy study. *Meteorit. Planet. Sci.* 33, 127–129.

Bland P. A., Alard O., Benedix G. K., et al. (2005) Volatile fractionation in the early solar system and chondrule/matrix complementarity. *Proc. Nat. Acad. Sci.* 102, 13755–13760.

Bland P. A., Artemieva N. A., Collins G. S., et al. (2008) Asteroids on the Moon: Projectile survival during low velocity impact (abstract #2045). *39th Lunar and Planetary Science Conference.*

Bland P. A., Spurný P., Towner M. C., et al. (2009) An anomalous basaltic meteorite from the innermost main belt. *Science* 325, 1525–1527.

Bland P. A., Collins G. S., Davison T. M., et al. (2014) Pressure-temperature evolution of primordial solar system solids during impact-induced compaction. *Nature Commun.* 5, 5451. https://doi.org/10.1038/ncomms6451.

Blander M. and Katz J. L. (1967) Condensation of primordial dust. *Geochim. Cosmochim. Acta* 31, 1025–1034.

Blood M. (2013) *Aspects of Meteorite Orientation*, Las Vegas, Nevada, 94 pp.

Blood M. (2019) *Meteorites and Sacred Stones on Coins of Ancient Greece & Rome*, San Bernardino, California, 85 pp.

Blum J. (2004) Grain growth and coagulation. *Astron. Soc. Pacific Conf. Ser.* 309, 369–391.

Boesenberg J. S. and Delaney J. S. (1997) A model composition of the basaltic achondrite planetoid. *Geochim. Cosmochim. Acta* 61, 3205–3225.

Bogard D. D. (1995) Impact ages of meteorites: A synthesis. *Meteorit. Planet. Sci.* 30, 244–268.

Bogard D. D. (2011) K-Ar ages of meteorites: Clues to parent-body thermal histories. *Chem. Der. Erde.* 71, 207–226.

Bogard D. D. and Garrison D. H. (2009) Ar–Ar and I–Xe ages and thermal histories of three unusual metal-rich meteorites. *Geochim. Cosmochim. Acta* 73, 6965–6983.

Bogard D. D., Husain L., and Wright R. J. (1976) 40Ar-39Ar dating of collisional events in chondrite parent bodies. *J. Geophys. Res.* 81, 5664–5678.

Bogard D. D., Garrison D. H., Norman M., Scott E. R. D., and Keil K. (1995) ^{39}Ar-^{40}Ar age and petrology of Chico: Large-scale melting on the L chondrite parent body. *Geochim. Cosmochim. Acta* 59, 1383–1399.

Bogard D. D., Garrison D. H., and McCoy T. J. (2000) Chronology and petrology of silicates from IIE iron meteorites: Evidence of a complex parent body evolution. *Geochim. Cosmochim. Acta* 64, 2133–2154.

Boley A. C., Morris M. A., and Desch S. J. (2013) High-temperature processing of solids through solar nebular bow shocks: 3D radiation hydrodynamics simulations with particles. *Astrophys. J.* 776, 23 pp.

Bollard J., Connelly J. N., Whitehouse M. J., et al. (2017) Early formation of planetary building blocks inferred from Pb isotopic ages of chondrules. *Sci. Adv.* 3, 9 pp.

Borovička J., Spurný P. and Brown P. (2015) Small Near-Earth Asteroids as a source of meteorites. In *Asteroids IV* (edited by Michel P., Demeo F. E. and Bottke W. F.), pp. 257–280, University of Arizona Press, Tucson.

Boss A. P. and Durisen, R. H. (2005) Chondrule-forming shock fronts in the solar nebula: A possible unified scenario for planet and chondrite formation. *Astrophys. J.* 621, L137–L140.

Boss A. P. and Graham J. A. (1993) Clumpy disk accretion and chondrule formation. *Icarus* 106, 168–178.

Bostock J. and Riley H. T. (1855) *The Natural History of Pliny. Translated with Copious Notes and Illustrations*, Henry G. Bohn, London, 542 pp.

Bottke W. F., Nolan M. C., Greenberg R., and Kolvoord R. A. (1994a) Collisional lifetimes and impact statistics of Near-Earth Asteroids. In *Hazards due to Comets and Asteroids* (edited by Gehrels T.), pp. 337–357, University of Arizona Press, Tucson.

Bottke W. F., Nolan M. C., Greenberg R., and Kolvoord R. A. (1994b) Velocity distributions among colliding asteroids. *Icarus* 107, 255–268.

Bottke W. F., Richardson D. C., Michel P., and Love S. G. (1999) 1620 Geographos and 433 Eros: Shaped by planetary tides. *Astron. J.* 117, 1921–1928.

Bottke W. F., Jedicke R., Morbidelli A., Petite J.-M., and Gladman B. (2000) Understanding the distribution of Near-Earth Asteroids. *Science* 288, 2190–2194.

Bottke W. F., Vokrouhlický D., Brož M., Nesvorný D., and Morbidelli A. (2001) Dynamical spreading of asteroid families by the Yarkovsky effect. *Science* 294, 1693–1696.

Bottke W. F., Vokrouhlický D., Rubincam D., and Brož M. (2002) The effect of Yarkovsky thermal forces on the dynamical evolution of asteroids and meteoroids. In *Asteroids III* (edited by Bottke W. F., Cellino A., Paolicchi P. and Binzel R. P.), pp. 395–408, University of Arizona Press, Tucson.

Bottke W. F., Durda D. D., Nesvorný D., et al. (2005) The fossilized size distribution of the main asteroid belt. *Icarus* 175, 111–140.

Bottke W. F., Vokrouhlický D., Rubincam D. P., and Nesvorný D. (2006) The Yarkovsky and YORP effects: Implications for asteroid dynamics. *Annu. Rev. Earth Planet. Sci.* 34, 157–191.

Bottke W., Vokrouhlický D., Nesvorný D., and Shrbeny L. (2010) (6) Hebe really is the H chondrite parent body (abstract). *Bull. Amer. Astronom. Soc.* 42, 1051.

Bottke W. F., Vokrouhlický D., Walsh K. J., et al. (2015) In search of the source of asteroid (101955) Bennu: applications of the stochastic YORP model. *Icarus* 247, 191–217. https://doi.org/10.1016/j.icarus.2014.09.046.

Bottke W. F., Vokrouhlický D., Ballouz R.-J., et al. (2020) Interpreting the cratering histories of Bennu, Ryugu, and other spacecraft-explored asteroids. *Astronom. J.* 160, 37 pp.

Bottomley R., Grieve R., York D., and Masaitis V. (1997) The age of the Popigai impact event and its relation to events at the Eocene/Oligocene boundary. *Nature* 388, 365–368.

Bournon J. L. C. de. (1802) Mineralogical descriptions of the various stones said to have fallen upon the Earth; and description of various kinds of native iron. *Philos. Trans. R. Soc.*,92, 180–187, 203–210.

Bouvier A. and Wadhwa M. (2010) The age of the solar system redefined by the oldest Pb-Pb age of a meteoritic inclusion. *Nat. Geosci.* 3, 637–641.

Bowden K. E. (2002) Effects of loading path on the shock metamorphism of porous quartz: An experimental study. Ph.D. thesis, University College London, 228 pp.

Boyd F. R. and England J. L. (1965) The rhombic enstatite-clinoenstatite inversion. In *Year Book – Carnegie Institution of Washington*, pp. 117–120. Carnegie Institution of Washington, Washington D. C.

Bradley J. P. (2005) Interplanetary dust particles. In *Meteorites, Comets, and Planets* (edited by Davis A. M.) Treatise on Geochemistry, vol. 1, pp. 689–711. Elsevier–Pergamon, Oxford.

Bradley J. P., Sandford S. A. and Walker R. M. (1988) Interplanetary dust particles. In *Meteorites and the Early Solar System* (edited by Kerridge J. F. and Matthews M. S.), pp. 861–895, University of Arizona Press, Tucson.

Branco W. and Fraas E. (1905) Das kryptovulcanische Becken von Steinheim. In *Aus den Abhandlungen der Königlich Preussischen Akademie der Wissenschaften*, pp. 1–64, Verlag G. Reimer, Berlin.

Brandes E. A. and Brook G. B., eds. (1998) *Smithells Metals Reference Book*, 7th ed. Butterworth-Heinemann.

Brandstätter F., Ivanov A. V., and Kurat G. (1996) An ordinary chondrite fragment (R3) in the Kaidun carbonaceous chondrite. *Meteorit. Planet. Sci.* 31, p. A20.

Brandstätter F., Konzett J., Koeberl C., and Ferrière L. (2013) The Ischgl meteorite, a new LL6 chondrite from Tyrol, Austria. *Ann. Naturhis. Mus. Wien, Series A* 115, 5–18.

Braukmüller N., Wombacher F., Hezel D. C., Escoube R., and Münker C. (2018) The chemical composition of carbonaceous chondrites: Implications for volatile element depletion, complementarity and alteration. *Geochim. Cosmochim. Acta* 239, 14–48.

Brearley A. J. (1990) Carbon-rich aggregates in type 3 ordinary chondrites: Characterization, origins, and thermal history. *Geochem. Cosmochim. Acta* 54, 831–850.

Brearley A. J. (1993) Matrix and fine-grained rims in the unequilibrated CO3 chondrite, ALHA77307: origins and evidence for diverse, primitive nebular dust components. *Geochim. Cosmochim. Acta* 57, 1521–1550.

Brearley A. J. (1997) Phyllosilicates in the matrix of the unique carbonaceous chondrite, LEW 85332 and possible implications for the aqueous alteration of CI chondrites. *Meteorit. Planet. Sci.* 32, 377–388

Brearley A. J. (2006) The action of water. In *Meteorites and the Early Solar System II* (edited by Lauretta, D. S. and McSween, H. Y.), pp. 587–624, University of Arizona Press, Tucson.

Brearley A. J. and Jones R. H. (1998) Chondritic meteorites. In *Planetary Materials* (edited by Papike J. J.), pp. 3–1–3–398, Mineralogical Society of America, Washington, D.C.

Brearley A. J., Bajt S., and Sutton S. R. (1995) Distribution of moderately volatile trace elements in fine-grained chondrule rims in the unequilibrated CO3 chondrite, ALH A77307. *Geochim. Cosmochim. Acta* 59, 4307–4316.

Brett R. and Sato M. (1984) Intrinsic oxygen fugacity measurements on seven chondrites, a pallasite, and a tektite and the redox state of meteorite parent bodies. *Geochim. Cosmochim. Acta* 48, 111–120.

Brezina A. (1885) Die Meteoritensammlung des k.k. mineralogischen Hofkabinetes in Wien am. 1 Mai 1885. *Jahrbuch der Geologischen Reichsanstalt*. 35, 151–276.

Brezina A. (1904) The arrangement of collections of meteorites. *Trans. Amer. Phil. Soc.* 43, 211–247.

Bridges J. C. and Hutchison R. (1997) A survey of clasts and large chondrules in ordinary chondrites. *Meteorit. Planet. Sci.* 32, 389–394.

Bridges J. C., Hutchison R., Franchi I. A., Alexander C. M. O'D., and Pillinger C. T. (1995) A feldspar-nepheline achondrite clast in Parnallee. *Proc. NIPR Symp. Antarct. Meteorites* 8, 195–203.

Bridges J. C., Alexander C. M. O'D., Hutchison R., Franchi I. A., and Pillinger C. T. (1997) Sodium-, chlorine-rich mesostases in Chainpur (LL3) and Parnallee (LL3) chondrules. *Meteorit. Planet. Sci.* 32, 555–565.

Brigham C. A., Yabuki H., Ouyang Z., et al. (1986) Silica-bearing chondrules and clasts in ordinary chondrites. *Geochim. Cosmochim. Acta* 50, 1655–1666.

Britt D. T. and Consolmagno G. J. (2001) Modeling the structure of high porosity asteroids. *Icarus* 152, 134–139.

Britt D. T. and Consolmagno G. J. (2003) Stony meteorite porosities and densities: A review of the data through 2001. *Meteorit. Planet. Sci.* 38, 1161–1180.

Britt D. T. and Pieters C. M. (1991) Black ordinary chondrites: An analysis of abundance and fall frequency. *Meteoritics* 26, 279–285.

Britt D. T. and Pieters C. M. (1994) Darkening in black and gas-rich ordinary chondrites: The spectral effects of opaque morphology and distribution. *Geochim. Cosmochim. Acta* 58, 3905–3919.

Brown P., Pack D., Edwards W. N., et al. (2004) The orbit, atmospheric dynamics, and initial mass of the Park Forest meteorite. *Meteorit. Planet. Sci.* 39, 1781–1796.

Brown P. G., Assink J. D., Astiz L., et al. (2013) A 500-kiloton airburst over Chelyabinsk and an enhanced hazard from small impactors. *Nature* 503, 238–241.

Brown P. G., McCausland P. J. A., Hildebrand A. R., et al. (2023) The Golden meteorite fall: Fireball trajectory, orbit, and meteorite characterization. *Meteorit. Planet. Sci.* 58, 1773–1807.

Brownlee D. E. and Rajan R. S. (1973) Micrometeorite craters discovered on chondrule-like objects from Kapoeta meteorite. *Science* 182, 1341–1344.

Brownlee D. E., Joswiak D. J., Schlutter D. J., et al. (1995) Identification of individual cometary IDPs by thermally stepped He release (abstract). *Lunar Planet. Sci.* 26, 183–184.

Brownlee D. E., Tsou P., Anderson J. D., et al. (2003) Stardust: Comet and interstellar dust sample return mission. *J. Geophys. Res. Planets* 108, 15 pp.

Brownlee D., Joswiak D., and Matrajt G. (2012) Overview of the rocky component of Wild 2 comet samples: Insight into the early solar system, relationship with meteoritic materials and the differences between comets and asteroids. *Meteorit. Planet. Sci.* 47, 453–470.

Brož M., Vernazza P., Marsset M., et al. (2024) Young asteroid families as the primary source of meteorites. arXiv:2403.08552 [astro-ph.EP]. https://doi.org/10.48550/arXiv.2403.08552.

Bryant D. (2018) *Spacerocks: A Collectors' Guide to Meteorites, Tektites and Impactites*, Heathland Books, United Kingdom, 168 pp.

Bryson J. F. J., Weiss B. P., Getzin B., et al. (2019) Paleomagnetic evidence for a partially differentiated ordinary chondrite parent asteroid. *J. Geophys. Res. Planets* 124, 1880–1898.

Buchwald V. F. (1967) Study of six iron meteorites: Babb's Mill, Föllinge, Indian Valley, Netschaëvo, Scottsville and Seeläegen. *Analecta Geol.* 2, 1–66.

Buchwald V. F. (1975) *Handbook of Iron Meteorites*, University of California Press, Berkeley, California, 1418 pp.

Budde G., Burkhardt C., Brennecka G. A., et al. (2016) Molybdenum isotopic evidence for the origin of chondrules and a distinct genetic heritage of carbonaceous and non-carbonaceous meteorites. *Earth Planet. Sci. Lett.* 454, 293–303.

Budde G., Burkhardt C., and Kleine T. (2017) The distinct genetics of carbonaceous and non-carbonaceous meteorites inferred from molybdenum isotopes. *Annu. Meet. Meteorit. Soc.* abstract #6271.

Budde G., Burkhardt C., and Kleine T. (2018) Early solar system dynamics inferred from molybdenum isotope anomalies in meteorites. *Lunar Planet. Sci.* 49, abstract #2353.

Buddhue J. D. (1957) The Oxidation and Weathering of Meteorites. In *Publications in Meteoritics No. 3 of the University of New Mexico*, University of New Mexico Press, Albuquerque, 161 pp.

Buhl S. and Wimmer K. (2013) Trajectory projection of the Chelyabinsk superbolide and location of recorded meteorite finds. OpenStreetMap contributors (openstreetmap.org).

Bukovanská M., Jakeš P., Pernicka E., and El Goresy A. (1983) Ustí Nad Orlici: A new L6 chondrite from Czechoslovakia. *Meteoritics* 18, 223–240.

Bunch T. E. and Keil K. (1971) Chromite and ilmenite in nonchondritic meteorites. *Amer. Mineral.* 56, 146–157.

Bunch T. E. and Olsen E. (1968) Potassium feldspar in Weekeroo Station, Kodaikanal, and Colomera iron meteorites. *Science* 160, 1223–1225.

Bunch T. E. and Quaide W. L. (1968) Shatter cones in the Danny Boy nuclear crater. In *Shock Metamorphism of Natural Materials* (edited by French B. M. and Short N. M.), pp. 267–285, Mono Book Corp., Baltimore.

Bunch T. E., Keil K., and Snetsinger K. G. (1967) Chromite composition in relation to chemistry and texture in ordinary chondrites. *Geochim. Cosmochim. Acta* 31, 1569–1582.

Bunch T. E., Keil K., and Olsen E. (1970) Mineralogy and petrology of silicate inclusions in iron meteorites. *Contrib. Mineral. Petrol.*, 25, 297–340.

Burbine T. H. (2017) *Asteroids: Astronomical and Geological Bodies*, Cambridge University Press, Cambridge, 367 pp.

Burbine T. H., Gaffey M. J., and Bell J. F. (1992) S-asteroids 387 Aquitania and 980 Anacostia: Possible fragments of the breakup of a spinel-bearing parent body with CO3/CV3 affinities. *Meteorit. Planet. Sci.* 27, 424–434.

Burbine T. H., Buchanan P. C., Binzel R. P., et al. (2001) Vesta, Vestoids, and the howardite, eucrite, diogenite group: Relationships and the origin of spectral differences. *Meteorit. Planet. Sci.* 36, 761–781.

Burbine T. H., McCoy T. J., Meibom A., Gladman B., and Keil K. (2002) Meteoritic parent bodies: Their number and identification. In *Asteroids III* (edited by Bottke W. F., Cellino A., Paolicchi P. and Binzel R. P.), pp. 653–667, University of Arizona Press, Tucson.

Burbine T. H., McCoy T. J., Jarosewich E., and Sunshine J. M. (2003) Deriving asteroid mineralogies from reflectance spectra: Implications for the MUSES-C target asteroid. *Antarct. Meteor. Res.* 16, 185–195.

Burchell M. J. and Leliwa-Kopystynski J. (2010) The large crater on the small Asteroid (2867) Steins. *Icarus* 210, 707–712.

Burke J. G. (1986) *Cosmic Debris: Meteorites in History*, University of California Press, Berkeley, 445 pp.

Burkhardt C., Kleine T., Oberli F., et al. (2011) Molybdenum isotope anomalies in meteorites: Constraints on solar nebula evolution and origin of the Earth. *Earth Planet. Sci. Lett.* 312, 390–400.

Buseck P. R. (1977) Pallasite meteorites – mineralogy, petrology and geochemistry. *Geochim. Cosmochim. Acta* 41, 711–740.

Buseck P. R. and Keil K. (1966) Meteoritic rutile. *Am. Mineral.* 51, 1506–1515.

Busemann H., Alexander C. M. O'D., and Nittler L. R. (2007) Characterization of insoluble organic matter in primitive meteorites by microRaman spectroscopy. *Meteorit. Planet. Sci.* 42, 1387–1416.

Busfield A., Gilmour J. D., Whitby J. A., and Turner G. (2004) Iodine-xenon analysis of ordinary chondrite halide: Implications for early solar system water. *Geochim. Cosmochim. Acta* 68, 195–202.

Butler R. F. (1972) Natural remanent magnetization and thermomagnetic properties of Allende meteorite. *Earth Planet. Sci. Lett.* 17, 120–128.

Caillet Komorowski C., El Goresy A., Miyahara M., Boudouma O., and Ma C. (2012) Discovery of Hg-Cu-bearing metal-sulfide assemblages in a primitive H-3 chondrite: Towards a new insight in early solar system processes. *Earth Planet. Sci. Lett.* 349–350, 261–271.

Cain P., McSween H. Y., and Woodward N. B. (1986) Structural deformation of the Leoville chondrite. *Earth Planet. Sci. Lett.* 77, 165–175.

Cameron A. G. W. (1966) The accumulation of chondritic material. *Earth Planet. Sci. Lett.* 1, 93–96.

Cameron A. G. W. (1995) The first ten million years in the solar nebula. *Meteoritics* 30, 133–161.

Cameron A. G. W. and Fegley M. B. (1982) Condensation events near the total evaporation front within the primitive solar nebula (abstract). *Lunar Planet. Sci.* 13, 81–82.

Cameron A. G. W., Benz W., and Wasson J. T. (1990) Heating during asteroidal collisions (abstract). *Lunar Planet. Sci.* 21, 155–156.

Cameron A. G. W., Benz W., and Wasson J. T. (1991) Heating during asteroidal collisions II (abstract). *Lunar Planet. Sci.* 22, 173.

Carapezza M., Nuccio P. M., and Valenza M. (1976) Piancaldoli meteorite: Chemistry, mineralogy and petrology. *Meteoritics* 11, 195–205.

Carman H. (1995) *Collecting Meteorites: Starting in Your Own Backyard. Hill of Content*, Australia, 78 pp.

Carporzen L., Weiss B. P., Elkins-Tanton L. T., et al. (2010) Magnetic evidence for a partially differentiated carbonaceous chondrite parent body. *Proc. Natl. Acad. Sci. USA* 42, 944.

Carter N. L., Raleigh C. B., and DeCarli P. S. (1968) Deformation of olivine in stony meteorites. *J. Geophys. Res.* 73, 5439–5461.

Cassidy W. A. (1983) The remarkably low surface density of meteorites at Allan Hills and implications in this for climate change. In *Antarctic Earth Science* (edited by Oliver R. L., James P. R., and Jago J. B.), pp. 623–625. Australian Academy of Sciences. Canberra.

Cassidy W. A. (2003) *Meteorites, Ice, and Antarctica: A Personal Account*, Cambridge University Press, Cambridge, 349 pp.

Cassidy W. A., Olsen E., and Yanai K. (1977) Antarctica: A deep-freeze storehouse for meteorites. *Science* 198, 727–731.

Cassidy W., Harvey R., Schutt J., Delisle G., and Yani K. (1992) The meteorite collection sites in Antarctica. *Meteoritics* 27, 490–525.

Ceplecha Z. (1961) Multiple fall of Pribram meteorites photographed. 1. Double-station photographs of the fireball and their relations to the found meteorites. *Bull. Astron. Inst. Czechosl.* 12, 21–47.

Ceplecha Z. (1994) Impacts of meteoroids larger than 1 m into the Earth's atmosphere. *Astron. Astrophys.* 286, 967–970.

Ceplecha Z. and McCrosky R. E. (1976) Fireball end heights: A diagnostic for the structure of meteoric material. *J. Geophys. Res.* 81, 6257–6275.

Ceplecha Z. and Revelle D. O. (2005) Fragmentation model of meteoroid motion, mass loss, and radiation in the atmosphere. *Meteorit. Planet. Sci.* 40, 35–54.

Ceplecha Z., Brown P., Hawkes R. L., et al. (1996) Video observations, atmospheric path, orbit and fragmentation record of the fall of the Peekskill meteorite. *Earth Moon Planets* 72, 395–404.

Ceplecha Z., Borovička J., Elford W. G., et al. (1998) Meteor phenomena and bodies. *Space Sci. Rev.* 84, 327–471.

Cerantola V., Walte N. P., and Rubie D. C. (2015) Deformation of a crystalline olivine aggregate containing two immiscible liquids: Implications for early core-mantle differentiation. *Earth Planet. Sci. Lett.* 417, 67–77.

Chabot N. L. (2004) Sulfur contents of the parental metallic cores of magmatic iron meteorites. *Geochim. Cosmochim. Acta* 68, 3607–3618.

Chakraborty S. (1997) Rates and mechanisms of Fe-Mg interdiffusion in olivine at 980–1300°C. *J. Geophys. Res.* 102, 12317–12331.

Chakraborty S. (2025) Nucleating ideas and crystals – To do or not to do. *Elements* 21, 3–4.

Chambers J. E. (2004) Planetary accretion in the inner solar system. *Earth Planet. Sci. Lett.* 223, 241–252.

Chan Q. H. S., Zolensky M. E., Kebukawa Y., et al. (2018a) Organic matter in extraterrestrial water-bearing salt crystals. *Sci. Adv.* 4. https://doi.org/10.1126/sciadv.aao3521.

Chan Q. H. S., Zolensky M. E., Kebukawa Y., et al. (2018b) Primitive O-, N-, and C-rich organic vein preserved in a xenolith hosted in the

metamorphosed Carancas meteorite. *Lunar Planet. Sci.* 49th, abstract#1191.
Chang C. and Wänke H. (1969) Beryllium-10 in iron meteorites, their cosmic-ray exposure and terrestrial ages. In *Meteorite Research* (edited by Millman P. M.), pp. 397–406. D. Riedel Publishing Company, Dordrecht, Holland.
Chao E. C. T., Shoemaker E. M., and Madsen B. M. (1960) First natural occurrence of coesite. *Science* 132, 220–221.
Chao E. C. T., Fahey J. J., Littler J., and Milton D. J. (1962) Stishovite, SiO_2, a very high-pressure new mineral from Meteor Crater, Arizona. *J. Geophys. Res.* 67, 419–421.
Chapman C. R. (1996) S-type asteroids, ordinary chondrites, and space weathering: The evidence from Galileo's fly-bys of Gaspra and Ida. *Meteorit. Planet. Sci.* 31, 699–725.
Chapman C. R. (2004a) The hazard of near-Earth asteroid impacts on Earth. *Earth Planet. Sci. Lett.* 222, 1–15.
Chapman C. R. (2004b) Space weathering of asteroid surfaces. *Annu. Rev. Earth Planet. Sci.* 32, 539–567.
Chapman C. R. (2007) The asteroid impact hazard and interdisciplinary issues. In *Comet/Asteroid Impacts and Human Society* (edited by Bobrowsky P. T. and Rickman H.), pp. 145–162, Springer, Berlin.
Chapman C. R., Morrison D., and Zellner B. (1975) Surface properties of asteroids: A synthesis of polarimetry, radiometry, and spectrophotometry. *Icarus* 25, 104–130.
Chaumard N., Devouard B., Bouvier A., and Wadhwa M. (2014) Metamorphosed calcium-aluminum-rich inclusions in CK carbonaceous chondrites. *Meteorit. Planet. Sci.* 49, 419–452.
Chaussidon M. and Robert F. (1995) Nucleosynthesis of ^{11}B-rich boron in the pre-solar cloud recorded in meteoritic chondrules. *Nature* 374, 337–339.
Chaussidon M. and Robert F. (1996) Boron and lithium isotope variations in chondrules: The signature of presolar nucleosynthesis (abstract). *Lunar Planet. Sci.* 27, 207–208.
Che S., Domanik K. J., and Zega T. J. (2023a) In-situ formation of halite in the Sidi El Habib 001 (H5) ordinary chondrite: Implications for hydrothermal alteration in ordinary chondrite parent bodies. *Geochim. Cosmochim. Acta* 348, 85–106.
Che S., Domanik K. J., Chang Y.-J., and Zega T. J. (2023b) The important role of fluid chemistry in the hydrothermal alteration of ordinary chondrites: Insights from halite and sphalerite in the Sidi El Habib 001 (H5) meteorite. *Earth Planet. Sci. Lett.* 621, 118374, 12 pp.
Chen C., Shu J., and Mao H. K. (2008) Xieite, a new mineral of high-pressure $FeCr_2O_4$ polymorph. *Chin. Sci. Bull.* 53, 3341–3345.
Chen M. and El Goresy A. (2000) The nature of maskelynite in shocked meteorites: Not diaplectic glass but a glass quenched from shock-induced dense melt at high pressures. *Earth Planet. Sci. Lett.* 179, 489–502.
Chen M., Sharp T. G., El Goresy A., Wopenka B., and Xie X. (1996) The majorite-pyrope + magnesiowüstite assemblage: Constraints on the history of shock veins in chondrites. *Science* 271, 1570–1573.
Chen M., Xie X., and El Goresy A. (2004) A shock-produced (Mg,Fe)SiO_3 glass in the Suizhou meteorite. *Meteorit. Planet. Sci.* 39, 1797–1808.
Chen T., Gong B., and Tang C. (2023) Origin and evolution of cracks in the glaze surface of a ceramic during the cooling process. *Materials* 16, 5508, 18 pp. https://doi.org/10.3390/ma16165508.
Cheng A. F. (2002) Near earth asteroid rendezvous: Mission summary. In *Asteroids III* (edited by Bottke W. F., Cellino A., Paolicchi P. and Binzel R. P.), pp. 351–366, University of Arizona Press, Tucson.
Cheng A. F. and Barnouin-Jha O. S. (1999) Giant craters on Mathilde. *Icarus* 140, 34–48.
Cheng A. F., Izenberg N., Chapman C. R., and Zuber M. T. (2002) Ponded deposits on asteroid 433 Eros. *Meteorit. Planet. Sci.* 37, 1095–1105.
Cheng A. F., Barnouin-Jha O. S., Hirata N., et al. (2007) Fundamentally distinct outcomes of asteroid collisional evolution: Itokawa and Eros. *J. Geophys. Res.* 34. https://doi.org/10.1029/2007GL029559.
Chennaoui-Aoudjehane H., Jambon A., Rjimati E., Jull A. J. T., and Leclerc-Giscard M. D. (2009) The Late Quaternary fall at Al Haggounia (Morocco): The ^{14}C evidence. *Meteorit. Planet. Sci.* 44, A100.
Chesley S. R., Ostro S. J., Vokrouhlicky D., et al. (2003) Direct detection of the Yarkovsky effect by radar ranging to asteroid 6489 Golevka. *Science* 302, 1739–1742.
Chiang E. I. (2004) Dust in protoplanetary disks. In *Astrophysics of Dust, ASP Conference Series 309* (edited by Witt A. N., Clayton G. C. and Draine B. T.), pp. 213–228, Astronomical Society of the Pacific, San Francisco.
Choe W. H., Huber H., Rubin A. E., Kallemeyn G. W., and Wasson J. T. (2010) Compositions and taxonomy of 15 unusual carbonaceous chondrites. *Meteorit. Planet. Sci.* 45, 531–554.
Choi B.-G., McKeegan K. D., Krot A. N., and Wasson J. T. (1998) Extreme oxygen-isotope compositions in magnetite from unequilibrated ordinary chondrites. *Nature* 392, 577–579.
Chou C.-L. and Cohen A. J. (1973) Gallium and germanium in the metal and silicates of L- and LL-chondrites. *Geochim. Cosmochim. Acta* 37, 315–327.
Chou C.-L., Baedecker P. A., and Wasson J. T. (1973) Distribution of Ni, Ga, Ge, and Ir between metal and silicate portion of H-group chondrites. *Geochim. Cosmochim. Acta* 37, 2159–2171.

Christophe Michel-Lévy M. (1976) La matrice noire et blanche de la chondrite de Tieschitz (H3). *Earth Planet. Sci. Lett.* 30, 143–150.

Christophe Michel-Lévy M., Robin E., and Blanc Ph. (1995) Occurrence of noble metals in aluminum- and chromium-rich objects in ordinary chondrites and of baddeleyite (ZrO_2) in a chromite inclusion. *Meteoritics* 30, 15–19.

Chung Y. and Lee C.-W. (2013) Electrochemistry of gallium. *J. Electrochem. Sci. Technol.* 4, 1–18.

Ciesla F. J. (2010) The distributions and ages of refractory objects in the solar nebula. *Icarus* 208, 455–467.

Ciesla F. J. and Charnley S. B. (2006) The physics and chemistry of nebular evolution. In *Meteorites and the Early Solar System II* (edited by Lauretta D. S. and McSween H. Y.), pp. 209–230, University of Arizona Press, Tucson.

Ciesla F. J. and Hood L. L. (2002) The nebular shock wave model for chondrule formation: Shock processing in a particle-gas suspension. *Icarus* 158, 281–293.

Ciesla F. J., Lauretta D. S., Cohen B. A., and Hood L. L. (2003) A nebular origin for chondritic fine-grained phyllosilicates. *Science* 299, 549–552.

Ciesla F. J., Hood L. L., and Weidenschilling S. J. (2004) Evaluating planetesimal bow shocks as sites for chondrule formation. *Meteorit. Planet. Sci.* 39, 1809–1821.

Ciesla F. J., Davison T. M., Collins G. S., and O'Brien D. P. (2013) Thermal consequences of impacts in the early solar system. *Meteorit. Planet. Sci.* 48, 2559–2576.

Clayton D. D. (1980) Chemical energy in cold-cloud aggregates: The origin of meteoritic chondrules. *Astrophys. J.* 239, L37–L41.

Clayton D. D. and Nittler L. R. (2004) Astrophysics with presolar stardust. *Annu. Rev. Astron. Astrophys.* 42, 39–78.

Clayton R. N. (1993) Oxygen isotopes in meteorites. *Annu. Rev. Earth Planet. Sci.* 21, 115–149.

Clayton R. N. (2005) Oxygen isotopes in meteorites. In *Treatise on Geochemistry, Meteorites, Comets, and Planets* (edited by Davis A. M.), pp. 129–142, Elsevier, Amsterdam.

Clayton R. N. and Mayeda T. K. (1983) Oxygen isotopes in eucrites, shergottites, nakhlites, and chassignites. *Earth Planet. Sci. Lett.* 62, 1–6.

Clayton R. N. and Mayeda T. K. (1989) Oxygen isotope studies of carbonaceous chondrites. *Geochim. Cosmochim. Acta* 63, 2089–2104.

Clayton R. N., Grossman L., and Mayeda T. K. (1973) A component of primitive nuclear composition in carbonaceous meteorites. *Science* 182, 485–488.

Clayton R. N., Mayeda T. K., Olsen E. J., and Prinz M. (1983) Oxygen isotope relationships in iron meteorites. *Earth Planet. Sci. Lett.* 65, 229–232.

Clayton R. N., Mayeda T. K., and Rubin A. E. (1984) Oxygen isotopic compositions of enstatite chondrites and aubrites. *Proc. Lunar Planet. Sci. Conf. 15th*, C245–C249.

Clayton R. N., Mayeda T. K., and Brownlee D. E. (1986) Oxygen isotopes in deep-sea spherules. *Earth Planet. Sci. Lett.* 79, 235–240.

Clayton R. N., Mayeda T. K., Goswami J. N., and Olsen E. J. (1991) Oxygen isotope studies of ordinary chondrites. *Geochim. Cosmochim. Acta* 55, 2317–2337.

Connelly J. N., Bizzarro M., Krot A. N., et al. (2012) The absolute chronology and thermal processing of solids in the solar protoplanetary disk. *Science* 338, 651–655.

Connolly H. C. and Hewins R. H. (1995) Chondrules as products of dust collisions with totally molten droplets within a dust-rich nebular enivornment: An experimental investigation. *Geochim. Cosmochim. Acta* 59, 3231–3246.

Connolly H. C. and Jones R. H. (2016) Chondrules: The canonical and noncanonical views. *J. Geophys. Res. Planets* 121, 1885–1899.

Connolly H. C. and Love S. G. (1998) The formation of chondrules: petrologic tests of the shock wave model. *Science* 280, 62–67.

Consolmagno G. J. and Drake, M. J. (1977) Composition and evolution of the eucrite parent body: Evidence from rare earth elements. *Geochim. Cosmochim. Acta* 41, 1271–1282.

Consolmagno G. J., Britt D. T., and Stoll C. P. (1998) The porosities of ordinary chondrites: Models and interpretation. *Meteorit. Planet. Sci.* 33, 1221–1230.

Consolmagno G. J., Macke R. J., Rochette P., Britt D. T., and Gattacceca J. (2006) Density, magnetic susceptibility, and the characterization of ordinary chondrite falls and showers. *Meteorit. Planet. Sci.* 41, 331–342.

Consolmagno G. J., Britt D. T., and Macke R. J. (2008) The significance of meteorite density and porosity. *Chem. der Erde* 68, 1–29.

Cook D. L., Walker R. J., Horan M. F., Wasson J. T., and Morgan J. W. (2004) Pt-Re-Os systematics of group IIAB and IIIAB iron meteorites. *Geochim. Cosmochim. Acta* 68, 1413–1431.

Corrigan C. M., Lunning N. G., and Ziegler K. (2015) An H chondrite melt clast in an LL chondrite: Evidence for mixing of ordinary chondrite parent bodies. *Lunar Planet. Sci.* 46th, abstract#2678.

Crabb J. and Schultz L. (1981) Cosmic ray exposure ages of the ordinary chondrites and their significance for parent body stratigraphy. *Geochim. Cosmochim. Acta* 45, 2151–2160.

Cressy F. (2016) *From Weston to Creston: A Compendium of Witnessed US Meteorite Falls 1807 to 2016*, Printed at Artistic Printing, Salt Lake City, Utah, 257 pp.

Croning R. A. (2018) *Meteorite Hunting and Collecting*, Self-published, 152 pp.

Cruikshank D. P., Tholen D. J., Hartmann W. K., Bell J. H., and Brown R. H. (1991) Three basaltic Earth-approaching asteroids and the source of basaltic meteorites. *Icarus* 89, 1–13.

Crutcher R. M. (2012) Magnetic fields in molecular clouds. *Annu. Rev. Astron. Astrophys.* 50, 29–53.

Cuzzi J. N. and Hogan R. C. (2003) Blowing in the wind. I. Velocities of chondrule-sized particles in a turbulent protoplanetary nebula. *Icarus* 164, 127–138.

Cuzzi J. N. and Olson D. M. (2017) Recovering 3D particle size distributions from 2D sections. *Meteorit. Planet. Sci.* 52, 532–545.

Cuzzi J. N., Hogan R. C., Paque J. M., and Dobrovolskis A. R. (2001) Size-selective concentration of chondrules and other small particles in protoplanetary nebula turbulence. *Astrophys. J.* 546, 496–508.

Daeschler E. B., Shubin N. H., and Jenkins F. A. (2006) A Devonian tetrapod-like fish and the evolution of the tetrapod body plan. *Nature* 440, 757–763.

Dai L. and Timmermann A. (2025) Climatic and ecological responses to Bennu-type asteroid collisions. *Sci. Adv.* 11. https://doi.org/10.1126/sciadv.adq5399.

Dauphas N., Davis A. M., Marty B., and Reisberg L. (2004) The cosmic molybdenum–ruthenium isotope correlation. *Earth Planet. Sci. Lett.* 226, 465–475.

Dauphas N., Chen J. H., Zhang J., et al. (2014a) Calcium-48 isotopic anomalies in bulk chondrites and achondrites: evidence for a uniform isotopic reservoir in the inner protoplanetary disk. *Earth Planet. Sci. Lett.* 407, 96–108.

Dauphas N., Burkhardt C., Warren P. H., and Teng F.-Z. (2014b) Geochemical arguments for an Earth-like Moon-forming impactor. *Phil. Trans. R. Soc. A*. 372, 20130244. http://dx.doi.org/10.1098/rsta.2013.0244.

Davis A. M., Grossman L., and Ganapathy R. (1977) Chemical characterization of a "mysterite"-bearing clast from the Supuhee chondrite. *Geochim. Cosmochim. Acta* 41, 853–856.

Davis A. M., editor (2005) *Meteorites, Comets and Planets*, Elsevier, Amsterdam, 737 pp.

Davis A. M. (2006) Volatile element evolution and loss. In *Meteorites and the Early Solar System II* (edited by Lauretta D. S. and McSween H. Y.), pp. 295–307, University of Arizona Press, Tucson.

Davis D. R., Chapman C. R., Greenberg R., and Weidenschilling S. J. (1979) Collisional evolution of asteroids: Populations, rotations, and velocities. In *Asteroids* (ed. Gehrels T.), pp. 528–557, University of Arizona Press, Tucson.

Davis D. R., Chapman C. R., Weidenschilliing S. J., and Greenberg R. (1985) Collisional history of asteroids: Evidence from Vesta and the Hirayama families. *Icarus* 62, 30–53.

Davison T. M., Collins G. S., and Ciesla F. J. (2010) Numerical modeling of heating in porous planetesimal collisions. *Icarus* 208, 468–481.

Davison T. M., Ciesla F. J., and Collins G. S. (2012) Post-impact thermal evolution of porous planetesimals. *Geochim. Cosmochim. Acta* 95, 252–269.

Davison T. M., O'Brien D. P., Ciesla F. J., and Collins G. S. (2013) The early impact histories of meteorite parent bodies. *Meteorit. Planet. Sci.* 48, 1849–1918.

Dawson K. R., Maxwell J. A., and Parsons D. E. (1960) A description of the meteorite which fell near Abee, Alberta, Canada. *Geochim. Cosmochim. Acta* 21, 127–144.

Debaille V., Gattacceca J., Roland J., et al. (2022) Northwest Africa 13188: A meteorite from Earth? *85th Annu. Meet. Meteorit. Soc.*, abstract#6294.

DeHart J. M., Lofgren G. E., Lu J., Benoit P. H., and Sears D. W. G. (1992) Chemical and physical studies of chondrules X. Cathodoluminescence and phase composition studies of metamorphism and nebular processes in chondrules of type 3 ordinary chondrites. *Geochim. Cosmochim. Acta* 56, 3791–3807.

Delaney J. S. (1994) A model composition for Mars derived from the oxygen isotopic ratios of martian/SNC meteorites (abstract). *Meteoritics* 29, 459.

Delbo M., Gayon-Markt J., Busso G., et al. (2012) Asteroid spectroscopy with Gaia. *Planet. Space Sci.* 73, 86–94.

Deloule E. and Robert F. (1996) Origin of water in meteorites: Ion-probe determinations of D/H ratios in chondrules (abstract). *Lunar Planet. Sci.* 27, 307–308.

de Luc G.-A. (1803) Nouvelles considérations sur la masse de Fer de Sibérie et sur les pierres supposées tombées du ciel sur la terre. *Bibliothèque Britannique* 23, 78–112.

Demidova S. I., Merle R., Kenny G. G., et al. (2020) Possible LL chondrite projectile in Luna-16 soil samples. *Lunar Planet. Sci.* 51, abstract#1368.

Demidova S. I., Whitehouse M. J., Merle R., et al. (2022) A micrometeorite from a stony asteroid identified in Luna 16 soil. *Nat. Astron.* 6, 560–567.

DeMeo F. E. and Carry B. (2013) The taxonomic distribution of asteroids from multi-filter all-sky photometric surveys. *Icarus* 226, 723–741.

DeMeo F. E. and Carry B. (2014) Solar System evolution from compositional mapping of the asteroid belt. *Nature* 505, 629–634.

DeMeo F. E., Binzel R. P., Slivan S. M., and Bus S. J. (2009) An extension of the Bus asteroid taxonomy into the near-infrared. *Icarus* 202, 160–180.

Dennison J. E. and Lipschutz M. E. (1987) Chemical studies of H chondrites. II: Weathering effects in the Victoria Land, Antarctica population and comparison of two Antarctic populations with non-Antarctic falls. *Geochim. Cosmochim. Acta* 51, 741–754.

Derkowski A. and Kuligiewicz A. (2023) Thermal analysis and thermal reactions of smectites: A review of methodology, mechanisms, and kinetics. *Clays Clay Miner.* 70, 946–972.

De Sanctis M. C., Ammannito E., Raponi A., et al. (2020) Fresh emplacement of hydrated sodium chloride on Ceres from ascending salty fluids. *Nat. Astron.* 4, 786–793.

Desch S. J. and Connolly H. C. (2002) A model of the thermal processing of particles in solar nebula shocks: Application to the cooling rates of chondrules. *Meteorit. Planet. Sci.* 37, 183–207.

Desch S. J. and Cuzzi J. N. (2000) The generation of lightning in the solar nebula. *Icarus* 143, 87–105.

Desch S. J. and Mouschovias T. C. (2001) The magnetic decoupling stage of star formation. *Astrophys. J.* 550, 314–333. https://doi.org/10.1086/319703.

Desch S. J., Morris M. A., Connolly H. C., and Boss A. P. (2012) The importance of experiments: Constraints on chondrule formation models. *Meteorit. Planet. Sci.* 47, 1139–1156.

Desch S. J., Kalyaan A., and Alexander C. M. O'. D. (2018) The effect of Jupiter's formation on the distribution of refractory elements and inclusions in meteorites. *Astrophys. J. Suppl. Ser.* 238 (11), 31 pp.

Deutsch A., Masaitis V. L., Langenhorst F., and Grieve R. A. F. (2000) Popigai, Siberia – Well preserved giant impact structure, national treasury, and world's geological heritage. *Episodes J. Int. Geosci.* 23, 3–12. https://doi.org/10.18814/epiiugs/2000/v23i1/002.

Devogèle M., Tanga P., Cellino A., et al. (2018) New polarimetric and spectroscopic evidence of anomalous enrichment in spinel-bnearing calcium-aluminum-rich inclusions among L-type asteroids. *Icarus* 304, 31–57.

Diaz A. (2024) A taste of space: New $200 vodka features an out-of-this-world ingredient. *New York Post*, online 2 May 2024.

Dietz R. S. (1946) The meteoritic impact origin of the Moon's surface features. *J. Geol.* 54, 359–375.

Dietz R. S. (1947) Meteorite impact suggested by the orientation of shatter-cones at the Kentland, Indiana disturbance. *Science* 105, 42–43.

Dietz R. S (1959) Shatter cones in cryptoexplosion structures (meteorite impact?). *J. Geol.* 67, 496–505.

Dietz R. S (1963) Cryptoexplosion structures: A discussion. *Amer. J. Sci.* 261, 650–664.

Dietz R. S. (1966) Striated surfaces on meteorites: Shock fractures, not slickensides. *Meteoritics* 3, 31–33.

Dietz R. S. (1968) Shatter cones in cryptoexplosion structures. In *Shock Metamorphism of Natural Materials* (edited by French B. M. and Short N. M.), pp. 267–285. Mono Book Corp., Baltimore.

Dirac P. A. M. (1938) A new basis for cosmology. *Proc. R. Soc. A.* 165 (921), 199–208.

Dixon E. T., Bogard D. D., Garrison D. H., and Rubin A. E. (2004) ^{39}Ar-^{40}Ar evidence for early impact events on the LL parent body. *Geochim. Cosmochim. Acta* 68, 3779–3790.

Dobrică E. and Brearley A. J. (2014) Widespread hydrothermal alteration minerals in the fine-grained matrices of the Tieschitz unequilibrated ordinary chondrite. *Meteorit. Planet. Sci.* 49, 1323–1349.

Dobrică E. and Brearley A. J. (2016) Microchondrules in two unequilibrated ordinary chondrites: Evidence for formation by splattering from chondrules during stoichastic collisions in the solar nebula. *Meteorit. Planet. Sci.* 51, 884–905.

Dobrică E. and Brearley A. J. (2020) Amorphous silicates in the matrix of Semarkona: The first evidence for the localized preservation of pristine matrix materials in the most unequilibrated ordinary chondrites. *Meteorit. Planet. Sci.* 55, 649–668.

Dobrică E., Krot A. N., and Brearley A. J. (2023) Phosphates – the role of aqueous fluids in the evolution of ordinary chondrite parent asteroids. *Microsc. Microanal.* 29, 828–829. https://doi.org/10.1093/micmic/ozad067.411.

Dobrică E., Megevand V., Krot A. N., and Brearley A. J. (2024) The role of apatite in understanding shock metamorphic processes. *86th Annu. Meet. Meteorit. Soc.*, abstract#6049.

Dodd R. T. (1965) Preferred orientation of chondrules in chondrites. *Icarus* 4, 308–316.

Dodd R. T. (1969) Metamorphism of the ordinary chondrites: A review. *Geochim. Cosmochim. Acta* 33, 161–203.

Dodd R. T. (1974) Petrology of the St. Mesmin chondrite. *Contrib. Mineral. Petrol.* 46, 129–145.

Dodd R. T. (1978) The composition and origin of large microporphyritic chondrules in the Manych (L-3) chondrite. *Earth Planet. Sci. Lett.* 39, 52–66.

Dodd R. T. (1981) *Meteorites – A Petrologic-Chemical Synthesis*, Cambridge University Press, Cambridge, 368 pp.

Dodd R. T. and Jarosewich E. (1979) Incipient melting in and shock classification of L-group chondrites. *Earth Planet. Sci. Lett.* 44, 335–340.

Dodd R. T. and Jarosewich E. (1982) The composition of incipient shock melts in L6 chondrites. *Earth Planet. Sci. Lett.* 59, 355–363.

Dodd R. T. and Van Schmus W. R. (1965) Significance of the unequilibrated ordinary chondrites. *J. Geophys. Res.* 70, 3801–3811.

Dodd R. T., Van Schmus W. R., and Koffman D. M. (1967) A survey of the unequilibrated ordinary chondrites. *Geochim. Cosmochim. Acta* 31, 921–951.

Dodd R. T., Grover J. E., and Brown G. E. (1975) Pyroxenes in the Shaw (L-7) chondrite. *Geochim. Cosmochim. Acta* 39, 1585–1594.

Dodd R. T., Wolf S. F., and Lipschutz M. E. (1993) An H chondrite stream: Identification and confirmation. *J. Geophys. Res.* 98, 15105–15118.

Dohmen R., Becker H.-W., and Chakraborty S. (2007) Fe-Mg diffusion in olivine I: Experimental determination between 700 and 1,200°C as a function of composition, crystal orientation and oxygen fugacity. *Phys. Chem. Miner.* 34, 389–407.

Dohnanyi J. S. (1972) Interplanetary objects in review: Statistics of their masses and dynamics. *Icarus* 17, 1–48.

Dominik C., Blum J., Cuzzi J. N., and Wurm G. (2006) Growth of dust as the initial step toward planet formation. In *Protostars and Planets V* (edited by Reipurth B., Jewitt D. and Keil K.), pp. 783–800, University of Arizona Press, Tucson.

Donaldson C. H. (1976) An experimental investigation of olivine morphology. *Contrib. Mineral. Petrol.* 57, 187–213.

Downes H. and Mittlefehldt D. W. (2006) Identification of a common R chondrite impactor on the ureilite parent body. *Meteorit. Planet. Sci.* 41, A46.

Downes H., Mittlefehldt D. W., Kita N. T., and Valley J. W. (2008) Evidence from polymict ureilite meteorites for a disrupted and re-accreted single ureilite parent asteroid gardened by several distinct impactors. *Geochim. Cosmochim. Acta* 72, 4825–4844.

Drążkowska J., Bitsch B., Lambrechts M., et al. (2023) Planet formation theory in the era of ALMA and Kepler: From pebbles to exoplanets. In *Protostars and Planets VII* (edited by Inutsuka S., Aikawa Y., Muto T., Tomida K. and Tamura M.), pp. 717–751, University of Arizona Press, Tucson.

Drolshagen E., Ott T., Koschny D., Drolshagen G., Schmitt A. K., and Poppe B. (2020) Velocity distribution of larger meteoroids and small asteroids impacting Earth. *Planet. Space Sci.* 184, 104869.

Drolshagen G., Koschny, D., Drolshagen S., Kretschmer J., and Poppe B. (2017) Mass accumulation of Earth from interplanetary dust, meteoroids, asteroids and comets. *Planet. Space Sci.* 143, 21–27.

Drouart A., Dubrulle B., Gautier D., and Robert F. (1999) Structure and transport in the solar nebula from constraints on deuterium enrichment and giant planets formation. *Icarus* 140, 129–155. https://doi.org/10.1006/icar.1999.6137.

Drozd R. J., Kennedy B. M., Morgan C. J., Podosek F. A., and Taylor G. J. (1976) The excess fission xenon problem in lunar samples, *Proc. Lunar Sci. Conf. 7th*, 599–623.

Du K., Li S., Leya I., et al. (2011) The Kumtag meteorite strewn field. *Adv. Space Res.* 67, 4089–4098.

Dudorov A. E. and Eretnova O. V. (2020) The rate of falls of meteorites and bolides. *Sol. Syst. Res.* 54, 223–235.

Dunham E. T., Volskis P., and McKeegan K. D. (2022) CAI size distributions in NCs and CCs. *85th Meeting of the Meteoritical Society*, abstract#6507.

Dunham E. T., Sheikh A., Opara D., et al. (2023) Calcium-aluminum-rich inclusions in non-carbonaceous chondrites: Abundances, sizes, and mineralogy. *Meteorit. Planet. Sci.* 58, 643–671.

Dunn T. L., Cressey G., McSween H. Y., and McCoy T. (2010a) Analysis of ordinary chondrites using powder X-ray diffraction: 1. Modal mineral abundances. *Meteorit. Planet. Sci.* 45, 123–134.

Dunn T. L., McCoy T. J., Sunshine J. M., and McSween H. Y. (2010b) A coordinated spectral, mneralogical, and compositional study of ordinary chondrites. *Icarus* 208, 789–797.

Dunn T. L., Bujrbine T. H., Bottke W. F., and Clkark J. P. (2013) Mineralogies and source regions of near-Earth asteroids. *Icarus* 222, 273–282.

Dye S. T., Huang Y., Lekic V., McDonough W. F., and Šrámek O. (2015) Geo-neutrinos and Earth models. *Phys. Procedia*, 61, 310–318.

Ebel D. S. (2006) Condensation of rocky material in astrophysical environments. In *Meteorites and the Early Solar System II* (edited by Lauretta D. S. and McSween H. Y.), pp. 253–277, University of Arizona Press, Tucson.

Ebel D. S. and Grossman L. (1997) Direct condensation of ferromagnesian liquids from cosmic gases (abstract). *Lunar Planet. Sci.* 28, 317–318.

Ebel D. S. and Grossman L. (1998) Effects of dust enrichment on solid and liquid compositions in equilibrium with cosmic gases. *Lunar Planet. Sci.* 29, abstract#1421.

Ebel D. S. and Grossman L. (2000) Condensation in dust enriched systems. *Geochim. Cosmochim. Acta* 64, 339–366.

Ebel D. S., Gemma M. E., Alpert S. P., et al. (2024) Abundance, sizes, and major element compositions of components in CR and LL chondrites: Formation from single reservoirs. *Meteorit. Planet. Sci.* 59, 2276–2295.

Ebel D. S., Gemma M. E., Weisberg M. K., and Friedrich J. M. (2025) Metal-forsterite condensate nodules and H-L-LL compositions. *Meteorit. Planet. Sci.* 60, 757–777.

Ebert S., Nagashima K., Bischoff A., Berndt J., and Krot A. N. (2022) Mineralogy, petrology, and oxygen isotopic compositions of aluminum-rich chondrules from unequilibrated ordinary and the Dar al Gani 083 (CO3.1) chondrite. *Geochim. Cosmochim. Acta* 336, 448–468.

Ebihara M., Shinonaga T., Nakahara H., et al. (1989) Depth-profiles of halogen abundance and integrated intensity of hydration band near 3 μm in ALH 77231, Antarctic L6 chondrite. In *Differences Between Antarctic and Non-Antarctic Meteorites*, pp. 32–37. LPI Tech. Rpt. 90-01, Lunar and Planetary Institute, Houston.

Edwards G. H. and Blackburn T. (2020) Accretion of a large LL parent planetesimal from a recently formed chondrule population. *Sci. Adv.* 6, eaay8641.

Ehlmann A. J. and Keil K. (1988) Classification of six ordinary chondrites from Texas. *Meteoritics* 23, 361–364.

Einstein A. (1936) *Ideas and Opinions*. Translated by Sonja Bargmann, Crown Publishers, New York (1954), 377 pp.

Eisenhour D. D. (1996) Determining chondrule size distributions from thin-section measurements. *Meteorit. Planet. Sci.* 31, 243–248.

Eisenhour D. D. and Buseck P. R. (1995) Chondrule formation by radiative heating: A numerical model. *Icarus* 117, 197–211.

Eisenhour D. D., Daulton T. L., and Buseck P. R. (1994) Electromagnetic heating in the early solar nebula and the formation of chondrules. *Science* 265, 1067–1070.

Elkins-Tanton L. T., Weiss B. P., and Zuber M. T. (2011) Chondrites as samples of differentiated planetesimals. *Earth Planet. Sci. Lett.* 305, 1–10.

Engrand C. and Maurette M. (1998) Carbonaceous micrometeorites from Antarctica. *Meteorit. Planet. Sci.* 33, 565–580.

Enzel Y., Brown W. J., Anderson R. Y., McFadden L. D., and Wells S. G. (1992) Short-duration Holocene lakes in the Mojave River drainage basin, Southern California. *Quatern. Res.* 38, 60–73.

Eugster O. (2003) Cosmic-ray exposure ages of meteorites and lunar rocks and their significance. *Chem. Der. Erde.* 63, 3–30.

Eugster O., Herzog G. F., Marti K., and Caffee M. W. (2006) Irradiation records, cosmic-ray exposure ages, and transfer times of meteorites. In *Meteorites and the Early Solar System II* (edited by Lauretta D. S. and McSween H. Y.), pp. 829–851. University of Arizona Press, Tucson.

Evatt G. W., Smedley A. R. D., Joy K. H., et al. (2020) The spatial flux of Earth's meteorite falls found via Antarctic data. *Geology* 48, 683–687.

Fang L., Moynier F., Chaussidon M., et al. (2025) The initial solar system abundance of ^{60}Fe and early core formation of the first asteroids. *Sci. Adv.* 11. https://doi.org/10.1126/sciadv.adp9381.

Farinella P., Froeschlé C., and Gonczi R. (1993a) Meteorites from the asteroid 6 Hebe. *Celest. Mech. Dyn. Astron.* 56, 287–305.

Farinella P., Gonczi R., Froschlé Ch., and Froeschlé C. (1993b) The injection of asteroid fragments into resonances. *Icarus* 101, 174–187.

Farinella P., Vokrouhlický D. and Hartmann W. K. (1998) Meteorite delivery via Yarkovsky orbital drift. *Icarus* 132, 378–387.

Farrington O. C. (1915) *Meteorites: Their Structure, Composition, and Terrestrial Relations*, Lakeside Press, Chicago, 233 pp.

Feige J., Airo A., Berger D., et al. (2024) Transport of dust across the Solar System: Constraints on the spatial origin of individual micrometeorites from cosmic-ray exposure. *Phil. Trans. R. Soc.* A382, 20230197. https://doi.org/10.1098/rsta.2023.0197.

Feldstein S. N., Jones R. H., and Papike J. J. (2001) Disequilibrium partial melting experiments on the Leedey L6 chondrite: Textural controls on melting processes. *Meteorit. Planet. Sci.* 36, 1421–1441.

Fernandes D., Rudraswami N. G., Pandey M., and Singh V. P. (2024) Chemical compositions of Fe-rich relict olivines from cosmic spherules, understanding their links with ordinary and carbonaceous chondrites. *Meteorit. Planet. Sci.* 59, 605–625.

Ferrière L., Brandstätter F., and Koeberl C. (2013) Finding shatter cones in meteorites from the Natural History Museum Vienna collection. *76th Annu. Meet. Meteorit. Soc.*, abstract#5193.

Ferrière L., Brandstätter F., and Koeberl C. (2025) Shatter cone occurrences in chondrite and achondrite meteorites – Implications for the disruption of meteoroids. *87th Annu. Meet. Meteorit. Soc.*, abstract#5321.

Finnegan J. A. (2015) "Hiss, clicks and pops" – The enigmatic sounds of meteors. *WGN, J. Int. Meteor Organi.* 43, (2): 47–57.

Fitoussi C. and Bourdon B. (2012) Silicon isotope evidence against an enstatite chondrite. *Earth. Sci.* 335, 1477–1480.

Florin G., Luais B., Rushmer T., and Alard O. (2020) Influence of redox processes on the germanium isotopic composition of ordinary chondrites. *Geochim. Cosmochim. Acta* 269, 270–291.

Florin G., Alard O., Luais B., and Rushmer T. (2023) Abundances of siderophile elements in H-chondrite metal grains: Implications for the origin of metal in unequilibrated ordinary chondrites. *Geochim. Cosmochim. Acta* 341, 1–15.

Floyd C. J., Benito S., Martin P.-E., et al. (2024) Chondrule sizes within the CM carbonaceous chondrites and measurement methodologies. *Meteorit. Planet. Sci.* 59, 1–20. https://doi.org/10.1111/maps.14250.

Flynn G. J., Consolmagno G. J., Brown P., and Macke R. J. (2018) Physical properties of the stone meteorites: Implications for the properties of their parent bodies. *Chem. der Erde* 78, 269–298.

Fodor R. V. (1976) *Meteorites: Stones from the Sky*, Dodd, Mead & Co., New York, 47 pp.

Fodor R. V. and Keil K. (1973) Composition and origin of lithic fragments in L- and H-group chondrites (abstract). *Meteoritics* 8, 366–367.

Fodor R. V. and Keil K. (1975) Implications of poikilitic textures in LL-group chondrites. *Meteoritics* 10, 325–339.

Fodor R. V. and Keil K. (1976a) Carbonaceous and non-carbonaceous lithic fragments in the Plainview, Texas, chondrite: Origin and history. *Geochim. Cosmochim. Acta* 40, 177–178, IN3-IN12, 179–189.

Fodor R. V. and Keil K. (1976b) A komatiite-like lithic fragment with spinifex texture in the Eva meteorite: Origin from a supercooled impact-melt of chondritic parentage. *Earth Planet. Sci. Lett.* 29, 1–6.

Fodor R. V. and Keil K. (1978) Catalog of lithic fragments in LL-group chondrites. *Spec. Publ. Univ. New Mexico Inst. Meteorit.* 19, 1–38.

Fodor R. V. and Taylor G. J. (1979) *Impact!*, Leisure Books, New York, 302 pp.

Fodor R. V., Keil K., Jarosewich E., and Huss G. I. (1971) Mineralogy, petrology, and chemistry of the Burdett, Kansas, chondrite. *Chem. Der. Erde.* 30, 103–113.

Fodor R. V., Keil K., and Jarosewich E. (1972) The Oro Grande, New Mexico, chondrite and its lithic inclusion. *Meteoritics* 7, 496–507.

Fodor R. V., Prinz M., and Keil K. (1974) Implications of K-rich lithic fragments and chondrules in the Bhola brecciated chondrite (abstract). *Prog. Geol. Sec. Amer. Meet. 87th*, 739–740.

Fodor R. V., Keil K., Wilkening L. L., Bogard D. D., and Gibson E. K. (1976) Origin and history of a meteorite parent body regolith breccia: carbonaceous and non-carbonaceous lithic fragments in the Abbott, New Mexico, chondrite. *Tectonics and Mineral Resources of Southwestern U.S., New Mexico Geol. Soc. Spec. Publ.* 6, 206–218.

Fodor R. V., Keil K., and Gomes C. B. (1977) Studies of Brazilian meteorites IV. Origin of a dark colored. unequilibrated lithic fragment in the Rio Negro chondrite. *Rev. Bras. Geosciênc.* 7, 45–67.

Folco L., Glass B. P., D'Orazio M., and Rochette P. (2018) Australasian microtektites: Impactor identification using Cr, Co and Ni ratios. *Geochim. Cosmochim. Acta* 222, 550–568.

Folco L., Rochette P., D'Orazio M., and Masotta M. (2023) The chondritic impactor origin of the Ni-rich component in Australasian tektites and microtektites. *Geochim. Cosmochim. Acta* 360, 231–240.

Foshag W. F. (1941) Problems in the study of meteorites. *Amer. Mineral.* 26, 137–144.

Fourcroy A.-F. de. (1803) Memoire on the stones which have fallen from the atmosphere, and particularly near L'Aigle in the Department of L'Orne, on the 26th of April last. *Philos. Mag. (Tilloch's)* 16, 299–305.

Frank D. R. (2023) Elemental and isotopic compositions of carbonaceous xenoliths in meteorites and of samples returned from comet 81P/Wild 2: An investigation into the diversity of early solar system materials. Ph.D. thesis, University of Hawaii at Manoa, 291 pp.

Frank D. R., Zolensky M. E., and Le L. (2014) Olivine in terminal particles of Stardust aerogel tracks and analogous grains in chondrite matrix. *Geochim. Cosmochim. Acta* 142, 240–259.

Fredriksson K. (1963) Chondrules and the meteorite parent bodies. *Trans. N.Y. Acad. Sci.* 25, 756–769.

Fredriksson K. and Keil K. (1963) The light-dark structure in the Pantar and Kapoeta stone meteorites. *Geochim. Cosmochim. Acta* 27, 717–739.

Fredriksson K. and Wlotzka F. (1985) Morro do Rocio: An unequilibrated H5 chondrite. *Meteoritics* 20, 467–478.

Fredriksson K., Nelen J., and Melson W. G. (1970) Petrography and origin of lunar breccias and glasses. *Proc. Apollo 11 Lunar Sci. Conf.*, 419–432.

Fredriksson K., Noonan A., and Nelen J. (1973) Meteoritic, lunar and Lonar impact chondrules. *Moon* 7, 475–482.

Fredriksson K., Miller J., Nelen J., and Darsa S. (1981) The Tambakwatu chondrite. *Meteoritics* 16, 77–81.

Freer R. (1981) Diffusion in silicate minerals and glasses: A data digest and guide to the literature. *Contrib. Mineral. Petrol.* 76, 440–454.

French B. M. (1998) *Traces of Catastrophe: A Handbook of Shock-metamorphic Effects in Terrestrial Meteorite Impact Structures*, LPI Contribution No. 954, Lunar and Planetary Institute, Houston, 120 pp.

Friedrich J. M., Ruzicka A., Rivers M. L., et al. (2013) Metal veins in the Kernouvé (H6 S1) chondrite: evidence for pre- or syn-metamorphic shear deformation. *Geochim. Cosmochim. Acta* 116, 71–83.

Friedrich J. M., Rubin A. E., Beard S. P., et al. (2014a) Ancient porosity preserved in ordinary chondrites: Shock, compaction and thermal metamorphism. *Meteorit. Planet. Sci.* 49, 1214–1231.

Friedrich J. M., Weisberg M. K., and Rivers M. L. (2014b) Multiple impact events recorded in the NWA 7298 H chondrite breccia and the dynamical evolution of an ordinary chondrite asteroid. *Earth Planet. Sci. Lett.* 394, 13–19.

Friedrich J. M., Weisberg M. K., Ebel D. S., et al. (2015) Chondrule size and related physical properties: A compilation and evaluation of current data across all meteorite groups. *Chem. Der. Erde.* 75, 419–443.

Friedrich J. M., Chen M. M., Giordano S. A., et al. (2022) Size-frequency distributions and physical properties of chondrules from x-ray computed microtomography and digital data extraction. *Microsc. Res. Tech.* 85, 1814–1824.

Friedrich J. M., Riveros E. M., Macke R. J., et al. (2025) Physical properties, internal structure, and the three-dimensional petrography of CI chondrites. *Meteorit. Planet. Sci.* 60, 632–645.

Fritz J., Wünnemann K., Greshake A., et al. (2011) Shock pressure calibration for lunar plagioclase. *Lunar Planet. Sci.* 42 (abstract#1196).

Frossard P., Bonnand P., Boyet M., and Bouvier A. (2024) Role of redox conditions and thermal metamorphism in the preservation of Cr isotopic anomalies in components of non-carbonaceous chondrites. *Geochim. Cosmochim. Acta* 367, 165–178.

Fu R. R., Weiss B. P., Lima E. A., et al. (2014a) Solar nebula magnetic fields recorded in the Semarkona meteorite. *Science* 346. https://doi.org/10.1126/science.1258022.

Fu R. R., Lima E. A., and Weiss B. P. (2014b) No nebular magnetization in the Allende CV carbonaceous chondrite. *Earth Planet. Sci. Lett.* 404, 54–66.

Fujimaki H., Matsu-Ura M., Sunagawa I., and Aoki K. (1981) Chemical compositions of chondrules and matrices in the ALHA-77015 chondrite (L3). *Mem. Nat. Inst. Polar Res.* 20, 161–174.

Fujita T., Kojima H., and Yanai K. (1999) Origin of metal-troilite aggregates in six ordinary chondrites. *Antarct. Meteorite Res.* 12, 19–35.

Fujiwara A., Kamimoto G., and Tsukamoto A. (1977) Destruction of basaltic bodies by high velocity impact. *Icarus* 31, 277–278.

Fujiwara A., Kawaguchi J., Yeomans D. K., et al. (2006) The rubble pile asteroid Itokawa as observed by Hayabusa. *Science* 312, 1330–1334.

Fukimoto K., Miyahara M., Sakai T., et al. (2020) Back-transformation mechanisms of ringwoodite and majorite in an ordinary chondrite. *Meteorit. Planet. Sci.* 55, 1–15.

Fukuda K., Hibiya Y., Kastelle C. R., et al. (2024) Radial transport and nebular thermal processing of millimeter-sized solids in the solar protoplanetary disk inferred from Cr-Ti-O isotope systematics of chondrules. *Meteorit. Planet. Sci.* 59, 3282–3304.

Funaki M. and Wasilewski P. (1999) A relation of magnetization and sulphidation in the parent body of Allende (CV3) carbonaceous chondrite. *Meteorit. Planet. Sci.* 34, A39.

Funk S. C., Bischoff A., and Schlüter J. (2011) Xenoliths in carbonaceous and ordinary chondrites. *Meteorit. Planet. Sci.* 46, A71, abstract#5318.

Gaffey M. J. (1976) Spectral reflectance characteristics of the meteorite classes. *J. Geophys. Res.* 81, 905–920.

Gaffey M. J. (1984) Rotational spectral variations of asteroid (8) Flora: Implications for the nature of the S-type asteroids and for the parent bodies of the ordinary chondrites. *Icarus* 60, 83–114.

Gaffey M. J. and Gilbert S. L. (1998) Asteroid 6 Hebe: The probable parent body of the H-type ordinary chondrites and the IIE iron meteorites. *Meteorit. Planet. Sci.* 33, 1281–1295.

Gaffey M. J., Bell J. F., Brown R. H., et al. (1993) Mineralogic variations within the S-type asteroid class. *Icarus* 106, 573–602.

Ganguly J., Tirone M., Chakraborty S., and Domanik K. (2013) H-chondrite parent asteroid: A multistage cooling, fragmentation and re-accretion history constrained by thermometric studies, diffusion kinetic modeling and geochronological data. *Geochim. Cosmochim. Acta* 105, 206–220.

Garai S., Olson P. L., and Sharp Z. D. (2025) Building Earth with pebbles made of chondritic components. *Geochim. Cosmochim. Acta* 390, 86–104.

Garlick G. F. J. (1970) *Luminescent Materials*, Clarendon, Oxford University Press.

Garrison D. H. and Bogard D. D. (2000) Noble gases in the Monahans chondrite and halite: ^{39}Ar-^{40}Ar age, space exposure age, trapped solar gases and neutron fluence. *Lunar Planet. Sci.* 31, abstract#1137.

Garrison D. H. and Bogard D. D. (2001) ^{39}Ar-^{40}Ar and space exposure ages of the unique Portales Valley H chondrite. *Lunar Planet. Sci.* XXXII, abstract#1137.

Gastineau-Lyons H. K., McSween H. Y., and Gaffey M. J. (2002) A critical evaluation of oxidation versus reduction during metamorphism of L and LL group chondrites, and implications for asteroid spectroscopy. *Meteorit. Planet. Sci.* 37, 75–89.

Gattacceca J., Eisenlohr P., and Rochette P. (2004) Calibration of in situ magnetic susceptibility measurements. *Geophys. J. Int.* 158, 42–49.

Gattacceca J., Rochette P., Denise M., Consolmagno G., and Folco L. (2005) An impact origin for the foliation of chondrites. *Earth Planet. Sci. Lett.* 234, 351–368.

Gattacceca J., Krzesińska A. M., Marrocchi Y., et al. (2017) Young asteroid mixing revealed in ordinary chondrites: The case of NWA 5764, a polymict LL breccia with L clasts. *Meteorit. Planet. Sci.* 62, 2289–2304.

Geer M. R. and Yancey H. F. (1938) Expression and interpretation of the size composition of coal. *Amer. Inst. Min. Met. Eng. Trans. Coal Div.* 130, 250–269.

Geiss J. and Gloeckler G. (2003) Isotopic Composition of H, He and Ne in the protosolar cloud. *Space Rev.* 106, 3–18. https://doi.org/10.1023/A:1024651232758.

Genge M. J. and Grady M. M. (1999) The fusion crusts of stony meteorites: Implications for the atmospheric reprocessing of extraterrestrial materials. *Meteorit. Planet. Sci.* 34, 341–356.

Genge M. J., van Ginneken M., and Suttle M. D. (2020) Micrometeorites: Insights into the flux, sources and atmospheric entry of extraterrestrial dust at Earth. *Planet. Space Sci.* 187, 104900.

Gerber S., Burkhardt C., Budde G., Metzler K., and Kleine T. (2017) Mixing and transport of dust in the early solar nebula as inferred from titanium isotope variations among chondrules. *Astrophys. J. Lett.* 841, L17.

Gessler N., Gessler P., Matson R. D., and Verish R. S. (2002) Implications of ice rafting for meteorite recovery on seasonally wet playas (abstract). *Meteorit. Planet. Sci.* 37, A51.

Ghanbarzadeh S., Hesse M. A., and Prodanovic M. (2017) Percolative core formation in planetesimals enabled by hysteresis in metal connectivity. *Proc. Natl. Acad. Sci.* 114, 13,406–13,411.

Gibbons R. V., Hörz, and Schaal R. B. (1976) The chemistry of some individual lunar soil agglutinates. *Proc. Lunar Sci. Conf. 7th*, 405–421.

Gibson E. K., Lange D. E., Keil K., Schmidt T. E., and Rhodes J. M. (1977) The Kramer Creek, Colorado meteorite: A new L4 chondrite. *Meteoritics* 12, 95–107.

Gibson R. L. (2002) Impact-induced melting of Archean granulites in the Vredefort Dome, South Africa. I: Anatexis of metapelitic granulites. *J. Metamorph. Geol.* 20, 57–70.

Gillet P., Chen M., Dubrovinsky L., and El Goresy A. (2000) Natural $NaAlSi_3O_8$-hollandite in the shocked Sixiangkou meteorite. *Science* 287, 1633–1636.

Gilmer M. (2020) About NWA meteorites: A brief history of the Saharan gold rush of meteorites. http://galactic-stone.com/about-nwa-meteorites/. Accessed November 17, 2020.

Gilmour I. (2005) Structural and isotopic analysis of organic matter in carbonaceous chondrites. In *Meteorites, Comets, and Planets* (edited by Davis A. M.) Treatise on Geochemistry, vol. 1. pp. 269–290, Oxford: Elsevier–Pergamon.

Gladman B., Michel P., and Froeschlé C. (2000) The Near-Earth Object population. *Icarus* 146, 176–189.

Glavin D. P., Dworkin J. P., Alexander C. M. O. et al. (2025) Abundant ammonia and nitrogen-rich soluble organic matter in samples from asteroid (101955) Bennu. *Nat. Astron.* 9, 199–210.

Gnos E., Wimmer K., Hofmann B., et al. (2025) The giant L5 strewn field Jiddat al Harasis 091, Sultanate of Oman. *87th Annu. Meet. Meteorit. Soc.*, abstract#5008.

Goetze C. and Kohlstedt D. L. (1973) Laboratory study of dislocation climb and diffusion in olivine. *J. Geophys. Res.* 78, 5961–5971.

Goldstein J. I. and Michael J. R. (2006) The formation of plessite in meteoritic metal. *Meteorit. Planet. Sci.* 41, 553–570.

Gomes C. B. and Keil K. (1980) *Brazilian Stone Meteorites*, University of New Mexico Press, Albuquerque, 162 pp.

Gooding J. L. (1979) Petrogenetic properties of chondrules in unequilibrated H-, L-, and LL-group chondritic meteorites. Ph.D. dissertation, University of New Mexico, Albuquerque, 392 pp.

Gooding J. L. and Keil K. (1981) Relative abundances of chondrule primary textural types in ordinary chondrites and their bearing on conditions of chondrule formation. *Meteoritics* 16, 17–43.

Goodrich C. A. (2023) Polymict ureilites: The answer to life, the universe, and everything. *85th Annu. Meet. Meteorit. Soc.*, abstract#6166.

Goodrich C. A., Fioretti A. M., Zolensky M., et al. (2018) The Almahata Sitta polymict ureilite from the University of Khartoum collection: Classification, distribution of clast types in the strewn field, new meteorite types, and implications for the structure of Asteroid 2008 TC3. *Lunar Planet. Sci.* 49, abstract#1321.

Goodrich C. A., Kring D. A., and Greenwood R. C. (2021) Xenoliths in ordinary chondrites and ureilites: Implications for early solar system dynamics. *Meteorit. Planet. Sci.* 56, 1949–1987.

Göpel C., Manhès G., and Allègre C. J. (1994) U-Pb systematics of phosphates from equilibrated ordinary chondrites. *Earth Planet. Sci. Lett.* 121, 153–171.

Gorshkov E. S. (1973) Elastic properties of meteorites. *Meteoritika* 32, 126–135.

Goryunov M. V., Petrova E. V., Chgukin A. V., et al. (2025) Comparison of the iron-bearing crystals and phases from Tamdakht H5 and Annama H5 ordinary chondrites by x-ray diffraction, magnetization measurements and Mössbauer spectroscopy. *Meteorit. Planet. Sci.* 60, 1520–1544.

Goudy S. P., Telus M., and Chapman B. (2023) Evidence for multiple early impacts on the H chondrite parent body from electron backscatter diffraction analysis. *Meteorit. Planet. Sci.* 58, 501–515.

Gould S. J. (1989) *Wonderful Life: The Burgess Shale and the Nature of History*, W. W. Norton, New York, 347 pp.

Gounelle M. (2006) The meteorite fall at L'Aigle and the Biot report: Exploring the cradle of meteoritics. In *The History of Meteoritics and Key Meteorite Collections: Fireballs, Falls and Finds* (edited by McCall G. J. H., Bowden A. J. and Howarth R. J.), pp. 73–89. Geological Society, London.

Grady M. M. (2000) *Catalogue of Meteorites*, Fifth edition, Cambridge University Press, Cambridge, UK, 689 pp.

Grady M., Pratesi G., and Moggi Cecchi V. (2014) *Atlas of Meteorites*, Cambridge University Press, Cambridge, 373 pp.

Graf T. and Marti K. (1994) Collisional records in LL-chondrites. *Meteoritics* 29, 643–648.

Graf T. and Marti K. (1995) Collisional history of H chondrites. *J. Geophys. Res.* 100, 21247–21263.

Graham A. L., Bevan A. W. R., and Hutchison R. (1985) *Catalogue of Meteorites*, Fourth edition, University of Arizona Press, Tucson, 460 pp.

Grant H., Tartèse R., Jones R., et al. (2023) Bulk minerlaogy, water abundance, and hydrogen isotopic composition of unequilibrated ordinary chondrites. *Meteorit. Planet. Sci.* 58, 1365–1381.

Granvik M., Morbidelli A., Jedicke R., Bolin B., Bottke W. F., Beshore E., Vokrouhlický D., Nesvorný D. and Michel P. (2018) Debiased orbit and absolute-magnitude distributions for near-Earth objects. *Icarus* 312, 181–207.

Greenberg R., Wacker J., Hartmann W. K., and Chapman C. R. (1978) Planetsimals to planets: Numerical simulation of collisional evolution. *Icarus* 35, 1–26.

Greenstreet S. and Gladman B. (2013) High-inclination Atens are indeed rare. *Astrophys. J. Let.* 767, L18. https://doi.org/10.1088/2041-8205/767/1/L18.

Greenstreet S., Gladman B., Ngo H., Granvik M. and Larson S. (2012a) Production of Near-Earth Asteroids on retrograde orbits. *Astrophys. J. Lett.* 749, L39. https://doi.org/10.1088/2041-8205/749/2/L39.

Greenstreet S., Ngo H. and Gladman B. (2012b) The orbital distribution of Near-Earth Objects inside Earth's orbit. *Icarus* 217, 355–366.

Greenwood J. P. and Hess P. C. (1996) Congruent melting kinetics: Constraints on chondrule formation. In *Chondrules and the Protoplanetary Disk* (edited by Hewins R. H., Jones R. H. and Scott E. R. D.), pp. 205–211, Cambridge University Press: Cambridge.

Greenwood J. P., Rubin A. E., and Wasson J. T. (2000) Oxygen-isotopes in R-chondrite magnetite and olivine: Links between R chondrites and ordinary chondrites. *Geochim. Cosmochim. Acta* 64, 3897–3911.

Greenwood N. N. (1963) The chemistry of gallium. *Adv. Inorganic Chem. Radiochem.* 5, 91–134.

Greenwood R. C., Barrat J.-A., Scott E. R. D., et al. (2015) Geochemistry and oxygen isotope composition of main-group pallasites and olivine-rich clasts in mesosiderites: Implications for the "Great Dunite Shortage" and HED-mesosiderite connection. *Geochim. Cosmochim. Acta* 169, 115–136.

Greenwood R. C., Burbine T. H., Miller M. F., and Franchi I. A. (2017) Melting and differentiation of early-formed asteroids: The perspective from high precision oxygen isotope studies. *Chem. Erde Geochem.* 77, 1–43.

Greg R. P. (1860) *A Catalogue of Meteorites and Fireballs*, Rep. Brit. Assoc.

Gregory T., Luu T.-H., Coath C. D., Russell S. S., and Elliott T. (2020) Primordial formation of major silicates in a protoplanetary disc with homogeneous $^{26}Al/^{27}Al$. *Sci. Adv.* 6, eaay9626 (8 pp.)

Greshake A. (1997) The primitive matrix components of the unique carbonaceous chondrite Acfer 094: A TEM study. *Geochim. Cosmochim. Acta* 61, 437–452.

Grier J. A., Kring D. A., Swindle T. D., et al. (2004) Analyses of the chondritic meteorite Orvinio (H6): Insight into the origins and evolution of shocked H chondrite material. *Meteorit. Planet. Sci.* 39, 1475–1493.

Grimm R. E. (1985) Penecontemporaneous metamorphism, fragmentation, and reassembly of ordinary chondrite parent bodies. *J. Geophys. Res.* 90, 2022–2028.

Grimm R. E. and McSween H. (1993) Heliocentric zoning of the asteroid belt by 26-Al heating. *Science* 259, 653–655.

Gritsevich M., Vinnikov V., Kohout T., et al. (2014) A comprehensive study of distribution laws for the fragments of Košice meteorite. *Meteorit. Planet. Sci.* 49, 328–345.

Gritsevich M., Moilanen J., Visuri J., et al. (2024) The fireball of November 24, 1970, as the most probable source of the Ischgl meteorite. *Meteorit. Planet. Sci.* 59, 1658–1691.

Grossman J. N. (1985) Semarkona: The least metamorphosed ordinary chondrite (abstract). *Meteoritics* 20, 656–657.

Grossman J. N. and Brearley A. J. (2005) The onset of metamorphism in ordinary and carbonaceous chondrites. *Meteorit. Planet. Sci.* 40, 87–122.

Grossman J. N. and Rubin A. E. (1986) The origin of chondrules and clasts bearing calcic plagioclase in ordinary chondrites (abstract). *Lunar Planet. Sci.* 17, 293–294.

Grossman J. N. and Rubin A. E. (2006) White paper report for the Nomenclature Committee on the composition of olivine and pyroxene in equilibrated ordinary chondrites. 6 pp.

Grossman J. N. and Wasson J. T. (1983) Refractory precursor components of Semarkona chondrules and the fractionation of refractory elements among chondrites. *Geochim. Cosmochim. Acta* 47, 759–771.

Grossman J. N. and Wasson J. T. (1985) The origin and history of the metal and sulfide components of chondrules. *Geochim. Cosmochim. Acta* 49, 925–939.

Grossman J. N., Rubin A. E., Rambaldi E. R., Rajan R. S., and Wasson J. T. (1985) Chondrules in the Qingzhen type-3 enstatite chondrite: Possible precursor components and comparison to ordinary chondrite chondrules. *Geochim. Cosmochim. Acta* 49, 1781–1795.

Grossman J. N., Alexander C. M. O'D., Wang J., and Brearley A. J. (2000) Bleached chondrules: Evidence for widespread aqueous processes on the parent asteroids of ordinary chondrites. *Meteorit. Planet. Sci.* 35, 467–486.

Grossman L. (1972) Condensation in the primitive solar nebula. *Geochim. Cosmochim. Acta* 36, 507–619.

Group G. (1981) Terrestrial chondrules, glass spherules and accretioinary lapilli from the suevite, Ries Crater, Germany. *Earth Planet. Sci. Lett.* 55, 407–418.

Guimon R. K., Keck D. B., Weeks S. K., DeHart J., and Sears D. W. G. (1985) Chemical and physical studies of type 3 chondrites – IV: Annealing studies of a type 3.4 ordinary chondrite and the metamorphic history of meteorites. *Geochim. Cosmochim. Acta* 49, 1515–1524.

Guimon R. K., Lofgren G. E., and Sears D. W. G. (1986) Devitrifcation and the thermoluminescence-metamorphism relationship in ordinary chondrites. *Geophys. Res. Lett.* 13, 969–972.

Guimon R. K., Lofgren G. E., and Sears D. W. G. (1988) Chemical and physical studies of type 3 chondrites. IX: Thermoluminescence and hydrothermal annealing experiments and their relationship to metamorphism and aqueous alteration in type <3.3 ordinary chondrites. *Geochim. Cosmochim. Acta* 52, 119–127.

Guo Z., Li Y., Liu S., et al. (2020) Discovery of nanophase iron particles and high pressure clinoenstatite in a heavily shocked ordinary chondrite: Implications for the decomposition of pyroxene. *Geochim. Cosmochim. Acta* 272, 276–286.

Gyngard F., Ávila J. N., Ireland T. R., and Zinner E. (2014) More interstellar exposure ages of large presolar SiC grains from the Murchison meteorite. *Lunar Planet. Sci.* 45, abstract#2348.

Gyngard F., Jadhav M., Nittler L. R., Stroud R. M., and Zinner E. (2018) Bonanza: An extremely large dust grain from a supernova. *Geochim. Cosmochim. Acta* 221, 60–86.

Gyollai I., Kereszturi A., Kereszty Z., Szabó M., and Chatzitheodoridis E. (2017) Shock-induced alteratioons in the recently found H chondrite Csátalja meteorite and its implications. *Central Eur. Geol.*, 1–28. https://doi .org/10.1556/24.60.2017.007.

Gyollai I., Kereszturi Á., Chatzitheodoridis E., et al. (2023) Formation and shock impact history of the Cstalja ordinary chondrite. *Meteorit. Planet. Sci.* 58, 2–24.

Haack H., Taylor G. J., Scott E. R. D., and Keil K. (1992) Thermal history of chondrites: Hot accretion vs. metamorphic heating. *Geophys. Res. Lett.* 19, 2235–2238.

Haggerty S. E. (1972) An enstatite chondrite from Hadley Rille (abstract). In *The Apollo 15 Lunar Samples* (ed. Chamberlain J. W. and Watkins C.), pp. 85–87, Lunar Science Institute, Houston, Texas.

Haghighipour N. (2004) Growth of dust particles and accumulation of centimeter-sized objects in the vicinity of a pressure-enhanced region of a solar nebula. *Lunar Planet. Sci.* 35, abstract#2001.

Hahn O. (1880) *Die Meteorite (Chondrite) und ihre Organismen mit 32 Tafeln photographischer Abbildungen*, Laupp'schen, Tübingen, 220 pp.

Haldane J. B. S. (1927) Possible Worlds and Other Essays. *Chatto and Windus*, London, 312 pp.

Halley E. (1714) An account of several extraordinary meteors or lights in the sky. *Philos. Trans. R. Soc. London* 29, 159–164.

Halliday I. (1964) The variation in the frequency of meteorite impact with geographic latitude. *Meteoritics* 2, 271–278.

Halliday I. A. (2001) The present day flux of meteorites to the Earth. In *Accretion of Extraterrestrial Matter Throughout Earth's History* (edited by Peuker-Ehrenbrink B. and Schmitz B.), pp. 305–318, Kluwer Academic/Plenum, New York.

Halliday I. and Griffin A. A. (1982) A study of the relative rates of meteorite falls on the Earth's surface. *Meteoritics* 17, 31–46.

Halliday I., Blackwell A. T., and Griffin A. A. (1978) The Innisfree meteorite and the Canadian camera network. *J. Roy. Astron. Soc. Can.* 72, 15–39.

Hamilton W. (1795) An account of the late eruption of Mount Vesuvius. *Philos. Trans. R. Soc.* 85, 73–116.

Han J. and Dunham E. T. (2025) Evidence for metasomatic alteration in ordinary chondrites. *Lunar Planet. Sci.* 56, abstract#1072.

Hansen M. (1958) *Constitution of Binary Alloys*, McGraw-Hill, New York, 1305 pp.

Harlan S., Jenniskens P., Zolensky M. E., et al. (2016) Meteorites found on Misfits Flat dry lake, Nevada. *Meteorit. Planet. Sci.* 51, 757–772.

Harris A. W. (1996) The rotation rates of very small asteroids: Evidence for "rubble pile" structure (abstract). *Lunar Planet. Sci.* 27, 493–494.

Harrison K. P. and Grimm R. E. (2010) Thermal constraints on the early history of the H-chondrite parent body reconsidered. *Geochim. Cosmochim. Acta* 74, 5410–5423.

Hart R. J., Cloete M., McDonald I., Carlson R. W., and Andreoli M. A. G. (2002) Siderophile-rich inclusions from the Morokweng impact melt sheet, South Africa: Possible fragments of a chondritic meteorite. *Earth Planet. Sci. Lett.* 198, 49–62.

Hartlep T. and Cuzzi J. N. (2020) Cascade model for planetesimal formation by turbulent clustering. *Astrophys. J.* 892, 120, (21 pp.)

Hartmann W. K. (1978) Planet formation: Mechanism of early growth. *Icarus* 33, 50–61.

Hartmann W. K. (1979) Diverse puzzling asteroids and a possible unifed explanation. In *Asteroids* (edited by Gehrels T.), pp. 466–479, University of Arizona Press, Tucson.

Harvey R. (2003) The origin and significance of Antarctic meteorites. *Chem. Der. Erde*. 63, 93–147.

Harvey R. P. and Cassidy W. A. (1989) A statistical comparison of Antarctic finds and modern falls: Mass frequency distributions and relative abundance by type. *Meteoritics* 24, 9–14.

Harvey R. P., Bennett M. L., and McSween H. Y. (1993) Pyroxene equilibration temperatures in metamorphosed ordinary chondrites (abstract). *Lunar Planet. Sci.* XXIV, 615–616.

Harvey R. P., Schutt J., and Karner J. (2015) Fieldwork methods of the US Antarctic Search for Meteorites Program. In *35 Seasons of U.S. Antarctic Meteorites (1976–2010): A Pictorial Guide to the Collection* (edited by Righter K., Corrigan C. M., McCoy T. J. and Harvey R. P.), pp. 23–41, American Geophysical Union, Washington, D. C.

Hasegawa Y., Wakita S., Matsumoto Y., and Oshino S. (2016) Chondrule formation via impact jetting triggered by planetary accretion. *Astrophysical J.* 816, 14 pp.

Heck P. R., Schmitz B., Baur H., Halliday A. N., and Wieler R. (2004) Fast delivery of meteorites to Earth after a major asteroid collision. *Nature* 430, 323–325.

Heck P. R., Schmitz B., Baur H., and Wieler R. (2008) Noble gases in fossil micrometeorites and meteorites from 470 Myr old sediments from southern Sweden, and new evidence for the Lchondrite parent body break-up event. *Meteorit. Planet. Sci.* 43, 517–528.

Heck P. R., Ushikubo T., Schmitz B., et al. (2010) A single asteroidal source for extraterrestrial Ordovician chromite grains from Sweden and China: High-precision oxygen three-isotope SIMS analysis. *Geochim. Cosmochim. Acta* 74, 487–509.

Heck P. R., Greer J., Kööp L., et al. (2020) Lifetimes of interstellar dust from cosmic ray exposure ages of presolar silicon carbide. *Proc. Natl. Acad. Sci.* https://doi.org/10.1073/pnas.1904573117.

Heck P. R., Schmitz B., Ritter X., et al. (2024) Unusual sources of fossil micrometeorites deduced from relict chromite in the small size fraction in ~467 Ma old limestone. *Meteorit. Planet. Sci.* 59, 502–513.

Heide F. (1964) *Meteorites*. University of Chicago Press, Chicago, 144 pp.

Heide F. and Brückner H.-P. (1968) Der olivine-hypersthen-Chondrite von Wolamo, Äthiopien. *Chem. Der. Erde.* 27, 265–338.

Heide F. and Wlotzka F. (1995) *Meteorites: Messengers from Space*, Springer, Berlin, 231 pp.

Heiken G. (1975) Petrology of lunar soils, *Rev. Geophys.* 13, 567–587.

Hellmann J. L., Kruijer T. S., Van Orman J. A., Metzler K., and Kleine T. (2019) Hf-W chronology of ordinary chondrites. *Geochim. Cosmochim. Acta* 258, 290–309.

Hellmann J. L., Kruijer T. S., Metzler K., et al. (2020) Hf-W chronology of a macrochondrule from the L5/6 chondrite Northwest Africa 8192. *Meteorit. Planet. Sci.* 55, 2241–2255.

Henke S., Gail H.-P., Trieloff M., Schwarz W., and Kleine T. (2012) Thermal history modelling of the H chondrite parent body. *Astron. Astrophys.* 545, 15. https://doi.org/10.1051/0004-6361/201219100.

Herbert F. (1989) Primordial electrical induction heating of asteroids. *Icarus* 78, 402–410.

Herbert F. and Sonett C. P. (1979) Electromagnetic heating of minor planets in the early solar system. *Icarus* 40, 484–496.

Herbst W. and Greenwood J. P. (2016) A new mechanism for chondrule formation: Radiative heating by hot planetesimals. *Icarus* 267, 364–367.

Herbst W. and Greenwood J. P. (2024) Chondrule formation during low-speed collisions of planetesimals: A hybrid splash-flyby framework. *Meteorit. Planet. Sci.* 59, 895–909.

Herd C. D. K., Friedrich J. M., Greenwood R. C., and Franchi I. A. (2013) An igneous-textured clast in the Peace Rive meteorite: Insights into accretion and metamorphism of asteroids in the early solar system. *Canadian J. Earth Sci.* 50, 14–25.

Herkenhoff K. E., Sullivan R. J., Newman C. E., et al. (2023) Comparison of ventifact orientations and recent wind direction indicators on the floor of Jezero Crater, Mars. *J. Geophys. Res. Planets* 128, e2022JE007599.

Herndon J. M. and Herndon M. A. (1977) Aluminum-26 as a planetoid heat source in the early solar system. *Meteoritics* 12, 459–465.

Herschel W. (1781) *Account of a Comet*, Philosophical Transactions of the Royal Society of London 71, 492–501.

Herzog G. F. (2005) Cosmic-ray exposure ages of meteorites. In *Meteorites, Comets, and Planets*, (edited by Davis A. M.) Treatise on Geochemistry, vol. 1. pp. 347–380, Elsevier–Pergamon, Oxford.

Hewins R. H. (1997) Chondrules. *Annu. Rev. Earth Planet. Sci.* 25, 61–83.

Hewins R. H., Klein L. C. and Fasano B. V. (1981) Conditions of formation of pyroxene excentroradial chondrules. *Proc. Lunar Planet. Sci. Conf.* 12B, 1123–1133.

Hewins R. H., Connolly H. C., Lofgren G. E., and Libourel G. (2005) Experimental constraints on chondrule formation. In *Chondrules and the Protoplanetary Disk* (edited by Krot A. N., Scott E. R. D. and Reipurth B.), pp. 286–316, ASP Conference Series 341, Astronomical Society of the Pacific, San Francisco.

Hey M. H. (1966) *Catalogue of Meteorites*, 3rd Revised and enlarged edition. British Museum (Natural History), London, 637 pp.

Heymann D. (1967) On the origin of hypersthene chondrites: ages and shock effects of black chondrites. *Icarus* 6, 189–221.

Heymann D. and Mazor E. (1968) Noble gases in unequilibrated ordinary chondrites. *Geochim. Cosmochim. Acta* 32, 1–19.

Heyse J. V. (1978) The metamorphic history of LL-group ordinary chondrites. *Earth Planet. Sci. Lett.* 40, 365–381.

Hezel D. C., Metzler K., and Hochstein M. (2025) Revealing the relationship between 2-D and 3-D chondrule size-frequency distribution in a meteorite. *Meteorit. Planet. Sci.* 60, 869–875.

Higuchi H., Ganapathy R., Morgan J. W., and Anders E. (1977) "Mysterite": A late condensate from the solar nebula. *Geochim. Cosmochim. Acta* 41, 843–852.

Hilton C. D., Bermingham K. R., Walker R. J., and McCoy T. J. (2019) Genetics, crystallization sequence, and age of the South Byron Trio iron meteorites: New insights to carbonaceous chondrite (CC) type parent bodies. *Geochim. Cosmochim. Acta* 251, 217–228.

Hirayama K. (1918) Groups of asteroids probably of common origin. *Astronom. J.* 31, 185–188.

Hirayama K. (1928) Families of asteroids. *Jpn. J. Astron. Geophys.* 5, 137–162.

Hirayama K. (1933) Present state of the families of asteroids. *Proc. Imperial Acad. Jpn.* 9, 482–485.

Hiroi T., Zolensky M. E., Pieters C. M., and Lipschutz M. E. (1996) Thermal metamorphism of the C, G, B, and F asteroids seen from the 0.7 μm, 3 μm, and UV absorption strengths in comparison with carbonaceous chondrites. *Meteorit. Planet. Sci.* 31, 321–327.

Hiyagon H. and Hashimoto A. (1999) ^{16}O excesses in olivine inclusions in Yamato-86009 and Murchison chondrites and their relation to CAIs. *Science* 283, 828–831.

Hollister L. S., Bindi L., Yao N., et al. (2014) Impact-induced shock and the formation of natural quasicrystals in the early solar system. *Nat. Commun.* 5, 4040.

Holmèn B. A. and Wood J. A. (1986) Chondrules that indent one another: Evidence for hot accretion? (abstract). *Meteoritics* 21, 399.

Hood L. L. (1998) Thermal processing of chondrule precursors in planetesimal bow shocks. *Meteorit. Planet. Sci.* 33, 97–107.

Hood L. L. and Horányi M. (1991) Gas dynamic heating of chondrule precursor grains in the solar nebula. *Icarus* 93, 259–269.

Hood L. L. and Horányi M. (1993) The nebular shock wave model for chondrule formation: One-dimensional calculations. *Icarus* 106, 179–189.

Hood L. L. and Kring D. A. (1996) Models for multiple heating mechanisms. In *Chondrules and the Protoplanetary Disk* (edited by Hewins R. H., Jones R. H. and Scott E. R. D.), pp. 265–276, Cambridge University Press, Cambridge.

Hooshanginejad A. and Lee S. (2022) Dynamics of a partially wetting droplet under wind and gravity. *Phys. Rev. Fluids* 7, 033601.

Hopp T., Dauphas N., Spitzer F., Burkhard T. C., and Kleine T. (2022) Earth's accretion inferred from iron isotopic anomalies of supernova nuclear statistical equilibrium origin. *Earth Planet. Sci. Lett.* 577, 117245.

Horan M. F., Walker R. J., Morgan J. W., Grossman J. N., and Rubin A. E. (2003) Highly siderophile elements in chondrites. *Chem. Geol.* 196, 5–20.

Horányi M. and Robertson S. (1996) Chondrule formation in lightning discharges: Status of theory and experiments. In *Chondrules and the Protoplanetary Disk* (edited by Hewins R. H., Jones R. H. and Scott E. R. D.), pp. 303–310, Cambridge University Press, Cambridge.

Horányi M., Morfill G., Goertz C. K., and Levy E. H. (1995) Chondrule formation in lightning discharges. *Icarus* 114, 174–185.

Horejsi M. (2011) The New Concord meteorite: Let's stop kicking a dead horse. *Meteorite Times Magazine*, May 1, 2011.

Horstmann M. and Bischoff A. (2010) Characterization of spectacular lithologies from the Almahata Sitta breccia. *41st Lunar and Planetary Science Conference*, abstract #1784.

Horstmann N. and Bischoff A. (2014) The Almahata Sitta polymict breccia and the late accretion of asteroid TC3. *Chem. Der. Erde*. 74, 149–183.

Horstmann M., Bischoff A., Pack A., and Laubenstein M. (2010) Almahata Sitta – fragment MS-CH: Characterization of a new chondrite type. *Meteorit. Planet. Sci.* 45, 1657–1667.

Hörz F. and Schaal R. B. (1981) Asteroidal agglutinate formation and implications for asteroidal surfaces. *Icarus* 46, 337–353.

Hörz F., Cintala M. J., See T. H., and Le L. (2005) Shock melting of ordinary chondrite powders and implications for asteroidal regoliths. *Meteorit. Planet. Sci.* 40, 1329–1346.

Housen K. R. and Holsapple K. A. (1999) Impact cratering on porous low-density bodies (abstract). *Lunar Planet. Sci.* 30, 1228.

Housen K. R., Wilkening L. L., Chapman C. R., and Greenberg R. (1979) Asteroidal regoliths. *Icarus* 39, 317–351.

Howard K. T., Alexander C. M. O'D., Schrader D. L., and Dyl K. A. (2015) Classification of hydrous meteorites (CR, CM and C2 ungrouped) by phyllosilicate fraction: PSD-XRD modal mineralogy and planetesimal environments. *Geochim. Cosmochim. Acta* 149, 206–222.

Howgego J. (2025) *The Meteorite Hunters*, Oneworld, London, 254 pp.

Hu H.-N. and Taylor L. A. (1977) Lack of chemical fractionation in major and minor elements during agglutinate formation. *Proc. Lunar Sci. Conf. 8th*, 3645–3656.

Hu J. and Sharp T. G. (2017) Back-transformation of high-pressure minerals in shocked chondrites: Low-pressure mineral evidence for strong shock. *Geochim. Cosmochim. Acta* 215, 277–294.

Hu J. and Sharp T. G. (2022) Formation, preservation and extinction of high-pressue minerals in meteorites: Temperature effects in shock metamorphism and shock classification. *Prog. Earth Planet. Sci.* 9, 1–22.

Hughes D. W. (1978a) Chondrule mass distribution and the Rosin and Weibull statistical functions. *Earth Planet. Sci. Lett.* 39, 371–376.

Hughes D. W. (1978b) A disaggregation and thin section analysis of size and mass distributions of the chondrules in the Bjurböle and Chainpur meteorites. *Earth Planet. Sci. Lett.* 38, 391–400.

Hughes D. W. (1981) Meteorite falls and finds: Some statistics. *Meteoritics* 16, 269–281.

Hugo R. C., Ruzicka A. M., and Rubin A. E. (2020) Mesoscale and microscale shock effects in the LL6 S4 chondrites Saint-Séverin and Elbert: A tale of two breccias. *Meteorit. Planet. Sci.* 55, 1418–1438.

Humayun M. and Campbell A. J. (2002) The duration of ordinary chondrite metamorphism inferred from tungsten microdistribution in metal. *Earth Planet. Sci. Lett.* 198, 225–243.

Humboldt A. V. (1849) Cosmos: A Sketch of a Physical Description of the Universe I. H. G. Bohn, pp. 97–212.

Hume D. (1748) *Philosophical Essays Concerning Human Understanding*, A Millar, London.

Hunt A. C., Benedix G. K., Hammond S. J., et al. (2017) A geochemical study of the winonaites: Evidence for limited partial melting and constraints on the precursor composition. *Geochim. Cosmochim. Acta* 199, 13–30.

Huss G. R. (1980) Heterogeneous shock effects in type 3 ordinary chondrites (abstract). *Meteoritics* 15, 305–306.

Huss G. R. (1988) The role of interstellar dust in the formation of the solar system. *Earth, Moon, Planets* 40, 165–211.

Huss G. R. (1990) Meteorite infall as a function of mass: Implications for the accumulation of meteorites on Antarctic ice. *Meteoritics* 25, 41–56.

Huss G. R. (1991) Meteorite mass distributions and differences between Antarctic and non-Antarctic meteorites. *Geochim. Cosmochim. Acta* 55, 105–111.

Huss G. R. (2004) Implications of isotopic anomalies and presolar grains for the formation of the solar system. *Antarctic Meteorit. Res.* 17, 132–152.

Huss G. R. and Alexander E. C. (1987) On the presolar origin of the "normal planetary" noble gas component in meteorites. *Proc. Lunar Planet. Sci. Conf. 17th*, E710–E716.

Huss G. R. and Lewis R. S. (1994) Noble gases in presolar diamonds II: Component abundances reflect thermal processing. *Meteoritics* 29, 811–829.

Huss G. R. and Lewis R. S. (1995) Presolar diamond, SiC, and graphite in primitive chondrites: Abundances as a function of meteorite class and petrologic type. *Geochim. Cosmochim. Acta* 59, 115–160.

Huss G. R., Keil K. and Taylor G. J. (1981) The matrices of unequilibrated ordinary chondrites: Implications for the origin and history of chondrites. *Geochim. Cosmochim. Acta* 45, 33–51.

Huss G. R., Lewis R. S. and Hemkin S. (1996) The "normal planetary" noble gas component in primitive chondrites: Compositions, carrier, and metamorphic history. *Geochim. Cosmochim. Acta* 60, 3311–3340.

Huss G. R., MacPherson G. J., Wasserburg G. J., Russell S. S. and Srinivasan G. (2001) Aluminum-26 in calcium-aluminum-rich inclusions and chondrules from unequilibrated ordinary chondrites. *Meteorit. Planet. Sci.* 36, 975–997.

Huss G. R., Meshik A. P., Smith J. B., and Hohenberg C. M. (2003) Presolar diamond, silicon carbide, and graphite in carbonaceous chondrites: Implications for thermal processing the solar nebula. *Geochim. Cosmochim. Acta* 67, 4823–4848.

Huss G. R., Rubin A. E., and Grossman J. N. (2006) Thermal metamorphism in chondrites. In *Meteorites and the Early Solar System II* (edited by Lauretta D. and McSween H. Y.), pp. 567–586. University of Arizona Press, Tucson.

Huston T. J. and Lipschutz M. E. (1984) Chemical studies of L chondrites, III. Mobile trace elements and $^{40}Ar/^{39}Ar$ ages. *Geochim. Cosmochim. Acta* 48, 1319–1329.

Hutchison R. (1982) Meteorites – Evidence for the interrelationships of materials in the solar system of 4.55 Ga ago. *Phys. Earth Planet. Inter.* 29, 199–208.

Hutchison R. (1987) Chromian-manganoan augite in the interchondrule matrix of the Tieschitz (H3) chondritic meteorite. *Mineral. Mag.* 51, 311–316.

Hutchison R. (1996) Hot accretion of the ordinary chondrites: The rocks don't lie (abstract). *Lunar Planet. Sci.* XXVII, 579–580.

Hutchison R. (2004) *Meteorites: A Petrologic, Chemical and Isotopic Synthesis*, Cambridge University Press, 506 pp.

Hutchison R. and Bevan A. W. R. (1983) Conditions and time of chondrule accretion. In *Chondrules and Their Origins* (edited by King E. A.), pp. 162–179. Lunar and Planetary Institute, Houston, TX.

Hutchison R. and Graham A. L. (1975) Significance of calcium-rich differentiates in chondritic meteorites. *Nature* 255, 471.

Hutchison R., Bevan A. W. R., and Hall J. M. (1977) *Appendix to the Catalogue of Meteorites*, British Museum of Natural History, London, 297 pp.

Hutchison R., Bevan A. W. R., Agrell S. O., and Ashworth J. R. (1979) Accretion temperature of the Tieschitz, H3, chondritic meteorite. *Nature* 280, 116–119.

Hutchison R., Bevan A. W. R., Agrell S. O., and Ashworth J. R. (1980) Thermal history of the H-group of chondritic meteorites. *Nature* 287, 787–790.

Hutchison R., Alexander C. M., and Barber D. J. (1987) The Semarkona meteorite: First occurrence of smectite in an ordinary chondrite and its implications. *Geochim. Cosmochim. Acta* 51, 1875–1882.

Hutchison R., Williams C. T., Din V. K., et al. 1988. A planetary, H-group pebble in the Barwell, L6, unshocked chondritic meteorite. *Earth Planet. Sci. Lett.* 90, 105–118.

Hutchison R., Alexander C. M. O'D., and Bridges J. C. (1998) Elemental redistribution in Tieschitz and the origin of white matrix. *Meteorit. Planet. Sci.* 33, 1169–1179.

Hutson M. L. (1989) Shock effects in H-group chondrites. *Lunar Planet. Sci.* 20, 436.

Hyde B. C., Tait K. T., Moser D. E., Rumble D., and Thompson M. S. (2020) Accretionary mixing of a eucrite impactor and the regolith of the L chondrite parent body. *Meteorit. Planet. Sci.* 55, 20–35.

Iida A., Nakamoto T., Susa H., and Nakagawa Y. (2001) A shock heating model for chondrule formation in a protoplanetary disk. *Icarus* 153, 430–450.

Ikeda Y. (1992) An overview of the research consortium, "Antarctic carbonaceous chondrites with CI affinities, Yamato-86720, Yamato-82162, and Belgica-7904. *Proc. NIPR Symp. Antarct. Meteorit.* 5, 49–73.

Ikeda Y. and Takeda H. (1979) Petrology of the Yamate-74442 chondrite. *Proc. Symp. Antarct. Meteorit. 4th*, 123–139.

Ikeda Y., Kimura M., Mori H., and Takeda H. (1981) Chemical composition of matrices of unequilibrated ordinary chondrites. *Mem. Natl. Inst. Polar Res. Sp. Is.* 20, 124–144.

Ikeda Y., Prinz M., and Nehru C. E. (2000) Lithic and mineral clasts in the Dar al Gani (DAG) 319 polymict ureilite. *Antarctic Met. Res.* 13, 177–221.

Ikeda Y., Kita N. T., Morishita Y., and Weisberg M. K. (2003) Primitive clasts in the Dar al Gani 319 polymict ureilite: Precursors of the ureilites. *Ant. Meteorite Res.* 16, 105–127.

Ikramuddin M., Matza S., and Lipschutz M. E. (1977) Thermal metamorphism of primitive meteorites. V. Ten trace elements in Tieschitz H3 chondrite heated at 400–1000°C. *Geochim. Cosmochim. Acta* 41, 1247–1256.

Isa J., Rubin A. E., and Wasson J. T. (2014) R-chondrite bulk-chemical compositions and diverse oxides: Implications for parent-body processes. *Geochim. Cosmochim. Acta* 124, 131–151.

Ito M., Tomioka N., Uesugi M., et al. (2022) A pristine record of outer Solar System materials from asteroid Ryugu's returned sample. *Nature Astron.* 6, 1163–1171. https://doi.org/10.1038/s41550-022-01745-5.

Itoh S., Russell S. S., and Yurimoto H. (2007) Oxygen and magnesium isotopic compositions of amoeboid olivine aggregates from the Semarkona LL3.0 chondrite. *Meteorit. Planet. Sci.* 42, 1241–1247.

Ivanova M. A., Lorenz C. A., Humayun M., et al. (2019) Properties of a new grouplet of G metal-rich chondrites. *82nd Annu. Meet. Meteorit. Soc.*, abstract#6143.

Ivanova M. A., Lorenz C. A., Humayun M., et al. (2020) Sierra Gorda 009: A new member of the metal-rich G chondrites grouplet. *Meteorit. Planet. Sci.* 55, 1764–1792.

Ivanova M. A., Britvin S. N., Gulyaeva R. I., et al. (2025) Cafeosite, $Ca_4Fe^{2+}{}_3Fe^{3+}{}_2\square O_6S_4$, a new meteoritic oxysulfide, a redox indicator of metamorphic alteration of carbonaceous asteroids. *Meteorit. Planet. Sci.* 60, 175–189.

Izawa M. R. M., Flemming R. L., Banerjee N. R., and McCausland P. J. A. (2011) Micro-X-ray difraction assessment of shock stage in enstatite chondrites. *Meteorit. Planet. Sci.* 46, 638–651.

Jacobsen B., Yin Q., Moynier F., et al. (2008) ^{26}Al-^{26}Mg and ^{207}Pb-^{206}Pb systematics of Allende CAIs: Canonical solar initial ^{26}Al/^{27}Al ratio reinstated. *Earth Planet. Sci. Lett.* 272, 353–364.

Jacquet E. (2014) The quasi-universality of chondrule size as a constraint for chondrule formation models. *Icarus* 232, 176–186.

Jacquet E. (2021) Collisions and compositional variability in chondrule-forming events. *Geochim. Cosmochim. Acta* 296, 18–37. https://doi.org/10.1016/j.gca.2020.12.025.

Jacquet E. and Doisneau B. (2024) The secondary classification of unequilibrated chondrites. *Meteorit. Planet. Sci.* 59, 3150–3180. https://doi.org/10.1111/maps.14270.

Jacquet E. and Robert F. (2013) Water transport in protoplanetary disks and the hydrogen isotopic composition of chondrites. *Icarus* 223, 722–732. https://doi.org/10.1016/j.icarus.2013.01.022.

Jamsja N. and Ruzicka A. (2010) Shock and thermal history of Northwest Africa 4859, an annealed impact-melt breccia of LL chondrite parentage containing unusual igenous features and pentlandite. *Meteorit. Planet. Sci.* 45, 828–849.

Jaques A. L. and Fitzgerald M. J. (1982) The Nilpena ureilite, an unusual polymict breccia: Implications for origin. *Geochim. Cosmochim. Acta* 46, 893–900.

Jaret S. J., Woerner W. R., Phillips B. L., et al. (2015) Maskelynite formation via solid-state transformation: Evidence of infrared and X-ray anisotropy. *J. Geophys. Res. Planets* 120, 570–587.

Jarosewich E. (1990) Chemical analyses of meteorites: A compilation of stony and iron meteorite analyses. *Meteoritics* 25, 323–337.

Javoy M. (1995) The integral enstatite chondrite model of the earth. *Geophys. Res. Lett.* 22, 2219–2222.

Javoy M., Kaminski E., Guyot F., et al. (2010) The chemical composition of the Earth: Enstatite chondrite models. *Earth Planet. Sci. Lett.* 293, 259–268.

Jenniskens P. and Devillepoix A. R. (2025) Review of asteroid, meteor, and meteorite-type links. *Meteorit. Planet. Sci.* 60, 928–973.

Jenniskens P., Betlem H., Betlem J., et al. (1994) The Mbale meteorite shower. *Meteoritics* 29, 246–254.

Jenniskens P., Shaddad M. H., Numan D., et al. (2009) The impact and recovery of asteroid 2008 TC_3. *Nature*, 458, 485–488.

Jenniskens P., Gabadirwe M., Yin Q.-Z., et al. (2021) The impact and recovery of asteroid 2018 LA. *Meteorit. Planet. Sci.* 56, 844–893.

Jin Z. and Bose M. (2019) New clues to ancient water on Itokawa. *Sci. Adv.*, eaav8106. https://doi.org/10.1126/sciadv.aav8106.

Johansen A. and Lambrechts M. (2017) Forming planets via pebble accretion. *Ann. Rev. Earth Planet. Sci.* 45, 359–387.

Johansen A. and Okuzumi S. (2018) Harvesting the decay energy of ^{26}Al to drive lightning discharge in protoplanetary disks. *Astron. Astrophys.* 609, 22 pp.

Johansen A. and Youdin A. (2007) Protoplanetary disk turbulence driven by the streaming instability: Nonlinear saturation and particle concentation. *Astrophys. J.* 662, 627–641.

Johansen A., Oishi J. S., Mac Low M.-M., et al. (2007) Rapid planetesimal formation in turbulent circumstellar disks. *Nature* 448, 1022–1025.

Johansen A., Youdin A., and Klahr H. (2009) Zonal flows and long-lived axisymmetric pressure bumps in magnetorotational turbuilence. *Astrophys. J.* 697, 1269–1289.

Johansen A., Blum J., Tanaka H., et al. (2014) The multifaceted planetesimal formation process. In *Protostars and Planets VI* (edited by Beuther H., Klessen R. S., Dullemond C. P. and Henning T.), pp. 547–570, University of Arizona Press, Tucson.

Johansen A., Jacquet E., Cuzzi J. N., Morbidelli A., and Gounelle M. (2015) New paradigms for asteroid formation. In *Asteroids IV* (edited by Michel P., Demeo F. E. and Bottke W. F.), pp. 471–492, University of Arizona Press, Tucson.

Johnson B. C., Minton D. A., Melosh H. J., and Zuber M. T. (2015) Impact jetting as the origin of chondrules. *Nature* 517, 339–341.

Johnson J. M., Zolensky M. E., Chan Q., and Kring D. A. (2016) Intriguing dehydrated phyllosilicates found in an unusual clast in the LL3.15 chondrite North West Africa 6925. *Lunar Planet. Sci.* 47, abstract#1608.

Johnson J. R., Bell J. F., Gasnault O., et al. (2014) First iron meteorites observed by the Mars Science Laboratory (MSL) rover Curiosity. *Amer. Geophys. Union, Fall Meet. 2014*, abstract id. P51E-3989.

Johnson S. (1755) *A Dictionary of the English Language*, W. Strahan, London, 2348 pp.

Johnson T. E., Benedix G. K. and Bland P. A. (2016) Metamorphism and partial melting of ordiunary chondrites: Calculated phase equilibria. *Earth Planet. Sci. Lett.* 433, 21–30.

Jones J. H. and Drake M. J. (1983) Experimental investigations of trace element fractionations in iron meteorites. II: The influence of sulfur. *Geochim. Cosmochim. Acta* 47, 1199–1209.

Jones R. H. (1990) Petrology and mineralogy of Type II, FeO-rich chondrules in Semarkona (LL3.0): Origin by closed-system fractional crystallization, with evidence for supercooling. *Geochim. Cosmochim. Acta* 54, 1785–1802.

Jones R. H. (1994) Petrology of FeO-poor, pórphyritic pyroxene chondrules in the Sermarkona chondrite. *Geochim. Cosmochim. Acta* 58, 5325–5340.

Jones R. H. (1996) FeO-rich, porphyritic pyroxene chondrules in unequilibrated ordinary chondrites. *Geochim. Cosmochim. Acta* 60, 3115–3138.

Jones R. H. (1996) Relcit grains in chondrules: Evidence for chondrule recycling. In *Chondrules and the Protoplanetary Disk* (edited by Hewins R. H., Jones R. H. and Scott E. R. D.), pp. 163–172. Cambridge University Press, Cambridge.

Jones R. H. and Danielson L. R. (1997) A chondrule origin for dusty relict olivine in unequilibrated chondrites. *Meteorit. Planet. Sci.* 32, 753–760.

Jones R. H. and Scott E. R. D. (1989) Petrology and thermal history of Type IA chondrules in the Semarkona (LL3.0) chondrite. *Proceedings Lunar and Planetary Science Conference 19th*, 523–536.

Jones R. H., McCubbin F. M., Dreeland L., et al. (2014) Phosphate minerals in LL chondrites: A record of the action of fluids during metamorphism on ordinary chondrite parent bodies. *Geochim. Cosmochim. Acta* 132, 120–140.

Jones R. H., McCubbin F. M. and Guan Y. (2016) Phosphate minerals in the H-group of ordinary chondrites, and fluid activity recorded by apatite heterogeneity in the Zag H3-6 regolith breccia. *Amer. Mineral.* 101, 2452–2467.

Jonker G., van Maldeghem F., van Ginneken M., Krämer Ruggiu L., and Goderis S. (2024) Oxygen isotopic compositions of fresh rooftop micrometeorites from the Budel collection – Insights into the contemporary cosmic dust flux. *Meteorit. Planet. Sci.* 59, 1236–1257.

Jopek T. J., Valsecchi G. B. and Froeschlé C. (2002) Asteroid meteoroid streams. In *Asteroids III* (edited by Bottke W. F., Cellino A., Paolicchi P. and Binzel R. P.), pp. 645–652. University of Arizona Press, Tucson.

Joswiak D. J., Brownlee D. E., Nguyen A. N., and Messenger S. (2017) Refractory maerials in comet samples. *Meteorit. Planet. Sci.* 52, 1612–1648.

Joswiak D. J., Brownlee D. E., Westphal A. J., et al. (2024) Compositional evidence for chondrule origins of low-Ca pyroxenes in comet Wild 2 and a giant cluster IDP. *Meteorit. Planet. Sci.* 59, 1790–1819.

Jourdan F., Timms N. E., Eroglu E., et al. (2017) Collisional history of asteroid Itokawa. *Geology* 45, 819–822.

Jourdan F., Timms N. E., Nakamura T., et al. (2023) Rubble pile asteroids are forever. *Proc. Nat. Acad. Sci.* 120, e2214353120.

Jurewicz A. J. G., Mittlefehldt D. W., and Jones J. H. (1995) Experimental partial melting of the St. Severin (LL) and Lost City (H) chondrites. *Geochim. Cosmochim. Acta* 59, 391–408.

Jurewicz A. J. G., Amarsi A. M., Burnett D. S., and Grevesse N. (2024) Differences in elemental abundances between CI chondrites and the solar photosphere. *Meteorit. Planet. Sci.* 59, 3193–3214.

Kadlag Y., Haberthür D., Leya I., Hlushchuk R., and Mezger K. (2023) Physical properties and average atomic numbers of chondrules using compuated tomography. *Planet. Space Sci.* 238, 105799, 11 pp.

Kaeter D., Ziemann M. A., Böttger U., et al. (2017) The Chelyabinsk meteorite: New insights from a comprehensive electron microscopy and Raman spectroscopy study with evidence for graphite in olivine of ordinary chondrites. *Meteorit. Planet. Sci.* 53, 416–432.

Kallemeyn G. W., Boynton W. V., Willis J., and Wasson J. T. (1978) Formation of the Bencubbin polymict meteoritic breccia. *Geochim. Cosmochim. Acta* 42, 507–515.

Kallemeyn G. W., Rubin A. E., Wang D., and Wasson J. T. (1989) Ordinary chondrites: Bulk compositions, classification, lithophile-element fractionations and composition-petrographic type relationships. *Geochim. Cosmochim. Acta* 53, 2747–2767.

Kallemeyn G. W., Rubin A. E., and Wasson J. T. (1991) The compositional classification of chondrites: V. The Karoonda (CK) group of carbonaceous chondrites. *Geochim. Cosmochim. Acta* 55, 881–892.

Kallemeyn G. W., Rubin A. E., and Wasson J. T. (1994) The compositional classification of chondrites: VI. The CR carbonaceous chondrite group. *Geochim. Cosmochim. Acta* 58, 2873–2888.

Kallemeyn G. W., Rubin A. E., and Wasson J. T. (1996) The compositional classification of chondrites VII. The R chondrite group. *Geochim. Cosmochim. Acta* 60, 2243–2256.

Kamp P. J. J. and Lowe D. J. (1982) Geology and terrestrial age of the Derrik Peak meteorite occurrence, Antarctica. *Meteoritics* 17, 119–127.

Kaneko H., Sato K., Ikeda C., and Nakamoto T. (2023) Cooling rates of chondrules after lightning discharege in solid-rich environments. *Astrophys. J.* 947, 22 pp.

Kasen D., Metzger B., Barnes J., Quataert E., and Ramirez-Ruiz E. (2017) Origin of the heavy elements in binary neutron-star mergers from a gravitational-wave event. *Nature* 551, 80–84.

Kato T. and Kumazawa M. (1985) Garnet phase of $MgSiO_3$ filling the pyroxene–ilmenite gap at very high temperature. *Nature* 316, 803–805.

Keay C. S. L. (1980) Anomalous sounds from the entry of meteor fireballs. *Science* 210, 11–15.

Keay C. S. L. (1992) Electrophonic sounds from large meteor fireballs. *Meteoritics* 27, 144–148.

Kegler P. and Holzheid A. (2011) Determination of the formal Ge-oxide species in silicate melts at oxygen fugacities applicable to terrestrial core formation scenarios. *Eur. J. Mineral.* 23, 369–378.

Keil K. (1962) Quantitative-erzmikroskopishe Integrationsanalyse der Chondrite. *Chem. der Erde* 22, 281–348.

Keil K. (1968) Mineralogical and chemical relationships among enstatite chondrites. *J. Geophys. Res.* 73, 6945–6976.

Keil K. (1969) Meteorite composition. In *Handbook of Geochemistry* (edited by Wedepohl K. H.), pp. 78–115, Springer, Cham, Switzerland.

Keil K. (1982) Composition and origin of chondritic breccias. In *Workshop on Lunar Breccias and Soils and Their Meteoritic Analogs*. LPI Tech. Rept. 82-02, pp. 65–83.

Keil K. and Fodor R. V. (1973) Composition and origin of lithic fragments in LL-group chondrites (abstract). *Meteoritics* 8, 394–396.

Keil K. and Fodor R. V. (1980) Origin and history of the polymict-brecciated Tysnes Island chondrite and its carbonaceous and non-carbonaceous lithic fragments. *Chem. der Erde* 39, 1–26.

Keil K. and Fredriksson K. (1964) The iron, magnesium, and calcium distribution in coexisting olivines and rhombic pyroxenes of chondrites. *J. Geophys. Res.* 69, 3487–3515.

Keil K. and McCoy T. J. (2018) Acapulcoite-lodranite meteorites: Ultramafic asteroidal partial melt residues. *Chem. er Erde* 78, 153–203.

Keil K. and Papike J. J. (1983) Meteorites from the Earth's Moon. Special session abstracts, *14th Lunar and Planetary Science Conference*, LPI Contribution 501. Lunar and Planetary Institute, Houston.

Keil K., Mason B., Wiik H. B., and Fredriksson K. (1964) The Chainpur meteorite. *Amer. Mus. Novitates* 2173, 1–28.

Keil K., Kurat G., Prinz M., and Green J. A. (1972) Lithic fragments, glasses and chondrules from Luna 16 fines. *Earth Planet. Sci. Lett.* 13, 243–256.

Keil K., Kirchner E., Gomes C. B., Jarosewich E., and Murta R. L. L. (1978) Studies of Brazilian meteorites XIV. Mineralogy, petrology, and chemistry of the Conquista, Minas Gerais, chondrite. *Meteoritics* 13, 177–187.

Keil K., Fodor R. V., Starzyk P. M., et al. (1980) A 3.6-b.y.old impact-melt rock fragment in the Plainview chondrite: Implications for the age of the H-group chondrite parent body regolith formation. *Earth Planet. Sci. Lett.* 51, 235–247.

Keil K., Conrad G. and Welsh J. E. (1981) The Beaver-Harison, Beaver County, Utah L6 chondrite. *Meteoritics* 16, 13–16.

Keil K., Conrad G. H., King E. A., and San Miguel A. (1986) Petrology and classification of the Garraf, Spain chondrite. *Meteoritics* 21, 125–129.

Keil K., Stöffler D., Love S. G., and Scott E. R. D. (1997) Constraints on the role of impact heating and melting in asteroids. *Meteorit. Planet. Sci.* 32, 349–363.

Keller L. P. (1998) A transmission electron microscope study of iron-nickel carbides in the matrix of the Semarkona unequilibrated ordinary chondrite. *Meteorit. Planet. Sci.* 33, 913–919.

Kelley M. C. and Price C. (2017) On the electrophonic generation of audio frequency sound by meteors. *Geophys. Res. Lett.* 44, 2987–2990.

Kempe W. and Müller O. (1969) The stony meteorite Krähenberg: Its chemical composition and the Rb-Sr age of the light and dark portions. In *Meteorite Research* (edited by Millman P. M.), pp. 418–428, D. Reidel Pub. Co., Dordrecht, Holland.

Kempton R. (1996) Abee – more questions than answers. *Meteorite* 2, 18–19.

Kenkmann T., Artemieva N. A., Wünnemann K., et al. (2009) The Carancas meteorite impact crater, Peru: Geologic surveying and modeling of crater formation and atmospheric passage. *Meteorit. Planet. Sci.* 44, 985–1000.

Kennedy A. K., Hutchison R., Hutcheon I. D., and Agrell S. O. (1992) A unique high Mn/Fe microgabbros in the Parnallee (LL3) ordinary chondrite: Nebular mixture or planetary differentiate from a previously unrecognized planetary body? *Earth Planet. Sci. Lett.* 113, 191–205.

Kerraouch I., Ebert S., Patzek M., et al. (2019) A light, chondritic xenolith in the Murchison (CM) chondrite – Formation by fluid-assisted percolation during metasomatism? *Geochemistry* 79, 125518.

Kerraouch I., Bischoff A., Krot A. N., et al. (2025) Hydrated chondritic clasts challenge secular evolution models of the non-carbonaceous reservoir. *Meteorit. Planet. Sci.* 60, abstract.

Kerridge J. F. (1979) Fractionation of refractory lithophile elements among chondritic meteorites. *Proc. Lunar Planet. Sci. Conf. 10th*, 989–996.

Kerridge J. F. and Kieffer S. W. (1977) A constraint on impact theories of chondrule formation. *Earth Planet. Sci. Lett.* 35, 35–42.

Kessel R., Beckett J. R., and Stolper E. M. (2007) The thermal history of equilibrated ordinary chondrites and the relationship between textural maturity and temperature. *Geochim. Cosmochim. Acta* 71, 1855–1881.

Kichinka K. (2005) The art of collecting meteorites, *Bookmasters*, Ashland, Ohio, 219 pp.

Kieffer S. W. (1975a) Droplet chondrules. *Science* 189, 333–340.

Kieffer S. W. (1975b) From regolith to rock by shock. *Moon* 13, 301–320.

Killgore K. and Killgore M. (2002) *Southwest Meteorite Collection: A Pictorial Catalog*, Southwest Meteorite Press, Payson, Arizona, 201 pp.

Kimura M., Hiyagon H., Palme H., et al. (2002) Yamato 792947, 793408 and 82038: The most primitive H chondrites, with abundant refractory inclusions. *Meteorit. Planet. Sci.* 37, 1417–1434.

Kimura M., Grossman J. N., and Weisberg M. K. (2008) Fe-Ni metal in primitive chondrites: Indicators of classification and metamorphic conditions for ordinary and CO chondrites. *Meteorit. Planet. Sci.* 43, 1161–1177.

Kimura M., Weisberg M. K., and Yamaguchi A. (2024) Subtype 3.0 chondrites: Petrologic classification criteria. *Meteorit. Planet. Sci.* 59, 858–877.

King A. J., Bates H. C., Krietsch D., et al. (2019) The Yamato-type (CY) carbonaceous chondrite group: Analogues for the surface of asteroid Ryugu? *Geochemistry* 79, 125531.

King E. A., Butler J. C., and Carman M. F. (1972a) Chondrules in Apollo 14 samples and size analyses of Apollo 14 and 15 fines. *Proc. Lunar Sci. Conf. 3rd*, 673–686.

King E. A., Carman M. F., and Butler J. C. (1972b) Chondrules in Apollo 14 samples: Implications for the origin of chondritic meteorites. *Science* 175, 59–60.

Kingsley C. (1855) *Westward Ho! Voyages and Adventures of Sir Amyas Leigh, Knight, of Burroughs, in the County of Devon, in the Reign of Her Most Glorious Majesty Queen Elizabeth*, A. L. Burt, New York, 634 pp.

Kirkwood D. (1866) On the theory of meteors. *Proceedings of American Association for the Advancement of Science for 1866*, 8–14.

Kirkwood D. (1867) *Meteoric Astronomy: A Treatise on Shooting Stars, Fire-balls and Aerolites*, J. B. Lippincott & Co., Philadelphia, 129 pp.

Kita N. T. and Ushikubo T. (2012) Evolution of protoplanetary disk inferred from ^{26}Al chronology of individual chondrules. *Meteorit. Planet. Sci.* 47, 1108–1119.

Kita N. T., Nagahara H., Togashi S., and Morishita Y. (2000). A short duration of chondrule formation in the solar nebula: evidence from ^{26}Al in Semarkona ferromagnesian chondrules. *Geochim. Cosmochim. Acta* 64, 3913–3922.

Kita N. T., Huss G. R., Tachibana S., et al. (2005) Constraints on the origin of chondrules and CAIs from short-lived radionuclides. In *Chondrites and the Protoplanetary Disk* (edited by Krot A. N., Scott E. R. D. and Reipurth B.), pp. 558–587. ASP Conference Series 341, Astronomical Society of the Pacific, San Francisco.

Kita N. T., Nagahara H., Tachibana S., et al. (2010) High precision SIMS oxygen three isotope study of chondrules in LL3 chondrites: Role of ambient gas during chondrule formation. *Geochim. Cosmochim. Acta* 74, 6610–6635.

Kleine T., Mezger K., Palme H., Scherer E., and Münker C. (2005) Early core formation in asteroids and late accretion of chondrite parent bodies: Evidence from ^{182}Hf-^{182}W in CAIs, metal-rich chondrites, and iron meteorites. *Geochim. Cosmochim. Acta* 69, 5805–5818.

Kleine T., Touboul M., Van Orman J. A., et al. (2008) Hf-W thermochronometry: Closure temperature and constraints on the accretion and cooling history of the H chondrite parent body. *Earth Planet. Sci. Lett.* 270, 106–118.

Klöck W., Thomas K. L., McKay D. S., and Palme H. (1989) Unusual olivine and pyroxene composition in interplanetary dust and unequilibrated ordinary chondrites. *Nature* 339, 126–128.

Kodolányi J., Hoppe P., Vollmer C., Berndt J., and Müller M. (2022) Iron-60 in the early solar system revisited: Insights from in situ isotope analysis of chondritic troilite. *Astrophys. J.* 929, 19 pp. https://doi.org/10.3847/1538-4357/ac5910.

Koeberl C., Reimold W. U., Horsch H. E., and Merkle R. K. W. (1990) New mineralogical and chemical data on the Machinga (L6) chondrite, Malawi. *Meteoritics* 25, 23–26.

Koeberl C., Armstrong R. A., and Reimold W. U. (1997) Morokweng, South Africa: A large impact structure of Jurassic-Cretaceous boundary age. *Geology* 25, 731–734.

Koeberl C., Shukolyukov A., and Lugmair G. W. (2007) Chromium isotopic studies of terrestrial impact craters: Identification of meteoritic components at Bosumtwi, Clearwater East, Lappajärvi, and Rochechouart. *Earth Planet. Sci. Lett.* 256, 534–546.

Koefoed P. and Wang K. (2025) Unravelling chondrule formation processes: Clues from the potassium isotopic composition of chondrules from unequilibrated ordinary chondrites. *Geochim. Cosmochim. Acta* 398, 163–177.

Kohout T., Kallonen A., Suuronen J.-P., et al. (2014) Density, porosity, mineralogy, and internal structure of cosmic dust and alteration of its properties during high-velocity atmospheric entry. *Meteorit. Planet. Sci.* 49, 1157–1170.

Kong P., Ebihara M., and Xie X. (1998) Revaluation of formation of metal nodules in ordinary chondrites. *Meteorit. Planet. Sci.* 33, 993–998.

Kong P., Deloule E., and Palme H. (2000) REE-bearing sulfide in Bishunpur (LL3.1), a highly unequilibrated ordinary chondrite. *Earth Planet. Sci. Lett.* 177, 1–7.

Konig H., Keil K., Hintenberger H., Wlotzka F., and Begemann F. (1961) Untersuchungen an Steinmeteoriten mit extrem hohem Edelgasgehalt. I. Der Chondrit Pantar. *Zeitschr. Naturforsch* 16a, 1124–1130.

Konig H., Keil K., and Hintenberger H. (1962) Untersuchungen an Steinmeteoriten mit extrem hohem Edelgasgehalt. II. Der Chondrit Tabor. *Zeitschr. Naturforsch* 17a, 357–358.

Korochantseva E. V., Trieloff M., Lorenz C. A., et al. (2007) L-chondrite asteroid breakup tied to Ordovician meteorite shower by multiple isochron ^{40}Ar-^{39}Ar dating. *Meteorit. Planet. Sci.* 42, 113–130.

Korotev R. (n.d.) Some Meteorite Information. https://sites.wustl.edu/meteoritesite/items/meteorite-testing/#XRFgun.

Korschinek G., Bergmaier A., Faestermann T., et al. (2010) A new value for the half-life of ^{10}Be by heavy-ion elastic recoil detection and liquid scintillation counting. *Nuclear Instrum. Methods Phys. Res. Sec. B: Beam Interact Mater. Atoms* 268, 187–191.

Kovach H. A. and Jones R. H. (2010) Feldspar in type 4–6 ordinary chondrites: Metamorphic processing on the H and LL chondrite parent bodies. *Meteorit. Planet. Sci.* 45, 246–264.

Kracher A., Keil K., Kallemeyn G. W., et al. (1985) The Leoville (CV3) accretionary breccia. *J. Geophys. Res.* 90 (S01), 123–135. https://doi.org/10.1029/JB090iS01p00123.

Kring D. A., Swindle T. D., Britt D. T., and Grier J. A. (1996) Cat Mountain: A meteoritic sample of an impact-melted asteroid regolith. *J. Geophys. Res.* 101, 29,353–29,371.

Kring D. A., Hill D. H., Gleason J. D., et al. (1999) Portales Valley: A meteoritic sample of the brecciated and metal-veined floor of an impact crater on an H chondrite asteroid. *Meteorit. Planet. Sci.* 34, 663–669.

Kring D. A., Jull A. J. T., McHargue L. R., et al. (2001) Gold Basin meteorite strewn field, Mojave Desert, northwestern Arizona: Relic of a small late Pleistocene impact event. *Meteorit. Planet. Sci.* 36, 1057–1066.

Kronk G. W. (1988) *Meteor Showers: A Descriptive Catalog*. Enslow Publishers, Hillside, New Jersey, 291 pp.

Krot A. N. and Rubin A. E. (1993) Chromite-rich mafic silicate chondrules in ordinary chondrites: Formation by impact melting. *Lunar Planet. Sci.* 24, 827–828.

Krot A. N. and Rubin A. E. (1994) Glass-rich chondrules in ordinary chondrites. *Meteoritics* 29, 697–706.

Krot A. N. and Rubin A. E. (1996) Microchondrule-bearing chondrule rims: Constraints on chondrule formation. In *Chondrules and the Protoplanetary Disk* (edited by R. H. Hewins, R. H. Jones and E. R. D. Scott), pp. 181–184, Cambridge University Press.

Krot A. N. and Wasson J. T. (1994) Silica-merrihueite/roedderite-bearing chondrules and clasts in ordinary chondrites: New occurrences and possible origin. *Meteorit. Planet. Sci.* 29, 707–718.

Krot A. N. and Wasson J. T. (1995) Igneous rims on low-FeO and high-FeO chondrules in ordinary chondrites. *Geochim. Cosmochim. Acta* 59, 4951–4966.

Krot A. N., Rubin A. E., and Kononkova N. N. (1993) First occurrence of pyrophanite ($MnTiO_3$) and baddeleyite (ZrO_2) in an ordinary chondrite. *Meteoritics* 28, 232–239.

Krot A. N., Rubin A. E., Keil K., and Wasson J. T. (1997a) Microchondrules in ordinary chondrites: Implications for chondrule formation. *Geochim. Cosmochim. Acta*, 61, 463–473.

Krot A. N., Zolensky M. E., Wasson J. T., et al. (1997b) Carbide-magnetite assemblages in type-3 ordinary chondrites. *Geochim. Cosmochim. Acta* 61, 219–237.

Krot A. N., Weber D., Greshake A., et al. (1999) Relic Ca, Al-rich inclusions in chondrules from the carbonaceous chondrites Acfer 182 and Acfer 094. *Lunar Planet. Sci.* 30, abstract#1511.

Krot A. N., Meibom A., and Keil K. (2000) A clast of Bali-like oxidized CV material in the reduced CV chondrite breccia Vigarano. *Meteorit. Planet. Sci.* 35, 817–825.

Krot A. N., Libourel G., Goodrich C. A., and Petaev M. I. (2004a) Silica-rich igneous rims around magnesian chondrules in CR carbonaceous chondrites: Evidence for condensation origin from fractionated nebular gas. *Meteorit. Planet. Sci.* 39, 1931–1955.

Krot A. N., Petaev M. I., Russell S. S., et al. (2004b) Amoeboid olivine aggregates and related objects in carbonaceous chondrites: Records of nebular and asteroid processes. *Chem. der Erde* 64, 185–239.

Krot A. N., Keil K., Goodrich C. A., and Scott E. R. D. (2005a) Classification of meteorites. *Meteorites, Comets, and Planets* (edited by Davis A. M.), pp. 83–128, Elsevier, Amsterdam.

Krot A. N., Amelin Y., Cassen P., and Meibom A. (2005b) Young chondrules in CB chondrites from a giant impact in the early Solar System. *Nature* 436, 989–992.

Krot A. N., Hutcheon I. D., Brearley A. J., et al. (2006) Timescales and settings for alteration of chondritic meteorites. In *Meteorites and the Early Solar System II* (edited by Lauretta D. S. and McSween H. Y.), pp. 525–553, University of Arizona Press, Tucson.

Krot A. N., Amelin Y., Bland P., et al. (2009) Origin and chronology of chondritic components: A review. *Geochim. Cosmochim. Acta* 73, 4963–4997.

Krot A. N., Nagashima K., Wasserburg G. J., et al. (2014a) Calcium-aluminum-rich inclusions with fractionation and unknown nuclear effects (FUN CAIs): I. Mineralogy, petrology, and oxygen isotopic compositions. *Geochim. Cosmochim. Acta* 145, 206–247.

Krot A. N., Keil K., Scott E. R. D., Goodrich C. A., and Weisberg, M. K. (2014b) Classification of meteorites and their genetic relationships. In *Treatise on Geochemistry* 2nd edition (edited by Davis A. M.), 63 pp., Volume 1: Meteorites and Cosmochemical Processes, Elsevier, Oxford.

Krot A. N., Nagashima K., Alexander C. M. O'D., et al. (2015) Sources of water and aqueous activity on the chondrite parent asteroids. In *Asteroids IV* (edited by Michel P., DeMeo F. E. and Bottke W. F.), pp. 635–660, University of Arizona Press, Tucson.

Krot A. N., Nagashima K., van Kooten E. M. M., and Bizzarro M. (2017) Calcium-aluminum-rich inclusions recycled during formation of porphyritic chondrules from CH carbonaceous chondrites. *Geochim. Cosmochim. Acta* 201, 185–223.

Kruijer T. S. and Kleine T. (2019) Age and origin of IIE iron meteorites inferred from Hf-W chronology. *Geochim. Cosmochim. Acta* 262, 92–103.

Kruijer T. S., Burkhardt C., Budde G., and Kleine T. (2017a) Age of Jupiter inferred from the distinct genetics andformation times of meteorites. *Proc. Natl. Acad. Sci.* 114, 6712–6716.

Kruijer T. S., Kleine T., Burkhardt C., and Budde G. (2017b) Dating the formation of Jupiter through W and Moisotope analyses of meteorites (abstract #1386). *48th Lunar and Planetary Science Conference*. CD-ROM.

Kruijer T. S., Burkhardt C., Budde G., and Kleine T. (2017c) Dating the formation of Jupiter using the distinct genetics and formation times of meteorites. *Annu. Meet. Meteorit. Soc.* abstract #6333.

Krumbein W. C. and Tisdel F. W. (1940) Size distributions of source rocks of sediments. *Amer. J. Sci.* 238, 296–305.

Kuebler K. E., McSween H. Y., Carlson W. D., and Hirsch D. (1999) Sizes and masses of chondrules and metal-troilite grains in ordinary chondrites: Possible implications for nebula sorting. *Icarus* 141, 96–106.

Kullerud G. (1963) The Fe-Ni-S system. *Carnegie Inst. Wash. Yr. Bk.* 62, 175–189.

Kunihiro T., Rubin A. E., McKeegan K. D., and Wasson J. T. (2004) Initial $^{26}Al/^{27}Al$ in carbonaceous-chondrite chondrules: Too little ^{26}Al to melt asteroids. *Geochim. Cosmochim. Acta* 68, 2947–2957.

Kunz J., Falter M., and Jessberger E. K. (1997) Shocked meteorites: Argon-40-argon-39 evidence for multiple impacts. *Meteorit. Planet. Sci.* 32, 647–670.

Kurahashi E., Kita N. T., Nagahara H., and Morishita Y. (2008) ^{26}Al-^{26}Mg systematics of chondrules in a primitive CO chondrite. *Geochim. Cosmochim. Acta* 72, 3865–3882.

Kurat G. (1969) The formation of chondrules and chondrites and some observations on chondrules from the Tieschitz meteorite. In *Meteorite Research* (edited by Millman P.), pp. 185–190. D. Reidel Publ. Co., Dordrecht, Netherlands.

Kurat G. (1982) Impact origin of chondrules (abstract). In *Papers Presented to the Conference on Chondrules and their Origins*, p. 36, Lunar and Planetary Institute, Houston.

Kushiro I. and Mysen B. O. (1979) Melting experiments on a Yamato chondrite. *Proc. 4th Symp. Antarctic Met., Mem. Natl. Inst. Polar Res.* 15, 165–170.

Kyte F. T. (1998) A meteorite from the Cretaceous/Tertiary boundary. *Nature* 396, 237–239.

Kyte F. T. (2002) Unmelted meteoritic debris collected from Eltanin ejecta in Polarstern cores from expedition ANT XII/4. *Deep-Sea Res. II* 49, 1063–1071.

Laity J. E. and Bridges N. T. (2009) Venifacts on Earth and Mars: Analytical, field, and laboratory studies supporting sand abrasion and windward feature development. *Geomorphology* 105, 202–217.

Lal D. (1980) Surface evolution records in chondrites and lunar regolith: Implications to early accretion in the solar system. In *Early Solar System Processes and the Present Solar System* (edited by Lal D.), pp. 219–238, North-Holland Publishing, Amsterdam.

Lal D. and Rajan R. S. (1969) Observations relating to space irradiation of individual crystals of gas-rich meteorites. *Nature* 223, 269–271.

Lamar D. L. and Romig M. F. (1964) Anomalous sounds and electromagnetic effects associated with fireball entry. *Meteoritics* 2, 127–136.

Lambert P., Lewis C., and Moore C. B. (1984) Repeated shock and thermal metamorphism of the Abernathy meteorite. *Meteorit. Planet. Sci.* 19, 29–48.

Lambrechts M., Johansen A., and Morbidelli A. (2014) Separating gas-giant and ice-giant planets by halting pebble accretion. *Astron. Astrophys.* 572, A35.

Lange D. E., Keil K., and Gomes C. B. (1979) The Mafra meteorite and its lithic clasts: A genomict L-group chondrite breccia (abstract). *Meteoritics* 14, 472–473.

Langenhorst F. (2002) Shock metamorphism of some minerals: Basic introduction and microstructural observations. *Bull. Czech Geol. Surv.* 77, 265–282.

Langenhorst F., Deutsch A., Stöffler D., and Hornemann U. (1992) Effect of temperature on shock metamorphism of single-crystal quartz. *Nature* 356, 507–509.

Langevin Y. and Maurette M. (1981) Grain size and maturity in lunar and asteroidal regoliths (abstract). *Lunar Planet. Sci.* 12, 595–597.

La Paz L. (1951) Injuries from falling meteorites. *Contrib. Meteorit. Soc.* [Meteorit. Planet. Sci.] 5, 75–82.

Lappe S.-C. L. L., Church N. S., Kasama T., et al. (2011) Mineral magnetism of dusty olivine: A credible recorder of pre-accretionary remanence. *Geochem. Geophys. Geosyst.* 12, Q12Z35. https://doi.org/10.1029/2011GC003811.

Larimer J. W. and Anders E. (1967) Chemical fractionation of meteorites, II. Abundance patterns and their interpretation. *Geochim. Cosmochim. Acta* 31, 1239–1270.

Larimer J. W. and Bartholomay H. A. (1983) Gas-dust fractionation and the origin of chondritic meteorites (abstract). *Lunar Planet. Sci.* 14, 423.

Laul J. C., Morgan J. W., Ganapathy R., and Anders E. (1971) Meteoritic material in lunar samples: Characterization from trace elements. *Proc. Lunar Sci, Conf. 2nd*, 1139–1158.

Laul J. C., Ganapathy R., Anders E., and Morgan J. W. (1973) Chemical fractionations in meteorites – VI. Accretion temperatures of H-, LL-, and E-chondrites from abundance of volatile trace elements. *Geochim. Cosmochim. Acta* 37, 329–357.

Lauretta D. S. (2024) *The Asteroid Hunter: A Scientist's Journey to the Dawn of our Solar System*, Grand Central Publishing, New York, 320 pp.

Lauretta D. S. and Killgore M. (2005) *A Color Atlas of Meteorites in Thin Section*, Golden Retriever Publications and Southwest Meteorite Press, Payson, Arizona, 301 pp.

Lauretta D. S., Lodders K., Fegley B., and Kremser D. T. (1997) The origin of sulfide-rimmed metal grains in ordinary chondrites. *Earth Planet. Sci. Lett.* 151, 289–301.

Lauretta D. S., Connolly H. C., Aebersold J. E., et al. (2024) Asteroid (101955) Bennu in the laboratory: Properties of the sample collected by OSIRIS-REx. *Meteorit Planet Sci.* 59, 2453–2486.

Lee D.-C. and Halliday A. N. (2000) Accretion of primitive planetesimals: Hf-W isotopic evidence from enstatite chondrites. *Science* 288, 1629–1631.

Lee M. R., Russell S. S., Arden J. W., and Pillinger C. T. (1995) Nierite (Si_3N_4), a new mineral from ordinary and enstatite chondrites. *Meteoritics* 30, 387–398.

Lee P. (1996) Dust levitation on asteroids. *Icarus* 124, 181–194.

Lee T., Papanastassiou D. A., and Wasserburg G. J. (1976) Demonstration of Mg-26 excess in Allende and evidence for Al-26. *Geophys. Res. Lett.* 3, 41–44.

Le Feuvre and Wieczorek M. A. (2008) Nonuniform cratering of the terrestrial planets. *Icarus* 197, 291–306.

Le Houérou H. N. (1980) The rangelands of the Sahel. *J. Range Manage.* 33, 41–46.

Leili M. H., Chennaoui Aoudjehane H., Devouard B., et al. (2024) Strewn field, mineralogy, and petrology of Al Haggounia 001: A unique enstatite chondrite. *Meteorit. Planet. Sci.* 59, 301–320.

Leonard F. C. (1953) Introducing meteoritics: The journal of the Meteoritical Society and the Institute of Meteoritics of the University of New Mexico. *Meteoritics* 1, 1–4.

Leonard F. C. and Slanin B. (1941) Statistical study of the meteorite falls of the world: 3. Their time distribution. *Popular Astron.* 49, 551–560.

Levin B. Ju. and Simonenko A. N. (1969) Meteorite radiants and orbits. In *Meteorite Research* (edited by Millman P. M.), pp. 552–558, D. Reidel Pub. Co., Dordrecht, Holland.

Levy E. H. and Araki S. (1989) Magnetic reconnection flares in the protoplanetary nebula and the possible origin of meteoritic chondrules. *Icarus* 81, 74–91.

Lewis C. S. (1946) The meteorite. In *The Collected Poems of C. S. Lewis: A Critical Edition* (edited by King D. W.), Kent State University Press, Kent, Ohio, 485 pp.

Lewis J. A. and Jones R. H. (2016) Phosphate and feldspar mineralogy of equilibrated L chondrites: The record of metasomatism during metamorphism in ordinary chondrite parent bodies. *Meteorit. Planet. Sci.* 51, 1886–1913.

Lewis J. A. and Jones R. H. (2019) Primary feldspar in the Semarkona LL3.00 chondrite: Constraints on chondrule formation and secondary alteration. *Meteorit. Planet. Sci.* 54, 72–89.

Lewis J. A. and Jones R. H. (2022) Exsolution in alkali feldspar in ordinary chondrites: Ubiquitous evidence for rapid cooling at high temperatures. *Geochim. Cosmochim. Acta* 321, 293–310.

Lewis J. A., Jones R. H., and Brearley A. J. (2022) Plagioclase alteration and equilibration in ordinary chondrites: Metasomatism during thermal metamorphism. *Geochim. Cosmochim. Acta* 316, 201–229.

Lewis J. S. (1972) Metal/silicate fractionation in the solar system. *Earth Planet. Sci. Lett.* 15, 286–290.

Li Y., Rubin A. E., Hsu W., and Ziegler K. (2020) Early impact events on chondritic parent bodies: Insights from petrography and geochemistry of the LL7 breccia NWA 11004. *J. Geophys. Res. Planets* 125, 16 pp. https://doi.org/10.1029/2019JE006360.

Li Y., Rubin A. E., and Hsu W. (2021) Formation of metallic-Cu-bearing mineral assemblages in type-3 ordinary and CO chondrites. *Amer. Mineral.* 106, 1751–1767.

Li Y., Wang Y., Jiang HJ., et al. (2025) The geochronology and cooling history of type 7 chondrites: Insights into the early impact events on chondritic parent body. *Geochim. Cosmochim. Acta* 389, 280–293.

Li Z., Xu Y.-K., Wang S.-J., et al. (2025) Ni isotopic compositions in shocked ordinary chondrites: Insights into the influence of shock processes. *Meteorit. Planet. Sci.* 60, 1545–1554.

Libourel G. and Portail M. (2018) Chondrules as direct thermochemical sensors of solar protoplanetary disk gas. *Sci. Adv.* 4, eaar3321, 12 pp.

Liffman K. (1992) The formation of chondrules via ablation. *Icarus* 100, 608–620.

Liffman K. and Brown M. J. I. (1996) The protostellar jet model of chondrule formation. In *Chondrules and the Protoplanetary Disk* (edited by Hewins R. H., Jones R. H. and Scott E. R. D.), pp. 285–302, Cambridge University Press, Cambridge.

Li H.-B. (2021) Magnetic fields in molecular clouds – Observation and interpretation. *Galaxies* 9, 41. https://doi.org/10.3390/galaxies9020041.

Lin C., Hollister L. S., MacPherson G. J., et al. (2017) Evidence of cross-cutting and redox reaction in Khatyrka meteorite reveals metallic-Al minerals formed in outer space. *Sci. Rep.*, 7. https://doi.org/10.1038/s41598-017-01445-5.

Lin T. C. and Qun P. (1986) On the formation of regmaglypts on meteorites. *Fluid Dyn. Res.* 1, 191–199.

Lindsley D. H. (1983) Pyroxene thermometry. *Amer. Mineral.* 68, 477–493.

Lingner D. W., Huston T. J., Hutson M., and Lipschutz M. E. (1987) Chemical studies of H chondrites. I: Mobile trace elements and gas retention ages. *Geochim. Cosmochim. Acta* 51, 727–739.

Lipschutz M. E. and Woolum D. S. (1988) Highly labile elements. In *Meteorites and the Early Solar System* (edited by J. F. Kerridge and M. S. Matthews), pp. 462–487. University of Arizona, Tucson.

Lipschutz M. E., Gaffey M. J., and Pellas P. (1989) Meteoritic parent bodies: Nature, number, size and relation to present-day asteroids. In *Asteroids II* (edited by Binzel R. P., Gehrels T. and Matthews M. S.), pp. 740–777, University of Arizona Press, Tucson.

Lipschutz M. E., Zolensky M. E., and Bell M. S. (1999) New petrographic and trace element data on thermally metamorphosed carbonaceous chondrites. *Antarct. Meteorite Res.* 12, 57–80.

Lipschutz M. E., Wolf S. F., Vogt S., et al. (1993) Consotrium study of the unusual H chondrite regolith breccia, Noblesville. *Meteoritics* 28, 528–537.

Llorca J., Trigo-Rodríguez J. M., Ortiz J. L., et al. (2005) The Villalbeto de la Peña meteorite fall: I. Fireball energy, meteorite recovery, strewn field and petrography. *Meteorit. Planet. Sci.* 40, 795–804.

Lodders K. (2000) An oxygen isotope mixing model for the accretion and composition of rocky planets. *Space Sci. Rev.* 92, 341–354.

Lodders K. (2003) Solar system abundances and condensation temperatures of the elements. *Astrophys. J.* 591, 1220–1247.

Lodders K. (2021) Relative atomic solar system abundances, mass fractions, and atomic masses of the elements and their isotopes, composition of the solar photosphere, and compositions of the major chondritic meteorite groups. *Space Sci. Rev.* 217, 1–33. https://doi.org/10/1007/s11214-021-00825-8.

Lodders K. and Amari S. (2005) Presolar grains from meteorites: Remnants from the early times of the solar system. *Chem. Der. Erde. – Geochem.* 65, 93–166.

Lodders K. and Fegley B. (1997) An oxygen isotope model for the composition of Mars. *Icarus* 126, 373–394.

Lofgren G. E. (1989) Dynamic crystallization of chondrule melts of porphyritic olivine composition: Textures experimental and natural. *Geochim. Cosmochim. Acta* 53, 461–470.

Lofgren G. E. (1996) A dynamic crystallization model for chondrule melts. In *Chondrules and the Protoplanetary Disk* (edited by Hewins R. H., Jones R. H. and Scott E. R. D.), pp. 187–196. Cambridge University Press, Cambridge.

Lonsdale J. T. (1937) The Plantersville meteorite, Grimes County, Texas. *Amer. Mineral.* 22, 877–888.

Lorin J. C. and Pellas P. (1979) Preirradiation history of Djermaia (H) chondritic breccia. *Icarus* 40, 502–509.

Lorin J. C., Pellas P., Schultz L., and Singer P. (1970) Evidence for different irradiation histories of xenoliths from gas-rich Djermaia chondrite. *Trans. Amer. Geophys. Union* 51, 340.

Love S. G. and Pettit D. R. (2004) Fast, repeatable clumping of solid particles in microgravity. *Lunar Planet. Sci.* 35., abstract#1119.

Love S. G., Hörz F., and Brownlee D. E. (1993) Target porosity effects in impact cratering and collisional disruption. *Icarus* 105, 216–224.

Lovering J. F. (1962) The evolution of the meteorites – evidence for the co-existence of chondritic, achondritic and iron meteorites in a typical parent meteorite body. In *Researches on Meteorites* (edited by Moore C. B.), pp. 179–197, Wiley & Sons, Hoboken, New Jersey.

Lucas M. P., Dygert N., Ren J., et al. (2020) Evidence for early fragmentation-reassembly of ordinary chondrite (H, L, and LL) parent bodies from REE-in-two-pyroxene thermometry. *Geochim. Cosochim. Acta* 290, 366–390.

Łuszczek K. and Krzesińska A. M. (2020) Copper in ordinary chondrites: Proxies for resource potential of asteroids and constraints for minimum-invasive and economically efficient exploitation. *Planet. Space Sci.* 194, 105092, 9 pp.

Lux G., Keil K. and Taylor G. J. (1980) Metamorphism of the H-group chondrites: Implications from compositional and textural trends in chondrules. *Geochim. Cosmochim. Acta* 44, 841–855.

Ma C., Lin C., Bindi L., and Steinhardt P. J. (2017) Hollisterite (Al_3Fe), kryachkoite $(Al,Cu)_6(Fe,Cu)$, and stolperite (AlCu): Three new minerals from the Khatyrka CV3 carbonaceous chondrite. *Amer. Mineral.* 102, 690–693.

Ma C., Tschauner O., Bindi L., Beckett J. R., and Xie X. (2019) A vacancy-rich, partially inverted spinelloid silicate, $(Mg,Fe,Si)_2(Si,\square)O_4$, as a major matrix phase in shock melt veins of the Tenham and Suizhou L6 chondrites. *Meteorit. Planet. Sci.* 54, 1907–1918.

Ma L., Williams D. B., and Goldstein J. I. (1998) Determination of the Fe-rich portion of the Fe-Ni-S phase diagram. *J. Phase Equilibria* 19, 299–309.

Macdougall D., Rajan R. S., Hutcheon I. D., and Price P. B. (1973) Irradiation history and accretionary processes in lunar and meteoritic breccias. *Proc. Lunar Sci. Conf. 4th*, 2319–2336.

Macdougall D., Rajan R. S., and Price P. B. (1974) Gas-rich meteorites: Possible evidence for origin on a regolith. *Science* 183, 73–74.

Macente A., Daly L., Griffin S., et al. (2025) The petrology of the Ozerki meteorite constrained by electron backscatter diffraction and x-ray computed tomography. *Meteorit. Planet. Sci.* 60, 1095–1118.

Macke R. J. (2010) Survey of meteorite physical properties: Density, porosity and magnetic susceptibility. *Electron. Theses Diss.* 1638. https://stars.library.ucf.edu/etd/1638.

MacLennan E. M., Emery J. P., Lucas M. P., McClure L. M., and Lindsay S. S. (2024a) Space weathering, grain size, and metamorphic heating effects on ordinary chondrite spectral reflectance parameters. *Meteorit. Planet. Sci.* 59, 1329–1352.

MacLennan E. M., Emery J. P., McClure L. M., et al. (2024b) Multifactor characterization of space weathering among siliceous asteroids. I: Grain size effects. *Meteorit. Planet. Sci.* 59, 1353–1384.

MacPherson G. J. (2005) Calcium-aluminum-rich inclusions in chondritic meteorites. In *Treatise on Geochemistry*, vol. 1 (edited by Davis A. M.), pp. 201–246. Elsevier, Amsterdam.

MacPherson G. J., Wark D. A., and Armstrong J. T. (1988) Primitive material surviving in chondrites: Refractory inclusions. In *Meteorites and the Early Solar System* (edited by Kerridge J. F. and Matthews M. S.), pp. 746–807, University of Arizona Press, Tucson.

MacPherson G. J., Davis A. M., and Zinner E. K. (1995) The distribution of aluminum-26 in the early solar system – A reappraisal. *Meteoritics* 30, 365–386.

MacPherson G. J., Huss G. R., and Davis A. M. (2003) Extinct ^{10}Be in Type A calcium-aluminum-rich inclusions from CV chondrites. *Geochim. Cosmochim. Acta* 67, 3165–3179.

MacPherson G. J., Simon S. B., Davis A. M., Grossman L., and Krot A. N. (2005) Calcium-aluminum-rich inclusions: Major unanswered questions. In *Chondrites and the Protoplanetary Disk* (edited by Krot A. N., Scott E. R. D. and Reipurth B.), pp. 225–250, ASP Conference 341, Astronomical Society of the Pacific, San Francisco.

MacPherson G. J., Lin C., Hollister L. S., et al. (2016) The Khatyrka meteorite: A summary of evidence for a natural origin of its remarkable Cu-Al alloys. *Lunar Planet. Sci.* 47, abstract#2655.

Maeda R., Goderis S., Yamaguchi A., et al. (2023) Fluid mobilization of rare earh elements, Th, and U during the terrestrial alteration of H chondrites. *Meteorit. Planet. Sci.* 58, 1018–1038.

Mai C., Desch S. J., Boley A. C., and Weiss B. P. (2018) Magnetic fields recorded by chondrules formed in nebular shocks. *Astrophys. J.* 857, 13 pp.

Maier W. D., Andreoli M. A. G., McDonald I., et al. (2006) Discovery of a 25-cm asteroid clast in the giant Morokweng imapct crater, South Africa. *Nature* 441, 203–206.

Mainzer A., Grav T., Bauer J., et al. (2011) NEOWISE observations of Near-Earth objects: Preliminary results. *Astrophys. J.* 743. https://doi.org/10.1088/0004-637X/743/2/156.

Maksimova A. A., Petrova E. V., Chukin A. V., et al. (2020) Characterization of the matrix and fusion crust of the recent meteorite fall Ozerki L6. *Meteorit. Planet. Sci.* 55, 231–244.

Mann C. R., Boley A. C., and Morris M. A. (2016) Planetary embryo bow shocks as a mechanism for chondrule formation. *Astrophys. J.* 818, 20 pp.

Marrocchi Y., Villeneuve J., Batanova V., Piani L., and Jacquet E. (2018) Oxygen isotopic diversity of chondrule precursors and the nebular origin of chondrules. *Earth Planet. Sci. Lett.* 496, 132–141.

Marrocchi Y., Bonal L., Gattacceca J., et al. (2020) The Piancaldoli meteorite: A forgotten primitive LL3.10 chondrite. *Meteorit. Planet. Sci.* 55, 1924–1935.

Marrocchi Y., Longeau A., Goupil R. L., et al. (2024) Isotopic evolution of the inner solar system revealed by size-dependent oxygen isotopic variations in chondrites. *Geochim. Cosmochim. Acta* 371, 52–64.

Marsset M., Carry B., Dumas C., et al. (2017) 3D shape of asteroid (6) Hebe from VLT/SPHERE imaging: Implications for the origin of ordinary H chondrites. *Astron. Astrophys.* 604, A64, 1–12.

Marsset M., Vernazza P., Brož M., et al. (2024) The Massalia asteroid family as the origin of ordinary L chondrites. arXiv:2403.08548v1 [astro-ph.EP]. https://doi.org/10.48550/arXiv.2403.08548.

Marti K. and Graf T. (1992) Cosmic-ray exposure history of ordinary chondrites. *Ann. Rev. Earth Planet. Sci.* 20, 221–243.

Martin P. M. and Hughes D. W. (1980) Meteoritic chondrules and the Weibull function. *Earth Planet. Sci. Lett.* 49, 175–180.

Martin P. M. and Mills A. A. (1977) Does the lunar regolith follow Rosin's law? *Moon* 16, 215–219.

Marvin U. B. (1996). Ernst Florens Friedrich Chladni (1756–1827) and the origins of modern meteorite research. *Meteorit. Planet. Sci.* 31, 545–588.

Marvin U. B. (2006) Meteorites in history: An overview from the Renaissance to the 20th century. In *The History of Meteoritics and Key Meteorite Collections: Fireballs, Falls and Finds* (ed. by McCall G. J. H., Bowden A. J., and Howarth R. J.), pp. 15–71, Geological Society of London, Special Publication 256, London.

Marvin U. B., Wood J. A., Taylor G. J., et al. (1971) Relative proportions and probable sources of rock fragments in the Apollo 12 lunar soil samples. *Proc. Lunar Sci. Conf. 2nd*, 679–699.

Mason B. (1962) *Meteorites*, John Wiley and Sons, Inc., New York, 274 pp.

Mason B. (1963) The carbonaceous chondrites. *Space Sci. Rev.* 1, 621–646.

Mason B. (1965) Feldspar in chondrites. *Science* 148, 943.

Mason B. and Wiik H. B. (1962) The Renazzo meteorite. *Amer. Museum Novit.* 2106, 1–11.

Mason B. and Wiik H. B. (1966) The composition of the Bath, Frankfort, Kakangari, Rose City and Tadjera meteorites. *Amer. Museum Novit.* 2272, 1–24.

Maurel C., Bryson J. F. J., Lyons R. L., et al. (2020) Meteorite evidence for partial differentiation and protracted accretion of planetesimals. *Sci. Adv.* 6. https://doi.org/10.1126/sciadv.aba1303.

McBreen B. and Hanlon L., (1999) Gamma-ray bursts and the origin of chondrules and planets. *Astro. Astrophys.* 351, 759–765.

McCall G. J. (1973) *Meteorites and Their Origins*, David & Charles, Devon, England, 352 pp.

McCall G. J. H. (2018) The Bencubbin meteorite: Further details, inclujding microscopic character of host material and two chondrite enclaves. *Mineral. Mag.* 36, 726–739.

McCall G. J. H. and De Laeter J. R. (1965) *Catalogue of Western Australian Meteorite Collections*, Spec. Publication No. 3, Western Australian Museum.

McCanta M. C., Treiman A. H., Dyar M. D., et al. (2008a) The LaPaz Icefield 04840 meteorite: Mineralogy, metamorphism, and origin of an amphibole- and biotite-bearing R chondrite. *Geochim. Cosmochim. Acta* 72, 5757–5780.

McCanta M. C., Beckett J. R., and Stolper E. M. (2008b) Zonation of phosphorus in olivine: Dynamic crystallization experiments and a study of chondrule olivine in unequilibrated ordinary chondrites. *Lunar Planet. Sci.* 39, abstract#1807.

McCanta M. C., Beckett J. R., and Stolper E. M. (2016) Correlations and zoning patterns of phosphorus and chromium in olivine from H chondrites and the LL chondrite Semarkona. *Meteorit. Planet. Sci.* 51, 520–546.

McConville P., Kelley S., and Turner G. (1988) Laser probe ^{40}Ar-^{39}Ar studies of the Peace River shocked L6 chondrite. *Geochim. Cosmochim. Acta* 52, 2487–2499.

McCord T. B., Adams J. B., and Johnson T. V. (1970) Asteroid Vesta: Spectral reflectivity and compositional implications. *Science* 168, 1445–1447.

McCoy T. J., Scott E. R. D., Jones R. H., Keil K., and Taylor G. J. (1991) Composition of chondrule silicates in LL3–5 chondrites and implications for their nebular history and parent body metamorphism. *Geochim. Cosmochim. Acta* 55, 601–619.

McCoy T. J., Keil K., Ash R. D., et al. (1993) Roosevelt County 075: A petrologic, chemical and isotopic study of the most unequilibrated known H chondrite. *Meteorit. Planet. Sci.* 28, 681–691.

McCoy T. J., Keil K., Clayton R. N., et al. (1996) A petrologic, chemical, and isotopic study of Monument Draw and comparison with other acapulcoites: Evidence for formation by incipient partial melting. *Geochim. Cosmochim. Acta* 60, 2681–2708.

McCoy T. J., Keil K., Muenow D. W., and Wilson L. (1997) Partial melting and melt migration in the acapulcoite-lodranite parent body. *Geochim. Cosmochim. Acta* 61, 639–650.

McCoy T. J., Mittlefehldt D. W., and Wilson L. (2006) Asteroid differentiation. In *Meteorites and the Early Solar System II* (edited by Lauretta D. S. and McSween H. Y.), pp. 733–745, University of Arizona Press, Tucson.

McCoy T. J., Corrigan C. M., Nagashima K., et al. (2019). The Milton pallasite and South Byron Trio Irons: Evidence for oxidation and core crystallization. *Geochim. Cosmochim. Acta* 259, 358–370.

McCoy T. J., Russell S. S., Zega T. J. et al. (2025) An evaporite sequence from ancient brine recorded in Bennu samples. *Nature* 637, 1072–1077. https://doi .org/10.1038/s41586-024-08495-6.

McCrosky R. E., Posen A., Schwartz G., and Shao C.-Y. (1971) Lost City meteorite – Its recovery and a comparison with other fireballs. *J. Geophys. Res. Solid Earth* 76, 4090–4108.

McDermott K. H., Greenwood R. C., Scott E. R. D., Franchi I. A., and Anand M. (2016) Oxygen isotope and petrological study of silicate inclusions in IIE iron meteorites and their relationship with H chondrites. *Geochim. Cosmochim. Acta* 173, 97–113.

McDonald I. (2002) Clearwater East impact structure: A re-interpretation of the projectile type using new platinum-group element data from meteorites. *Meteorit. Planet. Sci.* 37, 459–464.

McDonald I., Andreoli M. A. G., Hart R. J., and Tredoux M. (2001) Platinum-group elements in the Morokweng impact structure, South Africa: Evidence for the impact of a large ordinary chondrite projectile at the Jurassic-Cretaceous boundary. *Geochim. Cosmochim. Acta* 65, 299–309.

McDonough W. F. and Sun S.-S. (1995) The composition of the Earth. *Chem. Geol.* 120, 223–253.

McHone J. F., Shoemaker C., Killgore M., and Killgore K. (2012) Two shatter-coned NWA meteorites. *Lunar Planet. Sci.* 43, abstract#2359.

McKay D. S. (1991) Lunar meteorites and lunar regolith breccia: Some similarities and differences (abstract). *NIPR Syrup. Antarct. Meteorit.* 16, 96–99.

McKay D. S. and Basu A. (1983) The production curve for agglutinates in planetary regoliths. *J. Geophys. Res.– Solid Earth* 88, B193–B199.

McKay D. S., Heiken G. H., Taylor R. M., et al. (1972a) Apollo 14 soils: Size distribution and particle types. *Proc. Lunar Sci. Conf. 3rd*, 983–994.

McKay D. S., Clanton U. S., Morrison D. A., and Ladle G. H. (1972b) Vapor phase crystallization in Apollo 14 breccia. *Proc. Third Lunar Sci. Conf.*, 739–752.

McKay D. S., Bogard D. D., Morris R. V., et al. (1986) Apollo 16 regolith breccias: Characterization and evidence for early formation in the mega-regolith. *Proc. Lunar Planet. Sci. Conf. 16th, Part 2, J. Geophys. Res.* 91, suppl., D277–D303.

McKay D. S., Bogard D. D., Morris R. V., et al. (1989) Apollo 15 regolith breccias: Window to a KREEP regolith. *Proc. Lunar Planet. Sci. Conf. 19th*, 19–41.

McKay D. S., Heiken G., Basu A., et al. (1991) The lunar regolith. In *Lunar Sourcebook: A User's Guide to the Moon* (edited by Heiken G. H., Vaniman D. T. and French B. M.), Cambridge University Press, Cambridge, pp. 285–356.

McKeegan K. D., Leshin L. A., Russell S. S., and MacPherson G. J. (1998) Oxygen isotopic abundances in calcium-aluminum-rich inclusions from ordinary chondrites: Implications for nebular heterogeneity. *Science* 280, 414–418.

McKeegan K. D., Chaussidon M., and Robert F. (2000) Incorporation of short-lived ^{10}Be in a calcium-aluminum-rich inclusion from the Allende meteorite. *Science* 289, 1334–1337.

McKinley S. G., Scott E. R. D., Taylor G. J., and Keil K. (1981) A unique type 3 ordinary chondrite containing graphite-magnetite aggregates – Allan Hills A77011. *Lunar Planet. Sci. Conf.* 12B, 1039–1048.

McNaughton N. J., Borthwick J., Fallick A. E., and Pillinger C. T. (1981) Deuterium/hydrogen ratios in unequilibrated ordinary chondrites. *Nature* 294, 639–641.

McSween H. Y. (1976) A new type of chondritic meteorite found in lunar soil. *Earth Planet. Sci. Lett.* 31, 193–199.

McSween H. Y. (1977) Chemical and petrographic constraints on the origin of chondrules and inclusions in carbonaceous chondrites. *Geochim. Cosmochim. Acta* 41, 1843–1860.

McSween H. Y. (1987) *Meteorites and Their Parent Planets*, Cambridge University Press, Cambridge, 237 pp.

McSween H. Y. and Huss G. R. (2010) *Cosmochemistry*, Cambridge University Press, Cambridge. 549 pp.

McSween H. Y. and Labotka T. C. (1993) Oxidation during metamorphism of the ordinary chondrites. *Geochim. Cosmochim. Acta* 57, 1108–1114.

McSween H. Y. and Lipschutz M. E. (1980) Origin of volatile-rich H chondrites with light/dark structures. *Proc. Lunar Planet. Sci. Conf. 11th*, 853–864.

McSween H. Y. and Patchen A. D. (1989) Pyroxene thermobarometry in LL-group chondrites and implications for parent body metamorphism. *Meteoritics* 24, 219–226.

McSween H. Y. and Richardson S. M. (1977) The composition of carbonaceous chondrite matrix. *Geochim. Cosmochim. Acta* 41, 1145–1161.

McSween H. Y. and Wilkening L. L. (1980) A note on the Allan Hills A77278 unequilibrated ordinary chondrite. *Meteoritics* 15, 193–199.

McSween H. Y., Sears D. W. G., and Dodd R. T. (1988) Thermal metamorphism. In *Meteorites and the Early Solar System* (edited by J. F. Kerridge and M. S. Matthews), pp. 102–113. University of Arizona Press, Tucson.

McSween H. Y., Ghosh A., Grimm R. E., Wilson L., and Young E. D. (2002) Thermal evolution models of asteroids. In *Asteroids III* (edited by Bottke W. F., Cellino A., Paolicchi P., and Binzel R. P.), pp. 559–571. University of Arizona Press, Tuscon.

McSween H. Y., Binzel R. P., De Sanctis M. C., et al. (2013) Dawn; the Vesta-HED connection; and the geologic context for eucrites, diogenites, and howardites. *Meteorit. Planet. Sci.* 48, 2090–2104.

Médard E. and Grove T. L. (2008) The effect of H_2O on the olivine liquidus of basaltic melts: Experiments and thermodynamic models. *Contrib. Mineral. Petrol.* 155, 417–432.

Meek N. (1994) The stratigraphy and geomorphology of Coyote Basin, central Mojave Desert, California. *San Bernardino Co. Mus. Assoc. Quarterly* 41, 5–13.

Meier M. M. M., Schmitz B., Lindskog A., Maden C., and Wieler R. (2014) Cosmic-ray exposure ages of fossil micrometeorites from mid-Ordovician sediments at Lynna River, Russia. *Geochim. Cosmochim. Acta* 125, 338–350.

Meisel R., Walker R. and Morgan J. (1996) The osmium isotopic composition of the Earth's primitive upper mantle. *Nature* 383, 517–520.

Melosh H. J. (1989) *Impact Cratering: A Geologic Process*, Oxford University Press, Oxford, U.K., 245 pp.

Melosh H. J. (2005) The mechanics of pseudotachylite formation in impact events. In *Impact Tectonics* (edited by Koeberl C. and Henkel H.), pp. 55–80, Springer, Cham, Switzerland.

Melosh H. J. (2011) *Planetary Surface Processes*, Cambridge University Press, Cambridge, 500 pp.

Merrill G. P. (1920) On chondrules and chondritic structure in meteorites. *Proc. National Acad. Sci.* 6, 449–472.

Merrill G. P. (1921) On metamorphism in meteorites. *Geol. Soc. Am. Bull.* 32, 395–416.

Merrill G. P. (1934) The story of meteorites. In *Minerals from Earth and Sky* (edited by Abbot C. G.), pp. 1–163, Smithsonian Institution, New York.

Messenger S. (2002) Opportunities for the stratospheric collection of dust from short-period comets. *Meteorit. Planet. Sci.* 37, 1491–1505.

Metzler K. (2012) Ultrarapid chondrite formation by hot chondrule accretion? Evidence from unequilibrated ordinary chondrites. *Meteorit. Planet. Sci.* 47, 2193–2217. https://doi.org/10.1111/maps.12009.

Metzler K. (2018) From 2D to 3D chondrule size data: Some empirical ground truths. *Meteorit. Planet. Sci.* 53, 1489–1499.

Michikami T., Moriguchi K., and Hasegawa Fujiwara A. (2007) Ejecta velocity distribution for impact cratering experiments on porous and low strength targets. *Planet. Space Sci.* 55, 70–88.

Migliorini F., Manara A., Scaltriti F., et al. (1997) Surface properties of (6) Hebe: A possible parent body of ordinary chondrites. *Icarus* 128, 104–113.

Mikouchi T., Makishima J., Koizumi E., and Zolensky M. E. (2005) Porphyritic olivine-pyroxene clast in Kaidun: First discovery of an ordinary chondrite clast? *Lunar Planet. Sci.* 36, abstract#1956.

Mill J. S. (1859) *On Liberty*, Reprinted in 2001 by Batoche Books, Kitchener, Ontario, Canada, 109 pp.

Milton D. J. and DeCarli P. S. (1963) Maskelynite: Formation by explosive shock. *Science* 140, 670–671.

Minster J.-F. and Allègre C. J. (1979) ^{87}Rb-^{87}Sr chronology of H chondrites: Constraint and speculations on the early evolution of their parent body. *Earth Planet. Sci. Lett.* 42, 333–347.

Misawa K. and Fujita T. (1994) A relict refractory inclusion in a ferromagnesian chondrule from the Allende meteorite. *Nature* 368, 723–726.

Misawa K., Watanabe S., Kitamura M., et al. (1992) A noritic clast from the Hedjaz chondrite breccia: Implications for melting events in the early solar system. *Geochem. J.* 26, 435–446.

Mittlefehldt D. W. and Lindstrom M. M. (1997) Magnesian basalt clasts from the EET 92014 and Kapoeta howardites and a discussion of alleged primary magnesian HED basalts. *Geochim. Cosmochim. Acta* 61, 453–462.

Mittlefehldt D. W. and Lindstrom M. M. (2001) Petrology and geochemistry of Patuxent Range 91501, a clast-poor impact melt from the L-chondrite parent body and Lewis Cliff 88663, an L7 chondrite. *Meteorit. Planet. Sci.* 36, 439–457.

Mittlefehldt D. W., Rubin A. E., and Davis A. M. (1992) Mesosiderite clasts with the most extreme positive Eu anomalies among solar system rocks. *Science* 257, 1096–1099.

Miura H., Nakamota T., and Doi M. (2008) Origin of three-dimensional shapes of chondrules I. Hydrodynamics simulations of rotating droplet exposed to high-velocity rarefied gas flow. *Icarus* 197, 269–281.

Miyahara M., Ohtani E., and Yamaguchi A. (2016) Dissociation from plagioclase into jadeite + coesite in a shocked LL7 chondrite (abstract). *Seventh Symposium on Polar Science*, NIPR, Japan.

Miyahara M., Yamaguchi A., Saitoh M., et al. (2020) Systematic investigations of high-pressure polymorphs in shocked ordinary chondrites. *Meteorit. Planet. Sci.* 55, 2619–2651.

Miyahara M., Noguchi T., Matsumoto T., et al. (2024) Microscopic slickenside as a record of weak shock metamorphism in the surface layer of asteroid Ryugu. *Meteorit. Planet. Sci.* 59. https://doi.org/10.1111/maps.14271.

Miyamoto H., Yano H., Scheeres D. et al. (2007) Regolith migration and sorting on asteroid Itokawa. *Science* 316, 1011–1014.

Miyamoto M., Fujii N., and Takeda H. (1981). Ordinary chondrite parent body: An internal heating model. *Proc. Lunar Planet. Sci. Conf. 12th*, 1145–1152.

Miyamoto M., McKay D. S., McKay G. A., and Duke M. B. (1986) Chemical zoning and homogenization of olivines in ordinary chondrites and implications for thermal histories of chondrules. *J. Geophys. Res.* 91, 12804–12816.

Moilanen J., Gritsevich M., and Lyytinen E. (2021) Determination of strewn fields for meteorite falls. *Mon. Notices R. Astron. Soc.* 503, 3337–3350.

Molin G. M., Tribaudino M., and Brizi E. (1994) Zaoyang chondrite cooling history from Fe^{2+}-Mg intracrystalline ordering in pyroxenes. *Mineral. Mag.* 58, 143–150.

Monari G., Famaey B., Carrillo I., et al. (2018) The escape speed curve of the Galaxy obtained from Gaia DR2 implies a heavy Milky Way. *Astron. Astrophys.* 616. https://doi.org/10.1051/0004-6361/201833748.

Monnereau M., Toplis M. J., Baratoux D., and Guignard J. (2013) Thermal history of the H-chondrite parent body: Implications for metamorphic grade and accretionary time-scales. *Geochim. Cosmochim. Acta* 119, 302–321.

Morbidelli A. and Gladman B. (1998) Orbital and temporal distributions of meteorites originating in the asteroid belt. *Meteorit. Planet. Sci.* 33, 999–1016.

Morbidelli A. and Vokrouhlický D. (2003) The Yarkovsky-driven origin of near-Earth asteroids. *Icarus* 163, 120–134.

Morbidelli A., Bottke W. F., Froeschlé Ch., and Michel P. (2002) Origin and evolution of near-earth objects. In *Asteroids III* (edited by Bottke W. F., Cellino A., Paolicchi P. and Binzel R. P.), pp. 409–422, University of Arizona Press, Tucson.

Morbidelli A., Bitsch B., Crida A., et al. (2016) Fossilized condensation lines in the solar system protoplanetary disk. *Icarus* 267, 368–376.

Morden S. J., and Collinson D. W. (1992) The implications of the magnetism of ordinary chondrite meteorites. *Earth Planet. Sci. Lett.* 109, 185–204.

Moreau J.-G., Kohout T., Wünnemann K., Halodova P., and Haloda J. (2019) Shock physics mesoscale modeling of shock stage 5 and 6 in ordinary and enstatite chondrites. *Icarus* 332, 50–65.

Moreau J.-G., Jõeleht A., Stojic A. N., et al. (2024) Experimentally induced troilite melt pervasion in chondritic analog materials: A study for FeNi-FeS darkening in chondrites. *Meteorit. Planet. Sci.* 59. https://doi.org/10.1111/maps.14274.

Morfill G., Spruit H., and Levy E. H. (1993) Physical processes and conditions associated with the formation of protoplanetary disks. In *Protostars and Planets III* (edited by Levy E. H. and Lunine J. I.), pp. 939–978, University of Arizona Press, Tucson.

Morgan J. W., Krähenbühl U., Ganapathy R., and Anders E. (1972) Trace elements in Apollo 15 samples: Implications for meteorite influx and volatile depletion on the Moon. *Proc. Lunar Sci. Conf. 3rd*, 1361–1376.

Morita T., Nakamura T., Koyama C., et al. (2025) Reproduction of complete rim texture of barred olivine chondrule. *87th Annual Meeting of the Meteoritical Society*, abstract#5179.

Morris M. A. and Boley A. C. (2018) Formation of chondrules by shock waves. In *Chondrules: Records of Protoplanetary Disk Processes* (edited by Russell S. S., Connolly H. C. and Krot A. N.), pp. 375–399, Cambridge University Press, Cambridge.

Morris M. A. and Garvie L. A. J. (2013) New constraints on the formation of igneous rims around chondrules. *Lunar Planet. Sci.* 44, abstract#2852.

Morris M. A., Boley A. C., Desch S. J., and Athanassiadou T. (2012) Chondrule formation in bow shocks around eccentric planetary embryos. *Astrophys. J.* 752, 17 pp.

Mostefaoui S., Kita N. T., Togashi S., et al. (2002) The relative formation ages of ferromagnesian chondrules inferred from their initial aluminum-26/aluminum-27 ratios. *Meteorit. Planet. Sci.* 37, 421–438.

Mostefaoui S., Lugmair G. W., Hoppe P., and El Goresy A. (2003) Evidence for live iron-60 in Semarkona and Chervony Kut: A nanoSIMS study. *Lunar Planet. Sci.* 34, 1585.

Mostefaoui S., Zinner E., Hoppe P., Stadermann F. J., and El Goresy A. (2005) In situ survey of graphite in unequilibrated chondrites: Morphologies, C, N, O, and H isotopic ratios. *Meteorit. Planet. Sci.* 40, 721–743.

Movshovitz N., Nimmo F., Korycansky D. G., Asphaug E., and Owen J. M. (2014) Disruption and reaccretion of midsized moons during an outer solar system Late Heavy Bombardment. *Geophys. Res. Lett.* 42, 256–263.

Moynier F., Day J. M. D., Okui W., et al. (2012) Planetary scale Sr isotopic heterogeneity and the age of volatile depletion of the early solar system materials. *Astrophys. J.* 758, 45–51.

Mueller R. F. and Olsen E. J. (1967) The olivine, pyroxene and metal content of chondritic meteorites as a consequence of Prior's rule. *Mineral Mag.* 36, 311–318.

Müller D. R., Altwegg K., Berthelier J. J., et al. (2022) High D/H ratios in water and alkanes in Comet 7P/Churyumov-Gerasimenko measured with Rosetta/ROSINA DFMS. *Astron. Astrophys.* 662, A69, 1–13. https://doi.org/10.1051/0004-6361/202142922.

Müller O. and Zähringer J. (1966) Chemische unterschiede bei uredelgashaltigen steinmeteoriten. *Earth Planet. Sci. Lett.* 1, 25–29.

Müller O., Baedecker P. A., and Wasson J. T. (1971) Relationship between siderophilic-element content and oxidation state of ordinary chondrites. *Geochim. Cosmochim. Acta* 35, 1121–1137.

Muranushi T. (2010) Dust-dust collisional charging and lightning in protoplanetary discs. *Mon. Not. R. Astron. Soc.* 401, 2641–2664.

Murayama J. K., Nakai S., Kato M., and Kumazawa M. (1986) A dense polymorph of $Ca_3(PO_4)_2$: a high-pressure phase of apatite decomposition and its geochemical significance. *Phys. Earth Planet. Inter.* 44, 293–303.

Murdoch N., Sánchez P. Schwartz S. R., and Miyamoto H. (2015) Asteroid surface geophysics. In *Asteroids IV* (edited by Michel P., DeMeo F. E. and Bottke W. F.), pp. 767–792, University of Arizona Press, Tucson.

Musolino A., Suttle M. D., Folco L., et al. (2024) Early fluid migration and alteration fronts in the CM chondrite Reckling Peak 17085. *Meteorit. Planet. Sci.* 59, 1–23. https://doi.org/10.1111/maps.14261.

Muxworthy A. R., Bland P. A., Davison T. M., et al. (2017) Evidence for an impact-induced magnetic fabric in Allende, and exogenous alternatives to the core dynamo theory for Allende magnetization. *Meteorit. Planet. Sci.* 52, 2132–2146.

Nagahara H. (1981) Evidence for secondary origin of chondrules. *Nature* 292, 135–136.

Nagahara H. (1984) Matrices of type 3 ordinary chondrites – primitive nebular records. *Geochim. Cosmochim. Acta* 48, 2581–2595.

Nagahara H., Kita N. T., Ozawa K., and Morishita Y. (1999) Condensation during chondrule formation: Elemental and Mg isotopic evidence. *Lunar Planet. Sci.* 30, abstract#1342.

Nagahara H., Kita N. T., Ozawa K., and Morishita Y. (2008) Condensation of major elements during chondrule formation and its implication to the origin of chondrules. *Geochim. Cosmochim. Acta* 72, 1442–1465.

Nagasawa M., Tanaka K. K., Tanaka H., et al. (2019) Shock-generating planetesimals perturbed by a giant planet in a gas disk. *Astrophys. J.* 871, 15 pp.

Nagata T. and Funaki M. (1983) Paleointensity of the Allende carbonaceous chondrite. *Mem. Natl. Inst. Polar Res.* Spec. Issue 30, 403–434.

Nakagawa Y., Sekiya M. and Hayashi C. (1986) Settling and growth of dust particles in a laminar phase of a low mass solar nebula. *Icarus* 67, 375–390.

Nakamura N., Morikawa N., Hutchison R., et al. (1994) Trace element and isotopic characteristics of inclusions in the Yamato ordinary chondrites Y-75097, Y-793241 and Y-794046. *Proc. NIPR Symp. Antarct. Meteorites* 7, 125–143.

Nakamura T., Tomeoka K., Takaoka K., Sekine T., and Takeda H. (2000) Impact-induced textural changes of CV carbonaceous chondrites: Experimental reproduction. *Icarus* 146, 289–300.

Nakamura T., Noguchi T., Tanaka M, et al. (2011) Itokawa dust particles: A direct link between S-type asteroids and ordinary chondrites. *Science* 333, 1113–1116.

Nakamura E., Makishima A., Moriguti T., et al. (2012) Space environment of an asteroid preserved on micrograins returned by the Hayabusa spacecraft. *Proc. Natl. Acad. Sci.* 109, E624–E629.

Nakamuta Y. and Motomura Y. (1999) Sodic plagioclase thermometry of type 6 ordinary chondrites: Implications for the thermal histories of parent bodies. *Meteorit. Planet. Sci.* 34, 763–772.

Nakashima D., Nakamura T., and Okazaki R. (2006) Cosmic-ray exposure age and heliocentric distance of the parent bodies of enstatite chondrites ALH 85119 and MAC 88136. *Meteorit. Planet. Sci.* 41, 851–862.

Nakashima D., Kita N. T., Ushikubo T., et al. (2013) Oxygen three-isotope ratios of silicate particles returned from asteroid Itokawa by the Hayabusa spacecraft: A strong link with equilibrated LL chondrites. *Earth Planet. Sci. Lett.* 379, 127–136.

Nanne J. A. M., Nimmo F., Cuzzi J. N., and Kleine T. (2019) Origin of the non-carbonaceous-carbonaceous meteorite dichotomy. *Earth Planet. Sci. Lett.* 511, 44–54.

Nath R., Mali S. K. D., Panwar N., et al. (2025) Classifying meteorites with MetNet: A deep learning approach using reflectance spectroscopy. *Meteorit. Planet. Sci.* 60, 1063–1072.

Neal C. W., Dodd R. T., Jarosewich E., and Lipschutz M. E. (1981) Chemical studies of L-chondrites-I. A study of possible chemical subgroups. *Geochim. Cosmochim. Acta* 45, 891–898.

Nelson V. E. and Rubin A. E. (2002) Size-frequency distributions of chondrules and chondrule fragments in LL3 chondrites: Implications for parent-body fragmentation of chondrules. *Meteorit. Planet. Sci.* 37, 1361–1376.

Nesvorný D., Vokrouhlický D., Morbidelli A., and Bottke W. F. (2009) Asteroidal source of L chondrite meteorites. *Icarus* 200, 698–701.

Nesvorný D., Deienno R., Bottke W. F., et al. (2023) NEOMOD: A new orbital distribution model for Near Earth Objects. *Astron. J.* 166. https://doi.org/10.3847/1538-3881/ace040.

Neukampf J., Marrocchi Y., Villeneuve J., and Roskosz M. (2023) Using Li concentration and isotopic composition to unravel the link between type I and II chondrule olivine. Goldschmidt 2023 conference, Lyon, France. https://conf.goldschmidt.info/goldschmidt/2023/meetingapp.cgi/Paper/17994.

Newcomb S. (1861). On the secular variations and mutual relations of the orbits of the asteroids. *Mem. Amer. Acad. Arts Sci.* 8, Part I, 123–152.

Newsom H. E., Mangold N., Kah L. C., et al. (2015) Gale crater and impact processes – Curiosity's first 364 Sols on Mars. *Icarus* 249, 108–128.

Newton H. A. (1888) Upon the relation which the former orbits of those meteorites that are in our collections and that were seen to fall had to the Earth's orbit. *Amer. J. Sci.* 36, 1–14.

Ni P., Chabot N. L., Ryan C. J., and Shahar A. (2020) Heavy iron isotope composition of iron meteorites explained by core crystallization. *Nat. Geosci.* 13, 611–615.

Nichols R. H., Hagee B. E., and Hohenberg C. M. (1991) Tieschitz chondrules; I-Xe systematics (abstract). *Lunar Planet. Sci.* 22, 975–976.

Nield T. (2011) *The Falling Sky*, Lyons Press, Guilford, Connecticut, 289 pp.

Niemeier M. and Bischoff A. (2006) Glanerbrug – An LL4-6 fragmental breccia with huge L chondrite clasts. *Lunar Planet. Sci.* 37, abstract#1625.

Niépce de Saint-Victor, A. (1861) Cinquième mémoire sur une nouvelle action de la lumière inconnu jusqu'ici (Fifth memoir on a new action of light). *Comptes Rendus Hebdomadaires Des Séances de l'Académie Des Sciences* 53, 33–35.

Nininger H. H. (1933) *Our Stone-pelted Planet*, Houghton Mifflin, Boston, 237 pp.

Nininger H. H. (1934) The great meteor of March 24, 1933. *Pop. Astron.* 42, 291–306.

Nininger H. H. (1936a) Importance of meteorite collections. *Mineralogist* 4, 9–10.

Nininger H. H. (1936b) Terminology in meteoritics. *Pop. Astron.* 44, 194–203.

Nininger H. H. (1938) Meteorite collecting among ancient Americans. *Amer. Antiq.* 4, 39–40.

Nininger H. H. (1939) Sound from ether waves. *Pop. Astron.* 47, 97–99.

Nininger H. H. (1944) The Puente-Ladron, Socorro County, New Mexico, aerolite. *Contrib. Soc. Res. Meteorit.* 3, 165–167.

Nininger H. H. (1948) Terrestrialization of meteorites. *Mineralogist* 16, 62–64.

Nininger H. H. (1952) *Out of the Sky: An Introduction to Meteoritics*, University of Denver Press, Denver, 336 pp.

Nininger H. H. (1972) *Find a Falling Star*, Eriksson, New York, 254 pp.

Nininger H. H. (1977) *Meteorites: A Photographic Study of Surface Features, Part 1 – Shapes,*

Publication No. 16, Center for Meteorite Studies, Arizona State University, Tempe, Arizona, 148 pp.

Nininger H. H. (1981) *Meteorites: A Photographic Study of Surface Features, Part 2 – Orientation*, Publication No. 19, Center for Meteorite Studies, Arizona State University, Tempe, Arizona, 74 pp.

Nininger H. H. and Nininger A. D. (1950) The nininger collection of meteorites: A catalog and a history. *American Meteorite Museum*, Winslow, Arizona, 144 pp.

Nittler L. R., Alexander C. M. O'D., Gao X., Walker R. M., and Zinner E. K. (1994) Interstellar oxide grains from the Tieschitz ordinary chondrite. *Nature* 370, 443–446.

Noda M. (2024) Valera – Cow killer meteorite. Meteorite Times Magazine, March 1, 2024.

Noguchi T., Tsuchiyama A., Hirata N., et al. (2010) Surface morphological features of boulders on Asteroid 25143 Itokawa. *Icarus* 206, 319–326.

Noguchi T., Kimura M., Hashimoto T., et al. (2014) Sylvite and halite on particles recovered from 25143 Itokawa: A preliminary report. *Meteorit. Planet. Sci.* 49, 1305–1314.

Noguchi T., Tominaga T., Takase M., Yamaguchi A., and Imae N. (2025) Collection of Antarctic micrometeorites stored in the surface snow near the Dome Fuji Station. *Meteorit. Planet. Sci.* 60, 701–716.

Noonan A. F. and Nelen J. A. (1976) A petrographic and mineral chemistry study of the Weston, Connecticut, chondrite. *Meteoritics* 11, 111–130.

Norris R. D., Norris J. M., Lorenz R. D., Ray J., and Jackson B. (2014) Sliding rocks on Racetrack Playa, Death Valley National Park: First observation of rocks in motion. *PLoS One* 9(8), e105948.

Norton O. R. (1998) *Rocks from Space* (2nd edition), Mountain Press, Missoula, Montana, 447 pp.

Norton O. R. (2002) *The Cambridge Encyclopedia of Meteorites*, Cambridge University Press, Cambridge, 354 pp.

Norton O. R. and Chitwood L. A. (2008) *Field Guide to Meteors and Meteorites*, Springer, London, 287 pp.

Notkin G. (2011) Meteorite hunting: How to find treasure from space. *Aerolite Meteorites*, Tucson, Arizona, 84 pp.

Nyström J. O. and Wickman F. E. (1991) The Ordovician chondrite from Brunflo, central Sweden, II. Secondary minerals. *Lithos* 27, 167–185.

Nyström J. O., Lindström M., and Wickman F. E. (1988) Discoverey of a second Ordovician meteorite using chromite as a tracer. *Nature* 336, 572–574.

O'Hara M. J. and Schairer J. F. (1963) The join diopside-pyrope at atmospheric pressure. *Carnegie Inst. Washington Yearb.* 62, 107–115.

Ohtani E., Kagawa N., and Fujino K. (1991) Stability of majorite (Mg,Fe)SiO_3 at high pressures and 1800°C. *Earth Planet. Sci. Lett.* 102, 158–166.

Okabayashi S., Yokoyama T., Nakanishi N. and Iwamori H. (2019) Fractionation of highly siderophile elements in metal grains from unequilibrated ordinary chondrites: Implications for the origin of chondritic metals. *Geochim. Cosmochim. Acta* 244, 197–215.

O'Keefe J. D. and Ahrens T. J. (1985) Impact and explosion crater ejecta, fragment size, and velocity. *Icarus* 62, 328–338.

Olmsted D. (1834) Observations of the meteors of Nov. 13th, 1833. *Amer. J. Sci. Arts* 25, 363–411.

Olsen E. (1981) Vugs in ordinary chondrites. *Meteoritics* 16, 43–59.

Olsen E. (1989) The Chondrule. *Geochim. Cosmochim. Acta* 53, 1674.

Olsen E. J. and Bunch T. E. (1984) Equilibration temperatures of the ordinary chondrites: A new evaluation. *Geochim. Cosmochim. Acta* 48, 1363–1365.

Olsen E. and Jarosewich E. (1970) The chemical composition of the silicate inclusions in the Weekeroo Station iron meteorite. *Earth Planet. Sci. Lett.* 8, 261–266.

Olsen E. and Jarosewich E. (1971) Chondrules: First occurrence in an iron meteorite. *Science* 174, 583–585.

Olsen E., Davis A., Clarke R. S., et al. (1994) Watson: A new link in the IIE iron chain. *Meteoritics* 29, 200–213.

Onuma N., Clayton R. N., and Mayeda T. K. (1972) Oxygen isotope cosmothermometer. *Geochimica et Cosmochimica Acta* 36, 169–188.

Opeil C. P., Consolmagno G. J., Safarik D. J., and Britt D. T. (2012) Stony meteorite thermal properties and their relationship to meteorite chemical and physical states. *Meteorit. Planet. Sci.* 47, 319–329.

Öpik E. J. (1951) Collision probabilities with the planets and the distribution of interplanetary matter. *Proc. R. Irish Acad.* 54A, 165–199.

Oriti R. A. (1965) Dry lakes and meteorites. *Griffith Obser.* 29, 150–151.

Ormel C. W. (2008) *The Early Stages of Planet Formation: How to Grow from Small to Large*. Ph.D. Dissertation, University of Groningen. 256 pp.

Ormö J., Sturkell E., Alwmark C., and Melosh H. J. (2014) The Lockne-Målingen Doublet, Sweden: Unique Impact of a Binary Asteroid from a Main Belt Breakup Event. *77th Annu. Meet. Meteorit. Soc.*, abstract#5044.

Osinski G. R. and Ferrière L. (2016) Shatter cones: (Mis)understood? *Sci. Adv.* 2(8). https://doi.org/10.1126/sciadv.1600616.

Ostro S. J., Hudson R. S., Jurgens R. F., et al. (1995) Radar images of asteroid 4179 Toutatis. *Science* 270, 80–83.

Ostrowski D. and Bryson K. (2019) The physical properties of meteorites. *Planet. Space Sci.* 165, 148–178.

Ovenden M. W. (1972). Bode's law and the missing planet. *Nature* 239, 508–509.

Ovenden M. W. (1973). Planetary distances and the missing planet. In *Recent Advances in Dynamical Astronomy* (edited by Tapley B. D. and Szebehely V.), pp. 319–332. Reidel, Dordrecht.

Pack A., Shelley J. M. G., and Palme H. (2004) Chondrules with peculiar REE patterns: Implications for solar nebula condensation at high C/O. *Science* 303, 997–1000.

Pack A., Höweling A., Hezel D. C., et al. (2017) Tracing the oxygen isotope composition of the upper Earth's atmosphere using cosmic spherules. *Nat. Commun.* 8, 15702. https://doi.org/10.1038/ncomms15702.

Palme H. and O'Neill H. (2014) Cosmochemical estimates of mantle composition. In *Planets, Asteroids, Comets and The Solar System*, pp. 149–211. volume 2 of Treatise on Geochemistry (edited by Davis A. M.), Elsevier–Pergamon, Oxford.

Palme H., Göbel E., and Grieve R. A. F. (1979) The distribution of volatile and siderophile elements in the impact melt of East Clearwater (Quebec). *Proc. Lunar Planet. Sci. Conf. 10th*, 2465–2492.

Paneth F. A. (1956) The frequency of meteorite falls throughout the ages. *Vistas Astron.* 2, 1681–1686.

Pape J., Mezger K., Bouvier A.-S., and Baumgartner L. P. (2019) Time and duration of chondrule formation: Constraints from ^{26}Al-^{26}Mg ages of individual chondrules. *Geochim. Cosmochim. Acta* 244, 416–436.

Parisi A. F., Catlos E. J., Brookfield M. E., et al. (2024) The Ordovician meteorite event in North America: Age of the Slate Islands impact structure, northern Lake Superior, Ontario, Canada. *Meteorit. Planet. Sci.* 59, 1751–1774.

Park J., Turrin B. D., Herzog G. F., et al. (2015) ^{40}Ar/^{39}Ar age of material returned from asteroid 25143 Itokawa. *Meteorit. Planet. Sci.* 50, 2087–2098.

Parnell J. (2009) Global mass wasting at continental margins during Ordovician high meteorite flux. *Nat. Geosci.* 2, 57–61.

Paszun D. and Dominik C. (2008) Numerical determination of the material properties of porous dust cakes. *Astron. Astrophys.* 484, 859–868.

Pauls A. and Gladman B. (2005) Decoherence time scales for "meteoroid streams." *Meteorit. Planet. Sci.* 40, 1241–1256.

Pedersen M. and Rasmussen C. M. Ø. (2019) The glacially induced Middle Ordovician rise of the Plectambonitoidea (Brachiopoda) and the phylogenetic placement of the Baltic genus Ukoa. *Palaeogr. Palaeoclimatol. Palaeoecol.* 534, 109319.

Pedersen S. G., Schiller M., Connelly J. N., and Bizzarro M. (2019) Testing accretion mechanisms of the H chondrite parent body utilizing nucleosynthetic anomalies. *Meteorit. Planet. Sci.* 54, 1215–1227.

Pedreschi F., Mery D., Bunger A, and Yanez V. (2011) Computer vision classification of potato chips by color. *J. Food Process Eng.* 34, 1714–1728.

Pejovic B. (1982) *Man and Meteorites*, Thames Head Limited, Avening Tetbury Gloucestershire, U.K., 119 pp.

Pellas P. (1972) Irradiation history of grain aggregates in ordinary chondrites. Possible clues to the advanced stages of accretion. In *From Plasma to Planet* (edited by Elvius A.), pp. 65–90, Almavist and Wiksell, Stockholm.

Pellas P. and Storzer D. (1981) ^{244}Pu fission track thermometry and its application to stony meteorites. *Proc. R. Soc. London, Ser. A* 374, 253–270.

Petaev M. I. and Wood J. A. (1998) The condensation with partial isolation CWPI model of condensation in the solar nebula. *Meteorit. Planet. Sci.* 33, 1123–1137.

Peterson C. (1976) A source mechanism for meteorites controlled by the Yarkovsky effect. *Icarus* 29, 91–111.

Petrova E. V., Chukin A. V., Varga G., et al. (2024) Characterization of bulk interior and fusion crust of Calama 009 L6 ordinary chondrite. *Meteorit. Planet. Sci.* 59, 2865–2879.

Pickering W. H. (1910) The orbits of meteorites. *Pop. Astron.* 18, 262–276.

Pillinger C. T. and Pillinger J. M. (1996) The Wold Cottage meteorite: Not just any ordinary chondrite. *Meteorit. Planet. Sci.* 31, 589–605.

Pillinger C. T., Pillinger J. M., Greenwood R. C., et al. (2011) The meteorite from Lake House. *74th Annu. Meet. Meteorit. Soc.*, abstract#5326.

Pillinger C. T., Pillinger J. M., Johnson D., et al. (2014) The Danebury Iron Age meteorite – An H5 ordinary chondrite "find" from Hampshire, England. *Meteorit. Planet. Sci.* 49, 946–957.

Pillinger J. M., Pillinger C. T., Powell M., and Butler M. (2011) Application of lichenometry to meteorite studies. *74th Annu. Meet. Meteorit. Soc.*, abstract#5277.

Piralla M., Villeneuve J., Batanova V., Jacquet E., and Marrocchi Y. (2021) Conditions of chondrule formation in ordinary chondrites. *Geochim. Cosmochim. Acta* 313, 298–312.

Pittarello L., Goderis S., Soens B., et al. (2019) Meteoroid atmospheric entry investigated with plasma flow experiments: Petrography and geochemistry of the recovered material. *Icarus* 331, 170–178.

Podolak M. and Cameron A. G. W. (1974) Possible formation of meteoritic chondrules and inclusions in the precollapse Jovian protoplanetary atmosphere. *Icarus* 23, 326–333.

Podolak M., Prialnik D., Bunch T. E., Cassen P., and Reynolds R. (1993) Secondary processing of chondrules and refractory inclusions CAIs by gas-dynamic heating. *Icarus* 104, 97–109.

Pohl L. and Britt D. T. (2020) Strengths of meteorites – An overview and analysis of available data. *Meteorit. Planet. Sci.* 55, 962–987.

Poole G. M., Rehkämper M., Coles B. J., and Goldberg T. (2017) Nucleosynthetic molybdenum isotope anomalies in iron meteorites – New evidence for thermal processing of solar nebula material. *Earth Planet. Sci. Lett.* 473, 215–226.

Popescu M., Vaduvescu O., de León J., et al. (2019) Near-Earth asteroids spectroscopic survey at Isaac Newton Telescope. *Astron. Astrophys.* 627, A124.

Popova O. P., Jenniskens P., Emel'yanenko V., et al. (2013) Chelyabinsk airburst, damage assessment, meteorite recovery, and characterization. *Science* 342, 1069–1073.

Popper K. (1963) *Conjectures and Refutations*, Routledge and Keagan Paul, London, London, 608 pp.

Pourkhorsandi H., Debaille V., Armytage R. M. G., et al. (2021) The effects of terrestrial weathering on samarium–neodymium isotopic composition of ordinary chondrites. *Chem. Geol.* 562, 120056.

Povenmire H. (1984) The Grayton Beach, Florida meteorite. *Meteoritics* 19, 89–90.

Prasad M. S., Rudraswami N. G., De Araujo A., Babu E. V. S. S. K., and Kumar T. V. (2015) Ordinary chondritic micrometeorites from the Indian Ocean. *Meteorit. Planet. Sci.* 50, 1013–1031.

Prince C. J. (2011) *A Professor, a President, and a Meteor: The Birth of American Science*, Prometheus Books, Amherst, New York, 254 pp.

Prinz M., Fodor R. V., and Keil K. (1977) Comparison of lunar rocks and meteorites: Implications to histories of the moon and parent meteorite bodies. *Soviet-American Conference on Cosmochemistry of the Moon and Planets* 1, NASA SP-370, 183–199.

Prinz M., Nehru C. E., Weisberg M. K., et al. (1984) H chondritic clasts in a Yamato L6 chondrite: Implications for metamorphism. *Meteoritics* 19, 292–293.

Prinz M., Weisberg M. K., Nehru C. E., and Delaney J. S. (1986) North Haig and Nilpena: Paired polymict ureilites with Angra dos Reis-related and other clasts (abstract). *Lunar Planet. Sci.* 17, 681–682.

Prinz M., Weisberg M. K., and Nehru C. E. (1988) Gunlock, a new type 3 ordinary chondrite with a golfball-sized chondrule (abstract). *Meteoritics* 23, 297.

Prinz M., Weisberg M. K., Clayton R. N., and Mayeda T. K. (1993) Ordinary and Carlisle Lakes-like chondrite clasts in the Weatherford chondrite breccia. *Meteoritics* 28, 419–420.

Prior G. T. (1916) On the genetic relationship and classification of meteorites. *Mineral. Mag.* 18, 26–44.

Prior G. T. and Hey M. H. (1953) *Catalogue of Meteorites*, British Museum (Natural History), Clowes and Sons, London, 432 pp.

Przylibski T. A., Długosz-Lisiecka M., Łuszczek K., and Blutstein K. (2023) Results of statistical analysis of primary components concentrations in bulk chemical composition of ordinary chondrite groups. *Meteorit. Planet. Sci.* 58, 1246–1259.

Qin L. and Carlson R. W. (2016) Nucleosynthetic isotopic anomalies and their cosmochemical significance. *Geochem. J.* 50, 43–65.

Qin L., Nittler L. R., Alexander C. M. O'D., et al. (2011) Extreme ^{54}Cr-rich nano-oxides in the CI chondrite Orgueil – Implication for a late supernova injection into the solar system. *Geochim. Cosmochim. Acta* 75, 629–644.

Quirico E., Raynal P.-I., and Bourot-Denise M. (2003) Metamorphic grade of organic matter in six unequilibrated ordinary chondrites. *Meteorit. Planet. Sci.* 38, 795–811.

Radomsky P. M., Turrin R. P., and Hewins, R. H. (1986) Dynamic crystallization experiments on a pyroxene-olivine chondrule composition (abstract). *Lunar Planet. Sci.* 17, 687–688.

Raducan S. D., Jutzi M., Cheng A. F., et al. (2024) Physical properties of asteroid Dimorphos as derived from the DART impact. *Nat. Astron.* 8, 445–455.

Raikes S. A, and Ahrens T. J. (1979) Post-shock temperatures in minerals. *Geophys. J. Int.* 58, 717–747.

Rajan R. S. (1974) On the irradiation history and origin of gas-rich meteorites. *Geochim. Cosmochim. Acta* 38, 777–788.

Rambaldi E. R. (1981) Relict grains in chondrules. *Nature* 293, 558–561.

Rambaldi E. R. and Wasson J. T. (1981) Metal and associated phases in Bishunpur, a highly unequilibrated ordinary chondrite. *Geochim. Cosmochim. Acta* 45, 1001–1015.

Rambaldi E. R. and Wasson J. T. (1982) Fine, nickel-poor Fe–Ni grains in the olivine of unequilibrated ordinary chondrites. *Geochim. Cosmochim. Acta* 46, 929–939.

Rambaldi E. R., Sears D. W., and Wasson J. T. (1980) Si-rich Fe-Ni grains in highly unequilibrated chondrites. *Nature* 287, 817–820.

Rambaldi E. R., Fredriksson B. J., and Fredriksson K. (1981) Primitive ultrafine matrix in ordinary chondrites. *Earth Planet. Sci. Lett.* 56, 107–126.

Ramdohr P. (1967a) Die Schmelzkruste der Meteoriten. *Earth Planet. Sci. Lett.* 2, 197–209.

Ramdohr P. (1967b) Chromite and chromite chondrules in meteorites – I. *Geochim. Cosmochim. Acta* 31, 1961–1967.

Ramdohr P. (1973) *The Opaque Minerals in Stony Meteorites*, Elsevier, Amsterdam, 245 pp.

Ramírez I., Bajkova A. T., Bobylev V. V., et al. (2014) Elemental abundances of solar sibling candidates. *Astrophys. J.* 787:154 (17 pp).

Raymond S. N. and Izidoro A. (2017) Origin of water in the inner solar system: Planetesimals scattered inward during Jupiter and Saturn's rapid gas accretion. *Icarus* 297, 134–148.

Raymond S. N. and Morbidelli A. (2014) The grand tack model: A critical review. In *Complex Planetary Systems, Proceedings IAU Symposium 310* (edited by Knežević Z. and Lemaitre A.), pp. 194–203.

Rasmussen C. M. Ø., Ullmann C. V., Jakobsen K. G., et al. (2016) Onset of main Phanerozoic marine radiation sparked by emerging Mid Ordovician icehouse. *Sci. Rep.* 6, p. 18884.

Recca S. J., Scott E. R. D., Keil K., et al. (1986) Ragland, an LL3.4 chondrite find from New Mexico. *Meteoritics* 21, 217–229.

Reeves H. and Audouze J. (1968) Early heat generation in meteorites. *Earth Planet. Sci. Lett.* 4, 135–141.

Regelous M., Elliott T., and Coath C. (2008) Nickel isotope heterogeneity in the early solar system. *Earth Planet. Sci. Lett.* 272, 330–338.

Reichenbach C. (1860) Meteoriten en Meteoriten. *Ann. Phys. und Chem. von J. C. Poggendorff*, 111.

Reisener R. J. and Goldstein J. I. (2003) Ordinary chondrite metallography: Part 2. Formation of zoned and unzoned metal particles in relatively unshocked H, L, and LL chondrites. *Meteorit. Planet. Sci.* 38, 1679–1696.

Reitmeijer F. J. M. and MacKinnon I. D. R. (1985) Poorly graphitized carbon as a new cosmothermometer for primitive extraterrestrial materials. *Nature* 315, 733–736.

Rendtel J. and Knöfel A, (1989) Analysis of annual and diurnal variation of fireball rates and the population index of fireballs from different compilations of visual observations. *Bull. Astron. Inst. Czechosl.* 40, 53–63.

Reuter K. B., Williams D. B., and Goldstein J. I. (1987) Determination of the Fe-Ni phase diagram below 400°C. *Metall. Trans. A* 20A, 719–725.

Reuter K. B., Williams D. B., and Goldstein J. I. (1988) Low temperature transformations in the metallic phases of iron and stony-iron meteorites. *Geochim. Cosmochim. Acta* 52, 617–626.

ReVelle D. O. (1979) A quasi-simple ablation model for large meteorite entry: theory vs. observations. *J. Atmos. Terres. Phys.* 41, 453–473.

Rieber M., Graf F., Hase M., Roth N., and Weigand B. (2000) Numerical simulations of moving spherical and strongly deformed droplets. Proc. 16th Int. Conf. on Liquid Atomization and Spray Systems.

Righter K. (2018) Curator's comments. In *Antarctic Meteorite Newsletter* (edited by Satterwhite C. and Righter K.), Houston, TX: NASA Johnson Space Center.

Righter K. and Drake M. J. (1997) A magma ocean on Vesta: Core formation and petrogenesis of eucrites and diogenites. *Meteorit. Planet. Sci.* 32, 929–944.

Righter K., Abell P., Agresti D., et al. (2015) Mineralogy, petrology, chronology, and exposure history of the Chelyabinsk meteorite and parent body. *Meteorit. Planet. Sci.* 50, 1790–1819.

Ringwood A. E. (1965) Origin of chondrites. *Nature* 207, 701–704.

Ringwood A. E. (1966) Genesis of chondritic meteorites. *Rev. Geophys. Space Phys.* 4, 113–175.

Rivkin A. S., Davies J. K., Clark B. E., Trilling D. E., and Brown R. H. (2001) Aqueous alteration of S asteroid 6 Hebe? *Lunar Planet Sci.* 32, abstract#1723.

Robertson D., Pokorný P., Granvik M., Wheeler L., and Rumpf C. (2021) Latitude variation of flux and impact angle of asteroid collisions with Earth and Moon. *Planetary Sci. J.* 2:88. https://doi.org/10.3847/PSJ/abefda.

Robinson M. S., Thomas P. C., Veverka J., Murchie S. L., and Wilcox B. B. (2002) The geology of 433 Eros. *Meteorit. Planet. Sci.* 37, 1651–1684.

Rochette P., Sagnotti L., Bourot-Denise M., et al. (2003) Magnetic classification of stony meteorites: 1. Ordinary chondrites. *Meteorit. Planet. Sci.* 38, 251–268.

Rochette P., Gattacceca J., Bonal L., et al. (2008a) Magnetic classification of stony meteorites: 2. Non-ordinary chondrites. *Meteorit. Planet. Sci.* 43, 959–980.

Rochette P., Folco L., Suavet C., et al. (2008b) Micrometeorites from the Transantarctic Mountains. *Proc. Natl. Acad. Sci.* 105, 18206–18211.

Rochette P., Gattacceca J., Bourot-Denise M., et al. (2009) Magnetic classification of stony meteorites: 3. Achondrites. *Meteorit. Planet. Sci.* 44, 405–427.

Rochette P., Gattacceca J., and Lewandowski M. (2012) Magnetic classification of meteorites and application to the Soltmany fall. *Meteorites* 2, 67–71. https://doi.org/10.5277/met120108.

Roddy D. J. and Davis L. K. (1977) Shatter cones formed in large-scale experimental explosion craters. In *Impact and Explosion Cratering: Planetary and Terrestrial Implications* (edited by Roddy D. J., Pepin R. O. and Merrill R. B.), pp. 715–750. Pergamon, New York.

Roller M. (1870) Schreiben des Herrn Mathias Roller, Assistent am Polytechnikum zu Ofen, an den Herausgeber. *Astron. Nachr.* 75, 331–332.

Romig A. D. and Goldstein J. I. (1980) Determination of the Fe-Ni-P phase diagrams at low temperatures (700 to 300°C). *Trans. Metallurg. Soc. AIME* 11A, 1151–1159.

Rosenshein E. B., Ivanova M. A., Dickinson T. L., et al. (2006) Oxide-bearing and FeO-rich clasts in aubrites. *Meteorit. Planet. Sci.* 41, 496–503.

Rosin P. and Rammler E. (1933) Laws governing the fineness of powdered coal. *J. Inst. Fuel* 7, 29–36.

Ross A. J., Downes H., Smith C. L., and Jones A. P. (2010) DaG 1047: A polymict ureilite containing exotic clasts including a chondrite. *Lunar Planet. Sci.* 41, abstract#2361.

Rout S. S., Heck P. R., Zaluzec N. J., et al. (2017) Shocked chromites in fossil L chondrites: A Raman spectroscopy and transmission electron microscopy study. *Meteorit. Planet. Sci.* 52, 1776–1796.

Roy S. K. (1957) The problems of the origin and structure of chondrules in stony meteorites. *Field.– Geol.* 10, 383–396.
Roy S. K. and Wyant R. K. (1951) The Benld meteorite. *Geol. Series Field Mus. Natural His.* 7, 145–157.
Roy S. K. and Wyant R. K. (1955) The Paragould meteorite. *Field. Geol.* 10, 283–304.
Rubin A. E. (1982) Petrology and Origin of Brecciated Chondritic Meteorites. Ph.D. thesis, University of New Mexico, Albuquerque, 220 pp.
Rubin A. E. (1983) The Adhi Kot breccia and implications for the origin of chondrules and silica-rich clasts in enstatite chondrites. *Earth Planet. Sci. Lett.* 64, 201–212.
Rubin A. E. (1984a) Manganiferous orthopyroxene and olivine in the Allende meteorite. *Amer. Mineral.* 69, 880–888.
Rubin A. E. (1984b) Coarse-grained chondrule rims in type 3 chondrites. *Geochim. Cosmochim. Acta* 48, 1779–1789.
Rubin A. E. (1985) Impact melt products of chondritic material. *Rev. Geophys.* 23, 277–300.
Rubin A. E. (1989) An olivine-microchondrule-bearing clast in the Krymka meteorite. *Meteoritics* 24, 191–192.
Rubin A. E. (1990) Kamacite and olivine in ordinary chondrites: Intergroup and intragroup relationships. *Geochim. Cosmochim. Acta* 54, 1217–1232.
Rubin A. E. (1992) A shock-metamorphic model for silicate darkening and compositionally variable plagioclase in CK and ordinary chondrites. *Geochim. Cosmochim. Acta* 56, 1705–1714.
Rubin A. E. (1994a) Metallic copper in ordinary chondrites. *Meteoritics* 29, 93–98.
Rubin A. E. (1994b) Euhedral tetrataenite in the Jelica meteorite. *Mineral. Mag.* 58, 215–221.
Rubin A. E. (1995a) Petrologic evidence for collisional heating of chondritic asteroids. *Icarus* 113, 156–167.
Rubin A. E. (1995b) Fractionation of refractory siderophile elements in metal from the Rose City meteorite. *Meteoritics* 30, 412–417.
Rubin A. E. (1997a) The Hadley Rille enstatite chondrite and its agglutinate-like rim: Impact melting during accretion to the Moon. *Meteorit. Planet. Sci.* 32, 135–141.
Rubin A. E. (1997b) Igneous graphite in chondritic meteorites. *Mineral. Mag.* 61, 699–703.
Rubin A. E. (1997c) The Galim LL/EH polymict breccia: Evidence for impact-induced exchange between reduced and oxidized meteoritic material. *Meteorit. Planet. Sci.* 32, 489–492.
Rubin A. E. (1999) Formation of large metal nodules in ordinary chondrites. *J. Geophys. Res. Planets* 104, 30799–30804.
Rubin A. E. (2000) Petrologic, geochemical and experimental constraints on models of chondrule formation. *Earth Sci. Rev.* 50, 3–27.
Rubin A. E. (2002a) The Smyer H-chondrite impact-melt breccia and evidence for sulfur vaporization. *Geochim. Cosmochim. Acta* 66, 683–695.
Rubin A. E. (2002b) Post-shock annealing of MIL99301 (LL6): Implications for impact heating of ordinary chondrites. *Geochim. Cosmochim. Acta* 66, 3327–3337.
Rubin A. E. (2002c) *Disturbing the Solar System: Impacts, Close Encounters, and Coming Attractions*, Princeton University Press, Princeton, New Jersey, 361 pp.
Rubin A. E. (2003) Chromite-plagioclase assemblages as a new shock indicator; Implications for the shock and thermal histories of ordinary chondrites. *Geochim. Cosmochim. Acta* 67, 2695–2709.
Rubin A. E. (2004a) Aluminian low-Ca pyroxene in a Ca-Al-rich chondrule from the Semarkona meteorite. *Amer. Mineral.* 89, 867–872.
Rubin A. E. (2004b) Post-shock annealing and post-annealing shock in equilibrated ordinary chondrites: Implications for the thermal and shock histories of chondritic asteroids. *Geochim. Cosmochim. Acta* 68, 673–689.
Rubin A. E. (2004c) Numismatic astronomy. *Griffith Obser.* 67, 2–19.
Rubin A. E. (2005a) Relationships among intrinsic properties of ordinary chondrites: Oxidation state, bulk chemistry, oxygen-isotopic composition, petrologic type and chondrule size. *Geochim. Cosmochim. Acta* 69, 4907–4918.
Rubin A. E. (2005b) What heated the asteroids? *Scient. Amer.* 292, 80–87.
Rubin A. E. (2006) Shock, post-shock annealing and post-annealing shock in ureilites. *Meteorit. Planet. Sci.* 41, 125–133.
Rubin A. E. (2007) Petrogenesis of acapulcoites and lodranites: A shock-melting model. *Geochim. Cosmochim. Acta* 71, 2383–2401.
Rubin A. E. (2010a) Physical properties of chondrules in different chondrite groups: Implications for multiple melting events in dusty environments. *Geochim. Cosmochim. Acta* 74, 4807–4828.
Rubin A. E. (2010b) Impact melting in the Cumberland Falls and Mayo Belwa aubrites. *Meteorit. Planet. Sci.* 45, 265–275.
Rubin A. E. (2011) Origin of the differences in refractory-lithophile-element abundances among chondrite groups. *Icarus* 213, 547–558.
Rubin A. E. (2013a) Multiple melting in a four-layered barred-olivine chondrule with compositionally heterogeneous glass from LL3.0 Semarkona. *Meteorit. Planet. Sci.* 48, 445–456.
Rubin A. E. (2013b) An amoeboid olivine inclusion (AOI) in CK3 NWA 1559, comparison to AOIs in CV3 Allende, and the origin of AOIs in CK and CV chondrites. *Meteorit. Planet. Sci.* 48, 432–444.
Rubin A. E. (2013c) Numismeteoritics. *Meteorite* 19(4), 24–27.

Rubin A. E. (2014) Shock and annealing in the amphibole- and mica-bearing R chondrites. *Meteorit. Planet. Sci.* 49, 1057–1075.

Rubin A. E. (2015a) Impact features of enstatite-rich meteorites. *Chem. Der. Erde.* 75, 1–28.

Rubin A. E. (2015b) An American on Paris: Extent of aqueous alteration of a CM chondrite and the petrography of its refractory and amoeboid olivine inclusions. *Meteorit. Planet. Sci.* 50, 1595–1612.

Rubin A. E. (2015c) Maskelynite in asteroidal, lunar and planetary basaltic meteorites: An indicator of shock pressure during impact ejection from their parent bodies. *Icarus* 257, 221–229.

Rubin A. E. (2016) Impact melting of the largest known enstatite meteorite: Al Haggounia 001, a fossil EL chondrite. *Meteorit. Planet. Sci.* 51, 1576–1587.

Rubin A. E. (2017a) Multiple indicators for multiple melting of chondrules. *Chondrules and Protoplanetary Disk 2017, LPI Contrib.* 1963. abstract#2006.

Rubin A. E. (2017b) Fourteen points – A Wilsonian approach to chondrule formation: Evidence for multiple melting of chondrules. *Meteorit. Planet. Sci.* 52, abstract#6039.

Rubin A. E. (2018a) Carbonaceous and non-carbonaceous iron meteorites: Differences in chemical, physical and collective properties. *Meteorit. Planet. Sci.* 53, 2357–2371.

Rubin A. E. (2018b) Mechanisms accounting for variations in the proportions of carbonaceous and ordinary chondrites in different mass ranges. *Meteorit. Planet. Sci.* 53, 2181–2192.

Rubin A. E. (2018c) Evaluation of petrologic evidence for high partial pressures of SiO(g) in the solar nebula. *Meteorit. Planet. Sci.* 53, 2596–2607.

Rubin A. E. (2021) Benford's Law: Applications to ordinary-chondrite mass distributions. *Meteorit. Planet. Sci.* 56, 379–392.

Rubin A. E. (2022a) IIE irons: Origin, relationship to ordinary chondrites, and evidence for siderophile-element fractionations caused by chondrule formation. *Meteorit. Planet. Sci.* 57, 163–184.

Rubin A. E. (2022b) The recognition of meteorites and ice ages. *Oxford Research Encyclopedia of Planetary Science*. https://doi.org/10.1093/acrefore/9780190647926.013.262.

Rubin A. E. (2023) *Surface/Volume: How Geometry Explains why Grain Elevators Explode, Hummingbirds Hover, and Asteroids are Colder than Ice*, Springer, Cham, Switzerland, 182 pp.

Rubin A. E. (2024) Differences in bulk Fe content and density between Type-I and Type-II ordinary-chondrite chondrules: Implications for parent body heterogeneities in oxidation state and O-isotopic composition. *Meteorit. Planet. Sci.* 59, 2403–2410.

Rubin A. E. (2025) Positive correlation between shock stage and petrologic type in ordinary chondrites: Implications for the internal structures and thermal histories of ordinary-chondrite parent asteroids. *Meteorit. Planet. Sci.* 60, 17–31.

Rubin A. E. and Baecker B. (2021) Evidence from phosphorus x-ray mapping for a multi-step process in the formation of olivine phenocrysts in FeO-rich porphyritic chondrules. *Meteorit. Planet. Sci.* 56, 1478–1501.

Rubin A. E. and Bottke W. F. (2009) On the origin of shocked and unshocked CM clasts in H-chondrite regolith breccias. *Meteorit. Planet. Sci.* 44, 701–724.

Rubin A. E. and Brearley A. J. (1996) A critical evaluation of the evidence for hot accretion. *Icarus* 124, 86–96.

Rubin A. E. and Grossman J. N. (1985) Phosphate-sulfide assemblages and Al/Ca ratios in type 3 chondrites. *Meteoritics* 20, 479–489.

Rubin A. E. and Grossman J. N. (2010) Meteorite and meteoroid: New comprehensive definitions. *Meteorit. Planet. Sci.* 45, 114–122.

Rubin A. E. and Jones R. H. (2003) Spade: An H-chondrite impact-melt breccia that experienced post-shock annealing. *Meteorit. Planet. Sci.* 38, 1507–1520.

Rubin A. E. and Kallemeyn G. W. (1990) Lewis Cliff 85332: A unique carbonaceous chondrite. *Meteoritics* 25, 215–225.

Rubin A. E. and Keil K. (1983) Mineralogy and petrology of the Abee enstatite chondrite breccia and its dark inclusions. *Earth Planet. Sci. Lett.* 62, 118–131.

Rubin A. E. and Krot A. N. (1996) Multiple heating of chondrules. In *Chondrules and the Protoplanetary Disk* (ed. Hewins R. H., Jones R. H. and Scott E. R. D.), pp. 173–180. Cambridge University Press, Cambridge.

Rubin A. E. and Li Y. (2019) Formation and destruction of magnetite in CO3 chondrites and other chondrite groups. *Chem. Erde. Geochem*, 125528.

Rubin A. E. and Ma C. (2021) *Meteorite Mineralogy*, Cambridge University Press, Cambridge. 404 pp.

Rubin A. E. and Matson R. D. (2008) Coincidental compositional and orbital correspondences among some ordinary chondrites: No strong evidence for meteoroid streams. *Earth Moon Planets* 103, 73–88.

Rubin A. E. and Mittlefehldt D. W. (1992) Classification of mafic clasts from mesosiderites: Implications for endogenous igneous processes. *Geochim. Cosmochim. Acta* 56, 827–840.

Rubin A. E. and Mittlefehldt D. W. (1993) Evolutionary history of the mesosiderite asteroid: A chronologic and petrologic synthesis. *Icarus*, 101, 232–252.

Rubin A. E. and Moore W. B. (2011) What's up? Preservation of gravitational direction in the LAR 06299 LL impact-melt breccia. *Meteorit. Planet. Sci.* 46, 737–747.

Rubin A. E. and Pernicka E. (1989) Chondrules in the Sharps H3 chondrite: Evidence for intergroup compositional differences among ordinary chondrite chondrules. *Geochim. Cosmochim. Acta* 53, 187–195.

Rubin A. E. and Read W. F. (1984) The Brownell and Ness County (1894) L6 chondrites: Further sorting-out of Ness County meteorites. *Meteoritics* 19, 153–160.

Rubin A. E. and Scott E. R. D. (1997) Abee and related EH chondrite impact-melt breccias. *Geochim. Cosmochim. Acta* 61, 425–435.

Rubin A. E. and Swindle T. D. (2011) Flattened chondrules in the LAP 04581 LL5 chondrite: Evidence for an oblique impact into LL3 material and subsequent collisional heating. *Meteorit. Planet. Sci.* 46, 587–600.

Rubin A. E. and Turrin B. D. (2024) Nature and timing of a significant reduction event on the L-chondrite parent asteroid. *Meteorit. Planet. Sci.* 59, 836–857.

Rubin A. E. and Wasson J. T. (1987) Chondrules, matrix and coarse-grained chondrule rims in the Allende meteorite: origin, interrelationships and possible precursor components. *Geochim. Cosmochim. Acta* 51, 1923–1937.

Rubin A. E. and Wasson J. T. (1988) Chondrules and matrix in the Ornans CO3 meteorite: Possible precursor components. *Geochim. Cosmochim. Acta* 52, 425–432.

Rubin A. E. and Wasson J. T. (2005) Non-spherical lobate chondrules in CO3.0 Y-81020: General implications for the formation of low-FeO porphyritic chondrules in CO chondrites. *Geochim. Cosmochim. Acta* 69, 211–220.

Rubin A. E., Taylor G. J., Keil K., and Nelson G. (1981a) The Correo and Suwanee Spring meteorites: Two new ordinary chondrite finds. *Meteoritics* 16, 9–12.

Rubin A. E., Keil K., Taylor G. J., et al. (1981b) Derivation of a heterogeneous lithic fragment in the Bovedy L-group chondrite from impact-melted porphyritic chondrules. *Geochim. Cosmochim. Acta* 45, 2213–2228.

Rubin A. E., Scott E. R. D., and Keil K. (1982a) Microchondrule-bearing clast in the Piancaldoli LL3 meteorite: A new kind of type 3 chondrite and its relevance to the history of chondrules. *Geochim. Cosmochim. Acta* 46, 1763–1776.

Rubin A. E., Taylor G. J., Scott E. R. D., and Keil K. (1982b) Petrologic insights into the fragmentation history of asteroids. In *Workshop on Lunar Breccias and Soils and Their Meteoritic Analogs* (edited by Taylor G. J. and Wilkening L. L.), pp. 107–110, LPI Tech. Rpt. 82-02, Lunar and Planetary Institute, Houston.

Rubin A. E., Scott E. R. D., Taylor G. J., et al. (1983a) Nature of the H chondrite parent body regolith: Evidence from the Dimmitt breccia. *Proc. Lunar Planet. Sci. Conf. 13th*, A741–A754.

Rubin A. E., Rehfeldt A., Peterson E., Keil K. and Jarosewich E. (1983b) Fragmental breccias and the collisional evolution of ordinary chondrite parent bodies. *Meteoritics* 18, 179–196.

Rubin A. E., Jerde E., Zong P., et al. (1986) Properties of the Guin ungrouped iron meteorite: The origin of Guin and of group-IIE irons. *Earth Planet. Sci. Lett.* 76, 209–226.

Rubin A. E., Wang D., Kallemeyn G. W. and Wasson J. T. (1988a) The Ningqiang meteorite: Classification and petrology of an anomalous CV chondrite. *Meteoritics* 23, 13–23.

Rubin A. E., Fegley B. and Brett R. (1988b) Oxidation state in chondrites. In *Meteorites and the Early Solar System* (edited by Kerridge J. F. and Matthews M. S.), pp. 488–511, University of Arizona Press, Tucson.

Rubin A. E., Scott E. R. D. and Keil K. (1997) Shock metamorphism of enstatite chondrites. *Geochim. Cosmochim. Acta* 61, 847–858.

Rubin A. E., Sailer A. L. and Wasson J. T. (1999) Troilite in the chondrules of type-3 ordinary chondrites: Implications for chondrule formation. *Geochim. Cosmochim. Acta* 63, 2281–2298.

Rubin A. E., Verish R. S., Moore C. B. and Oriti R. A. (2000) Numerous unpaired meteorites exposed on a deflating playa lake at Lucerne Valley, California. *Meteorit. Planet. Sci.* 35, A181–A183.

Rubin A. E., Ulff-Møller F., Wasson J. T. and Carlson W. D. (2001) The Portales Valley meteorite breccia: Evidence for impact-induced melting and metamorphism of an ordinary chondrite. *Geochim. Cosmochim. Acta* 65, 323–342.

Rubin A. E., Zolensky M. E. and Bodnar R. J. (2002) The halite-bearing Zag and Monahans (1998) meteorite breccias: Shock metamorphism, thermal metamorphism and aqueous alteration on the H-chondrite parent body. *Meteorit. Planet. Sci.* 37, 125–141.

Rubin A. E., Trigo-Rodríguez J. M., Kunihiro T., Kallemeyn G. W., and Wasson J. T. (2005) Carbon-rich chondritic clast PV1 from the Plainview H-chondrite regolith breccia: Formation from H3 chondrite material by possible cometary impact. *Geochim. Cosmochim. Acta* 69, 3419–3430.

Rubin A. E., Ziegler K. and Young E. D. (2008) Size scales over which ordinary chondrites and their parent asteroids are homogeneous in oxidation state and oxygen-isotopic composition. *Geochim. Cosmochim. Acta* 72, 948–958.

Rubin A. E., Griset C. D., Choi B.-G., and Wasson J. T. (2009) Clastic matrix in EH3 chondrites. *Meteorit. Planet. Sci.* 44, 589–601.

Rubin A. E., Breen J. P., Isa J. and Tutorow S. (2017) NWA 10214 – An LL3 chondrite breccia with an assortment of metamorphosed, shocked, and unique chondrite clasts. *Meteorit. Planet. Sci.* 52, 372–390.

Rubin A. E., Dunn T., Garner K., Cecchi M., and Hernandez M. (2024) Barred olivine chondrules in ordinary chondrites: Constraints on chondrule formation. *Meteorit. Planet. Sci.* 59, 670–884.

Rubincam D. P. (1995) Asteroid orbit evolution due to thermal drag. *J. Geophys. Res.* 100, 1585–1594.

Rubincam D. P. (1998) Yarkovsky drag on small asteroids and Mars-Earth delivery. *J. Geophys. Res.* 103, 1725–1732.

Rudnick R. L., Ashwal L. D., Henry D. J., et al. (1985) Fluid inclusions in stony meteorites – A cautionary note. *Proc. Lunar Planet. Sci. Conf. 15th*, C669–C675.

Rudraswami N. G. and Goswami J. N. (2007) ^{26}Al in chondrules from unequilibrated L chondrites: Onset and duration of chondrule formation in the early solar system. *Earth Planet. Sci. Lett.* 257, 231–244.

Rudraswami N. G., Marrocchi Y., Prasad M. S., et al. (2019) Oxygen isotopic and chemical composition of chromites in micrometeorites: Evidence of ordinary chondrite precursors. *Meteorit. Planet. Sci.* 54, 1347–1361.

Russell B. (1935) *Religion and Science*, Oxford University Press, London, 256 pp.

Russell S. S., Lee M. R., Arden J. W., and Pillinger C. T. (1995) The isotopic composition and origins of silicon nitride from ordinary and enstatite chondrites. *Meteoritics* 30, 399–404.

Russell S. S., Srinivasan G., Huss G. R., Wasserburg G. J., and MacPherson G. J. (1996) Evidence for widespread 26Al in the solar nebula and constraints for nebula time scales. *Science* 273, 757–762.

Russell S. S., McCoy T. J., Jarosewich E., and Ash R. D. (1998) The Burnwell, Kentucky, low iron oxide chondrite fall: Description, classification and origin. *Meteorit. Planet. Sci.* 33, 853–856.

Russell S. S., MacPherson G. J., Leshin L. A., and McKeegan K. D. (2000) 16O enrichments in aluminum-rich chondrules from ordinary chondrites. *Earth Planet. Sci. Lett.* 184, 57–74.

Russell S. S., Connolly H. C. and Krot A. N. (editors) (2018) *Chondrules: Records of Protoplanetary Disk Processes*, Cambridge University Press, Cambridge, 450 pp.

Ruzicka A. (2014) Silicate-bearing iron meteorites and their implications for the evolution of asteroidal parent bodies. *Chem. Der. Erde. Geochem.* 74, 3–48.

Ruzicka A. M. and Hugo R. C. (2018) Electron backscatter diffraction (EBSD) study of seven heavily metamorphosed chondrites: Deformation systematics and variations in pre-shock temperature and post-shock annealing. *Geochim. Cosmochim. Acta* 234, 115–147.

Ruzicka A. and Hutson M. (2010) Comparative petrology of silicates in the Udei Station (IAB) and Miles (IIE) iron meteorites: Implications for the origin of silicate-bearing irons. *Geochim. Cosmochim. Acta* 74, 394–433.

Ruzicka A., Kring D. A., Hill D. H., et al. (1995) Silica-rich orthopyroxenite in the Bovedy chondrite. *Meteoritics* 30, 57–70.

Ruzicka A., Fowler G. W., Snyder G. A., et al. (1999) Petrogenesis of silicate inclusions in the Weekeroo Station IIE iron meteorite: Differentiation, remelting, and dynamic mixing. *Geochim. Cosmochim. Acta* 63, 2123–2143.

Ruzicka A., McHone J. F., and Killgore M. (2000) Portales Valley: Discovery of a large graphite nodule (abstract). *Meteorit. Planet. Sci.* 35, A140.

Ruzicka A., Killgore M., Mittlefehldt D. W., and Fries M. D. (2005) Portales Valley: Petrology of a metallic-melt meteorite breccia. *Meteorit. Planet. Sci.* 40, 261–295.

Ruzicka A., Floss C., and Hutson M. (2008) Relict olivine grains, chondrule recycling, and implications for the chemical, thermal, and mechanical processing of nebular materials. *Geochim. Cosmochim. Acta* 72, 5530–5557.

Ruzicka A., Hugo R., and Hutson M. (2015) Deformation and thermal histories of ordinary chondrites: Evidence for post-deformation annealing and syn-metamorphic shock. *Geochim. Cosmochim. Acta* 163, 219–233.

Ruzicka A. M., Friedrich J. M., Hutson M. L., et al. (2020) Shock compaction heating and collisional processes in the production of type 3 ordinary chondrites: Lessons from the (nearly) unique L3 chondrite melt breccia Northwest Africa 8709. *Meteorit. Planet. Sci.* 55, 2117–2140.

Ruzicka A. M., Hugo R. C., Friedrich J. M., and Ream M. T. (2024) Accretion of warm chondrules in weakly metamorphosed ordinary chondrites and their subsequent reprocessing. *Geochim. Cosmochim. Acta* 378, 1–35.

Ruzmaikina T. V. and Ip W. H. (1994) Chondrule formation in the accretional shock. *Icarus* 112, 430–447.

Ryan E. V. and Melosh H. J. (1998) Impact fragmentation: From the laboratory to asteroids. *Icarus* 133, 1–24.

Sadaka C., Gattacceca J., Gounelle M., et al. (2025) Systematic meteorite collection in the Catalina Dense Collection area (Chile): Description and statistics. *Meteorit. Planet. Sci.* 60, 308–323.

Sahagian D. L. and Proussevitch A. A. (1998) 3D particle size distributions from 2D observations: Stereology for natural applications. *J. Volcanol. Geotherm. Res.* 84, 173–196.

Salafia O. S., Ghisellini G., Ghirlanda G. and Colpi M. (2018) Interpreting GRB170817A as a giant flare from a jet-less double neutron star merger. *Astron. Astrophys.* 619, 5 pp. https://doi.org/10.1051/0004-6361/201732259.

Sambridge M., Tkalčić H., and Jackson A. (2010) Benford's Law in the natural sciences. *Geophys. Res. Lett.* 37, L22301. https://doi.org/10.1029/2010GL044830.

Sanchez J. A., Reddy V., Thirouin A., et al. (2024) The population of small near-Earth objects: Composition, source regions, and rotational properties. *Planet. Sci. J.* 5, 45 pp. https://doi.org/10.3847/PSJ/ad445f.

Sanders I. S. and Scott E. R. D. (2012) The origin of chondrules and chondrites: Debris from low-velocity impacts between molten planetesimals? *Meteorit. Planet. Sci.* 47, 2170–2192.

Sano T. and Yamashita S. (2004) Experimental petrology of basement lavas from Ocean Drilling Program Leg 192: Implications for differentiation processes in Ontong Java Plateau magmas. In *Origin and Evolution of the Ontong Java Plateau* (edited by Fitton J. G., Mahoney J. J., Wallace P. J. and Saunders A. D.), pp. 185–218 Geological Society Special Publications 229, London.

Sasaki S., Nakamura K., Hamabe Y., Kurahashi Y., and Hiroi T. (2001) Production of iron nanoparticles by laser irradiation in a simulation of lunar-like space weathering. *Nature* 410, 555–557.

Sauers E. (2024) What does a meteorite taste like? Someone found out and bottled it. *Mashable*, posted online June 1, 2024.

Saunier G., Poitrasson F., Moine B., Gregoire M., and Seddiki A. (2010) Effect of hot desert weathering on the bulk-rock iron isotope composition of L6 and H5 ordinary chondrites. *Meteorit. Planet. Sci.* 45, 195–209.

Savage P. S. and Moynier F. (2013) Silicon isotopic variation in enstatite meteorites: clues to their origin and Earth-forming material. *Earth Planet. Sci. Lett.* 361, 487–496.

Schaal R. B., Hörz F., Thompson T. D., and Bauer J. F. (1979) Shock metamorphism of granulated lunar basalt. *Proc. Lunar Planet. Sci. Conf. 10th*, 2547–2571.

Scheeres D. J. (2005) Solar radiation pressure and transient flows on asteroid surfaces. *Lunar Planet. Sci.* XXXVI, abstract#1919.

Scherer P., Schultz L., Knauer M., et al. (1997). Allan Hills 88019: An Antarctic H-chondrite with a very long terrestrial age. *Meteorit. Planet. Sci.* 32, 769–773.

Scherstén A., Elliot T., Hawkesworth C., Russell S., and Masarik J. (2006) Hf-W evidence for rapid differentiation of iron meteorite parent bodies. *Earth Planet. Sci. Lett.* 241, 530–542.

Schlüter J., Schultz L., Thiedig F., Al-Mahdi B. O., and Abu Aghreb A. E. (2002) The Dar al Gani meteorite field (Libyan Sahara): Geological setting, pairing of meteorites, and recovery density. *Meteorit. Planet. Sci.* 37, 1079–1093.

Schmidt T. E. (1973) The third rock garden – a meteorite! *Earth Sci.* 26, 63–65.

Schmieder M., Schwarz W. H., Trieloff M., et al. (2015) New ^{40}Ar/^{39}Ar dating of the Clearwater Lake impact structures (Québec, Canada) – Not the binary asteroid impact it seems? *Geochim. Cosmochim. Acta* 148, 304–324.

Schmitt R. T. (2000) Shock experiments with the H6 chondrite Kernouvé: Pressure calibration of microscopic shock effects. *Meteorit. Planet. Sci.* 35, 545–560.

Schmitz B. (2013) Extraterrestrial spinels and the astronomical perspective on Earth's geological record and evolution of life. *Chem. Der Erde. Geochem.* 73, 117–145.

Schmitz B., Lindström M., Asaro F., and Tassinari M. (1996) Geochemistry of meteorite-rich marine limestones strata and fossil meteorites from the lower Ordovician at Kinnekulle, Sweden. *Earth Planet. Sci. Lett.* 145, 31–48.

Schmitz B., Peucker-Ehrenbrink B., Lindström M., and Tassinari M. (1997) Accretion rates of meteorites and cosmic dust in the Early Ordovician. *Science* 278, 88–90.

Schmitz B., Tassinari M., and Peucker-Ehrenbrink B. (2001) A rain of ordinary chondritic meteorites in the Early Ordovician. *Earth Planet. Sci. Lett.* 194, 1–15.

Schmitz B., Häggström T., and Tassinari M. (2003) Sediment-dispersed extraterrestrial chromite traces a major asteroid disruption event. *Science* 300, 961–964.

Schmitz B., Harper D. A. T., Peucker-Ehrenbrink B., et al. (2008) Asteroid breakup linked to the Great Ordovician Biodiversification Event. *Nat. Geosci.* 1, 49–53.

Schmitz B., Farley K. A., Goderis S., et al. (2019) An extraterrestrial trigger for the mid-Ordovician ice age: Dust from the breakup of the L-chondrite parent body. *Sci. Adv.* 5, eaax4184.

Schmitz B., Schmieder M., Liao S., Martin E., and Terfelt F. (2022) Impact-crater ages and micrometeorite paleofluxes compared: Evidence for the importance of ordinary chondrites in the flux of meteorites and asteroids to Earth over the past 500 million years. In *From the Guajira Desert to the Apennines, and from Mediterranean Microplates to the Mexican Killer Asteroid: Honoring the Career of Walter Alvarez* (edited by Koeberl C., Claeys P. and Montanari A.), pp. 557, Geological Society of America, Boulder, Colorado. https://doi.org/10.1130/SPE557.

Schneider J. M., Burkhardt C., Marrocchi Y., Brennecka G. A., and Kleine T. (2020) Early evolution of the solar accretion disk inferred from Cr-Ti-O isotopes in individual chondrules. *Earth Planet. Sci. Lett.* 551, 116585.

Schrader D. L., McCoy T. J., and Davidson J. (2016) Widespread evidence for high-temperature formation of pentlandite in chondrites. *Geochim. Cosmochim. Acta* 189, 359–376.

Schrader D. L., McCoy T. J., and Gardner-Vandy K. (2017a) Relict chondrules in primitive achondrites: Remnants from their precursor parent bodies. *Geochim. Cosmochim. Acta* 205, 295–312.

Schrader D. L., Nagashima K., Krot A. N., et al. (2017b) Distribution of ^{26}Al in the CR chondrite chondrule-forming region of the protoplanetary disk. *Geochim. Cosmochim. Acta* 201, 275–302.

Schrader D. L., Nagashima K., Davidson J., et al. (2020) Outward migration of chondrule fragments in the early solar system: O-isotope evidence for rocky material crossing the Jupiter Gap? *Geochim. Cosmochim. Acta* 282, 133–155.

Schrader D. L., Zega T. J., Davidson J., McCoy T. J., and Domanik K. J. (2022) Pentlandite in chondrites: A compositional indicator of oxygen fugacity. *85th Annu. Meet. Meteorit. Soc.*, abstract#6110.

Schrader D. L., Torrano Z. A., Foustoukos D. I., et al. (2025) Reassessing the proposed "CY chondrites": Evidence for multiple meteorite types and parent bodies from Cr-Ti-H-C-N isotopes and bulk elemental compositions. *Geochim. Cosmochim. Acta* 390, 24–37.

Schramm L. S., Brownlee D. E., and Wheelock M. M. (1989) Major element composition of stratospheric micro-meteorites. *Meteorit. Planet. Sci.* 24, 99–112.

Schultz L. and Kruse H. (1978) Light noble gases in stony meteorites – A compilation. *Nucl. Track Detection* 2, 65–103.

Schultz L. and Kruse H. (1989) Helium, neon, and argon in meteorites – A data compilation. *Meteoritics* 24, 155–172.

Schultz L. and Weber H. W. (1996) Noble gases and H chondrite meteoroid streams: No confirmation. *J. Geophys. Res.* 101, 21177–21181.

Schultz L., Signer P., Pellas P., and Poupeau G. (1971) Assam: A gas-rich hypersthene chondrite. *Earth Planet. Sci. Lett.* 12, 119–123.

Schutt J., Schultz L., Zinner E., and Zolensky M. E. (1986) Search for meteorites in the Allan Hills region, 1985–1986. *Antarc. J. United States* 21, 81–82.

Schwerer F. C., Nagata T., and Fisher R. M. (1971) Electrical conductivity of lunar surface rocks and chondritic meteorites. *Moon* 2, 408–422.

Scott E. R. D. (1973) The nature of dark-etching rims in meteoritic taenite. *Geochim. Cosmochim. Acta* 37, 2283–2294.

Scott E. R. D. (1982) Origin of rapidly solidified metal-troilite grains in chondrules and iron meteorites. *Geochim. Cosmochim. Acta* 46, 813–823.

Scott E. R. D. (1984) Pairing of meteorites found in Victoria Land, Antarctica. *Mem. Natl. Inst. Polar Res.* 35, 102–125.

Scott E. R. D. (2007) Chondrites and the protoplanetary disk. *Ann. Rev. Earth Planet. Sci.* 35, 577–620.

Scott E. R. D. and Krot A. N. (2005) Chondrites and their components. In *Treatise on Geochemistry*, vol. 1 (edited by A. M. Davis), pp. 143–200, Elsevier, Amsterdam.

Scott E. R. D. and Rajan R. S. (1979) Thermal history of the Shaw chondrite. *Proc. Lunar Planet. Sci. Conf. 10th*, 1031–1043.

Scott E. R. D. and Rajan R. S. (1981) Metallic minerals, thermal histories and parent bodies of some xenolithic, ordinary chondrite meteorites. *Geochim. Cosmochim, Acta* 45, 53–67.

Scott E. R. D. and Taylor G. J. (1982) Primitive breccias among type 3 ordinary chondrites – Origin and relation to regolith breccias. In *Workshop on Lunar Breccias and Soils and their Meteoritic Analogs* (edited by Taylor G. J. and Wilkening L. L.), pp. 130–134, LPI Technical Report 82-02, Lunar and Planetary Institute, Houston.

Scott E. R. D. and Taylor G. J. (1983) Chondrules and other components in C, O, and E chondrites: Similarities in their properties and origins. *Proc. Lunar Planet. Sci. Conf. 14th*, B275–B286.

Scott E. R. D., Taylor G. J., Rubin A. E., Okada A., and Keil K. (1981a) Graphite-magnetite aggregates in ordinary chondritic meteorites. *Nature* 291, 544–546.

Scott E. R. D., Rubin A. E., Taylor G. J., and Keil K. (1981b) New kind of type 3 chondrite with a graphite-magnetite matrix. *Earth Planet. Sci. Lett.* 56, 19–31.

Scott E. R. D., Taylor G. J., and Maggiore P. (1982) A new LL3 chondrite, Allan Hills A79003, and observations on matrices in ordinary chondrites. *Meteoritics* 17, 65–75.

Scott E. R. D., Rubin A. E., Taylor G. J., and Keil K. (1984) Matrix material in type 3 chondrites – occurrence, heterogeneity and relationship with chondrules. *Geochim. Cosmochim, Acta* 48, 1741–1757.

Scott E. R. D., Lusby D., and Keil K. (1985) Ubiquitous brecciation after metamorphism in equilibrated ordinary chondrites. *Proc. Lunar Planet. Sci. Conf.* 16, D137–D148.

Scott E. R. D., Taylor G. J., and Keil K. (1986) Accretion, metamorphism, and brecciation of ordinary chondrites: Evidence from petrologic studies of meteorites from Roosevelt County, New Mexico. *Proc. Lunar Planet. Sci. Conf. 17th*, E115–E123.

Scott E. R. D., Brearley A. J., Keil K., et al. (1988a) Nature and origin of C-rich ordinary chondrites and chondritic clasts. *Proc. Lunar Planet. Sci. Conf. 18th*, 513–523.

Scott E. R. D., Barber D. J., Alexander C. M. O'D., Hutchison R., and Peck J. A. (1988b) Primitive material surviving in chondrites; matrix. In *Meteorites, the Early Solar System* (edited by J. F. Kerridge and M. S. Matthews), pp. 718–745, University of Arizona, Tuscon.

Scott E. R. D., Taylor G. J., Newsom H., et al. (1989) Chemical, thermal and impact processing of asteroids. In *Asteroids II* (edited by Binzel R. P., Gehrels T. and Matthews M. S.), pp. 701–739, University of Arizona Press, Tucson.

Scott E. R. D., Keil K., and Stöffler D. (1992) Shock metamorphism of carbonaceous chondrites. *Geochim. Cosmochim. Acta* 56, 4281–4293.

Scott E. R. D., Jones R. H., and Rubin A. E. (1994) Classification, metamorphic history and pre-metamorphic composition of chondrules. *Geochim. Cosmochim. Acta* 58, 1203–1209.

Scott E. R. D., Love S. G., and Krot A. N. (1996) Formation of chondrules and chondrites in the protoplanetary nebula. In *Chondrules and the Protoplanetary Disk* (edited by Hewins R. H., Jones R. H. and Scott E. R. D.), pp. 87–96, Cambridge University Press, Cambridge.

Scott E. R. D., Krot T. V., Goldstein J. I., and Taylor G. J. (2011) Thermal and impact history of H chondrites: Was the onion shell structure punctured by impacts during metamorphism? *74th Annual Meeting of the Meteoritical Society*, abstract #5516.

Scott E. R. D., Krot T. V., Goldstein J. I., and Wakita S. (2014) Thermal and impact history of the H chondrite parent asteroid during metamorphism: Constraints from metallic Fe-Ni. *Geochim. Cosmochim. Acta* 136, 13–37.

Scott E. R. D., Krot A. N., and Sanders I. S. (2017) Constraints from isotopic dichotomy of meteorites and from a trans-Jovian origin for carbonaceous chondrites on the origin of chondritic components. *80th Annu. Meet. Meteorit. Soc.*, abstract #6395.

Scott E. R. D., Krot A. N., and Sanders I. S. (2018) Isotopic dichotomy among meteorites and its bearing on the protoplanetary disk. *Astrophys. J.* 854, 12 pp.

Scott J. M., Negrini M., Faure K., et al. (2023) Multi-zone fusion crust formation and classification of the 2004 Auckland meteorite (L6, S5, and W0). *Meteorit. Planet. Sci.* 58, 328–340.

Sears D. W. (1975) Sketches in the history of meteoritics 1: The birth of the science. *Meteoritics* 10, 215–225.

Sears D. W. (1978) *The Nature and Origin of Meteorites*, Oxford University Press, New York, 187 pp.

Sears D. W. G. (1980) Formation of E chondrites and aubrites – a thermodynamic model. *Icarus* 43, 184–202.

Sears D. W. G. (2004) *The Origin of Chondrules and Chondrites*, Cambridge University Press, Cambridge, 209 pp.

Sears D. W. and Axon H. J. (1976) Metal of high Co content in LL chondrites. *Meteoritics* 11, 97–101.

Sears D. W. and Mills A. A. (1974) Thermoluminescence and the terrestrial age of meteorites. *Meteoritics* 9, 47–67.

Sears D. W., Grossman J. N., Melcher C. L., Ross L. M., and Mills A. A. (1980) Measuring metamorphic history of unequilibrated ordinary chondrites. *Nature* 287, 791–795.

Sears D. W., Grossman J. N., and Melcher C. L. (1982) Chemical and physical studies of type 3 chondrites – I: Metamorphism related studies of Antarctic and other type 3 ordinary chondrites. *Geochim. Cosmochim. Acta* 46, 2471–2481.

Sears D. W. G., DeHart J. M., Hasan F. A., and Lofgren G. E. (1990) Induced thermoluminescence and cathodoluminescence studies of meteorites. In *Spectroscopic Characterization of Minerals and Their Surfaces* (edited by Coyné L. M. et al.), pp. 190–222, American Chemical Society, Washington, DC.

Sears D. W. G., Hasan F. A., Batchelor J. D., and Lu J. (1991) Chemical and physical studies of type 3 chondrites – XI: Metamorphism, pairing, and brecciation of ordinary chondrites. *Proc. Lunar Planet. Sci.* 21, 493–512.

Sears D. W. G., Huang S., and Benoit P. H. (1996) Open-system behavior during chondrule formation. In *Chondrules and the Protoplanetary Disk* (edited by Hewins R. H., Jones R. H. and Scott E. R. D.), pp. 221–231, Cambridge University Press, Cambridge.

Seifert L. B., Haenecour P., Ramprasad T., and Zega T. J. (2023) An in situ study of presolar grains and the fine-grained matrices of the Meteorite Hills 00526 and Queen Alexandra Range 97008 unequilibrated ordinary chondrites. *Meteorit. Planet. Sci.* 58, 1079–1098.

Sekiya M. (1997) Do the chondrule sizes indicate the maximum size of dust aggregations in the solar nebula? *Antarctic Meteorit. XXII* 22, 168–170.

Sellards E. H. (1929) Sounds reported accompanying the fall of a meteor. *Science* 64, 297–298.

Selsis F., Lemmon M. T., Vaubaillon J., and Bell J. F. (2005) A martian meteor and its parent comet. *Nature* 435, 581.

Semenenko V. P. and Girich A. L. (1995) Mineralogy of a unique graphite-containing fragment in the Krymka chondrite (LL3). *Mineral. Mag.* 59, 445–456.

Semenenko V. P., Bischoff A., Weber I., Perron C., and Girich A. L. (2001) Mineralogy of fine-grained material in the Krymka (LL3.1) chondrite. *Meteorit. Planet. Sci.* 36, 1067–1085.

Semenenko V. P., Girich A. L., and Nittler L. R. (2004) An exotic kind of cosmic material: Graphite-containing xenoliths from the Krymka (LL3.1) chondrite. *Geochim. Cosmochim. Acta* 68, 455–475.

Servais T., Cascales-Miñana B., and Harper D. A. T. (2021) The Great Ordovician Biodiversification Event (GOBE) is not a single event. *Paleontol. Res.* 25, 315–328.

Sharp T. G. and DeCarli P. S. (2006) Shock effects in meteorites. In *Meteorites and the Early Solar System II* (edited by Lauretta D. and McSween H. Y.), pp. 653–677. University of Arizona Press, Tucson.

Sharpton V. L., Dressler B. O., Herrick R. R., Schneiders B., and Scott J. (1996) New constraints on the Slate Islands impact structure, Ontario, Canada. *Geology* 24, 851–854.

Shaw P. A. and Thomas D. S. G. (1997) Pans, playas and salt lakes. In *Arid Zone Geomorphology: Process, Form and Change in Drylands*, 2nd ed. (edited by Thomas D. S. G.), pp. 293–317. Wiley & Sons, Chichester, U.K.

Shen D., Li S., and Qin L., et al. (2025) Dsicovery of a distinct chondritic reservoir in a xenolith from an ordinary chondrite. *87th Annu. Meet. Meteorit. Soc.*, abstract#5088.

Shepard C. U. (1846) Report on Meteorites. *Amer. J. Sci. Arts (1820–1879)* 2, 377–392.

Shermer M. (2001) Colorful pebbles and Darwin's dictum. *Scient. Amer.* 248, 38.

Shima M., Murayama S., Wakabayashi F., Okada A., and Yabuki H. (1986) Two new chondrite-falls in Japan. *Meteoritics* 21, 59–78.

Shober P. M., Devillepoix H. A. R., Sansom E. K., et al. (2022) Arpu Kuilpu: An H5 from the outer main belt. *Meteorit. Planet. Sci.* 57, 1146–1157.

Shoemaker E. M. (1962) Interpretation of lunar craters. In *Physics and Astronomy of the Moon* (edited by Kopal Z.), pp. 283–359, Academic Press, New York.

Shu F. H., Shang H., and Lee T. (1996) Toward an astrophysical theory of chondrites. *Science* 271, 1545–1552.

Shu F. H., Shang H., Glassgold A. E., and Lee T. (1997) X-rays and fluctuating x-winds from protostars. *Science* 277, 1475–1479.

Shukolyukov A. and Lugmair G. W. (1993) 60Fe in eucrites. *Earth Planet. Sci. Lett.* 119, 159–166.

Shukolyukov A. and Lugmair G. W. (1998) Isotopic evidence for the Cretaceous-Tertiary impactor and its type. *Science* 282, 927–929.

Shukolyukov A. and Lugmair G. W. (2000) Extraterrestrial matter on Earth: Evidence from the Cr isotopes. In *Catastrophic Events and Mass Extinction: Impacts and Beyond*, pp. 197–198. LPI Contribution No. 1053, Lunar and Planetary Institute, Houston.

Shukolyukov A. and Lugmair G. W. (2006) Manganese-chromium systematics of carbonaceous chondrites. *Earth Planet. Sci. Lett.* 250, 200–213.

Sikdar J. and Rai V. K. (2020) Si-Mg isotopes in enstatite chondrites and accretion of reduced planetary bodies. *Sci. Rep.* 10, 1273. https://doi.org/10.1038/s41598-020-57635-1.

Silliman B. and Kingsley J. L. (1809) Memoir on the origin and composition of the meteoric stones which fell from the atmosphere, in the county of Fairfield, and state of Connecticut, on the 14th of December, 1807. *Trans. Amer. Philos. Soc.* 6, 335–345.

Simms M. J. (2021) The extraordinary story of the L chondrite meteorites. *Geol. Today* 37, 225–230.

Simon S. B., Grossman L., Clayton R. N., et al. (2004) The fall, recovery, and classification of the Park Forest meteorite. *Meteorit. Planet. Sci.* 39, 625–634.

Simon S. B., Sutton S. R., and Grossman L. (2008) Constraints on the oxidation state of chondrule precursors from titanium XANES analysis of Semarkona chondrules. *Lunar Planet. Sci.* 39, abstract#1352.

Simon J. B., Beckwith K., and Armitage P. J. (2012) Emergent mesoscale phenomena in magnetized accretion disc turbulence. *Monthly Notices R. Astronom. Soc.* 422, 2685–2700.

Simon J. I., Cuzzi J. N., McCain K. A., et al. (2018) Particle size distributions in chondritic meteorites: Evidence for pre-planetesimal histories. *Earth Planet. Sci. Lett.* 494, 69–82.

Singerling S. A., Nittler L. R., Barosch J., et al. (2022) TEM analyses of in situ presolar grains from unequilibrated ordinary chondrite LL3.0 Semarkona. *Geochim. Cosmochim. Acta* 328, 130–152.

Singerling S. A., Nittler L. R., Barosch J., et al. (2023) Tracing the history of an unusual compound presolar grain from progenitor star to asteroid parent body host. *Geochim. Cosmochim. Acta* 344, 230–243.

Sipiera P. P., Tarter J., Moore C. B., Dod B. D., and Johnston R. A. (1980) Gomez, Terry County, Texas: A new meteorite find. *Meteoritics* 15, 201–210.

Sipiera P. S., Olsen E. J., Eatough D. L., and Dod B. D. (1983) Summary of several recent chondrite finds from the Texas panhandle. *Meteoritics* 18, 63–75.

Sipiera P. P., Becker M. J., and Kawachi Y. (1987) Classification of twenty-six chondrites from Roosevelt County, New Mexico. *Meteoritics* 22, 151–155.

Siron G., Fukuda K., Kimura M., and Kita N. T. (2021) New constraints from ^{26}Al-^{26}Mg chronology of anorthite bearing chondrules in unequilibrated ordinary chondrites. *Geochim. Cosmochim. Acta* 293, 103–126.

Skinner W. R. (1990) Bipolar outflows and a new model of the early solar system: Part II. The origins of chondrules (abstract). *Lunar Planet. Sci.* 21, 1168–1169.

Skinner W. R., McSween H. Y., and Patchen A. D. (1989) Bleached chondrules and the diagenetic histories of ordinary chondrite parent bodies (abstract). *Meteoritics* 29, 326.

Skirak P., Opiła G., Piestrzyński A., Kozub-Budzyń G., and Kapusta C. (2025) Microtextures and structures in metal-sulfide grains in NWA 869 meteorite and their relation to thermal history of L-chondrite parent body. *Meteorit. Planet. Sci.* 60, 616–631.

Slater-Reynolds V. and McSween H. Y. (2005) Peak metamorphic temperatures in type 6 ordinary chondrites: An evaluation of pyroxene and plagioclase geothermometry. *Meteorit. Planet. Sci.* 40, 745–754.

Smith B. A. and Goldstein J. I. (1977) The metallic microstructures and thermal histories of severely reheated chondrites. *Geochim. Cosmochim. Acta* 41, 1061–1072.

Smith C., Russell S., and Almedia N. (2018) *Meteorites: The Story of Our Solar System*, Natural History Museum, London, 128 pp.

Smith D. G. W. (1980) The mineral chemistry of the Innisfree meteorite. *Canadian Mineralogist* 18, 433–442.

Smith D. G. W. and Launspach S. (1991) The composition of metal phases in Bruderheim (L6) and implications for the thermal histories of ordinary chondrites – Erratum. *Earth Planet. Sci. Lett.* 102, 79–93.

Smith J. V. (1979) Mineralogy of the planets: A voyage in space and time. *Mineral. Mag.* 43, 1–89.

Smyth J. R. (1974) Experimental study on the polymorphism of enstatite. *Amer. Mineral.* 59, 345–352.

Snetsinger K. G. and Keil K. (1969) Ilmenite in ordinary chondrites. *Amer. Mineral.* 54, 780–786.

Snetsinger K. G., Bunch T. E., and Keil K. (1968) Electron microprobe analysis of vanadium in the presence of titanium. *Amer. Mineral.* 53, 1770–1773.

Sneyd D. S., McSween H. Y., Sugiura N., Strangway D. W., and Nord G. L. (1988) Origin of petrofabrics and magnetic anisotropy in ordinary chondrites. *Meteoritics* 23, 139–149.

Soldani A. (1794) (with seven letters from G. Wm. Thomson) Sopra Piogetta di Sassi Accaduta Nella Serf de 16. Giugno 1794 in Lusignan d'Asso Nel Sanese (p. 288) [Dissertation]. Siena.

Sonett C. P. (1979) On the origin of chondrules. *Geophys. Res. Lett.* 6, 677–680.

Sonett C. P., Colburn D. S., Schwartz K., and Keil K. (1970) The melting of asteroidal-sized bodies by unipolar dynamo induction from a primordial T Tauri sun. *Astrophys. Space Phys.* 7, 446–488.

Sorby H. C. (1864) On the microscopical structure of meteorites. *Proc. Royal Soc. London* 13, 333–334.

Sorby H. C. (1877) On the structure and origin of meteorites. *Nature* 15, 495–498.

Spargur C. S. and Gooding J. L. (1992) Calorimetric "weatherometer" for stony meteorites (abstract). *Lunar Planet. Sci.* 23, 1337–1338.

Spencer J. R., Stern S. A., Moore J. M., et al. (2020) The geology and geophysics of Kuiper Belt object (486958) Arrokoth. *Science* 367, eaay3999.

Spray J. G. (2010) Frictional melting processes in planetary materials: From hypervelocity impact to earthquakes. *Ann. Rev. Earth Planet. Sci.* 38, 221–254.

Spurný P., Borovička J., Haloda J., Shrbený L., and Heinlein D. (2016) Two very precisely instrumentally documented meteorite falls: Žďár nad Sázavou and Stubenberg – Prediction and Reality. In *79th Annu. Meet. Meteorit. Soc.*, abstract#6221.

Spurný P., Borovička J., and Shrbený L. (2020) The Žďár nad Sázavou meteorite fall: Fireball trajectory, photometry, dynamics, fragmentation, orbit, and meteorite recovery. *Meteorit. Planet. Sci.* 55, 376–401.

Stanley G. M. (1955) Origin of playa stone tracks, Racetrack Playa, Inyo County, California. *Bull. Geol. Soc. Amer.* 66, 1329–1350.

Stauffer H. (1961) Primordial argon and neon in carbonaceous chondrites and ureilites. *Geochim. Cosmochim. Acta* 24, 70–82.

Steele I. M. (1986) Compositions and textures of relic forsterite in carbonaceous and unequilibrated ordinary chondrites. *Geochim. Cosmochim. Acta* 50, 1379–1395.

Steller T., Burkhardt C., Yang C., and Kleine T. (2022) Nucleosynthetic zinc isotope anomalies reveal a dual origin of terrestrial volatiles. *Icarus* 386, 115171.

Stelzner Th., Heide K., Bischoff A., et al. (1999) An interdisciplinary study of weathering effects in ordinary chondrites from the Acfer region, Algeria. *Meteorit. Planet. Sci.* 34, 787–794.

Stephant A., Carli C., Anand M., et al. (2023) Tissemouminites: A new group of primitive achondrites spanning the transition between acapulcoites and winonaites. *Meteorit. Planet. Sci.* 58, 111–134.

Stephens I. W., Looney L. W., Kwon W., et al. (2014) Spatially resolved magnetic field structure in the disk of a T Tauri star. *Nature* 514, 597–599. Medline https://doi.org/10.1038/nature13850.

Stewart S. T., Lock S. J., Carter P. J., et al. (2025) Planetesimal impact vapor plumes and nebular shocks form chondritic mixtures. *Planet. Sci. J.* 6. https://doi.org/10.48550/arXiv.2503.05636.

Stinchcomb B. L. (2011) Meteorites. *Schiffer*, Atglen, PA, USA, 160 pp.

Stöffler D., Ostertag R., Jammes C., et al. (1986) Shock metamorphism and petrography of the Shergotty achondrite. *Geochim. Cosmochim. Acta* 50, 889–903.

Stöffler D., Keil K., and Scott E. R. D. (1991) Shock metamorphism of ordinary chondrites. *Geochim. Cosmochim. Acta* 55, 3845–3867.

Stöffler D., Hamann C., and Metzler K. (2018) Shock metamorphism of planetary silicate rocks and sediments: Proposal for an updated classification system. *Meteorit. Planet. Sci.* 53, 5–49.

Story-Maskelyne M. H. N. (1870) On the mineral constituents of meteorites. *Proc. R. Soc. London* 18, 114–122.

Storz J., Ludwig T., Bischoff A., Schwarz W. H., and Trieloff M. (2021) Graphite in ureilites, enstatite chondrites, and unique clasts in ordinary chondrites – Insights from the carbon-isotope composition. *Geochim. Cosmochim. Acta* 307, 86–104.

Strickland A. (2018) Ingredients for life found in meteorites that crashed to Earth. CNN report, January 11, 2018.

Struempler A. W. and Kaelin R. (1987) The L5 Bayard meteorite: A 1982 find. *Meteoritics* 22, 105–108.

Stuart J. S. and Binzel R. P. (2004) Bias-corrected population, size distribution, and impact hazard for the near-Earth objects. *Icarus* 170, 295–311.

Suavet C., Cordier C., Rochette P., et al. (2011) Ordinary chondrite-related giant (>800 μm) cosmic spherules from the Transantarctic Mountains, Antarctica. *Geochim. Cosmochim. Acta* 75, 6200–6210.

Suess H. E. (1949) Zur Chemie der Planeten und Meteoritenbindung. *Z. Electrochem.* 53, 237–241.

Sugiura N. and Fujiya W. (2014) Correlated accretion ages and ε^{54}Cr of meteorite parent bodies and the evolution of the solar nebula. *Meteorit. Planet. Sci.* 49, 772–787.

Sunshine J. M. and Pieters C. M. (1998) Determining the composition of olivine from reflectance spectroscopy. *J. Geophys. Res.* 103, 13,675–13,688.

Sunshine J. M., Connolly H. C., McCoy T. J., Bus S. J., and La Croix L. M. (2008) Ancient asteroids enriched in refractory inclusions. *Science* 320, 514–517.

Suttle M. D., Greshake A., King A. J., et al. (2021) The alteration history of the CY chondrites, investigated through analysis of a new member: Dhofar 1988. *Geochim. Cosmochim. Acta* 295, 286–309.

Suttle M. D., King A. J., Harrison C. S., et al. (2023) The mineralogy and alteration history of the Yamato-type (CY) carbonaceous chondrites. *Geochim. Cosmochim. Acta* 361, 245–264.

Swindel G. W. and Jones W. B. (1954) The Sylacauga, Talladega County, Alabama, aerolite: A recent meteoritic fall that injured a human being. *Meteoritics* 1, 125–132.

Swindle T. D. and Kring D. A. (2008) Chronological evidence for the late heavy bombardment in ordinary chondrite meteorites. In Workshop on Early Solar System Impact Bombardment. Lunar & Planetary Institute, abstract#3004.

Swindle T. D., Caffee M. W., Hohenberg C. M., Lindstrom M. M., and Taylor G. J. (1991a) Iodine-xenon studies of petrographically and chemically characterized Chainpur chondrules. *Geochim. Cosmochim. Acta* 55, 861–880.

Swindle T. D., Grossman J. N., Olinger C. T., and Garrison D. H. (1991b) Iodine-xenon, chemical, and petrographic studies of Semarkona chondrules: Evidence for the timing of aqueous alteration. *Geochim. Cosmochim. Acta* 55, 3723–3734.

Swindle T. D., Kring D. A., and Weirich J. R. (2014) ^{40}Ar/^{39}Ar ages of impacts involving ordinary chondrite meteorites. *Geol. Soc. Lond. Special Publ.* 378, 333–347. https://doi.org/10.1144/SP378.6.

Symes S. J. K., Sears D. W. G., Akridge D. G., Huang S., and Benoit P. H. (1998) The crystalline lunar spherules: Their formation and implications for the origin of meteoritic chondrules. *Meteorit. Planet. Sci.* 33, 13–29.

Tachibana S. and Huss G. R. (2003) The initial abundance of ^{60}Fe in the solar system. *Astrophys. J. Lett.* 588, L41–L44.

Tachibana S., Nagahara H., Mostefaoui S., and Kita N. T. (2003) Correlation between relative ages inferred from ^{26}Al and bulk compositions of ferromagnesian chondrules in least equilibrated ordinary chondrites. *Meteorit. Planet. Sci.* 38, 939–962.

Tachibana S., Huss G. R., Kita N. T., Shimoda H., and Morishita Y. (2005) The abundances of iron-60 in pyroxene chondrules from unequilibrated ordinary chondrites (abstract). *Lunar Planet. Sci.* XXXVI, abstract#1529.

Tagle R. and Claeys P. (2005) An ordinary chondrite impactor for the Popigai crater, Siberia. *Geochim. Cosmochim. Acta* 69, 2877–2889.

Tait A. W., Tomkins A. G., Godel B. M., Wilson S., and Hasalova P. (2014) Investigation of the H7 ordinary chondrite, Watson 012: Implications for recognition and classification of Type 7 meteorites. *Geochim. Cosmochim. Acta* 134, 175–196.

Takagi M. (2005) Compositional distributions of olivine and low-Ca pyroxene in chondrules from LL3.0 Semarkona: Development of a numerical computer model to simulate the olivine distribution and implications for chondrule formation. M.S. Thesis, University of California, Los Angeles, 149 pp.

Takahashi E. (1983) Melting of a Yamato L3 chondrite (Y-74191) up to 30 kbar. *Proc. 8th Symp. Antarctic Met., Mem. Natl. Inst. Polar Res.* 30, 168–180.

Takeda H., Huston T. J., and Lipschutz M. E. (1984) On the chondrite-achondrite transition: Mineralogy and chemistry of Yamato 74160 (LL7). *Earth Planet. Sci. Lett.* 71, 329–339.

Takenouchi A., Mikouchi T., Yamaguchi A., and Zolensky M. E. (2015) Mineralogical comparison of olivine in shergottites and a shocked L chondrite: Implications for shock histories of brown olivine. *Lunar Planet. Sci.* 46th, abstract#1650.

Takigawa A., Tachibana S., Huss G. R., et al. (2014) Morphology and crystal structures of solar and presolar Al_2O_3 in unequilibrated ordinary chondrites. *Geochim. Cosmochim. Acta* 124, 309–327.

Tancredi G., Roland S., and Bruzzone S. (2015) Distribution of boulders and the gravity potential on asteroid Itokawa. *Icarus* 247, 279–290.

Tandon S. N. and Wasson J. T. (1968) Gallium, germanium, indium and iridium variations in a suite of L-group chondrites. *Geochim. Cosmochim. Acta* 32, 1087–1110.

Taylor G. J. (1982) Petrologic comparison of lunar and meteoritic breccias. In *Workshop on Lunar Breccias and Soils and Their Meteoritic Analogs*, LPI Tech. Rept. 82-02 (edited by Taylor G. J. and Wilkening L. L.), pp. 153–167, Lunar and Planetary Institute, Houston.

Taylor G. J. (2004) Asteroid heating: A shocking view. *Planet. Sci. Res. Discov.*, www.psrd.hawaii.edu/April04/asteroidHeating.html.

Taylor G. J. and Heymann D. (1969) Shock, reheating, and the gas retention ages of chondrites. *Earth Planet. Sci. Lett.* 7, 151–161.

Taylor G. J. and Heymann D. (1971) Postshock thermal histories of reheated chondrites. *J. Geophys. Res.* 76, 1879–1893.

Taylor G. J., Marvin U. B., Reid J. B., and Wood J. A. (1972) Noritic fragments in the Apollo 14 and 12 soils and the origin of Oceanus Procellarum. *Proc. Lunar Sci. Conf. 3rd*, 995–1014.

Taylor G. J., Wentworth S., Warner R. D., and Keil K. (1978) Agglutinates as recorders of fossil soil compositions. *Proc. Lunar Planet. Sci. Conf. 9th*, 1959–1967.

Taylor G. J., Keil K., Berkley J. L., et al. (1979) The Shaw meteorite: History of a chondrite consisting of impact-melted and metamorphic lithologies. *Geochim. Cosmochim. Acta* 43, 323–337.

Taylor G. J., Okada A., Scott E. R. D., et al. (1981) The occurrence and implications of carbide-magnetite assemblages in unequilibrated ordinary chondrites. *Lunar Planet. Sci.* 12, 1076–1078.

Taylor G. J., Scott E. R. D. and Keil K. (1983) Cosmic setting for chondrule formation. In *Chondrules and Their Origins* (edited by King E. A.), pp. 262–278, Lunar and Planetary Institute, Houston.

Taylor G. J., Maggiore P., Scott E. R. D., Rubin A. E., and Keil K. (1987) Original structures, and fragmentation and reassembly histories of asteroids: Evidence from meteorites. *Icarus* 69, 1–13.

Terfelt F. and Schmitz B. (2021) Asteroid break-ups and meteorite delivery to Earth the past 500 million years. *Proc. Natl. Acad. Sci.* 118, e2020977118.

Testi L., Natta A., Shepherd D. S., and Wilner D. J. (2003) Large grains in the disk of CQ Tau. *Astron. Astrophys.* 403, 323–328.

Thaisen K. G. and Taylor L. A. (2009) Meteorite fusion crust variability. *Meteorit. Planet. Sci.* 44, 871–878.

Thomas P. C., Veverka J., Robinson M. S., and Murchie S. (2001) Shoemaker crater as the source of most ejecta blocks on the asteroid 433 Eros. *Nature* 413, 394–396.

Thorslund P. and Wickman F. (1981) Middle Ordovician chondrite in fossiliferous limestone from Brunflo, central Sweden. *Nature* 289, 285–286.

Thorslund P., Wickman F. E., and Nyström J. O. (1984) The Ordovician chondrite from Brunflo, central Sweden, I. General description and primary minerals. *LIthos* 17, 87–100.

Tissandier L., Libourel G., and Robert F. (2002) Gas-melt interactions and their bearing on chondrule formation. *Meteorit. Planet. Sci.* 37, 1377–1389.

Tobin J. (2019) Fifty years of cutting meteorites. Meteorite Times Magazine, November 1, 2019 issue.

Tobin J. P. (2023) The Darker Path: Stories of Light and Shadow. *Independently published*, 190 pp.

Tomioka N. and Miyahara M. (2017) High-pressure minerals in shocked meteorites. *Meteorit. Planet. Sci.* 52, 2017–2039.

Tomioka N., Miyahara M., and Ito M. (2016) Discovery of natural $MgSiO_3$ tetragonal garnet in a shocked chondritic meteorite. *Sci. Adv.* 2, e1501725.

Tomkins A. G. (2009) What metal-troilite textures can tell us about post-impact metamorphism in chondrite meteorites. *Meteorit. Planet. Sci.* 44, 1133–1149.

Tomkins A. G., Johnson T. E., and Mitchell J. T. (2020) A review of the chondrite-achondrite transition, and a metamorphic facies series for equilibrated primitive stony meteorites. *Meteorit. Planet. Sci.* 55, 857–885.

Tomkins A. G., Martin E. L., and Cawood P. A. (2024) Evidence suggesting that earth had a ring in the Ordovician. *Earth Planet. Sci. Lett.* 646, 118991, 1–11.

Trappitsch R., Boehnke P., Stephan T., et al. (2018) New constraints on the abundance of ^{60}Fe in the early solar system. *Astrophys. J. Lett.* 857, L15, 6 pp.

Treiman A. H., Gleason J. D., and Bogard D. D. (2000) The SNC meteorites are from Mars. *Planet. Space Sci.* 48, 1213–1230.

Trieloff M., Jessberger E. K., Herrwerth I., et al. (2003) Structure and thermal history of the H-chondrite parent asteroid revealed by thermochronometry. *Nature* 422, 502–506.

Trigo-Rodríguez J. M. and Blum J. (2022) Learning about comets from the study of mass distributions and fluxes of meteoroid streams. *Month. Not. R. Astron. Soc.* 512, 2277–2289.

Trinquier A., Birck J.-L., and Allègre C. J. (2007) Widespread ^{54}Cr heterogeneity in the inner solar system. *Astrophys. J.* 655, 1179–1185.

Trinquier A., Elliott T., Ulfbeck D., et al. (2009) Origin of nucleosynthetic isotope heterogeneity in the solar protoplanetary disk. *Science* 324, 374–376.

Trombka J. I., Squyres S. W., Brückner J., et al. (2000) The elemental composition of asteroid 433 Eros: Results of the NEAR-Shoemaker x-ray spectrometer. *Science* 289, 2101–2105.

Tschauner O., Ma C., Beckett J. R., et al. (2014) Discovery of bridgmanite, the most abundant mineral in Earth, in a shocked meteorite. *Science* 346, 1100–1102.

Tschermak G. (1872) Die Meteoriten von Shergotty und Gopalpur. *Sitzber. Akad. Wiss. Wien, Math.-Naturwiss. Kl. Abt. I.* 65, 122–146.

Tschermak G. (1883) Beitrag zur Classification der Meteoriten. *Sitzber. Akad. Wiss. Wien, Math.-Naturwiss. Kl. Abt. I.* 88, 347–371.

Tschermak G. (1895) *Die mikroscopische Beschaffenheit der Meteoriten erläutert durch photographische Abbildungen*, Schweitzerbart, E., Stuttgart.

Tsuchiyama A. (2014) Asteroid Itokawa: A source of ordinary chondrites and a laboratory for surface processes. *Elements* 10, 45–50.

Tsuchiyama A., Osada Y., Nakano T., and Uesugi K. (2004) Experimental reproduction of classic barred olivine chondrules: Open-system behavior of chondrule formation. *Geochim. Cosmochim. Acta* 68, 653–672.

Turner G. (1988) Dating of secondary events. In *Meteorites and the Early Solar System* (edited by Kerridge J. F. and Matthews M. S.), pp. 276–288, University of Arizona Press, Tucson.

Turner G., Enright M. C., and Cadogan P. H. (1978) The early history of chondrite parent bodies inferred from ^{40}Ar-^{39}Ar ages. *Proc. Lunar Planet. Sci. Conf. 9th*, 989–1025.

Udden J. A. (1917) The Texas meteor of October 1, 1917. *University of Texas Bulletin*, no. 1772, 46–47.

Udden J. A. (1921) The Troup, Texas, meteorite. *Proc. United States National Museum* 59, 471–476.

Uehara M. and Gattacceca J. (2023) The "Meteorite meter": A handheld instrument for the combined measurement of magnetic susceptibility and electrical conductivity, with application to meteorite identification and classification. *Meteorit. Planet. Sci.* 58, 1629–1641.

Uehara M. and Nakamura N. (2006) Experimental constraints on magnetic stability of chondrules and the paleomagnetic significance of dusty olivines. *Earth Planet. Sci. Lett.* 250, 292–305. https://doi.org/10.1016/j.epsl.2006.07.042.

Uehara M., Gattacceca J., Leroux H., Jacob D. and van der Beek C. J. (2011) Magnetic microstructures of metal grains in equilibrated ordinary chondrites and implications for paleomagnetism of meteorites. *Earth Planet. Sci. Lett.* 306, 241–252.

Uhlig H. H. (1955) Contribution of metallurgy to the origin of meteorites. Part II – The significance of Neumann bands in meteorites. *Geochim. Cosmochim. Acta* 7, 34–42.

Ulrich G. E., Hodges C. A., and Muehlberger W. R. (1981) Geology of the Apollo 16 area, Central Lunar Highlands. *Geological Survey Professional Paper 1048*, U.S. Government Printing Office, Washington D.C., 539 pp.

Utas, J. A., Rubin, A. E., and Ziegler, K. (2017) Northwest Africa 10085: An equilibrated Kakangari chondrite. *Lunar Planet. Sci.* 48, abstract#2906.

Urey H. C. (1952) *The Planets, Their Origins and Development*, Yale University Press, New Haven, 245 pp.

Urey H. C. (1965) Meteorites and the moon: Cosmic-ray ages and contaminants provide evidence that meteorites may come from the moon. *Science* 147, 1262–1265.

Urey H. C. (1967) On the origin of chondrules and chondrites. *Meteoritics* 3, 127–128.

Urey H. C. and Craig H. (1953) The composition of the stone meteorites and the origin of the meteorites. *Geochim. Cosmochim. Acta* 4, 36–82.

Usselman T. M. (1975) Experimental approach to the state of the core. Part 1: The liquidus relations of the Fe-Ni-S system from 30 to 100 kb. *Amer. J. Sci.* 275, 278–290.

Vacher L. G., Eschrig J., Bonal L., et al. (2025) Thermal metamorphism and volatile evolution in unequilibrated ordinary chondrites: Implications for the delivery of hydrogen to terrestrial planets. *Geochim. Cosmochim. Acta* 391, 106–126.

Van de Berg W. J., Van den Broeke M. R., Reijmer C. H., and Van Meijgaard E. (2006) Reassessment of the Antarctic surface mass balance using calibrated output of a regional atmospheric climate model. *J. Geophys. Res.* 111, D11104. https://doi.org/10.1029/2005JD006495.

van der Bogert C. H., Schultz P. H., and Spray J. G. (2003) Impact-induced frictional melting in ordinary chondrites: A mechanism for deformation, darkening, and vein formation. *Meteorit. Planet. Sci.* 38, 1521–1531.

Van Flandern T. C. (1978) A former asteroidal planet as the origin of comets. *Icarus* 36, 51–74.

van Ginneken M., Folco L., Cordier C. and Rochettte P. (2012) Chondritic micrometeorites from the Transantarctic Mountains. *Meteorit. Planet. Sci.* 47, 228–247.

van Ginneken M., Gattacceca J., Rochette P., et al. (2017) The parent body controls on cosmic spherule texture: Evidence from the oxygen isotopic compositions of large micrometeorites. *Geochim. Cosmochim. Acta* 212, 196–210.

Van Roosbroek N., Debaille V., Pittarello L., et al. (2015) The formation of IIE iron meteorites investigated by the chondrule-bearing Mont Dieu meteorite. *Meteorit. Planet. Sci.* 50, 1173–1196.

Van Schmus W. R. (1969) The mineralogy and petrology of chondritic meteorites. *Earth Sci. Rev.* 5, 145–184.

Van Schmus W. R. and Koffman D. M. (1967) Equilibration temperatures of iron and magnesium in chondritic meteorites. *Science* 155, 1009–1011.

Van Schmus W. R. and Ribbe P. H. (1968) The composition and structural state of feldspar from chondritic meteorites. *Geochim. Cosmochim. Acta* 32, 1327–1342.

Van Schmus W. R. and Wood J. A. (1967) A chemical-petrologic classification for the chondritic meteorites. *Geochim. Cosmochim. Acta* 31, 747–765.

Varela M. E., Roszjar J., Sylvester P., and Garcia L. (2025) Bulk major and trace element abundances of barred olivine chondrules: Evidence about the physicochemical conditions in their region of nebular formation. *Meteorit. Planet. Sci.* 60, 1365–1383.

Vaubaillon J. (2022) Hyperbolic meteors: Is CNEOS 2014-01-08 interstellar? WGN, *J. IMO* 50, 140–143.

Vauquelin L.-N. (1802) Mémoire sur les pierres dites tombées du ciel. *J. Mines* 13, 308–322.

Velbel M. A. (2014) Terrestrial weathering of ordinary chondrites in nature and continuing during laboratory storage and processing: Review and implications for Hayabusa sample integrity. *Meteorit. Planet. Sci.* 49, 154–171.

Verish R. (2003a) Dale Dry Lake – New classification, revised recovery information. *Lunar Planet. Sci.* 34, abstract#1801.

Verish R. (2003b) Reconstructing fragmented meteorite finds. Meteorite Times Magazine, July 1, 2003.

Verish R. (2012) Holbrook – 100th anniversary (2012) – meteorite hunt. *Meteorite Times Magazine*, August 1, 2012.

Vernazza P., Binzel R. P., Thomas C. A., et al. (2008) Compositional differences between meteorites and near-Earth asteroids. *Nature* 454, 858–860.

Vernazza P., Zanda B., Binzel R. P., et al. (2014) Multiple and fast: The accretion of ordinary chondrite parent bodies. *Astrophys. J.* 791:120, 22 pp.

Veverka J., Thomas P., Harch A., et al. (1997) NEAR's flyby of 253 Mathilde: Images of a C asteroid. *Science* 278, 2109–2114.

Veverka J., Thomas P., Harch A., et al. (1999) NEAR encounter with asteroid 253 Mathilde: Overview. *Icarus* 140, 3–16.

Villeneuve J., Chaussidon M. and Libourel G. (2009). Homogeneous distribution of ^{26}Al in the Solar System from the Mg isotopic composition of chondrules. *Science* 325, 985–988. https://doi .org/10.1126/science.1173907.

Vogt G. (1882) Les Prétendus Organismes des Météorites. *Geol. Mag.* 9, 424–429.

Vokrouhlický D. and Bottke W. F. (2001) The Yarkovsky thermal force on small asteroids and their fragments. Choosing the right albedo. *Astron. Astrophys.* 371, 350–353.

Vokrouhlický D., Milani A., and Chesley S. R. (2000) Yarkovsky effect on small near-Earth asteroids: Mathematical formulation and examples. *Icarus* 148, 118–138.

Wahl W. (1910) Beiträge zur Chemie der Meteoriten. *Z. Anorg. Allgem. Chem.* 69, 52–96.

Wai C. M. and Wasson J. T. (1977) Nebular condensation of moderately volatile elements and their abundances in ordinary chondrites. *Earth Planet. Sci. Lett.* 36, 1–13.

Wai C. M. and Wasson J. T. (1979) Nebular condensation of Ga, Ge and Sb and the chemical classification of iron meteorites. *Nature* 282, 790–793.

Wakaki S., Itoh S., Tanaka T., and Yurimoto H. (2013) Petrology, trace element abundances and oxygen isotopic compositions of a compound CAI-chondrule object from Allende. *Geochim. Cosmochim. Acta* 102, 261–279.

Walker R. J., McDonough W. F., Honesto J., et al. (2008) Modeling fractional crystallization of group IVB iron meteorites. *Geochim. Cosmochim. Acta* 72, 2198–2216.

Walsh T. M. and Lipschutz M. E. (1982) Chemical studies of L chondrites-II. Shock-induced trace element mobilization. *Geochim. Cosmochim. Acta* 46, 2491–2500.

Walsh K. J., Morbidelli A., Raymond S. N., O'Brien D. P., and Mandell A. M. (2011) A low mass for Mars from Jupiter's early gas-driven migration. *Nature* 475, 206–209.

Walsh K. J., Morbidelli A., Raymond S. N., O'Brien, and Mandell A. M. (2012) Populating the asteroid belt from two parent source regions due to the migration of giant planets – "The Grand Tack." *Meteorit. Planet. Sci.* 47, 1941–1947.

Walter J., Kurat G., Brandstätter F., Koeberl C., and Maurette M. (1995) The abundance of ordinary chondrite debris among Antarctic micrometeorites (abstract). *Meteoritics* 30, 592–593.

Walton C. R., Jeon H., Černok A., et al. (2023) In-situ phosphate U-Pb ages of the L chondrites. *Geochim. Cosmochim. Acta* 359, 191–204.

Walton C. R., Anand M., and Schönbächler M. (2025) Microfaults: Abundant shear deformation and frictional melting in chondrites. *Meteorit. Planet. Sci.* 60, 858–868.

Wang D. and Rubin A. E. (1987) Petrology of nine ordinary chondrite falls from China. *Meteoritics* 22, 97–104.

Wang H., Weiss B. P., Bai X.-N., et al. (2017) Lifetime of the solar nebula constrained by meteorite paleomagnetism. *Science* 355, 623–627.

Wang Y. and Hsu W. (2009) Petrology and mineralogy of the Ningqiang carbonaceous chondrite. *Meteorit. Planet. Sci.* 44, 763–780.

Wänke H., Dreibus G., Jagoutz E., Palme H., and Rammensee W. (1981) Chemistry of the Earth and the significance of primary and secondary objects for the formation of planets and meteorite parent bodies (abstract). *Lunar Planet. Sci.* 12, 1139–1141.

Wänke H., Dreibus G., and Jagoutz E. (1984) Mantle chemistry and accretion history of the Earth. In *Archaean Geochemistry* (edited by Kröner A., Hanson G. N., and Goodwin A. M.), pp. 1–24, Springer Verlag, Berlin.

Ward A. W. and Greeley E. R. (1984) Evolution of yardangs at Rogers Lake, California. *Geol. Soc. Am. Bull.* 95, 829–837.

Wark D. A. (1979) Birth of the presolar nebula: The sequence of condensation revealed in the Allende meteorite. *Astrophys. Space Sci.* 65, 275–295.

Wark D. A. and Lovering J. F. (1977) Marker events in the early evolution of the solar system: Evidence from rims on Ca-Al-rich inclusions in carbonaceous chondrites. *Proc. Lunar Sci. Conf.* 8, 95–112.

Warner J. L., Ashwal L. D., Bergman S. C., et al. (1983) Fluid inclusions in stony meteorites. Proc. Lunar Planet. Sci. Conf. 13th, A731–A735.

Warren P. H. (2001) Porosities of lunar meteorites: Strength, porosity, and petrologic screening during the meteorite delivery process. *J. Geophys. Res.* 106, 10,101–10,111.

Warren P. H. (2011a) Stable-isotope anomalies and the accretionary assemblage of the Earth and Mars:

A subordinate role for carbonaceous chondrites. *Earth Planet. Sci. Lett.* 331, 93–100.

Warren P. H. (2011b) Stable isotopes and the noncarbonaceous derivation of ureilites, in common with nearly all differentiated planetary materials. *Geochim. Cosmochim. Acta* 75, 6912–6926.

Warren P. H. and Rubin A. E. (2010) Pigeonite-selective impact smelting in ureilites. *Geochim. Cosmochim. Acta* 74, 5109–5133.

Warren P. H. and Rubin A. E. (2020) Trace element and textural evidence favoring lunar, not terrestrial, origin of the mini-granite in Apollo sample 14321. *Icarus* 347. https://doi.org/10.1016/j.icarus.2020.113771.

Warren P. H., Taylor G. J., and Keil K. (1983) Regolith breccia Allan Hills A81005: Evidence of lunar origin, and petrography of pristine and nonpristine clasts. *Geophys. Res. Lett.* 10, 779–782.

Wasserburg G. J., Sanz H. G., and Bence A. E. (1968) Potassium feldspar phenocrysts in the surface of the Colomera, an iron meteorite. *Science* 161, 684–687.

Wasserburg G. J., Lee T., and Papanastassiou D. A. (1977) Correlated O and Mg isotopic anomalies in Allende inclusions. II – magnesium. *Geophys. Res. Lett.* 4, 299–302.

Wasson J. T. (1972) Formation of ordinary chondrites. *Rev. Geophys. Space Phys.* 10, 711–759.

Wasson J. T. (1974) *Meteorites: Classification and Properties*, Springer, Berlin, 316 pp.

Wasson J. T. (1985) *Meteorites: Their Record of Early Solar-System History*, Freeman, New York, 267 pp.

Wasson J. T. (1988) The building stones of the planets. In *Mercury* (edited by Vilas F., Chapman C. R. and Matthews M. S.), pp. 622–650, University of Arizona Press, Tucson.

Wasson J. T. (1991) Layered tektites: A multiple impact origin for the Australasian tektites. *Earth Planet. Sci. Lett.* 102, 95–109.

Wasson J. T. (1992) Planetesimal heating by FU-Orionis-Type events (abstract). *Eos Trans. AGU* 73, 336.

Wasson J. T. (2003) Large aerial bursts: An important class of terrestrial accretionary events. *Astrobiology* 3, 163–179.

Wasson J. T. (2008) Evaporation of nebular fines during chondrule formation. *Icarus* 195, 895–907.

Wasson J. T. and Choe W.-H. (2009) The IIG iron meteorites: Probable formation in the IIAB core. *Geochim. Cosmochim. Acta* 73, 4879–4890.

Wasson J. T. and Kallemeyn G. W. (1988) Compositions of chondrites. *Phil. Trans. Royal Soc. London A* 325, 535–544.

Wasson J. T. and Kallemeyn G. W. (2002) The IAB iron-meteorite complex: A group, five subgroups, numerous grouplets, closely related, mainly formed by crystal segregation in rapidly cooling melts. *Geochim. Cosmochim. Acta* 66, 2445–2473.

Wasson J. T. and Krot A. N. (1994) Fayalite-silica association in unequilibrated ordinary chondrites: Evidence for aqueous alteration on a parent body. *Earth Planet. Sci. Lett.* 122, 403–416.

Wasson J. T. and Rasmussen K. L. (1994) The fine nebula dust component: A key to chondrule formation by lightning (abstract). *Papers Presented to Chondrules and the Protoplanetary Disk*, LPI Contribution No. 844, p. 43.

Wasson J. T. and Rubin A. E. (2009) Composition of matrix in the CR chondrite LAP 02342. *Geochim. Cosmochim. Acta* 73, 1436–1460.

Wasson J. T. and Rubin A. E. (2010) Metal in CR chondrites. *Geochim. Cosmochim. Acta* 74, 2212–2230.

Wasson J. T. and Wang J. (1986) A nonmagmatic origin of group-IIE iron meteorites. *Geochim. Cosmochim. Acta* 50, 725–732.

Wasson J. T. and Wang S. (1991) The histories of ordinary chondrite parent bodies: U,Th-He age distributions. *Meteoritics* 26, 161–167.

Wasson J. T. and Wetherill G. W. (1979) Dynamical, chemical and isotopic evidence regarding the formation locations of asteroids and meteorites. In *Asteroids* (edited by Gehrels T.), pp. 926–974, University of Arizona Press, Tucson.

Wasson J. T., Boynton W. V., Chou C.-L., and Baedecker P. A. (1975) Compositional evidence regarding the influx of interplanetary materials onto the lunar surface. *Moon* 13, 121–141.

Wasson J. T., Willis J., and Wai C. M. (1980) Origin of iron meteorite groups IAB and IIICD. *Zeitsch. Naturfor.* A35a, 781–795.

Wasson J. T., Benz W. and Rubin A. E. (1987) Heating of primitive, asteroid-size bodies by large impacts (abstract). *Meteoritics* 22, 525–526.

Wasson J. T., Rubin A. E., and Kallemeyn G. W. (1993) Reduction during metamorphism of four ordinary chondrites. *Geochim. Cosmochim. Acta* 57, 1865–1878.

Wasson J. T., Kallemeyn G. W., and Rubin A. E. (1994) Equilibration temperatures of EL chondrites: A major downward revision in the ferrosilite contents of enstatite. *Meteoritics* 29, 658–661.

Wasson J. T., Krot A. N., Lee M. S., and Rubin A. E. (1995) Compound chondrules. *Geochim. Cosmochim. Acta* 59, 1847–1869.

Wasson J. T., Kallemeyn G. W., and Rubin A. E. (2000) Chondrules in the LEW85332 ungrouped carbonaceous chondrite: Fractionation processes in the solar nebula. *Geochim. Cosmochim. Acta* 64, 1279–1290.

Wasson J. T., Huber H., and Malvin D. J. (2007) Formation of IIAB iron meteorites. *Geochim. Cosmochim. Acta* 71, 760–781.

Watanabe S., Kitamura M., and Morimoto N. (1985) A transmission electron microscope study of pyroxene chondrules in equilibrated L-group chondrites. *Earth Planet. Sci. Lett.* 72, 87–98.

Weber D. and Bischoff A. (1997) Refractory inclusions in the CR chondrite Acfer 059-El Djouf 001: Petrology, chemical composition, and relationship to inclusion populations in other types of carbonaceous chondrites. *Chem. Erde* 57, 1–24.

Weber H. W., Hintenberger H., and Begemann F. (1971) Noble gases in the Haverö ureilite. *Earth Planet. Sci. Lett.* 13, 205–209.

Wedepohl K. H. (1995) The composition of the continental crust. *Geochim. Cosmochim. Acta* 59, 1217–1232.

Wegener A. (1917) Das Dotenierende Meteor vom April 1916, 3-1/2 Uhr Nachmittags in Kurhessen, Schriften der Gesellschaft zur Beförderung der gasamten *Naturwissen. Marburg* 14, 83 pp.

Weibull W. (1951) A statistical distribution function of wide applicability. *J. Appl. Mech.* 18, 293–297.

Weidenschilling S. J. (1977) Aerodynamics of solid bodies in the solar nebula. *Month. Not. R. Astron. Soc.* 180, 57–70.

Weidenschilling S. J. (1980) Dust to planetesimals: Settling and coagulation in the solar nebula. *Icarus* 44, 172–189.

Weidenschilling S. J. and Cuzzi J. N. (2006) Accretion dynamics and timescales: Relation to chondrites. *Meteorites and the Early Solar System II* (edited by McSween H. Y. and Lauretta D. S.), pp. 473–485, University of Arizona Press, Tucson.

Weidenschilling S. J., Marzari F., and Hood L. L. (1998) The origin of chondrules at jovian resonances. *Science* 279, 681–684.

Weinbruch S. and Müller W. F. (1995) Constraints on the cooling rates of chondrules from the microstructure of clinopyroxene and plagioclase. *Geochim. Cosmochim. Acta* 59, 3221–3230.

Weinbruch S., Büttner H., Holzheid A., Rosenhauer M., and Hewins R. H. (1998) On the lower limit of chondrule cooling rates: The significance of iron loss in dynamic crystallization experiments. *Meteorit. Planet. Sci.* 33, 65–74.

Weinland D. F. (1881a) Korallen in Meteorsteinen. *Das Ausland* 54, 301–303.

Weinland D. F. (1881b) Weiteres über die Tiereste in Meteoriten. *Das Ausland* 54, 501–508.

Weinland D. F. (1882) *Ueber die in Meteoriten entdecken Thierreste*, G. Frohner, Esslingen, 12 pp.

Weirich J. R., Wittman A., Isachsen C. E., et al. (2011) The Ar-Ar age and petrology of Miller Range 05029: Evidence for a large impact in the very early solar system. *Meteorit. Planet. Sci.* 45, 1868–1888.

Weisberg M. K. and Kimura M. (2012) The unequilibrated enstatite chondrites. *Chem. Erde* 72, 101–115.

Weisberg M. K., Prinz M., and Nehru C. E. (1988) Macrochondrules in ordinary chondrites: Constraints on chondrule forming processes (abstract). *Meteoritics* 23, 309–310.

Weisberg M. K., Prinz M., and Nehru C. E. (1990) The Bencubbin chondrite breccia and its relationship to CR chondrites and the ALH85085 chondrite. *Meteoritics* 25, 269–279.

Weisberg M. K., Prinz M., Clayton R. N., and Mayeda T. K. (1992) Formation of layered chondrules in CR2 chondrites: A petrologic and oxygen isotopic study (abstract). *Meteoritics* 27, 306.

Weisberg M. K., Prinz M., Clayton R. N., and Mayeda T. K. (1993) The CR (Renazzo-type) carbonaceous chondrite group and its implications. *Geochim. Cosmochim. Acta* 57, 1567–1586.

Weisberg M. K., Prinz M., Clayton R. N., et al. (1996) The K (Kakangari) chondrite grouplet. *Geochim. Cosmochim. Acta* 60, 4253–4263.

Weisberg M. K., McCoy T. J., and Krot A. N. (2006) Systematics and Evaluation of Meteorite Classification. In *Meteorites and the Early Solar System II* (edited by Lauretta D. S. and McSween H. Y.), pp. 19–52, University of Arizona Press, Tucson.

Weiss B. P., Carporzen L., Elkins-Tanton L. T., et al. (2010) A partially differentiated parent body for CV chondrites. *Proc. Lunar Planet. Sci. Conf.* 41, 1688.

Weiss B. P., Bai X.-N., and Fu R. R. (2021) History of the solar nebula from meteorite paleomagnetism. *Sci. Adv.* 7, eaba5967 (19 pp).

Welten K. C., Alderliesten C., Van Der Borg K., Lindner L., Loeken T. and Schultz L. (1997) Lewis Cliff 86360: An Antarctic L-chondrite with a terrestrial age of 2.35 million years. Meteorit. *Planet. Sci.* 32, 775–780.

Welten K. C., Nishiizumi K., Finkel R. C., et al. (2004) Exposure history and terrestrial ages of ordinary chondrites from the Dar al Gani region, Libya. *Meteorit. Planet. Sci.* 39, 481–498.

Welten K. C., Caffee M. W., Hillegonds D. J., et al. (2011) Cosmogenic radionuclides in L5 and LL5 chondrites from Queen Alexandra Range, Antarctica: Identification of a large L/LL5 chondrite shower with a preatmospheric mass of approximately 50,000 kg. *Meteorit. Planet. Sci.* 46, 177–198.

Welten K. C., Caffee M. W., Kress M. E., et al. (2024) Cosmogenic radionuclides in meteorites from the Otway Massif blue ice area, Antarctica: An unusual, well-preserved H5 chondrite strewn field. *Meteorit. Planet. Sci.* 59, 3056–3071.

Westphal A. J., Fakra S. C., Gainsforth Z., et al. (2009) Mixing fraction of inner solar system material in comet 81P/Wild 2. *Astrophys. J.* 694, 18–28.

Wetherill G. W. (1968) Time of fall and origin of stone meteorites. *Science* 159, 79–82.

Wetherill G. W. (1979) Apollo objects. *Scient. Amer.* 240, 54–65.

Wetherill G. W. (1985) Asteroidal source of ordinary chondrites. *Meteoritics* 20, 1–22.

Wetherill G. W. (1987) Dynamical relationships between asteroids, meteorites, and Apollo-Amor objects. *Phil. Trans. R. Soc. London A* 323, 323–337.

Wetherill G. W. and Chapman C. R. (1988) Asteroids and meteorites. In *Meteorites and the Early Solar*

System (edited by Kerridge J. F. and Matthews M. S.), pp. 35–67, University of Arizona Press, Tucson.

Weyrauch M. and Bischoff A. (2012) Macrochondrules in chondrites – Formation by melting of mega-sized dust aggregates and/or by rapid collisions at high temperatures? *Meteorit. Planet. Sci.* 47, 2237–2250.

Whipple F. L. (1966) Chondrules: Suggestions concerning their origin. *Science* 153, 54–56.

Whitby J., Burgess R., Turner G., Gilmour J., and Bridges J. (2000) Extinct ^{129}I in halite from a primitive meteorite: Evidence for evaporite formation in the early solar system. *Science* 288, 1819–1821.

White C. M. (1989) The Wilder H5 chondrite: A 1982 find from Idaho. *Meteoritics* 24, 56.

Widom E., Rubin A. E., and Wasson J. T. (1986) Composition and formation of metal nodules and veins in ordinary chondrites. *Geochim. Cosmochim. Acta* 50, 1989–1995.

Wieler R., Baur H., Pedroni A., Signer P., and Pellas P. (1989a) Exposure history of the regolithic chondrite Fayetteville: I. Solar-gas-rich matrix. *Geochim. Cosmochim. Acta* 53, 1441–1448.

Wieler R., Graf T., Pedroni A., et al. (1989b) Exposure history of the regolithic chondrite Fayetteville: II. Solar-gas-free light inclusions. *Geochim. Cosmochim. Acta* 53, 1449–1459.

Wilkening L. L. (1971) Particle track studies and the origin of gas-rich meteorites. *Publication 11, Center for Meteorite Studies*, Arizona State University, 1–36.

Wilkening L. L. (1977) Meteorites in meteorites: Evidence for mixing among the asteroids. In *Comets, Asteroids, Meteorites* (edited by Delsemme A. H.), pp. 389–395, University of Toledo Press, Toledo, Ohio.

Wilkening L. L. (1978) Tysnes Island: An unusual clast composed of solidified, immiscible, Fe-FeS and silicate melts. *Meteoritics* 13, 1–9.

Wilkening L. L. and Marti K. (1976) Rare gases and fossil particle tracks in the Kenna ureilite. *Geochim. Cosmochim. Acta* 40, 1465–1473.

Williams C. V., Rubin A. E., Keil K., and San Miguel A. (1985) Petrology of the Cangas de Onis and Nulles regolith breccias: Implications for parent body history. *Meteoritics* 20, 331–345.

Willis J. and Goldstein J. I. (1981) A revision of metallographic cooling rate curves for chondrites. *Proc. Lunar Planet. Sci. Conf. 12th*, 1135–1143.

Willis J. and Goldstein J. I. (1982) The effects of C, P, and S on trace element partitioning during solidification of the Fe-Ni system. *Proceedings Lunar and Planetary Science Conference 13th*. pp. a435–a445.

Wilner D. J., D'Alessio P., Calvet N., Claussen M. J., and Hartmann L. (2005) Toward planetesimals in the disk around TW Hydrae: 3.5 centimeter dust emission. *Astrophys. J.* 626, L109–112.

Wisdom J. (1985) Meteorites may follow a chaotic route to Earth. *Nature* 315, 731–733.

Wittgenstein L. (1922) *Tractatus Logico-Philosophicus* (translated by Ramsey F. P. and Ogden C. K.), Kegan Paul, Trench, Trübner & Co., Ltd., London, 75 pp.

Wlotzka F. (1969) On the formation of chondrules and metal particles, by 'shock melting.' In *Meteorite Research* (edited by Millman P. M.), pp. 174–184, Reidel, Dordrecht.

Wlotzka F. (1985) Olivine-spinel and olivine-ilmenite thermometers in chondrites of different petrologic type (abstract). *Lunar Planet. Sci.* 16, 918–919.

Wlotzka F. (1987) Equilibration temperatures and cooling rates of chondrites: A new approach (abstract). *Meteoritics* 22, 529–531.

Wlotzka F. (1993) A weathering scale for the ordinary chondrites (abstract). *Meteoritics* 28, 460.

Wlotzka F. (2005) Cr spinel and chromite as petrogenetic indicators in ordinary chondrites: Equilibration temperatures of petrologic types 3.7 to 6. *Meteorit. Planet. Sci.* 40, 1673–1702.

Wlotzka F., Palme H., Spettel B., et al. (1983) Alkali differentiation in LL-chondrites. *Geochim. Cosmochim. Acta* 47, 743–757.

Wolf S. F. and Lipschutz M. E. (1995) Meteoroid streams: Evidence from meteorites recovered on Earth. *Earth, Moon Planet.* 68, 605–637.

Wood B. J., Smythe D. J., and Harrison T. (2019) The condensation temperatures of the elements: A reappraisal. *Amer. Mineral.* 104, 844–856.

Wood J. A. (1962a) Chondrules and the origin of the terrestrial planets. *Nature* 197, 127–130.

Wood J. A. (1962b) Metamorphism in chondrites. *Geochim. Cosmochim. Acta* 26, 739–749.

Wood J. A. (1963) On the origin of chondrules and chondrites. *Icarus* 2, 152–180.

Wood J. A. (1964) The cooling rates and parent planets of several iron meteorites. *Icarus* 3, 429–459.

Wood J. A. (1967) Chondrites: Their metallic minerals, thermal histories, and parent bodies. *Icarus* 6, 1–49.

Wood J. A. (1984) On the formation of meteoritic chondrules by aerodynamic drag heating in the solar nebula. *Earth Planet. Sci. Lett.* 70, 11–26.

Wood J. A. (1996) Processing of chondritic and planetary material in spiral density waves in the nebula. *Meteorit. Planet. Sci.* 31, 641–645.

Wood J. A. and Hashimoto A. (1993) Mineral equilibrium in fractionated nebular systems. *Geochim. Cosmochim. Acta* 57, 2377–2388.

Wood J. A. and Pellas P. (1991) What heated the parent meteorite planets? In *The Sun in Time* (edited by Sonett C. P., Giampapa M. S., and Matthews M. S.), pp. 740–760. University of Arizona Press, Tuscon.

Wood J. A., Marvin U. B., Reed J. B., et al. (1971) Mineralogy and petrology of the Apollo 12 lunar sample. *Smithson. Astrophys. Observ. Special Rep.* 333, 177.

Wyatt S. P. and Whipple F. L. (1950) The Poynting-Robertson effect on meteor orbits. *Astrophys. J.* 111, 134–141.

Xian H., Zhu J., Yang Y., et al. (2024) Formation of nanophase metallic iron through charge disproportionation of ferrous iron during micrometeoroid impact-induced splash melting. *Meteorit. Planet. Sci.* 59, 1692–1704.

Xiang J.-W. and Jin Z. (2024) Trace element systematics in unequilibrated ordinary chondrites. *86th Annu. Meet. Meteorit. Soc.*, abstract#3036.

Xie X. and Gu X. (2023) The breakdown of diopside to (Ca,Mg)SiO_3 perovskite – (Mg,Ca,Fe)SiO_3 glass – (Mg,Ca)SiO_3 glass – (Mg,Ca)SiO_3 majorite in a melt vein the Suizhou L6 chondrite. *Acta Geochim.* 42, 183–194.

Xie X. and Wang D. (1992) The behavior of Fe-Ni metal during thermal metamorphism of the Jilin chondrite. *Chin. J. Geochem.* 11, 23–31.

Xie X., Minitti M. E., Chen M., et al. (2003) Tuite, γ-$Ca_3(PO_4)_2$: A new mineral from the Suizhou L6 chondrite. *Eur. J. Mineral.* 15, 1001–1005.

Xie X., Zhai S., Chen M. and Yang H. (2013) Tuite, γ-$Ca_3(PO_4)_2$, formed by chloroapatite decomposition in a shock vein of the Suizhou L6 chondrite. *Meteorit. Planet. Sci.* 48, 1515–1523.

Xu Y.-K., Li Z., Li S.-J., et al. (2023) Shock-induced potassium and zinc isotope fractionation in ordinary chondrites and its implications. *Geochim. Cosmochim. Acta* 349, 1–10.

Yamaguchi A., Scott E. R. D., and Keil K. (1999) Origin of a unique impact-melt rock – the L-chondrite Ramsdorf. *Meteorit. Planet. Sci.* 34, 49–59.

Yamaguchi A., Kimura M., Barrat J.-A., and Greenwood R. (2019) Compositional diversity of ordinary chondrites inferred from petrology, bulk chemical, and oxygen isotopic compositions of the lowest FeO ordinary chondrites, Yamato 982717. *Meteorit. Planet. Sci.* 54, 1919–1929.

Yanai K. and Kojima H. (1991) Yamato-74063: chondritic meteorite classified between E and H chondrite groups. *Proc. NIPR Symp. Antarct. Met.* 4, 118–130.

Yang J. and Goldstein J. I. (2005) The formation mechanism of the Widmanstätten structure in meteorites. *Meteorit. Planet. Sci.* 40, 239–253.

Yang J., Goldstein J. I. and Scott E. R. D. (2010) Main-group pallasites: Thermal history, relationship to IIIAB irons, and origin. *Geochim. Cosmochim. Acta* 74, 4471–4492.

Yang L., Ciesla F. J. and Alexander C. M. O'D. (2013) The D/H ratio of water in the solar nebula during its formation and evolution. *Icarus* 226, 256–267. https://doi.org/10.1016/j.icarus.2013.05.027.

Yanai K. (1981) *Photographic Catalog of the Selected Antarctic Meteorites*, National Institute of Polar Research, Tokyo, 104 pp.

Yanai K. and Kojima H. (1987) *Photographic Catalog of the Selected Antarctic Meteorites*, National Institute of Polar Research, Tokyo, 298 pp.

Yang J. and Epstein S. (1983) Interstellar organic matter in meteorites. *Geochim. Cosmochim. Acta* 47, 2199–2216.

Yang L. and Ciesla F. J. (2012) The effects of disk building on the distributions of refractory materials in the solar nebula. *Meteorit. Planet. Sci.* 47, 99–119.

Yano H., Kubota T., Miyamoto H., et al. (2006) Touchdown of the Hayabusa spacecraft at the Muses Sea on Itokawa. *Science* 312, 1350–1353.

Yaroshevsky A. A. and Ivanov A. V. (2010) Geochemical diversity of meteorite minerals: Systematics and nomenclature of phosphides and silicides. *Geochem. Int.* 48, 808–814.

Yeomans D. K., Barriot J.-P., Dunham D. W., et al. (1997) Estimating the mass of asteroid 253 Mathilde from tracking data during the NEAR flyby. *Science* 278, 2106–2109.

Yin Q.-Z. (2005) From dust to planets: The tale told by moderately volatile elements. In *Chondrites and the Protoplanetary Disk, ASP Conference Series 341* (edited by Krot A. N., Scott E. R. D. and Reipurth B.), pp. 632–644, Astronomical Society of the Pacific, San Francisco.

Yokoyama T., Nagashima K., Nakai I., et al. (2022) Samples returned from the asteroid Ryugu are similar to Ivuna-type carbonaceous meteorites. *Science* 379. https://doi.org/10.1126/science.abn7850.

Yolcubal I., Sack R. O., Wang M.-S., and Lipschutz M. E. (1997) Formation conditions of igneous regions in ordinary chondrites: Chico, Rose City, and other heavily shocked H and L chondrites. *J. Geophys. Res. Planets* 102, 21589–21611.

Yomogida K. and Matsui T. (1983) Physical properties of ordinary chondrites. *J. Geophys. Res. Solid Earth* 88, 9513–9533.

Yoneda S. and Grossman L. (1994) Advanced generation condensation calculations: compositions of fassaite, spinel and plagioclase in the solar nebula (abstract). *Lunar Planet. Sci.* 25, 1533–1534.

Yoshino T., Walter M. J., and Katsura T. (2003) Core formation in planetesimals triggered by permeable flow. *Nature* 422, 154–157.

Youdin A. N. and Chiang E. I. (2004) Particle pileups and planetesimal formation. *Astrophys. J.* 601, 1109–1119.

Youdin A. N. and Goodman J. (2005) Streaming instabilities in protoplanetary disks. *Astrophys. J.* 620, 459–469.

Young E. D. and Russell S. S. (1998) Oxygen reservoirs in the early solar system inferred from an Allende CAI. *Science* 282, 452–455.

Young E. D., Shahar A, and Schlichting H. E. (2023) Earth shaped by primordial H_2 atmospheres. *Nature* 616, 306–311.

Yu Y., Hewins R. H., and Zanda B. (1996) Sodium and sulfur in chondrules: Heating times and cooling curves. In *Chondrules and the Protoplanetary Disk* (edited by Hewins R. H.,

Jones R. H. and Scott E. R. D.), pp. 213–219, Cambridge University Press, Cambridge.
Yu Y., Hewins R. H., Alexander C. M. O.'D. and Wang J. (2003) Experimental study of evaporation and isotopic mass fractionation of potassium in silicate. *Geochim. Cosmochim. Acta* 67, 773–786.
Yurimoto H., Abe K., Abe M., et al. (2011) Oxygen isotopic compositions of asteroidal materials returned from Itokawa by the Hayabusa mission. *Science*, 333, 1116–1119.
Zanda B., Bouret-Denise M., Perron C. and Hewins R. H. (1994) Origin and metamorphic redistribution of silicon, chromium, and phosphorus in the metal of chondrites. *Science* 265, 1846–1849.
Zanda B., Hewins R. H., Bourot-Denise M., Bland P. A., and Albarède F. (2006) Formation of solar nebula reservoirs by mixing chondritic components. *Earth Planet. Sci. Lett.* 248, 650–660.
Zavaritzky A. N. (1948) Structure of crystalline chondrites. *Meteoritika* 1.
Zavaritzky A. N. (1950) On some special features of meteorites. *Meteoritika* 3.
Zavaratskii A. N. and Kvasha L. G. (1952) *Meteorites of the USSR*, Kollektsia Akad., Nauka, (in Russian), Moscow.
Zellner B. (1973) Polarimetric albedos of asteroids (abstract). *Bull. Amer. Astron. Soc.* 5, 388.
Zellner B. and Gradie J. (1976) Minor planets and related objects. XX. Polarimetric evidence for the albedos and compositions of 94 asteroids. *Astron. J.* 81, 262–280.
Zennouri L., Chennaoui Aoudjehane H., Folco L., et al. (2025) Tamdakht: A Moroccan ordinary chondrite fall with an uncommon fusion crust. *Meteorit. Planet. Sci.* 60, 1216–1226.
Zhai S., Akaogi M., Kojitani H., Xue W., and Ito E. (2014) Thermodynamic investigation on β- and γ-$Ca_3(PO_4)_2$ and the phase equilibria. *Phys. Earth Planet. Interi.* 228, 144–149.
Zhang B., Chabot N. L., and Rubin A. E. (2022) Compositions of carbonaceous-type asteroidal cores in the early solar system. *Sci. Adv.* 8, eabo5781.
Zhang B., Chabot N. L., and Rubin A. E. (2024) Compositions of iron-meteorite parent bodies constrain the structure of the protoplanetary disk. *Proc. Natl. Acad. Sci.*, 121(23):e2306995121. https://doi.org/10.1073/pnas.2306995121.
Zhang C., Leyva V., Wang J., et al. (2024) Ionizing radiation exposure on Arrokoth shapes a sugar world. *Proc. Natl. Acad. Sci.*, 121 (24) e2320215121. https://doi.org/10.1073/pnas.2320215121.
Zhang J., Willimas D. B., and Goldstein J. I. (1994) Decomposition of Fe-Ni martensite: Implications for the low-temperature (≤500°C) Fe-Ni phase diagram. *Metall. Mater. Trans. A* 25, 1627–1637.
Zhang M., Brownlee D. E., Joswiak D. J., and Kita N. T. (2023) Rocky fragments in the largest track of comet 81P/Wild 2 – Chondrule debris formed in the icy outer solar system? *Lunar Planet. Sci. 54th*, abstract#1711.
Zhang M., Fukuda K., Tappa M. J., et al. (2024) Forging inner-disk Al-rich chondrules by interactions of CAI-like melt and ambient gas. *Geochim. Cosmochim. Acta* 379, 89–110.
Zhang T. A., Liao S. Y., Wu R. C., and Schmitz B. (2024) L-chondrite body breakup in Ordovician strata in China – A time tie point globally and across the inner solar system. *Earth Planet. Sci. Lett.* 643, 118891.
Zinner E. (1988) Interstellar cloud material in meteorites. In *Meteorites and the Early Solar System* (edited by Kerridge J. F. and Matthews M. S.), pp. 956–983, University of Arizona Press, Tucson.
Zinner E. K. (2005) Presolar grains. In *Meteorites, Comets, and Planets* (edited by Davis A. M.), pp. 17–39. Elsevier, Oxford.
Zinner E. (2014) Presolar grains. In *Meteorites and Cosmochemical Processes* (edited by Holland, H. D. and Turekian, K. K.), pp. 181–210, Treatise on Geochemistry, vol. 1. Elsevier.
Zolensky M. E. (1997) Structural water in the Bench Crater chondrite returned from the Moon. *Meteorit. Planet. Sci.* 32, 15–18.
Zolensky M. and Ivanov A. (2003) The Kaidun microbreccia meteorite: A harvest from the inner and outer Asteroid belt. *Chem. Erde Geochem.* 63, 185–246.
Zolensky M. E., Bodnar R. J., Gibson E. K., et al. (1999a) Asteroidal water within fluid inclusion-bearing halite in an H5 chondrite, Monahans. *Science* 285, 1377–1379.
Zolensky M. E., Bodnar R. J., and Rubin A. E. (1999b) Asteroidal water within fluid-inclusion-bearing halite in ordinary chondrites. *Meteorit. Planet. Sci.* 34, A124.
Zolensky M. E., Lee R., and Le L. (2004) Samples of asteroid surface ponded deposits in chondritic meteorites. *Lunar Planet. Sci.* 35, abstract#1332.
Zolensky M. E., Zega T. J., Yano H., et al. (2006) Mineralogy and petrology of comet 81P/Wild 2 nucleus samples. *Science* 314, 1735–1739.
Zolensky M. E., Fries M. D., Chan Q. H.-S., et al. (2015) Survival of organic materials in ancient cryovolcanically-produced halite crystals. Workshop on the Potential for Finding Life in a Europa Plume, Mofft Field, CA.
Zolensky M. E., Bodnar R. J., Fries M., et al. (2016) Ceres revealed in a grain of salt (abstract). *26th Goldschmidt Conference*, Yokohama, Japan.
Zolensky M. E., Abreu N. M., Velbel M. A., et al. (2018a) Primitive meteorites and asteroids. In *Physical, Chemical and Spectroscopic Observations Paving the Way to Exploration* (edited by Abreu N.), pp. 59–204, Elsevier, Amsterdam, 558 pp.
Zolensky M., Fries M., Chan Q. H.-S., et al. (2018b) Outer solar system material in inner solar system regolith breccias. 81st Meeting of the Meteoritical Society, abstract#6103.

Zolensky M., Johnson J., Ziegler K., et al. (2018c) Meteoritic evidence for injection of Trans-Neptuanian orbjects into the Inner Solar System. *Lunar Planet. Sci.* 49th, abstract#1789.

Zolensky M., Mikouchi T., Hagiya K., et al. (2022) Evidence for impact shock and regolith transportation on CM, CI, and CV chondrite parent asteroids. *Meteorit. Planet. Sci.* 57, 1902–1919.

Zolotov M. Yu. (2009) On the composition and differentiation of Ceres. *Icarus* 204, 183–193.

Zook H. A. (1980) A new impact model for the generation of ordinary chondrites (abstract). *Meteoritics* 15, 390–391.

Zook H. A. (1981) On a new model for the generation of chondrites (abstract). *Lunar Planet. Sci.* 12, 1242–1244.

Zsom A., Ormel C. W., Güttler C., Blum J., and Dullemond C. P. (2010) The outcome of protoplanetary dust growth: Pebbles, boulders, or planetesimals? II. Introducing the bouncing barrier. *Astron. Astrophys.* 513, A57.

Zurfluh F. J., Hofmann B. A., Gnos E., et al. (2012) Weathering and strontium contamination of meteorites recovered in the Sultanate of Oman. *Meteorite* 18, no. 1, 34–38.

Zurfluh F. J., Hofmann B. A., Gnos E., Eggenberger U., and Jull A. J. (2016) Weathering of ordinary chondrites from Oman: Correlation of weathering parameters with ^{14}C terrestrial ages and a refined weathering scale. *Meteorit. Planet. Sci.* 51, 1685–1700.

Index

Printed by Integrated Books International,
United States of America